List of the Elements with their Symbols and Atomic Masses*

Element	Symbol	Atomic number	Atomic mass†
Actinium	Ac	89	(227)
Aluminum	Al	13	26.98
Americium	Am	95	(243)
Antimony	Sb	51	121.8
Argon	Ar	18	39.95
Arsenic	As	33	74.92
Astatine	At	85	(210)
Barium	Ba	56	137.3
Berkelium	Bk	97	(247)
Beryllium	Be	4	9.012
Bismuth	Bi	83	209.0
Boron	B	5	10.81
Bromine	Br	35	79.90
Cadmium	Cd	48	112.4
Calcium	Ca	20	40.08
Californium	Cf	98	(249)
Carbon	C	6	12.01
Cerium	Ce	58	140.1
Cesium	Cs	55	132.9
Chlorine	Cl	17	35.45
Chromium	Cr	24	52.00
Cobalt	Co	27	58.93
Copper	Cu	29	63.55
Curium	Cm	96	(247)
Dysprosium	Dy	66	162.5
Einsteinium	Es	99	(254)
Erbium	Er	68	167.3
Europium	Eu	63	152.0
Fermium	Fm	100	(253)
Fluorine	F	9	19.00
Francium	Fr	87	(223)
Gadolinium	Gd	64	157.3
Gallium	Ga	31	69.72
Germanium	Ge	32	72.59
Gold	Au	79	197.0
Hafnium	Hf	72	178.5
Helium	He	2	4.003
Holmium	Ho	67	164.9
Hydrogen	H	1	1.008
Indium	In	49	114.8
Iodine	I	53	126.9
Iridium	Ir	77	192.2
Iron	Fe	26	55.85
Krypton	Kr	36	83.80
Lanthanum	La	57	138.9
Lawrencium	Lr	103	(257)
Lead	Pb	82	207.2
Lithium	Li	3	6.941
Lutetium	Lu	71	175.0
Magnesium	Mg	12	24.31
Manganese	Mn	25	54.94
Mendelevium	Md	101	(256)
Mercury	Hg	80	200.6
Molybdenum	Mo	42	95.94
Neodymium	Nd	60	144.2
Neon	Ne	10	20.18
Neptunium	Np	93	(237)
Nickel	Ni	28	58.69
Niobium	Nb	41	92.91
Nitrogen	N	7	14.01
Nobelium	No	102	(253)
Osmium	Os	76	190.2
Oxygen	O	8	16.00
Palladium	Pd	46	106.4
Phosphorus	P	15	30.97
Platinum	Pt	78	195.1
Plutonium	Pu	94	(242)
Polonium	Po	84	(210)
Potassium	K	19	39.10
Praseodymium	Pr	59	140.9
Promethium	Pm	61	(147)
Protactinium	Pa	91	(231)
Radium	Ra	88	(226)
Radon	Rn	86	(222)
Rhenium	Re	75	186.2
Rhodium	Rh	45	102.9
Rubidium	Rb	37	85.47
Ruthenium	Ru	44	101.1
Samarium	Sm	62	150.4
Scandium	Sc	21	44.96
Selenium	Se	34	78.96
Silicon	Si	14	28.09
Silver	Ag	47	107.9
Sodium	Na	11	22.99
Strontium	Sr	38	87.62
Sulfur	S	16	32.07
Tantalum	Ta	73	180.9
Technetium	Tc	43	(99)
Tellurium	Te	52	127.6
Terbium	Tb	65	158.9
Thallium	Tl	81	204.4
Thorium	Th	90	232.0
Thulium	Tm	69	168.9
Tin	Sn	50	118.7
Titanium	Ti	22	47.88
Tungsten	W	74	183.9
Unnilennium	Une	109	(266)
Unnilhexium	Unh	106	(263)
Unniloctium	Uno	108	(265)
Unnilpentium	Unp	105	(260)
Unnilquadium	Unq	104	(257)
Unnilseptium	Uns	107	(262)
Uranium	U	92	238.0
Vanadium	V	23	50.94
Xenon	Xe	54	131.3
Ytterbium	Yb	70	173.0
Yttrium	Y	39	88.91
Zinc	Zn	30	65.39
Zirconium	Zr	40	91.22

* All atomic masses have four significant figures. These values are recommended by the Committee on Teaching of Chemistry, International Union of Pure and Applied Chemistry.

† Approximate values of atomic masses for radioactive elements are given in parentheses.

ESSENTIAL CHEMISTRY

ESSENTIAL CHEMISTRY

RAYMOND CHANG

Williams College

THE MCGRAW-HILL COMPANIES, INC.

New York St. Louis San Francisco Auckland Bogotá Caracas Lisbon London Madrid Mexico City Milan Montreal New Delhi San Juan Singapore Sydney Tokyo Toronto

McGraw-Hill
*A Division of The **McGraw-Hill** Companies*

ESSENTIAL CHEMISTRY

This book is printed on acid-free paper.

1 2 3 4 5 6 7 8 9 0 VNH VNH 9 0 9 8 7 6 5

ISBN 0-07-011207-X

This book was set in New Aster by York Graphic Services, Inc.
The editors were Judith Kromm, Karen J. Allanson, and Jack Maisel;
the designers were Joan Greenfield and Joseph A. Piliero;
the production supervisor was Janelle S. Travers.
The photo editor was Kathy Bendo;
the photo researcher was Elyse Rieder.
Von Hoffmann Press, Inc., was printer and binder.

Library of Congress Cataloging-in-Publication Data

Chang, Raymond.
Essential chemistry / Raymond Chang.
p. cm.
Includes index.
ISBN 0-07-011207-X
1. Chemistry. I. Title.
QD33.C434 1996
540—dc20 95-41428

INTERNATIONAL EDITION

When ordering this title, use ISBN 0-07-114095-6.

About the Author

Raymond Chang was born in Hong Kong and grew up in Shanghai, China, and Hong Kong. He received his B.Sc. degree in chemistry from London University, England, and his Ph.D. in chemistry from Yale University. After doing postdoctoral research at Washington University and teaching for a year at Hunter College, he joined the chemistry department at Williams College, where he has taught since 1968. Professor Chang has written books on spectroscopy, physical chemistry, and industrial chemistry and has coauthored books on the Chinese language, a novel for juvenile readers, and children's picture books.

For relaxation, Professor Chang maintains a forest garden, plays tennis, and practices the violin.

CONTENTS IN BRIEF

CONTENTS

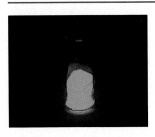

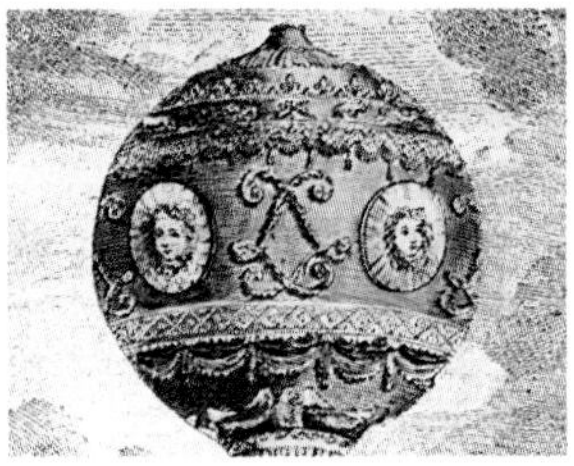

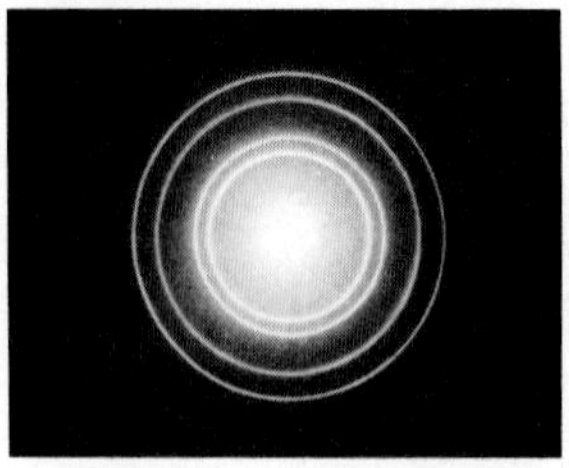

CHAPTER 8 THE PERIODIC TABLE 214

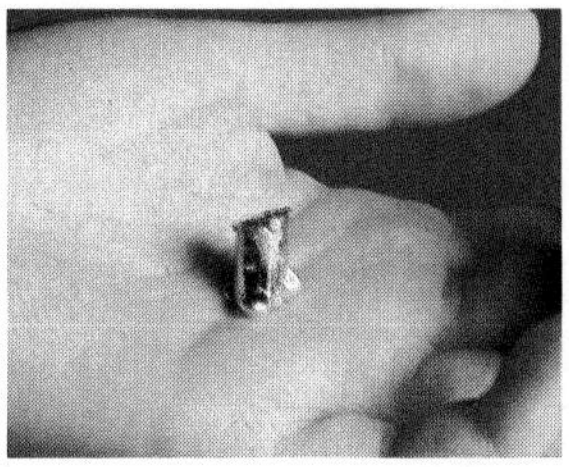

CHAPTER 9 CHEMICAL BONDING I: THE COVALENT BOND 248

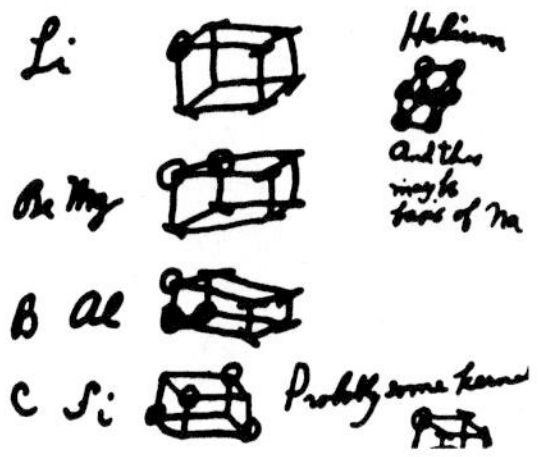

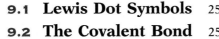

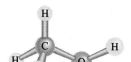

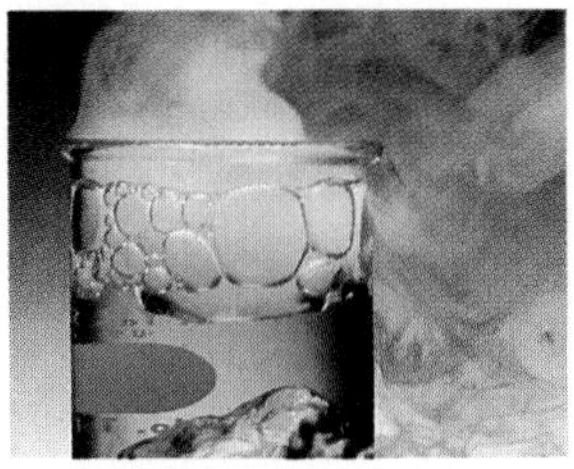

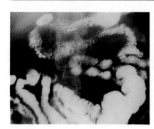

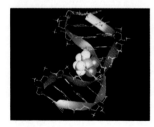

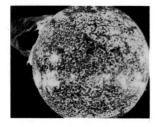

PREFACE

With a few simple images, the painting that is reproduced on the cover of this book—titled *Peace Through Chemistry*—captures the essence of chemistry. This art by Roy Lichtenstein serves as a colorful metaphor for this text, which presents only the material that is essential for a one-year general chemistry course. I do not presume to have created a masterpiece, but I offer this text as an alternative to the traditional 1100-page volumes for instructors who want to be able to cover every chapter in two semesters or three quarters.

My goal in writing this book was to include all the core topics that are necessary for a good foundation in general chemistry without sacrificing clarity and comprehension. Doing so in a brief form required a certain amount of selectivity. The selection process involved asking myself each time I started a new chapter, "What is essential for students to know about this area of chemistry?" Although some people may not agree totally with my choices, I feel that this approach has resulted in a book that students and instructors alike will be comfortable with.

ORGANIZATION

The organization of this text is fairly conventional and flexible. Chapters 1 to 12 follow the normal sequence for easy coordination with lab work. Organic chemistry and polymer chemistry, two topics that usually come at the end of a general chemistry text, are presented here in Chapters 13 and 14. This placement allows the use of organic compounds as examples and in problems for subsequent chapters and exposes students to the chemistry of natural and synthetic polymers. However, these chapters can be covered later in the course with no breaks in continuity. Chemical kinetics (Chapter 21) also can be introduced in a more traditional sequence, just before or after Chapter 15.

There are no discrete chapters on descriptive chemistry in this book because most instructors lack the time necessary for an extensive survey of the chemistry of the periodic groups. Instead, descriptive chemistry is integrated throughout the text to show how chemical principles are applied in the real world. Chapters 2, 3, 4, 8, and 18, in particular, have a descriptive orientation, and many of the examples and chapter problems include descriptive material.

Each chapter opens with a short historical vignette or contemporary story that is relevant to the subject of the chapter. By emphasizing chemistry as a human endeavor, the fruits of which are all around us, these introductions provide a real context for chemistry. I hope they will also stimulate the reader's interest in the content that follows.

Within the chapter, verbal explanations of concepts, theory, and mathematical relationships are reinforced with text and pictorial illustrations, worked examples, and practice exercises. A chapter summary and key words list are provided to help students review the chapter. Review questions and follow-up problems conclude each chapter.

PROBLEM-SOLVING APPROACH

The ability to analyze and solve problems is crucial for the successful completion of general chemistry. Consequently, this text includes 161 examples with full solutions to show how to approach the type of problem described in the example heading. The placement of these examples within the text demonstrates problem-solving strategies in context and the similar practice exercise that follows each example allows students to check their general understanding of the chemical concepts and principles discussed in the text while applying them to a specific question. Answers to the practice exercises appear at the end of a chapter.

Review questions and problems at the end of each chapter provide additional opportunities for concept review and practice in problem solving. The review questions check students' understanding of the conceptual side—the "why" of chemistry. The 880 problems cover the quantitative, experimental side of chemistry that yields numerical results. They explore "how" chemistry works and test the ability to apply conceptual logic and solve problems. The review questions and problems are grouped in sections that parallel the chapter discussion, and each problem is paired with another similar problem for additional practice. A set of "Miscellaneous Problems" at the end of the exercise section features more challenging problems as well as problems involving two or more concepts. These give students experience in identifying concepts and techniques needed to solve real problems. The answers to all even-numbered problems are provided in the back of the book.

SUPPLEMENTS

A number of excellent ancillary publications are available for use with this text. They are designed to make general chemistry more enjoyable for instructors and students alike.

- **Problem-Solving Workbook with Solutions,** by Brandon Cruickshank (Northern Arizona University) and Raymond Chang, is a success guide written for use with ***Essential Chemistry.*** It aims to help students hone their analytical and problem-solving skills by presenting detailed discussion of different types of problems and approaches to solving chemical problems, tutorial solutions for many of the end-of-chapter problems in the text, and several new problems for each chapter, along with strategies for solving them. The focus of the new problems is on real applications of chemical concepts both in everyday life and in related fields, such as biology. The solutions for all even-numbered end-of-chapter problems are included in this book.

- **Instructor's Resource Kit with Solutions,** by Vicky Ellis (Gulf Coast Community College) and Raymond Chang, is a complete manual for teaching a general chemistry course based on ***Essential Chemistry.*** This unique guide includes demonstrations that can be done in any classroom or assigned for homework, accompanied by discussion questions and tips to ensure success; information on relevant applications, along with references to additional articles and multimedia resources; chapter overviews and outlines; and annotated cross-references to other elements of the text package. In addition, this guide provides complete solutions to all end-of-chapter problems in the text.
- **Test Bank,** by Gary Wolf (Spokane Community College), contains over 2000 multiple choice, short answer, and true-false exam questions. The questions, which are graded in difficulty, are comparable to the problems in the text and include multistep problems that require conceptual analysis.
- **Computerized Test Bank with Algorithms** contains all of the questions in the print Test Bank plus algorithms and over 200 algorithm-based questions that instructors can edit to create their own test templates. This supplement is available in DOS and Windows versions for PC as well as for use with Macintosh computers.
- **Overhead Transparencies** include 200 full-color acetates of important illustrations from this text.
- **Chemistry at Work Videodisc** is a valuable teaching tool that provides access to 50 filmed laboratory demonstrations, as well as tables, illustrations, and photos.
- **Cooperative Chemistry Laboratory Manual,** by Melanie M. Cooper (Clemson University), is an innovative guide featuring open-ended problems designed to simulate experience in a real chemistry lab. Working in groups, students research one problem over a period of weeks, so that they might complete three or four projects during the semester, rather than one pre-programmed experiment per class. The emphasis here is on experimental design, analysis, problem solving, and communication.
- **Primis LabBase,** edited by Joseph Lagowski (The University of Texas at Austin), is a data-base collection of general chemistry lab experiments culled from the *Journal of Chemical Education* and experiments that Professor Lagowski has used at The University of Texas. Instructors can choose from over 40 experiments to design a customized lab manual.

Raymond Chang

ACKNOWLEDGMENTS

I would like to thank the following individuals, who read the manuscript for ***Essential Chemistry,*** for their helpful comments:

Lavoir Banks, Elgin Community College

Coran L. Cluff, Brigham Young University

Leslie DiVerdi, Colorado State University

Daryl Doyle, GMI Engineering and Management Institute

Anthony Guzzo, University of Wyoming

Anne Harmon, Lamar University

Grant N. Holder, Appalachian State University

Wyatt R. Murphy, Jr., Seton Hall University

Paul Poskozim, Northeastern Illinois University

David P. Richardson, Williams College

Chris D. Spindler, North Harris College

James B. Wood, Palm Beach Community College

It is also a pleasure to acknowledge the enthusiastic support and assistance given to me by the following people at McGraw-Hill's College Division: Karen Allanson, Cristene Burr, Robert Christie, Christopher Fitzpatrick, Jane Mackarell, Jennifer Speer, and Suzanne Thibodeau. In particular, I would like to acknowledge Kathy Bendo, Jack Maisel, and Janelle Travers, whose high professional standards and skill are responsible for the smooth production of the book, and Joseph Piliero for the pleasing and functional design. Finally, my thanks go to Denise Schanck, for her guidance, encouragement, and supervision at every stage of the writing of this book, and Judith Kromm, a superb development editor whose comments and suggestions have improved the text in so many ways.

Raymond Chang

ESSENTIAL CHEMISTRY

CHAPTER 1
INTRODUCTION

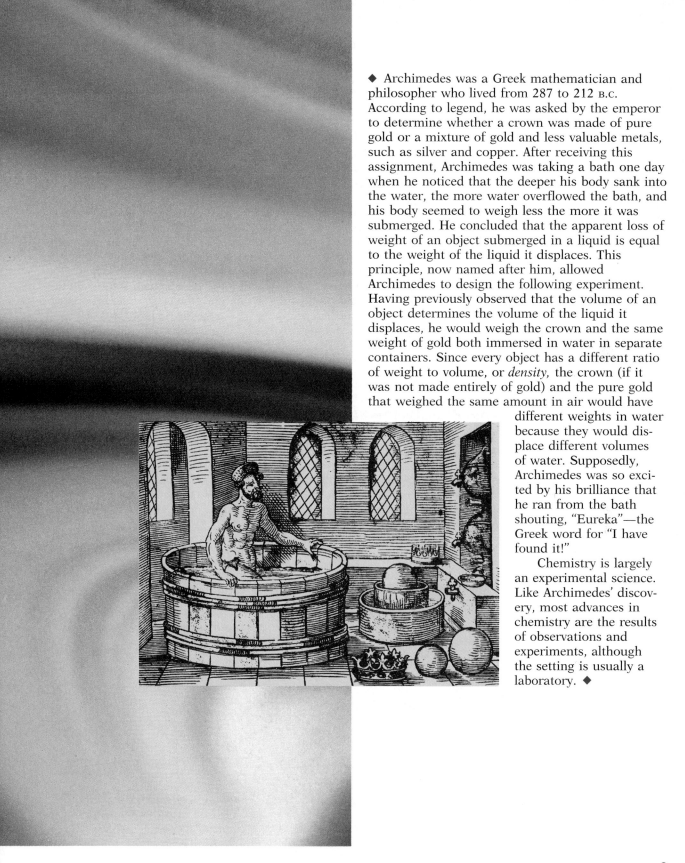

◆ Archimedes was a Greek mathematician and philosopher who lived from 287 to 212 B.C. According to legend, he was asked by the emperor to determine whether a crown was made of pure gold or a mixture of gold and less valuable metals, such as silver and copper. After receiving this assignment, Archimedes was taking a bath one day when he noticed that the deeper his body sank into the water, the more water overflowed the bath, and his body seemed to weigh less the more it was submerged. He concluded that the apparent loss of weight of an object submerged in a liquid is equal to the weight of the liquid it displaces. This principle, now named after him, allowed Archimedes to design the following experiment. Having previously observed that the volume of an object determines the volume of the liquid it displaces, he would weigh the crown and the same weight of gold both immersed in water in separate containers. Since every object has a different ratio of weight to volume, or *density,* the crown (if it was not made entirely of gold) and the pure gold that weighed the same amount in air would have different weights in water because they would displace different volumes of water. Supposedly, Archimedes was so excited by his brilliance that he ran from the bath shouting, "Eureka"—the Greek word for "I have found it!"

Chemistry is largely an experimental science. Like Archimedes' discovery, most advances in chemistry are the results of observations and experiments, although the setting is usually a laboratory. ◆

1.1 THE STUDY OF CHEMISTRY

Whether or not this is your first course in chemistry, you undoubtedly have some preconceived ideas about the nature of this science and about what chemists do. Most likely, you think chemistry is practiced in a laboratory by someone in a white coat who studies things in test tubes. This description is fine, up to a point. Chemistry is largely an experimental science, and a great deal of knowledge comes from laboratory research. In addition, however, today's chemist may use a computer to study the microscopic structure and chemical properties of substances or employ sophisticated electronic equipment to analyze pollutants from auto emissions or toxic substances in the soil. Many frontiers in biology and medicine are currently being explored at the level of atoms and molecules—the structural units on which the study of chemistry is based. Chemists participate in the development of new drugs and in agricultural research. What's more, they are seeking solutions to the problem of environmental pollution, along with replacements for energy sources. And most industries, whatever their products, have a basis in chemistry. For example, chemists developed the polymers (very large molecules) that manufacturers use to make a wide variety of goods, including clothing, cooking utensils, artificial organs, and toys. Indeed, because of its diverse applications, chemistry is often called the "central science."

How to Study Chemistry

Compared with other subjects, chemistry is commonly perceived to be more difficult, at least at the introductory level. There is some justification for this perception. For one thing, chemistry has a very specialized vocabulary. At first, studying chemistry is like learning a new language. Furthermore, some of the concepts are abstract. Nevertheless, with diligence you can complete this course successfully—and perhaps even pleasurably. Listed below are some suggestions to help you form good study habits and master the material:

- Attend classes regularly and take careful notes.
- If possible, always review the topics you learned in class the *same* day the topics are covered in class. Use this book to supplement your notes.
- Think critically. Ask yourself if you really understand the meaning of a term or the use of an equation. A good way to test your understanding is for you to explain a concept to a classmate or some other person.
- Do not hesitate to ask your instructor or your teaching assistant for help.

You will find that chemistry is much more than numbers, formulas, and abstract theories. It is a logical discipline brimming with interesting ideas and applications.

1.2 CLASSIFICATIONS OF MATTER

Matter is *anything that occupies space and has mass,* and ***chemistry*** is *the study of matter and the changes it undergoes.* All matter, at least in principle, can exist in three states: solid, liquid, and gas. Solids are rigid objects with definite shapes. Liquids are less rigid than solids and are fluid—they are able to flow and assume

FIGURE 1.1

The three states of matter. A hot poker changes ice into water and steam.

the shape of their containers. Like liquids, gases are fluid, but unlike liquids, they can expand indefinitely.

The three states of matter can be interconverted—they can be changed from one state into another. Upon heating, a solid will melt and become a liquid. Further heating will convert the liquid into a gas. On the other hand, cooling a gas will condense it into a liquid. Cooled further, the liquid will solidify. Figure 1.1 shows the three states of water.

Scientists also distinguish among several subcategories of matter based on composition and properties. The classifications of matter include substances, mixtures, elements, and compounds, as well as the fundamental units that make up elements and compounds—atoms and molecules—which we will consider in Chapter 2.

Substances and Mixtures

A ***substance*** is *matter that has a definite or constant composition and distinct properties.* Examples are water, silver, ethanol, table salt (sodium chloride), and carbon dioxide. Substances differ from one another in composition and can be identified by their appearance, smell, taste, and other properties. At present, over 8 million substances are known, and the list is growing rapidly.

A ***mixture*** is *a combination of two or more substances in which the substances retain their distinct identities.* Some examples are air, soft drinks, milk, and cement. Mixtures do not have constant composition. Therefore, samples of air collected in different cities would probably differ in composition because of differences in altitude, pollution, and so on.

Mixtures are either homogeneous or heterogeneous. When a spoonful of sugar dissolves in water, *the composition of the mixture,* after sufficient stirring, *is the same throughout the solution.* This solution is a ***homogeneous mixture.*** If sand is mixed with iron filings, however, the sand grains and the iron filings remain visible and separate (Figure 1.2). This type of mixture, in which *the composition is not*

(a) (b)

FIGURE 1.2

Separating iron filings from a heterogeneous mixture. The same technique is used on a larger scale to separate iron and steel from nonmagnetic objects such as aluminum, glass, and plastics.

uniform, is called a ***heterogeneous mixture.*** Adding oil to water creates another heterogeneous mixture because the liquid does not have a constant composition.

Any mixture, whether homogeneous or heterogeneous, can be created and then separated by physical means into pure components without changing the identities of the components. Thus, sugar can be recovered from a water solution by heating the solution and evaporating it to dryness. Condensing the water vapor will give us back the water component. To separate the iron-sand mixture, we can use a magnet to remove the iron filings from the sand, since sand is not attracted to the magnet [see Figure 1.2(b)]. After separation, the components of the mixture will have the same composition and properties as they did to start with.

Elements and Compounds

A substance can be either an element or a compound. An ***element*** is *a substance that cannot be separated into simpler substances by chemical means.* At present, 109 elements have been positively identified. (See the list inside the front cover of the book.) Eighty-three of them occur naturally on Earth. The others have been created by scientists.

Chemists use alphabetical symbols to represent the names of the elements. The first letter of the symbol for an element is *always* capitalized, but the second and third letters are *never* capitalized. For example, Co is the symbol for the element cobalt, whereas CO is the formula for carbon monoxide, which is made up of the elements carbon and oxygen. Table 1.1 shows some of the more common elements. The symbols for some elements are derived from their Latin names—for example, Au from *aurum* (gold), Fe from *ferrum* (iron), and Na from *natrium* (sodium)—while most of them are abbreviated forms of their English names. Special three-letter element symbols have been proposed for the most recently synthesized elements.

Figure 1.3 shows the most abundant elements in Earth's crust and in the human body. As you can see, only five elements (oxygen, silicon, aluminum, iron, and calcium) comprise over 90 percent of Earth's crust. Of these five elements, only oxygen is among the most abundant elements in living systems.

Most elements can interact with one or more other elements to form compounds. We define a ***compound*** as *a substance composed of atoms of two or more*

TABLE 1.1
Some Common Elements and Their Symbols

Name	Symbol	Name	Symbol	Name	Symbol
Aluminum	Al	Fluorine	F	Oxygen	O
Arsenic	As	Gold	Au	Phosphorus	P
Barium	Ba	Hydrogen	H	Platinum	Pt
Bromine	Br	Iodine	I	Potassium	K
Calcium	Ca	Iron	Fe	Silicon	Si
Carbon	C	Lead	Pb	Silver	Ag
Chlorine	Cl	Magnesium	Mg	Sodium	Na
Chromium	Cr	Mercury	Hg	Sulfur	S
Cobalt	Co	Nickel	Ni	Tin	Sn
Copper	Cu	Nitrogen	N	Zinc	Zn

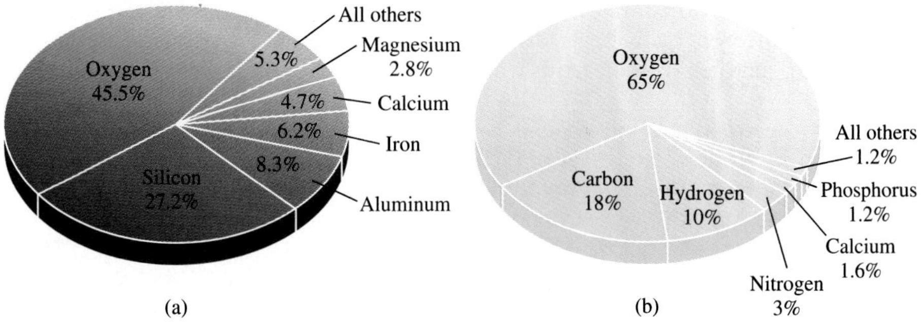

FIGURE 1.3

(a) Natural abundance of elements in Earth's crust in percent by mass. For example, oxygen's abundance is 45.5 percent. This means that in a 100-g sample of Earth's crust there are, on the average, 45.5 g of the element oxygen.
(b) Abundance of elements in the human body in percent by mass.

elements chemically united in fixed proportions. Hydrogen gas, for example, burns in oxygen gas to form water, a compound whose properties are distinctly different from those of the starting materials. Water is made up of two parts of hydrogen and one part of oxygen. This composition does not change, regardless of whether the water comes from a faucet in the United States, the Yangtze River in China, or the ice caps on Mars. Unlike mixtures, compounds can be separated only by chemical means into their pure components.

The relationships among elements, compounds, and other categories of matter are summarized in Figure 1.4.

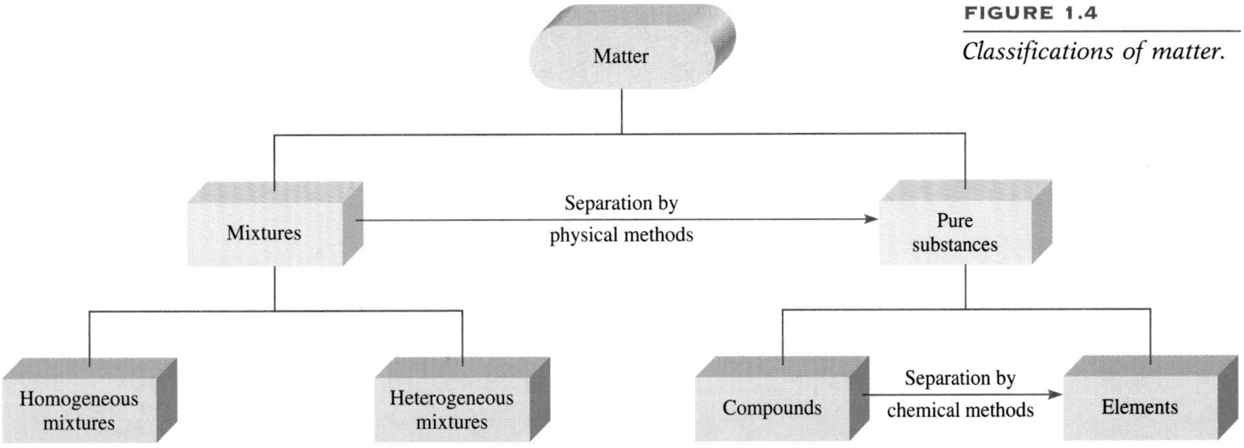

FIGURE 1.4

Classifications of matter.

1.3 PHYSICAL AND CHEMICAL PROPERTIES OF MATTER

Substances are identified by their properties as well as by their composition. Color, melting point, boiling point, and density are physical properties. A ***physical property*** *can be measured and observed without changing the composition or identity of a substance.* For example, we can measure the melting point of ice by heating a block of ice and recording the temperature at which the ice is converted to water. Water differs from ice only in appearance and not in composition, so this is a physical change; we can freeze the water to recover the original ice. Therefore, the melting point of a substance is a physical property. Similarly, when we say that helium gas is lighter than air, we are referring to a physical property.

On the other hand, the statement "Hydrogen gas burns in oxygen gas to form water" describes a ***chemical property*** of hydrogen because *in order to observe this property we must carry out a chemical change,* in this case burning. After the change, the original substances, hydrogen and oxygen gas, will have vanished and a chemically different substance—water—will have taken their place. We *cannot* recover hydrogen and oxygen from water by a physical change such as boiling or freezing.

One way to decompose water into hydrogen and oxygen is by electrolysis, which brings about a chemical change.

Every time we hard-boil an egg, we bring about a chemical change. When subjected to a temperature of about 100°C, the yolk and the egg white undergo reactions that alter not only their physical appearance but their chemical makeup as well. When eaten, the egg is changed again, by substances in the body called *enzymes.* This digestive action is another example of a chemical change. What happens during such a process depends on the chemical properties of the specific enzymes and of the food involved.

All measurable properties of matter fall into two categories: extensive properties and intensive properties. The measured value of an ***extensive property*** *depends on how much matter is being considered.* Mass, length, and volume are extensive properties. More matter means more mass. Values of the same extensive property can be added together. For example, two pieces of copper wire will have a combined mass that is the sum of the masses of each wire, and the volume occupied by the water in two beakers is the sum of the volumes of the water in each of the beakers.

The measured value of an ***intensive property*** *does not depend on the amount of matter being considered.* Temperature is an intensive property. Suppose that we have two beakers of water at the same temperature. If we combine them to make a single quantity of water in a larger beaker, the temperature of the larger amount of water will be the same as it was in two separate beakers. Unlike mass and volume, temperature and other intensive properties such as melting point, boiling point, and density are not additive.

1.4 MEASUREMENT

The study of chemistry depends heavily on measurement. For instance, chemists use measurements to compare the properties of different substances and to assess changes resulting from an experiment. A number of common devices enable us to make simple measurements of a substance's properties: The meter stick measures length; the buret, the pipet, the graduated cylinder, and the volumetric flask mea-

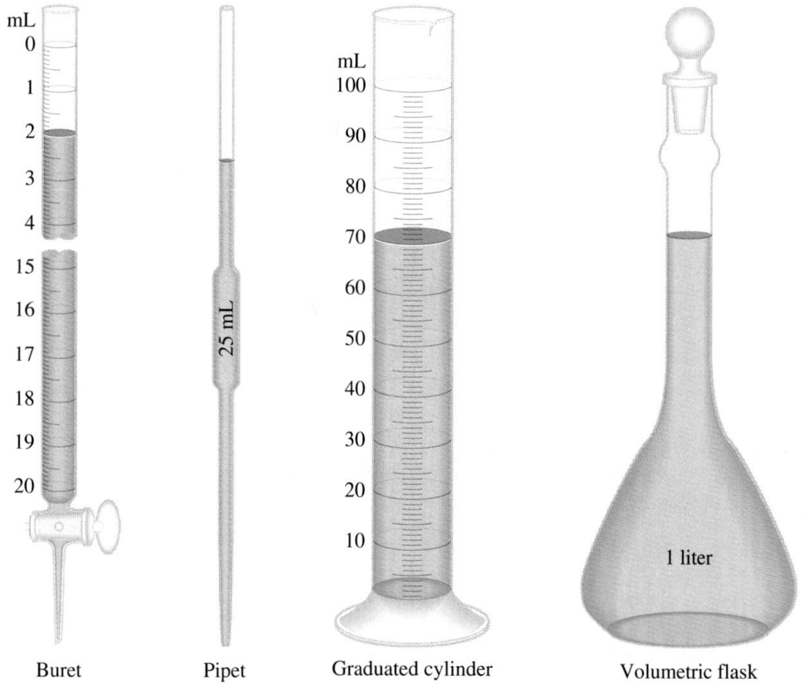

FIGURE 1.5

Common measuring devices found in a chemistry laboratory. These devices are not drawn to scale relative to one another. We will discuss the uses of these measuring devices in Chapter 4.

sure volume (Figure 1.5); the balance measures mass; the thermometer measures temperature. These instruments provide measurements of ***macroscopic properties,*** which *can be determined directly.* ***Microscopic properties,*** *on the atomic or molecular scale, must be determined by an indirect method,* as we will see in the next chapter.

A measured quantity is usually written as a number with an appropriate unit. To say that the distance between New York and San Francisco by car along a certain route is 5166 is meaningless. We must specify that the distance is 5166 kilometers. In science, units are essential to stating measurements correctly.

SI Units

For many years scientists recorded measurements in *metric units,* which are related decimally, that is, by powers of 10. In 1960, however, the General Conference of Weights and Measures, the international authority on units, proposed a revised metric system called the ***International System of Units*** (abbreviated *SI,* from the French *S*ystem *I*nternational d'Unites). Table 1.2 shows the seven SI base units. All other SI units of measurement can be derived from these base units. Like metric units, SI units are modified in decimal fashion by a series of prefixes, as shown in Table 1.3. We will use both metric and SI units in this book.

Measurements that we will utilize frequently in our study of chemistry include time, mass, volume, density, and temperature.

Mass and Weight

Mass is *a measure of the quantity of matter in an object.* The terms "mass" and "weight" are often used interchangeably, although, strictly speaking, they refer to

TABLE 1.2
SI Base Units

Base quantity	Name of unit	Symbol
Length	meter	m
Mass	kilogram	kg
Time	second	s
Electrical current	ampere	A
Temperature	kelvin	K
Amount of substance	mole	mol
Luminous intensity	candela	cd

TABLE 1.3
Prefixes Used with SI and Metric Units

Prefix	Symbol	Meaning	Example
Tera-	T	1,000,000,000,000, or 10^{12}	1 terameter (Tm) = 1×10^{12} m
Giga-	G	1,000,000,000, or 10^9	1 gigameter (Gm) = 1×10^9 m
Mega-	M	1,000,000, or 10^6	1 megameter (Mm) = 1×10^6 m
Kilo-	k	1,000, or 10^3	1 kilometer (km) = 1×10^3 m
Deci-	d	1/10, or 10^{-1}	1 decimeter (dm) = 0.1 m
Centi-	c	1/100, or 10^{-2}	1 centimeter (cm) = 0.01 m
Milli-	m	1/1,000, or 10^{-3}	1 millimeter (mm) = 0.001 m
Micro-	μ	1/1,000,000, or 10^{-6}	1 micrometer (μm) = 1×10^{-6} m
Nano-	n	1/1,000,000,000, or 10^{-9}	1 nanometer (nm) = 1×10^{-9} m
Pico-	p	1/1,000,000,000,000, or 10^{-12}	1 picometer (pm) = 1×10^{-12} m

An astronaut on the surface of the moon.

different quantities. In scientific terms, ***weight*** is *the force that gravity exerts on an object.* An apple that falls from a tree is pulled downward by Earth's gravity. The mass of the apple is constant and does not depend on its location, but its weight does. For example, on the surface of the moon the apple would weigh only one-sixth what it does on Earth, because the moon's gravity is only one-sixth that of Earth. This is why astronauts are able to jump about rather freely on the moon's surface despite their bulky suits and equipment. The mass of an object can be determined readily with a balance, and this process, oddly, is called weighing.

The SI base unit of mass is the *kilogram* (kg), but in chemistry the smaller gram (g) is more convenient:

$$1 \text{ kg} = 1000 \text{ g} = 1 \times 10^3 \text{ g}$$

VOLUME

Volume is *length (m) cubed,* so its SI-derived unit is the cubic meter (m^3). Generally, however, chemists work with much smaller volumes, such as the cubic centimeter (cm^3) and the cubic decimeter (dm^3):

$$1 \text{ cm}^3 = (1 \times 10^{-2} \text{ m})^3 = 1 \times 10^{-6} \text{ m}^3$$

$$1 \text{ dm}^3 = (1 \times 10^{-1} \text{ m})^3 = 1 \times 10^{-3} \text{ m}^3$$

Another common unit of volume is the liter (L). A ***liter*** is *the volume occupied* *by one cubic decimeter.* The liter is a metric unit, but chemists generally use L and mL for liquid volume. One liter is equal to 1000 milliliters (mL) or 1000 cubic centimeters:

$$\begin{aligned} 1\ \text{L} &= 1000\ \text{mL} \\ &= 1000\ \text{cm}^3 \\ &= 1\ \text{dm}^3 \end{aligned}$$

and one milliliter is equal to one cubic centimeter:

$$1\ \text{mL} = 1\ \text{cm}^3$$

Figure 1.6 compares the relative sizes of two volumes.

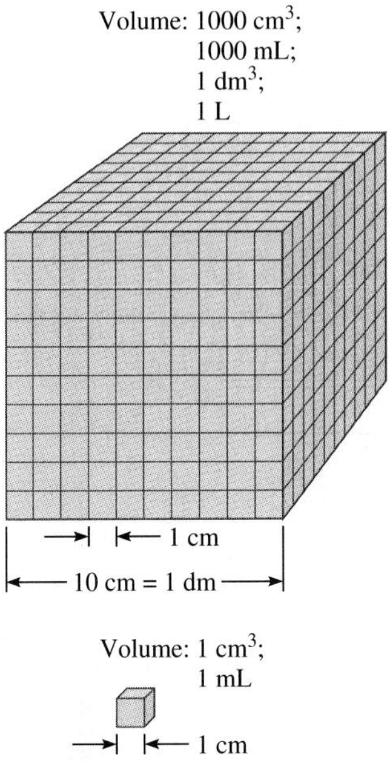

FIGURE 1.6

Comparison of two volumes, 1 mL and 1000 mL.

DENSITY

Density is *the mass of an object divided by its volume:*

$$\text{density} = \frac{\text{mass}}{\text{volume}}$$

or

$$d = \frac{m}{V}$$

where d, m, and V denote density, mass, and volume, respectively. Note that density is an intensive property that does not depend on the quantity of mass present. The reason is that V increases as m does, so the ratio of the two quantities always remains the same for a given material.

The SI-derived unit for density is the kilogram per cubic meter (kg/m^3). This unit is awkwardly large for most chemical applications. Therefore, grams per cubic centimeter (g/cm^3) and its equivalent, grams per milliliter (g/mL), are more commonly used for solid and liquid densities. Because gas densities are often very low, we express them in units of grams per liter (g/L):

$$1\ \text{g/cm}^3 = 1\ \text{g/mL} = 1000\ \text{kg/m}^3$$

$$1\ \text{g/L} = 0.001\ \text{g/mL}$$

EXAMPLE 1.1

Calculating Density

Gold is a precious metal that is chemically unreactive. It is used mainly in jewelry, dentistry, and electronic devices. A piece of gold ingot with a mass of 301 g has a volume of 15.6 cm^3. Calculate the density of gold.

Answer: The density of the gold metal is given by

$$\begin{aligned} d &= \frac{m}{V} \\ &= \frac{301\ \text{g}}{15.6\ \text{cm}^3} \\ &= 19.3\ \text{g/cm}^3 \end{aligned}$$

Gold bars.

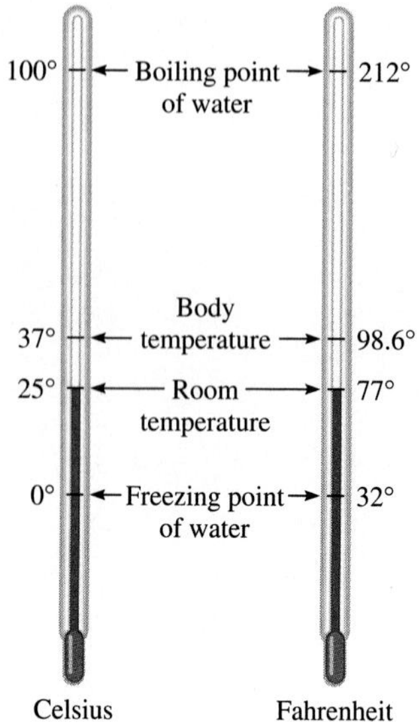

FIGURE 1.7

Comparison of the Celsius and Fahrenheit temperature scales.

PRACTICE EXERCISE*

A piece of platinum metal with a density of 21.5 g/cm^3 has a volume of 4.49 cm^3. What is its mass?

TEMPERATURE

Three temperature scales are currently in use. Their units are K (kelvin), °C (degree Celsius), and °F (degree Fahrenheit). The Fahrenheit scale, which is the most commonly used scale in the United States outside the laboratory, defines the normal freezing and boiling points of water to be exactly 32°F and 212°F, respectively. The Celsius scale divides the range between the freezing point (0°C) and boiling point (100°C) of water into 100 degrees (Figure 1.7). The degree Celsius is not an SI unit, but it may be used with SI units, and we will use it often.

The size of a degree on the Fahrenheit scale is only 100/180, or 5/9, times the degree Celsius. To convert degrees Fahrenheit to degrees Celsius, we write

$$?^\circ C = (^\circ F - 32^\circ F) \times \frac{5^\circ C}{9^\circ F}$$

To convert degrees Celsius to degrees Fahrenheit, we write

$$?^\circ F = \frac{9^\circ F}{5^\circ C} \times (^\circ C) + 32^\circ F$$

The SI unit of temperature is K (kelvin)—never "degrees kelvin." The Kelvin temperature scale is discussed in Chapter 5.

EXAMPLE 1.2

Converting between Degrees Celsius and Degrees Fahrenheit

(a) Solder is an alloy made of tin and lead that is used in electronic circuits. A certain solder has a melting point of 224°C. What is its melting point in degrees Fahrenheit? (b) Helium has the lowest boiling point of all the elements at −452°F. Convert this temperature to degrees Celsius.

Answer: (a) This conversion is carried out by writing

$$\frac{9^\circ F}{5^\circ C} \times (224^\circ C) + 32^\circ F = 435^\circ F$$

(b) Here we have

$$(-452^\circ F - 32^\circ F) \times \frac{5^\circ C}{9^\circ F} = -269^\circ C$$

PRACTICE EXERCISE

(a) Cesium has a low melting point of 28.4°C. What is the temperature in degrees Fahrenheit? (b) Convert 172.9°F (the boiling point of ethanol) to degrees Celsius.

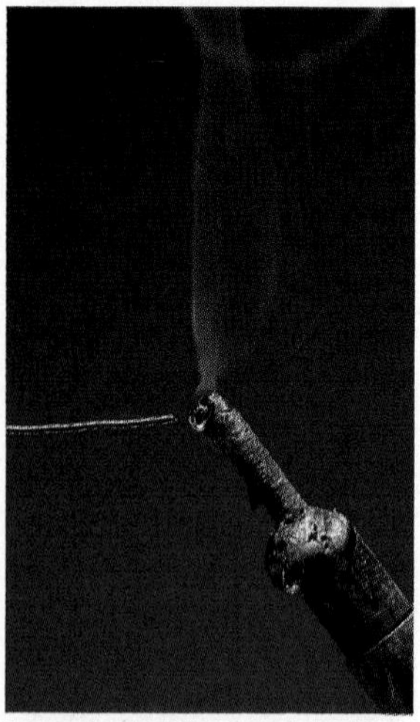

Solder is used extensively in the construction of electronic circuits.

*Answers to Practice Exercises appear at the end of the chapter.

1.5 HANDLING NUMBERS

Having surveyed some of the units used in chemistry, we now turn to conventions for handling numbers associated with measurements: scientific notation and significant figures.

SCIENTIFIC NOTATION

In chemistry, we often deal with numbers that are either extremely large or extremely small. For example, in 1 g of the element hydrogen there are roughly

$$602,200,000,000,000,000,000,000$$

hydrogen atoms. Each hydrogen atom has a mass of only

$$0.00000000000000000000000166 \text{ g}$$

These numbers are cumbersome to handle, and it is easy to make mistakes when using them in arithmetic computations. Consider the following multiplication problem:

$$0.0000000056 \times 0.00000000048 = 0.000000000000000002688$$

It would be easy to miss one zero or add one more zero after the decimal point. To handle very large and very small numbers, we use a system called *scientific notation.* Regardless of their magnitude, all numbers can be expressed in the form

$$N \times 10^n$$

where N is a number between 1 and 10 and n is an *exponent* that can be a positive or negative *integer* (whole number). Any number expressed in this way is said to be written in scientific notation.

Suppose that we are given a certain number and asked to express it in scientific notation. Basically, this amounts to finding n. We count the number of places that the decimal point must be moved to give the number N (which is between 1 and 10). If the decimal point has to be moved to the left, then n is a positive integer; if it has to be moved to the right, n is a negative integer. The following examples illustrate the use of scientific notation:

(a) Express 568.762 in scientific notation:

$$568.762 = 5.68762 \times 10^2$$

Note that the decimal point is moved to the left by two places and $n = 2$.

(b) Express 0.00000772 in scientific notation:

$$0.00000772 = 7.72 \times 10^{-6}$$

Note that the decimal point is moved to the right by six places and $n = -6$.

Next we will consider how scientific notation is handled in arithmetic operations.

Addition and Subtraction

To add or subtract using scientific notation, we first write each quantity with the same exponent n. Then we add or subtract the N parts of the numbers; the exponent parts remain the same. Consider the following examples:

$$(7.4 \times 10^3) + (2.1 \times 10^3) = 9.5 \times 10^3$$

$$\begin{aligned}(4.31 \times 10^4) + (3.9 \times 10^3) &= (4.31 \times 10^4) + (0.39 \times 10^4)\\ &= 4.70 \times 10^4\end{aligned}$$

$$\begin{aligned}(2.22 \times 10^{-2}) - (4.10 \times 10^{-3}) &= (2.22 \times 10^{-2}) - (0.41 \times 10^{-2})\\ &= 1.81 \times 10^{-2}\end{aligned}$$

Multiplication and Division

To multiply numbers expressed in scientific notation, we multiply the N parts of the numbers in the usual way, but *add* the exponents n together. To divide using scientific notation, we divide the N parts of the numbers as usual and *subtract* the exponents n. The following examples show how these operations are performed:

$$\begin{aligned}(8.0 \times 10^4) \times (5.0 \times 10^2) &= (8.0 \times 5.0)(10^{4+2})\\ &= 40 \times 10^6\\ &= 4.0 \times 10^7\end{aligned}$$

$$\begin{aligned}(4.0 \times 10^{-5}) \times (7.0 \times 10^3) &= (4.0 \times 7.0)(10^{-5+3})\\ &= 28 \times 10^{-2}\\ &= 2.8 \times 10^{-1}\end{aligned}$$

$$\begin{aligned}\frac{8.5 \times 10^4}{5.0 \times 10^9} &= \frac{8.5}{5.0} \times 10^{4-9}\\ &= 1.7 \times 10^{-5}\end{aligned}$$

$$\begin{aligned}\frac{6.9 \times 10^7}{3.0 \times 10^{-5}} &= \frac{6.9}{3.0} \times 10^{7-(-5)}\\ &= 2.3 \times 10^{12}\end{aligned}$$

SIGNIFICANT FIGURES

Except when all the numbers involved are integers (for example, in counting the number of students in a class), it is often impossible to obtain the exact value of the quantity under investigation. For this reason, it is important to indicate the margin of error in a measurement by clearly indicating the number of ***significant figures,*** which are *the meaningful digits in a measured or calculated quantity.* When significant figures are counted, the last digit is understood to be uncertain. For example, we might measure the volume of a given amount of liquid using a graduated cylinder (see Figure 1.5) with a scale that gives an uncertainty of 1 mL in the measurement. If the volume is found to be 6 mL, then the actual volume is in the range of 5 mL to 7 mL. We represent the volume of the liquid as (6 ± 1) mL. In this case, there is only one significant figure (the digit 6) that is uncertain by either plus or minus 1 mL. For greater accuracy, we might use a graduated cylinder that has finer divisions, so that the volume we measure is now uncertain by only 0.1 mL. If the volume of the liquid is now found to be 6.0 mL, we may express the quantity as (6.0 ± 0.1) mL, and the actual value is somewhere between 5.9 mL and 6.1 mL. We can further improve the measuring device and obtain more significant figures, but in each case, the last digit is always uncertain. The amount of this uncertainty depends on the particular measuring device we use.

Figure 1.8 shows a modern balance like the ones available in many general

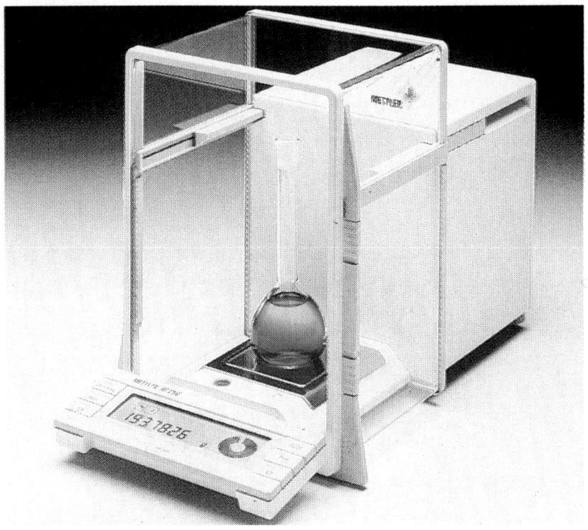

FIGURE 1.8

A single-pan balance.

chemistry laboratories. This balance readily measures the mass of objects to four decimal places. This means that the measured mass will have four significant figures (for example, 0.8642 g) or more (for example, 3.9745 g). Keeping track of the number of significant figures in a measurement means that calculations involving the data will reflect the precision of the measurement.

Guidelines for Using Significant Figures

The previous discussion shows that we must always be careful to write the proper number of significant figures. In general, it is fairly easy to determine how many significant figures are present in a number by following these rules:

- Any digit that is not zero is significant. Thus 845 cm has three significant figures, 1.234 kg has four significant figures, and so on.
- Zeros between nonzero digits are significant. Thus 606 m contains three significant figures, 40,501 kg contains five significant figures, and so on.
- Zeros to the left of the first nonzero digit are not significant. Their purpose is to indicate the placement of the decimal point. Thus 0.08 L contains one significant figure, 0.0000349 g contains three significant figures, and so on.
- If a number is greater than 1, then all the zeros written to the right of the decimal point count as significant figures. Thus 2.0 mg has two significant figures, 40.062 mL has five significant figures, and 3.040 dm has four significant figures. If a number is less than 1, then only the zeros that are at the end of the number and the zeros that are between nonzero digits are significant. Thus 0.090 kg has two significant figures, 0.3005 L has four significant figures, 0.00420 min has three significant figures, and so on.
- For numbers that do not contain decimal points, the trailing zeros (that is, zeros after the last nonzero digit) may or may not be significant. Thus 400 cm may have one significant figure (the digit 4), two significant figures (40), or three significant figures (400). We cannot know which is correct without more information. By using scientific notation, however, we avoid this ambiguity. In this particular case, we can express the number 400 as 4×10^2 for one significant figure, 4.0×10^2 for two significant figures, or 4.00×10^2 for three significant figures.

EXAMPLE 1.3

Determining Significant Figures

Determine the number of significant figures in the following measured quantities: (a) 478 cm, (b) 6.01 g, (c) 0.825 m, (d) 0.043 kg, (e) 1.310×10^{22} atoms, (f) 7000 mL.

Answer: (a) Three, (b) three, (c) three, (d) two, (e) four, (f) four possible answers. The number of significant figures in (f) may be four (7.000×10^3), three (7.00×10^3), two (7.0×10^3), or one (7×10^3).

PRACTICE EXERCISE

Determine the number of significant figures in each of the following measurements: (a) 24 mL, (b) 3001 g, (c) 0.0320 m^3, (d) 6.4×10^4 molecules, (e) 560 kg.

A second set of rules specifies how to handle significant figures in calculations:

- In addition and subtraction, the number of significant figures to the right of the decimal point in the final sum or difference is determined by the lowest number of significant figures to the right of the decimal point in any of the original numbers. Consider these examples:

$$\begin{array}{r l} 89.332 & \\ +\ 1.1 & \longleftarrow \text{one significant figure after the decimal point} \\ \hline 90.432 & \longleftarrow \text{round off to 90.4} \end{array}$$

$$\begin{array}{r l} 2.097 & \\ -0.12 & \longleftarrow \text{two significant figures after the decimal point} \\ \hline 1.977 & \longleftarrow \text{round off to 1.98} \end{array}$$

  The rounding-off procedure is as follows. To round off a number at a certain point we simply drop the digits that follow if the first of them is less than 5. Thus 8.724 rounds off to 8.72 if we want only two figures after the decimal point. If the first digit following the point of rounding off is equal to or greater than 5, we add 1 to the preceding digit. Thus 8.727 rounds off to 8.73, and 0.425 rounds off to 0.43.

- In multiplication and division, the number of significant figures in the final product or quotient is determined by the original number that has the smallest number of significant figures. The following examples illustrate this rule:

$$2.8 \times 4.5039 = 12.61092 \longleftarrow \text{round off to 13}$$

$$\frac{6.85}{112.04} = 0.0611388789 \longleftarrow \text{round off to 0.0611}$$

- Keep in mind that *exact numbers* obtained from definitions or by counting numbers of objects can be considered to have an infinite number of significant figures. If an object has the mass 0.2786 g, then the mass of eight such objects is

$$0.2786\ \text{g} \times 8 = 2.229\ \text{g}$$

We do *not* round off this product to one significant figure, because the number 8 is actually 8.00000 . . . , by definition. Similarly, to take the average of the two measured lengths 6.64 cm and 6.68 cm, we write

$$\frac{6.64 \text{ cm} + 6.68 \text{ cm}}{2} = 6.66 \text{ cm}$$

because the number 2 is actually 2.00000 . . . , by definition.

EXAMPLE 1.4

Handling Significant Figures in Calculations

Carry out the following arithmetic operations: (a) 11,254.1 + 0.1983, (b) 66.59 − 3.113, (c) 8.16 × 5.1355, (d) 0.0154 ÷ 883, (e) $2.64 \times 10^3 + 3.27 \times 10^2$.

Answer:

(a)
$$\begin{array}{r} 11{,}254.1 \\ +\quad 0.1983 \\ \hline 11{,}254.2983 \end{array} \longleftarrow \text{round off to } 11{,}254.3$$

(b)
$$\begin{array}{r} 66.59 \\ -\ 3.113 \\ \hline 63.477 \end{array} \longleftarrow \text{round off to } 63.48$$

(c) $8.16 \times 5.1355 = 41.90568 \longleftarrow$ round off to 41.9

(d) $\dfrac{0.0154}{883} = 0.0000174405436 \longleftarrow$ round off to 0.0000174, or 1.74×10^{-5}

(e) First we change 3.27×10^2 to 0.327×10^3 and then carry out the addition $(2.64 + 0.327) \times 10^3$. Following the procedure in (a), we find the answer is 2.97×10^3.

PRACTICE EXERCISE

Carry out the following arithmetic operations: (a) 26.5862 + 0.17, (b) 9.1 − 4.682, (c) $7.1 \times 10^4 \times 2.2654 \times 10^2$, (d) 6.54 ÷ 86.5542, (e) $(7.55 \times 10^4) - (8.62 \times 10^3)$.

The above rounding-off procedure applies to one-step calculations. In *chain calculations,* that is, calculations involving more than one step, we use a modified procedure. Consider the following two-step calculation:

first step: $A \times B = C$
second step: $C \times D = E$

Let us suppose that $A = 3.66$, $B = 8.45$, and $D = 2.11$. Depending on whether we round off C to three or four significant figures, we obtain a different number for E:

Method 1	*Method 2*
3.66 × 8.45 = 30.9	3.66 × 8.45 = 30.93
30.9 × 2.11 = 65.2	30.93 × 2.11 = 65.3

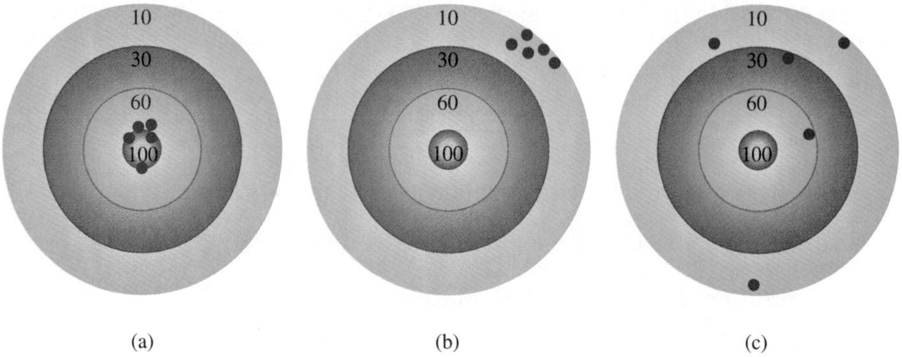

FIGURE 1.9

The distribution of darts on a dart board shows the difference between precise and accurate. (a) Good accuracy and good precision. (b) Poor accuracy and good precision. (c) Poor accuracy and poor precision.

However, if we had carried out the calculation as $3.66 \times 8.45 \times 2.11$ on a calculator without rounding off the intermediate result, we would have obtained 65.3 as the answer for E. The procedure of carrying the answers for all the intermediate calculations to *one more significant figure* and only rounding off the final answer to the correct number of significant figures will be used in some of the worked examples in this book. In general, we will show the correct number of significant figures in each step of the calculation.

Accuracy and Precision

In discussing measurements and significant figures it is useful to distinguish between accuracy and precision. ***Accuracy** tells us how close a measurement is to the true value of the quantity that was measured.* ***Precision** refers to how closely two or more measurements of the same quantity agree with one another* (Figure 1.9). Suppose that three students are asked to determine the mass of a piece of copper wire that their instructor knows has a mass of 2.000 g. The results of two successive weighings by each student are

	Student A	*Student B*	*Student C*
	1.964 g	1.972 g	2.000 g
	1.978 g	1.968 g	2.002 g
Average value	1.971 g	1.970 g	2.001 g

Student B's results are more *precise* than those of Student A (1.972 g and 1.968 g deviate less from 1.970 g than 1.964 g and 1.978 g from 1.971 g). Neither set of results is very *accurate,* however. Student C's results are not only *precise* but also the most *accurate,* since the average value is closest to the true value. Highly accurate measurements are usually precise too. On the other hand, highly precise measurements do not necessarily guarantee accurate results. For example, an improperly calibrated meter stick or a faulty balance may give precise readings that are in error.

1.6 THE FACTOR-LABEL METHOD OF SOLVING PROBLEMS

Careful measurements and the proper use of significant figures, along with correct calculations, will yield accurate numerical results. But to be meaningful, the answers also must be expressed in the desired units. The procedure we will use to convert between units in solving chemistry problems is called the *factor-label method* (also called *dimensional analysis*). A simple technique requiring little memorization, the factor-label method is based on the relationship between different units that express the same physical quantity.

We know, for example, that the unit "dollar" for money is different from the unit "penny." However, we say that 1 dollar is *equivalent* to 100 pennies because they both represent the same amount of money. This equivalence allows us to write

$$1 \text{ dollar} = 100 \text{ pennies}$$

Because 1 dollar is equal to 100 pennies, it follows that their ratio has a value of 1; that is,

$$\frac{1 \text{ dollar}}{100 \text{ pennies}} = 1$$

This ratio can be read as 1 dollar per 100 pennies. This fraction is called a *unit factor* (equal to 1) because the numerator and denominator describe the same amount of money.

We could also have written the ratio as

$$\frac{100 \text{ pennies}}{1 \text{ dollar}} = 1$$

This ratio reads as 100 pennies per dollar. The fraction 100 pennies/1 dollar is also a unit factor. We see that the reciprocal of any unit factor is also a unit factor. The usefulness of unit factors is that they allow us to carry out conversions between different units that measure the same quantity. Suppose that we want to convert 2.46 dollars into pennies. This problem may be expressed as

$$? \text{ pennies} = 2.46 \text{ dollars}$$

Since this is a dollar-to-penny conversion, we choose the unit factor that has the unit "dollar" in the denominator (to cancel the "dollars" in 2.46 dollars) and write

$$2.46 \cancel{\text{dollars}} \times \frac{100 \text{ pennies}}{1 \cancel{\text{dollar}}} = 246 \text{ pennies}$$

Note that the unit factor 100 pennies/1 dollar contains exact numbers, so it does not affect the number of significant figures in the final answer.

Next let us consider the conversion of 57.8 meters to centimeters. This problem may be expressed as

$$? \text{ cm} = 57.8 \text{ m}$$

By definition,

$$1 \text{ cm} = 1 \times 10^{-2} \text{ m}$$

Since we are converting "m" to "cm," we choose the unit factor that has meters in the denominator,

$$\frac{1 \text{ cm}}{1 \times 10^{-2} \text{ m}} = 1$$

and write the conversion as

$$? \text{ cm} = 57.8 \cancel{\text{m}} \times \frac{1 \text{ cm}}{1 \times 10^{-2} \cancel{\text{m}}}$$

$$= 5780 \text{ cm}$$
$$= 5.78 \times 10^3 \text{ cm}$$

Note that scientific notation is used to indicate that the answer has three significant figures. The unit factor 1 cm/1 $\times$ 10^{-2} m contains exact numbers; therefore, it does not affect the number of significant figures.

In the factor-label method the units are carried through the entire sequence of calculations. Therefore, if the equation is set up correctly, then all the units will cancel except the desired one. If this is not the case, then an error must have been made somewhere, and it can usually be spotted by reviewing the solution.

Conversion factors for some of the English system units commonly used in the United States for nonscientific measurements (for example, pounds and inches) are provided inside the back cover of this book.

EXAMPLE 1.5
Using the Factor-Label Method

A man weighs 162 pounds (lb). What is his mass in milligrams (mg)?

Answer: The problem can be expressed as

$$? \text{ mg} = 162 \text{ lb}$$

The conversion factors are

$$1 \text{ lb} = 453.6 \text{ g}$$

so the unit factor is

$$\frac{453.6 \text{ g}}{1 \text{ lb}} = 1$$

and

$$1 \text{ mg} = 1 \times 10^{-3} \text{ g}$$

so the unit factor is

$$\frac{1 \text{ mg}}{1 \times 10^{-3} \text{ g}} = 1$$

Thus

$$? \text{ mg} = 162 \cancel{\text{lb}} \times \frac{453.6 \cancel{\text{g}}}{1 \cancel{\text{lb}}} \times \frac{1 \text{ mg}}{1 \times 10^{-3} \cancel{\text{g}}} = 7.35 \times 10^7 \text{ mg}$$

PRACTICE EXERCISE

Convert 53.5 kilograms (kg) to pounds.

Note that unit factors may be squared or cubed, because $1^2 = 1^3 = 1$. The use of such factors is illustrated in Example 1.6.

EXAMPLE 1.6
Using the Factor-Label Method

The density of gold is 19.3 g/cm^3. Convert the density units to kg/m^3.

Answer: The problem can be stated as

$$? \text{ kg/m}^3 = 19.3 \text{ g/cm}^3$$

We need two unit factors—one to convert g to kg and the other to convert cm^3 to m^3. We know that

$$1 \text{ kg} = 1000 \text{ g}$$

so

$$\frac{1 \text{ kg}}{1000 \text{ g}} = 1$$

The second unit factor is

$$\left(\frac{1 \text{ cm}}{1 \times 10^{-2} \text{ m}}\right)^3 = 1$$

Thus we write

$$? \text{ kg/m}^3 = \frac{19.3 \text{ g}}{1 \text{ cm}^3} \times \frac{1 \text{ kg}}{1000 \text{ g}} \times \left(\frac{1 \text{ cm}}{1 \times 10^{-2} \text{ m}}\right)^3 = 19{,}300 \text{ kg/m}^3$$

$$= 1.93 \times 10^4 \text{ kg/m}^3$$

PRACTICE EXERCISE

The density of the lightest metal, lithium (Li), is 5.34×10^2 kg/m^3. Convert the density to g/cm^3.

SUMMARY

Chemists study matter and the substances of which it is composed. All substances, in principle, can exist in three states: solid, liquid, and gas. The interconversion between these states can be effected by a change in temperature.

The simplest substances in chemistry are elements. Compounds are formed by the combination of atoms of different elements. Substances have both unique physical properties that can be observed without changing the identity of the substances and unique chemical properties that, when they are demonstrated, do change the identity of the substances.

SI units are used to express physical quantities in all sciences, including chemistry. Numbers expressed in scientific notation have the form $N \times 10^n$, where N is between 1 and 10 and n is a positive or negative integer. Scientific notation helps us handle very large and very small quantities. Most measured quantities are inexact to some extent. The number of significant figures indicates the exactness of the measurement.

In the factor-label method of solving problems the units are multiplied together, divided into each other, or canceled like algebraic quantities. Obtaining the correct units for the final answer ensures that the calculation has been carried out properly.

KEY WORDS

Accuracy, p. 18
Chemical property, p. 8
Chemistry, p. 4
Compound, p. 6
Density, p. 11
Element, p. 6
Extensive property, p. 8
Heterogeneous mixture, p. 6
Homogeneous mixture, p. 5
Intensive property, p. 8
International System of Units (SI), p. 9
Liter, p. 10
Macroscopic property, p. 9
Mass, p. 9
Matter, p. 4
Microscopic property, p. 9
Mixture, p. 5
Physical property, p. 8
Precision, p. 18
Significant figures, p. 14
Substance, p. 5
Volume, p. 10
Weight, p. 10

QUESTIONS AND PROBLEMS

BASIC DEFINITIONS

Review Questions

1.1 Define the following terms: (a) matter, (b) mass, (c) weight, (d) substance, (e) mixture.

1.2 Which of the following statements is scientifically correct?
"The mass of the student is 56 kg."
"The weight of the student is 56 kg."

1.3 Give an example of a homogeneous mixture and an example of a heterogeneous mixture.

1.4 What is the difference between a physical property and a chemical property?

1.5 Give an example of an intensive property and an example of an extensive property.

1.6 Define the following terms: (a) element, (b) compound.

Problems

1.7 Do the following statements describe chemical or physical properties? (a) Oxygen gas supports combustion. (b) Fertilizers help to increase agricultural production. (c) Water boils below 100°C on top of a mountain. (d) Lead is denser than aluminum. (e) Sugar tastes sweet.

1.8 Does each of the following describe a physical change or a chemical change? (a) The helium gas inside a balloon tends to leak out after a few hours. (b) A flashlight beam slowly gets dimmer and finally goes out. (c) Frozen orange juice is reconstituted by adding water to it. (d) The growth of plants depends on the sun's energy in a process called photosynthesis. (e) A spoonful of table salt dissolves in a bowl of soup.

1.9 Which of the following properties are intensive and which are extensive? (a) length, (b) volume, (c) temperature, (d) mass.

1.10 Which of the following properties are intensive and which are extensive? (a) area, (b) color, (c) density.

1.11 Classify each of the following substances as an element or a compound: (a) hydrogen, (b) water, (c) gold, (d) sugar.

1.12 Classify each of the following as an element or a compound: (a) sodium chloride (table salt), (b) helium, (c) alcohol, (d) platinum.

UNITS

Review Questions

1.13 Give the SI units for expressing the following: (a) length, (b) area, (c) volume, (d) mass, (e) time, (f) force, (g) energy, (h) temperature.

1.14 Write the numbers for the following prefixes: (a) mega-, (b) kilo-, (c) deci-, (d) centi-, (e) milli-, (f) micro-, (g) nano-, (h) pico-.

1.15 Define density. What units do chemists normally use for density? Is density an intensive or extensive property?

1.16 Write the equations for converting degrees Celsius to degrees Fahrenheit and degrees Fahrenheit to degrees Celsius.

Problems

1.17 A lead sphere has a mass of 1.20×10^4 g, and its volume is 1.05×10^3 cm^3. Calculate the density of lead.

1.18 Mercury is the only metal that is a liquid at room temperature. Its density is 13.6 g/mL. How many grams of mercury will occupy a volume of 95.8 mL?

1.19 Calculate the temperature in degrees Celsius of the following: (a) a hot summer day of 95°F, (b) a cold winter day of 12°F, (c) a fever of 102°F, (d) a furnace operating at 1852°F.

1.20 (a) Normally the human body can endure a temperature of 105°F for only short periods of time without permanent damage to the brain and other vital organs. What is the temperature in degrees Celsius? (b) Ethylene glycol is a liquid organic compound that is used as an antifreeze in car radiators. It freezes at −11.5°C. Calculate its freezing temperature in degrees Fahrenheit. (c) The temperature on the surface of our sun is about 6.3×10^3°C. What is this temperature in degrees Fahrenheit?

SCIENTIFIC NOTATION

Problems

1.21 Express the following numbers in scientific notation: (a) 0.000000027, (b) 356, (c) 0.096.

1.22 Express the following numbers in scientific notation: (a) 0.749, (b) 802.6, (c) 0.000000621.

1.23 Convert the following to nonscientific notation: (a) 1.52×10^4, (b) 7.78×10^{-8}.

1.24 Convert the following to nonscientific notation: (a) 3.256×10^{-5}, (b) 6.03×10^6.

1.25 Express the answers to the following in scientific notation:
(a) $145.75 + (2.3 \times 10^{-1})$
(b) $79{,}500 \div (2.5 \times 10^2)$
(c) $(7.0 \times 10^{-3}) - (8.0 \times 10^{-4})$
(d) $(1.0 \times 10^4) \times (9.9 \times 10^6)$

1.26 Express the answers to the following in scientific notation:
(a) $0.0095 + (8.5 \times 10^{-3})$
(b) $653 \div (5.75 \times 10^{-8})$
(c) $850{,}000 - (9.0 \times 10^5)$
(d) $(3.6 \times 10^{-4}) \times (3.6 \times 10^6)$

SIGNIFICANT FIGURES

Problems

1.27 What is the number of significant figures in each of the following measured quantities? (a) 4867 mi, (b) 56 mL, (c) 60,104 tons, (d) 2900 g.

1.28 What is the number of significant figures in each of the following measured quantities? (a) 40.2 g/cm^3, (b) 0.0000003 cm, (c) 70 min, (d) 4.6×10^{19} atoms.

1.29 Carry out the following operations as if they were calculations of experimental results, and express each answer in the correct units and with the correct number of significant figures:
(a) 5.6792 m + 0.6 m + 4.33 m
(b) 3.70 g − 2.9133 g
(c) 4.51 cm × 3.6666 cm

1.30 Carry out the following operations as if they were calculations of experimental results, and express each answer in the correct units and with the correct number of significant figures:
(a) 7.310 km ÷ 5.70 km
(b) $(3.26 \times 10^{-3}$ mg$) - (7.88 \times 10^{-5}$ mg$)$
(c) $(4.02 \times 10^6$ dm$) + (7.74 \times 10^7$ dm$)$

THE FACTOR-LABEL METHOD

Problems

1.31 Carry out the following conversions: (a) 22.6 m to decimeters, (b) 25.4 mg to kilograms.

1.32 Carry out the following conversions: (a) 242 lb to milligrams, (b) 68.3 cm^3 to cubic meters.

1.33 The price of gold on a certain day in 1994 was $327 per ounce. How much did 1.00 g of gold cost that day? (1 ounce = 28.4 g.)

1.34 How many seconds are in a solar year (365.24 days)?

1.35 How many minutes does it take light from the sun to reach Earth? (The distance from the sun to Earth is 93 million mi; the speed of light = 3.00×10^8 m/s.)

1.36 A slow jogger runs a mile in 13 min. Calculate the speed in (a) in/s, (b) m/min, (c) km/h. (1 mi = 1609 m; 1 in = 2.54 cm.)

1.37 Carry out the following conversions: (a) A 6.0-ft person weighs 168 lb. Express this person's height in meters and weight in kilograms. (1 lb = 453.6 g; 1 m = 3.28 ft.) (b) The current speed limit in some states in the U.S. is 55 miles per hour. What is the speed limit in kilometers per hour? (c) The speed of light is 3.0×10^{10} cm/s. How many miles does light travel in 1 hour? (d) Lead is a toxic substance. The "normal" lead content in human blood is about 0.40 part per million (that is, 0.40 g of lead per million grams of blood). A value of 0.80 part per million (ppm) is considered to be dangerous. How many grams of lead are contained in 6.0×10^3 g of blood (the amount in an average adult) if the lead content is 0.62 ppm?

1.38 Carry out the following conversions: (a) 1.42 light-years to miles (a light-year is an astronomical measure of distance—the distance traveled by light in a year, or 365 days), (b) 32.4 yd to centimeters, (c) 3.0×10^{10} cm/s to ft/s, (d) 47.4°F to degrees Celsius, (e) −273.15°C (the lowest attainable temperature) to degrees Fahrenheit, (f) 71.2 cm^3 to m^3, (g) 7.2 m^3 to liters.

1.39 Aluminum is a lightweight metal (density = 2.70 g/cm^3) used in aircraft construction, high-voltage transmission lines, and foils. What is its density in kg/m^3?

1.40 The density of ammonia gas under certain conditions is 0.625 g/L. Calculate its density in g/cm^3.

MISCELLANEOUS PROBLEMS

1.41 Which of the following describe physical and which describe chemical properties? (a) Iron has a tendency to rust. (b) Rainwater in industrialized regions tends to be acidic. (c) Hemoglobin molecules have a red color. (d) When a glass of water is left out in the sun, the water gradually disappears. (c) Carbon dioxide in air is converted to more complex molecules by plants during photosynthesis.

1.42 In 1994 86.6 billion pounds of sulfuric acid were produced in the U.S. Convert this quantity to tons.

1.43 Suppose that a new temperature scale has been devised on which the melting point of ethanol (−117.3°C) and the boiling point of ethanol (78.3°C) are taken as 0°S and 100°S, respectively, where S is the symbol for the new temperature scale. Derive an equation relating a reading on this scale to a reading on the Celsius scale. What would this thermometer read at 25°C?

1.44 In the determination of the density of a rectangular metal bar, a student made the following measurements: length, 8.53 cm; width, 2.4 cm; height, 1.0 cm; mass, 52.7064 g. Calculate the density of the metal to the correct number of significant figures.

1.45 Calculate the mass of each of the following: (a) a sphere of gold of radius 10.0 cm [the volume of a sphere of radius r is $V = (\frac{4}{3})\pi r^3$; the density of gold = 19.3 g/cm^3], (b) a cube of platinum of edge length 0.040 mm (the density of platinum = 21.4 g/cm^3), (c) 50.0 mL of ethanol (the density of ethanol = 0.798 g/mL).

1.46 A cylindrical glass tube 12.7 cm in length is filled with mercury. The mass of mercury needed to fill the tube is found to be 105.5 g. Calculate the inner diameter of the tube. (The density of mercury = 13.6 g/mL.)

1.47 The following procedure was carried out to determine the volume of a flask. The flask was weighed dry and then filled with water. If the masses of the empty flask and filled flask were 56.12 g and 87.39 g, respectively, and the density of water is 0.9976 g/cm^3, calculate the volume of the flask in cubic centimeters.

1.48 A silver (Ag) object weighing 194.3 g is placed in a graduated cylinder containing 242.0 mL of water. The volume of water now reads 260.5 mL. From these data calculate the density of silver.

1.49 The experiment described in Problem 1.48 is a crude but convenient way to determine the density of some solids. Describe a similar experiment that would allow you to measure the density of ice. Specifically, what would be the requirements for the liquid used in your experiment?

1.50 The speed of sound in air at room temperature is about 343 m/s. Calculate this speed in miles per hour (mph).

1.51 The medicinal thermometer commonly used in homes can be read to ±0.1°F, while those in the doctor's office may be accurate to ±0.1°C. In degrees Celsius, express the percent error expected from each of these thermometers in measuring a person's body temperature of 38.9°C.

1.52 A thermometer gives a reading of 24.2°C ± 0.1°C. Calculate the temperature in degrees Fahrenheit. What is the uncertainty?

1.53 Vanillin (used to flavor vanilla ice cream and other foods) is the substance whose aroma the human nose detects in the smallest amount. The threshold limit is 2.0×10^{-11} g per liter of air. If the current price of 50 g of vanillin is $112, determine the cost to supply enough vanillin so that the aroma could be detectable in a large aircraft hangar of volume 5.0×10^7 ft^3.

1.54 A resting adult requires about 240 mL of pure oxygen/min and breathes about 12 times every minute. If inhaled air contains 20 percent oxygen by volume and exhaled air 16 percent, what is the volume of air per breath? (Assume that the volume of inhaled air is equal to that of exhaled air.)

1.55 The total volume of seawater is 1.5×10^{21} L. Assume that seawater contains 3.1 percent sodium chloride by mass and that its density is 1.03 g/mL. Calculate the total mass of sodium chloride in kilograms and in tons. (1 ton = 2000 lb; 1 lb = 453.6 g.)

1.56 Magnesium (Mg) is a valuable metal used in alloys, in batteries, and in chemical synthesis. It is obtained mostly from seawater, which contains about 1.3 g of Mg for every kilogram of seawater. Calculate the volume of seawater (in liters) needed to extract 8.0×10^4 tons of Mg, which is roughly the annual production in the U.S. (Density of seawater = 1.03 g/mL.)

1.57 A student is given a crucible and asked to prove whether it is made of pure platinum. She first weighs the crucible in air and then weighs it suspended in water (density = 0.9986 g/cm^3). The readings are 860.2 g and 820.2 g, respectively. Given that the density of platinum is 21.45 g/cm^3, what should her conclusion be based on these measurements? (*Hint:* An object suspended in a fluid is buoyed up by the mass of the fluid displaced by the object. Neglect the buoyancy of air.)

1.58 At what temperature does the numerical reading on a Celsius thermometer equal that on a Fahrenheit thermometer?

1.59 The surface area and average depth of the Pacific Ocean are 1.8×10^8 km^2 and 3.9×10^3 m, respectively. Calculate the volume of water in the ocean in liters.

1.60 Percent error is often expressed as the absolute value of the difference between the true value and the experimental value, divided by the true value:

$$\text{Percent error} = \frac{|\text{true value} - \text{experimental value}|}{|\text{true value}|} \times 100\%$$

where the vertical lines indicate absolute value. Calculate the percent error for the following measurements: (a) The density of alcohol (ethanol) is found to be 0.802 g/mL. (True value: 0.798 g/mL.) (b) The mass of gold in an earring is analyzed to be 0.837 g. (True value: 0.864 g.)

Answers to Practice Exercises: 1.1 96.5 g; **1.2** (a) 83.1°F, (b) 78.3°C; **1.3** (a) Two, (b) four, (c) three, (d) two, (e) three or two; **1.4** (a) 26.76, (b) 4.4, (c) 1.6×10^7, (d) 0.0756, (e) 6.69×10^4; **1.5** 118 lb; **1.6** 0.534 g/cm^3.

CHAPTER 2

ATOMS, MOLECULES, AND IONS

◆ A trip to the dentist is, for some people, a traumatic experience. Even thinking about the prospect may make your palms sweat and your heart beat faster. Chances are, however, that you don't worry about having your teeth X-rayed. This benign procedure, in which the technician first covers your torso with a lead blanket, places a film in your mouth, then focuses a narrow beam of X ray on your jaw, and leaves the room to throw the switch, lasts just a couple of minutes. Your mouth is actually exposed to the radiation for a second or so, and the blanket shields most of the rest of your body from the rays. In short, the whole process is carefully designed to limit your exposure to harmful radiation so that its diagnostic benefits can be utilized.

A century ago, decades before X-ray technology made its way into the dentist's office, very little was known about X rays and other types of high-energy radiation. Among the scientists interested in this phenomenon was a young graduate student named Marie Sklodowska Curie. With Henri Becquerel as her mentor at the Sorbonne in Paris and her husband, Pierre Curie, as her coworker, Marie Curie chose as the subject of her doctoral research radioactive uranium ore, which gives off radiation similar to X rays. In 1898 the Curies isolated a new element, which they named polonium after Marie's native country of Poland. Four months later, they discovered radium, another radioactive element.

In their makeshift laboratory in a shed, Marie and Pierre processed tons of pitchblende (a uranium oxide ore) to obtain just 0.1 g of radium chloride. They kept a sample of the fluorescent blue substance at their bedside, and they loved to visit the shed at night to look at the luminous test tubes

Marie Curie in her laboratory.

glowing like fairy lights. The unusual properties of radium made it the most important element to be discovered since oxygen. Marie Curie's thesis, which was based on the radium research, was hailed by many scientists as "the greatest contribution to science ever made by a doctoral student." Despite the potential medical applications of radium, the Curies refused to apply for patents on the process for extracting radium from uranium ore, because they believed that research should be pursued for its own sake, not for material rewards.

Along with Becquerel (the discoverer of radioactivity), the Curies were awarded the Nobel Prize in physics in 1903—the same year in which Marie received her doctoral degree. By then, Marie Curie was a household name, but the world's most famous scientist nevertheless encountered hostility and discrimination in the scientific community: She was denied admission to the French Academy of Sciences by one vote because she was a woman!

Marie Curie was the first woman to win a Nobel Prize, the first person to win two Nobel Prizes (the second in chemistry in 1911), and the first Nobel Prize winner to have a child go on to win a Nobel Prize. Her daughter, Iréne, and son-in-law, Frédéric Joliot, received the Nobel Prize in chemistry in 1935 for creating artificial radioactivity. Unfortunately, she did not live long enough to witness the occasion. After years of handling radioactive materials without protection, Marie died of leukemia in 1934. Her laboratory notebooks have been preserved, but they are unavailable for examination because they are still dangerously radioactive. Today almost everyone knows the value of safeguards against overexposure to radiation, be it from X rays or nuclear waste, largely because of Marie Curie. ◆

Photograph of 2.7 g radium bromide ($RaBr_2$), taken by its own light in 1922.

2.1 THE ATOMIC THEORY

In the fifth century B.C. the Greek philosopher Democritus expressed the belief that all matter consists of very small, indivisible particles, which he named *atomos* (meaning uncuttable or indivisible). Although Democritus' idea was not accepted by many of his contemporaries (notably Plato and Aristotle), somehow it endured. Experimental evidence from early scientific investigations provided support for the notion of "atomism" and gradually gave rise to the modern definitions of elements and compounds. It was in 1808 that an English scientist and schoolteacher, John Dalton, formulated a precise definition of the indivisible building blocks of matter that we call atoms.

John Dalton (1766–1844)

Dalton's work marked the beginning of the modern era of chemistry. The hypotheses about the nature of matter on which Dalton's atomic theory is based can be summarized as follows:

- Elements are composed of extremely small particles, called atoms. All atoms of a given element are identical, having the same size, mass, and chemical properties. The atoms of one element are different from the atoms of all other elements.
- Compounds are composed of atoms of more than one element. In any compound, the ratio of the numbers of atoms of any two of the elements present is either an integer or a simple fraction.
- A chemical reaction involves only the separation, combination, or rearrangement of atoms; it does not result in their creation or destruction.

Figure 2.1 is a schematic representation of the first two hypotheses.

Dalton's concept of an atom was far more detailed and specific than Democritus'. The first hypothesis states that atoms of one element are different from atoms of all other elements. Dalton made no attempt to describe the structure or composition of atoms—he had no idea what an atom is really like. But he did realize that the different properties shown by elements such as hydrogen and oxygen can be explained by assuming that hydrogen atoms are not the same as oxygen atoms.

The second hypothesis suggests that, in order to form a certain compound, we need not only atoms of the right kinds of elements, but the specific numbers of these atoms as well. This idea is an extension of a law published in 1799 by Joseph Proust, a French chemist. Proust's ***law of definite proportions*** states that *different samples of the same compound always contain its constituent elements in the same*

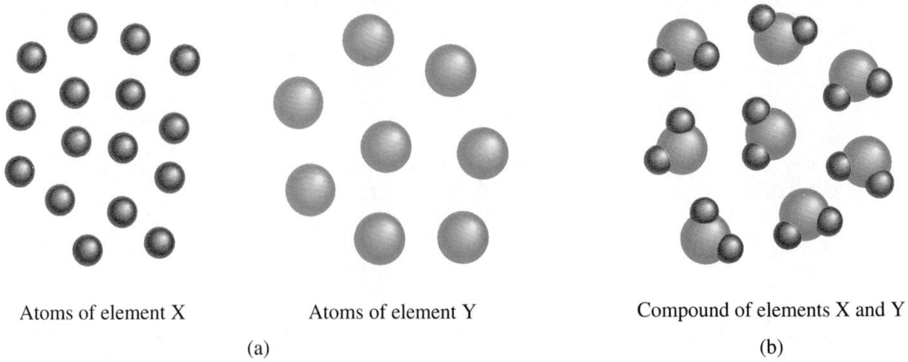

FIGURE 2.1

(a) According to Dalton's atomic theory, atoms of the same element are identical, but atoms of one element are different from atoms of other elements. (b) Compound formed from atoms of elements X and Y. In this case, the ratio of the atoms of element X to the atoms of element Y is 2:1.

proportion by mass. Thus, if we were to analyze samples of carbon dioxide gas obtained from different sources, we would find in each sample the same ratio by mass of carbon to oxygen. It stands to reason, then, that if the ratio of the masses of different elements in a given compound is fixed, the ratio of the atoms of these elements in the compound also must be constant.

Dalton's second hypothesis also supports another important law, the ***law of multiple proportions.*** According to this law, *if two elements can combine to form more than one compound, the masses of one element that combine with a fixed mass of the other element are in ratios of small whole numbers.* Dalton's theory explains the law of multiple proportions quite simply: Compounds differ in the number of atoms of each kind that combine. For example, carbon forms two stable compounds with oxygen, namely, carbon monoxide and carbon dioxide. Modern measurement techniques indicate that one atom of carbon combines with one atom of oxygen in carbon monoxide and that one atom of carbon combines with two oxygen atoms in carbon dioxide. Thus, the ratio of oxygen in carbon monoxide to oxygen in carbon dioxide is 1:2. This result is consistent with the law of multiple proportions.

Dalton's third hypothesis is another way of stating the ***law of conservation of mass,*** which is that *matter can be neither created nor destroyed.* Since matter is made of atoms that are unchanged in a chemical reaction, it follows that mass must be conserved as well. Dalton's brilliant insight into the nature of matter was the main stimulus for the rapid progress of chemistry during the nineteenth century.

2.2 THE STRUCTURE OF THE ATOM

On the basis of Dalton's atomic theory, we can define an ***atom*** as *the basic unit of an element that can enter into chemical combination.* Dalton imagined an atom that was both extremely small and indivisible. However, a series of investigations that begin the 1850s and extended into the twentieth century clearly demonstrated that atoms actually possess internal structure; that is, they are made up of even smaller particles, which are called *subatomic particles.* This research led to the discovery of three such particles—electrons, protons, and neutrons.

THE ELECTRON

In the 1890s many scientists became caught up in the study of ***radiation,*** *the emission and transmission of energy through space in the form of waves.* Information gained from this research contributed greatly to our understanding of atomic structure. One device used to investigate this phenomenon was a cathode ray tube, the forerunner of the television tube (Figure 2.2). It is a glass tube from which most of the air has been evacuated. When the two metal plates are connected to a high-voltage source, the negatively charged plate, called the *cathode,* emits an invisible ray. The cathode ray is drawn to the positively charged plate, called the *anode,* where it passes through a hole and continues traveling to the other end of the tube. When the ray strikes the specially coated surface, the cathode ray produces a strong fluorescence, or bright light.

In some experiments two electrically charged plates and a magnet were added to the *outside* of the cathode ray tube (see Figure 2.2). When the magnetic field is on and the electric field is off, the cathode ray strikes point A. When only the

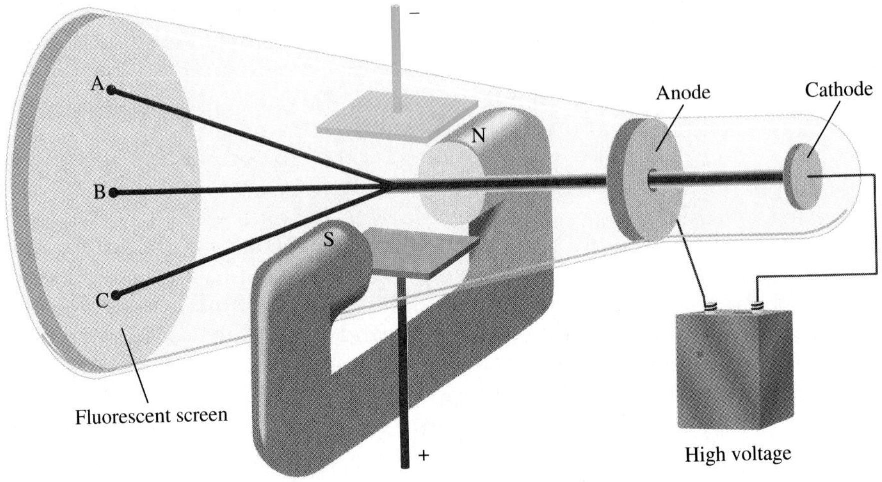

FIGURE 2.2

A cathode ray tube with an electric field perpendicular to the direction of the cathode rays and an external magnetic field. The symbols N and S denote the north and south poles of the magnet. The cathode rays will strike the end of the tube at A in the presence of a magnetic field, at C in the presence of an electric field, and at B when no external fields are present or when the effects of the electric field and magnetic field cancel each other.

electric field is on, the ray strikes point C. When both the magnetic and the electric fields are off or when they are both on but balanced so that they cancel each other's influence, the ray strikes point B. According to electromagnetic theory, a moving charged body behaves like a magnet and can interact with electric and magnetic fields through which it passes. Since the cathode ray is attracted by the plate bearing positive charges and repelled by the plate bearing negative charges, it must consist of negatively charged particles. We know these *negatively charged particles* as ***electrons.*** Figure 2.3 shows the effect of a bar magnet on the cathode ray.

An English physicist, J. J. Thomson, used a cathode ray tube and his knowledge of electromagnetic theory to determine the ratio of electric charge to the mass of an individual electron. The number he came up with is -1.76×10^8 C/g, where C stands for *coulomb,* which is the unit of electric charge. Thereafter, in a series of

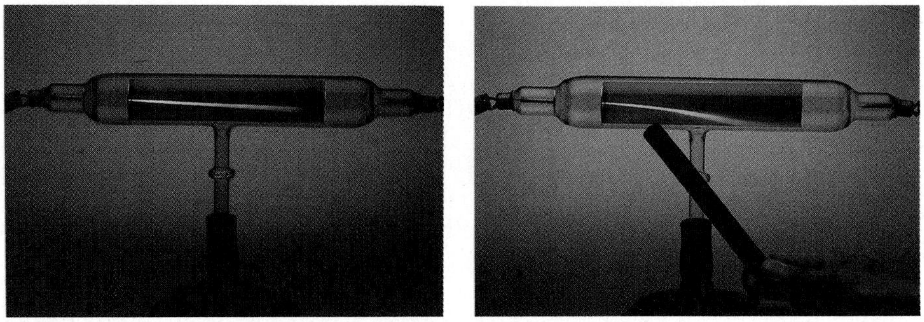

(a) (b)

FIGURE 2.3

(a) A cathode ray produced in a discharge tube. The ray itself has no color; the green color is due to the fluorescence of zinc sulfide coated on the screen when struck by the ray. (b) The cathode ray is bent in the presence of a magnet.

experiments carried out between 1908 and 1917, R. A. Millikan, an American physicist, found the charge of an electron to be -1.60×10^{-19} C. From these data he calculated the mass of an electron:

$$\text{mass of an electron} = \frac{\text{charge}}{\text{charge/mass}} = \frac{-1.60 \times 10^{-19}\ \text{C}}{-1.76 \times 10^{8}\ \text{C/g}} = 9.09 \times 10^{-28}\ \text{g}$$

which is an exceedingly small mass.

J. J. Thomson (1856–1940)

Radioactivity

In 1895, the German physicist Wilhelm Röntgen noticed that cathode rays caused glass and metals to emit very unusual rays. This highly energetic radiation penetrated matter, darkened covered photographic plates, and caused a variety of substances to fluoresce. Since these rays could not be deflected by a magnet, they could not contain charged particles as cathode rays do. Röntgen called them X rays.

Not long after Röntgen's discovery, Antoine Becquerel, a professor of physics in Paris, began to study fluorescent properties of substances. Purely by accident, he found that exposing thickly wrapped photographic plates to a certain uranium compound caused them to darken, even without the stimulation of cathode rays. Like X rays, the rays from the uranium compound were highly energetic and could not be deflected by a magnet, but they differed from X rays because they were generated spontaneously. One of Becquerel's students, Marie Curie, suggested the name ***radioactivity*** to describe this *spontaneous emission of particles and/or radiation.* Consequently, any element that spontaneously emits radiation is said to be *radioactive.*

Further investigation revealed that three types of rays are produced by the *decay,* or breakdown, of radioactive substances such as uranium. Two of the three kinds are deflected by oppositely charged metal plates (Figure 2.4). ***Alpha (α) rays***

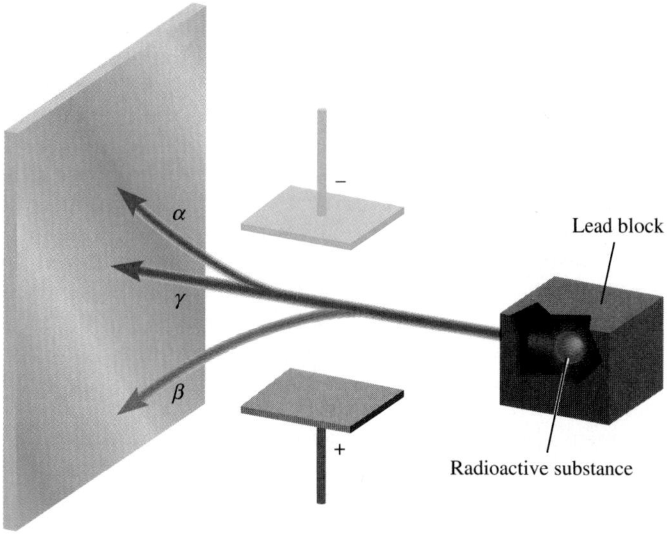

FIGURE 2.4

Three types of rays emitted by radioactive elements. β Rays consist of negatively charged particles (electrons) and are therefore attracted by the positively charged plate. The opposite holds true for α rays—they are positively charged and are drawn to the negatively charged plate. Because γ rays do not consist of charged particles, their movement is unaffected by an external electric field.

consist of *positively charged particles,* called ***α* particles,** and therefore are deflected by the positively charged plate. ***Beta (β) rays,*** or ***β* particles,** are *electrons* and are deflected by the negatively charged plate. The third type of radioactive radiation consists of *high-energy rays* called ***gamma (γ) rays.*** Like X rays, γ rays have no charge and are not affected by an external electric or magnetic field.

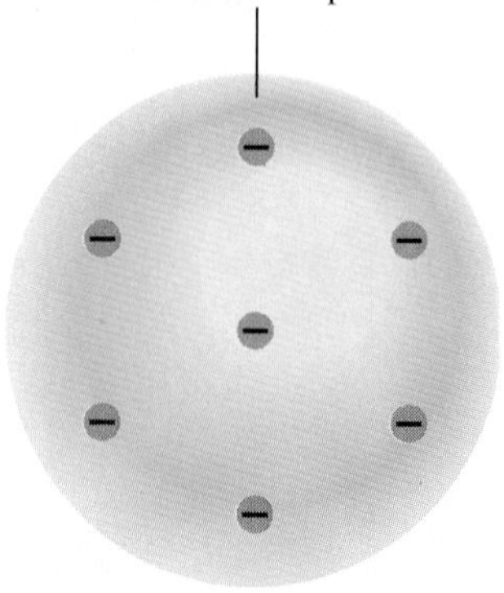

FIGURE 2.5

Thomson's model of the atom, sometimes described as the "plum-pudding" model, by analogy to a traditional English dessert containing raisins. The electrons are embedded in a uniform, positively charged sphere.

The Proton and the Nucleus

By the early 1900s, two features of atoms had become clear: They contain electrons, and they are electrically neutral. To maintain electrical neutrality, an atom must contain an equal number of positive and negative charges. On the basis of this information, Thomson proposed that an atom could be thought of as a uniform, positive sphere of matter in which electrons are embedded (Figure 2.5). Thomson's so-called plum-pudding model was the accepted theory for a number of years.

In 1910 the New Zealand physicist Ernest Rutherford, who had earlier studied with Thomson at Cambridge University, decided to use α particles to probe the structure of atoms. Together with his associate Hans Geiger and an undergraduate named Ernest Marsden, Rutherford carried out a series of experiments using very thin foils of gold and other metals as targets for α particles from a radioactive source (Figure 2.6). They observed that the majority of particles penetrated the foil either undeflected or with only a slight deflection. They also noticed that every now and then an α particle was scattered (or deflected) at a large angle. In some instances, an α particle actually bounced back in the direction from which it had come! This was a most surprising finding, for in Thomson's model the positive charge of the atom was so diffuse (spread out) that the positive α particles were expected to pass through with very little deflection. To quote Rutherford's initial reaction when told of this discovery: "It was as incredible as if you had fired a 15-inch shell at a piece of tissue paper and it came back and hit you."

To explain the results of the α-scattering experiment, Rutherford devised a new model of atomic structure, suggesting that most of the atom must be empty space. This structure would allow most of the α particles to pass through the gold foil with little or no deflection. The atom's positive charges, Rutherford proposed, are all concentrated in the ***nucleus,*** *a dense central core within the atom.* Whenever an α particle came close to a nucleus in the scattering experiment, it experienced a large repulsive force and therefore a large deflection. Moreover, an α particle traveling directly toward a nucleus would experience an enormous repulsion that could completely reverse the direction of the moving particle.

Ernest Rutherford (1871–1937)

The positively charged particles in the nucleus are called ***protons.*** In separate experiments, it was found that the charge of each proton has the same *magnitude* as that of an electron and that the mass of the proton is 1.67252×10^{-24} g—about 1840 times the mass of the oppositely charged electron.

At this stage of investigation, scientists perceived the atom as follows. The mass of a nucleus constitutes most of the mass of the entire atom, but the nucleus occupies only about $1/10^{13}$ of the volume of the atom. We express atomic (and molecular) dimensions in terms of the SI unit called the *picometer (pm),* where

$$1 \text{ pm} = 1 \times 10^{-12} \text{ m}$$

A typical atomic radius is about 100 pm, whereas the radius of an atomic nucleus is only about 5×10^{-3} pm. You can appreciate the relative sizes of an atom and its nucleus by imagining that if an atom were the size of the Houston Astrodome, the

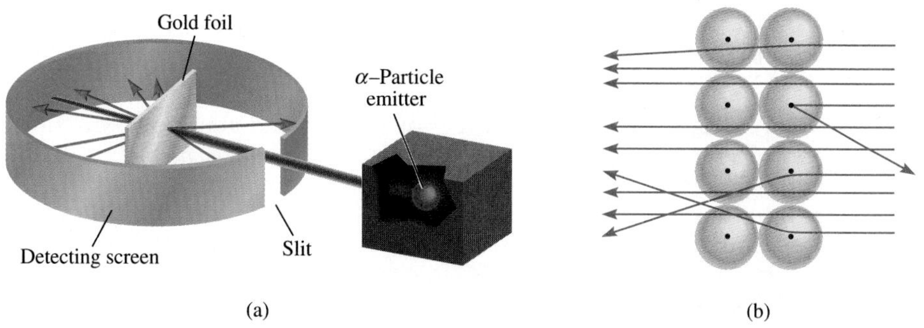

FIGURE 2.6

(a) Rutherford's experimental design for measuring the scattering of α particles by a piece of gold foil. Most of the α particles passed through the gold foil with little or no deflection. A few were deflected at wide angles. Occasionally an α particle was turned back. (b) Magnified view of α particles passing through and being deflected by nuclei.

volume of its nucleus would be comparable to that of a small marble. While the protons are confined to the nucleus of the atom, the electrons are conceived of as being spread out about the nucleus at some distance from it.

THE NEUTRON

Rutherford's model of atomic structure left one major problem unsolved. It was known that hydrogen, the simplest atom, contains only one proton and that the helium atom contains two protons. Therefore, the ratio of the mass of a helium atom to that of a hydrogen atom should be 2:1. (Because electrons are much lighter than protons, their contribution can be ignored.) In reality, however, the ratio is 4:1.

Rutherford and others postulated that there must be another type of subatomic particle in the atomic nucleus; the proof was provided by another English physicist, James Chadwick, in 1932. When Chadwick bombarded a thin sheet of beryllium with α particles, a very high energy radiation similar to γ rays was emitted by the metal. Later experiments showed that the rays actually consisted of *electrically neutral particles having a mass slightly greater than that of protons.* Chadwick named these particles ***neutrons.***

The mystery of the mass ratio could now be explained. In the helium nucleus there are two protons and two neutrons, but in the hydrogen nucleus there is only one proton and no neutrons; therefore, the ratio is 4:1.

There are other subatomic particles, but the electron, the proton, and the neutron are the three fundamental components of the atom that are important in chemistry. Table 2.1 shows the masses and charges of these elementary particles.

TABLE 2.1
Mass and Charge of Subatomic Particles

		Charge	
Particle	**Mass (g)**	**Coulomb**	**Charge unit**
Electron*	9.1095×10^{-28}	-1.6022×10^{-19}	−1
Proton	1.67252×10^{-24}	$+1.6022 \times 10^{-19}$	+1
Neutron	1.67495×10^{-24}	0	0

*More refined experiments have given us a more accurate value of an electron's mass than Millikan's.

2.3 ATOMIC NUMBER, MASS NUMBER, AND ISOTOPES

All atoms can be identified by the number of protons and neutrons they contain. *The number of protons in the nucleus of each atom of an element* is called the ***atomic number (Z).*** In a neutral atom the number of protons is equal to the number of electrons, so the atomic number also indicates the number of electrons present in the atom. The chemical identity of an atom can be determined solely by its atomic number. For example, the atomic number of nitrogen is 7; this means that each neutral nitrogen atom has 7 protons and 7 electrons. Or viewed another way, every atom in the universe that contains 7 protons is correctly named "nitrogen."

The ***mass number (A)*** is *the total number of neutrons and protons present in the nucleus of an atom of an element.* Except for the most common form of hydrogen, which has one proton and no neutrons, all atomic nuclei contain both protons and neutrons. In general, the mass number is given by

$$\begin{aligned} \text{mass number} &= \text{number of protons} + \text{number of neutrons} \\ &= \text{atomic number} + \text{number of neutrons} \end{aligned}$$

The number of neutrons in an atom is equal to the difference between the mass number and the atomic number, or $(A - Z)$. For example, the mass number of fluorine is 19, and the atomic number is 9 (indicating 9 protons in the nucleus). Thus the number of neutrons in an atom of fluorine is $19 - 9 = 10$. Note that all three quantities (atomic number, number of neutrons, and mass number) must be positive integers, or whole numbers.

In most cases atoms of a given element do not all have the same mass. *Atoms that have the same atomic number but different mass numbers* are called ***isotopes.*** For example, there are three isotopes of hydrogen. One, simply known as hydrogen, has one proton and no neutrons. The deuterium isotope has one proton and one neutron, and tritium has one proton and two neutrons. The accepted way to denote the atomic number and mass number of an atom of element X is as follows:

mass number → $^{A}_{Z}\text{X}$ ← atomic number

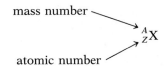

Thus, for the isotopes of hydrogen, we write

$^{1}_{1}\text{H}$	$^{2}_{1}\text{H}$	$^{3}_{1}\text{H}$
hydrogen	deuterium	tritium

As another example, consider two common isotopes of uranium with mass numbers of 235 and 238, respectively:

$$^{235}_{92}\text{U} \qquad ^{238}_{92}\text{U}$$

The first isotope is used in nuclear reactors and atomic bombs, whereas the second isotope lacks the properties necessary for these applications. With the exception of hydrogen, isotopes of elements are identified by their mass numbers. Thus the above two isotopes are called uranium-235 (pronounced "uranium two thirty-five") and uranium-238 (pronounced "uranium two thirty-eight").

The chemical properties of an element are determined primarily by the protons and electrons in its atoms; neutrons do not take part in chemical changes under normal conditions. Therefore, isotopes of the same element have similar chemistries, forming the same types of compounds and displaying similar reactivities.

EXAMPLE 2.1

Calculating the Number of Protons, Neutrons, and Electrons

Give the number of protons, neutrons, and electrons in each of the following species: (a) $^{17}_{8}O$, (b) $^{199}_{80}Hg$, (c) $^{200}_{80}Hg$.

Answer: (a) The atomic number is 8, so there are 8 protons. The mass number is 17, so the number of neutrons is $17 - 8 = 9$. The number of electrons is the same as the number of protons, that is, 8.
(b) The atomic number is 80, so there are 80 protons. The mass number is 199, so the number of neutrons is $199 - 80 = 119$. The number of electrons is 80.
(c) Here the number of protons is the same as in (b), or 80. The number of neutrons is $200 - 80 = 120$. The number of electrons is also the same as in (b), 80. The species in (b) and (c) are two isotopes of mercury that are chemically similar.

PRACTICE EXERCISE

How many protons, neutrons, and electrons are in the following isotope of copper: $^{63}_{29}Cu$?

2.4 THE PERIODIC TABLE

More than half of the elements known today were discovered between 1800 and 1900. During this period, chemists noted that many elements show very strong similarities to one another. Recognition of periodic regularities in physical and chemical behavior and the need to organize the large volume of available information about the structure and properties of elemental substances led to the development of the ***periodic table***—*a chart in which elements having similar chemical and physical properties are grouped together.* Figure 2.7 shows the modern periodic table, in which the elements are arranged by atomic number (shown above the element symbol) in *horizontal rows* called ***periods*** and in *vertical columns* known as ***groups*** or ***families,*** according to similarities in their chemical properties.

The elements can be divided into three categories—metals, nonmetals, and metalloids. A ***metal*** is *a good conductor of heat and electricity,* while a ***nonmetal*** is usually *a poor conductor of heat and electricity.* A ***metalloid*** *has properties that are intermediate between those of metals and nonmetals.* Figure 2.7 shows that the majority of known elements are metals; only seventeen elements are nonmetals, and eight elements are metalloids. From left to right across any period, the physical and chemical properties of the elements change gradually from metallic to nonmetallic. The periodic table is a handy tool that correlates the properties of the elements in a systematic way and helps us to make predictions about chemical behavior. We will take a closer look at this keystone of chemistry in Chapter 8.

Elements are often referred to collectively by their periodic table group num-

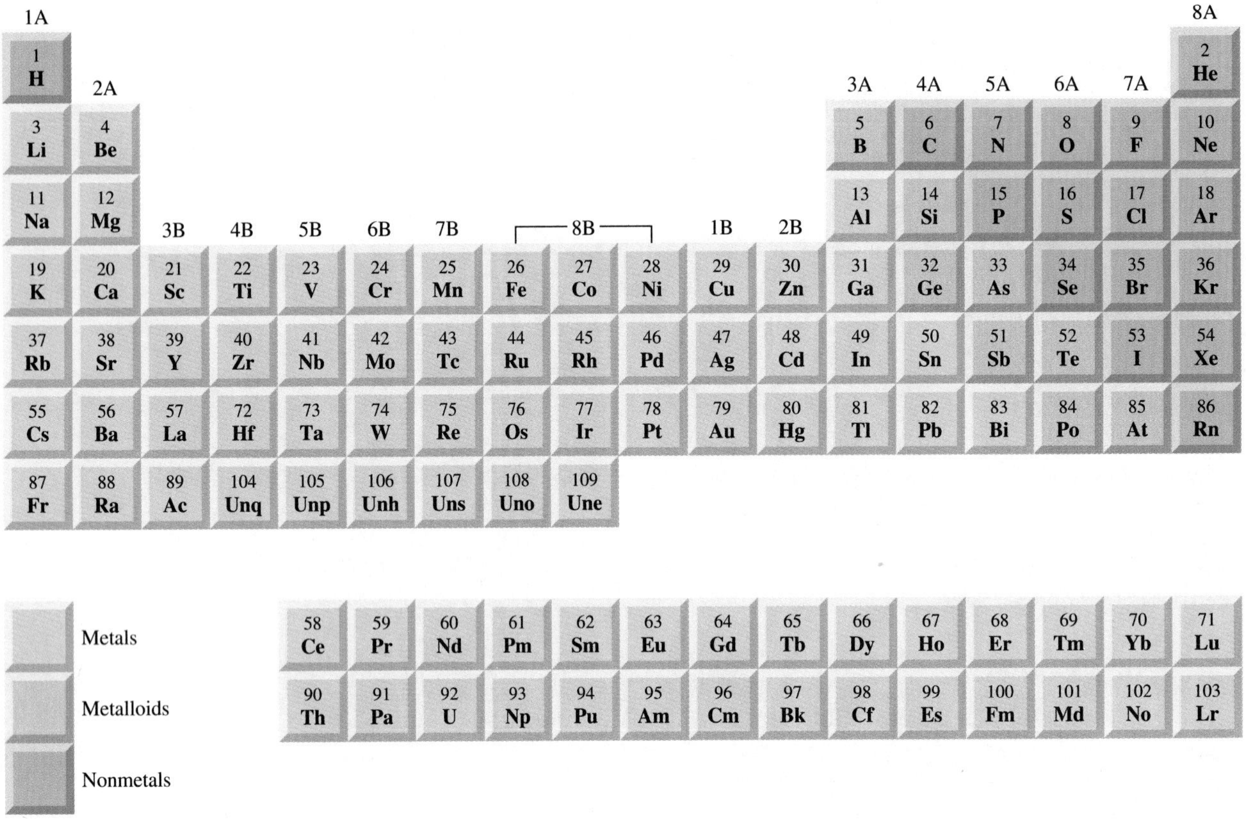

FIGURE 2.7

The modern periodic table. With the exception of hydrogen (H), nonmetals appear at the far right of the table. The elements are arranged according to their atomic numbers shown above their symbols. The two rows of metals beneath the main body of the table are conventionally set apart to avoid making the table too long. Actually, cerium (Ce) should follow lanthanum (La), and thorium (Th) should come right after actinium (Ac).

ber (Group 1A, Group 2A, and so on). However, for convenience, some element groups have special names. *The Group 1A elements (Li, Na, K, Rb, Cs, and Fr)* are called ***alkali metals,*** and *the Group 2A elements (Be, Mg, Ca, Sr, Ba, and Ra)* are called ***alkaline earth metals.*** *Elements in Group 7A (F, Cl, Br, I, and At)* are known as ***halogens,*** and *those in Group 8A (He, Ne, Ar, Kr, Xe, and Rn)* are called ***noble gases*** (or ***rare gases***). The names of other groups or families will be introduced later.

2.5 MOLECULES AND IONS

Of all the elements, only the six noble gases in Group 8A of the periodic table (He, Ne, Ar, Kr, Xe, and Rn) exist in nature as single atoms. Most matter is composed of molecules or ions formed by atoms.

MOLECULES

A ***molecule*** is *an aggregate of at least two atoms in a definite arrangement held together by chemical forces.* A molecule may contain atoms of the same element or atoms of two or more elements joined in a fixed ratio, in accordance with the law of definite proportions stated in Section 2.1. Thus, a molecule is not necessarily a compound, which, by definition, is made up of two or more elements (see Section 1.2). Hydrogen gas, for example, is a pure element, but it consists of molecules

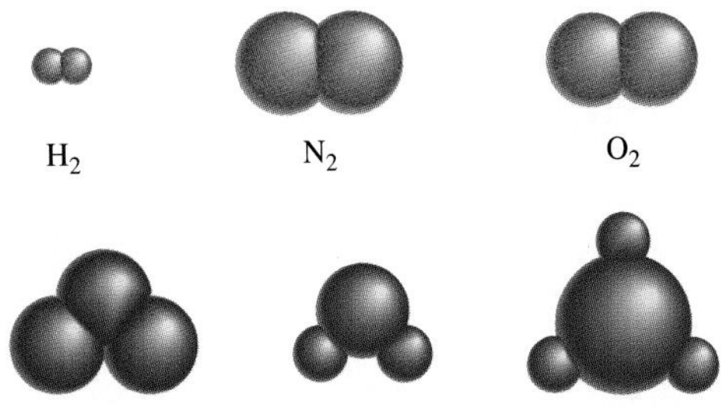

FIGURE 2.8

Common diatomic and polyatomic molecules.

made up of two H atoms each. Water, on the other hand, is a molecular compound that contains hydrogen and oxygen in a ratio of two H atoms and one O atom. Like atoms, molecules are electrically neutral.

The hydrogen molecule, symbolized as H_2, is called a ***diatomic molecule*** because it *contains only two atoms.* Other elements that normally exist as diatomic molecules are nitrogen (N_2) and oxygen (O_2), as well as the Group 7A elements—fluorine (F_2), chlorine (Cl_2), bromine (Br_2), and iodine (I_2). Of course, a diatomic molecule can contain atoms of different elements. Examples are hydrogen chloride (HCl) and carbon monoxide (CO).

The vast majority of molecules contain more than two atoms. They can be atoms of the same element, as in ozone (O_3), which is made up of three atoms of oxygen. Or they can be combinations of two or more different elements. *Molecules containing more than two atoms* are called ***polyatomic molecules.*** In addition to ozone, water (H_2O) and ammonia (NH_3) are polyatomic molecules. Figure 2.8 illustrates some common diatomic and polyatomic molecules.

IONS

An ***ion*** is *a charged species formed from a neutral atom or molecule when electrons are gained or lost as the result of a chemical change.* The number of positively charged protons in the nucleus of an atom remains the same during ordinary chemical changes (called chemical reactions), but negatively charged electrons may be lost or gained. The loss of one or more electrons from a neutral atom results in *an ion with a net positive charge,* which is called a ***cation.*** For example, a sodium atom (Na) can readily lose an electron to become a sodium cation, which is represented by Na^+:

Na atom	*Na^+ ion*
11 protons	11 protons
11 electrons	10 electrons

On the other hand, *an ion whose net charge is negative* due to an increase in the number of electrons is called an ***anion.*** A chlorine atom (Cl), for instance, can gain an electron to become the chloride ion Cl^-:

Cl atom	*Cl^- ion*
17 protons	17 protons
17 electrons	18 electrons

Sodium chloride (NaCl), ordinary table salt, is an ***ionic compound*** because it is *formed from cations and anions.*

An atom can lose or gain more than one electron, as in Mg^{2+}, Fe^{3+}, S^{2-}, and

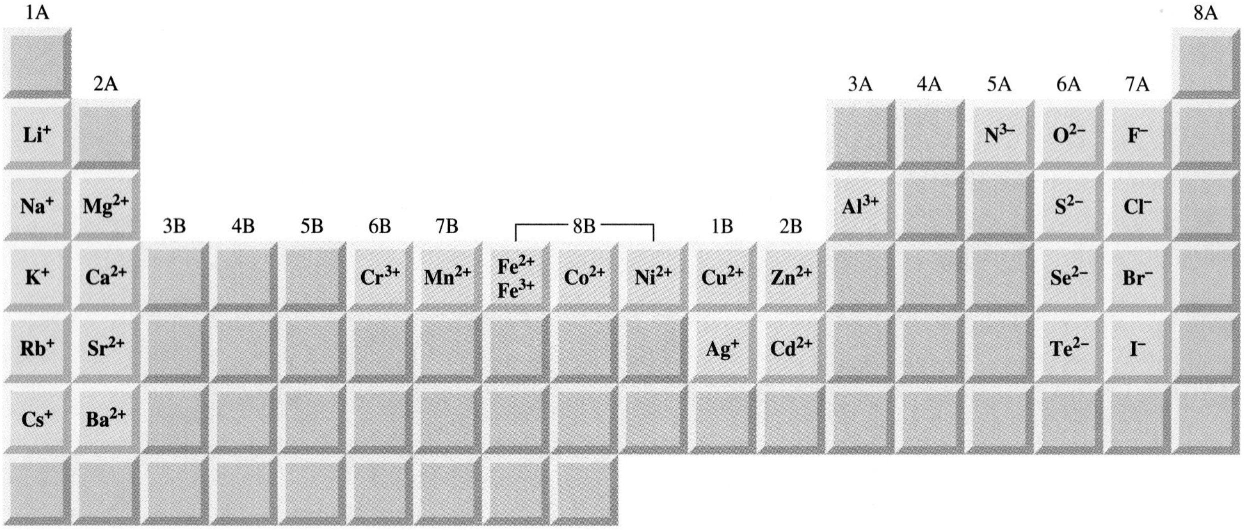

FIGURE 2.9

Common monatomic ions arranged according to their positions in the periodic table.

N^{3-}. These ions, as well as Na^+ and Cl^-, are called ***monatomic ions*** because they *contain only one atom.* Figure 2.9 shows the charges of some familiar monatomic ions. With very few exceptions, metals tend to form cations and nonmetals form anions.

In addition, two or more atoms can combine to form an ion that has a net positive or net negative charge, as in OH^- (hydroxide ion), CN^- (cyanide ion), and NH_4^+ (ammonium ion). These and other *ions containing more than one atom* are called ***polyatomic ions.***

2.6 CHEMICAL FORMULAS

Chemists use ***chemical formulas*** to *express the composition of molecules and ionic compounds in terms of chemical symbols.* By composition we mean not only the elements present but also the ratios in which the atoms are combined. Here we are concerned with two types of formulas: molecular formulas and empirical formulas.

MOLECULAR FORMULAS

A ***molecular formula*** *shows the exact number of atoms of each element in the smallest unit of a substance.* In our discussion of molecules, each example was given with its molecular formula in parentheses. Thus H_2 is the molecular formula for hydrogen, O_2 is oxygen, O_3 is ozone, and H_2O is water. For each of these molecules the subscript numeral indicates the number of atoms of an element present. There is no subscript for O in H_2O because there is only one atom of oxygen in a molecule of water, and the number "one" is omitted. Note that oxygen (O_2) and ozone (O_3) are both elemental forms of oxygen. *Different forms of the same element* are called ***allotropes.*** Two allotropic forms of the element carbon—diamond and graphite—are dramatically different not only in properties but also in their relative cost.

EMPIRICAL FORMULAS

The molecular formula of hydrogen peroxide, a substance used as an antiseptic and as a bleaching agent for textiles and hair, is H_2O_2. This formula indicates that each hydrogen peroxide molecule consists of two hydrogen atoms and two oxygen atoms. The ratio of hydrogen to oxygen atoms in this molecule is 2:2, or 1:1. The empirical formula of hydrogen peroxide is written as HO. Thus the ***empirical formula*** *tells us which elements are present and the simplest whole-number ratio of their atoms,* but not necessarily the actual number of atoms present in the molecule. As another example, consider the compound hydrazine (N_2H_4), which is used as a rocket fuel. The empirical formula of hydrazine is NH_2. Although the ratio of nitrogen to hydrogen is 1:2 in both the molecular formula (N_2H_4) and the empirical formula (NH_2), only the molecular formula tells us the actual number of N atoms (two) and H atoms (four) present in a hydrazine molecule.

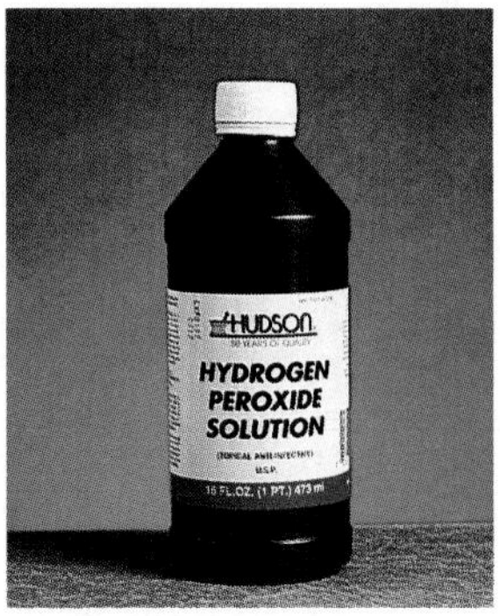

A bottle of hydrogen peroxide from the drugstore.

Empirical formulas are therefore the *simplest* chemical formulas; they are always written so that the subscripts in the molecular formulas are converted to the smallest possible whole numbers. Molecular formulas are the *true* formulas of molecules. As we will see in Chapter 3, when chemists analyze an unknown compound, the first step is usually the determination of the compound's empirical formula.

For many molecules, the molecular formula and the empirical formula are the same. Some examples are water (H_2O), ammonia (NH_3), carbon dioxide (CO_2), and methane (CH_4).

EXAMPLE 2.2
Relationship between Empirical Formulas and Molecular Formulas

Write the empirical formulas for the following molecules: (a) acetylene (C_2H_2), which is used for welding metals; (b) glucose ($C_6H_{12}O_6$), a substance known as blood sugar; and (c) nitrous oxide (N_2O), which is used as an anesthetic gas ("laughing gas") and as an aerosol propellant for whipped cream.

Answer: (a) There are two carbon atoms and two hydrogen atoms in acetylene. Dividing the subscripts by 2, we obtain the empirical formula CH.
(b) In glucose there are six carbon atoms, twelve hydrogen atoms, and six oxygen atoms. Dividing the subscripts by 6, we obtain the empirical formula CH_2O. Note that if we had divided the subscripts by 3, we would have obtained the formula $C_2H_4O_2$. Although the ratio of carbon to hydrogen to oxygen atoms in $C_2H_4O_2$ is the same as that in $C_6H_{12}O_6$ (1:2:1), $C_2H_4O_2$ is not the simplest formula because its subscripts are not in the smallest whole-number ratio.
(c) Since the subscripts in N_2O are already the smallest whole numbers, the empirical formula for nitrous oxide is the same as its molecular formula.

PRACTICE EXERCISE

Write the empirical formula for caffeine ($C_8H_{10}N_4O_2$), a substance found in tea and coffee.

The formulas of ionic compounds are always the same as their empirical formulas because ionic compounds do not consist of discrete molecular units. For

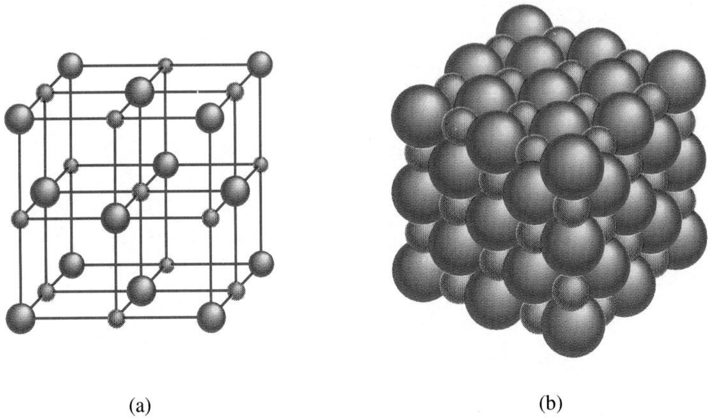

FIGURE 2.10

(a) Structure of solid NaCl. (b) In reality, the cations are in contact with the anions. In both (a) and (b), the smaller spheres represent Na^+ ions and the larger spheres Cl^- ions.

example, a solid sample of sodium chloride (NaCl) consists of equal numbers of Na^+ and Cl^- ions arranged in a three-dimensional network (Figure 2.10). In such compounds there is a 1:1 ratio of cations to anions so that the compounds are electrically neutral. As you can see in Figure 2.10, no Na^+ ion in NaCl is associated with just one particular Cl^- ion. In fact, each Na^+ ion is equally held by six Cl^- ions surrounding it, and vice versa. Thus NaCl is the empirical formula for sodium chloride. In other ionic compounds the actual structure may be different, but the arrangement of cations and anions is such that the compounds are all electrically neutral. Note that the charges on the cation and anion are not shown in the formula for an ionic compound.

In order for ionic compounds to be electrically neutral, the sum of the charges on the cation and anion in each formula unit must add up to zero. If the charges on the cation and anion are numerically different, we apply the following rule to make the formula electrically neutral: *The subscript of the cation is numerically equal to the charge on the anion, and the subscript of the anion is numerically equal to the charge on the cation.* If the charges are numerically equal, then no subscripts will be necessary. This rule follows from the fact that because the formulas of ionic compounds are empirical formulas, the subscripts must always be reduced to the smallest ratios. Let us consider some examples (see Figure 2.9).

- *Potassium bromide.* The potassium cation K^+ and the bromine anion Br^- combine to form the ionic compound potassium bromide. The sum of the charges is $+1 + (-1) = 0$, so no subscripts are necessary.
- *Zinc iodide.* The zinc cation Zn^{2+} and the iodine anion I^- combine to form zinc iodide. The sum of the charges is $+2 + 1(-1) = +1$. To make the charges add up to zero we multiply the -1 charge of the anion by 2 and add the subscript 2 to the symbol for iodine. Therefore the formula for zinc iodide is ZnI_2.
- *Aluminum oxide.* The cation is Al^{3+}, and the oxygen anion is O^{2-}. The following diagram helps us determine the subscripts for the compound formed by the cation and the anion:

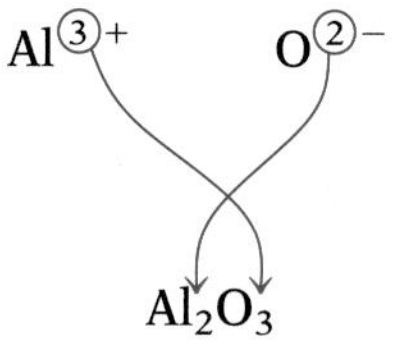

The sum of the charges is $2(+3) + 3(-2) = 0$. Thus the formula for aluminum oxide is Al_2O_3.

2.7 NAMING COMPOUNDS

In addition to using formulas to show the composition of molecules and compounds, chemists have developed a system for naming substances on the basis of their composition. First we divide them into three categories: ionic compounds, molecular compounds, and acids and bases. Then we apply certain rules to derive the scientific name for a given substance.

IONIC COMPOUNDS

In Section 2.5 we learned that ionic compounds are made up of cations (positive ions) and anions (negative ions). With the important exception of the ammonium ion, NH_4^+, all cations of interest to us are derived from metal atoms. Metal cations take their names from the elements. For example:

Element		*Name of cation*	
Na	sodium	Na^+	sodium ion (or sodium cation)
K	potassium	K^+	potassium ion (or potassium cation)
Mg	magnesium	Mg^{2+}	magnesium ion (or magnesium cation)
Al	aluminum	Al^{3+}	aluminum ion (or aluminum cation)

Many ionic compounds are ***binary compounds,*** or *compounds formed from just two elements.* For binary compounds the first element named is the metal cation, followed by the nonmetallic anion. Thus NaCl is sodium chloride. The anion is named by taking the first part of the element name (chlorine) and adding "-ide." Potassium bromide (KBr), zinc iodide (ZnI_2), and aluminum oxide (Al_2O_3) are also binary compounds. Table 2.2 shows the "-ide" nomenclature of some common monatomic anions according to their positions in the periodic table.

The "-ide" ending is also used for certain anion groups containing different elements, such as hydroxide (OH^-) and cyanide (CN^-). Thus the compounds LiOH and KCN are named lithium hydroxide and potassium cyanide. These and a number of other such ionic substances are called ***ternary compounds,*** meaning *com-*

TABLE 2.2
The "-ide" Nomenclature of Some Common Monatomic Anions According to Their Positions in the Periodic Table

Group 4A	Group 5A	Group 6A	Group 7A
C Carbide (C^{4-})*	N Nitride (N^{3-})	O Oxide (O^{2-})	F Fluoride (F^-)
Si Silicide (Si^{4-})	P Phosphide (P^{3-})	S Sulfide (S^{2-})	Cl Chloride (Cl^-)
		Se Selenide (Se^{2-})	Br Bromide (Br^-)
		Te Telluride (Te^{2-})	I Iodide (I^-)

*The word "carbide" is also used for the anion C_2^{2-}.

TABLE 2.3
Names and Formulas of Some Common Inorganic Cations and Anions

Cation	Anion
Aluminum (Al^{3+})	Bromide (Br^-)
Ammonium (NH_4^+)	Carbonate (CO_3^{2-})
Barium (Ba^{2+})	Chlorate (ClO_3^-)
Cadmium (Cd^{2+})	Chloride (Cl^-)
Calcium (Ca^{2+})	Chromate (CrO_4^{2-})
Cesium (Cs^+)	Cyanide (CN^-)
Chromium(III) or chromic (Cr^{3+})	Dichromate ($Cr_2O_7^{2-}$)
Cobalt(II) or cobaltous (Co^{2+})	Dihydrogen phosphate ($H_2PO_4^-$)
Copper(I) or cuprous (Cu^+)	Fluoride (F^-)
Copper(II) or cupric (Cu^{2+})	Hydride (H^-)
Hydrogen (H^+)	Hydrogen carbonate or bicarbonate (HCO_3^-)
Iron(II) or ferrous (Fe^{2+})	Hydrogen phosphate (HPO_4^{2-})
Iron(III) or ferric (Fe^{3+})	Hydrogen sulfate or bisulfate (HSO_4^-)
Lead(II) or plumbous (Pb^{2+})	Hydroxide (OH^-)
Lithium (Li^+)	Iodide (I^-)
Magnesium (Mg^{2+})	Nitrate (NO_3^-)
Manganese(II) or manganous (Mn^{2+})	Nitride (N^{3-})
Mercury(I) or mercurous (Hg_2^{2+})*	Nitrite (NO_2^-)
Mercury(II) or mercuric (Hg^{2+})	Oxide (O^{2-})
Potassium (K^+)	Permanganate (MnO_4^-)
Silver (Ag^+)	Peroxide (O_2^{2-})
Sodium (Na^+)	Phosphate (PO_4^{3-})
Strontium (Sr^{2+})	Sulfate (SO_4^{2-})
Tin(II) or stannous (Sn^{2+})	Sulfide (S^{2-})
Zinc (Zn^{2+})	Sulfite (SO_3^{2-})
	Thiocyanate (SCN^-)

*Mercury(I) exists as a pair as shown.

pounds consisting of three elements. Table 2.3 lists alphabetically the names of a number of common cations and anions.

Certain metals, especially the *transition metals,* can form more than one type of cation. Take iron as an example. Iron can form two cations: Fe^{2+} and Fe^{3+}. The accepted procedure for designating different cations of the *same* element is to use Roman numerals. The Roman numeral I is used for one positive charge, II for two positive charges, and so on. This is called the *Stock system.* In this system, the above ions are called iron(II) and iron(III), and the compounds $FeCl_2$ (containing the Fe^{2+} ion) and $FeCl_3$ (containing the Fe^{3+} ion) are called iron-two chloride and iron-three chloride, respectively. As another example, manganese (Mn) atoms can assume several different positive charges:

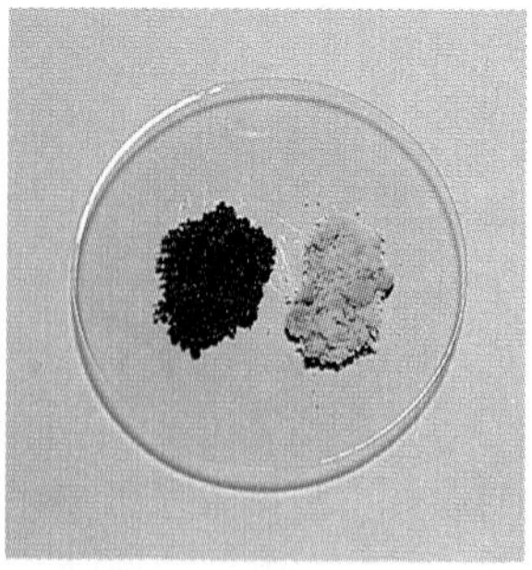

$FeCl_2$ (left) and $FeCl_3$ (right).

Mn^{2+}:	MnO	manganese(II) oxide
Mn^{3+}:	Mn_2O_3	manganese(III) oxide
Mn^{4+}:	MnO_2	manganese(IV) oxide

These compound names are pronounced manganese-two oxide, manganese-three oxide, and manganese-four oxide.

EXAMPLE 2.3
Naming Ionic Compounds

Name the following ionic compounds: (a) $Cu(NO_3)_2$, (b) KH_2PO_4, (c) NH_4ClO_3.

Answer: (a) Since the nitrate ion (NO_3^-) bears one negative charge (see Table 2.3), the copper ion must have two positive charges. Therefore, the compound is copper(II) nitrate.
(b) The cation is K^+, and the anion is $H_2PO_4^-$ (dihydrogen phosphate). The compound is potassium dihydrogen phosphate.
(c) The cation is NH_4^+ (ammonium ion), and the anion is ClO_3^-. The compound is ammonium chlorate.

PRACTICE EXERCISE

Name the following compounds: (a) PbO, (b) Li_2SO_3.

EXAMPLE 2.4
Writing Formulas Based on Names of Ionic Compounds

Write chemical formulas for the following compounds: (a) mercury(I) nitrite, (b) cesium sulfide, (c) calcium phosphate.

Answer: (a) The mercury(I) ion is diatomic, namely, Hg_2^{2+} (see Table 2.3), and the nitrite ion is NO_2^-. Therefore, the formula is $Hg_2(NO_2)_2$.
(b) Each sulfide ion bears two negative charges, and each cesium ion bears one positive charge (cesium is in Group 1A, as is sodium). Therefore, the formula is Cs_2S.
(c) Each calcium ion (Ca^{2+}) bears two positive charges, and each phosphate ion (PO_4^{3-}) bears three negative charges. To make the sum of charges equal to zero, the numbers of cations and anions must be adjusted:

$$3(+2) + 2(-3) = 0$$

Thus the formula is $Ca_3(PO_4)_2$.

PRACTICE EXERCISE

Write formulas for the following ionic compounds: (a) rubidium sulfate, (b) barium hydride.

Molecular Compounds

Unlike ionic compounds, molecular compounds contain discrete molecular units. They are usually composed of nonmetallic elements (see Figure 2.7). Many molecular compounds are binary compounds. Naming binary molecular compounds is similar to naming binary ionic compounds. We place the name of the first element in the formula first, and the second element is named by adding "-ide" to the root of the element name. Some examples are

HCl	Hydrogen chloride	SiC	Silicon carbide
HBr	Hydrogen bromide		

TABLE 2.4
Greek Prefixes Used in Naming Molecular Compounds

Prefix	Meaning
Mono-	1
Di-	2
Tri-	3
Tetra-	4
Penta-	5
Hexa-	6
Hepta-	7
Octa-	8
Nona-	9
Deca-	10

It is quite common for one pair of elements to form several different compounds. In these cases, confusion in naming the compounds is avoided by the use of Greek prefixes to denote the number of atoms of each element present (see Table 2.4). Consider the following examples:

CO	Carbon monoxide	SO_3	Sulfur trioxide
CO_2	Carbon dioxide	NO_2	Nitrogen dioxide
SO_2	Sulfur dioxide	N_2O_4	Dinitrogen tetroxide

The following guidelines are helpful when you are naming compounds with prefixes:

- The prefix "mono-" may be omitted for the first element. For example, PCl_3 is named phosphorus trichloride, not monophosphorus trichloride. Thus the absence of a prefix for the first element usually means that only one atom of that element is present in the molecule.
- For oxides, the ending "a" in the prefix is sometimes omitted. For example, N_2O_4 may be called dinitrogen tetroxide rather than dinitrogen tetraoxide.

Exceptions to the use of Greek prefixes are molecular compounds containing hydrogen. Traditionally, many of these compounds are called either by their common, nonsystematic names or by names that do not specifically indicate the number of H atoms present:

B_2H_6	Diborane	PH_3	Phosphine
CH_4	Methane	H_2O	Water
SiH_4	Silane	H_2S	Hydrogen sulfide
NH_3	Ammonia		

Note that even the order of writing the elements in the formulas is irregular. The above examples show that H is written first in water and hydrogen sulfide, whereas H is written last in the other compounds.

Writing formulas for molecular compounds is usually straightforward. Thus the name arsenic trifluoride means that there are one As atom and three F atoms in each molecule and the molecular formula is AsF_3. Note that the order of elements in the formula is the same as that in its name.

EXAMPLE 2.5
Naming Molecular Compounds

Name the following molecular compounds: (a) $SiCl_4$, (b) P_4O_{10}.

Answer: (a) Since four chlorine atoms are present, the compound is silicon tetrachloride.
(b) Four phosphorus atoms and ten oxygen atoms are present, so the compound is tetraphosphorus decoxide. Note that the "a" is omitted in "deca."

PRACTICE EXERCISE

Name the following molecular compounds: (a) NH_3, (b) Cl_2O_7.

EXAMPLE 2.6

Writing Formulas from Names of Molecular Compounds

Write chemical formulas for the following molecular compounds: (a) carbon disulfide, (b) disilicon hexabromide.

Answer: (a) Since one carbon atom and two sulfur atoms are present, the formula is CS_2.
(b) Two silicon atoms and six bromine atoms are present, so the formula is Si_2Br_6.

PRACTICE EXERCISE

Write chemical formulas for the following molecular compounds: (a) sulfur tetrafluoride, (b) dinitrogen pentoxide.

ACIDS AND BASES

Naming Acids

An ***acid*** can be described as *a substance that yields hydrogen ions (H^+) when dissolved in water.* Formulas for acids contain one or more hydrogen atoms as well as an anionic group. Anions whose names end in "-ide" have associated acids with a "hydro-" prefix and an "-ic" ending, as shown in Table 2.5. In some cases two different names are assigned to the same chemical formula. For instance, HCl is known as both hydrogen chloride and hydrochloric acid. The name used for this compound depends on its physical state. In the gaseous or pure liquid state, HCl is a molecular compound called hydrogen chloride. When it is dissolved in water, the molecules break up into H^+ and Cl^- ions; in this condition, the substance is called hydrochloric acid.

Acids that contain hydrogen, oxygen, and another element (the central element) are called ***oxoacids.*** The formulas of oxoacids are usually written with the H first, followed by the central element and then O, as illustrated by the following series of oxoacids:

H_2CO_3	Carbonic acid	H_2SO_4	Sulfuric acid
HNO_3	Nitric acid	$HClO_3$	Chloric acid

Often two or more oxoacids have the same central atom but a different num-

TABLE 2.5

Some Simple Acids

Anion	Corresponding acid
F^- (fluoride)	HF (hydrofluoric acid)
Cl^- (chloride)	HCl (hydrochloric acid)
Br^- (bromide)	HBr (hydrobromic acid)
I^- (iodide)	HI (hydroiodic acid)
CN^- (cyanide)	HCN (hydrocyanic acid)
S^{2-} (sulfide)	H_2S (hydrosulfuric acid)

ber of O atoms. Starting with the oxoacids whose names end with "-ic," we use the following rules to name these compounds.

- Addition of one O atom to the "-ic" acid: The acid is called "per . . . -ic" acid. Thus adding an O atom to $HClO_3$ changes chloric acid to perchloric acid, $HClO_4$.
- Removal of one O atom from the "-ic" acid: The acid is called "-ous" acid. Thus nitric acid, HNO_3, becomes nitrous acid, HNO_2.
- Removal of two O atoms from the "-ic" acid: The acid is called "hypo . . . -ous" acid. Thus when $HBrO_3$ is converted to HBrO, the acid is called hypobromous acid.

The rules for naming *anions of oxoacids,* called ***oxoanions,*** are as follows.

- When all the H ions are removed from the "-ic" acid, the anion's name ends with "-ate." For example, the anion CO_3^{2-} derived from H_2CO_3 is called carbonate.
- When all the H ions are removed from the "-ous" acid, the anion's name ends with "-ite." Thus the anion ClO_2^- derived from $HClO_2$ is called chlorite.
- The names of anions in which one or more but not all of the hydrogen ions have been removed must indicate the number of H ions present. For example, consider the anions derived from phosphoric acid:

H_3PO_4	Phosphoric acid	HPO_4^{2-}	Hydrogen phosphate
$H_2PO_4^-$	Dihydrogen phosphate	PO_4^{3-}	Phosphate

Note that we usually omit the prefix "mono-" when there is only one H in the anion. Table 2.6 gives the names of the oxoacids and oxoanions that contain chlorine, and Figure 2.11 summarizes the nomenclature for the oxoacids and oxoanions.

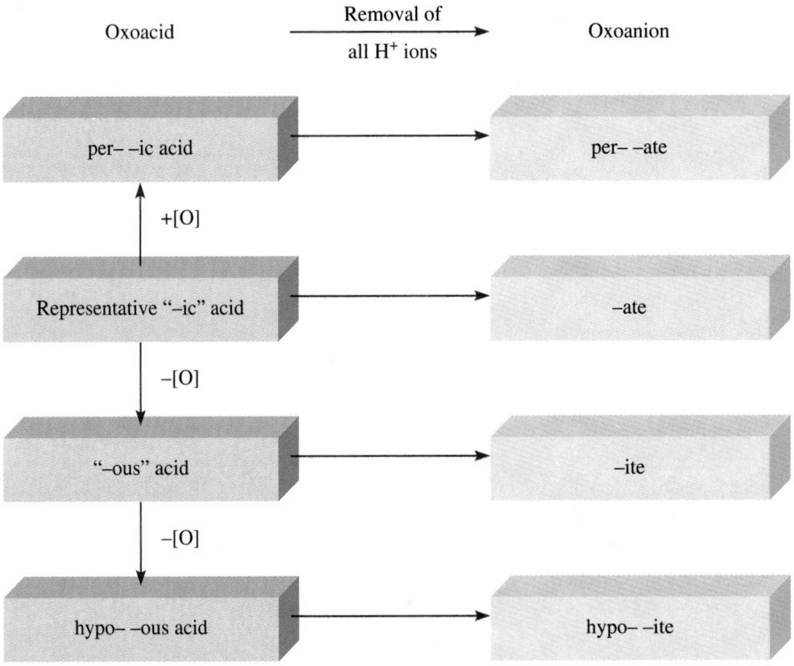

FIGURE 2.11

Naming oxoacids and oxoanions.

TABLE 2.6
Names of Oxoacids and Oxoanions That Contain Chlorine

Acid	Anion
$HClO_4$ (perchloric acid)	ClO_4^- (perchlorate)
$HClO_3$ (chloric acid)	ClO_3^- (chlorate)
$HClO_2$ (chlorous acid)	ClO_2^- (chlorite)
$HClO$ (hypochlorous acid)	ClO^- (hypochlorite)

EXAMPLE 2.7
Naming an Oxoacid and an Oxoanion

Name the following oxoacid and oxoanion: (a) H_3PO_3, (b) IO_4^-.

Answer: (a) We start with our reference acid, phosphoric acid (H_3PO_4). Since H_3PO_3 has one fewer O atom, it is called phosphorous acid.
(b) The parent acid is HIO_4. Since the acid has one more O atom than our reference iodic acid (HIO_3), it is called periodic acid. Therefore, the anion derived from HIO_4 is called periodate.

PRACTICE EXERCISE

Name the following oxoacid and oxoanion: (a) $HBrO_3$, (b) HSO_4^-.

Naming Bases

A ***base*** can be described as *a substance that yields hydroxide ions (OH^-)* when dissolved in water. Some examples are:

NaOH	Sodium hydroxide	$Ba(OH)_2$	Barium hydroxide
KOH	Potassium hydroxide		

Ammonia (NH_3), a molecular compound in the gaseous or pure liquid state, is also classified as a common base. At first glance this may seem to be an exception to the definition of a base. But note that as long as a substance *yields* hydroxide ions when dissolved in water, it need not contain hydroxide ions in its structure to be considered a base. In fact, when ammonia dissolves in water, NH_3 reacts partially with water to yield NH_4^+ and OH^- ions. Thus it is properly classified as a base.

SUMMARY

Modern chemistry began with Dalton's atomic theory, which states that all matter is composed of tiny, indivisible particles called atoms; that all atoms of the same element are identical; that compounds contain atoms of different elements combined in whole-number ratios; and that atoms are neither created nor destroyed in chemical reactions (the law of conservation of mass).

Atoms of constituent elements in a particular compound are always combined in the same proportions by mass (law of definite proportions). When two elements can combine to form more than one type of compound, the masses of one element that combine with a fixed mass of the other element are in a ratio of small whole numbers (law of multiple proportions).

An atom consists of a very dense central nucleus made up of protons and neutrons, plus electrons that move about the nucleus at a relatively large distance from it. Protons are positively charged, neutrons have no charge, and electrons are negatively charged. Protons and neutrons have roughly the same mass, which is about 1840 times greater than the mass of an electron.

The atomic number of an element is the number of protons in the nucleus of an atom of the element; it determines the identity of an element. The mass number is the sum of the number of protons and the number of neutrons in the nucleus. Isotopes are atoms of the same element that have the same number of protons but different numbers of neutrons.

Chemical formulas combine the symbols for the constituent elements with whole-number subscripts to show the type and number of atoms contained in the smallest unit of a compound. The molecular formula conveys the specific number and types of atoms combined in each molecule of a compound. The empirical formula shows the simplest ratios of the atoms in a molecule.

Chemical compounds are either molecular compounds (in which the smallest units are discrete, individual molecules) or ionic compounds (in which positive and negative ions are held together by mutual attraction). Ionic compounds are made up of cations and anions, formed when atoms lose electrons and gain electrons, respectively.

The names of many inorganic compounds can be deduced from a set of simple rules. The formulas can be written from the names of the compounds.

KEY WORDS

Acid, p. 45
Alkali metals, p. 36
Alkaline earth metals, p. 36
Allotrope, p. 38
Alpha (α) particles, p. 32
Alpha (α) rays, p. 31
Anion, p. 37
Atom, p. 29
Atomic number (Z), p. 34
Base, p. 47
Beta (β) particles, p. 32
Beta (β) rays, p. 32
Binary compound, p. 41
Cation, p. 37
Chemical formula, p. 38
Diatomic molecule, p. 37
Electron, p. 30
Empirical formula, p. 39
Families, p. 35
Gamma (γ) rays, p. 32
Groups, p. 35
Halogens, p. 36
Ion, p. 37
Ionic compound, p. 37
Isotope, p. 34
Law of conservation of mass, p. 29
Law of definite proportions, p. 28
Law of multiple proportions, p. 29
Mass number (A), p. 34
Metal, p. 35
Metalloid, p. 35
Molecular formula, p. 38
Molecule, p. 36
Monatomic ion, p. 38
Neutron, p. 33
Noble gases, p. 36
Nonmetal, p. 35
Nucleus, p. 32
Oxoacid, p. 45
Oxoanion, p. 46
Periodic table, p. 35
Periods, p. 35
Polyatomic ion, p. 38
Polyatomic molecule, p. 37
Proton, p. 32
Radiation, p. 29
Radioactivity, p. 31
Rare gases, p. 36
Ternary compound, p. 41

QUESTIONS AND PROBLEMS

STRUCTURE OF THE ATOM

Review Questions

2.1 Define the following terms: (a) α particle, (b) β particle, (c) γ ray, (d) X ray.

2.2 List the types of radiation that are known to be emitted by radioactive elements.

2.3 Compare the properties of the following: α particles, cathode rays, protons, neutrons, electrons. What is meant by the term "fundamental particle"?

2.4 Describe the contributions of the following scientists to our knowledge of atomic structure: J. J. Thomson, R. A. Millikan, Ernest Rutherford, James Chadwick.

2.5 A sample of a radioactive element is found to be losing mass gradually. Explain what is happening to the sample.

2.6 Describe the experimental basis for believing that the nucleus occupies a very small fraction of the volume of the atom.

Problems

2.7 The diameter of a neutral helium atom is about 1×10^2 pm. Suppose that we could line up helium atoms side by side in contact with one another. Approximately how many atoms would it take to make the distance from end to end 1 cm?

2.8 Roughly speaking, the radius of an atom is about 10,000 times greater than that of its nucleus. If an atom were magnified so that the radius of its nucleus became 10 cm, what would be the radius of the atom in miles? (1 mi = 1609 m.)

ATOMIC NUMBER, MASS NUMBER, AND ISOTOPES

Review Questions

2.9 Define the following terms: (a) atomic number, (b) mass number. Why does a knowledge of atomic number enable us to deduce the number of electrons present in an atom?

2.10 Why do all atoms of an element have the same atomic number, although they may have different mass numbers? What do we call atoms of the same elements with different mass numbers? Explain the meaning of each term in the symbol ${}^{A}_{Z}X$.

Problems

2.11 What is the mass number of an iron atom that has 28 neutrons?

2.12 Calculate the number of neutrons of Pu-239.

2.13 For each of the following species, determine the number of protons and the number of neutrons in the nucleus:

$${}^{3}_{2}He,\ {}^{4}_{2}He,\ {}^{24}_{12}Mg,\ {}^{25}_{12}Mg,\ {}^{48}_{22}Ti,\ {}^{79}_{35}Br,\ {}^{195}_{78}Pt$$

2.14 Indicate the number of protons, neutrons, and electrons in each of the following species:

$${}^{15}_{7}N,\ {}^{33}_{16}S,\ {}^{63}_{29}Cu,\ {}^{84}_{38}Sr,\ {}^{130}_{56}Ba,\ {}^{186}_{74}W,\ {}^{202}_{80}Hg$$

2.15 Write the appropriate symbol for each of the following isotopes: (a) $Z = 11, A = 23$; (b) $Z = 28, A = 64$.

2.16 Write the appropriate symbol for each of the following isotopes: (a) $Z = 74$, $A = 186$; (b) $Z = 80$; $A = 201$.

THE PERIODIC TABLE

Review Questions

2.17 What is the periodic table, and what is its significance in the study of chemistry? What are groups and periods in the periodic table?

2.18 Give two differences between a metal and a nonmetal.

2.19 Write the names and symbols for four elements in each of the following categories: (a) nonmetal, (b) metal, (c) metalloid.

2.20 Define, with two examples, the following terms: (a) alkali metals, (b) alkaline earth metals, (c) halogens, (d) noble gases.

2.21 Elements whose names end with "-ium" are usually metals; sodium is one example. Identify a nonmetal whose name also ends with "-ium."

2.22 Describe the changes in properties (from metals to nonmetals or from nonmetals to metals) as we move (a) down a periodic group and (b) across the periodic table.

2.23 Consult a handbook of chemical and physical data (ask your instructor where you can locate a copy of the handbook) to find (a) two metals less dense than water, (b) two metals more dense than mercury, (c) the densest known solid metallic element, (d) the densest known solid nonmetallic element.

2.24 Group the following elements in pairs that you would expect to show similar chemical properties: K, F, P, Na, Cl, and N.

MOLECULES AND CHEMICAL FORMULAS

Review Questions

2.25 What is the difference between an atom and a molecule?

2.26 What are allotropes?

2.27 What does a chemical formula represent? What is the ratio of the atoms in the following molecular formulas? (a) NO, (b) NCl_3, (c) N_2O_4, (d) P_4O_6.

2.28 Define molecular formula and empirical formula. What are the similarities and differences between the empirical formula and molecular formula of a compound?

2.29 Give an example of a case in which two molecules have different molecular formulas but the same empirical formula.

2.30 What does P_4 signify? How does it differ from 4P?

2.31 What is the difference between a molecule and a compound? Write a formula for a molecule that is also a compound and a formula for a molecule that is not a compound.

2.32 Give two examples of each of the following: (a) a diatomic molecule containing atoms of the same element, (b) a diatomic molecule containing atoms of different elements, (c) a polyatomic molecule containing atoms of the same element, (d) a polyatomic molecule containing atoms of different elements.

Problems

2.33 What are the empirical formulas of the following compounds? (a) C_2N_2, (b) C_6H_6, (c) C_9H_{20}, (d) P_4O_{10}, (e) B_2H_6.

2.34 What are the empirical formulas of the following compounds? (a) Al_2Br_6, (b) $Na_2S_2O_4$, (c) N_2O_5, (d) $K_2Cr_2O_7$.

IONIC COMPOUNDS

Review Questions

2.35 What is an ionic compound? How is electrical neutrality maintained in an ionic compound? Explain why the chemical formulas of ionic compounds are always the same as their empirical formulas.

2.36 Compare and contrast the properties of molecular compounds and ionic compounds.

Problems

2.37 Give the number of protons and electrons in each of the following ions: Na^+, Ca^{2+}, Al^{3+}, Fe^{2+}, I^-, F^-, S^{2-}, O^{2-}, N^{3-}.

2.38 Give the number of protons and electrons in each of the following ions: K^+, Mg^{2+}, Fe^{3+}, Br^-, Mn^{2+}, C^{4-}, Cu^{2+}.

2.39 Which of the following compounds are likely to be ionic? Which are likely to be molecular? $SiCl_4$, LiF, $BaCl_2$, B_2H_6, KCl, C_2H_4.

2.40 Which of the following compounds are likely to be ionic? Which are likely to be molecular? CH_4, NaBr, BaF_2, CCl_4, ICl, CsCl, NF_3.

NAMING INORGANIC COMPOUNDS

Problems

2.41 Name the following compounds: (a) KH_2PO_4, (b) K_2HPO_4, (c) HBr (gas), (d) HBr (in water), (e) Li_2CO_3, (f) $K_2Cr_2O_7$, (g) NH_4NO_2, (h) PF_3, (i) PF_5, (j) P_4O_6, (k) CdI_2, (l) $SrSO_4$, (m) $Al(OH)_3$.

2.42 Name the following compounds: (a) KClO, (b) Ag_2CO_3, (c) $FeCl_2$, (d) $KMnO_4$, (e) $CsClO_3$, (f) KNH_4SO_4, (g) FeO, (h) Fe_2O_3, (i) $TiCl_4$, (j) NaH, (k) Li_3N, (l) Na_2O, (m) Na_2O_2.

2.43 Write the formulas for the following compounds: (a) rubidium nitrite, (b) potassium sulfide, (c) sodium hydrogen sulfide, (d) magnesium phosphate, (e) calcium hydrogen phosphate, (f) potassium dihydrogen phosphate, (g) iodine heptafluoride, (h) ammonium sulfate, (i) silver perchlorate, (j) iron(III) chromate.

2.44 Write the formulas for the following compounds: (a) copper(I) cyanide, (b) strontium chlorite, (c) perbromic acid, (d) hydroiodic acid, (e) disodium ammonium phosphate, (f) lead(II) carbonate, (g) tin(II) fluoride, (h) tetraphosphorus decasulfide, (i) mercury(II) oxide, (j) mercury(I) iodide.

MISCELLANEOUS PROBLEMS

2.45 One isotope of a metallic element has a mass number of 65 and has 35 neutrons in the nucleus. The cation derived from the isotope has 28 electrons. Write the symbol for this cation.

2.46 In which one of the following pairs do the two species resemble each other most closely in chemical properties? (a) ${}^1_1H$ and ${}^1_1H^+$, (b) ${}^{14}_{7}N$ and ${}^{14}_{7}N^{3-}$, (c) ${}^{12}_{6}C$ and ${}^{13}_{6}C$.

2.47 The following table gives numbers of electrons, protons, and neutrons in atoms or ions of a number of elements. Answer the following: (a) Which of the species are neutral? (b) Which are negatively charged? (c) Which are positively charged? (d) What are the conventional symbols for all the species?

Atom or ion or element	A	B	C	D	E	F	G
Number of electrons	5	10	18	28	36	5	9
Number of protons	5	7	19	30	35	5	9
Number of neutrons	5	7	20	36	46	6	10

2.48 What is wrong or ambiguous about the following statements? (a) 1 mole of hydrogen, (b) four molecules of NaCl.

2.49 The following phosphorus sulfides are known:

P_4S_3, P_4S_7, and P_4S_{10}. Do these compounds obey the law of multiple proportions?

2.50 Which of the following are elements, which are molecules but not compounds, which are compounds but not molecules, and which are both compounds and molecules? (a) SO_2, (b) S_8, (c) Cs, (d) N_2O_5, (e) O, (f) O_2, (g) O_3, (h) CH_4, (i) KBr, (j) S, (k) P_4, (l) LiF.

2.51 Why is magnesium chloride ($MgCl_2$) not called magnesium(II) chloride?

2.52 Some compounds are better known by their common names than by their systematic chemical names. Consult a handbook, a dictionary, or your instructor for the chemical formulas of the following substances: (a) dry ice, (b) table salt, (c) laughing gas, (d) marble (chalk, limestone), (e) quicklime, (f) slaked lime, (g) baking soda, (h) milk of magnesia.

2.53 Fill in the blanks in the following table:

Symbol		$^{54}_{26}Fe^{2+}$			
Protons	5			79	86
Neutrons	6		16	117	136
Electrons	5		18	79	
Net charge			−3		0

2.54 (a) Which elements are most likely to form ionic compounds? (b) Which metallic elements are most likely to form cations with different charges?

2.55 Most ionic compounds contain either aluminum (a Group 3A metal) or a metal from Group 1A or Group 2A and a nonmetal-oxygen, nitrogen, or a halogen (Group 7A). Write the chemical formulas and names of all the binary compounds that can result from such combinations.

2.56 Which of the following symbols provides more information about the atom: ^{23}Na or $_{11}Na$? Explain.

2.57 Write the chemical formulas and names of acids that contain Group 7A elements. Do the same for elements in Groups 3A, 4A, 5A, and 6A.

2.58 Of the 109 elements known, only two are liquids at room temperature (25°C). What are they? (*Hint:* One element is a familiar metal and the other element is in Group 7A.)

2.59 For the noble gases (the Group 8A elements), $^{4}_{2}He$, $^{20}_{10}Ne$, $^{40}_{18}Ar$, $^{84}_{36}Kr$, and $^{132}_{54}Xe$, (a) determine the number of protons and neutrons in the nucleus of each atom and (b) determine the ratio of neutrons to protons in the nucleus of each atom. Describe any general trend you discover in the way this ratio changes with increasing atomic number.

2.60 List the elements that exist as gases at room temperature. (*Hint:* These elements can be found in Groups 5A, 6A, 7A, and 8A.)

2.61 The Group 1B metals, Cu, Ag, and Au, are called coinage metals. What chemical properties make them specially suitable for making coins and jewels?

2.62 The elements in Group 8A of the periodic table are called noble gases. Can you guess the meaning of "noble" in this context?

Answers to Practice Exercises: 2.1 29 protons, 34 neutrons, and 29 electrons; **2.2** $C_4H_5N_2O$; **2.3** (a) Lead(II) oxide, (b) lithium sulfite; **2.4** (a) Rb_2SO_4, (b) BaH_2; **2.5** (a) Nitrogen trifluoride, (b) dichlorine heptoxide; **2.6** (a) SF_4, (b) N_2O_5; **2.7** (a) Bromic acid, (b) hydrogen sulfate ion.

CHAPTER 3

STOICHIOMETRY

◆ When wood, paper, and wax are burned, they appear to lose mass. The decrease in mass that results from these combustion reactions was once attributed to the release of "phlogiston" into the air. For most of the eighteenth century scientists accepted the phlogiston theory. Then, in August 1774, the English chemist and clergyman Joseph Priestley isolated oxygen—which he called "dephlogisticated air"—as a product of the decomposition of mercury(II) oxide, HgO. The French chemist Antoine Lavoisier had noticed that nonmetals like phosphorus actually *gain* mass when they burn in air. He concluded that these nonmetals must combine with something in the air. This substance turned out to be Priestley's dephlogisticated air. Lavoisier named the new element "oxygen" (from the Greek word meaning "to form acid"), because he knew that it is also a constituent of all acids.

Born in 1743 Lavoisier is generally regarded as the father of modern chemistry. He was noted for his carefully controlled experiments and for the use of quantitative measurements. By carrying out chemical reactions, such as the decomposition of mercury(II) oxide, in a closed container, he showed that the total mass of the products equals the total mass of the reactants. In other words, the quantity of matter is not changed by chemical reactions. This observation is the basis of the law of conservation of mass and the principle underlying stoichiometry.

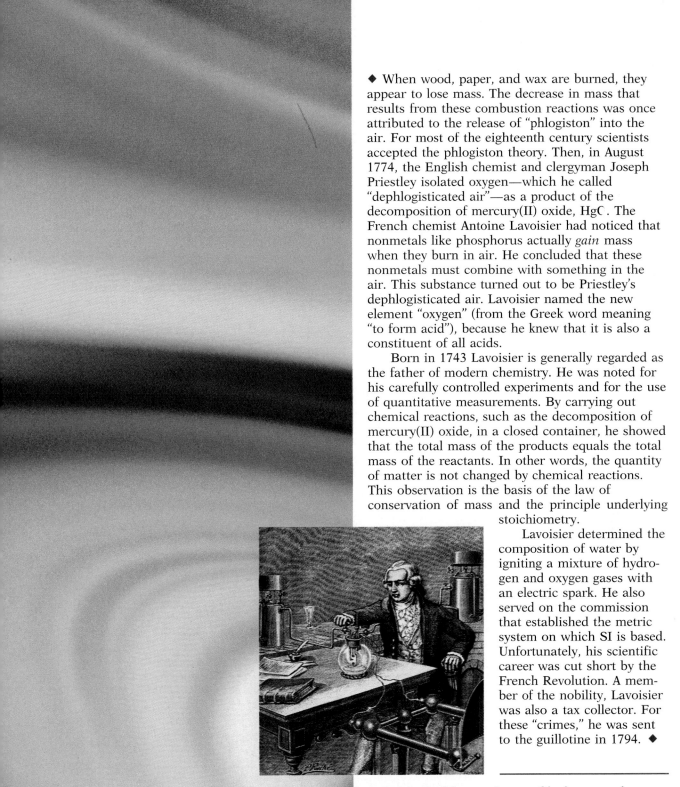

Lavoisier determined the composition of water by igniting a mixture of hydrogen and oxygen gases with an electric spark. He also served on the commission that established the metric system on which SI is based. Unfortunately, his scientific career was cut short by the French Revolution. A member of the nobility, Lavoisier was also a tax collector. For these "crimes," he was sent to the guillotine in 1794. ◆

Lavoisier igniting a mixture of hydrogen and oxygen gases.

3.1 ATOMIC MASS

In this chapter we will use what we have learned about chemical structure and formulas in studying the mass relationships of atoms and molecules. These relationships in turn will help us to explain the composition of compounds and the ways in which the composition changes.

The mass of an atom is related to the number of electrons, protons, and neutrons it has. Knowledge of an atom's mass is important in laboratory work. But atoms are extremely small particles—even the smallest speck of dust that our unaided eyes can detect contains as many as 1×10^{16} atoms! Clearly we cannot weigh a single atom, but it is possible to determine the mass of one atom *relative* to another experimentally. The first step is to assign a value to the mass of one atom of a given element so that it can be used as a standard.

By international agreement, an atom of the carbon isotope (called carbon-12) that has six protons and six neutrons has a mass of exactly 12 ***atomic mass units (amu).*** This carbon-12 atom serves as the standard, so one atomic mass unit is defined as *a mass exactly equal to one-twelfth the mass of one carbon-12 atom:*

$$\text{mass of one carbon-12 atom} = 12 \text{ amu}$$

$$1 \text{ amu} = \frac{\text{mass of one carbon-12 atom}}{12}$$

Experiments have shown that, on average, a hydrogen atom is only 8.400 percent as massive as the standard carbon-12 atom. If we accept the mass of one carbon-12 atom to be exactly 12 amu, then the ***atomic mass*** (that is, *the mass of the atom in atomic mass units*) of hydrogen must be 0.08400×12.00 amu, or 1.008 amu. (Note that atomic mass is also called atomic weight.) Similar calculations show that the atomic mass of oxygen is 16.00 amu and that of iron is 55.85 amu. Thus although we do not know just how much an average iron atom's mass is, we know that is approximately 56 times as massive as a hydrogen atom.

AVERAGE ATOMIC MASS

When you look up the atomic mass of carbon in a table such as the one on the inside front cover of this book, you will find that its value is not 12.00 amu but 12.01 amu. The reason for the difference is that most naturally occurring elements (including carbon) have more than one isotope. This means that when we measure the atomic mass of an element, we must generally settle for the *average* mass of the naturally occurring mixture of isotopes. For example, the natural abundances of carbon-12 and carbon-13 are 98.89 percent and 1.11 percent, respectively. The atomic mass of carbon-13 has been determined to be 13.00335 amu. Thus the average atomic mass of carbon can be calculated as follows:

$$\begin{aligned}\text{average atomic mass of natural carbon} \\ &= (0.9890)(12.00000 \text{ amu}) + (0.0110)(13.0335 \text{ amu}) \\ &= 12.0 \text{ amu}\end{aligned}$$

A more accurate determination gives the atomic mass of carbon as 12.01 amu. Note that in calculations involving percentages, we need to convert percentages to fractions. For example, 98.90 percent becomes 98.90/100, or 0.9890. Because there are many more carbon-12 atoms than carbon-13 atoms in naturally occurring carbon, the average atomic mass is much closer to 12 amu than 13 amu.

It is important to understand that when we say that the atomic mass of carbon is 12.01 amu, we are referring to the *average* value. If carbon atoms could be examined individually, we would find either an atom of atomic mass 12.00000 amu or one of 13.00335 amu, but never one of 12.01 amu.

EXAMPLE 3.1
Calculating Average Atomic Mass

Copper, a metal known since ancient times, is used in electrical cables and pennies, among other things. The atomic mass of its two stable isotopes, $^{63}_{29}Cu$ (69.09%) and $^{65}_{29}Cu$ (30.91%), are 62.93 amu and 64.9278 amu, respectively. Calculate the average atomic mass of copper. The percentages in parentheses denote the relative abundances.

Answer: Converting the percentages to fractions, we calculate the average atomic mass as follows:

$$(0.6909)(62.93 \text{ amu}) + (0.3091)(64.9278 \text{ amu}) = 63.55 \text{ amu}$$

PRACTICE EXERCISE

The atomic masses of the two stable isotopes of boron, $^{10}_{5}B$ (19.78%) and $^{11}_{5}B$ (80.22%), are 10.0129 amu and 11.0093 amu, respectively. Calculate the average atomic mass of boron.

Copper.

The atomic masses of many elements have been accurately determined to five or six significant figures. However, for purposes of calculation in this text, we will normally use atomic masses accurate only to four significant figures (see the table of atomic masses on the inside front cover).

3.2 MOLAR MASS OF AN ELEMENT AND AVOGADRO'S NUMBER

Atomic mass units provide a relative scale for the masses of the elements. But since atoms have such small masses, no usable scale can be devised to weigh them in calibrated units of atomic mass units. In any real situation we deal with samples containing enormous numbers of atoms. Therefore chemists have a special unit to describe a very large number of atoms. The idea of a unit to describe a particular number of objects is not new. For example, the pair (2 items), the dozen (12 items), and the gross (144 items) are all familiar units.

The unit defined by the SI system is the ***mole (mol),*** which is *the amount of substance that contains as many elementary entities (atoms, molecules, or other particles) as there are atoms in exactly 12 grams (or 0.012 kilogram) of the carbon-12 isotope.* Notice that this definition specifies only the *method* by which the number of elementary entities in a mole may be found. The actual *number* is determined experimentally. The currently accepted value is

$$1 \text{ mole} = 6.022045 \times 10^{23} \text{ particles}$$

Amedeo Avogadro (1776–1856)

This number is called ***Avogadro's number,*** in honor of the Italian scientist Amedeo Avogadro. For most calculations, we round this number to 6.022×10^{23}. Thus, just as 1 dozen oranges contains 12 oranges, 1 mole of hydrogen atoms contains 6.022×10^{23} H atoms. Figure 3.1 shows 1 mole each of several common elements.

We have seen that 1 mole of carbon-12 atoms has a mass of exactly 12 g and contains 6.022×10^{23} atoms. This mass of carbon-12 is its ***molar mass,*** defined as *the mass (in grams or kilograms) of 1 mole of units* (such as atoms or molecules) of a substance. Note that the molar mass of carbon-12 (in grams) is numerically equal to its atomic mass in amu. Likewise, the atomic mass of sodium (Na) is 22.99 amu, and its molar mass is 22.99 g; the atomic mass of phosphorus is 30.97 amu, and its molar mass is 30.97 g; and so on. If we know the atomic mass of an element, we also know its molar mass.

We can now calculate the mass (in grams) of a single carbon-12 atom. From our discussion we know that 1 mole of carbon-12 atoms weighs exactly 12 g. This means that

$$12.00 \text{ g carbon-12} = 1 \text{ mol carbon-12 atoms}$$

Therefore, we can write the unit factor as

$$\frac{12.00 \text{ g carbon-12}}{1 \text{ mol carbon-12 atoms}} = 1$$

(Note that we have used the unit "mol" to represent mole.) Similarly, since there are 6.022×10^{23} atoms in 1 mole of carbon-12 atoms, we have

$$1 \text{ mol carbon-12 atoms} = 6.022 \times 10^{23} \text{ carbon-12 atoms}$$

and the unit factor is

$$\frac{1 \text{ mol carbon-12 atoms}}{6.022 \times 10^{23} \text{ carbon-12 atoms}} = 1$$

We can now calculate the mass (in grams) of 1 carbon-12 atom as follows:

FIGURE 3.1

One mole each of several common elements. Clockwise from top left: carbon (black charcoal powder), sulfur (yellow powder), iron (as nails), copper (as pennies), and, in the center, mercury (shiny liquid metal).

$$1\ \cancel{\text{carbon-12 atom}} \times \frac{1\ \cancel{\text{mol carbon-12 atoms}}}{6.022 \times 10^{23}\ \cancel{\text{carbon-12 atoms}}} \times \frac{12.00\ \text{g carbon-12}}{1\ \cancel{\text{mol carbon-12 atoms}}}$$
$$= 1.993 \times 10^{-23}\ \text{g carbon-12}$$

Furthermore, we can find the relationship between atomic mass units and grams by noting that since the mass of every carbon-12 atom is exactly 12 amu, the number of grams equivalent to 1 amu is

$$\frac{\text{gram}}{\text{amu}} = \frac{1.993 \times 10^{-23}\ \text{g}}{1\ \cancel{\text{carbon-12 atom}}} \times \frac{1\ \cancel{\text{carbon-12 atom}}}{12\ \text{amu}}$$
$$= 1.661 \times 10^{-24}\ \text{g/amu}$$

Thus

$$1\ \text{amu} = 1.661 \times 10^{-24}\ \text{g}$$

and

$$1\ \text{g} = 6.022 \times 10^{23}\ \text{amu}$$

This example shows that Avogadro's number can be used to convert from the atomic mass unit to the mass in grams and vice versa.

Avogadro's number and the concept of molar mass enable us to carry out conversions between mass and moles of atoms and between the number of atoms and mass and to calculate the mass of a single atom. Examples 3.2 to 3.4 show how to carry out these conversions. We will employ the following unit factors in the calculations:

$$\frac{1\ \text{mol X}}{\text{molar mass of X}} = 1$$

$$\frac{1\ \text{mol X}}{6.022 \times 10^{23}\ \text{X atoms}} = 1$$

where X represents the symbol of an element.

EXAMPLE 3.2

Converting between Moles of Atoms and Mass of Atoms

Zinc (Zn) is a silvery metal that is used to form brass (with copper) and plate iron to prevent corrosion. (a) How many grams of Zn are there in 0.356 mole of Zn? (b) How many moles of Zn are in 668 g of Zn?

Answer (a) Since the molar mass of Zn is 65.39 g, the mass of Zn in grams is given by

$$0.356\ \cancel{\text{mol Zn}} \times \frac{65.39\ \text{g Zn}}{1\ \cancel{\text{mol Zn}}} = 23.3\ \text{g Zn}$$

Thus there are 23.3 g of Zn in 0.356 mole of Zn.

(b) From the known molar mass of Zn, we can convert the amount of Zn to moles:

$$668\ \cancel{\text{g Zn}} \times \frac{1\ \text{mol Zn}}{65.39\ \cancel{\text{g Zn}}} = 10.2\ \text{mol Zn}$$

Zinc.

PRACTICE EXERCISE

(a) Calculate the number of grams of helium (He) in 1.61 moles of helium.
(b) Calculate the number of moles of magnesium (Mg) in 0.317 g of Mg.

EXAMPLE 3.3
Calculating the Mass of a Single Atom

Silver.

Silver (Ag) is a precious metal used mainly in jewelry. What is the mass (in grams) of one Ag atom?

Answer: The molar mass of silver is 107.9 g. Since there are 6.022×10^{23} Ag atoms in 1 mole of Ag, the mass of 1 Ag atom is

$$1 \text{ Ag atom} \times \frac{1 \text{ mol Ag atoms}}{6.022 \times 10^{23} \text{ Ag atoms}} \times \frac{107.9 \text{ g Ag}}{1 \text{ mol Ag atoms}} = 1.792 \times 10^{-22} \text{ g Ag}$$

PRACTICE EXERCISE

What is the mass (in grams) of one iodine (I) atom?

EXAMPLE 3.4
Converting Mass in Grams to Number of Atoms

Sulfur.

Sulfur (S) is a nonmetallic element. Its presence in coal gives rise to the acid rain phenomenon. How many atoms are in 16.3 g of S?

Answer: Solving this problem requires two steps. First, we need to find the number of moles of S in 16.3 g of S (as in Example 3.2). Next, we need to calculate the number of S atoms from the known number of moles of S. We can combine the two steps as follows:

$$16.3 \text{ g S} \times \frac{1 \text{ mol S}}{32.07 \text{ g S}} \times \frac{6.022 \times 10^{23} \text{ S atoms}}{1 \text{ mol S}} = 3.06 \times 10^{23} \text{ S atoms}$$

PRACTICE EXERCISE

Calculate the number of atoms in 0.551 g of potassium (K).

3.3 MOLECULAR MASS

If we know the atomic masses, we can calculate the masses of molecules. The ***molecular mass*** is *the sum of the atomic masses (in amu) in the molecule.* For example, the molecular mass of H_2O is

$$2(\text{atomic mass of H}) + \text{atomic mass of O}$$

or $$2(1.008\text{ amu}) + 16.00\text{ amu} = 18.02\text{ amu}$$

EXAMPLE 3.5
Calculating Molecular Mass

Calculate the molecular masses of the following compounds: (a) sulfur dioxide (SO_2), which is mainly responsible for acid rain; (b) ascorbic acid, or vitamin C ($C_6H_8O_6$).

Answer: (a) From the atomic masses of S and O we get

$$\text{molecular mass of } SO_2 = 32.07\text{ amu} + 2(16.00\text{ amu})$$
$$= 64.07\text{ amu}$$

(b) From the atomic masses of C, H, and O we get

$$\text{molecular mass of } C_6H_8O_6 = 6(12.01\text{ amu}) + 8(1.008\text{ amu}) + 6(16.00\text{ amu})$$
$$= 176.12\text{ amu}$$

PRACTICE EXERCISE

What is the molecular mass of methanol (CH_4O)?

From the molecular mass we can determine the molar mass of a molecule or compound. The molar mass of a compound (in grams) is numerically equal to its molecular mass (in atomic mass units). For example, the molecular mass of water is 18.02 amu, so its molar mass is 18.02 g. Note that 1 mole of water weighs 18.02 g and contains 6.022×10^{23} H_2O *molecules,* just as 1 mole of elemental carbon contains 6.022×10^{23} carbon *atoms.*

As the following two examples show, a knowledge of the molar mass enables us to calculate the numbers of moles and individual atoms in a given quantity of a compound.

EXAMPLE 3.6
Calculating the Number of Moles in a Given Amount of a Compound

Methane (CH_4) is the principal component of natural gas. How many moles of CH_4 are present in 6.07 g of CH_4?

Answer: First we calculate the molar mass of CH_4:

$$\text{molar mass of } CH_4 = 12.01\text{ g} + 4(1.008\text{ g})$$
$$= 16.04\text{ g}$$

We then follow the procedure in Example 3.2:

$$6.07\text{ }\cancel{\text{g CH}_4} \times \frac{1\text{ mol CH}_4}{16.04\text{ }\cancel{\text{g CH}_4}} = 0.378\text{ mol CH}_4$$

Methane gas burning on a cooking range.

PRACTICE EXERCISE

Calculate the number of moles of chloroform ($CHCl_3$) in 198 g of chloroform.

EXAMPLE 3.7

Calculating the Number of Atoms in a Given Amount of a Compound

Urea.

How many hydrogen atoms are present in 25.6 g of urea [$(NH_2)_2CO$], which is used as fertilizer, in animal feed, and in the manufacture of polymers? The molar mass of urea is 60.06 g.

Answer: There are 4 hydrogen atoms in every urea molecule; therefore, the total number of hydrogen atoms is

$$25.6 \text{ g } (NH_2)_2CO \times \frac{1 \text{ mol } (NH_2)_2CO}{60.06 \text{ g } (NH_2)_2CO} \times \frac{6.022 \times 10^{23} \text{ molecules } (NH_2)_2CO}{1 \text{ mol } (NH_2)_2CO} \times \frac{4 \text{ H atoms}}{1 \text{ molecule } (NH_2)_2CO} = 1.03 \times 10^{24} \text{ H atoms}$$

We could calculate the number of nitrogen, carbon, and oxygen atoms by the same procedure. However, there is a shortcut. Note that the ratio of nitrogen to hydrogen atoms in urea is $\frac{2}{4}$, or $\frac{1}{2}$, and that of oxygen (and carbon) to hydrogen atom is $\frac{1}{4}$. Therefore the number of nitrogen atoms in 25.6 g of urea is $(\frac{1}{2})(1.03 \times 10^{24})$, or 5.15×10^{23} atoms. The number of oxygen atoms (and carbon atoms) is $(\frac{1}{4})(1.03 \times 10^{24})$, or 2.58×10^{23} atoms.

PRACTICE EXERCISE

How many H atoms are in 72.5 g of isopropanol, commonly called rubbing alcohol, C_3H_8O?

3.4 THE MASS SPECTROMETER

The most direct and most accurate method for determining atomic and molecular masses is mass spectrometry, which is depicted in Figure 3.2. In a *mass spectrometer,* a gaseous sample is bombarded by a stream of high-energy electrons. Collisions between the electrons and the gaseous atoms (or molecules) produce positive ions by dislodging an electron from each atom or molecule. These positive ions (of mass *m* and charge *e*) are accelerated by two oppositely charged plates as they pass through them. The emerging ions are deflected into a circular path by a magnet.

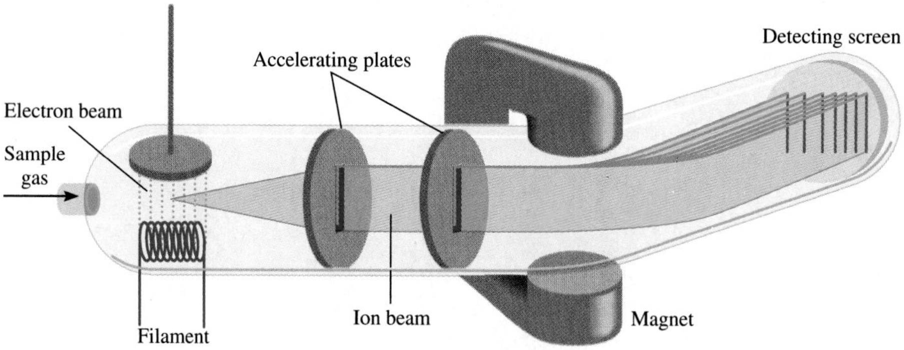

FIGURE 3.2

Schematic diagram of one type of mass spectrometer.

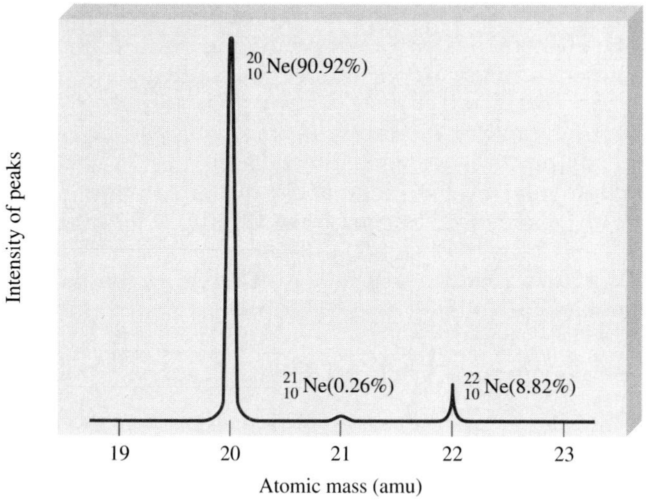

FIGURE 3.3

The mass spectrum of the three isotopes of neon.

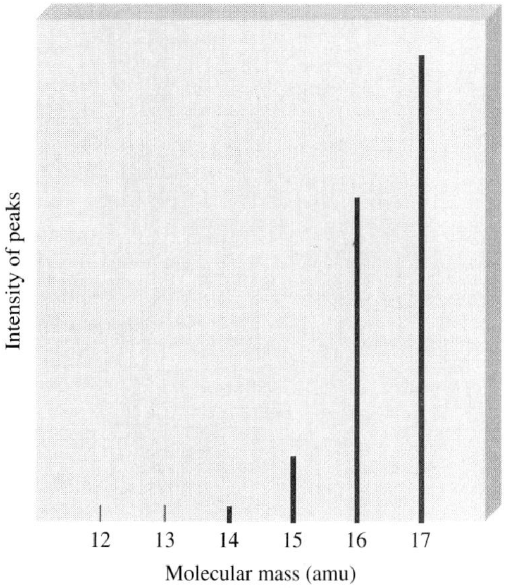

FIGURE 3.4

The mass spectrum of ammonia (NH_3) in bar graph form. The tallest peak at 17 amu corresponds to the NH_3^+ ion; the other peaks are fragment ions derived from the parent ion.

The radius of the path depends on the charge-to-mass ratio (that is, e/m). Ions of smaller e/m ratio trace a wider curve than those having a larger e/m ratio, so that ions with equal charges but different masses are separated from one another. Eventually they arrive at the detector, which registers a current for each type of ion. The amount of current generated is directly proportional to the number of ions, so it enables us to determine the relative abundance of isotopes.

The first mass spectrometer, developed in the 1920s by the English physicist F. W. Aston, was crude by today's standards. Nevertheless, it provided indisputable evidence of the existence of isotopes—neon-20 (atomic mass 19.9924 amu and natural abundance 90.92 percent) and neon-22 (atomic mass 21.9914 amu and natural abundance 8.82 percent). When more sophisticated and sensitive mass spectrometers became available, scientists were surprised to discover that neon has a third stable isotope with an atomic mass of 20.9940 amu and natural abundance 0.257 percent (Figure 3.3). This example illustrates how very important experimental accuracy is to a quantitative science like chemistry. Early experiments failed to detect neon-21 because its natural abundance is just 0.257 percent. In other words, only 26 in 10,000 Ne atoms are neon-21.

The masses of molecules are determined in a similar manner. Figure 3.4 shows the mass spectrum of ammonia (NH_3). The highest peak (at 17.03 amu) corresponds to the positively charged ammonia ion, NH_3^+. The other peaks represent cations produced by the breakdown of the NH_3^+ ion.

3.5 PERCENT COMPOSITION OF COMPOUNDS

As we have seen, the formula of a compound tells us the relative numbers of atoms of each element in a compound. However, suppose we had a sample of a compound in the laboratory and we needed to verify its purity for use in an experiment.

From the formula we could calculate what percent of the total mass of the compound is contributed by each element. Then, by comparing the result to the percent composition obtained experimentally for our sample, we could determine the purity of the sample.

The ***percent composition by mass*** is *the percent by mass of each element the compound contains.* Percent composition is obtained by dividing the mass of each element in 1 mole of the compound by the molar mass of the compound and multiplying by 100 percent. In 1 mole of hydrogen peroxide (H_2O_2), for example, there are 2 moles of H atoms and 2 moles of O atoms. The molar masses of H_2O_2, H, and O are 34.02 g, 1.008 g, and 16.00 g, respectively. Therefore, the percent composition of H_2O_2 is calculated as follows:

$$\%H = \frac{2 \times 1.008\ g}{34.02\ g} \times 100\% = 5.926\%$$

$$\%O = \frac{2 \times 16.00\ g}{34.02\ g} \times 100\% = 94.06\%$$

The sum of the percentage is 5.926 percent + 94.06 percent = 99.99 percent. The small discrepancy from 100 percent is due to the way we rounded off the molar masses of the elements. Note that the empirical formula (HO) would give us the same results.

EXAMPLE 3.8
Calculating Percent Composition of a Compound

Phosphoric acid (H_3PO_4) is used in detergents, fertilizers, and toothpastes. It is also the ingredient responsible for the "tangy" flavor of carbonated beverages. Calculate the percent composition by mass of H, P, and O in this compound.

Answer: The molar mass of H_3PO_4 is 97.99 g/mol. Therefore, the percent by mass of the elements in H_3PO_4 is

$$\%H = \frac{3(1.008\ g)}{97.99\ g} \times 100\% = 3.086\%$$

$$\%P = \frac{30.97\ g}{97.99\ g} \times 100\% = 31.61\%$$

$$\%O = \frac{4(16.00\ g)}{97.99\ g} \times 100\% = 65.31\%$$

The sum of the percentages is (3.086% + 31.61% + 65.31%) = 100.01%. The discrepancy from 100 percent is due to the way we rounded off.

PRACTICE EXERCISE

Calculate the percent composition by mass of all the elements in sulfuric acid (H_2SO_4).

The procedure used in the example can be reversed if necessary. Given the percent composition by mass of a compound, we can determine the empirical formula of the compound.

EXAMPLE 3.9

Determining Empirical Formula from Percent Composition

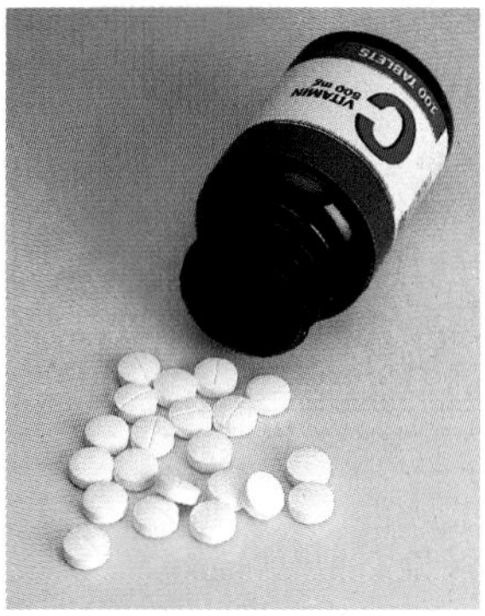

Vitamin C tablets.

Ascorbic acid (vitamin C) cures scurvy and may help prevent the common cold. It is composed of 40.92 percent carbon (C), 4.58 percent hydrogen (H), and 54.50 percent oxygen (O) by mass. Determine its empirical formula.

Answer: Because the sum of all the percentages is 100 percent, it is convenient to solve this type of problem by considering exactly 100 g of the substance. In 100 g of ascorbic acid there will be 40.92 g of C, 4.58 g of H, and 54.50 g of O. Next, we need to calculate the number of moles of each element in the compound. Let n_C, n_H, and n_O be the number of moles of elements present. Using the molar masses of these elements, we write

$$n_C = 40.92\ \cancel{g\ C} \times \frac{1\ \text{mol C}}{12.01\ \cancel{g\ C}} = 3.407\ \text{mol C}$$

$$n_H = 4.58\ \cancel{g\ H} \times \frac{1\ \text{mol H}}{1.008\ \cancel{g\ H}} = 4.54\ \text{mol H}$$

$$n_O = 54.50\ \cancel{g\ O} \times \frac{1\ \text{mol O}}{16.00\ \cancel{g\ O}} = 3.406\ \text{mol O}$$

Thus we arrive at the formula $C_{3.407}H_{4.54}O_{3.406}$, which gives the identity and the ratios of atoms present. However, since chemical formulas are written with whole numbers, we cannot have 3.407 C atoms, 4.54 H atoms, and 3.406 O atoms. We can make some of the subscripts whole numbers by dividing all the subscripts by the smallest subscript (3.406):

$$\text{C: } \frac{3.407}{3.406} = 1 \qquad \text{H: } \frac{4.54}{3.406} = 1.33 \qquad \text{O: } \frac{3.406}{3.406} = 1$$

This gives us $CH_{1.33}O$ as the formula for ascorbic acid. Next, we need to convert 1.33, the subscript for H, into an integer. This can be done by a trial-and-error procedure:

$$1.33 \times 1 = 1.33$$

$$1.33 \times 2 = 2.66$$

$$1.33 \times 3 = 3.99 \approx 4$$

Because 1.33×3 gives us an integer (4), we can multiply all the subscripts by 3 and obtain $C_3H_4O_3$ as the empirical formula for ascorbic acid.

PRACTICE EXERCISE

Determine the empirical formula of a compound having the following percent composition by mass: K: 24.75%; Mn: 34.77%; O: 40.51%.

Chemists often want to know the actual mass of an element in a certain mass of a compound. Since the percent composition by mass of the element in the substance can be readily calculated, such a problem can be solved in a rather direct way.

EXAMPLE 3.10

Calculating the Mass of an Element from Percent Composition

Chalcopyrite.

Chalcopyrite ($CuFeS_2$) is a principal ore of copper. Calculate the number of kilograms of Cu contained in 3.71×10^3 kg of chalcopyrite.

Answer: The molar masses of Cu and $CuFeS_2$ are 63.55 g and 183.5 g, respectively, so the percent composition by mass of Cu is

$$\%\text{Cu} = \frac{63.55\ \text{g}}{183.5\ \text{g}} \times 100\% = 34.63\%$$

To calculate the mass of Cu in a 3.71×10^3-kg sample of $CuFeS_2$, we need to convert the percentage to a fraction (that is, convert 34.63% to 0.3463) and write

$$\text{mass of Cu in CuFeS}_2 = 0.3463 \times 3.71 \times 10^3\ \text{kg} = 1.28 \times 10^3\ \text{kg}$$

This calculation can be simplified by combining the above two steps as follows:

$$\text{mass of Cu in CuFeS}_2 = 3.71 \times 10^3\ \text{kg CuFeS}_2 \times \frac{63.55\ \text{g Cu}}{183.5\ \text{g CuFeS}_2}$$

$$= 1.28 \times 10^3\ \text{kg Cu}$$

PRACTICE EXERCISE

Calculate the number of grams of Al in 371 g of Al_2O_3.

3.6 EXPERIMENTAL DETERMINATION OF EMPIRICAL FORMULAS

The fact that we can determine the empirical formula of a compound if we know the percent composition allows us to identify compounds experimentally. The procedure is as follows. First, chemical analysis determines the number of grams of each element present in a given amount of a compound. Then, the quantities in grams are converted to numbers of moles of the elements. Finally, the empirical formula of the compound is determined by the method shown in Example 3.9.

As a specific example let us consider the compound ethanol. When ethanol is burned in an apparatus such as that shown in Figure 3.5, carbon dioxide (CO_2) and

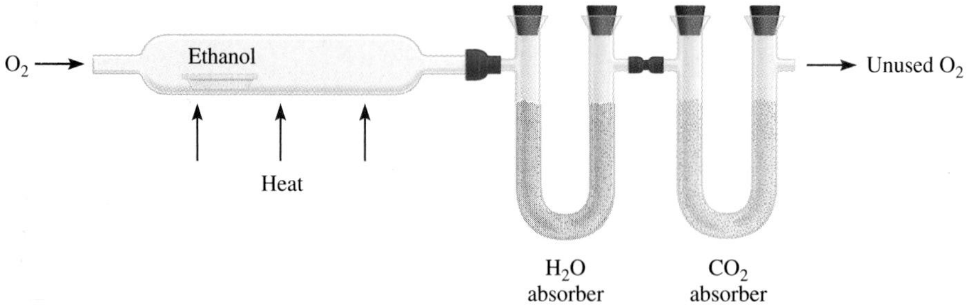

FIGURE 3.5

Apparatus for determining the empirical formula of ethanol. The absorbers are substances that can retain water and carbon dioxide, respectively.

water (H_2O) are given off. Since neither carbon nor hydrogen was in the inlet gas, we can conclude that both carbon (C) and hydrogen (H) were present in ethanol and that oxygen (O) may also be present. (Molecular oxygen was added in the combustion process, but some of the oxygen may also have come from the original ethanol sample.)

The mass of CO_2 and of H_2O produced can be determined by measuring the increase in mass of the CO_2 and H_2O absorbers, respectively. Suppose that in one experiment the combustion of 11.5 g of ethanol produced 22.0 g of CO_2 and 13.5 g of H_2O. We can calculate the masses of carbon and hydrogen in the original 11.5-g sample of ethanol as follows:

$$\text{mass of C} = 22.0\ \text{g}\ \cancel{CO_2} \times \frac{1\ \cancel{\text{mol}\ CO_2}}{44.01\ \cancel{\text{g}\ CO_2}} \times \frac{1\ \cancel{\text{mol}\ C}}{1\ \cancel{\text{mol}\ CO_2}} \times \frac{12.01\ \text{g C}}{1\ \cancel{\text{mol}\ C}}$$

$$= 6.00\ \text{g C}$$

$$\text{mass of H} = 13.5\ \text{g}\ \cancel{H_2O} \times \frac{1\ \cancel{\text{mol}\ H_2O}}{18.02\ \cancel{\text{g}\ H_2O}} \times \frac{2\ \cancel{\text{mol}\ H}}{1\ \cancel{\text{mol}\ H_2O}} \times \frac{1.008\ \text{g H}}{1\ \cancel{\text{mol}\ H}}$$

$$= 1.51\ \text{g H}$$

Thus, in 11.5 g of ethanol, 6.00 g are carbon and 1.51 g are hydrogen. The remainder must be oxygen, whose mass is

$$\begin{aligned} \text{mass of O} &= \text{mass of sample} - (\text{mass of C} + \text{mass of H}) \\ &= 11.5\ \text{g} - (6.00\ \text{g} + 1.51\ \text{g}) \\ &= 4.0\ \text{g} \end{aligned}$$

The number of moles of each element present in 11.5 g of ethanol is

$$\text{moles of C} = 6.00\ \cancel{\text{g C}} \times \frac{1\ \text{mol C}}{12.01\ \cancel{\text{g C}}} = 0.500\ \text{mol C}$$

$$\text{moles of H} = 1.51\ \cancel{\text{g H}} \times \frac{1\ \text{mol H}}{1.008\ \cancel{\text{g H}}} = 1.50\ \text{mol H}$$

$$\text{moles of O} = 4.0\ \cancel{\text{g O}} \times \frac{1\ \text{mol O}}{16.00\ \cancel{\text{g O}}} = 0.25\ \text{mol O}$$

The formula for ethanol is therefore $C_{0.50}H_{1.5}O_{0.25}$ (we round off the number of moles to two significant figures). Since the number of atoms must be an integer, we divide the subscripts by 0.25, the smallest subscript, and obtain for the empirical formula C_2H_6O.

Now we can better understand the word "empirical," which literally means "based only on observation and measurement." The empirical formula of ethanol is determined from analysis of the compound in terms of its component elements. No knowledge of how the atoms are linked together in the compound is required.

Determination of Molecular Formulas

The formula calculated from data about the percent composition by mass is always the empirical formula because the coefficients in the formula are always reduced to the smallest whole numbers. To calculate the actual or molecular formula we must know the *approximate* molar mass of the compound in addition to its empirical formula.

EXAMPLE 3.11

Finding the Molecular Formula of a Compound

A compound of nitrogen (N) and oxygen (O) has the composition 1.52 g of N and 3.47 g of O. The molar mass of this compound is known to be between 90 g and 95 g. Determine the molecular formula and the molar mass of the compound to four significant figures.

Answer: We first determine the empirical formula as outlined in Example 3.9. Let n_N and n_O be the number of moles of nitrogen and oxygen. Then

$$n_N = 1.52 \cancel{\text{g N}} \times \frac{1 \text{ mol N}}{14.01 \cancel{\text{g N}}} = 0.108 \text{ mol N}$$

$$n_O = 3.47 \cancel{\text{g O}} \times \frac{1 \text{ mol O}}{16.00 \cancel{\text{g O}}} = 0.217 \text{ mol O}$$

Thus the formula of the compound is $N_{0.108}O_{0.217}$. As in Example 3.9, we divide the subscripts by the smaller subscript, 0.108. After rounding off, we obtain NO_2 as the empirical formula. The molecular formula will be equal to the empirical formula or to some integral multiple of it (for example, two, three, four, or more times the empirical formula). The molar mass of the empirical formula NO_2 is

$$\text{empirical molar mass} = 14.01 \text{ g} + 2(16.00 \text{ g}) = 46.02 \text{ g}$$

Next we determine the number of (NO_2) units present in the molecular formula. This number is found by taking the ratio

$$\frac{\text{molar mass}}{\text{empirical molar mass}} = \frac{95 \text{ g}}{46.02 \text{ g}} = 2.1 \approx 2$$

Thus there are two NO_2 units in each molecule of the compound, so the molecular formula is $(NO_2)_2$, or N_2O_4. The molar mass of the compound is 2(46.02 g), or 92.04 g.

PRACTICE EXERCISE

A compound containing boron (B) and hydrogen (H) consists of 6.444 g of B and 1.803 g of H. The molar mass of the compound is about 30 g. What is its molecular formula?

3.7 CHEMICAL REACTIONS AND CHEMICAL EQUATIONS

Having discussed the masses of atoms and molecules, we turn next to what happens to atoms and molecules when chemical changes occur. A *chemical change* is called a ***chemical reaction.*** In order to communicate with one another about chemical reactions, chemists have devised a standard way to represent them using chemical equations. A ***chemical equation*** *uses chemical symbols to show what happens during a chemical reaction.* In this section we will learn how to write chemical equations and balance them.

WRITING CHEMICAL EQUATIONS

Consider what happens when hydrogen gas (H_2) burns in air (which contains oxygen, O_2) to form water (H_2O). This reaction can be represented by the chemical equation

$$H_2 + O_2 \longrightarrow H_2O \qquad (3.1)$$

where the + sign means "reacts with" and the → sign means "to yield." Thus this symbolic expression can be read: "Molecular hydrogen reacts with molecular oxygen to yield water." The reaction is assumed to proceed from left to right, as the arrow indicates.

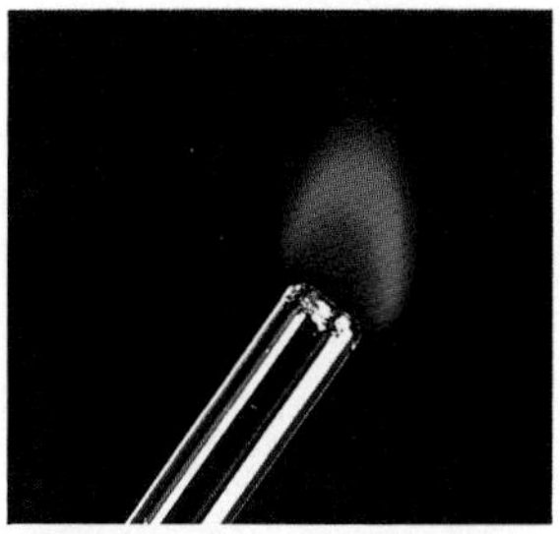

Hydrogen gas burning in air.

Expression (3.1) is not complete, however, because twice as many oxygen atoms are on the left side of the arrow (two) as on the right side (one). To conform with the law of conservation of mass, the same number of each type of atom must be on both sides of the arrow; that is, we must have as many atoms after the reaction ends as we did before it started. We can *balance* Equation (3.1) by placing the appropriate coefficient (2 in this case) in front of H_2 and H_2O:

$$2H_2 + O_2 \longrightarrow 2H_2O$$

This *balanced chemical equation* shows that "two hydrogen molecules can combine or react with one oxygen molecule to form two water molecules" (Figure 3.6). (Note that when the coefficient is 1, as in the case of O_2 in this equation, it is not shown.) Since the ratio of the number of molecules is equal to the ratio of the number of moles, the equation can also be read as "2 moles of hydrogen molecules react with 1 mole of oxygen molecules to produce 2 moles of water molecules." We know the mass of a mole of each of these substances, so we can also interpret the equation as "4.04 g of H_2 react with 32.00 g of O_2 to give 36.04 g of H_2O." These three ways of reading the equation are summarized in Table 3.1.

We refer to H_2 and O_2 in Equation (3.1) as ***reactants,*** which are *the starting materials in a chemical reaction.* Water is the ***product,*** which is *the substance*

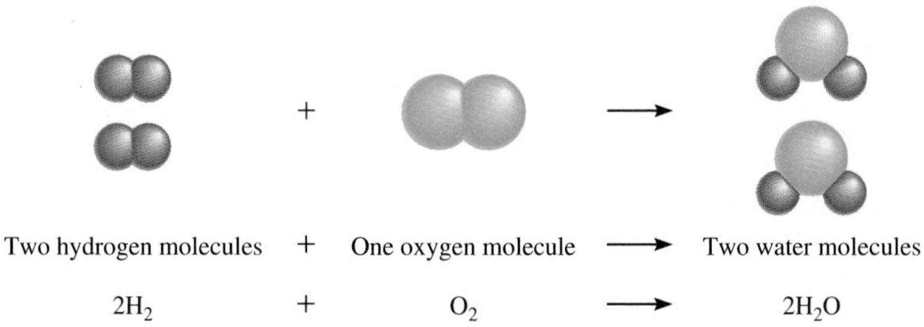

FIGURE 3.6

Three ways of representing the combustion of hydrogen. In accordance with the law of conservation of mass, the number of each type of atom must be the same on both sides of the equation.

TABLE 3.1

Interpretation of a Chemical Equation

$2H_2$	$+ O_2$	⟶	$2H_2O$
Two molecules	+ one molecule	⟶	two molecules
2 moles	+ 1 mole	⟶	2 moles
2(2.02 g) = 4.04 g	+ 32.00 g	⟶	2(18.02 g) = 36.04 g
36.04 g reactants			36.04 g product

formed as a result of a chemical reaction. A chemical equation, then, is just the chemist's shorthand description of a reaction. In a chemical equation the reactants are conventionally written on the left and the products on the right of the arrow:

$$\text{reactants} \longrightarrow \text{products}$$

To provide additional information, chemists often indicate the physical states of the reactants and products by using the letters *g*, *l*, and *s* to denote the gas, liquid, and solid state, respectively. For example,

$$2CO(g) + O_2(g) \longrightarrow 2CO_2(g)$$

$$2HgO(s) \longrightarrow 2Hg(l) + O_2(g)$$

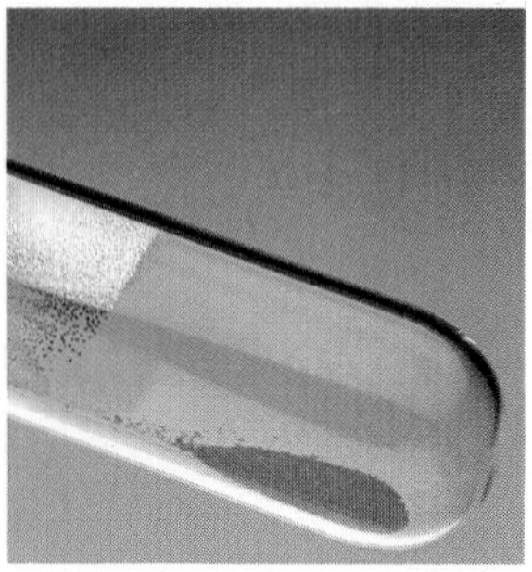

On heating, mercury(II) oxide (HgO) decomposes to form mercury and oxygen.

To represent what happens when sodium chloride (NaCl) is added to water, we write

$$NaCl(s) \xrightarrow{H_2O} NaCl(aq)$$

where *aq* denotes the aqueous (that is, water) environment. Writing H_2O above the arrow symbolizes the physical process of dissolving a substance in water, although it is sometimes left out for simplicity.

Balancing Chemical Equations

Suppose we want to write an equation to describe a chemical reaction that we have just carried out in the laboratory. How should we go about doing this? Because we know the identities of the reactants, we can write the chemical formulas for them. The identities of products are more difficult to establish. For simple reactions it is often possible to guess the product(s). For more complicated reactions involving three or more products, chemists may need to perform further tests to establish the presence of specific compounds. Thus we can conclude that a gaseous product is formed if we see bubbles appearing in water during a reaction carried out in the aqueous environment. Color change is another indication that a chemical reaction has occurred.

Once we have identified all the reactants and products and have written the correct formulas for them, we assemble them in the conventional sequence—reactants on the left separated from products on the right by an arrow. The equation written at this point is likely to be *unbalanced;* that is, the number of each type of atom on both sides of the equation differs. In general, we can balance a chemical equation by the following steps:

- Identify all reactants and products and write their correct formulas on the left side and right side of the equation, respectively.
- Begin balancing the equation by trying different coefficients to make the number of atoms of each element the same on both sides of the equation. We can change the coefficients (the numbers preceding the formulas) but not the subscripts (the numbers within formulas). Changing the subscripts could change the identity of the substance. For example, $2NO_2$ means "two molecules of nitrogen dioxide," but if we double the subscripts, we have N_2O_4, which is the formula of dinitrogen tetroxide, a completely different compound.
- Look for elements that appear only once on each side of the equation with the same number of atoms on each side: The formulas containing these elements must have the same coefficient. Next, look for elements that appear only once

on each side of the equation but in unequal numbers of atoms. Balance these elements. Finally, balance elements that appear in two or more formulas on the same side of the equation.

- Check your balanced equation to be sure that you have the same total number of each type of atom on both sides of the equation arrow.

Let's consider a specific example. In the laboratory, small amounts of oxygen gas can be conveniently prepared by heating potassium chlorate ($KClO_3$). The products are oxygen gas (O_2) and potassium chloride (KCl). From this information, we write

$$KClO_3 \longrightarrow KCl + O_2$$

(For simplicity, we omit the physical states of reactants and products.) We see that all three elements (K, Cl, and O) appear only once on each side of the equation, but only K and Cl appear in equal numbers of atoms on both sides. Thus $KClO_3$ and KCl must have the same coefficient. The next step is to make the number of O atoms the same on both sides of the equation. Since there are three O atoms on the left and two O atoms on the right of the equation, we can balance the O atoms by placing a 2 in front of $KClO_3$ and a 3 in front of O_2.

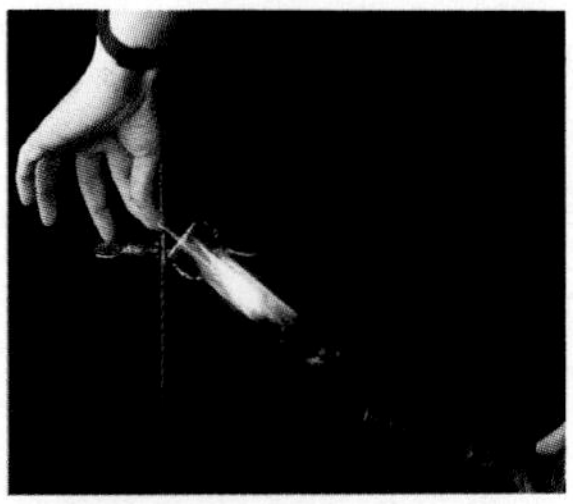

Heating potassium chlorate ($KClO_3$) produces oxygen, which supports the combustion of a wood splint.

$$2KClO_3 \longrightarrow KCl + 3O_2$$

Finally, we balance the K and Cl atoms by placing a 2 in front of KCl:

$$2KClO_3 \longrightarrow 2KCl + 3O_2 \qquad (3.2)$$

As a final check, we can draw up a balance sheet for the reactants and products where the number in parentheses indicates the number of atoms of each element:

Reactants	*Products*
K (2)	K (2)
Cl (2)	Cl (2)
O (6)	O (6)

Note that this equation could also be balanced with coefficients that are multiples of 2 (for $KClO_3$), 2 (for KCl), and 3 (for O_2); for example,

$$4KClO_3 \longrightarrow 4KCl + 6O_2$$

However, it is common practice to use the *simplest* possible set of whole-number coefficients to balance the equation. Equation (3.2) conforms to this convention.

Now let us consider the combustion (that is, burning) of the natural gas ethane (C_2H_6) in oxygen or air, which yields carbon dioxide (CO_2) and water. We write

$$C_2H_6 + O_2 \longrightarrow CO_2 + H_2O$$

We see that the number of atoms is not the same on both sides of the equation for any of the elements (C, H, and O). In addition, C and H appear only once on each side of the equation; O appears in two compounds on the right side (CO_2 and H_2O). To balance the C atoms, we place a 2 in front of CO_2:

$$C_2H_6 + O_2 \longrightarrow 2CO_2 + H_2O$$

To balance the H atoms, we place a 3 in front of H_2O:

$$C_2H_6 + O_2 \longrightarrow 2CO_2 + 3H_2O$$

At this stage, the C and H atoms are balanced, but the O atoms are not balanced, because there are seven O atoms on the right-hand side and only two O atoms on the left-hand side of the equation. This inequality of O atoms can be eliminated by writing $\frac{7}{2}$ in front of the O_2 on the left-hand side:

$$C_2H_6 + \tfrac{7}{2}O_2 \longrightarrow 2CO_2 + 3H_2O$$

The "logic" of the $\frac{7}{2}$ factor can be clarified as follows. There were seven oxygen atoms on the right-hand side of the equation, but only a pair of oxygen atoms (O_2) on the left. To balance them we ask how many *pairs* of oxygen atoms are needed to equal seven oxygen atoms. Just as 3.5 pairs of shoes equal seven shoes, $\frac{7}{2}O_2$ molecules equal seven O atoms. As the following tally shows, the equation is now completely balanced:

Reactants	*Products*
C (2)	C (2)
H (6)	H (6)
O (7)	O (7)

However, we normally prefer to express the coefficients as whole numbers rather than as fractions. Therefore, we multiply the entire equation by 2 to convert $\frac{7}{2}$ to 7:

$$2C_2H_6 + 7O_2 \longrightarrow 4CO_2 + 6H_2O$$

The final tally is

Reactants	*Products*
C (4)	C (4)
H (12)	H (12)
O (14)	O (14)

Note that the coefficients used in balancing the equation are the smallest possible set of whole numbers.

EXAMPLE 3.12

Balancing Chemical Equations

This beverage can is made of aluminum, which resists corrosion.

When aluminum metal is exposed to air, a protective layer of aluminum oxide (Al_2O_3) forms on its surface. This layer prevents further reaction between aluminum and oxygen, and this is the reason that aluminum beverage cans do not corrode. [In the case of iron, the rust, or iron(III) oxide, that forms is too porous to protect the iron metal underneath, so rusting continues.] Write a balanced equation for this process.

Answer: The unbalanced equation is

$$Al + O_2 \longrightarrow Al_2O_3$$

We see that both Al and O appear only once on each side of the equation but in unequal numbers. To balance the Al atoms, we place a 2 in front of Al:

$$2Al + O_2 \longrightarrow Al_2O_3$$

There are two O atoms on the left-hand side and three O atoms on the right-hand side of the equation. This inequality of O atoms can be eliminated by writing $\frac{3}{2}$ in front of the O_2 on the left-hand side of the equation:

$$2Al + \tfrac{3}{2}O_2 \longrightarrow Al_2O_3$$

As in the case of ethane described earlier, we multiply the entire equation by 2 to convert $\frac{3}{2}$ to 3:

$$4Al + 3O_2 \longrightarrow 2Al_2O_3$$

The final tally is

Reactants	*Products*
Al (4)	Al (4)
O (6)	O (6)

PRACTICE EXERCISE

Balance the equation representing the reaction between iron(III) oxide (Fe_2O_3) and carbon monoxide (CO) to yield iron (Fe) and carbon dioxide (CO_2).

3.8 AMOUNTS OF REACTANTS AND PRODUCTS

Having studied the writing and balancing of chemical equations, we are now ready to study the quantitative aspects of chemical reactions. *The mass relationships among reactants and products in a chemical reaction* represent the ***stoichiometry*** of the reaction. To interpret a reaction quantitatively, we need to apply our knowledge of molar masses and the mole concept.

The basic question posed in many stoichiometric calculations is, If we know the quantities of the starting materials (that is, the reactants) in a reaction, how much product will be formed? Or in some cases we may have to ask the reverse question: How much starting material must be used to obtain a specific amount of product? In practice, the units used for reactants (or products) may be moles, grams, liters (for gases), or other units. Regardless of which units are used, *the approach to determine the amount of product formed in a reaction* is called the ***mole method.*** It is based on the fact that *the stoichiometric coefficients in a chemical equation can be interpreted as the number of moles of each substance.* To illustrate the mole method, let us consider the combustion of carbon monoxide in air to form carbon dioxide:

$$2CO(g) + O_2(g) \longrightarrow 2CO_2(g)$$

The equation and the stoichiometric coefficients can be read as "2 moles of carbon monoxide gas combine with 1 mole of oxygen gas to form 2 moles of carbon dioxide gas."

The mole method consists of the following steps:

1. Write correct formulas for all reactants and products, and balance the resulting equation.
2. Convert the quantities of some or all given or known substances (usually reactants) into moles.
3. Use the coefficients in the balanced equation to calculate the number of moles of the sought or unknown quantities (usually products) in the problem.

4. Using the calculated numbers of moles and the molar masses, convert the unknown quantities to whatever units are required (typically grams).
5. Check that your answer is reasonable in physical terms.

Step 1 is a prerequisite to any stoichiometric calculation. We must know the identities of the reactants and products, and the mass relationships among them must not violate the law of conservation of mass (that is, we must have a balanced equation). Step 2 is the critical step of converting grams (or other units) of substances to number of moles. This conversion then allows us to analyze the actual reaction in terms of moles only.

To complete step 3, we need the balanced equation, furnished by step 1. The key point here is that the coefficients in a balanced equation tell us the ratio in which moles of one substance react with or form moles of another substance. Step 4 is similar to step 2, except that it deals with the quantities sought in the problem. Step 5 is often underestimated but is very important: Chemistry is an experimental science, and your answer must make sense in terms of real species in the real world. If you have set up the problem incorrectly or made a computational error, it will often become obvious when your answer turns out to be much too large or much too small for the amounts of materials you started with. Figure 3.7 shows three common types of stoichiometric calculations.

In stoichiometry we use the symbol ≎, which means "stoichiometrically equivalent to," or simply "equivalent to." In the balanced equation for the formation of carbon dioxide, 2 moles of CO react with 1 mole of O_2, so 2 moles of CO are equivalent to 1 mole of O_2:

$$2 \text{ mol CO} \approx 1 \text{ mol } O_2$$

In terms of the factor-label method, we can write the unit factor as

$$\frac{2 \text{ mol CO}}{1 \text{ mol } O_2} = 1 \quad \text{or} \quad \frac{1 \text{ mol } O_2}{2 \text{ mol CO}} = 1$$

Similarly, since 2 moles of CO (or 1 mole of O_2) produce 2 moles of CO_2, we can say that 2 moles of CO (or 1 mole of O_2) are equivalent to 2 moles of CO_2:

$$2 \text{ mol CO} \approx 2 \text{ mol } CO_2$$

$$1 \text{ mol } O_2 \approx 2 \text{ mol } CO_2$$

The following example illustrates the use of the five-step method in solving some typical stoichiometry problems.

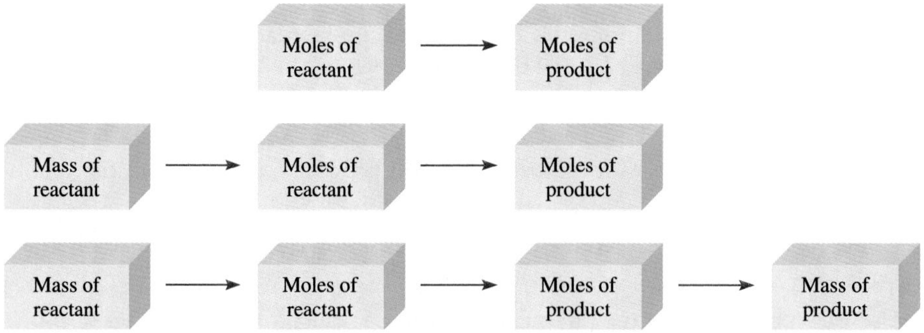

FIGURE 3.7

Three types of stoichiometric calculations based on the mole method.

EXAMPLE 3.13

Calculating the Amount of Product

All alkali metals react with water to produce hydrogen gas and the corresponding alkali metal hydroxide. A typical reaction is that between lithium and water:

$$2Li(s) + 2H_2O(l) \longrightarrow 2LiOH(aq) + H_2(g)$$

(a) How many moles of H_2 will be formed by the complete reaction of 6.23 moles of Li with water? (b) How many grams of H_2 will be formed by the complete reaction of 80.57 g of Li with water?

Lithium reacting with water.

Answer: (a)

Step 1: The balanced equation is given in the problem.

Step 2: No conversion is needed because the amount of the starting material, Li, is given in moles.

Step 3: Since 2 moles of Li produce 1 mole of H_2, or 2 mol Li $\Leftrightarrow$ 1 mol H_2, we calculate moles of H_2 produced as follows:

$$\text{moles of } H_2 \text{ produced} = 6.23\ \cancel{\text{mol Li}} \times \frac{1 \text{ mol } H_2}{2\ \cancel{\text{mol Li}}}$$

$$= 3.12 \text{ mol } H_2$$

Step 4: This step is not required.

Step 5: We began with 6.23 moles of Li and produced 3.12 moles of H_2. Since 2 moles of Li produce 1 mole of H_2, 3.12 moles is a reasonable quantity.

(b)

Step 1: The reaction is the same as in (a).

Step 2: The number of moles of Li is given by

$$\text{moles of Li} = 80.57\ \cancel{\text{g Li}} \times \frac{1 \text{ mol Li}}{6.941\ \cancel{\text{g Li}}} = 11.61 \text{ mol Li}$$

Step 3: Since 2 moles of Li produce 1 mole of H_2, or 2 mol Li $\Leftrightarrow$ 1 mol H_2, we calculate the number of moles of H_2 as follows:

$$\text{moles of } H_2 \text{ produced} = 11.61\ \cancel{\text{mol Li}} \times \frac{1 \text{ mol } H_2}{2\ \cancel{\text{mol Li}}} = 5.805 \text{ mol } H_2$$

Step 4: From the molar mass of H_2 (2.016 g), we calculate the mass of H_2 produced:

$$\text{mass of } H_2 \text{ produced} = 5.805\ \cancel{\text{mol } H_2} \times \frac{2.016 \text{ g } H_2}{1\ \cancel{\text{mol } H_2}} = 11.70 \text{ g } H_2$$

Step 5: The amount 11.70 g H_2 is a reasonable quantity.

PRACTICE EXERCISE

The reaction between nitric oxide (NO) and oxygen to form nitrogen dioxide (NO_2) is a key step in photochemical smog formation:

$$2NO(g) + O_2(g) \longrightarrow 2NO_2(g)$$

(a) How many moles of NO_2 are formed by the complete reaction of 0.254 mole of O_2? (b) How many grams of NO_2 are formed by the complete reaction of 1.44 g of NO?

After some practice, you may find it convenient to combine steps 2, 3, and 4 in a single equation, as the following example shows.

EXAMPLE 3.14

Calculating the Amount of Product

The food we eat is degraded, or broken down, in our bodies to provide energy for growth and function. A general overall equation for this very complex process represents the degradation of glucose ($C_6H_{12}O_6$) to carbon dioxide (CO_2) and water (H_2O):

$$C_6H_{12}O_6 + 6O_2 \longrightarrow 6CO_2 + 6H_2O$$

If 856 g of $C_6H_{12}O_6$ are consumed by the body over a certain period, what is the mass of CO_2 produced?

Answer:

Step 1: The balanced equation is given.

Steps 2, 3, and 4: From the balanced equation we see that 1 mol $C_6H_{12}O_6 \simeq$ 6 mol CO_2. The molar masses of $C_6H_{12}O_6$ and CO_2 are 180.2 g and 44.01 g, respectively. We combine all these data into one equation:

$$\text{mass of } CO_2 \text{ produced} = 856\,\cancel{\text{g } C_6H_{12}O_6} \times \frac{1\,\cancel{\text{mol } C_6H_{12}O_6}}{180.2\,\cancel{\text{g } C_6H_{12}O_6}} \times \frac{6\,\cancel{\text{mol } CO_2}}{1\,\cancel{\text{mol } C_6H_{12}O_6}} \times \frac{44.01\text{ g } CO_2}{1\,\cancel{\text{mol } CO_2}}$$

$$= 1.25 \times 10^3 \text{ g } CO_2$$

Step 5: The mass of CO_2 produced is 1.25×10^3 g, a reasonable amount of carbon dioxide to be produced when 856 g of $C_6H_{12}O_6$ are consumed.

PRACTICE EXERCISE

Methanol (CH_3OH) burns in air according to the equation

$$2CH_3OH + 3O_2 \longrightarrow 2CO_2 + 4H_2O$$

If 209 g of methanol are used up in a combustion process, what is the mass of H_2O produced?

3.9 LIMITING REAGENTS AND YIELDS OF REACTIONS

When a chemist carries out a reaction, the reactants are usually not present in exact ***stoichiometric amounts,*** that is, *in the proportions indicated by the balanced*

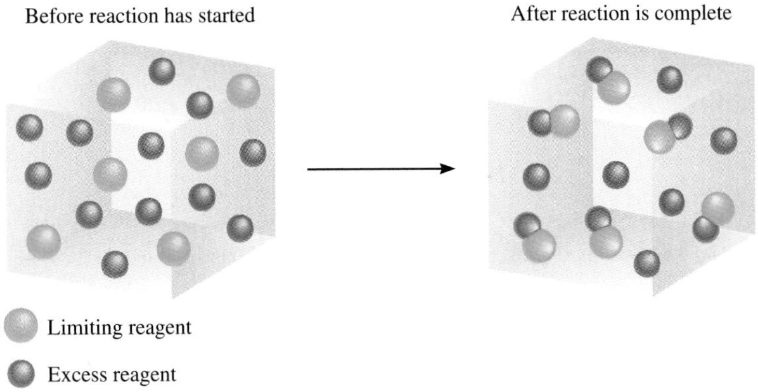

FIGURE 3.8

A limiting reagent is completely used up in a reaction.

equation. The reactant used up first in a reaction is called the ***limiting reagent,*** since the maximum amount of product formed depends on how much of this reactant was originally present (Figure 3.8). When this reactant is used up, no more product can be formed. *Other reactants, present in quantities greater than those needed to react with the quantity of the limiting reagent,* are called ***excess reagents.***

The concept of the limiting reagent is analogous to the relationship between men and women in a dance contest at a club. If there are fourteen men and only nine women, then only nine female/male pairs can compete. Five men will be left without partners. The number of women thus limits the number of men that can dance in the contest, and there is an excess of men.

Now consider what happens when there is a limiting reagent in a chemical reaction. Sulfur hexafluoride (SF_6), a colorless, odorless, and extremely stable compound, is formed by burning sulfur in an atmosphere of fluorine:

$$S(l) + 3F_2(g) \longrightarrow SF_6(g)$$

This equation tells us that 1 mole of S reacts with 3 moles of F_2 to produce 1 mole of SF_6. Suppose that 4 moles of S are added to 20 moles of F_2. Since 1 mol S $\simeq$ 3 mol F_2, the number of moles of F_2 needed to react with 4 moles of S is

$$4 \cancel{\text{mol S}} \times \frac{3 \text{ mol } F_2}{1 \cancel{\text{mol S}}} = 12 \text{ mol } F_2$$

But there are 20 moles of F_2 available, more than what is needed to completely react with S. Thus S must be the limiting reagent and F_2 the excess reagent. The amount of SF_6 produced depends only on how much S was originally present.

We can also calculate the number of moles of S needed to react with 20 moles of F_2. In this case we write

$$20 \cancel{\text{mol } F_2} \times \frac{1 \text{ mol S}}{3 \cancel{\text{mol } F_2}} = 6.7 \text{ mol S}$$

Since there are only 4 moles of S present, we arrive at the same conclusion that S is the limiting reagent and F_2 is the excess reagent.

In stoichiometric calculations involving limiting reagents, the first step is to decide which reactant is the limiting reagent. After the limiting reagent has been identified, the rest of the problem can be solved as outlined in Section 3.8. The following example illustrates this approach. We will not include step 5 in the calculations, but you should always examine the reasonableness of *any* chemical calculation.

EXAMPLE 3.15

Calculating Limiting Reagent and Excess Reagent

Urea [$(NH_2)_2CO$] is prepared by the reaction between ammonia and carbon dioxide:

$$2NH_3(g) + CO_2(g) \longrightarrow (NH_2)_2CO(aq) + H_2O(l)$$

In one process, 637.2 g of NH_3 are allowed to react with 1142 g of CO_2. (a) Which of the two reactants is the limiting reagent? (b) Calculate the mass of $(NH_2)_2CO$ formed. (c) How much of the excess reagent (in grams) is left at the end of the reaction?

Answer: (a) Since we cannot tell by inspection which of the two reactants is the limiting reagent, we have to proceed by first converting their masses into numbers of moles. The molar masses of NH_3 and CO_2 are 17.03 g and 44.01 g, respectively. Thus

$$\text{moles of } NH_3 = 637.2\ \cancel{\text{g } NH_3} \times \frac{1 \text{ mol } NH_3}{17.03\ \cancel{\text{g } NH_3}}$$

$$= 37.42 \text{ mol } NH_3$$

$$\text{moles of } CO_2 = 1142\ \cancel{\text{g } CO_2} \times \frac{1 \text{ mol } CO_2}{44.01\ \cancel{\text{g } CO_2}}$$

$$= 25.95 \text{ mol } CO_2$$

From the balanced equation we see that 2 mol $NH_3 \Leftrightarrow$ 1 mol CO_2; therefore, the number of moles of NH_3 needed to react with 25.95 moles of CO_2 is given by

$$25.95\ \cancel{\text{mol } CO_2} \times \frac{2 \text{ mol } NH_3}{1\ \cancel{\text{mol } CO_2}} = 51.90 \text{ mol } NH_3$$

Since there are only 37.42 moles of NH_3 present, not enough to react completely with the CO_2, NH_3 must be the limiting reagent and CO_2 the excess reagent.

(b) The amount of $(NH_2)_2CO$ produced is determined by the amount of limiting reagent present. Thus we write

$$\text{mass of } (NH_2)_2CO = 37.42\ \cancel{\text{mol } NH_3} \times \frac{1\ \cancel{\text{mol } (NH_2)_2CO}}{2\ \cancel{\text{mol } NH_3}} \times \frac{60.06 \text{ g } (NH_2)_2CO}{1\ \cancel{\text{mol } (NH_2)_2CO}}$$

$$= 1124 \text{ g } (NH_2)_2CO$$

(c) The number of moles of the excess reagent (CO_2) left is

$$25.95 \text{ mol } CO_2 - 37.42\ \cancel{\text{mol } NH_3} \times \frac{1 \text{ mol } CO_2}{2\ \cancel{\text{mol } NH_3}} = 7.24 \text{ mol } CO_2$$

and

$$\text{mass of } CO_2 \text{ left over} = 7.24\ \cancel{\text{mol } CO_2} \times \frac{44.01 \text{ g } CO_2}{1\ \cancel{\text{mol } CO_2}}$$

$$= 319 \text{ g } CO_2$$

PRACTICE EXERCISE

The reaction between aluminum and iron(III) oxide can generate temperatures approaching 3000°C and is used in welding metals:

$$2Al + Fe_2O_3 \longrightarrow Al_2O_3 + 2Fe$$

In one process 124 g of Al are reacted with 601 g of Fe_2O_3. (a) Calculate the mass (in grams) of Al_2O_3 formed. (b) How much of the excess reagent is left at the end of the reaction?

Example 3.15 brings out an important point. In practice, chemists usually choose the more expensive chemical as the limiting reagent so that all or most of it will be consumed in the reaction. In the synthesis of urea, NH_3 is invariably the limiting reagent because it is much more expensive than CO_2.

Yields of Reactions

The amount of limiting reagent present at the start of a reaction determines the ***theoretical yield*** of the reaction, that is, *the amount of product that would result if all the limiting reagent reacted.* The theoretical yield, then, is the *maximum* obtainable yield, predicted by the balanced equation. In practice, the *amount of product obtained* is almost always less than the theoretical yield. Therefore chemists define the ***actual yield*** as *the quantity of product that actually results from a reaction.* There are many reasons for the difference between actual and theoretical yields. For instance, many reactions are reversible, and so they do not proceed 100 percent from left to right. Even when a reaction is 100 percent complete, it may be difficult to recover all of the product from the reaction medium (say, from an aqueous solution). Some reactions are complex in the sense that the products formed may react further among themselves or with the reactants to form still other products. These additional reactions will reduce the yield of the first reaction.

To determine how efficient a given reaction is, chemists often figure the ***percent yield,*** which describes *the proportion of the actual yield to the theoretical yield.* It is calculated as follows:

$$\%\ \text{yield} = \frac{\text{actual yield}}{\text{theoretical yield}} \times 100\%$$

Percent yields may range from a fraction of 1 percent to 100 percent. Chemists strive to maximize the percent yield of product in a reaction. Other factors that can affect the percent yield of a reaction include temperature and pressure. We will study these effects later.

EXAMPLE 3.16

Calculating the Percent Yield of a Reaction

Titanium is a strong, lightweight, corrosion-resistant metal that is used in rockets, aircraft, and jet engines. It is prepared by the reaction of titanium(IV) chloride with molten magnesium between 950°C and 1150°C:

$$TiCl_4(g) + 2Mg(l) \longrightarrow Ti(s) + 2MgCl_2(l)$$

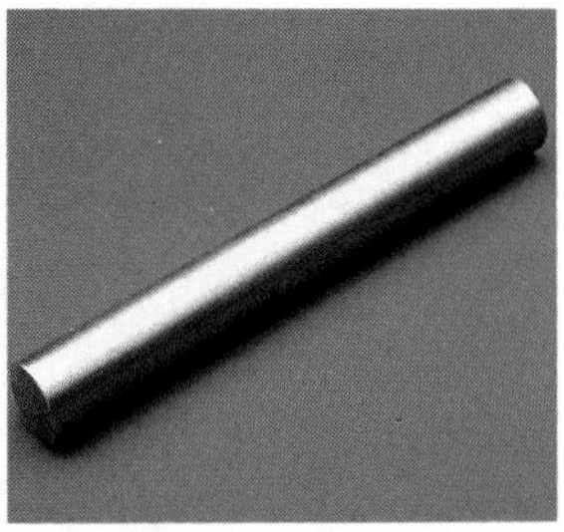

Titanium.

In a certain operation 3.54×10^7 g of $TiCl_4$ are reacted with 1.13×10^7 g of Mg. (a) Calculate the theoretical yield of Ti in grams. (b) Calculate the percent yield if 7.91×10^6 g of Ti are actually obtained.

Answer: (a) First we calculate the number of moles of $TiCl_4$ and Mg:

$$\text{moles of } TiCl_4 = 3.54 \times 10^7 \, \cancel{\text{g } TiCl_4} \times \frac{1 \text{ mol } TiCl_4}{189.7 \, \cancel{\text{g } TiCl_4}}$$

$$= 1.87 \times 10^5 \text{ mol } TiCl_4$$

$$\text{moles of Mg} = 1.13 \times 10^7 \, \cancel{\text{g Mg}} \times \frac{1 \text{ mol Mg}}{24.31 \, \cancel{\text{g Mg}}}$$

$$= 4.65 \times 10^5 \text{ mol Mg}$$

Next, we must determine which of the two substances is the limiting reagent. From the balanced equation we see that 1 mol $TiCl_4 \simeq 2$ mol Mg; therefore, the number of moles of Mg needed to react with 1.87×10^5 moles of $TiCl_4$ is

$$1.87 \times 10^5 \, \cancel{\text{mol } TiCl_4} \times \frac{2 \text{ mol Mg}}{1 \, \cancel{\text{mol } TiCl_4}} = 3.74 \times 10^5 \text{ mol Mg}$$

Since 4.65×10^5 moles of Mg are present, more than is needed to react with the amount of $TiCl_4$ we have, Mg must be the excess reagent and $TiCl_4$ the limiting reagent.

Since 1 mol $TiCl_4 \simeq 1$ mol Ti, the theoretical amount of Ti formed is

$$\text{mass of Ti formed} = 3.54 \times 10^7 \, \cancel{\text{g } TiCl_4} \times \frac{1 \, \cancel{\text{mol } TiCl_4}}{189.7 \, \cancel{\text{g } TiCl_4}} \times \frac{1 \, \cancel{\text{mol Ti}}}{1 \, \cancel{\text{mol } TiCl_4}} \times \frac{47.88 \text{ g Ti}}{1 \, \cancel{\text{mol Ti}}}$$

$$= 8.93 \times 10^6 \text{ g Ti}$$

(b) To find the percent yield, we write

$$\% \text{ yield} = \frac{\text{actual yield}}{\text{theoretical yield}} \times 100\%$$

$$= \frac{7.91 \times 10^6 \text{ g}}{8.93 \times 10^6 \text{ g}} \times 100\% = 88.6\%$$

PRACTICE EXERCISE

Industrially, vanadium metal, which is used in steel alloys, can be obtained by reacting vanadium(V) oxide with calcium at high temperatures:

$$5Ca + V_2O_5 \longrightarrow 5CaO + 2V$$

In one process 1.54×10^3 g of V_2O_5 react with 1.96×10^3 g of Ca. (a) Calculate the theoretical yield of V. (b) Calculate the percent yield if 803 g of V are obtained.

SUMMARY

Atomic masses are measured in atomic mass units (amu), a relative unit based on a value of exactly 12 for the carbon-12 isotope. The atomic mass given for the atoms of a particular element is usually the average of the naturally occurring isotope distribution of that element. The molecular mass of a molecule is the sum of the atomic masses of the atoms in the molecule. Both atomic mass and molecular mass can be accurately determined with a mass spectrometer.

A mole is an Avogadro's number (6.022×10^{23}) of atoms, molecules, or other particles. The molar mass (in grams) of an element or a compound is numerically equal to the mass of the atom, molecule, or formula unit (in amu) and contains an Avogadro's number of atoms (in the case of elements), molecules, or simplest formula units (in the case of ionic compounds).

The percent composition by mass of a compound is the percent by mass of each element present. If we know the percent composition by mass of a compound, we can deduce the empirical formula of the compound and also the molecular formula of the compound if the approximate molar mass is known.

Chemical changes, called chemical reactions, are represented by chemical equations. Substances that undergo change—the reactants—are written on the left and substances formed—the products—appear on the right of the arrow. Chemical equations must be balanced, in accordance with the law of conservation of mass. The number of atoms of each type of element in the reactants and products must be equal.

Stoichiometry is the quantitative study of products and reactants in chemical reactions. Stoichiometric calculations are best done by expressing both the known and unknown quantities in terms of moles and then converting to other units if necessary. A limiting reagent is the reactant that is present in the smallest stoichiometric amount. It limits the amount of product that can be formed. The amount of product obtained in a reaction (the actual yield) may be less than the maximum possible amount (the theoretical yield). The ratio of the two is expressed as the percent yield.

KEY WORDS

Actual yield, p. 77
Atomic mass, p. 54
Atomic mass unit (amu), p. 54
Avogadro's number, p. 56
Chemical equation, p. 66
Chemical reaction, p. 66
Excess reagent, p. 75
Limiting reagent, p. 75
Molar mass, p. 56
Mole (mol), p. 55
Mole method, p. 71
Molecular mass, p. 58
Percent composition by mass, p. 62
Percent yield, p. 77
Product, p. 67
Reactant, p. 67
Stoichiometric amount, p. 74
Stoichiometry, p. 71
Theoretical yield, p. 77

QUESTIONS AND PROBLEMS

ATOMIC MASS AND AVOGADRO'S NUMBER

Review Questions

3.1 What is an atomic mass unit?

3.2 What is the mass (in amu) of a carbon-12 atom?

3.3 When we look up the atomic mass of carbon, we find that its value is 12.01 amu rather than 12.00 amu as defined. Why?

3.4 Define the term "mole." What is the unit for mole in calculations? What does the mole have in common with the pair, the dozen, and the gross?

3.5 What does Avogadro's number represent?

3.6 Define molar mass. What are the commonly used units for molar mass?

3.7 Calculate the charge (in coulombs) and mass (in grams) of 1 mole of electrons.

3.8 Explain clearly what is meant by the statement "The atomic mass of gold is 197.0 amu."

Problems

3.9 The atomic masses of $^{35}_{17}Cl$ (75.53%) and $^{37}_{17}Cl$ (24.47%) are 34.968 amu and 36.956 amu, respectively. Calculate the average atomic mass of chlorine. The percentages in parentheses denote the relative abundances.

3.10 The atomic masses of $^{6}_{3}Li$ and $^{7}_{3}Li$ are 6.0151 amu and 7.0160 amu, respectively. Calculate the natural abundances of these two isotopes. The average atomic mass of Li is 6.941 amu.

3.11 Earth's population is about 5.5 billion. Suppose that every person on Earth participates in a process of counting identical particles at the rate of two particles per second. How many years would it take to count 6.0×10^{23} particles? Assume that there are 365 days in a year.

3.12 The thickness of a piece of paper is 0.0036 in. Suppose a certain book has an Avogadro's number of pages; calculate the thickness of the book in light-years. (*Hint:* See Problem 1.38 for the definition of light-year.)

3.13 What is the mass in grams of 13.2 amu?

3.14 How many amu are there in 8.4 g?

3.15 How many atoms are there in 5.10 moles of sulfur (S)?

3.16 How many moles of cobalt atoms are there in 6.00×10^9 (6 billion) Co atoms?

3.17 How many moles of calcium (Ca) atoms are in 77.4 g of Ca?

3.18 How many grams of gold (Au) are there in 15.3 moles of Au?

3.19 What is the mass in grams of a single atom of each of the following elements? (a) Hg, (b) Ne.

3.20 What is the mass in grams of a single atom of each of the following elements? (a) As, (b) Ni.

3.21 What is the mass in grams of 1.00×10^{12} lead (Pb) atoms?

3.22 How many atoms are present in 3.14 g of copper (Cu)?

3.23 Which of the following has more atoms: 1.10 g of hydrogen atoms or 14.7 g of chromium atoms?

3.24 Which of the following has a greater mass: 2 atoms of lead or 5.1×10^{-23} mole of helium?

MOLECULAR MASS

Problems

3.25 Calculate the molecular mass (in amu) of each of the following substances: (a) CH_4, (b) H_2O, (c) H_2O_2, (d) C_6H_6, (e) PCl_5.

3.26 Calculate the molar mass of the following substances: (a) S_8, (b) CS_2, (c) $CHCl_3$ (chloroform), (d) $C_6H_8O_6$ (ascorbic acid, or vitamin C).

3.27 Calculate the molar mass of a compound if 0.372 mole of it has a mass of 152 g.

3.28 How many moles of ethane (C_2H_6) are present in 0.334 g of C_2H_6?

3.29 Calculate the numbers of C, H, and O atoms in 1.50 g of glucose ($C_6H_{12}O_6$), a sugar.

3.30 Urea [$(NH_2)_2CO$] is a compound used for fertilizer and many other things. Calculate the number of N, C, O, and H atoms in 1.68×10^4 g of urea.

3.31 Pheromones are a special type of compound secreted by the females of many insect species to attract the males for mating. One pheromone has the molecular formula $C_{19}H_{38}O$. Normally, the amount of this pheromone secreted by a female insect is about 1.0×10^{-12} g. How many molecules are there in this quantity?

3.32 The density of water is 1.00 g/mL at 4°C. How many water molecules are present in 2.56 mL of water at this temperature?

MASS SPECTROMETRY

Review Questions

3.33 Describe the operation of a mass spectrometer.

3.34 Describe how you would determine the isotopic abundance of an element from its mass spectrum.

Problems

3.35 Carbon has two stable isotopes, $^{12}_{6}C$ and $^{13}_{6}C$, and fluorine has only one stable isotope, $^{19}_{9}F$. How many peaks would you observe in the mass spectrum of the positive ion of CF_4^+? Assume no decomposition of the ion into smaller fragments.

3.36 Hydrogen has two stable isotopes, $^{1}_{1}H$ and $^{2}_{1}H$, and sulfur has four stable isotopes, $^{32}_{16}S$, $^{33}_{16}S$, $^{34}_{16}S$, $^{36}_{16}S$. How many peaks would you observe in the mass spectrum of the positive ion of hydrogen sulfide, H_2S^+? Assume no decomposition of the ion into smaller fragments.

PERCENT COMPOSITION AND CHEMICAL FORMULAS

Review Questions

3.37 Define percent composition by mass of a compound.

3.38 Describe how the knowledge of the percent composition by mass of an unknown compound of high purity can help us identify the compound.

3.39 What does the word "empirical" in empirical formula mean?

3.40 If we know the empirical formula of a compound, what additional information do we need in order to determine its molecular formula?

Problems

3.41 Tin (Sn) exists in Earth's crust as SnO_2. Calculate the percent composition by mass of Sn and O in SnO_2.

3.42 Sodium, a member of the alkali metal family (Group 1A), is very reactive and thus is never found in nature in its elemental state. It forms ionic compounds with members of the halogen family (Group 7A). Calculate the percent composition by mass of all the elements in each of these compounds: (a) NaF, (b) NaCl, (c) NaBr, (d) NaI.

3.43 For many years the organic compound chloroform ($CHCl_3$) was used as an inhalation anesthetic in spite of the fact that it is also a toxic substance that may cause severe liver, kidney, and heart damage. Calculate the percent composition by mass of this compound.

3.44 Allicin is the compound responsible for the characteristic smell of garlic. An analysis of the compound gives the following percent composition by mass: C: 44.4%; H: 6.21%; S: 39.5%; O: 9.86%. Calculate its empirical formula. What is its molecular formula given that its molar mass is about 162 g?

3.45 Cinnamic alcohol is used mainly in perfumery, particularly soaps and cosmetics. Its molecular formula is $C_9H_{10}O$. (a) Calculate the percent composition by mass of C, H, and O in cinnamic alcohol. (b) How many molecules of cinnamic alcohol are contained in a sample of mass 0.469 g?

3.46 All the substances listed below are fertilizers that contribute nitrogen to the soil. Which of these is the richest source of nitrogen on a mass percentage basis?
(a) Urea, $(NH_2)_2CO$
(b) Ammonium nitrate, NH_4NO_3
(c) Guanidine, $HNC(NH_2)_2$
(d) Ammonia, NH_3

3.47 The formula for rust can be represented by Fe_2O_3. How many moles of Fe are present in 24.6 g of the compound?

3.48 How many grams of sulfur (S) are needed to combine with 246 g of mercury (Hg) to form HgS?

3.49 Calculate the mass in grams of iodine (I_2) that will react completely with 20.4 g of aluminum (Al) to form aluminum iodide (AlI_3).

3.50 Tin(II) fluoride (SnF_2) is often added to toothpaste as an ingredient to prevent tooth decay. What is the mass of F in grams in 24.6 g of the compound?

3.51 What are the empirical formulas of the compounds with the following compositions? (a) 2.1% H, 65.3% O, 32.6% S; (b) 20.2% Al, 79.8% Cl; (c) 40.1% C, 6.6% H, 53.3% O; (d) 18.4% C, 21.5% N, 60.1% K.

3.52 Peroxyacylnitrate (PAN) is one of the components of smog. It is a compound of C, H, N, and O. Determine the percent composition of oxygen and the empirical formula from the following percent composition by mass: 19.8% C, 2.50% H, 11.6% N.

3.53 The molar mass of caffeine is 194.19 g. Is the molecular formula of caffeine $C_4H_5N_2O$ or $C_8H_{10}N_4O_2$?

3.54 Monosodium glutamate (MSG), a food-flavor enhancer, has been blamed for "Chinese restaurant syndrome," the symptoms of which are headaches and chest pains. MSG has the following composition by mass: 35.51% C, 4.77% H, 37.85% O, 8.29% N, and 13.60% Na. What is its molecular formula if its molar mass is 169 g?

CHEMICAL REACTIONS AND CHEMICAL EQUATIONS

Review Questions

3.55 Define the following terms: chemical reaction, reactant, product.

3.56 What is the difference between a chemical reaction and a chemical equation?

3.57 Why must a chemical equation be balanced? What law is obeyed by a balanced chemical equation?

3.58 Write the symbols used to represent gas, liquid, solid, and the aqueous phase in chemical equations.

Problems

3.59 Balance the following equations using the method outlined in Section 3.7:
(a) $C + O_2 \rightarrow CO$
(b) $CO + O_2 \rightarrow CO_2$
(c) $H_2 + Br_2 \rightarrow HBr$
(d) $K + H_2O \rightarrow KOH + H_2$
(e) $Mg + O_2 \rightarrow MgO$
(f) $O_3 \rightarrow O_2$
(g) $H_2O_2 \rightarrow H_2O + O_2$
(h) $N_2 + H_2 \rightarrow NH_3$
(i) $Zn + AgCl \rightarrow ZnCl_2 + Ag$
(j) $S_8 + O_2 \rightarrow SO_2$

3.60 Balance the following equations using the method outlined in Section 3.7:
(a) $KClO_3 \rightarrow KCl + O_2$
(b) $KNO_3 \rightarrow KNO_2 + O_2$
(c) $NH_4NO_3 \rightarrow N_2O + H_2O$
(d) $NH_4NO_2 \rightarrow N_2 + H_2O$
(e) $NaHCO_3 \rightarrow Na_2CO_3 + H_2O + CO_2$
(f) $P_4O_{10} + H_2O \rightarrow H_3PO_4$
(g) $HCl + CaCO_3 \rightarrow CaCl_2 + H_2O + CO_2$
(h) $Al + H_2SO_4 \rightarrow Al_2(SO_4)_3 + H_2$
(i) $CO_2 + KOH \rightarrow K_2CO_3 + H_2O$
(j) $CH_4 + O_2 \rightarrow CO_2 + H_2O$

AMOUNTS OF REACTANTS AND PRODUCTS

Review Questions

3.61 Define the following terms: stoichiometry, stoichiometric amounts, limiting reagent, excess reagent, theoretical yield, actual yield, percent yield.

3.62 On what law is stoichiometry based?

3.63 Describe the basic steps involved in the mole method.

3.64 Why is it essential to use balanced equations in solving stoichiometric problems?

Problems

3.65 Consider the combustion of carbon monoxide (CO) in oxygen gas:

$$2CO(g) + O_2(g) \longrightarrow 2CO_2(g)$$

Starting with 3.60 moles of CO, calculate the number of moles of CO_2 produced if there is enough oxygen gas to react with all of the CO.

3.66 Silicon tetrachloride ($SiCl_4$) can be prepared by heating Si in chlorine gas:

$$Si(s) + 2Cl_2(g) \longrightarrow SiCl_4(l)$$

In one reaction, 0.507 mole of $SiCl_4$ is produced. How many moles of molecular chlorine were used in the reaction?

3.67 The annual production of sulfur dioxide from burning coal and fossil fuels, auto exhaust, and other sources is about 26 million tons. The equation for the reaction is

$$S(s) + O_2(g) \longrightarrow SO_2(g)$$

How much sulfur, present in the original materials, would result in that quantity of SO_2?

3.68 When baking soda (sodium bicarbonate or sodium hydrogen carbonate, $NaHCO_3$) is heated, it releases carbon dioxide gas, which is responsible for the rising of cookies, doughnuts, and bread. (a) Write a balanced equation for the decomposition of the compound (one of the products is Na_2CO_3). (b) Calculate the mass of $NaHCO_3$ required to produce 20.5 g of CO_2.

3.69 When potassium cyanide (KCN) reacts with acids, a deadly poisonous gas, hydrogen cyanide (HCN), is given off. Here is the equation:

$$KCN(aq) + HCl(aq) \longrightarrow KCl(aq) + HCN(g)$$

If a sample of 0.140 g of KCN is treated with an excess of HCl, calculate the amount of HCN formed, in grams.

3.70 Fermentation is a complex chemical process of wine making in which glucose is converted into ethanol and carbon dioxide:

$$\underset{\text{glucose}}{C_6H_{12}O_6} \longrightarrow \underset{\text{ethanol}}{2C_2H_5OH} + 2CO_2$$

Starting with 500.4 g of glucose, what is the maximum amount of ethanol in grams and in liters that can be obtained by this process? (Density of ethanol = 0.789 g/mL.)

3.71 Each copper(II) sulfate unit is associated with five water molecules in crystalline copper(II) sulfate pentahydrate ($CuSO_4 \cdot 5H_2O$). When this compound is heated in air above 100°C, it loses the water molecules and also its blue color:

$$CuSO_4 \cdot 5H_2O \longrightarrow CuSO_4 + 5H_2O$$

If 9.60 g of $CuSO_4$ are left after heating 15.01 g of the blue compound, calculate the number of moles of H_2O originally present in the compound.

3.72 For many years the recovery of gold (that is, the separation of gold from other materials) involved the treatment of gold by isolation from other substances, using potassium cyanide:

$$4Au + 8KCN + O_2 + 2H_2O \longrightarrow 4KAu(CN)_2 + 4KOH$$

What is the minimum amount of KCN in moles needed in order to extract 29.0 g (about an ounce) of gold?

3.73 Limestone ($CaCO_3$) is decomposed by heating to quicklime (CaO) and carbon dioxide. Calculate how many grams of quicklime can be produced from 1.0 kg of limestone.

3.74 Nitrous oxide (N_2O) is also called "laughing gas." It can be prepared by the thermal decomposition of ammonium nitrate (NH_4NO_3). The other product is H_2O. (a) Write a balanced equation for this reaction. (b) How many grams of N_2O are formed if 0.46 mole of NH_4NO_3 is used in the reaction?

3.75 The fertilizer ammonium sulfate [$(NH_4)_2SO_4$] is prepared by the reaction between ammonia (NH_3) and sulfuric acid:

$$2NH_3(g) + H_2SO_4(aq) \longrightarrow (NH_4)_2SO_4(aq)$$

How many kilograms of NH_3 are needed to produce 1.00×10^5 kg of $(NH_4)_2SO_4$?

3.76 A common laboratory preparation of oxygen gas is the thermal decomposition of potassium chlorate ($KClO_3$). Assuming complete decomposition, calculate the number of grams of O_2 gas that can be obtained starting with 46.0 g of $KClO_3$. (The products are KCl and O_2.)

LIMITING REAGENTS

Review Questions

3.77 Define limiting reagent and excess reagent. What is the significance of the limiting reagent in predicting the amount of the product obtained in a reaction?

3.78 Give an everyday example that illustrates the limiting reagent concept.

Problems

3.79 Nitric oxide (NO) reacts instantly with oxygen gas to give nitrogen dioxide (NO_2), a dark-brown gas:

$$2NO(g) + O_2(g) \longrightarrow 2NO_2(g)$$

In one experiment 0.886 mole of NO is mixed with 0.503 mole of O_2. Calculate which of the two reactants is the limiting reagent. Calculate also the number of moles of NO_2 produced.

3.80 The depletion of ozone (O_3) in the stratosphere has been a matter of great concern among scientists in recent years. It is believed that ozone can react with nitric oxide (NO) that is discharged from the high-altitude jet plane, the SST. The reaction is

$$O_3 + NO \longrightarrow O_2 + NO_2$$

If 0.740 g of O_3 reacts with 0.670 g of NO, how many grams of NO_2 would be produced? Which compound is the limiting reagent? Calculate the number of moles of the excess reagent remaining at the end of the reaction.

3.81 Propane (C_3H_8) is a component of natural gas and is used in domestic cooking and heating. (a) Balance the following equation representing the combustion of propane in air:

$$C_3H_8 + O_2 \longrightarrow CO_2 + H_2O$$

(b) How many grams of carbon dioxide can be produced by burning 3.65 moles of propane? Assume that oxygen is the excess reagent in this reaction.

3.82 Consider the reaction

$$MnO_2 + 4HCl \longrightarrow MnCl_2 + Cl_2 + 2H_2O$$

If 0.86 mole of MnO_2 and 48.2 g of HCl react, which reagent will be used up first? How many grams of Cl_2 will be produced?

REACTION YIELD

Review Questions

3.83 Why is the yield of a reaction determined only by the amount of the limiting reagent?

3.84 Why is the actual yield of a reaction almost always smaller than the theoretical yield?

Problems

3.85 Hydrogen fluoride is used in the manufacture of Freons (which destroy ozone in the stratosphere) and in the production of aluminum metal. It is prepared by the reaction

$$CaF_2 + H_2SO_4 \longrightarrow CaSO_4 + 2HF$$

In one process 6.00 kg of CaF_2 are treated with an excess of H_2SO_4 and yield 2.86 kg of HF. Calculate the percent yield of HF.

3.86 Nitroglycerin ($C_3H_5N_3O_9$) is a powerful explosive. Its decomposition may be represented by

$$4C_3H_5N_3O_9 \longrightarrow 6N_2 + 12CO_2 + 10H_2O + O_2$$

This reaction generates a large amount of heat

and many gaseous products. It is the sudden formation of these gases, together with their rapid expansion, that produces the explosion. (a) What is the maximum amount of O_2 in grams that can be obtained from 2.00×10^2 g of nitroglycerin? (b) Calculate the percent yield in this reaction if the amount of O_2 generated is found to be 6.55 g.

3.87 Titanium(IV) oxide (TiO_2) is a white substance produced by the action of sulfuric acid on the mineral ilmenite ($FeTiO_3$):

$$FeTiO_3 + H_2SO_4 \longrightarrow TiO_2 + FeSO_4 + H_2O$$

Its opaque and nontoxic properties make it suitable as a pigment in plastics and paints. In one process 8.00×10^3 kg of $FeTiO_3$ yielded 3.67×10^3 kg of TiO_2. What is the percent yield of the reaction?

3.88 Ethylene (C_2H_4), an important industrial organic chemical, can be prepared by heating hexane (C_6H_{14}) at 800°C:

$$C_6H_{14} \longrightarrow C_2H_4 + \text{other products}$$

If the yield of ethylene production is 42.5 percent, what mass of hexane must be reacted to produce 481 g of ethylene?

MISCELLANEOUS PROBLEMS

3.89 The atomic mass of element X is 33.42 amu. A 27.22-g sample of X combines with 84.10 g of another element Y to form a compound XY. Calculate the atomic mass of Y.

3.90 How many moles of O are needed to combine with 0.212 mole of C to form (a) CO and (b) CO_2?

3.91 A research chemist used a mass spectrometer to study the two isotopes of an element. Over time, she recorded a number of mass spectra of these isotopes. On analysis, she noticed that the ratio of the taller peak (the more abundant isotope) to the shorter peak (the less abundant isotope) gradually increased with time. Assuming that the mass spectrometer was functioning normally, what do you think was causing this change?

3.92 The aluminum sulfate hydrate [$Al_2(SO_4)_3 \cdot xH_2O$] contains 8.20 percent of Al by mass. Calculate x, that is, the number of water molecules associated with each $Al_2(SO_4)_3$ unit.

3.93 Mustard gas ($C_4H_8Cl_2S$) is a poisonous gas that was used in World War I and banned afterward. It causes general destruction of body tissues, resulting in the formation of large water blisters. There is no effective antidote. Calculate the percent composition by mass of the elements in mustard gas.

3.94 The carat is the unit of mass used in jewelry, and 1 carat is exactly 200 mg. How many carbon atoms are present in a 24-carat diamond?

3.95 An iron bar weighed 664 g. After the bar had been standing in moist air for a month, exactly one-eighth of the iron turned to rust (Fe_2O_3). Calculate the final mass.

3.96 A certain metal oxide has the formula MO. A 39.46-g sample of the compound is strongly heated in an atmosphere of hydrogen to remove oxygen as water molecules. At the end, 31.70 g of the metal M is left over. If O has an atomic mass of 16.00 amu, calculate the atomic mass of M and identify the element.

3.97 An impure sample of zinc (Zn) is treated with an excess of sulfuric acid (H_2SO_4) to form zinc sulfate ($ZnSO_4$) and molecular hydrogen (H_2). (a) Write a balanced equation for the reaction. (b) If 0.0764 g of H_2 is obtained from 3.86 g of the sample, calculate the percent purity of the sample. (c) What assumptions must you make in (b)?

3.98 One of the reactions that occurs in a blast furnace, where iron ore is converted to cast iron, is

$$Fe_2O_3 + 3CO \longrightarrow 2Fe + 3CO_2$$

Suppose that 1.64×10^3 kg of Fe are obtained from a 2.62×10^3-kg sample of Fe_2O_3. Assuming that the reaction goes to completion, what is the percent purity of Fe_2O_3 in the original sample?

3.99 Carbon dioxide (CO_2) is the gas that is mainly responsible for global warming (the so-called greenhouse effect). The burning of fossil fuels is a major cause of the increased concentration of CO_2 in the atmosphere. Carbon dioxide is also the end product of metabolism (see Example 3.14). Using glucose as an example of food, calculate the annual production of CO_2 in grams, assuming that each person consumes 5.0×10^2 g of glucose per day. The world's population is 5.5 billion, and there are 365 days in a year.

3.100 Carbohydrates are compounds containing carbon, hydrogen, and oxygen in which the hydrogen to oxygen ratio is 2 : 1. A certain carbohydrate contains 40.0% of carbon by mass. Calculate the empirical and molecular formulas of the compound if the approximate molar mass is 178 g.

Answers to Practice Exercises: **3.1** 10.81 amu. **3.2** (a) 6.44 g He, (b) 0.0130 mol Mg. **3.3** 2.107×10^{-22} g. **3.4** 8.49×10^{21} K atoms. **3.5** 32.04 amu. **3.6** 1.66 moles. **3.7** 5.81×10^{24} H atoms. **3.8** H: 2.055%; S: 32.69%; O: 65.25%. **3.9** $KMnO_4$ (potassium permanganate). **3.10** 196 g. **3.11** B_2H_6. **3.12** $Fe_2O_3 + 3CO \longrightarrow 2Fe + 3CO_2$. **3.13** (a) 0.508 mole, (b) 2.21 g. **3.14** 235 g. **3.15** (a) 234 g, (b) 234 g. **3.16** (a) 863 g, (b) 93.0%.

CHAPTER 4
REACTIONS IN AQUEOUS SOLUTIONS

A lock of Napoleon's hair.

◆ After his defeat at Waterloo in 1815, Napoleon was exiled to St. Helena, a small island in the Atlantic, where he spent the last six years of his life. In the 1960s, samples of his hair were analyzed and found to contain a high level of arsenic, suggesting that he might have been poisoned. The prime suspects are the governor of St. Helena, with whom Napoleon did not get along, and the French royal family, who wanted to prevent his return to France.

Elemental arsenic is not that harmful. The commonly used poison is actually arsenic(III) oxide, As_2O_3. This white compound dissolves in water, is tasteless, and, if administered over a period of time, is hard to detect. It was once known as the "inheritance powder": adding As_2O_3 to grandfather's wine would hasten his demise so the grandson could inherit the estate!

In 1832 the English chemist James Marsh devised a procedure for detecting arsenic that now bears his name. In the Marsh test, hydrogen, formed by the reaction between zinc and sulfuric acid, is allowed to react with a sample of the suspected poison. If As_2O_3 is present, it reacts with hydrogen to form a toxic gas, arsine (AsH_3). When arsine gas is heated, it decomposes to form arsenic, which is recognized by its metallic luster. The Marsh test is an effective deterrent to murder by As_2O_3. But it was invented too late to do Napoleon any good, if, in fact, he was a victim of deliberate arsenic poisoning.

Doubts about the conspiracy theory of Napoleon's death developed in the early 1990s, when a sample of the wallpaper from his drawing room was found to contain copper arsenite, a green pigment that was commonly used at the time Napoleon lived. It has been suggested that the damp climate on St. Helena promoted the growth of molds on the wallpaper. To rid themselves of arsenic, the molds could have converted it to trimethyl arsine [$(CH_3)_3As$], a volatile and highly poisonous compound. Prolonged exposure to this vapor could have ruined Napoleon's health and would also account for the presence of arsenic in his body, though it may not have been the primary cause of his death. This provocative theory is supported by the fact that Napoleon's regular guests suffered from gastrointestinal disturbances and other symptoms of arsenic poisoning and that their health all seemed to improve whenever they spent hours working outdoors in the garden, their main hobby on the island.

We will probably never know whether Napoleon died from arsenic poisoning, intentional or accidental, but the exercise in historical sleuthing provides a fascinating example of the use of chemical analysis. It plays an essential part in a range of endeavors from pure research to practical applications like quality control of commercial products, disease diagnosis, and forensic science. ◆

4.1 GENERAL PROPERTIES OF AQUEOUS SOLUTIONS

Many chemical reactions and virtually all biological processes take place in an aqueous environment. Therefore it is important to understand the properties of different substances in solution with water. To start with, what exactly is a solution? A ***solution*** is *a homogeneous mixture of two or more substances. The substance present in a smaller amount* is called the ***solute,*** while *the substance present in a larger amount* is called the ***solvent.*** A solution may be gaseous (such as air), solid (such as an alloy), or liquid (seawater, for example). In this section we will discuss only ***aqueous solutions,*** in which *the solute initially is a liquid or a solid and the solvent is water.*

ELECTROLYTES VERSUS NONELECTROLYTES

All solutes that dissolve in water fit into one of two categories: electrolytes and nonelectrolytes. An ***electrolyte*** is *a substance that, when dissolved in water, results in a solution that can conduct electricity.* A ***nonelectrolyte*** *does not conduct electricity when dissolved in water.* Figure 4.1 shows an easy and straightforward method of distinguishing between electrolytes and nonelectrolytes. A pair of platinum electrodes is immersed in a beaker of water. To light the bulb, electric current must flow from one electrode to the other, thus completing the circuit. Pure water is a very poor conductor of electricity. However, if we add a small amount of sodium chloride (NaCl), the bulb will glow as soon as the salt dissolves in the water. Solid NaCl, an ionic compound, breaks up into Na^+ and Cl^- ions when it dissolves in water. The Na^+ ions are attracted to the negative electrode and the Cl^- ions to the positive electrode. This movement sets up an electrical current that is equivalent to

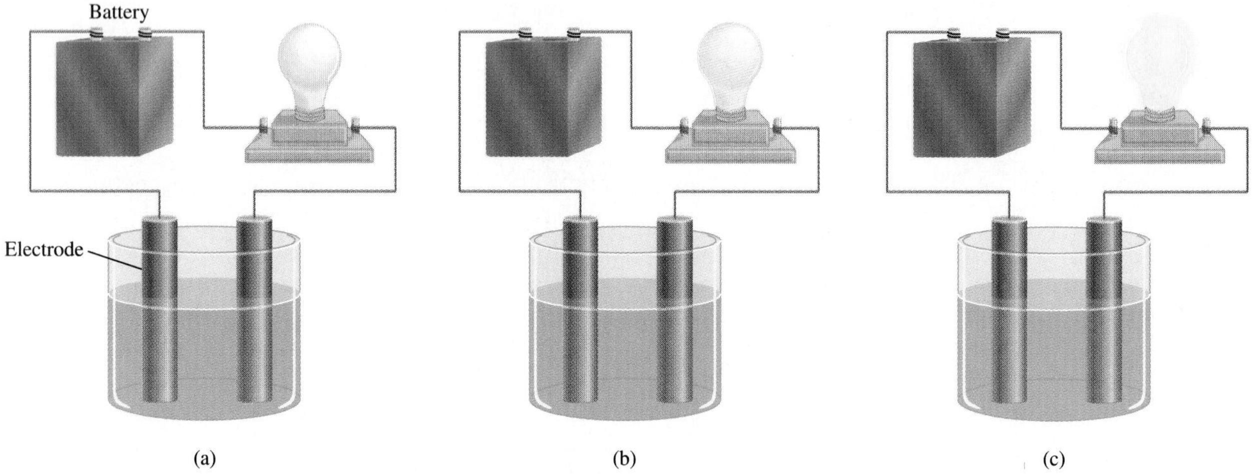

FIGURE 4.1

An arrangement for distinguishing between electrolytes and nonelectrolytes. (a) A nonelectrolyte solution does not contain ions, and the light bulb is not lit. (b) A weak electrolyte solution contains a small number of ions, and the light bulb is dimly lit. (c) A strong electrolyte solution contains a large number of ions, and the light bulb is brightly lit.

TABLE 4.1
Classification of Solutes in Aqueous Solution

Strong electrolyte	Weak electrolyte	Nonelectrolyte
HCl	CH_3COOH	$(NH_2)_2CO$ (urea)
HNO_3	HF	CH_3OH (methanol)
$HClO_4$	HNO_2	C_2H_5OH (ethanol)
H_2SO_4*	NH_3	$C_6H_{12}O_6$ (glucose)
NaOH	H_2O†	$C_{12}H_{22}O_{11}$ (sucrose)
$Ba(OH)_2$		
Ionic compounds		

*H_2SO_4 has two ionizable H^+ ions.
†Pure water is an extremely weak electrolyte.

the flow of electrons along a metal wire. Because the NaCl solution conducts electricity, we say that NaCl is an electrolyte. Pure water contains very few ions, so it cannot conduct electricity.

Comparing the light bulb's brightness for the *same molar amounts* of dissolved substances helps us distinguish between strong and weak electrolytes. A characteristic of strong electrolytes is that the solute is assumed to be 100 percent dissociated into ions in solution. (By *dissociation* we mean the breaking up of the compound into cations and anions.) Thus we can represent sodium chloride dissolving in water as

$$NaCl(s) \xrightarrow{H_2O} Na^+(aq) + Cl^-(aq)$$

What this equation says is that all the sodium chloride that enters the solution ends up as Na^+ and Cl^- ions; there are no undissociated NaCl units in solution.

Table 4.1 lists examples of strong electrolytes, weak electrolytes, and nonelectrolytes. Ionic compounds, such as sodium chloride, potassium iodide (KI), and calcium nitrate [$Ca(NO_3)_2$], are strong electrolytes. It is interesting to note that human body fluids contain many strong and weak electrolytes.

Water is a very effective solvent for ionic compounds. Although water is an electrically neutral molecule, it has a positive end (the H atoms) and a negative end (the O atom), or positive and negative "poles"; for this reason it is often referred to as a *polar* solvent. When an ionic compound such as sodium chloride dissolves in water, the three-dimensional network of the ions in the solid is destroyed, and the Na^+ and Cl^- ions are separated from each other. In solution, each Na^+ ion is surrounded by a number of water molecules orienting their negative ends toward the cation. Similarly, each Cl^- ion is surrounded by water molecules with their positive ends oriented toward the anion (Figure 4.2). *The process in which an ion is*

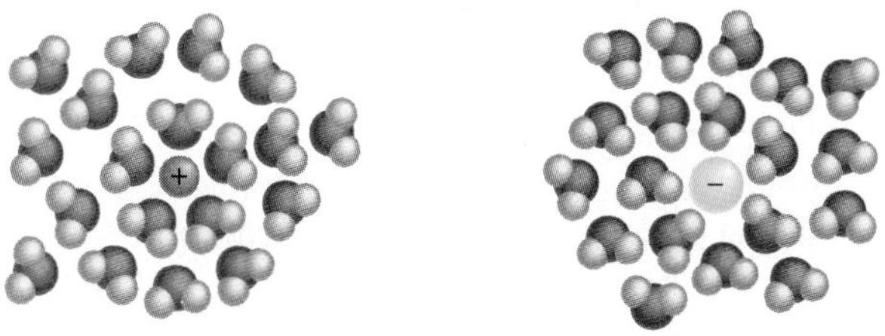

FIGURE 4.2
Hydration of Na^+ and Cl^- ions.

surrounded by water molecules arranged in a specific manner is called ***hydration.*** Hydration helps to stabilize ions in solution and prevents cations from combining with anions.

Acids and bases are also electrolytes. Some acids, including hydrochloric acid (HCl) and nitric acid (HNO_3), are strong electrolytes. These acids ionize completely in water; for example, when hydrogen chloride gas dissolves in water, it forms hydrated H^+ and Cl^- ions:

$$HCl(g) \xrightarrow{H_2O} H^+(aq) + Cl^-(aq)$$

In other words, *all* the dissolved HCl molecules separate into hydrated H^+ and Cl^- ions in solution. Thus when we write HCl(*aq*), it is understood that it is a solution of only $H^+(aq)$ and $Cl^-(aq)$ ions and there are no hydrated HCl molecules present. On the other hand, certain acids, such as acetic acid (CH_3COOH), which is found in vinegar, ionize to a much less extent. We represent the ionization of acetic acid as

$$CH_3COOH(aq) \rightleftharpoons CH_3COO^-(aq) + H^+(aq)$$

where CH_3COO^- is called the acetate ion. (In this book we will use the term "dissociation" for ionic compounds and "ionization" for acids and bases.) By writing the formula of acetic acid as CH_3COOH we indicate that the ionizable proton is in the COOH group.

The double arrow $\rightleftharpoons$ in a reaction means that the reaction is ***reversible;*** that is, *the reaction can occur in both directions.* Initially, a number of CH_3COOH molecules break up to yield CH_3COO^- and H^+ ions. As time goes on, some of the CH_3COO^- and H^+ ions recombine to form CH_3COOH molecules. Eventually, a state is reached in which the acid molecules break up as fast as the ions recombine. Such *a chemical state, in which no net change can be observed* (although continuous activity is taking place on the molecular level), is called ***chemical equilibrium.*** Acetic acid, then, is a weak electrolyte because its ionization in water is incomplete. By contrast, in a hydrochloric acid solution the H^+ and Cl^- ions have no tendency to recombine to form molecular HCl. We use the single arrow to represent complete ionizations.

4.2 PRECIPITATION REACTIONS

Having discussed the general properties of aqueous solutions, we can study some common and important reactions in this medium. In this section we will consider ***precipitation reactions,*** which are *characterized by the formation of an insoluble product, or precipitate.* A ***precipitate*** is *an insoluble solid that separates from the solution.* Precipitation reactions usually involve ionic compounds. For example, when an aqueous solution of lead nitrate [$Pb(NO_3)_2$] is added to sodium iodide (NaI), a yellow precipitate of lead iodide (PbI_2) is formed (Figure 4.3):

$$Pb(NO_3)_2(aq) + 2NaI(aq) \longrightarrow PbI_2(s) + 2NaNO_3(aq)$$

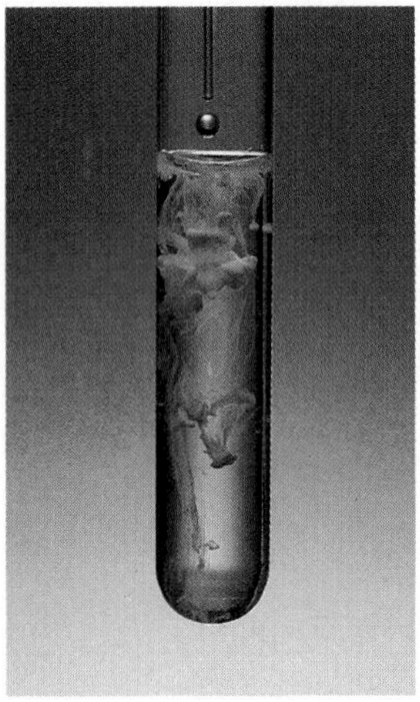

FIGURE 4.3

Formation of yellow PbI_2 precipitate as a solution of $Pb(NO_3)_2$ is added to a solution of NaI.

SOLUBILITY

How can we predict whether a precipitate will form when two solutions are mixed or when a compound is added to a solution? We need to know the ***solubility*** of the solute, that is, *the maximum amount of solute that will dissolve in a given quantity of solvent at a specific temperature.* All ionic compounds are strong electrolytes, but

TABLE 4.2
Solubility Characteristics of Ionic Compounds in Water at 25°C

1. All alkali metal (Group 1A) compounds are soluble.
2. All ammonium (NH_4^+) compounds are soluble.
3. All compounds containing nitrate (NO_3^-), chlorate (ClO_3^-), and perchlorate (ClO_4^-) are soluble.
4. Most hydroxides (OH^-) are insoluble. The exceptions are the alkali metal hydroxides and barium hydroxide [$Ba(OH)_2$]. Calcium hydroxide [$Ca(OH)_2$] is slightly soluble.
5. Most compounds containing chlorides (Cl^-), bromides (Br^-), or iodides (I^-) are soluble. The exceptions are those containing Ag^+, Hg_2^{2+}, and Pb^{2+}.
6. All carbonates (CO_3^{2-}), phosphates (PO_4^{3-}), and sulfides (S^{2-}) are insoluble; the exceptions are those of alkali metals and the ammonium ion.
7. Most sulfates (SO_4^{2-}) are soluble. Calcium sulfate ($CaSO_4$) and silver sulfate (Ag_2SO_4) are slightly soluble. Barium sulfate ($BaSO_4$), mercury(II) sulfate ($HgSO_4$), and lead sulfate ($PbSO_4$) are insoluble.

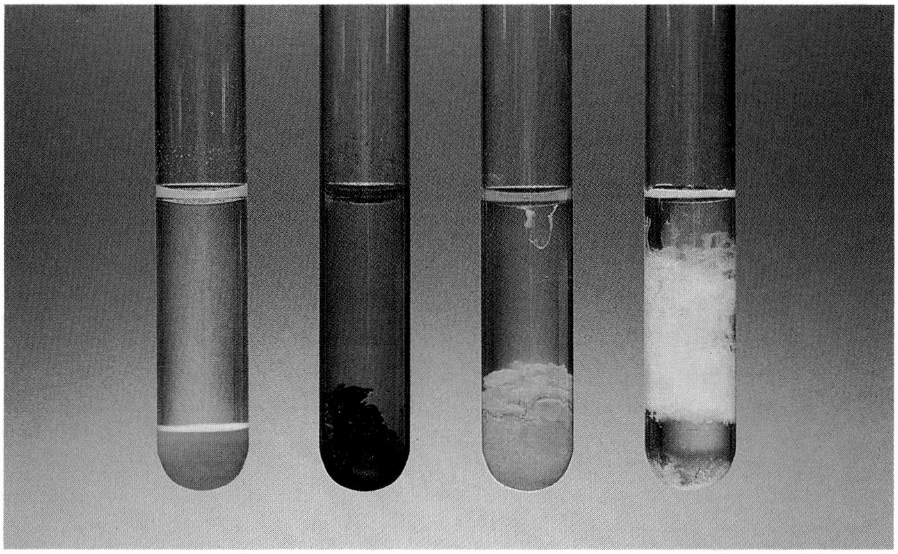

FIGURE 4.4

Appearance of several precipitates. From left to right: CdS, PbS, Ni(OH)$_2$, Al(OH)$_3$.

they are not equally soluble. We classify ionic compounds as "soluble," "slightly soluble," or "insoluble." Table 4.2 provides solubility rules that help us determine how a given compound will behave in aqueous solution. Figure 4.4 shows several insoluble substances.

EXAMPLE 4.1
Applying Solubility Rules

Classify the following ionic compounds as soluble, slightly soluble, or insoluble: (a) silver sulfate (Ag_2SO_4), (b) calcium carbonate ($CaCO_3$), (c) sodium phosphate (Na_3PO_4).

Answer: (a) According to rule 7 in Table 4.2, Ag_2SO_4 is slightly soluble. (b) According to rule 6, $CaCO_3$ is insoluble. (c) Sodium is an alkali metal (Group 1A). According to rule 1, Na_3PO_4 is soluble.

PRACTICE EXERCISE

Classify the following ionic compounds as soluble, slightly soluble, or insoluble: (a) CuS, (b) $Ca(OH)_2$, (c) $Zn(NO_3)_2$.

MOLECULAR EQUATIONS AND IONIC EQUATIONS

The equation describing the precipitation of lead iodide on page 90 is called a ***molecular equation*** because *the formulas of the compounds are written as though all species existed as molecules or whole units.* A molecular equation is useful because it tells us the identity of the reactants (that is, lead nitrate and sodium iodide). If we wanted to carry out this reaction, the molecular equation would be the one to use. However, a molecular equation does not accurately describe what actually happens at the microscopic level. As pointed out earlier, when ionic compounds dissolve in water, they break apart completely into their component cations and anions. To be more faithful to reality, the equations should show the dissociation of dissolved ionic compounds into ions. Therefore, returning to the reaction between sodium iodide and lead nitrate, we would write

$$Pb^{2+}(aq) + 2NO_3^-(aq) + 2Na^+(aq) + 2I^-(aq) \longrightarrow PbI_2(s) + 2Na^+(aq) + 2NO_3^-(aq)$$

Such an equation, which *shows dissolved ionic compounds in terms of their free ions,* is called an ***ionic equation.*** *Ions that are not involved in the overall reaction,* in this case the Na^+ and NO_3^- ions, are called ***spectator ions.*** Since the spectator ions appear on both sides of the equation and are unchanged in the chemical reaction, they can be canceled. To focus on the change that actually occurs, we write the ***net ionic equation,*** that is, *the equation that shows only the species that actually take part in the reaction:*

$$Pb^{2+}(aq) + 2I^-(aq) \longrightarrow PbI_2(s)$$

Similarly, when an aqueous solution of barium chloride ($BaCl_2$) is added to an aqueous solution of sodium sulfate (Na_2SO_4), a white precipitate of barium sulfate ($BaSO_4$) is formed (Figure 4.5):

$$BaCl_2(aq) + Na_2SO_4(aq) \longrightarrow BaSO_4(s) + 2NaCl(aq)$$

The ionic equation for the reaction is

$$Ba^{2+}(aq) + 2Cl^-(aq) + 2Na^+(aq) + SO_4^{2-}(aq) \longrightarrow BaSO_4(s) + 2Na^+(aq) + 2Cl^-(aq)$$

Canceling the spectator ions (Na^+ and Cl^-) on both sides of the equation gives us the net ionic equation

$$Ba^{2+}(aq) + SO_4^{2-}(aq) \longrightarrow BaSO_4(s)$$

The following steps summarize the procedure for writing ionic and net ionic equations.

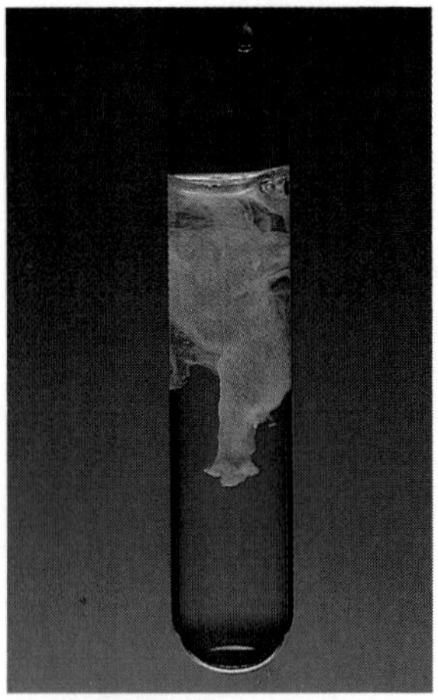

FIGURE 4.5

Formation of $BaSO_4$ precipitate.

- Write a balanced molecular equation for the reaction.
- Rewrite the equation to indicate which substances are in ionic form in solution. Remember that all strong electrolytes, when dissolved in solution, are completely dissociated into cations and anions. This procedure gives us the ionic equation.
- Identify and cancel spectator ions on both sides of the equation to arrive at the net ionic equation.

EXAMPLE 4.2

Using Solubility Rules to Predict the Products of Reactions in Solution

Predict the products of the following reaction and write a net ionic equation for the reaction

$$K_3PO_4(aq) + Ca(NO_3)_2(aq) \longrightarrow ?$$

Answer: Both reactants are soluble salts, but according to solubility rule 6, calcium ions (Ca^{2+}) and phosphate ions (PO_4^{3-}) can form an insoluble compound, calcium phosphate [$Ca_3(PO_4)_2$]. Therefore this is a precipitation reaction. The other product, potassium nitrate (KNO_3), is soluble and remains in solution. The molecular equation is

$$2K_3PO_4(aq) + 3Ca(NO_3)_2(aq) \longrightarrow 6KNO_3(aq) + Ca_3(PO_4)_2(s)$$

and the ionic equation is

$$6K^+(aq) + 2PO_4^{3-}(aq) + 3Ca^{2+}(aq) + 6NO_3^-(aq) \longrightarrow 6K^+(aq) + 6NO_3^-(aq) + Ca_3(PO_4)_2(s)$$

Canceling the spectator K^+ and NO_3^- ions, we obtain the net ionic equation

$$3Ca^{2+}(aq) + 2PO_4^{3-}(aq) \longrightarrow Ca_3(PO_4)_2(s)$$

Note that because we balanced the molecular equation first, the net ionic equation is balanced as to the number of atoms on each side, and the number of positive and negative charges on the left-hand side is the same.

Formation of $Ca_3(PO_4)_2$ precipitate.

PRACTICE EXERCISE

Predict the precipitate produced by the following reaction and write a net ionic equation for the reaction

$$Al(NO_3)_3(aq) + 3NaOH(aq) \longrightarrow ?$$

4.3 ACID-BASE REACTIONS

Acids and bases are as familiar as aspirin and milk of magnesia, although many people do not know their chemical names—acetylsalicylic acid (aspirin) and magnesium hydroxide (milk of magnesia). In addition to being the basis of many medicinal and household products, acid-base chemistry is important in industrial processes and essential in sustaining biological systems.

General Properties of Acids and Bases

In Section 2.7 we defined acids as substances that ionize in water to produce H^+ ions and bases as substances that ionize in water to produce OH^- ions. These definitions were formulated in the late nineteenth century by the Swedish chemist Svante Arrhenius to classify substances whose properties in aqueous solutions were well known.

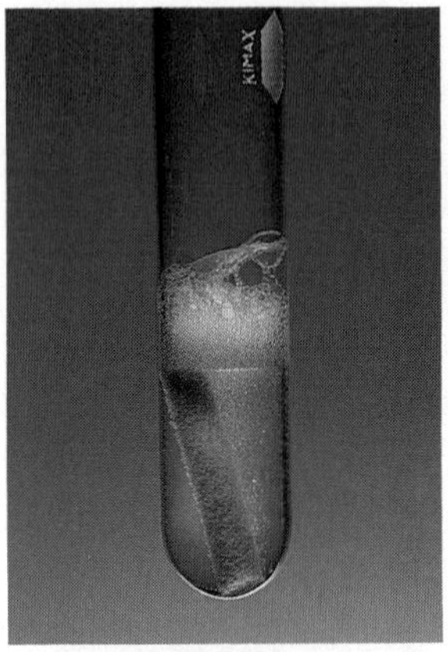

FIGURE 4.6

A piece of blackboard chalk, which is mostly $CaCO_3$, reacts with hydrochloric acid.

Acids

- Acids have a sour taste; for example, vinegar owes its taste to acetic acid, and lemons and other citrus fruits contain citric acid.
- Acids cause color changes in plant dyes; for example, they change the color of litmus from blue to red.
- Acids react with certain metals such as zinc, magnesium, and iron to produce hydrogen gas. A typical reaction is that between hydrochloric acid and magnesium:

$$2HCl(aq) + Mg(s) \longrightarrow MgCl_2(aq) + H_2(g)$$

- Acids react with carbonates and bicarbonates such as Na_2CO_3, $CaCO_3$, and $NaHCO_3$ to produce carbon dioxide gas (Figure 4.6). For example,

$$2HCl(aq) + CaCO_3(s) \longrightarrow CaCl_2(aq) + H_2O(l) + CO_2(g)$$

$$HCl(aq) + NaHCO_3(s) \longrightarrow NaCl(aq) + H_2O(l) + CO_2(g)$$

- Aqueous acid solutions conduct electricity.

Bases

- Bases have a bitter taste.
- Bases feel slippery; for example, soaps, which contain bases, exhibit this property.
- Bases cause color changes in plant dyes; for example, they change the color of litmus from red to blue.
- Aqueous base solutions conduct electricity.

Johannes Nicolaus Brønsted (1879–1947).

Brønsted Acids and Bases

Arrhenius's definitions of acids and bases are limited in that they apply only to aqueous solutions. Broader definitions, which were proposed by the Danish chemist Johannes Brønsted in 1932, describe an acid as *a proton donor* and a base as *a proton acceptor.* Substances that behave according to this definition are called ***Brønsted acids*** and ***Brønsted bases.***

Nitric acid is a Brønsted acid since it donates a proton in water:

$$HNO_3(aq) \longrightarrow H^+(aq) + NO_3^-(aq)$$

Because hydrogen ions (or protons) are hydrated in solution, they are often represented as H_3O^+, which is called a ***hydronium ion.*** Thus the ionization of nitric acid is better represented as

$$HNO_3(aq) + H_2O(l) \longrightarrow H_3O^+(aq) + NO_3^-(aq)$$

In this text we will use both $H^+(aq)$ and H_3O^+ to represent the hydrated proton. The $H^+(aq)$ notation is for convenience, and the H_3O^+ notation is closer to reality. Keep in mind that they represent the same species in aqueous solution.

Among the acids commonly used in the laboratory are hydrochloric acid (HCl), nitric acid (HNO_3), acetic acid (CH_3COOH), sulfuric acid (H_2SO_4), and phosphoric acid (H_3PO_4). The first three acids are ***monoprotic acids;*** that is, *each unit of the acid yields one hydrogen ion:*

$$HCl(aq) \longrightarrow H^+(aq) + Cl^-(aq)$$

$$HNO_3(aq) \longrightarrow H^+(aq) + NO_3^-(aq)$$

$$CH_3COOH(aq) \rightleftharpoons CH_3COO^-(aq) + H^+(aq)$$

As mentioned earlier, because the ionization of acetic acid is incomplete (note the double arrows), it is a weak electrolyte. For this reason it is called a weak acid. On the other hand, both HCl and HNO_3 are strong acids because they are strong electrolytes, so they are completely ionized in solution (note the use of single arrows).

Sulfuric acid (H_2SO_4) is called a ***diprotic acid*** because *each unit of the acid gives up two H^+ ions,* in two separate steps:

$$H_2SO_4(aq) \longrightarrow H^+(aq) + HSO_4^-(aq)$$

$$HSO_4^-(aq) \rightleftharpoons H^+(aq) + SO_4^{2-}(aq)$$

H_2SO_4 is a strong electrolyte/strong acid (the first step of ionization is complete), but HSO_4^- is a weak acid and we need a double arrow to represent its incomplete ionization.

Triprotic acids, which *yield three H^+ ions,* are relatively few in number. The best known triprotic acid is phosphoric acid, whose ionizations are

$$H_3PO_4(aq) \rightleftharpoons H^+(aq) + H_2PO_4^-(aq)$$

$$H_2PO_4^-(aq) \rightleftharpoons H^+(aq) + HPO_4^{2-}(aq)$$

$$HPO_4^{2-}(aq) \rightleftharpoons H^+(aq) + PO_4^{3-}(aq)$$

Note that all three species (H_3PO_4, $H_2PO_4^-$, and HPO_4^{2-}) are weak acids, and we use the double arrows to represent each ionization step. Anions such as $H_2PO_4^-$ and HPO_4^{2-} are generated when ionic compounds such as NaH_2PO_4 and Na_2HPO_4 dissolve in water.

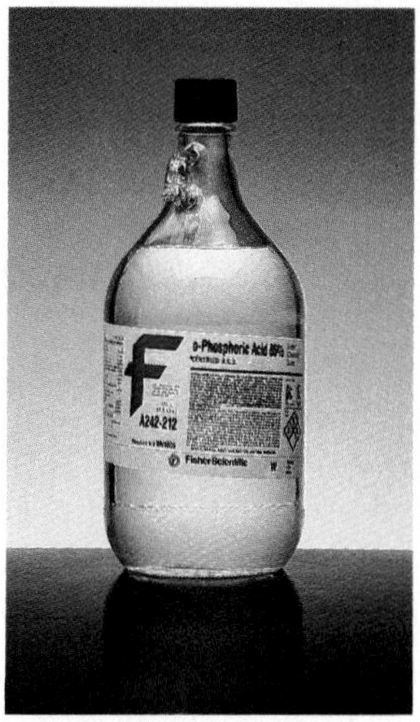

A bottle of phosphoric acid (H_3PO_4).

Table 4.1 shows that sodium hydroxide (NaOH) and barium hydroxide [$Ba(OH)_2$] are strong electrolytes. This means that they are completely ionized in solution:

$$NaOH(s) \xrightarrow{H_2O} Na^+(aq) + OH^-(aq)$$

$$Ba(OH)_2(s) \xrightarrow{H_2O} Ba^{2+}(aq) + 2OH^-(aq)$$

The OH^- ion can accept a proton as follows:

$$H^+(aq) + OH^-(aq) \longrightarrow H_2O(l)$$

Thus OH^- is a Brønsted base.

Ammonia (NH_3) is classified as a Brønsted base because it can accept a H^+ ion, as shown in the following equation:

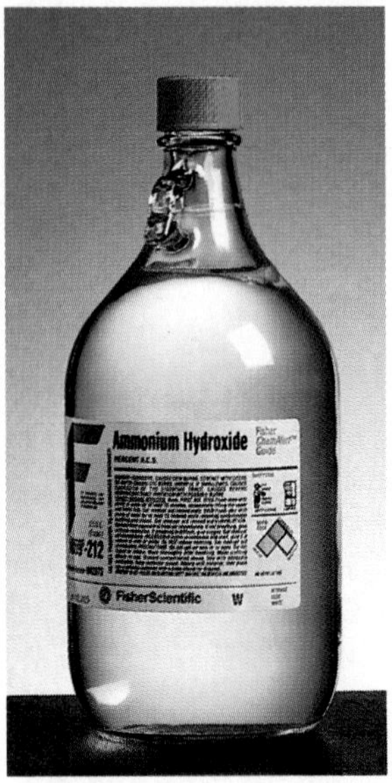

A bottle of NH_4OH.

$$NH_3(aq) + H_2O(l) \rightleftharpoons NH_4^+(aq) + OH^-(aq)$$

Ammonia is a weak electrolyte (and therefore classified as a weak base) because only a small fraction of dissolved NH_3 molecules react with water to form NH_4^+ and OH^- ions.

The most commonly used strong base in the laboratory is sodium hydroxide, because it is cheap and soluble. (In fact, all of the alkali metal hydroxides are soluble.) The most commonly used weak base is aqueous ammonia solution, which is sometimes erroneously called ammonium hydroxide. There is no evidence that the species NH_4OH actually exists. All of the Group 2A elements form hydroxides of the type $M(OH)_2$, where M denotes an alkaline earth metal. Of these hydroxides, only $Ba(OH)_2$ is soluble. Magnesium and calcium hydroxides are used in medicine and industry. Hydroxides of other metals, such as $Al(OH)_3$ and $Zn(OH)_2$, are insoluble and are less commonly used.

EXAMPLE 4.3

Classifying Brønsted Acids and Brønsted Bases

Classify each of the following species as a Brønsted acid or base: (a) HBr, (b) NO_2^-, (c) HCO_3^-.

Answer: (a) HBr dissolves in water to yield H^+ and Br^- ions:

$$HBr(aq) \longrightarrow H^+(aq) + Br^-(aq)$$

Therefore HBr is a Brønsted acid.

(b) In solution the nitrite ion can accept a proton to form nitrous acid:

$$NO_2^-(aq) + H^+(aq) \longrightarrow HNO_2(aq)$$

This property makes NO_2^- a Brønsted base.

(c) The bicarbonate ion is a Brønsted acid because it ionizes in solution as follows:

$$HCO_3^-(aq) \rightleftharpoons H^+(aq) + CO_3^{2-}(aq)$$

It is also a Brønsted base because it can accept a proton:

$$HCO_3^-(aq) + H^+(aq) \rightleftharpoons H_2CO_3(aq)$$

PRACTICE EXERCISE

Classify each of the following species as a Brønsted acid or base: (a) SO_4^{2-}, (b) HI.

Acid-Base Neutralization

An acid-base reaction, also called a ***neutralization reaction,*** is *a reaction between an acid and a base.* Aqueous acid-base reactions are generally characterized by the following equation:

$$\text{acid} + \text{base} \longrightarrow \text{salt} + \text{water}$$

A ***salt*** is *an ionic compound made up of a cation other than H^+ and an anion other* *than OH^- or O^{2-}*. All salts are strong electrolytes. The substance we know as table salt, NaCl, is a familiar example. It is a product of the following acid-base reaction:

$$HCl(aq) + NaOH(aq) \longrightarrow NaCl(aq) + H_2O(l)$$

However, since both the acid and the base are strong electrolytes, they are completely ionized in solution. The ionic equation is

$$H^+(aq) + Cl^-(aq) + Na^+(aq) + OH^-(aq) \longrightarrow Na^+(aq) + Cl^-(aq) + H_2O(l)$$

Therefore, the reactions can be represented by the net ionic equation

$$H^+(aq) + OH^-(aq) \longrightarrow H_2O(l)$$

Both Na^+ and Cl^- are spectator ions.

If we had started the above reaction with equal molar amounts of the acid and the base, at the end of the reaction we would have only a salt and no leftover acid or base. This is a characteristic of acid-base neutralization reactions.

The following are further examples of acid-base neutralization reactions, represented by molecular equations:

$$HF(aq) + KOH(aq) \longrightarrow KF(aq) + H_2O(l)$$

$$H_2SO_4(aq) + 2NaOH(aq) \longrightarrow Na_2SO_4(aq) + 2H_2O(l)$$

$$HNO_3(aq) + NH_3(aq) \longrightarrow NH_4NO_3(aq)$$

If we had expressed $NH_3(aq)$ as $NH_4^+(aq)$ and $OH^-(aq)$, then the last reaction would yield a salt (NH_4NO_3) and H_2O.

4.4 OXIDATION-REDUCTION REACTIONS

Whereas acid-base reactions can be characterized as proton-transfer processes, the class of reactions called ***oxidation-reduction*** (or ***redox***) ***reactions*** are considered *electron-transfer reactions*. Oxidation-reduction reactions are very much a part of the world around us. They range from the combustion of fossil fuels to the action of household bleach. In addition, most metallic and nonmetallic elements are obtained from their ores by the process of oxidation or reduction.

Consider the formation of calcium oxide (CaO) from calcium and oxygen:

$$2Ca(s) + O_2(g) \longrightarrow 2CaO(s)$$

Calcium oxide (CaO) is an ionic compound made up of Ca^{2+} and O^{2-} ions. In this reaction, two Ca atoms give up or transfer four electrons to two O atoms (in O_2). For convenience, this process can be considered as two separate steps, one involving the loss of four electrons by the two Ca atoms and the other being the gain of four electrons by an O_2 molecule:

$$2Ca \longrightarrow 2Ca^{2+} + 4e^-$$

$$O_2 + 4e^- \longrightarrow 2O^{2-}$$

Each of these steps is called a ***half-reaction,*** which *explicitly shows the electrons involved.* The sum of the half-reactions gives the overall reaction:

$$2Ca + O_2 + 4e^- \longrightarrow 2Ca^{2+} + 2O^{2-} + 4e^-$$

or if we cancel the electrons that appear on both sides of the equation,

$$2Ca + O_2 \longrightarrow 2Ca^{2+} + 2O^{2-}$$

Finally, the Ca^{2+} and O^{2-} ions combine to form CaO:

$$2Ca^{2+} + 2O^{2-} \longrightarrow 2CaO$$

By convention, we do not show the charges in the formula of an ionic compound, so that calcium oxide is normally represented as CaO rather than $Ca^{2+}O^{2-}$.

The half-reaction that involves loss of electrons is called an ***oxidation reaction.*** The term "oxidation" was originally used by chemists to denote combinations of elements with oxygen. However, it now has a broader meaning that includes reactions not involving oxygen. *The half-reaction that involves gain of electrons* is called a ***reduction reaction.*** In the formation of calcium oxide, calcium is oxidized. It is said to act as a ***reducing agent*** because it *donates electrons* to oxygen and causes oxygen to be reduced. Oxygen is reduced and acts as an ***oxidizing agent*** because it *accepts electrons* from calcium, causing calcium to be oxidized. Note that the extent of oxidation in a redox reaction must be equal to the extent of reduction; that is, the number of electrons lost by a reducing agent must be equal to the number of electrons gained by an oxidizing agent.

A common type of redox reaction is the reaction between metals and acids, represented by

$$\text{metal} + \text{acid} \longrightarrow \text{salt} + \text{molecular hydrogen}$$

Consider the following reactions (Figure 4.7):

$$Fe(s) + 2HCl(aq) \longrightarrow FeCl_2(aq) + H_2(g)$$

$$Zn(s) + 2HCl(aq) \longrightarrow ZnCl_2(aq) + H_2(g)$$

$$Mg(s) + 2HCl(aq) \longrightarrow MgCl_2(aq) + H_2(g)$$

In these reactions, the metal acts as the reducing agent by transferring two electrons to the H^+ ions supplied by HCl. In the case of iron, the half-reactions are

$$Fe \longrightarrow Fe^{2+} + 2e^-$$

$$2H^+ + 2e^- \longrightarrow H_2$$

and we can represent the overall reaction as a net ionic equation:

$$Fe + 2H^+ \longrightarrow Fe^{2+} + H_2$$

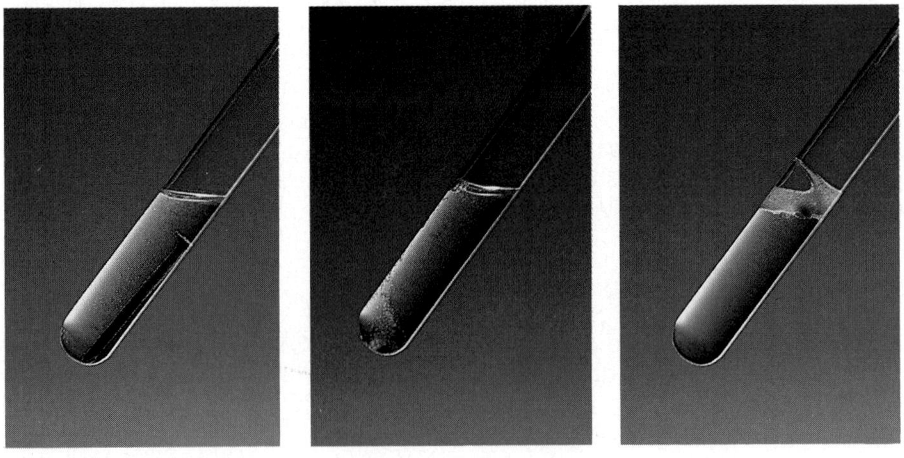

FIGURE 4.7

Left to right: Reactions of iron (Fe), zinc (Zn), and magnesium (Mg) with hydrochloric acid to form hydrogen gas and the metal chlorides ($FeCl_2$, $ZnCl_2$, $MgCl_2$). The reactivity of these metals is reflected in the rate of hydrogen gas evolution, which is slowest for the least reactive metal, Fe, and fastest for the most reactive metal, Mg.

Oxidation Number

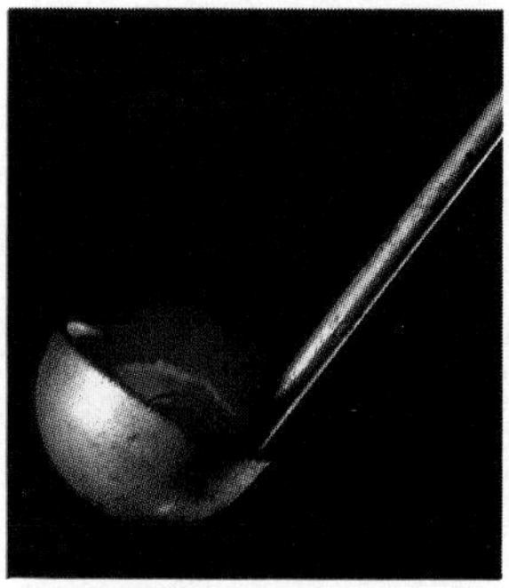
Sulfur burning in air to form sulfur dioxide.

The definitions of oxidation and reduction in terms of loss and gain of electrons apply to the formation of ionic compounds such as CaO. However, these definitions do not accurately characterize the formation of hydrogen chloride (HCl) and sulfur dioxide (SO_2):

$$H_2(g) + Cl_2(g) \longrightarrow 2HCl(g)$$
$$S(s) + O_2(g) \longrightarrow SO_2(g)$$

Because HCl and SO_2 are not ionic but molecular compounds, no electrons are actually transferred in the formation of these compounds, as they are in the case of CaO. Nevertheless, chemists find it convenient to treat these reactions as redox reactions because experimental measurements show that there is a *partial* transfer of electrons (from H to Cl in HCl and from S to O in SO_2).

To keep track of electrons in redox reactions, it is useful to assign oxidation numbers to the reactants and products. ***Oxidation number*** (also called ***oxidation state***) refers to *the number of charges an atom would have in a molecule (or an ionic compound) if electrons were transferred completely*. For example, we can rewrite the above equations for the formation of HCl and SO_2 as follows:

$$\overset{0}{H_2}(g) + \overset{0}{Cl_2}(g) \longrightarrow 2\overset{+1}{H}\overset{-1}{Cl}(g)$$
$$\overset{0}{S}(s) + \overset{0}{O_2}(g) \longrightarrow 2\overset{+4}{S}\overset{-2}{O_2}(g)$$

The numbers above the element symbols are the oxidation numbers. In both of the reactions shown, there is no charge on the atoms in the reactant molecules. Thus their oxidation number is zero. For the product molecules, however, it is assumed that complete electron transfer has taken place and atoms have gained or lost electrons. The oxidation numbers reflect the number of electrons "transferred."

This format allows us to identify elements that are oxidized (oxidation number increase) and reduced (oxidation number decrease) at a glance. The elements that show an increase in oxidation number, hydrogen and sulfur in the above examples, are oxidized in the reactions. Chlorine and oxygen are reduced, so their oxidation numbers show a decrease from their initial values. Note that the sum of the oxidation numbers of H and Cl in HCl (+1 and −1) is zero. Likewise, if we add the charges on S (+4) and two atoms of O [$2 \times (-2)$], the total is zero. The reason is that the HCl and SO_2 molecules are neutral, so the charges must cancel.

We use the following rules to assign oxidation numbers:

1. In free elements (that is, in the uncombined state), each atom has an oxidation number of zero. Thus each atom in H_2, Br_2, Na, Be, K, O_2, and P_4 has the same oxidation number: zero.

2. For ions composed of only one atom, the oxidation number is equal to the charge on the ion. Thus Li^+ has an oxidation number of +1; Ba^{2+} ion, +2; Fe^{3+}, +3; I^- ion, −1; O^{2-} ion, −2; and so on. All alkali metals have an oxidation number of +1, and all alkaline earth metals have an oxidation number of +2 in their compounds. Aluminum has an oxidation number of +3 in all its compounds.

3. The oxidation number of oxygen in most compounds (for example, MgO and H_2O) is −2, but in hydrogen peroxide (H_2O_2) and peroxide ion (O_2^{2-}), its oxidation number is −1.

4. The oxidation number of hydrogen is +1, except when it is bonded to metals in binary compounds. In these cases (for example, LiH, NaH, and CaH_2), its oxidation number is −1.

5. Fluorine has an oxidation number of −1 in *all* its compounds. Other halogens (Cl, Br, and I) have negative oxidation numbers when they occur as halide ions in their compounds. When combined with oxygen—for example in oxoacids and oxoanions (see Section 2.7)—they have positive oxidation numbers.

6. In a neutral molecule, the sum of the oxidation numbers of all the atoms must be zero. In a polyatomic ion, the sum of oxidation numbers of all the elements in the ion must be equal to the net charge of the ion. For example, in the ammonium ion, NH_4^+, the oxidation number of N is −3 and that of H is +1. Thus the sum of the oxidation numbers is $-3 + 4(+1) = +1$, which is equal to the net charge of the ion.

EXAMPLE 4.4

Assigning Oxidation Numbers

Assign oxidation numbers to all the elements in the following compounds and ion: (a) Li_2O, (b) HNO_3, (c) $Cr_2O_7^{2-}$.

Answer: (a) By rule 2 we see that lithium has an oxidation number of +1 (Li^+) and oxygen an oxidation number of −2 (O^{2-}).

(b) This is the formula for nitric acid, which yields a H^+ ion and a NO_3^- ion in solution. From rule 4 we see that H has an oxidation number of +1. Thus the other group (the nitrate ion) must have a net oxidation number of −1. Since oxygen has an oxidation number of −2, the oxidation number of nitrogen (labeled x) is given by

$$-1 = x + 3(-2)$$

or

$$x = +5$$

(c) From rule 6 we see that the sum of the oxidation numbers in $Cr_2O_7^{2-}$ must be −2. We know that the oxidation number of O is −2, so all that remains is to determine the oxidation number of Cr, which we call y. The sum of the oxidation numbers in the ion is

$$-2 = 2(y) + 7(-2)$$

or

$$y = +6$$

PRACTICE EXERCISE

Assign oxidation numbers to all the elements in the following compound and ion: (a) PF_3, (b) MnO_4^-.

Figure 4.8 shows the known oxidation numbers of the more familiar elements,

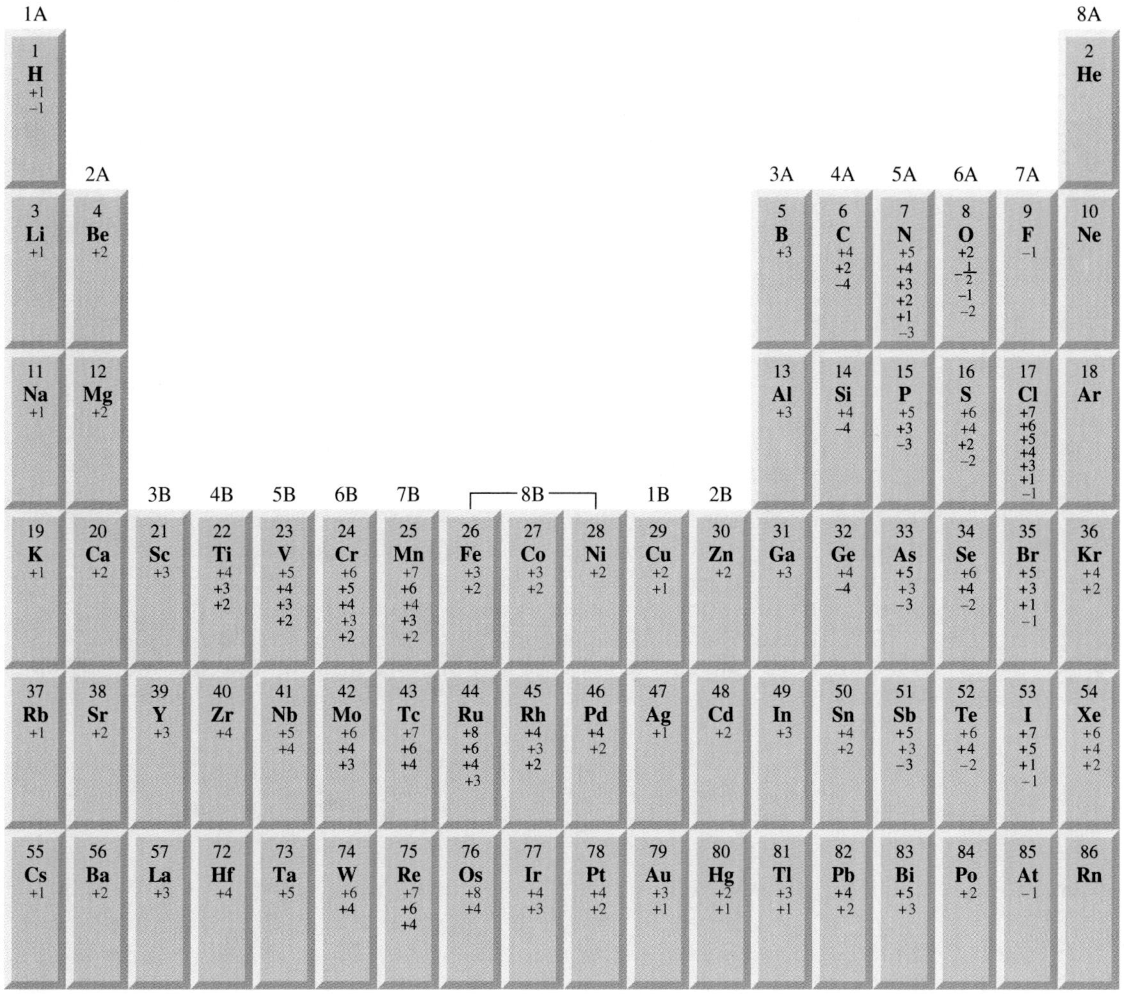

1A
1 H +1 −1
2A
3 Li +1
4 Be +2
11 Na +1
12 Mg +2
3B 4B 5B 6B 7B 8B 1B 2B
19 K +1
20 Ca +2
21 Sc +3
22 Ti +4 +3 +2
23 V +5 +4 +3 +2
24 Cr +6 +5 +4 +3 +2
25 Mn +7 +6 +4 +3 +2
26 Fe +3 +2
27 Co +3 +2
28 Ni +2
29 Cu +2 +1
30 Zn +2
37 Rb +1
38 Sr +2
39 Y +3
40 Zr +4
41 Nb +5 +4
42 Mo +6 +4 +3
43 Tc +7 +6 +4
44 Ru +8 +6 +4 +3
45 Rh +4 +3 +2
46 Pd +4 +2
47 Ag +1
48 Cd +2
55 Cs +1
56 Ba +2
57 La +3
72 Hf +4
73 Ta +5
74 W +6 +4
75 Re +7 +6 +4
76 Os +8 +4
77 Ir +4 +3
78 Pt +4 +2
79 Au +3 +1
80 Hg +2 +1
3A 4A 5A 6A 7A 8A
5 B +3
6 C +4 +2 −4
7 N +5 +4 +3 +2 +1 −3
8 O +2 −1/2 −1 −2
9 F −1
2 He
10 Ne
13 Al +3
14 Si +4 −4
15 P +5 +3 −3
16 S +6 +4 +2 −2
17 Cl +7 +6 +5 +4 +3 +1 −1
18 Ar
31 Ga +3
32 Ge +4 −4
33 As +5 +3 −3
34 Se +6 +4 −2
35 Br +5 +3 +1 −1
36 Kr +4 +2
49 In +3
50 Sn +4 +2
51 Sb +5 +3 −3
52 Te +6 +4 −2
53 I +7 +5 +1 −1
54 Xe +6 +4 +2
81 Tl +3 +1
82 Pb +4 +2
83 Bi +5 +3
84 Po +2
85 At −1
86 Rn

FIGURE 4.8

The oxidation numbers of elements in their compounds. The more common oxidation numbers are in color.

arranged according to their positions in the periodic table. This arrangement highlights the following general features of oxidation numbers:

- Metallic elements have only positive oxidation numbers, whereas nonmetallic elements may have either positive or negative oxidation numbers.
- The highest oxidation number a representative element can have is its group number in the periodic table. For example, the halogens are in Group 7A, so their highest possible oxidation number is +7.
- The transition metals (Groups 1B, 3B–8B) usually have several possible oxidation numbers (Figure 4.9).

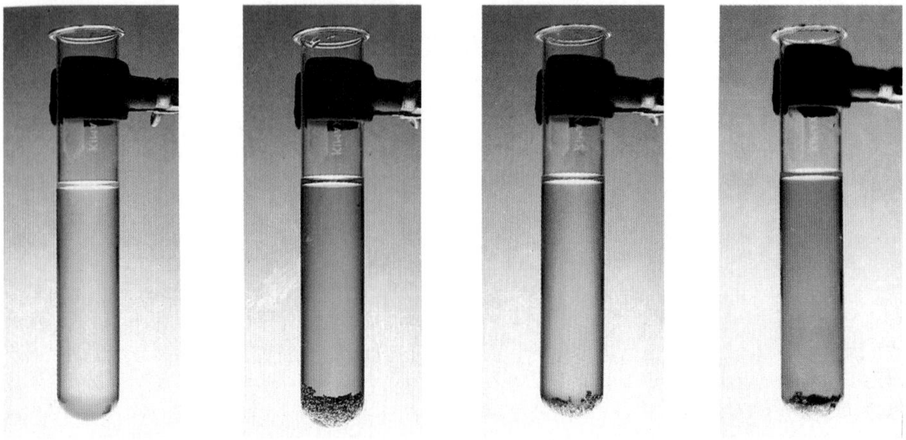

FIGURE 4.9

Left to right: Colors of aqueous solutions of compounds containing vanadium in four oxidation states (V, VI, III, and II).

Activity Series

The reactions between metals and hydrochloric acid described in the preceding section are examples of ***displacement reactions,*** so called because *one ion or atom in the reaction is replaced (displaced) by another ion (or atom) of another element.* In the metal-HCl reaction, the H^+ ions are displaced by a metal ion.

Displacement reactions are among the most common redox reactions. A metal in a compound can also be displaced by another metal in the uncombined state. For example, when metallic zinc is added to a solution containing copper sulfate ($CuSO_4$), it displaces Cu^{2+} ions from the solution (Figure 4.10):

$$\overset{0}{\mathrm{Zn}}(s) + \overset{+2}{\mathrm{Cu}}\mathrm{SO_4}(aq) \longrightarrow \overset{+2}{\mathrm{Zn}}\mathrm{SO_4}(aq) + \overset{0}{\mathrm{Cu}}(s)$$

or

$$\overset{0}{\mathrm{Zn}}(s) + \overset{+2}{\mathrm{Cu}}{}^{2+}(aq) \longrightarrow \overset{+2}{\mathrm{Zn}}{}^{2+}(aq) + \overset{0}{\mathrm{Cu}}(s)$$

Similarly, metallic copper displaces silver ions from a solution containing silver nitrate ($AgNO_3$) (also shown in Figure 4.10):

FIGURE 4.10

Left to right: When a piece of Zn metal is placed in an aqueous $CuSO_4$ solution, Zn atoms enter the solution as Zn^{2+} ions, and Cu^{2+} ions are converted to solid Cu (first beaker). In time, the blue color of the $CuSO_4$ solution disappears (second beaker). When a piece of Cu wire is placed in an aqueous $AgNO_3$ solution, Cu atoms enter the solution as Cu^{2+} ions, and Ag^+ ions are converted to solid Ag (third beaker). Gradually, the solution acquires the blue color characteristic of hydrated Cu^{2+} ions (fourth beaker).

$$\overset{0}{Cu}(s) + 2\overset{+1}{Ag}NO_3(aq) \longrightarrow \overset{+2}{Cu}(NO_3)_2(aq) + 2\overset{0}{Ag}(s)$$

or

$$\overset{0}{Cu}(s) + 2\overset{+1}{Ag^+}(aq) \longrightarrow \overset{+2}{Cu^{2+}}(aq) + 2\overset{0}{Ag}(s)$$

Reversing the roles of the metals would result in no reaction. In other words, copper metal will not displace zinc ions from zinc sulfate, and silver metal will not displace copper ions from copper nitrate.

An easy way to predict whether a metal or hydrogen displacement reaction will actually occur is to refer to an ***activity series,*** like the one shown in Figure 4.11. Basically, an activity series is *a convenient summary of the results of many possible displacement reactions,* similar to those already discussed. According to this series, any metal above hydrogen will displace it from water or from an acid, but metals below hydrogen will not react with either water or an acid. In fact, any species listed in the series will react with a compound containing any species listed below it. For example, Zn is above Cu, so zinc metal will displace copper ions from copper sulfate.

Using this series, we can predict the reaction of all alkali metals (Group 1A) and some alkaline metals (Group 2A) with cold water. For example, sodium (Group 1A) and calcium (Group 2A) react with water as follows (Figure 4.12):

$$2\overset{0}{Na}(s) + 2\overset{+1}{H_2}O(l) \longrightarrow 2\overset{+1}{Na}O\overset{+1}{H}(aq) + \overset{0}{H_2}(g)$$

$$\overset{0}{Ca}(s) + 2\overset{+1}{H_2}O(l) \longrightarrow \overset{+2}{Ca}(O\overset{+1}{H})_2(s) + \overset{0}{H_2}(g)$$

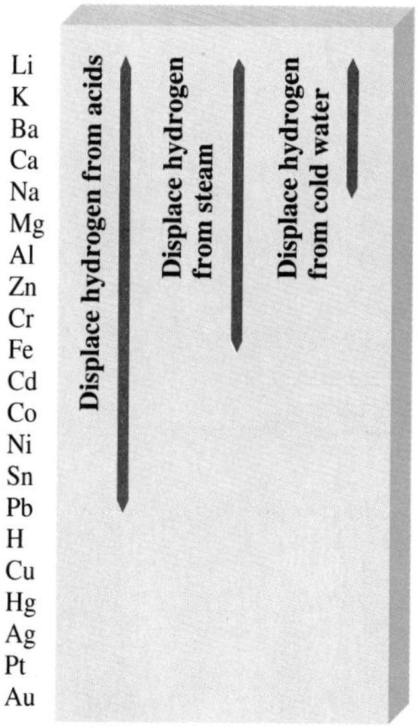

FIGURE 4.11

The activity series for metals. The metals are arranged according to their ability to displace hydrogen from an acid or water. Li (lithium) is the most reactive metal, and Au (gold) is the least reactive.

4.5 CONCENTRATION AND DILUTION OF SOLUTIONS

To study solution stoichiometry we must know how much of the reactants are present in a solution and also how to control the amounts of reactants used to bring about a reaction in aqueous solution.

(a) (b)

FIGURE 4.12

Reactions of (a) sodium (Na) and (b) calcium (Ca) with cold water. Note that the reaction is more vigorous with Na than with Ca.

Concentration of Solutions

The ***concentration of a solution*** is *the amount of solute present in a given quantity of solution.* (For this discussion we will assume the solute is a liquid or a solid and the solvent is a liquid.) One of the most common units of concentration in chemistry is ***molarity (M)***, or ***molar concentration.*** Molarity is the *number of moles of solute in 1 liter of solution;* it is defined by the equation

$$M = \text{molarity} = \frac{\text{moles of solute}}{\text{liters of soln}} \tag{4.1}$$

where "soln" denotes "solution." Thus, a 1.46 molar glucose ($C_6H_{12}O_6$) solution, expressed as 1.46 M $C_6H_{12}O_6$, contains 1.46 moles of the solute ($C_6H_{12}O_6$) in 1 liter of the solution; a 0.52 molar urea [$(NH_2)_2CO$] solution, expressed as 0.52 M $(NH_2)_2CO$, contains 0.52 mole of $(NH_2)_2CO$ (the solute) in 1 liter of solution; and so on.

Of course, we do not always need (or desire) to work with solutions of exactly 1 liter (L). This is not a problem as long as we remember to convert the volume of the solution to liters. Thus, a 500-mL solution containing 0.730 moles of $C_6H_{12}O_6$ also has a concentration of 1.46 M:

$$M = \text{molarity} = \frac{0.730 \text{ mol}}{0.500 \text{ L}}$$

$$= 1.46 \text{ mol/L} = 1.46\ M$$

As you can see, the unit of molarity is moles per liter, so a 500-mL solution containing 0.730 mole of $C_6H_{12}O_6$ is equivalent to 1.46 mol/L or 1.46 M. Note that concentration, like density, is an intensive property, so its value does not depend on how much of the solution is present.

The procedure for preparing a solution of known molarity is as follows. First, the solute is accurately weighed and transferred to a volumetric flask through a funnel (Figure 4.13). Next, water is added to the flask, which is carefully swirled to dissolve the solid. After *all* the solid has dissolved, more water is added slowly to

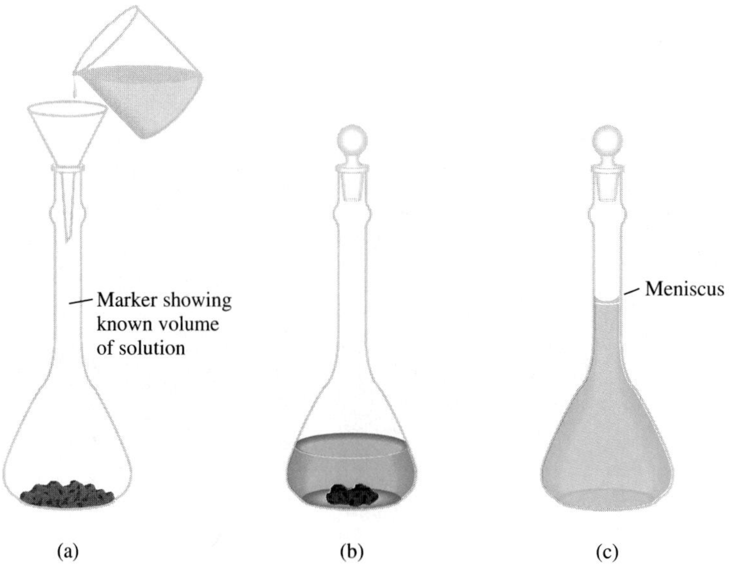

FIGURE 4.13

Procedure for preparing a solution of known molarity. (a) A known amount of a solid solute is put into the volumetric flask; then water is added through a funnel. (b) The solid is slowly dissolved by gently shaking the flask. (c) After the solid has completely dissolved, more water is added to bring the level of solution to the mark. Knowing the volume of the solution and the amount of solute dissolved in it, we can calculate the molarity of the prepared solution.

bring the level of solution exactly to the volume mark. Knowing the volume of solution (which is the volume of the flask) and the quantity of compound (the number of moles) dissolved, we can calculate the molarity of the solution using Equation (4.1). Note that this procedure does not require knowing the amount of water added, as long as the volume of the final solution is known.

EXAMPLE 4.5
Preparing a Solution of Known Molarity

How many grams of potassium dichromate ($K_2Cr_2O_7$) are required to prepare a 250-mL solution whose concentration is 2.16 *M*?

Answer: The first step is to determine the number of moles of $K_2Cr_2O_7$ in 250 mL of solution:

$$\text{moles of } K_2Cr_2O_7 = 250 \cancel{\text{mL soln}} \times \frac{2.16 \text{ mol } K_2Cr_2O_7}{1000 \cancel{\text{mL soln}}}$$

$$= 0.540 \text{ mol } K_2Cr_2O_7$$

The molar mass of $K_2Cr_2O_7$ is 294.2 g, so we write

$$\text{grams of } K_2Cr_2O_7 \text{ needed} = 0.540 \cancel{\text{mol } K_2Cr_2O_7} \times \frac{294.2 \text{ g } K_2Cr_2O_7}{1 \cancel{\text{mol } K_2Cr_2O_7}}$$

$$= 159 \text{ g } K_2Cr_2O_7$$

A $K_2Cr_2O_7$ solution.

PRACTICE EXERCISE

What is the molarity of an 85.0-mL ethanol (C_2H_5OH) solution containing 1.77 g of ethanol?

DILUTION OF SOLUTIONS

Concentrated solutions are often stored in the chemical stockroom and used as needed in the laboratory. Frequently we dilute these "stock" solutions before working with them. *The procedure for preparing a less concentrated solution from a more concentrated one* is called ***dilution.***

Suppose that we want to prepare 1 L of 0.400 *M* $KMnO_4$ solution from a solution of 1.00 *M* $KMnO_4$. For this purpose we need 0.400 mole of $KMnO_4$. Since there is 1.00 mole of $KMnO_4$ in 1 L, or 1000 mL, of a 1.00 *M* $KMnO_4$ solution, there is 0.400 mole of $KMnO_4$ in 0.400 × 1000 mL, or 400 mL of the same solution:

$$\frac{1.00 \text{ mol}}{1000 \text{ mL soln}} = \frac{0.400 \text{ mol}}{400 \text{ mL soln}}$$

Therefore, we must withdraw 400 mL from the 1.00 *M* $KMnO_4$ solution and dilute it to 1000 mL by adding water (in a 1-L volumetric flask). This method gives us 1 L of the desired solution of 0.400 *M* $KMnO_4$.

In carrying out a dilution process, it is useful to remember that adding more solvent to a given amount of the stock solution changes (decreases) the concentration of the solution without changing the number of moles of solute present in the solution (Figure 4.14); that is,

$$\text{moles of solute before dilution} = \text{moles of solute after dilution}$$

Two $KMnO_4$ solutions of different concentrations.

(a) (b)

FIGURE 4.14

The dilution of a more concentrated solution to a less concentrated one does not change the total number of moles of solute.

Since molarity is defined as moles of solute in 1 L of solution, we see that the number of moles of solute is given by

$$\underbrace{\frac{\text{moles of solute}}{\text{liters of solution}}}_{M} \times \underbrace{\text{volume of soln (in liters)}}_{V} = \text{moles of solute}$$

or

$$MV = \text{moles of solute}$$

Because all the solute comes from the original stock solution, we can conclude that

$$\underbrace{M_{\text{initial}}V_{\text{initial}}}_{\substack{\text{moles of solute}\\ \text{before dilution}}} = \underbrace{M_{\text{final}}V_{\text{final}}}_{\substack{\text{moles of solute}\\ \text{after dilution}}} \quad (4.2)$$

where M_{initial} and M_{final} are the initial and final concentrations of the solution in molarity and V_{initial} and V_{final} are the initial and final volumes of the solution, respectively. Of course, the units of V_{initial} and V_{final} must be the same (milliliters or liters) for the calculation to work. To check the reasonableness of your results, be sure that $M_{\text{initial}} > M_{\text{final}}$ and $V_{\text{final}} > V_{\text{initial}}$.

EXAMPLE 4.6

Diluting a Solution

Describe how you would prepare 5.00×10^2 mL of 1.75 M H_2SO_4 solution, starting with an 8.61 M stock solution of H_2SO_4.

Answer: Since the concentration of the final solution is less than that of the original one, this a dilution process. We prepare for the calculation by tabulating our data:

$$M_{\text{initial}} = 8.61\ M \qquad M_{\text{final}} = 1.75\ M$$

$$V_{\text{initial}} = ? \qquad V_{\text{final}} = 5.00 \times 10^2\ \text{mL}$$

Substituting in Equation (4.2),

$$(8.61\ M)(V_{\text{initial}}) = (1.75\ M)(5.00 \times 10^2\ \text{mL})$$

$$V_{\text{initial}} = \frac{(1.75\ M)(5.00 \times 10^2\ \text{mL})}{8.61\ M}$$

$$V_{initial} = 102 \text{ mL}$$

Thus we must dilute 102 mL of the 8.61 *M* H_2SO_4 solution with sufficient water to give a final volume of 5.00×10^2 mL in a 500-mL volumetric flask to obtain the desired concentration.

PRACTICE EXERCISE

How would you prepare 2.00×10^2 mL of an 0.866 *M* NaOH solution, starting with a 5.07 *M* stock solution?

4.6 SOLUTION STOICHIOMETRY

In Chapter 3 we studied stoichiometric calculations in terms of the mole method, which treats the coefficients in a balanced equation as the number of moles of reactants and products. In working with solutions of known molarity, we have to use the relationship MV = moles of solute. We will examine two types of common solution stoichiometry here: gravimetric analysis and titration.

GRAVIMETRIC ANALYSIS

Gravimetric analysis is *an analytical technique based on the measurement of mass.* One type of gravimetric analysis experiment involves the formation, isolation, and mass determination of a precipitate. Generally this procedure is applied to ionic compounds. A sample substance of unknown composition is dissolved in water and allowed to react with another substance to form a precipitate. The precipitate is filtered off, dried, and weighed. Knowing the mass and chemical formula of the precipitate formed, we can calculate the mass of a particular chemical component (that is, the anion or cation) of the original sample. From the mass of the component and the mass of the original sample, we can determine the percent composition by mass of the component in the original compound.

A reaction that is often studied in gravimetric analysis is

$$AgNO_3(aq) + NaCl(aq) \longrightarrow NaNO_3(aq) + AgCl(s)$$

or, expressed as a net ionic equation,

$$Ag^+(aq) + Cl^-(aq) \longrightarrow AgCl(s)$$

The precipitate is silver chloride (see solubility rule 5 in Table 4.2). As an example, let us say that we wanted to determine *experimentally* the percent by mass of Cl in NaCl. We would first accurately weigh out a sample of NaCl and dissolve it in water. Next, we would add enough $AgNO_3$ solution to the NaCl solution to cause the precipitation of all the Cl^- ions present in solution as AgCl. In this procedure NaCl is the limiting reagent and $AgNO_3$ the excess reagent. The AgCl precipitate is separated from the solution by filtration, dried, and weighed. From the measured mass of AgCl, we can calculate the mass of Cl (using the percent by mass of Cl in AgCl). Since this same amount of Cl was present in the original NaCl sample, we can calculate the percent by mass of Cl in NaCl. Figure 4.15 shows how this procedure is done.

Gravimetric analysis is a highly accurate technique, since the mass of a sample can be measured accurately. However, this procedure is applicable only to reac-

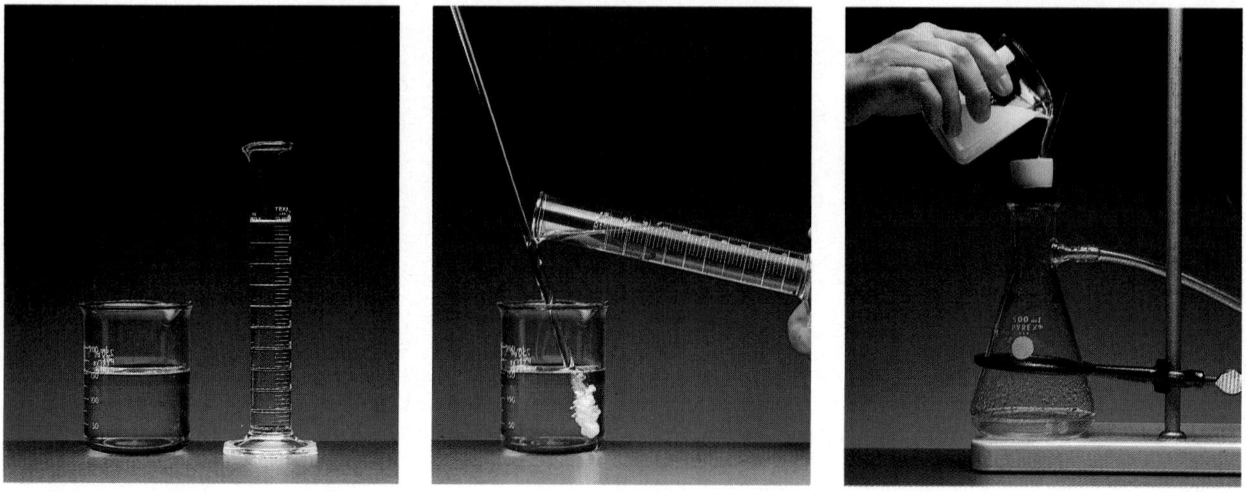

(a) (b) (c)

FIGURE 4.15

Basic steps for gravimetric analysis. (a) A solution containing a known amount of NaCl in a beaker. (b) The precipitation of AgCl upon the addition of $AgNO_3$ solution from a measuring cylinder. (c) The solution containing the AgCl precipitate is filtered through a preweighed sintered-disk crucible, which allows the liquid (but not the precipitate) to pass through. The crucible is then removed from the apparatus, dried in an oven, and weighed again. The difference between this mass and that of the empty crucible gives the mass of the AgCl precipitate.

tions that go to completion, or have nearly 100 percent yield. Thus, if AgCl were slightly soluble instead of being insoluble, it would not be possible to remove all the Cl^- ions from the NaCl solution and the subsequent calculation would be in error.

EXAMPLE 4.7

Determining Percent by Mass by Gravimetric Analysis

A 0.5662-g sample of an ionic compound containing chloride ions and an unknown metal is dissolved in water and treated with an excess of $AgNO_3$. If the mass of the AgCl precipitate that forms is 1.0882 g, what is the percent by mass of Cl in the original compound?

Answer: The mass of Cl in the AgCl precipitate is given by

$$\text{mass of Cl} = 1.0882\,\cancel{\text{g AgCl}} \times \frac{1\,\cancel{\text{mol AgCl}}}{143.4\,\cancel{\text{g AgCl}}} \times \frac{1\,\cancel{\text{mol Cl}}}{1\,\cancel{\text{mol AgCl}}} \times \frac{35.45\text{ g Cl}}{1\,\cancel{\text{mol Cl}}}$$

$$= 0.2690\text{ g Cl}$$

We calculate the percent by mass of Cl in the unknown sample as

$$\%\text{Cl by mass} = \frac{\text{mass of Cl}}{\text{mass of sample}} \times 100\%$$

$$= \frac{0.2690\text{ g}}{0.5662\text{ g}} \times 100\%$$

$$= 47.51\%$$

PRACTICE EXERCISE

A sample of 0.3220 g of an ionic compound containing the bromide ion (Br^-) is dissolved in water and treated with an excess of $AgNO_3$. If the mass of the AgBr precipitate that forms is 0.6964 g, what is the percent of Br in the original compound?

TITRATIONS

Quantitative studies of acid-base neutralization reactions are most conveniently carried out using a procedure known as ***titration.*** In a titration experiment, *a solution of accurately known concentration,* called a ***standard solution,*** *is added gradually to another solution of unknown concentration, until the chemical reaction between the two solutions is complete.* If we know the volumes of the standard and unknown solutions used in the titration, along with the concentration of the standard solution, we can calculate the concentration of the unknown solution.

Sodium hydroxide is one of the bases commonly used in the laboratory. However, because it is difficult to obtain solid sodium hydroxide in a pure form, a solution of sodium hydroxide must be *standardized* before it can be used in accurate analytical work. We can standardize the sodium hydroxide solution by titrating it against an acid solution of accurately known concentration. The acid often chosen for this task is a monoprotic acid called potassium hydrogen phthalate (KHP), for which the molecular formula is $KHC_8H_4O_4$. KHP is a white, soluble solid that is commercially available in highly pure form. The reaction between KHP and sodium hydroxide is

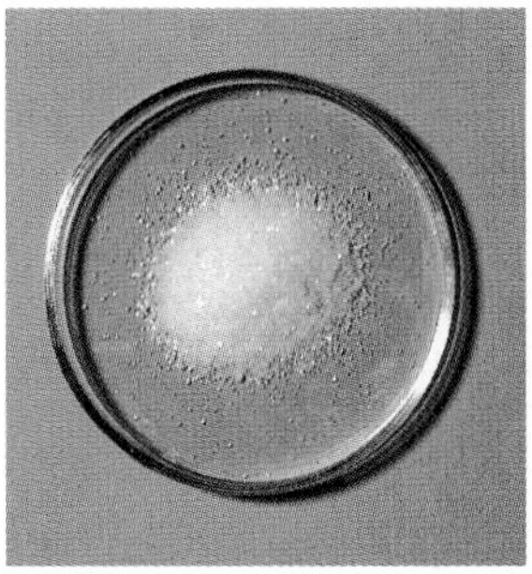

Potassium hydrogen phthalate.

$$KHC_8H_4O_4(aq) + NaOH(aq) \longrightarrow KNaC_8H_4O_4(aq) + H_2O(l)$$

The net ionic equation is

$$HC_8H_4O_4^-(aq) + OH^-(aq) \longrightarrow C_8H_4O_4^{2-}(aq) + H_2O(l)$$

The procedure for the titration is shown in Figure 4.16. First, a known amount of KHP is transferred to an Erlenmeyer flask and some distilled water is added to make up a solution. Next, NaOH solution is carefully added to the KHP solution from a buret until we reach the ***equivalence point,*** that is, *the point at which the acid has completely reacted with or been neutralized by the base.* The point is usually

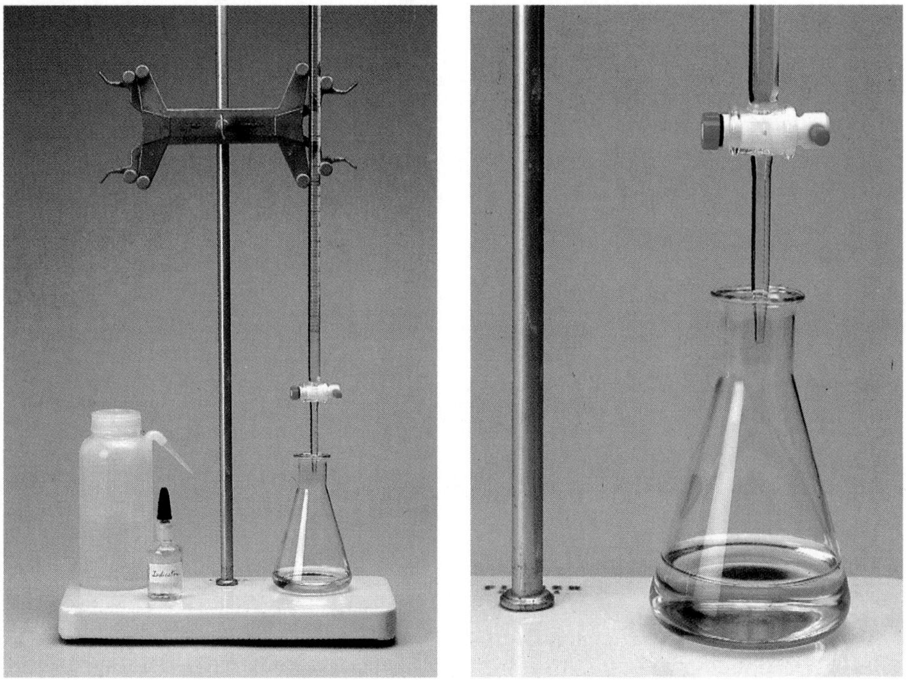

(a) (b)

FIGURE 4.16

(a) Apparatus for acid-base titration. A NaOH solution from the buret is added to a KHP solution in an Erlenmeyer flask. (b) A reddish-pink color is observed when the equivalence point is reached. The color here has been intensified for visual display.

signaled by a sharp color change of an indicator that has been added to the acid solution. In acid-base titrations, ***indicators*** are *substances that have distinctly different colors in acidic and basic media.* One commonly used indicator is phenolphthalein, which is colorless in acidic and neutral solutions but reddish pink in basic solutions. At the equivalence point, all the KHP has been neutralized by the NaOH and the solution is still colorless. However, if we add just one more drop of NaOH solution from the buret, the solution will immediately turn pink because the solution is now basic. Knowing the mass (and hence the number of moles) of KHP reacted, we can calculate the concentration of the NaOH solution.

The acid-base neutralization reaction between NaOH and KHP is one of the simplest types of neutralization known. Suppose, though, that instead of KHP, a diprotic acid such as H_2SO_4 is used. The reaction is represented by

$$2\text{NaOH}(aq) + \text{H}_2\text{SO}_4(aq) \longrightarrow \text{Na}_2\text{SO}_4(aq) + 2\text{H}_2\text{O}(l)$$

Since 2 mol NaOH $\Leftrightarrow$ 1 mol H_2SO_4, we need twice the amount of NaOH to react completely with a H_2SO_4 solution as needed for the *same* concentration of KHP solution.

EXAMPLE 4.8

Acid-Base Titration

How many milliliters (mL) of a 0.610 *M* NaOH solution are needed to completely neutralize 20.0 mL of a 0.245 *M* H_2SO_4 solution?

Answer: The equation for the neutralization reaction is shown above. First we calculate the number of moles of H_2SO_4 in 20.0 mL solution:

$$\text{moles H}_2\text{SO}_4 = \frac{0.245 \text{ mol H}_2\text{SO}_4}{1 \cancel{\text{L soln}}} \times \frac{1 \cancel{\text{L soln}}}{1000 \cancel{\text{mL soln}}} \times 20.0 \cancel{\text{mL soln}}$$

$$= 4.90 \times 10^{-3} \text{ mol H}_2\text{SO}_4$$

From the stoichiometry we see that 1 mol H_2SO_4 $\Leftrightarrow$ 2 mol NaOH. Therefore the number of moles of NaOH reacted must be $2(4.90 \times 10^{-3})$ mol, or 9.80×10^{-3} mol. From the definition of molarity [see Equation (4.1)] we have

$$\text{liters of soln} = \frac{\text{moles of solute}}{\text{molarity}}$$

or

$$\text{volume of NaOH} = \frac{9.80 \times 10^{-3} \text{ mol NaOH}}{0.610 \text{ mol/L soln}}$$

$$= 0.0161 \text{ L, or } 16.1 \text{ mL}$$

PRACTICE EXERCISE

How many milliliters of a 1.28 *M* H_2SO_4 solution are needed to neutralize 60.2 mL of a 0.427 *M* KOH solution?

SUMMARY

Aqueous solutions are electrically conducting if the solutes are electrolytes. If the solutes are nonelectrolytes, the solutions do not conduct electricity.

From general rules about solubilities of ionic compounds, we can predict whether a precipitate will form in a reaction.

Arrhenius acids ionize in water to give H^+ ions, and Arrhenius bases ionize in water to give OH^- ions. Brønsted acids donate protons, and Brønsted bases accept protons. The reaction of an acid with a base is called neutralization.

In redox reactions, oxidation and reduction always occur simultaneously. Oxidation is characterized by the loss of electrons, reduction by the gain of electrons. Oxidation numbers help us keep track of charge distribution and are assigned to all atoms in a compound or ion according to a specific set of rules. Oxidation can be defined as an increase in oxidation number; reduction can be defined as a decrease in oxidation number. The activity series helps us predict the outcome of a displacement reaction, which is a type of redox reaction.

The concentration of a solution is the amount of solute present in a given amount of solution. Molarity expresses concentration as the number of moles of solute in 1 L of solution. Adding a solvent to a solution, a process known as dilution, decreases the concentration (molarity) of the solution without changing the total number of moles of solute present in the solution.

Gravimetric analysis is a technique for determining the identity of a compound and/or the concentration of a solution by measuring mass. Gravimetric experiments often involve precipitation reactions.

In acid-base titration, a solution of known concentration (say, a base) is added gradually to a solution of unknown concentration (say, an acid) with the goal of determining the unknown concentration. The point at which the reaction in the titration is complete is called the equivalence point.

KEY WORDS

Activity series, p. 103
Aqueous solution, p. 88
Brønsted acid, p. 94
Brønsted base, p. 94
Chemical equilibrium, p. 90
Concentration of a solution, p. 104
Dilution, p. 105
Diprotic acid, p. 95
Displacement reaction, p. 102
Electrolyte, p. 88
Equivalence point, p. 109
Gravimetric analysis, p. 107
Half-reaction, p. 97
Hydration, p. 90
Hydronium ion, p. 94
Indicator, p. 110
Ionic equation, p. 92
Molar concentration, p. 104
Molarity (*M*), p. 104
Molecular equation, p. 92
Monoprotic acid, p. 95
Net ionic equation, p. 92
Neutralization reaction, p. 96
Nonelectrolyte, p. 88
Oxidation number, p. 99
Oxidation reaction, p. 98
Oxidation-reduction reaction, p. 97
Oxidation state, p. 99
Oxidizing agent, p. 98
Precipitate, p. 90
Precipitation reaction, p. 90
Redox reaction, p. 97
Reducing agent, p. 98
Reduction reaction, p. 98
Reversible reaction, p. 90
Salt, p. 96
Solubility, p. 90
Solute, p. 88
Solution, p. 88
Solvent, p. 88
Spectator ion, p. 92
Standard solution, p. 109
Titration, p. 109
Triprotic acid, p. 95

QUESTIONS AND PROBLEMS

AQUEOUS SOLUTIONS

Review Questions

4.1 Define the following terms: solute, solvent, solution, electrolyte, nonelectrolyte, hydration, reversible reaction, chemical equilibrium.

4.2 What is the difference between the following symbols used in chemical equations: $\rightarrow$ and $\rightleftharpoons$?

4.3 Water, as we know, is an extremely weak electrolyte and therefore cannot conduct electricity. Yet we are often cautioned not to operate electrical appliances when our hands are wet. Why?

4.4 LiF is a strong electrolyte. When we write LiF(*aq*), what species are actually present in the solution?

Problems

4.5 Identify each of the following substances as a strong electrolyte, a weak electrolyte, or a nonelectrolyte: (a) H_2O, (b) KCl, (c) HNO_3, (d) CH_3COOH, (e) $C_{12}H_{22}O_{11}$.

4.6 Identify each of the following substances as a strong electrolyte, a weak electrolyte, or a nonelectrolyte: (a) $Ba(NO_3)_2$, (b) Ne, (c) NH_3, (d) NaOH.

4.7 The passage of electricity through an electrolyte solution is caused by the movement of (a) electrons only, (b) cations only, (c) anions only, (d) both cations and anions.

4.8 Predict and explain which of the following systems are electrically conducting: (a) solid NaCl, (b) molten NaCl, (c) an aqueous solution of NaCl.

4.9 You are given a water-soluble compound X. Describe how you would determine whether it is an electrolyte or a nonelectrolyte. If it is an electrolyte, how would you determine whether it is strong or weak?

4.10 Explain why a solution of HCl in benzene does not conduct electricity but in water it does.

PRECIPITATION REACTIONS

Review Questions

4.11 What is the difference between an ionic equation and a molecular equation?

4.12 What is the advantage of writing net ionic equations for precipitation reactions?

Problems

4.13 Characterize the following compounds as soluble or insoluble in water: (a) $Ca_3(PO_4)_2$, (b) $Mn(OH)_2$, (c) $AgClO_3$, (d) K_2S.

4.14 Characterize the following compounds as soluble or insoluble in water: (a) $CaCO_3$, (b) $ZnSO_4$, (c) $Hg(NO_3)_2$, (d) $HgSO_4$, (e) NH_4ClO_4.

4.15 Write ionic and net ionic equations for the following reactions:
(a) $2AgNO_3(aq) + Na_2SO_4(aq) \rightarrow$
(b) $BaCl_2(aq) + ZnSO_4(aq) \rightarrow$
(c) $(NH_4)_2CO_3(aq) + CaCl_2(aq) \rightarrow$

4.16 Write ionic and net ionic equations for the following reactions:
(a) $Na_2S(aq) + ZnCl_2(aq) \rightarrow$
(b) $2K_3PO_4(aq) + 3Sr(NO_3)_2(aq) \rightarrow$
(c) $Mg(NO_3)_2(aq) + 2NaOH(aq) \rightarrow$

4.17 Which of the following processes will likely result in a precipitation reaction? (a) Mixing a $NaNO_3$ solution with a $CuSO_4$ solution. (b) Mixing a $BaCl_2$ solution with a K_2SO_4 solution. Write the net ionic equation.

4.18 On the basis of the solubility rules given in this chapter, suggest one method by which you might separate (a) K^+ from Ag^+, (b) Ag^+ from Pb^{2+}, (c) NH_4^+ from Ca^{2+}, (d) Ba^{2+} from Cu^{2+}. All cations are assumed to be in aqueous solution, and the common anion is the nitrate ion.

ACID-BASE REACTIONS

Review Questions

4.19 List the general properties of acids and bases.

4.20 Give Arrhenius's and Brønsted's definitions of an acid and a base.

4.21 Give an example each of a monoprotic acid, a diprotic acid, and triprotic acid.

4.22 What are the characteristics of an acid-base neutralization reaction?

4.23 Give four examples of a salt.

4.24 Identify the following as a weak or strong acid or base: (a) NH_3, (b) H_3PO_4, (c) LiOH, (d) HCOOH (formic acid), (e) H_2SO_4, (f) HF, (g) $Ba(OH)_2$.

Problems

4.25 Identify each of the following species as a Brønsted acid or base, or both: (a) HI, (b) CH_3COO^-, (c) $H_2PO_4^-$.

4.26 Identify each of the following species as a Brønsted acid or base, or both: (a) PO_4^{3-}, (b) ClO_2^-, (c) NH_4^+, (d) HCO_3^-.

4.27 Balance the following equations and write the corresponding ionic and net ionic equations (if appropriate):
(a) $HBr(aq) + NH_3(aq) \rightarrow$ (HBr is a strong acid)
(b) $Ba(OH)_2(aq) + H_3PO_4(aq) \rightarrow$
(c) $HClO_4(aq) + Mg(OH)_2(s) \rightarrow$

4.28 Balance the following equations and write the corresponding ionic and net ionic equations (if appropriate):
(a) $CH_3COOH(aq) + KOH(aq) \rightarrow$
(b) $H_2CO_3(aq) + NaOH(aq) \rightarrow$
(c) $HNO_3(aq) + Ba(OH)_2(aq) \rightarrow$

OXIDATION-REDUCTION REACTIONS

Review Questions

4.29 Define the following terms: half-reaction, oxidation reaction, reduction reaction, reducing agent, oxidizing agent, redox reaction.

4.30 Is it possible to have a reaction in which oxidation occurs and reduction does not? Explain.

Problems

4.31 For the complete redox reactions given below, (i) break down each reaction into its half-reactions; (ii) identify the oxidizing agent; (iii) identify the reducing agent.
(a) $2Sr + O_2 \rightarrow 2SrO$
(b) $2Li + H_2 \rightarrow 2LiH$
(c) $2Cs + Br_2 \rightarrow 2CsBr$
(d) $3Mg + N_2 \rightarrow Mg_3N_2$

4.32 For the complete redox reactions given below, write the half-reactions and identify the oxidizing and reducing agents:
(a) $4Fe + 3O_2 \rightarrow 2Fe_2O_3$
(b) $Cl_2 + 2NaBr \rightarrow 2NaCl + Br_2$
(c) $Si + 2F_2 \rightarrow SiF_4$
(d) $H_2 + Cl_2 \rightarrow 2HCl$

OXIDATION NUMBERS

Review Questions

4.33 Define oxidation number. Explain why, except for ionic compounds, oxidation number does not have any physical significance.

4.34 (a) Without referring to Figure 4.8, give the oxidation numbers of the alkali and alkaline earth metals in their compounds. (b) Give the highest oxidation numbers that the Groups 3A–7A elements can have.

Problems

4.35 Arrange the following species in order of increasing oxidation number of the sulfur atom: (a) H_2S, (b) S_8, (c) H_2SO_4, (d) S^{2-}, (e) HS^-, (f) SO_2, (g) SO_3.

4.36 Indicate the oxidation number of phosphorus in each of the following compounds: (a) HPO_3, (b) H_3PO_2, (c) H_3PO_3, (d) H_3PO_4, (e) $H_4P_2O_7$, (f) $H_5P_3O_{10}$.

4.37 Give the oxidation number of the underlined atom in the following molecules and ions: (a) $\underline{Cl}F$, (b) $\underline{I}F_7$, (c) $\underline{C}H_4$, (d) $\underline{C}_2H_2$, (e) $\underline{C}_2H_4$, (f) $K_2\underline{Cr}O_4$, (g) $K_2\underline{Cr}_2O_7$, (h) $K\underline{Mn}O_4$, (i) $NaH\underline{C}O_3$, (j) $\underline{Li}_2$, (k) $Na\underline{I}O_3$, (l) $K\underline{O}_2$, (m) $\underline{P}F_6^-$, (n) $K\underline{Au}Cl_4$.

4.38 Give the oxidation number of the underlined atom in the following molecules and ions: (a) $\underline{Cs}_2O$, (b) $Ca\underline{I}_2$, (c) $\underline{Al}_2O_3$, (d) $H_3\underline{As}O_3$, (e) $\underline{Ti}O_2$, (f) $\underline{Mo}O_4^{2-}$, (g) $\underline{Pt}Cl_4^{2-}$, (h) $\underline{Pt}Cl_6^{2-}$, (i) $\underline{Sn}F_2$, (j) $\underline{Cl}F_3$, (k) $\underline{Sb}F_6^-$.

4.39 Give the oxidation number for each of the following species: H_2, Se_8, P_4, O, U, As_4, B_{12}.

4.40 Give the oxidation number of the underlined atom in the following molecules and ions: (a) $Mg_3\underline{N}_2$, (b) $Cs\underline{O}_2$, (c) $Ca\underline{C}_2$, (d) $\underline{C}O_3^{2-}$, (e) $\underline{C}_2O_4^{2-}$, (f) $Zn\underline{O}_2^{2-}$, (g) $Na\underline{B}H_4$, (h) $\underline{W}O_4^{2-}$.

4.41 Nitric acid is a strong oxidizing agent. State which of the following species is *least* likely to be produced when nitric acid reacts with a strong reducing agent such as zinc metal, and explain why: N_2O, NO, NO_2, N_2O_4, N_2O_5, NH_4^+.

4.42 On the basis of oxidation number considerations, one of the following oxides would not react with molecular oxygen: NO, N_2O, SO_2, SO_3, P_4O_6. Which one is it? Why?

4.43 Predict the outcome of the reactions represented by the following equations by using the activity series, and balance the equations:
(a) $Cu(s) + HCl(aq) \rightarrow$
(b) $I_2(s) + NaBr(aq) \rightarrow$
(c) $Mg(s) + CuSO_4(aq) \rightarrow$
(d) $Cl_2(g) + KBr(aq) \rightarrow$

4.44 Which of the following metals can react with water? (a) Au, (b) Li, (c) Hg, (d) Ca, (e) Pt.

CONCENTRATION OF SOLUTIONS

Review Questions

4.45 Define molarity. Why is molarity a convenient concentration unit?

4.46 Describe steps involved in preparing a solution of known molar concentration using a volumetric flask.

Problems

4.47 Calculate the mass of NaOH in grams required to prepare a 5.00×10^2-mL solution of concentration 2.80 *M*.

4.48 A quantity of 5.25 g of NaOH is dissolved in a sufficient amount of water to make up exactly 1 L of solution. What is the molarity of the solution?

4.49 How many moles of $MgCl_2$ are present in 60.0 mL of a 0.100 *M* $MgCl_2$ solution?

4.50 How many grams of KOH are present in 35.0 mL of a 5.50 *M* solution?

4.51 Calculate the molarity of each of the following solutions: (a) 29.0 g of ethanol (C_2H_5OH) in 545 mL of solution, (b) 15.4 g of sucrose ($C_{12}H_{22}O_{11}$) in 74.0 mL of solution, (c) 9.00 g of sodium chloride (NaCl) in 86.4 mL of solution.

4.52 Calculate the molarity of each of the following solutions: (a) 6.57 g of methanol (CH_3OH) in 1.50×10^2 mL of solution, (b) 10.4 g of calcium chloride ($CaCl_2$) in 2.20×10^2 mL of solution, (c) 7.82 g of naphthalene ($C_{10}H_8$) in 85.2 mL of benzene solution.

4.53 Calculate the volume in milliliters of a solution required to provide the following: (a) 2.14 g of sodium chloride from a 0.270 *M* solution, (b) 4.30 g of ethanol from a 1.50 *M* solution, (c) 0.85 g of acetic acid (CH_3COOH) from a 0.30 *M* solution.

4.54 Determine how many grams of each of the following solutes would be needed to make 2.50×10^2 mL of a 0.100 *M* solution: (a) cesium iodide (CsI), (b) sulfuric acid (H_2SO_4), (c) sodium carbonate (Na_2CO_3), (d) potassium dichromate ($K_2Cr_2O_7$), (e) potassium permanganate ($KMnO_4$).

DILUTION OF SOLUTIONS

Review Questions

4.55 Describe the basic steps involved in the dilution of a solution of known concentration.

4.56 Write the equation that enables you to calculate the concentration of a diluted solution. Define all the terms.

Problems

4.57 Describe how to prepare 1.00 L of a 0.646 *M* HCl solution, starting with a 2.00 *M* HCl solution.

4.58 A quantity of 25.0 mL of a 0.866 *M* KNO_3 solution is poured into a 500-mL volumetric flask, and water is added until the volume of the solution is exactly 500 mL. What is the concentration of the final solution?

4.59 How would you prepare 60.0 mL of 0.200 *M* HNO_3 from a stock solution of 4.00 *M* HNO_3?

4.60 You have 505 mL of a 0.125 *M* HCl solution, and you want to dilute it to exactly 0.100 *M*. How much water should you add?

4.61 A 35.2-mL, 1.66 *M* $KMnO_4$ solution is mixed with a 16.7-mL, 0.892 *M* $KMnO_4$ solution. Calculate the concentration of the final solution.

4.62 A 46.2-mL, 0.568 *M* calcium nitrate [$Ca(NO_3)_2$] solution is mixed with an 80.5-mL, 1.396 *M* calcium nitrate solution. Calculate the concentration of the final solution.

GRAVIMETRIC ANALYSIS

Review Questions

4.63 Define gravimetric analysis. Describe the basic steps involved in gravimetric analysis. How does such an experiment help us determine the identity of a compound or the purity of a compound if its formula is known?

4.64 Distilled water must be used in the gravimetric analysis of chlorides. Why?

Problems

4.65 The volume 30.0 mL of 0.150 *M* $CaCl_2$ is added to 15.0 mL of 0.100 *M* $AgNO_3$. What is the mass in grams of AgCl precipitate formed?

4.66 A sample of 0.6760 g of an unknown compound containing barium ions (Ba^{2+}) is dissolved in water and treated with an excess of Na_2SO_4. If the mass of the $BaSO_4$ precipitate formed is 0.4105 g, what is the percent by mass of Ba in the original unknown compound?

4.67 How many grams of NaCl are required to precipitate practically all the Ag^+ ions from 2.50×10^2 mL of 0.0113 *M* $AgNO_3$ solution? Write the net ionic equation for the reaction.

4.68 The concentration of Cu^{2+} ions in the water (which also contains sulfate ions) discharged from a certain industrial plant is determined by adding excess sodium sulfide (Na_2S) solution to 0.800 L of the water. The molecular equation is

$$Na_2S(aq) + CuSO_4(aq) \longrightarrow Na_2SO_4(aq) + CuS(s)$$

Write the net ionic equation and calculate the molar concentration of Cu^{2+} in the water sample if 0.0177 g of solid CuS is formed.

ACID-BASE TITRATIONS

Review Questions

4.69 Define acid-base titration, standard solution, equivalence point.

4.70 Describe the basic steps involved in acid-base titration. Why is this technique of great practical value? How does an acid-base indicator work?

Problems

4.71 Calculate the volume in milliliters of a 1.420 *M*

NaOH solution required to titrate the following solutions:
(a) 25.00 mL of a 2.430 *M* HCl solution
(b) 25.00 mL of a 4.500 *M* H_2SO_4 solution
(c) 25.00 mL of a 1.500 *M* H_3PO_4 solution

4.72 What volume of a 0.50 *M* KOH solution is needed to neutralize completely each of the following?
(a) 10.0 mL of a 0.30 *M* HCl solution
(b) 10.0 mL of a 0.20 *M* H_2SO_4 solution
(c) 15.0 mL of a 0.25 *M* H_3PO_4 solution

MISCELLANEOUS PROBLEMS

4.73 Classify the following reactions according to the types discussed in the chapter:
(a) $Cl_2 + 2OH^- \rightarrow Cl^- + ClO^- + H_2O$
(b) $Ca^{2+} + CO_3^{2-} \rightarrow CaCO_3$
(c) $NH_3 + H^+ \rightarrow NH_4^+$
(d) $2CCl_4 + CrO_4^{2-} \rightarrow 2COCl_2 + CrO_2Cl_2 + 2Cl^-$
(e) $Ca + F_2 \rightarrow CaF_2$
(f) $2Li + H_2 \rightarrow 2LiH$
(g) $Ba(NO_3)_2 + Na_2SO_4 \rightarrow 2NaNO_3 + BaSO_4$
(h) $CuO + H_2 \rightarrow Cu + H_2O$
(i) $Zn + 2HCl \rightarrow ZnCl_2 + H_2$
(j) $2FeCl_2 + Cl_2 \rightarrow 2FeCl_3$

4.74 Using the apparatus shown in Figure 4.1, a student found that the light bulb was brightly lit when the electrodes were immersed in a sulfuric acid solution. However, after the addition of a certain amount of barium hydroxide [$Ba(OH)_2$] solution, the light began to dim even though $Ba(OH)_2$ is also a strong electrolyte. Explain.

4.75 Someone gave you a colorless liquid. Describe three chemical tests you would perform on the liquid to show that it is water.

4.76 You are given two colorless solutions, one containing NaCl and the other sucrose ($C_{12}H_{22}O_{11}$). Suggest a chemical and a physical test that would distinguish between these two solutions.

4.77 Chlorine (Cl_2) is used to purify drinking water. Too much chlorine is harmful to humans. The excess chlorine is often removed by treatment with sulfur dioxide (SO_2). Balance the following equation that represents this procedure:

$$Cl_2 + SO_2 + H_2O \longrightarrow Cl^- + SO_4^{2-} + H^+$$

4.78 Before aluminum was obtained by electrolytic reduction from its ore (Al_2O_3), the metal was produced by chemical reduction. Which metals would you use to reduce Al^{3+} to Al?

4.79 Oxygen (O_2) and carbon dioxide (CO_2) are colorless and odorless gases. Suggest two chemical tests that would allow you to distinguish between them.

4.80 Based on oxidation number, explain why carbon monoxide (CO) is flammable but carbon dioxide (CO_2) is not.

4.81 Which of the following aqueous solutions would you expect to be the best conductor of electricity at 25°C? Explain your answer.
(a) 0.20 *M* NaCl
(b) 0.60 *M* CH_3COOH
(c) 0.25 *M* HCl
(d) 0.20 *M* $Mg(NO_3)_2$

4.82 A 5.00×10^2-mL sample of 2.00 *M* HCl solution is treated with 4.47 g of magnesium. Calculate the concentration of the acid solution after all the metal has reacted. Assume that the volume remains unchanged.

4.83 Calculate the volume (in liters) of a 0.156 *M* $CuSO_4$ solution that would react with 7.89 g of zinc.

4.84 Sodium carbonate (Na_2CO_3) can be obtained in very pure form and can be used to standardize acid solutions. What is the molarity of a HCl solution if 28.3 mL of the solution is required to react with 0.256 g of Na_2CO_3?

4.85 A 3.664-g sample of a monoprotic acid was dissolved in water and required 20.27 mL of a 0.1578 *M* NaOH solution for neutralization. Calculate the molar mass of the acid.

4.86 Acetic acid (CH_3COOH) is an important ingredient of vinegar. A sample of 50.0 mL of a commercial vinegar is titrated against a 1.00 *M* NaOH solution. What is the concentration (in *M*) of acetic acid present in the vinegar if 5.75 mL of the base were required for the titration?

4.87 Calculate the mass of precipitate formed when 2.27 L of 0.0820 *M* $Ba(OH)_2$ are mixed with 3.06 L of 0.0664 *M* Na_2SO_4.

4.88 Milk of magnesia is an aqueous suspension of magnesium hydroxide [$Mg(OH)_2$] used to treat acid indigestion. Calculate the volume of a 0.035 *M* HCl solution (a typical acid concentration in an upset stomach) needed to react with two spoonfuls of milk of magnesia [approximately 10.0 mL at 0.080 g $Mg(OH)_2$/mL].

Answers to Practice Exercises: 4.1 (a) Insoluble, (b) slightly soluble, (c) soluble. **4.2** $Al^{3+}(aq) + 3OH^-(aq) \longrightarrow Al(OH)_3(s)$. **4.3** (a) Brønsted base, (b) Brønsted acid. **4.4** (a) P: +3, F: −1; (b) Mn: +7, O: −2. **4.5** 0.452 *M*. **4.6** Dilute 34.2 mL of the stock solution to 200 mL. **4.7** 92.02%. **4.8** 10.0 mL.

CHAPTER 5

GASES

◆ The first manned balloon flight was made by the Montgolfier brothers in June 1783 in France, using a crude version of today's hot-air balloon. Two months later, the French scientist Jacques Charles carried out another balloon trial in Paris. Exploiting his knowledge of recent discoveries in the study of gases, Charles decided to inflate the balloon with hydrogen, which s lighter than air. He made his balloon out of silk and coated it with a rubber solution to prevent the gas from escaping into the atmosphere. The hydrogen gas was prepared by reacting 500 pounds of sulfuric acid (H_2SO_4) with 1000 pounds of iron. It took several days to inflate the balloon to a diameter of 13 ft. The flight lasted 45 minutes and landed 15 miles away from the launch site. The local inhabitants were so terrified by the balloon that they tore it to shreds.

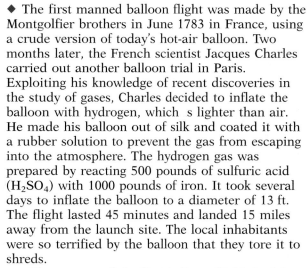

The success of the first balloon flights had far-reaching effects. Around the world their potential value for travel and in warfare was recognized quickly. The invention of the balloon also allowed scientists to reach altitudes greater than 10,000 ft for the purpose of measuring the temperature and pressure of the atmosphere and collecting air samples. Their studies provided a further experimental basis for understanding gas behavior and gave support to the work that led to the overthrow of the phlogiston theory (see p. 53). Thus the first manned balloon flights were partly responsible for the birth of modern chemistry. ◆

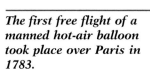

The first free flight of a manned hot-air balloon took place over Paris in 1783.

5.1 SUBSTANCES THAT EXIST AS GASES

We live at the bottom of an ocean of air whose composition by volume is roughly 78 percent N_2, 21 percent O_2, and 1 percent other gases, including CO_2. In the 1990s, the chemistry of this vital mixture of gases has become a source of great interest because of the detrimental effects of environmental pollution. Here we will focus generally on the behavior of substances that exist as gases under normal atmospheric conditions, which are defined as 25°C and 1 atmosphere (atm) pressure (see Section 5.2).

NO_2 gas.

Only 11 elements are gases under normal atmospheric conditions. Table 5.1 lists these, along with a number of gaseous compounds. Note that the elements hydrogen, nitrogen, oxygen, fluorine, and chlorine exist as gaseous diatomic molecules. Another form of oxygen, ozone (O_3), is also a gas at room temperature. All the elements in Group 8A, the noble gases, are monatomic gases: He, Ne, Ar, Kr, Xe, and Rn.

Of the gases listed in Table 5.1, only O_2 is essential for our survival. Hydrogen cyanide (HCN) is a deadly poison. Carbon monoxide (CO), hydrogen sulfide (H_2S), nitrogen dioxide (NO_2), O_3, and sulfur dioxide (SO_2) are somewhat less toxic. The gases He, Ne, and Ar are chemically inert; that is, they do not react with any other substance. Most gases are colorless. Exceptions are F_2, Cl_2, and NO_2. The dark-brown color of NO_2 is sometimes visible in polluted air. All gases have the following physical characteristics:

- Gases assume the volume and shape of their containers.
- Gases are the most compressible of the states of matter.
- Gases will mix evenly and completely when confined to the same container.
- Gases have much lower densities than liquids and solids.

TABLE 5.1
Some Substances Found as Gases at 1 atm and 25°C

Elements	Compounds
H_2 (molecular hydrogen)	HF (hydrogen fluoride)
N_2 (molecular nitrogen)	HCl (hydrogen chloride)
O_2 (molecular oxygen)	HBr (hydrogen bromide)
O_3 (ozone)	HI (hydrogen iodide)
F_2 (molecular fluorine)	CO (carbon monoxide)
Cl_2 (molecular chlorine)	CO_2 (carbon dioxide)
He (helium)	NH_3 (ammonia)
Ne (neon)	NO (nitric oxide)
Ar (argon)	NO_2 (nitrogen dioxide)
Kr (krypton)	N_2O (nitrous oxide)
Xe (xenon)	SO_2 (sulfur dioxide)
Rn (radon)	H_2S (hydrogen sulfide)
	HCN (hydrogen cyanide)*

*The boiling point of HCN is 26°C, but it is close enough to qualify as a gas at ordinary atmospheric conditions.

At this point, it is useful to distinguish between a "gas" and a "vapor," two terms that are often used interchangeably but do not have exactly the same meaning. A *gas* is a substance that is *normally* in the gaseous state at ordinary temperatures and pressures; a *vapor* is the gaseous form of any substance that is a liquid or a solid at normal temperatures and pressures. Thus, at 25°C and 1 atm pressure, we speak of water *vapor* and oxygen *gas*.

5.2 PRESSURE OF A GAS

Gases exert pressure on any surface with which they come in contact, because gas molecules are constantly in motion. We humans have adapted so well physiologically to the pressure of the air around us that we are usually unaware of its presence, perhaps as fish are not conscious of the water's pressure on them.

It is easy to demonstrate the existence of atmospheric pressure. One everyday example is the ability to drink a liquid through a straw. Sucking air out of the straw reduces the pressure inside. The greater atmospheric pressure on the liquid pushes it up into the straw to replace the air that has been sucked out.

SI Units of Pressure

Pressure is one of the most readily measurable properties of a gas. In order to understand how we measure the pressure of a gas, it is helpful to know how the units of measurement are derived. We begin with velocity and acceleration.

Velocity is defined as the change in distance with elapsed time; that is,

$$\text{velocity} = \frac{\text{distance moved}}{\text{elapsed time}}$$

The SI unit for velocity is meters per second (m/s), although we also use centimeters per second (cm/s).

Acceleration is the change in velocity with time, or

$$\text{acceleration} = \frac{\text{change in velocity}}{\text{elapsed time}}$$

Acceleration is measured in meters per second squared (m/s^2) or centimeters per second squared (cm/s^2).

The second law of motion formulated by Sir Isaac Newton in the late seventeenth century defines another term, from which the units of pressure are derived, namely, *force*. According to this law,

$$\text{force} = \text{mass} \times \text{acceleration}$$

In this context, force has *the SI units kg* m/s^2, or more conveniently, ***newton (N)***, where

$$1\text{ N} = 1\text{ kg m/s}^2$$

(It is interesting to note that 1 N is roughly equivalent to the force of gravity exerted on an apple.)

Finally, we define ***pressure*** as *force applied per unit area:*

$$\text{pressure} = \frac{\text{force}}{\text{area}}$$

The SI unit of pressure is the ***pascal (Pa),*** honoring the French mathematician and physicist Blaise Pascal, where

$$1 \text{ Pa} = 1 \text{ N/m}^2$$

Atmospheric Pressure

The atoms and molecules of the gases in the atmosphere, like those of all other matter, are subject to Earth's gravitational pull. As a consequence, the atmosphere is much denser near the surface of Earth than at high altitudes, and it exerts pressure on the surface. The force experienced by any area exposed to Earth's atmosphere is equal to *the weight of the column of air above it.* It is *the pressure exerted by this column of air* that we refer to as ***atmospheric pressure.*** The actual value of atmospheric pressure depends on location, temperature, and weather conditions.

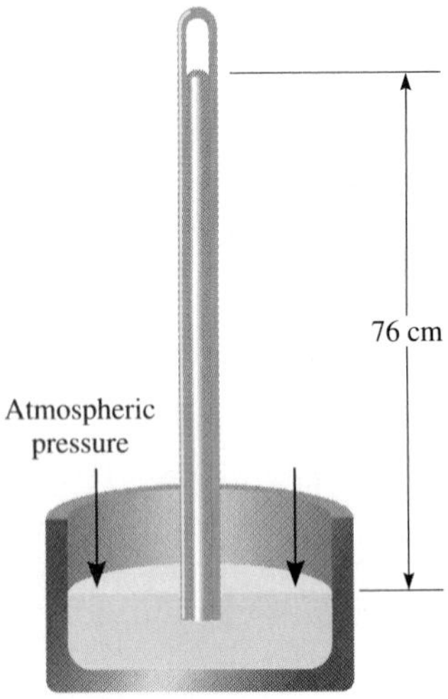

FIGURE 5.1

A barometer for measuring atmospheric pressure. Above the mercury in the tube is a vacuum. The column of mercury is supported by the atmospheric pressure.

One instrument that measures atmospheric pressure is the ***barometer.*** A simple barometer can be constructed by filling a long glass tube, closed at one end, with mercury, and then carefully inverting the tube in a dish of mercury, making sure that no air enters the tube. Some mercury in the tube will flow down into the dish, creating a vacuum at the top (Figure 5.1). The weight of the mercury column remaining in the tube is supported by atmospheric pressure acting on the surface of the mercury in the dish. The ***standard atmospheric pressure (1 atm)*** is equal to *the pressure that supports a column of mercury exactly 760 mm (or 76 cm) high at 0°C at sea level.* The standard atmosphere thus equals a pressure of 760 mmHg, where mmHg represents the pressure exerted by a column of mercury 1 mm high. The mmHg unit is also called the *torr,* after the Italian scientist Evangelista Torricelli, who invented the barometer. Thus

$$1 \text{ torr} = 1 \text{ mmHg}$$

and

$$1 \text{ atm} = 760 \text{ mmHg}$$
$$= 760 \text{ torr}$$

EXAMPLE 5.1

Converting between mmHg and Atmospheres

Passenger jet flying at high altitude.

The pressure outside a jet plane flying at high altitude falls considerably below atmospheric pressure at sea level. The air inside the cabin must therefore be pressurized to protect the passengers. What is the pressure in atmospheres in the cabin if the barometer reading is 688 mmHg?

Answer: The pressure in atmospheres is calculated as follows:

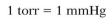

$$\text{pressure} = 688 \cancel{\text{mmHg}} \times \frac{1 \text{ atm}}{760 \cancel{\text{mmHg}}}$$
$$= 0.905 \text{ atm}$$

PRACTICE EXERCISE

Convert 749 mmHg to atmospheres.

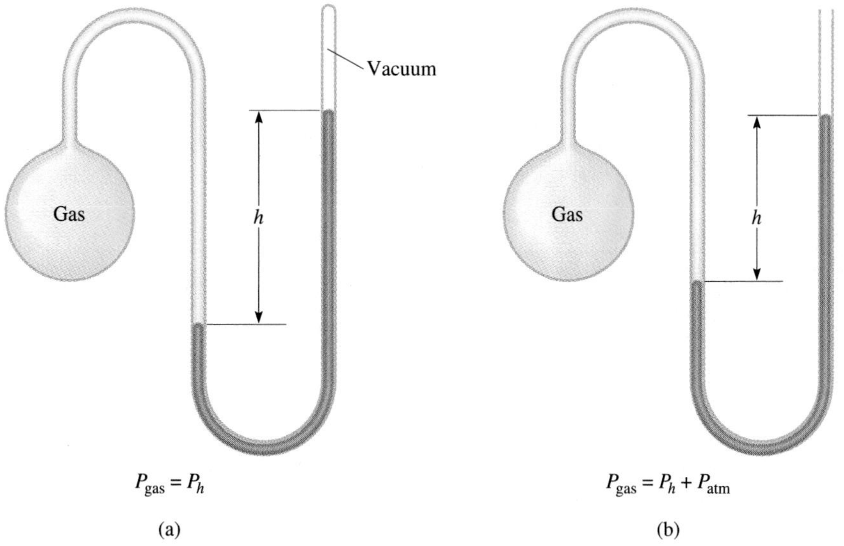

FIGURE 5.2

Two types of manometers used to measure gas pressures. (a) Gas pressure is less than atmospheric pressure. (b) Gas pressure is greater than atmospheric pressure.

The relation between atmospheres and pascals (see Appendix 1) is

$$1 \text{ atm} = 101{,}325 \text{ Pa}$$

and since 1000 Pa = 1 kPa (kilopascal),

$$1 \text{ atm} = 1.01325 \times 10^2 \text{ kPa}$$

The above equation gives the SI definition of 1 atm.

The device scientists use to measure the pressure of gases other than the atmosphere is called a *manometer.* The principle of operation of a manometer is similar to that of a barometer. There are two types of manometers, shown in Figure 5.2. The closed-tube manometer is normally used to measure pressures below atmospheric pressure [Figure 5.2(a)]. The open-tube manometer is more suited for measuring pressures equal to or greater than atmospheric pressure [Figure 5.2(b)].

Nearly all barometers and most manometers use mercury as the working fluid, despite the fact that it is a toxic substance with a harmful vapor. The reason is that mercury has a very high density (13.6 g/mL) compared with most other liquids. Since the height of the liquid in a column is inversely proportional to the liquid's density, we can construct manageably small barometers and manometers.

5.3 THE GAS LAWS

The gas laws we will study in this chapter summarize several centuries' worth of research on the physical properties of gases. These generalizations about the macroscopic behavior of gaseous substances have played a major role in the development of chemistry.

THE PRESSURE-VOLUME RELATIONSHIP: BOYLE'S LAW

When a helium balloon ascends, it expands because the pressure outside gradually decreases. Conversely, when the volume of a gas is compressed, the pressure of the

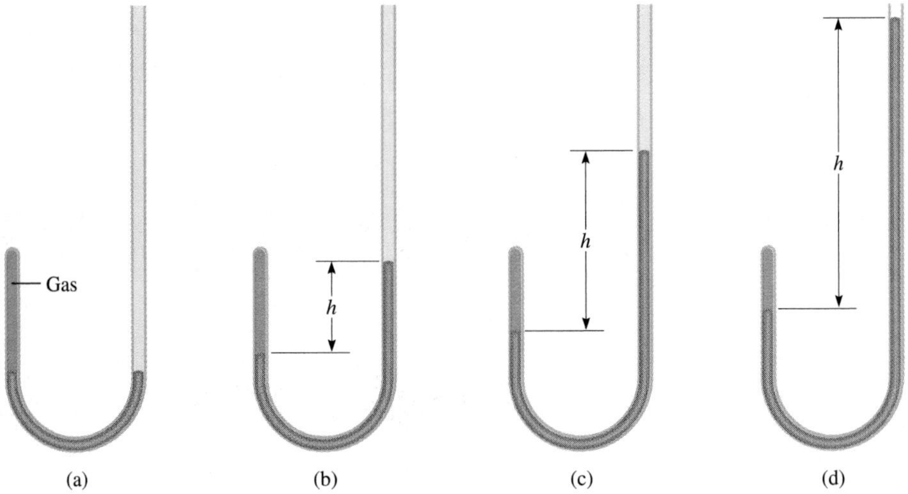

FIGURE 5.3

Apparatus for studying the relationship between pressure and volume of a gas. In (a) the pressure of the gas is equal to the atmospheric pressure. The pressure exerted on the gas increases from (a) to (d) as mercury is added, and the volume of the gas decreases, as predicted by Boyle's law. The extra pressure exerted on the gas is shown by the difference in the mercury levels (h mmHg). The temperature of the gas is kept constant.

Robert Boyle (1627–1691)

gas increases. In the seventeenth century, the English chemist Robert Boyle studied the behavior of gases systematically and quantitatively, using an apparatus like that shown in Figure 5.3. In Figure 5.3(a) the pressure exerted on the gas by the mercury added to the tube is equal to atmospheric pressure. In Figure 5.3(b) an increase in pressure due to the addition of more mercury is indicated by the unequal levels of mercury in the two columns; consequently, the volume of the gas decreases. Boyle noticed that when temperature is held constant, the volume V of a given amount of a gas decreases as the total applied pressure P—atmospheric pressure plus the pressure due to the added mercury—is increased. This relationship between pressure and volume is shown pictorially in Figure 5.3(b), (c), and (d). On the other hand, if the applied pressure is decreased, the gas volume becomes larger.

Boyle's law, which summarizes his observations of the pressure-volume relationship, states that *the volume of a fixed amount of gas maintained at constant temperature is inversely proportional to the gas pressure;* that is,

$$V \propto \frac{1}{P}$$

where the symbol $\propto$ means *proportional to.* To change $\propto$ to an equals sign, we must write

$$V = k_1 \times \frac{1}{P} \tag{5.1a}$$

where k_1 is a constant called the *proportionality constant.* Rearranging Equation (5.1a), we obtain

$$PV = k_1 \tag{5.1b}$$

Equation (5.1b) is another form of Boyle's law. It means that the product of the pressure and volume of a gas at constant temperature is a constant.

Figure 5.4 shows two conventional ways of expressing Boyle's findings graphically. Figure 5.4(a) is a graph of the equation $PV = k_1$; Figure 5.4(b) is a graph of the equivalent equation $P = k_1 \times 1/V$. Note that the latter equation takes the form of a linear equation $y = mx + b$, where $b = 0$.

Although the individual values of pressure and volume can vary greatly for a

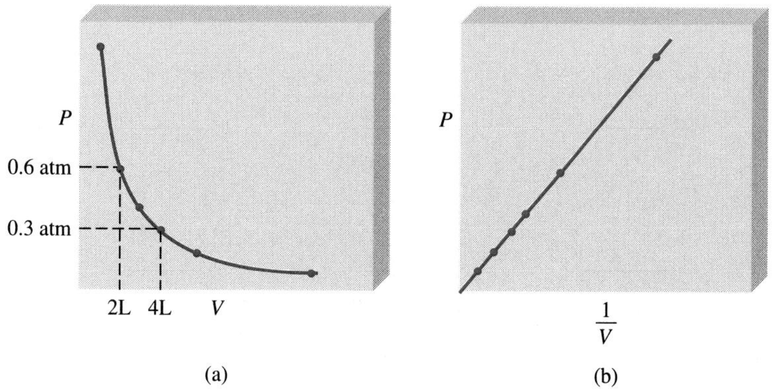

FIGURE 5.4

Graphs showing variation of the volume of a gas sample with the pressure exerted on the gas, at constant temperature. (a) P versus V. Note that the volume of the gas doubles as the pressure is halved. (b) P versus 1/V.

given sample of gas, as long as the temperature is held constant and the amount of the gas does not change, $P \times V$ is always equal to the same constant. Therefore, for a given sample of gas under two different sets of conditions at constant temperature, we can write

$$P_1V_1 = k_1 = P_2V_2$$

or

$$P_1V_1 = P_2V_2 \tag{5.2}$$

where V_1 and V_2 are the volumes at pressures P_1 and P_2, respectively.

THE TEMPERATURE-VOLUME RELATIONSHIP: CHARLES' AND GAY-LUSSAC'S LAW

Boyle's law depends on the temperature of the system remaining constant. But suppose the temperature changes: How does a change in temperature affect the volume and pressure of a gas? Let us begin with the effect of temperature on the volume of a gas. The earliest investigators of this relationship were French scientists Jacques Charles and Joseph Gay-Lussac. Their studies showed that, at constant pressure, the volume of a gas sample expands when heated and contracts when cooled (Figure 5.5). In fact, the quantitative relations involved in these changes in temperature and volume turn out to be remarkably consistent. For example, we observe an interesting phenomenon when we study the temperature-volume relationship at various pressures. At any given pressure, the plot of volume versus temperature yields a straight line. By extending the line to zero volume, we find the intercept on the temperature axis to be −273.15°C. At any other pressure, we obtain a different straight line for the volume-temperature plot, but we get the *same* zero-volume temperature intercept at −273.15°C (Figure 5.6). (In practice, we can measure the volume of a gas over only a limited temperature range, because all gases condense at low temperatures to form liquids.)

In 1848 the Scottish physicist Lord Kelvin (William Thomson) realized the significance of this behavior. He identified −273.15°C as *theoretically the lowest attainable temperature,* called ***absolute zero.*** With *absolute zero as a starting point,* he set up an ***absolute temperature scale,*** now called the ***Kelvin temperature scale.*** One degree Celsius is equal *in magnitude* to one kelvin (K). (Note that the absolute temperature scale has no degree sign, so 25 K is called twenty-five kelvins.) The only difference between the absolute temperature scale and the Celsius

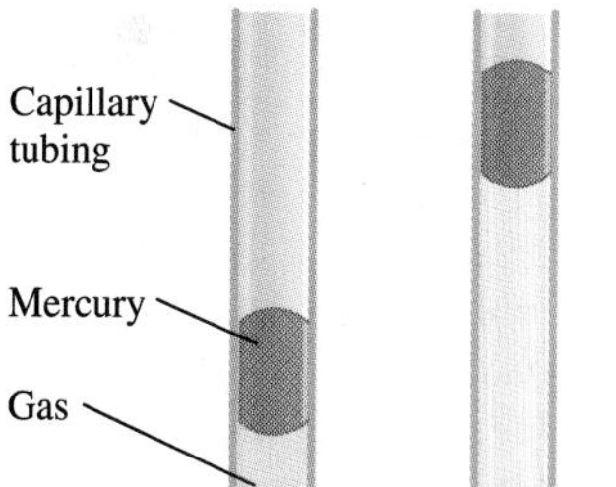

FIGURE 5.5

Variation of the volume of a gas sample with temperature, at constant pressure. The pressure exerted on the gas is the sum of the atmospheric pressure and the pressure due to the weight of the mercury.

FIGURE 5.6

Plots of the volume of a gas versus temperature at different pressures. (The pressure increases from P_1 to P_4.) When these lines are extrapolated, or extended (the dashed portions), they all intersect at the point representing zero volume and a temperature of −273.15°C.

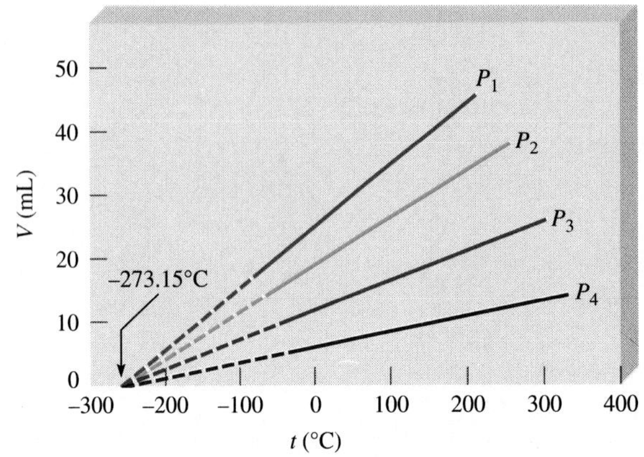

scale is that the zero position is shifted. Important points on the scales match up as follows:

Absolute zero: 0 K = −273.15°C

Freezing point of water: 273.15 K = 0°C

Boiling point of water: 373.15 K = 100°C

The relationship between degrees Celsius and kelvins is

$$T(\text{K}) = t(°\text{C}) + 273.15°\text{C} \tag{5.3}$$

In most calculations we will use 273 instead of 273.15 as the term relating kelvins and degrees Celsius. By convention we use T to denote absolute (kelvin) temperature and t to indicate temperature on the Celsius scale. Figure 5.7 shows corresponding temperatures on the Celsius and Kelvin scales for a number of systems.

FIGURE 5.7

Relationship between the Celsius temperature scale and the Kelvin temperature scale. The distances between the temperature marks are not drawn to scale.

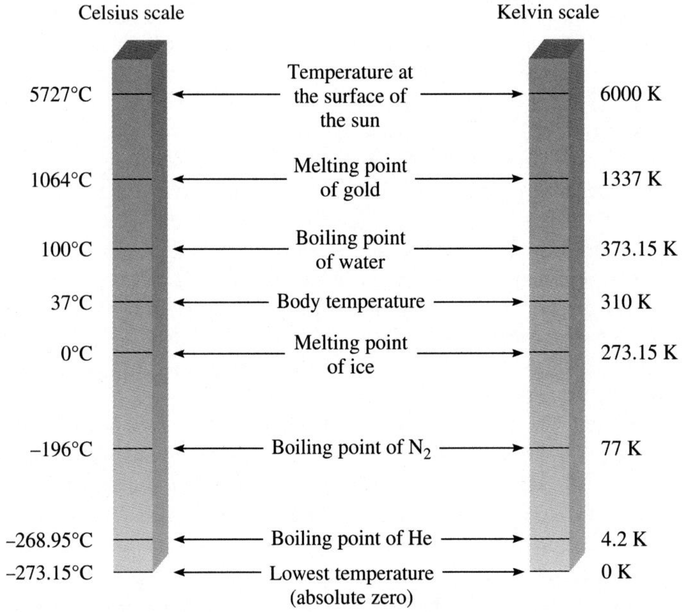

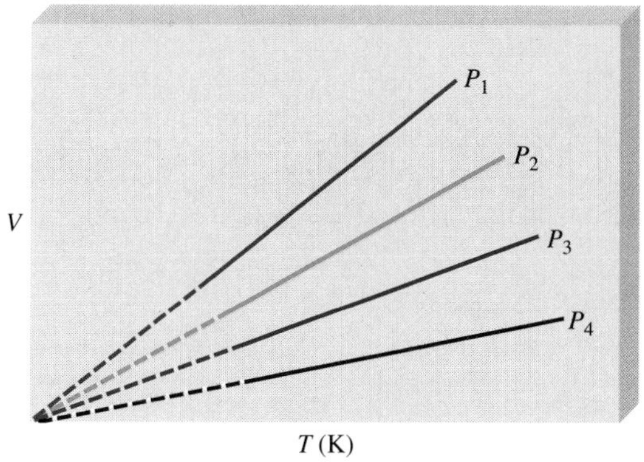

FIGURE 5.8

Volume of a gas sample versus temperature at constant pressure. The graph is similar to Figure 5.6, except that the temperature is in kelvins and the line goes through the origin.

The dependence of volume on temperature is given by

$$V \propto T$$

$$V = k_2T$$

or

$$\frac{V}{T} = k_2 \qquad (5.4)$$

where k_2 is the proportionality constant. Equation (5.4) is known as ***Charles' and Gay-Lussac's law,*** or simply ***Charles' law.*** Charles' law states that *the volume of a fixed amount of gas maintained at constant pressure is directly proportional to the absolute temperature of the gas.*

Just as we did for pressure-volume relationships at constant temperature, we can compare two sets of conditions for a given sample of gas at constant pressure. From Equation (5.4) we can write

$$\frac{V_1}{T_1} = k_2 = \frac{V_2}{T_2}$$

or

$$\frac{V_1}{T_1} = \frac{V_2}{T_2} \qquad (5.5)$$

Jacques Charles (1746–1832)

where V_1 and V_2 are the volumes of the gases at temperatures T_1 and T_2 (both in kelvins), respectively. Figure 5.8 shows straight-line plots of V versus T at different pressures. This graph is similar to Figure 5.6, although the Kelvin scale is used for temperature.

THE VOLUME-AMOUNT RELATIONSHIP: AVOGADRO'S LAW

The work of the Italian scientist Amedeo Avogadro (see Section 3.2) complemented the studies of Boyle, Charles, and Gay-Lussac. In 1811 he published a hypothesis proposing that at the same temperature and pressure, equal volumes of different gases contain the same number of molecules (or atoms if the gas is monatomic). It follows that the volume of any given gas must be proportional to the number of molecules present; that is,

$$V \propto n$$

$$V = k_3n \qquad (5.6)$$

where n represents the number of moles and k_3 is the proportionality constant.

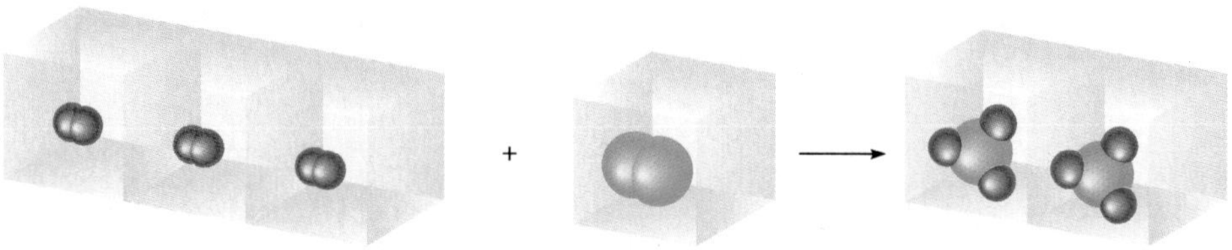

$3H_2\ (g)$	+	$N_2\ (g)$	⟶	$2NH_3\ (g)$
3 molecules	+	1 molecule	⟶	2 molecules
3 moles	+	1 mole	⟶	2 moles
3 volumes	+	1 volume	⟶	2 volumes

FIGURE 5.9

Volume relationship of gases in a chemical reaction. The ratio of the volumes of molecular hydrogen to molecular nitrogen is 3:1, and that of ammonia (the product) to molecular hydrogen and molecular nitrogen combined (the reactants) is 2:4, or 1:2.

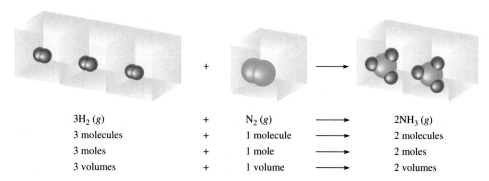

Equation (5.6) is the mathematical expression of ***Avogadro's law,*** which states that *at constant pressure and temperature, the volume of a gas is directly proportional to the number of moles of the gas present.* From Avogadro's law we learn that when two gases react with each other, their reacting volumes have a simple ratio to each other. If the product is a gas, its volume is related to the volume of the reactants by a simple ratio (a fact demonstrated earlier by Gay-Lussac). For example, consider the synthesis of ammonia from molecular hydrogen and molecular nitrogen:

$$\underset{3\text{ mol}}{3H_2(g)} + \underset{1\text{ mol}}{N_2(g)} \longrightarrow \underset{2\text{ mol}}{2NH_3(g)}$$

Since, at the same temperature and pressure, the volumes of gases are directly proportional to the number of moles of the gases present, we can now write

$$\underset{3\text{ volumes}}{3H_2(g)} + \underset{1\text{ volume}}{N_2(g)} \longrightarrow \underset{2\text{ volumes}}{2NH_3(g)}$$

The volume ratio of molecular hydrogen to molecular nitrogen is 3:1, and that of ammonia (the product) to molecular hydrogen and molecular nitrogen (the reactants) is 2:4, or 1:2 (Figure 5.9).

5.4 THE IDEAL GAS EQUATION

Let us summarize the gas laws we have discussed so far:

$$\text{Boyle's law: } V \propto \frac{1}{P} \qquad \text{(at constant } n \text{ and } T)$$

$$\text{Charles' law: } V \propto T \qquad \text{(at constant } n \text{ and } P)$$

$$\text{Avogadro's law: } V \propto n \qquad \text{(at constant } P \text{ and } T)$$

We can combine all three expressions to form a single master equation for the behavior of gases:

$$V \propto \frac{nT}{P}$$

$$= R\frac{nT}{P}$$

or

$$PV = nRT \tag{5.7}$$

where R, *the proportionality constant,* is called the ***gas constant.*** Equation (5.7), which *describes the relationship among* the four experimental variables P, V, T, and n, is called the ***ideal gas equation.*** An ***ideal gas*** is *a hypothetical gas whose pressure-volume-temperature behavior can be completely accounted for by the ideal gas equation.* The molecules of an ideal gas do not attract or repel one another, and their volume is negligible compared with the volume of the container. Although there is no such thing in nature as an ideal gas, discrepancies in the behavior of real gases over reasonable temperature and pressure ranges do not significantly affect calculations. Thus we can safely use the ideal gas equation to solve many gas problems.

Before we can apply Equation (5.7) to a real system, we must evaluate the gas constant R. At 0°C (273.15 K) and 1 atm pressure, many real gases behave like an ideal gas. Experiments show that under these conditions, 1 mole of an ideal gas occupies 22.414 L. Figure 5.10 compares this molar volume with that of a basketball. *The conditions 0°C and 1 atm are called* ***standard temperature and pressure,*** often abbreviated ***STP.*** From Equation (5.7) we can write

$$\begin{aligned} R &= \frac{PV}{nT} \\ &= \frac{(1\ \text{atm})(22.414\ \text{L})}{(1\ \text{mol})(273.15\ \text{K})} = 0.082057 \frac{\text{L}\cdot\text{atm}}{\text{K}\cdot\text{mol}} \\ &= 0.082057\ \text{L}\cdot\text{atm/K}\cdot\text{mol} \end{aligned}$$

The dots between L and atm and between K and mol remind us that both L and atm are in the numerator and both K and mol are in the denominator. For most calculations, we will round off the value of R to three significant figures (0.0821 L · atm/K · mol) and use 22.41 L for the molar volume of a gas at STP. In addition, we will assume that temperatures given in degrees Celsius are exact, so that they do not affect the number of significant figures.

EXAMPLE 5.2

Applying the Ideal Gas Equation

Sulfur hexafluoride (SF_6) is a colorless, odorless, very unreactive gas. Calculate the pressure (in atmospheres) exerted by 1.82 moles of the gas in a steel vessel of volume 5.43 L at 69.5°C.

FIGURE 5.10

A comparison of the molar volume at STP (which is approximately 22.4 L) with a basketball.

Answer: From Equation (5.7) we write

$$\begin{aligned} P &= \frac{nRT}{V} \\ &= \frac{(1.82\ \text{mol})(0.0821\ \text{L} \cdot \text{atm/K} \cdot \text{mol})(69.5 + 273)\ \text{K}}{5.43\ \text{L}} \\ &= 9.42\ \text{atm} \end{aligned}$$

PRACTICE EXERCISE

Calculate the volume (in liters) occupied by 2.12 moles of nitric oxide (NO) at 6.54 atm and 76°C.

EXAMPLE 5.3

Finding the Molar Volume of a Gas at STP

Calculate the volume (in liters) occupied by 7.40 g of CO_2 at STP.

Answer: Recognizing that 1 mole of an ideal gas occupies 22.41 L at STP, we write

$$\begin{aligned} V &= 7.40\ \cancel{\text{g CO}_2} \times \frac{1\ \cancel{\text{mol CO}_2}}{44.01\ \cancel{\text{g CO}_2}} \times \frac{22.41\ \text{L}}{1\ \cancel{\text{mol CO}_2}} \\ &= 3.77\ \text{L} \end{aligned}$$

PRACTICE EXERCISE

What is the volume occupied by 49.8 g of HCl at STP?

Equation (5.7) is useful for problems that do not involve changes in P, V, T, and n for a gas sample. At times, however, we need to deal with changes in pressure, volume, and temperature, or even in the amount of a gas. When conditions change, we must employ a modified form of Equation (5.7) that is derived as follows. From Equation (5.7),

$$R = \frac{P_1V_1}{n_1T_1} \quad \text{(before change)}$$

and

$$R = \frac{P_2V_2}{n_2T_2} \quad \text{(after change)}$$

so that

$$\frac{P_1V_1}{n_1T_1} = \frac{P_2V_2}{n_2T_2}$$

Often the amount of the gas does not change in a process. In this case, $n_1 = n_2$ and the above equation becomes

$$\frac{P_1V_1}{T_1} = \frac{P_2V_2}{T_2} \tag{5.8}$$

Note that from Equation (5.8) we can generate Equation (5.2) by setting $T_1 = T_2$ or Equation (5.5) by setting $P_1 = P_2$.

EXAMPLE 5.4

Applying Equation (5.8)

A small bubble rises from the bottom of a lake, where the temperature and pressure are 8°C and 6.4 atm, to the water's surface, where the temperature is 25°C and pressure is 1.0 atm. Calculate the final volume (in milliliters) of the bubble if its initial volume was 2.1 mL.

Answer: We start by writing

Initial conditions	*Final conditions*
$P_1 = 6.4$ atm	$P_2 = 1.0$ atm
$V_1 = 2.1$ mL	$V_2 = ?$
$T_1 = (8 + 273)$ K	$T_2 = (25 + 273)$ K

The amount of the gas in the bubble remains constant, so that $n_1 = n_2$. To calculate the final volume, V_2, we rearrange Equation (5.8) as follows:

$$\begin{aligned} V_2 &= V_1 \times \frac{P_1}{P_2} \times \frac{T_2}{T_1} \\ &= 2.1 \text{ mL} \times \frac{6.4 \text{ atm}}{1.0 \text{ atm}} \times \frac{298 \text{ K}}{281 \text{ K}} \\ &= 14 \text{ mL} \end{aligned}$$

Thus the bubble's volume increases from 2.1 mL to 14 mL because of the decrease in water pressure and the increase in temperature.

PRACTICE EXERCISE

A sample of radioactive radon gas initially at 4.0 L, 1.2 atm, and 66°C undergoes a change so that its final volume and temperature become 1.7 L and 42°C. What is its final pressure? Assume the number of moles remains constant.

Density and Molar Mass of a Gaseous Substance

The ideal gas equation can be applied to determine the density or molar mass of a gaseous substance. Rearranging Equation (5.7), we write

$$\frac{n}{V} = \frac{P}{RT}$$

The number of moles of the gas, n, is given by

$$n = \frac{m}{\mathcal{M}}$$

FIGURE 5.11

An apparatus for measuring the density of a gas.

where m is the mass of the gas in grams and $\mathcal{M}$ is its molar mass. Therefore

$$\frac{m}{\mathcal{M}V} = \frac{P}{RT}$$

Since density, d, is mass per unit volume, we can write

$$d = \frac{m}{V} = \frac{P\mathcal{M}}{RT} \tag{5.9}$$

Equation (5.9) allows us to calculate the density of a gas (given in units of grams per liter). More often, the density of a gas can be measured, so this equation can be rearranged for us to calculate the molar mass of a gaseous substance:

$$\mathcal{M} = \frac{dRT}{P} \tag{5.10}$$

In a typical experiment, a bulb of known volume is filled with the gaseous substance under study. The temperature and pressure of the gas sample are recorded, and the total mass of the bulb plus gas sample is determined (Figure 5.11). The bulb is then evacuated (emptied) and weighed again. The difference in mass is the mass of the gas. The density of the gas is equal to its mass divided by the volume of the bulb. Then we can calculate the molar mass of the substance using Equation (5.10).

EXAMPLE 5.5

Determining Molar Mass from Gas Density

A chemist has synthesized a greenish-yellow gaseous compound of chlorine and oxygen and finds that its density is 7.71 g/L at 36°C and 2.88 atm. Calculate the molar mass of the compound and determine its molecular formula.

Answer: We substitute in Equation (5.10)

$$\begin{aligned}\mathcal{M} &= \frac{dRT}{P}\\ &= \frac{(7.71\text{ g/L})(0.0821\text{ L}\cdot\text{atm/K}\cdot\text{mol})(36+273)\text{ K}}{2.88\text{ atm}}\\ &= 67.9\text{ g/mol}\end{aligned}$$

We can determine the molecular formula of the compound by trial and error, using only the knowledge of the molar masses of chlorine (35.45 g) and oxygen (16.00 g). We know that a compound containing one Cl atom and one O atom would have a molar mass of 51.45 g, which is too low, while the molar mass of a compound made up of two Cl atoms and one O atom is 86.90 g, which is too high. Thus the compound must contain one Cl atom and two O atoms, and the formula must be ClO_2, which has a molar mass of 67.45 g.

PRACTICE EXERCISE

The density of a gaseous organic compound is 3.38 g/L at 40°C and 1.97 atm. What is its molar mass?

5.5 DALTON'S LAW OF PARTIAL PRESSURES

Thus far we have concentrated on the behavior of pure gaseous substances, but experimental studies very often concern mixtures of gases, such as air. For a gaseous mixture, we need to understand how the total gas pressure is related to *the pressures of individual gas components in the mixture,* called ***partial pressures.*** In 1801 Dalton formulated a law, now known as ***Dalton's law of partial pressures,*** which states that *the total pressure of a mixture of gases is just the sum of the pressures that each gas would exert if it were present alone.*

Consider a case in which two gases, A and B, are in a container of volume V. The pressure exerted by gas A, according to Equation (5.7), is

$$P_A = \frac{n_A RT}{V}$$

where n_A is the number of moles of A present. Similarly, the pressure exerted by gas B is

$$P_B = \frac{n_B RT}{V}$$

In a mixture of gases A and B, the total pressure P_T is the result of the collisions of both types of molecules, A and B, with the walls of the container. Thus, according to Dalton's law,

$$\begin{aligned} P_T &= P_A + P_B \\ &= \frac{n_A RT}{V} + \frac{n_B RT}{V} \\ &= \frac{RT}{V}(n_A + n_B) \\ &= \frac{nRT}{V} \end{aligned}$$

where n, the total number of moles of gases present, is given by $n = n_A + n_B$, and P_A and P_B are the partial pressures of gases A and B, respectively. In general, the total pressure of a mixture of gases is given by

$$P_T = P_1 + P_2 + P_3 + \cdots \tag{5.11}$$

where $P_1, P_2, P_3, \ldots$ are the partial pressures of components 1, 2, 3, ...

To see how each partial pressure is related to the total pressure, consider again the case of a mixture of gases A and B. Dividing P_A by P_T, we obtain

$$\begin{aligned} \frac{P_A}{P_T} &= \frac{n_A RT/V}{(n_A + n_B)RT/V} \\ &= \frac{n_A}{n_A + n_B} \\ &= X_A \end{aligned}$$

where X_A is called the mole fraction of gas A. The ***mole fraction*** is *a dimensionless quantity that expresses the ratio of the number of moles of one component to the* *number of moles of all components present.* It is always smaller than 1, except when A is the only component present. In that case, $n_B = 0$ and $X_A = n_A/n_A = 1$. We can now express the partial pressure of A as

$$P_A = X_A P_T$$

Similarly,

$$P_B = X_B P_T$$

Note that the sum of mole fractions must be unity. If only two components are present, then

$$X_A + X_B = \frac{n_A}{n_A + n_B} + \frac{n_B}{n_A + n_B} = 1$$

If a system contains more than two gases, then the partial pressure of the ith component is related to the total pressure by

$$P_i = X_i P_T \qquad (5.12)$$

where X_i is the mole fraction of substance i.

EXAMPLE 5.6

Applying Dalton's Law of Partial Pressures

A mixture of noble gases contains 4.46 moles of neon (Ne), 0.74 mole of argon (Ar), and 2.15 moles of xenon (Xe). Calculate the partial pressures of the gases if the total pressure is 2.00 atm at a certain temperature.

Answer: The mole fraction of Ne is

$$X_{Ne} = \frac{n_{Ne}}{n_{Ne} + n_{Ar} + n_{Xe}} = \frac{4.46 \text{ mol}}{4.46 \text{ mol} + 0.74 \text{ mol} + 2.15 \text{ mol}}$$
$$= 0.607$$

From Equation (5.12),

$$\begin{aligned} P_{Ne} &= X_{Ne} P_T \\ &= 0.607 \times 2.00 \text{ atm} \\ &= 1.21 \text{ atm} \end{aligned}$$

Similarly,

$$\begin{aligned} P_{Ar} &= 0.10 \times 2.00 \text{ atm} \\ &= 0.20 \text{ atm} \end{aligned}$$

and

$$\begin{aligned} P_{Xe} &= 0.293 \times 2.00 \text{ atm} \\ &= 0.586 \text{ atm} \end{aligned}$$

PRACTICE EXERCISE

A sample of natural gas contains 8.24 moles of methane (CH_4), 0.421 mole of ethane (C_2H_6), and 0.116 mole of propane (C_3H_8). If the total pressure of the gases is 1.37 atm, what are the partial pressures of the gases?

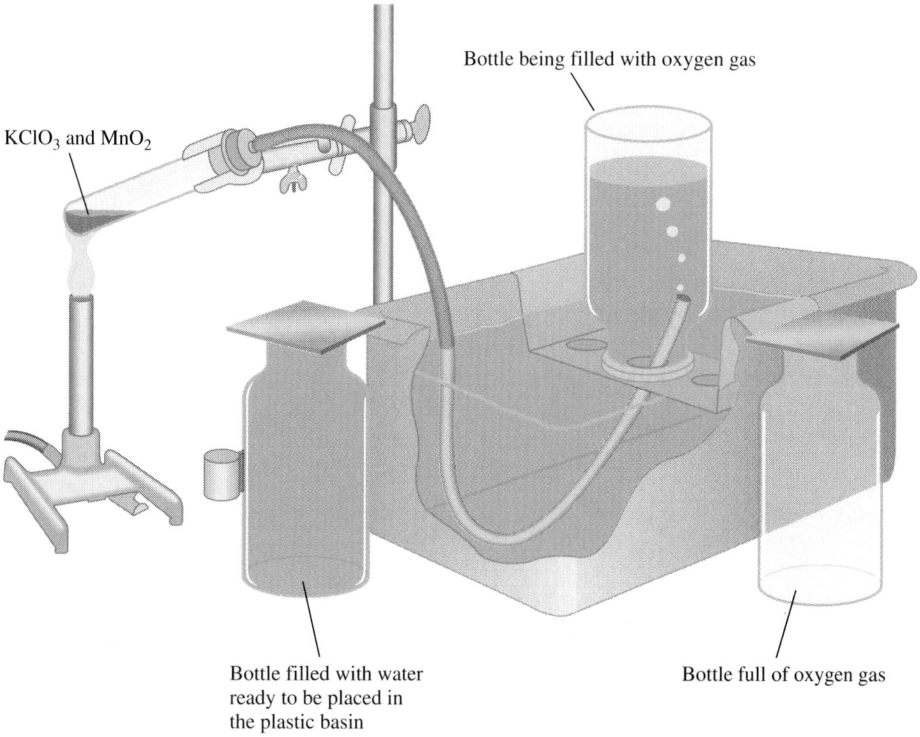

FIGURE 5.12

An apparatus for collecting gas over water. The oxygen generated by heating potassium chlorate ($KClO_3$) in the presence of a small amount of manganese dioxide (MnO_2), to speed up the reaction, is bubbled through water and collected in a bottle as shown. Water originally present in the bottle is pushed into the trough by the oxygen gas.

Dalton's law of partial pressures is useful for calculating volumes of gases collected over water. For example, oxygen gas is conveniently prepared in the laboratory by heating potassium chlorate to produce KCl and O_2:

$$2KClO_3(s) \longrightarrow 2KCl(s) + 3O_2(g)$$

The evolved oxygen gas can be collected over water, as shown in Figure 5.12. Initially, the inverted bottle is completely filled with water. As oxygen gas is generated, the gas bubbles rise to the top and water is pushed out of the bottle. This method of collecting a gas is based on the assumptions that the gas does not react with water and that it is not appreciably soluble in it. These assumptions are valid for oxygen gas, but not for gases such as NH_3, which dissolves readily in water. The oxygen gas collected in this way is not pure, however, because water vapor is also present in the bottle. The total gas pressure is equal to the sum of the pressures exerted by the oxygen gas and the water vapor:

$$P_T = P_{O_2} + P_{H_2O}$$

Consequently, we must allow for the pressure caused by the presence of water vapor when we calculate the amount of O_2 generated. Table 5.2 shows the pressure of water vapor at various temperatures. These data are plotted in Figure 5.13.

TABLE 5.2
Pressure of Water Vapor at Various Temperatures

t (°C)	Water vapor pressure (mmHg)
0	4.58
5	6.54
10	9.21
15	12.79
20	17.54
25	23.76
30	31.82
35	42.18
40	55.32
45	71.88
50	92.51
55	118.04
60	149.38
65	187.54
70	233.7
75	289.1
80	355.1
85	433.6
90	525.76
95	633.90
100	760.00

EXAMPLE 5.7
Calculating the Mass of a Gas Collected over Water

Oxygen gas generated in the decomposition of potassium chlorate is collected as shown in Figure 5.12. The volume of the gas collected at 24°C and atmo-

FIGURE 5.13

The pressure of water vapor as a function of temperature. Note that at the boiling point of water (100°C) the pressure is 760 mmHg, which is exactly equal to 1 atm.

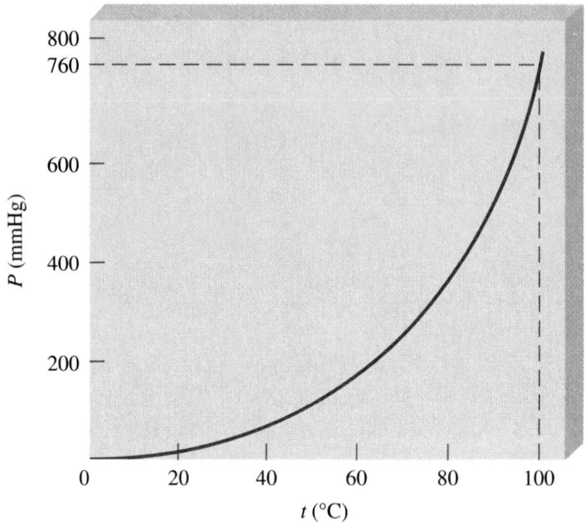

spheric pressure of 762 mmHg is 128 mL. Calculate the mass (in grams) of oxygen gas obtained. The pressure of the water vapor at 24°C is 22.4 mmHg.

Answer: Our first step is to calculate the partial pressure of O_2. We know that

$$P_T = P_{O_2} + P_{H_2O}$$

Therefore

$$\begin{aligned} P_{O_2} &= P_T - P_{H_2O} \\ &= 762 \text{ mmHg} - 22.4 \text{ mmHg} \\ &= 740 \text{ mmHg} \\ &= 740 \cancel{\text{mmHg}} \times \frac{1 \text{ atm}}{760 \cancel{\text{mmHg}}} \\ &= 0.974 \text{ atm} \end{aligned}$$

From the ideal gas equation we write

$$PV = nRT = \frac{m}{\mathcal{M}}RT$$

where m and $\mathcal{M}$ are the mass of O_2 collected and the molar mass of O_2, respectively. Rearranging the equation we get

$$\begin{aligned} m &= \frac{PV\mathcal{M}}{RT} = \frac{(0.974 \text{ atm})(0.128 \text{ L})(32.00 \text{ g/mol})}{(0.0821 \text{ L} \cdot \text{atm/K} \cdot \text{mol})(273 + 24) \text{ K}} \\ &= 0.164 \text{ g} \end{aligned}$$

PRACTICE EXERCISE

Hydrogen gas generated when calcium metal reacts with water is collected as shown in Figure 5.12. The volume of gas collected at 30°C and pressure of 988 mmHg is 641 mL. What is the mass (in grams) of the hydrogen gas obtained? The pressure of water vapor at 30°C is 31.82 mmHg.

5.6 THE KINETIC MOLECULAR THEORY OF GASES

The gas laws help us to predict the behavior of gases, but they do not explain what happens at the molecular level to cause the changes we observe in the macroscopic world. For example, why does the volume of a gas expand upon heating?

In the nineteenth century, a number of physicists, notably Ludwig Boltzmann in Germany and James Clerk Maxwell in England, found that the physical properties of gases can be explained in terms of the motions of individual molecules. This molecular movement is a form of *energy*, which we define as the capacity to do work or to produce change. In mechanics, *work* is defined as force times distance. Since energy can be measured as work, we can write

$$\begin{aligned} \text{energy} &= \text{work done} \\ &= \text{force} \times \text{distance} \end{aligned}$$

The SI unit of energy is the ***joule (J):***

$$\begin{aligned} 1\text{ J} &= 1\text{ kg m}^2/\text{s}^2 \\ &= 1\text{ N m} \end{aligned}$$

Alternatively, energy can be expressed in kilojoules (kJ):

$$1\text{ kJ} = 1000\text{ J}$$

As we will see in Chapter 6, there are many different kinds of energy. The type of energy possessed by a moving object is called kinetic energy. In other words, ***kinetic energy (KE)*** is *energy of motion.*

The findings of Maxwell, Boltzmann, and others resulted in a number of generalizations about gas behavior that have since become known as the ***kinetic molecular theory of gases*** (or simply the kinetic theory of gases). Central to the kinetic theory are the following assumptions:

1. A gas is composed of molecules that are separated from one another by distances far greater than their own dimensions. The molecules can be considered to be "points"; that is, they possess mass but have negligible volume.

2. Gas molecules are in constant motion in random directions, and they frequently collide with one another. Collisions among molecules are perfectly elastic. Although energy may be transferred from one molecule to another as a result of a collision, the total energy of all the molecules in a system remains the same.

3. Gas molecules exert neither attractive nor repulsive forces on one another.

4. The average kinetic energy of the molecules is proportional to the temperature of the gas in kelvins. Any two gases at the same temperature will have the same average kinetic energy. The average kinetic energy of a molecule is given by

$$\text{KE} = \tfrac{1}{2}m\overline{u^2}$$

where m is the mass of the molecule and u is its speed. The horizontal bar denotes an average value. The quantity $\overline{u^2}$ is called *mean-square speed;* it is the average of the square of the speeds of all the molecules (N) present:

$$\overline{u^2} = \frac{u_1^2 + u_2^2 + \cdots + u_N^2}{N}$$

From assumption 4 we write

$$\text{KE} \propto T$$

$$\tfrac{1}{2}m\overline{u^2} \propto T$$

$$\tfrac{1}{2}m\overline{u^2} = CT \quad (5.13)$$

where C is the proportionality constant and T the absolute temperature.

According to the kinetic molecular theory, gas pressure is the result of collisions between molecules and the walls of the container. The gas pressure depends on the frequency of collision per unit area and how "hard" the molecules strike the wall. The theory also provides a molecular interpretation of temperature. According to Equation (5.13), the absolute temperature of a gas is a measure of the average kinetic energy of the molecules. In other words, the absolute temperature is an index of the random motion of the molecules—the higher the temperature, the more energetic the motion. Because it is related to the temperature of the gas sample, random molecular motion is sometimes referred to as thermal motion.

Distribution of Molecular Speeds

The kinetic theory of gases allows us to investigate molecular motion in more detail. Suppose we have a large number of molecules of a gas, say, 1 mole, in a container. As you might expect, the motion of the molecules is totally random and unpredictable. As long as we hold the temperature constant, however, the average kinetic energy and the mean-square speed will remain unchanged as time passes. The characteristic that is of interest to us here is the spread, or distribution, of molecular speeds. At a given instant, how many molecules are moving at a particular speed? An equation to solve this problem was formulated by Maxwell in 1860. The equation, which incorporates the above assumptions, is based on a statistical analysis of molecular behavior.

Figure 5.14 shows typical *Maxwell speed distribution curves* for an ideal gas at two different temperatures. At a given temperature, the distribution curve tells us the number of molecules moving at a certain speed. The peak of each curve gives the *most probable speed,* that is, the speed of the largest number of molecules. Note that at the higher temperature [Figure 5.14(b)] the most probable speed is greater than that at the lower temperature [Figure 5.14(a)]. Moreover, as temperature increases, not only does the peak shift toward the right, but the curve also flattens out, indicating that larger numbers of molecules are moving at greater speeds.

The kinetic molecular theory conveniently enables us to relate macroscopic

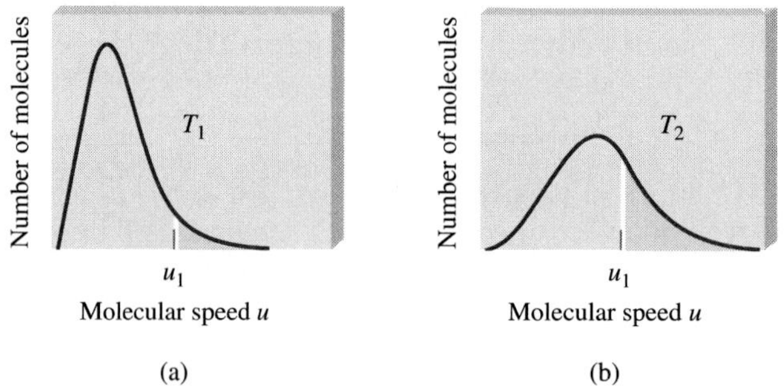

FIGURE 5.14

Maxwell's speed distribution for a gas at (a) temperature T_1 and (b) a higher temperature T_2. The shaded areas represent the number of molecules traveling at a speed equal to or greater than a certain speed of u_1.

quantities P and V to molecular parameters such as molar mass ($\mathcal{M}$) and mean-square speed. Using an equation derived from the assumptions of the theory:

$$PV = \tfrac{1}{3}n\mathcal{M}\overline{u^2}$$

From the ideal gas equation we have

$$PV = nRT$$

Combining the above two equations, we obtain

$$\tfrac{1}{3}n\mathcal{M}\overline{u^2} = nRT$$

or

$$\overline{u^2} = \frac{3RT}{\mathcal{M}}$$

Taking the square root of both sides gives

$$\sqrt{\overline{u^2}} = u_{rms} = \sqrt{\frac{3RT}{\mathcal{M}}} \qquad (5.14)$$

where u_{rms} is the root-mean-square speed; that is, it is the square root of the mean-square speed. Equation (5.14) shows that the root-mean-square speed of a gas increases with the square root of its temperature (in kelvins). Because $\mathcal{M}$ appears in the denominator, it follows that the heavier the gas, the more slowly its molecules move. If we use 8.314 J/K · mol for R (see Appendix 1) and convert the molar mass to kilograms per mole, then u_{rms} will be in meters per second (m/s).

EXAMPLE 5.8

Calculating Root-Mean-Square Speeds

Calculate the root-mean-square speeds of helium atoms and nitrogen molecules in meters per second at 25°C.

Answer: We need Equation (5.14) for this calculation. For He the molar mass is 4.003 g/mol, or 4.003×10^{-3} kg/mol:

$$\begin{aligned} u_{rms} &= \sqrt{\frac{3RT}{\mathcal{M}}} \\ &= \sqrt{\frac{3(8.314\ \text{J/K}\cdot\text{mol})(298\ \text{K})}{4.003 \times 10^{-3}\ \text{kg/mol}}} \\ &= \sqrt{1.86 \times 10^{6}\ \text{J/kg}} \end{aligned}$$

Using the conversion factor,

$$1\ \text{J} = 1\ \text{kg m}^2/\text{s}^2$$

we get

$$\begin{aligned} u_{rms} &= \sqrt{1.86 \times 10^{6}\ \text{kg m}^2/\text{kg}\cdot\text{s}^2} \\ &= \sqrt{1.86 \times 10^{6}\ \text{m}^2/\text{s}^2} \\ &= 1.36 \times 10^{3}\ \text{m/s} \end{aligned}$$

Similarly for N_2 (molar mass 2.802×10^{-2} kg/mol)

$$u_{rms} = \sqrt{\frac{3(8.314 \text{ J/K} \cdot \text{mol})(298 \text{ K})}{2.802 \times 10^{-2} \text{ kg/mol}}}$$

$$= \sqrt{2.65 \times 10^5 \text{ m}^2/\text{s}^2}$$

$$= 515 \text{ m/s}$$

Because of its smaller mass, a helium atom, on the average, moves about 2.6 times faster than a nitrogen molecule at the same temperature (1360 ÷ 515 = 2.64).

PRACTICE EXERCISE

Calculate the root-mean-square speed of molecular chlorine in meters per second at 20°C.

Jupiter. The interior of this massive planet consists mainly of liquid hydrogen.

The calculation in Example 5.8 has an interesting relationship to the composition of Earth's atmosphere. Earth, unlike, say, Jupiter, does not have appreciable amounts of hydrogen or helium in its atmosphere. Why is this the case? A much smaller planet than Jupiter, Earth has a weaker gravitational attraction for these lighter molecules. A fairly straightforward calculation shows that to escape Earth's gravitational field, a molecule must possess a speed equal to or greater than 1.1×10^3 m/s. This is usually called the *escape velocity.* Because the average speed of helium is considerably greater than that of molecular nitrogen or molecular oxygen, more helium atoms escape from Earth's atmosphere into outer space. Consequently, only a trace amount of helium is present in our atmosphere. On the other hand, Jupiter, with a mass about 320 times greater than that of Earth, retains both heavy and light gases in its atmosphere.

5.7 DEVIATION FROM IDEAL BEHAVIOR

So far our discussion has assumed that molecules in the gaseous state do not exert any force, either attractive or repulsive, on one another. Further, we have assumed that the volume of the molecules is negligibly small compared with that of the container. A gas that satisfies these two conditions is said to exhibit *ideal behavior.*

These would seem to be fair assumptions, although we cannot expect them to hold under all conditions. For example, without intermolecular forces, gases could not condense to form liquids. The important question is: Under what conditions will gases most likely exhibit nonideal behavior?

Figure 5.15 shows the PV/RT versus P plots for three real gases at a given temperature. These plots offer a test of ideal gas behavior. According to the ideal gas equation (for 1 mole of the gas), PV/RT equals 1, regardless of the actual gas pressure. For real gases, this is true only at moderately low pressures (<5 atm); significant deviations are observed as pressure increases. The attractive forces among molecules operate at relatively short distances. For a gas at atmospheric pressure, the molecules are relatively far apart and these attractive forces are negli-

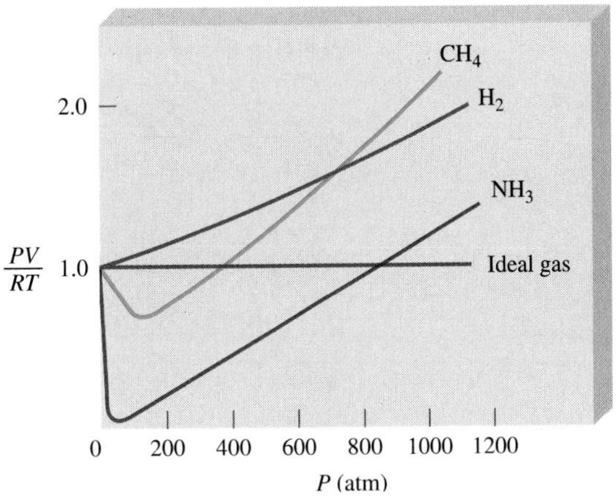

FIGURE 5.15

Plot of PV/RT versus P of 1 mole of a gas at 0°C. For 1 mole of an ideal gas, PV/RT is equal to 1, no matter what the pressure of the gas is. Real gases, on the other hand, deviate from ideality at high pressures. At very low pressures, all gases exhibit ideal behavior; that is, their PV/RT values all converge to 1 as P approaches zero.

gible. At high pressures, the density of the gas increases; the molecules are much closer to one another. Then intermolecular forces can be significant enough to affect the motion of the molecules, and the gas will not behave ideally.

Another way to observe the nonideality of gases is to lower the temperature. Cooling a gas decreases the molecules' average kinetic energy, which in a sense deprives molecules of the drive they need to break away from their mutual attractive influences.

The nonideality of gases can be dealt with mathematically by modifying Equation (5.7), taking into account intermolecular forces and finite molecular volumes. Such an analysis was first made by the Dutch physicist J. D. van der Waals in 1873. Besides being mathematically simple, van der Waals's treatment provides us with an interpretation of real gas behavior at the molecular level.

Consider the approach of a particular molecule toward the wall of a container (Figure 5.16). The intermolecular attractions exerted by its neighbors tend to soften the impact made by this molecule against the wall. The overall effect is a lower gas pressure than we would expect for an ideal gas. Van der Waals suggested that the pressure exerted by an ideal gas, P_{ideal}, is related to the experimentally measured pressure, P_{real}, by

$$P_{\text{ideal}} = \underset{\substack{\uparrow \\ \text{observed} \\ \text{pressure}}}{P_{\text{real}}} + \underset{\substack{\uparrow \\ \text{correction} \\ \text{term}}}{\frac{an^2}{V^2}}$$

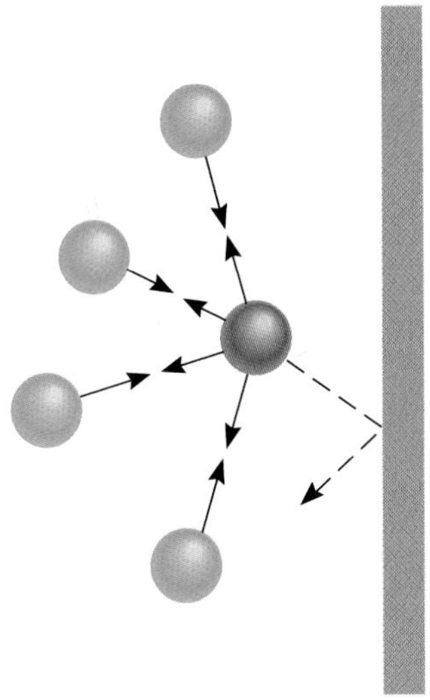

FIGURE 5.16

The speed of a molecule that is moving toward the container wall (red sphere) is reduced by the attractive forces exerted by its neighbors (green spheres).

where a is a constant and n and V are the number of moles and volume of the gas, respectively. The correct term for pressure (an^2/V^2) can be understood as follows. The interaction between molecules that gives rise to nonideal behavior depends on how frequently any two molecules approach each other closely. The number of such "encounters" increases with the square of the number of molecules per unit volume, $(n/V)^2$, because the presence of each of the two molecules in a particular region is proportional to n/V. The quantity P_{ideal} is the pressure we would measure if there were no intermolecular attractions. The quantity a, then, is just a proportionality constant in the correction term for pressure.

Another correction concerns the volume occupied by the gas molecules. The quantity V in Equation (5.7) represents the volume of the container. However,

Johannes van der Waals (1837–1923)

TABLE 5.3
Van der Waals Constants of Some Common Gases

Gas	a (atm · L^2/mol^2)	b (L/mol)
He	0.034	0.0237
Ne	0.211	0.0171
Ar	1.34	0.0322
Kr	2.32	0.0398
Xe	4.19	0.0266
H_2	0.244	0.0266
N_2	1.39	0.0391
O_2	1.36	0.0318
Cl_2	6.49	0.0562
CO_2	3.59	0.0427
CH_4	2.25	0.0428
CCl_4	20.4	0.138
NH_3	4.17	0.0371
H_2O	5.46	0.0305

each molecule does occupy a finite, although small, intrinsic volume, so the effective volume of the gas becomes $(V - nb)$, where n is the number of moles of the gas and b is a constant. The term nb represents the volume occupied by n moles of the gas.

Having taken into account the corrections for pressure and volume, we can rewrite Equation (5.7) as

$$\underbrace{\left(P + \frac{an^2}{V^2}\right)}_{\text{corrected pressure}}\underbrace{(V - nb)}_{\text{corrected volume}} = nRT \tag{5.15}$$

Equation (5.15) is known as the ***van der Waals equation.*** The van der Waals constants a and b are selected for each gas to give the best possible agreement between the equation and actually observed behavior.

Table 5.3 lists the values of a and b for a number of gases. The value of a is an expression of how strongly molecules of a given type of gas attract one another. We see that the helium atoms have the weakest attraction for one another, because helium has the smallest a value. There is also a rough correlation between molecular size and b. Generally, the larger the molecule (or atom), the greater b is, but the relationship between b and molecular (or atomic) size is not a simple one.

EXAMPLE 5.9
Comparing Pressures Using the Ideal Gas Equation and the van der Waals Equation

A quantity of 3.50 moles of NH_3 occupies 5.20 L at 47°C. Calculate the pressure of the gas (in atmospheres) using (a) the ideal gas equation and (b) the van der Waals equation.

Answer: (a) We have the following data:

$$V = 5.20 \text{ L}$$
$$T = (47 + 273) \text{ K} = 320 \text{ K}$$
$$n = 3.50 \text{ mol}$$
$$R = 0.0821 \text{ L} \cdot \text{atm/K} \cdot \text{mol}$$

which we substitute in the ideal gas equation:

$$P = \frac{nRT}{V}$$
$$= \frac{(3.50 \text{ mol})(0.0821 \text{ L} \cdot \text{atm/K} \cdot \text{mol})(320 \text{ K})}{5.20 \text{ L}}$$
$$= 17.7 \text{ atm}$$

(b) From Table 5.3, we have

$$a = 4.17 \text{ atm} \cdot \text{L}^2/\text{mol}^2$$
$$b = 0.0371 \text{ L/mol}$$

It is convenient to calculate the correction terms first, for Equation (5.15). They are

$$\frac{an^2}{V^2} = \frac{(4.17 \text{ atm} \cdot \text{L}^2/\text{mol}^2)(3.50 \text{ mol})^2}{(5.20 \text{ L})^2} = 1.89 \text{ atm}$$
$$nb = (3.50 \text{ mol})(0.0371 \text{ L/mol}) = 0.130 \text{ L}$$

Finally, substituting in the van der Waals equation, we write

$$(P + 1.89 \text{ atm})(5.20 \text{ L} - 0.130 \text{ L}) = (3.50 \text{ mol})(0.0821 \text{ L} \cdot \text{atm/K} \cdot \text{mol})(320 \text{ K})$$
$$P = 16.2 \text{ atm}$$

PRACTICE EXERCISE

Using the data in Table 5.3, calculate the pressure exerted by 4.37 moles of molecular chlorine confined in a volume of 2.45 L at 38°C. What is the pressure using the ideal gas equation?

SUMMARY

Under atmospheric conditions, a number of elemental substances are gases: H_2, N_2, O_2, O_3, F_2, Cl_2, and the Group 8A elements (the noble gases).

Gases exert pressure because their molecules move freely and collide with any surface in their paths. Gas pressure units include millimeters of mercury (mmHg), torr, pascals, and atmospheres. One atmosphere equals 760 mmHg, or 760 torr.

The pressure-volume relationships of ideal gases are governed by Boyle's law: Volume is inversely proportional to pressure (at constant T and n). The temperature-volume relationships of ideal gases are described by Charles' and Gay-Lussac's law: Volume is directly proportional to temperature (at constant P and n).

Absolute zero (−273.15°C) is the lowest theoretically attainable temperature. On the Kelvin temperature scale, 0 K is absolute zero. In all gas law calculations, temperature must be expressed in kelvins. The amount-volume relationships of ideal gases are described by Avogadro's law: Equal volumes of gases contain equal numbers of molecules (at the same T and P).

The ideal gas equation, $PV = nRT$, combines the laws of Boyle, Charles, and Avogadro. This equation describes the behavior of an ideal gas.

Dalton's law of partial pressure states that in a mixture of gases each gas exerts the same pressure as it would if it were alone and occupied the same volume.

The kinetic molecular theory, a mathematical way of describing the behavior of gas molecules, is based on the following assumptions: Gas molecules are separated by distances far greater than their own dimensions, they possess mass but have negligible volume, they are in constant motion, and they frequently collide with one another. The molecules neither attract nor repel one another. A Maxwell speed distribution curve shows how many gas molecules are moving at various speeds at a given temperature. As temperature increases, more molecules move at greater speeds.

The van der Waals equation is a modification of the ideal gas equation that takes into account the nonideal behavior of real gases. It corrects for two facts: Real gas molecules do exert forces on each other and they do have volume. The van der Waals constants are determined experimentally for each gas.

KEY WORDS

Absolute temperature scale, p. 123
Absolute zero, p. 123
Atmospheric pressure, p. 120
Avogadro's law, p. 126
Barometer, p. 120
Boyle's law, p. 122
Charles' and Gay-Lussac's law, p. 125
Charles' law, p. 125
Dalton's law of partial pressures, p. 131
Gas constant, p. 127
Ideal gas, p. 127
Ideal gas equation, p. 127
Joule (J), p. 135
Kelvin temperature scale, p. 123
Kinetic energy (KE), p. 135
Kinetic molecular theory of gases, p. 135
Mole fraction, p. 131
Newton (N), p. 119
Partial pressure, p. 131
Pascal (Pa), p. 120
Pressure, p. 119
Standard atmospheric pressure (1 atm), p. 120
Standard temperature and pressure (STP), p. 127
van der Waals equation, p. 140

QUESTIONS AND PROBLEMS

SUBSTANCES THAT EXIST AS GASES

Review Questions

5.1 Name five elements and five compounds that exist as gases at room temperature.

5.2 Describe the physical characteristics of gases.

PRESSURE

Review Questions

5.3 Define pressure and give the common units for pressure.

5.4 Describe how a barometer and a manometer are used to measure pressure.

5.5 Why is mercury a more suitable substance to use in a barometer than water?

5.6 Explain why the height of mercury in a barometer is independent of the cross-sectional area of the tube.

5.7 Would it be easier to drink water with a straw on the top or at the foot of Mt. Everest? Explain.

5.8 Is the atmospheric pressure in a mine that is 500 m below sea level greater or less than 1 atm?

5.9 What is the difference between a gas and a vapor? At 25°C, which of the following substances in the gas phase should be properly called a gas and which should be called a vapor: molecular nitrogen (N_2) or mercury?

5.10 If the maximum distance that water may be brought up a well by using a suction pump is 34 ft (10.3 m), explain how it is possible to obtain water and oil from hundreds of feet below the surface of Earth.

5.11 Why is it that if the barometer reading falls in one part of the world, it must rise somewhere else?

5.12 Why do astronauts have to wear protective suits when they are on the surface of the moon?

Problems

5.13 Convert 562 mmHg to kPa and 2.0 kPa to mmHg.

5.14 The atmospheric pressure at the summit of Mt. McKinley is 606 mmHg on a certain day. What is the pressure in atm?

THE GAS LAWS

Review Questions

5.15 State the following gas laws in words and also in the form of an equation: Boyle's law, Charles' law, Avogadro's law. In each case, indicate the conditions under which the law is applicable and also give the units for each quantity in the equation.

5.16 Explain why a helium weather balloon expands as it rises in the air. Assume that the temperature remains constant.

Problems

5.17 A gas occupying a volume of 725 mL at a pressure of 0.970 atm is allowed to expand at constant temperature until its pressure becomes 0.541 atm. What is its final volume?

5.18 At 46°C a sample of ammonia gas exerts a pressure of 5.3 atm. What is the pressure when the volume of the gas is reduced to one-tenth (0.10) of the original value at the same temperature?

5.19 The volume of a gas is 5.80 L, measured at 1.00 atm. What is the pressure of the gas in mmHg if the volume is changed to 9.65 L? (The temperature remains constant.)

5.20 A sample of air occupies 3.8 L when the pressure is 1.2 atm. (a) What volume does it occupy at 6.6 atm? (b) What pressure is required in order to compress it to 0.075 L? (The temperature is kept constant.)

5.21 Convert the following temperatures to Kelvin: 0°C, 37°C, 100°C, −225°C.

5.22 Convert the following temperatures to degrees Celsius: 77 K (boiling point of nitrogen), 4.2 K (boiling point of helium), 6.0×10^3 K (surface temperature of the sun).

5.23 A 36.4-L volume of methane gas is heated from 25°C to 88°C at constant pressure. What is the final volume of the gas?

5.24 Under constant-pressure conditions a sample of hydrogen gas initially at 88°C and 9.6 L is cooled until its final volume is 3.4 L. What is its final temperature?

5.25 Ammonia burns in oxygen gas to form nitric oxide (NO) and water vapor. How many volumes of NO are obtained from one volume of ammonia at the same temperature and pressure?

5.26 Molecular chlorine and molecular fluorine combine to form a gaseous product. Under the same conditions of temperature and pressure it is found that one volume of Cl_2 reacts with three volumes of F_2 to yield two volumes of the product. What is the formula of the product?

THE IDEAL GAS EQUATION

Review Questions

5.27 Describe the characteristics of an ideal gas.

5.28 Write the ideal gas equation and also state it in words. Give the units for each term in the equation.

5.29 What are standard temperature and pressure (STP)? What is the significance of STP in relation to the volume of 1 mole of an ideal gas?

5.30 Why is the density of a gas much lower than that of a liquid or solid under atmospheric conditions? What units are normally used to express the density of gases?

Problems

5.31 A sample of nitrogen gas kept in a container of volume 2.3 L and at a temperature of 32°C exerts a pressure of 4.7 atm. Calculate the number of moles of gas present.

5.32 A sample of 6.9 moles of carbon monoxide gas is present in a container of volume 30.4 L. What is the pressure of the gas (in atm) if the temperature is 62°C?

5.33 What volume would 5.6 moles of sulfur hexafluoride (SF_6) gas occupy if the temperature and pressure of the gas are 128°C and 9.4 atm?

5.34 A certain amount of gas at 25°C and at a pressure of 0.800 atm is contained in a glass vessel. Suppose that the vessel can withstand a pressure of 2.00 atm. How high can you increase the temperature of the gas without bursting the vessel?

5.35 A gas-filled balloon having a volume of 2.50 L at 1.2 atm and 25°C is allowed to rise to the stratosphere (about 30 km above the surface of Earth), where the temperature and pressure are −23°C and 3.00×10^{-3} atm, respectively. Calculate the final volume of the balloon.

5.36 The temperature of 2.5 L of a gas initially at STP is increased to 250°C at constant volume. Calculate the final pressure of the gas in atm.

5.37 The pressure of 6.0 L of an ideal gas in a flexible container is decreased to one-third of its original pressure, and its absolute temperature is decreased by one-half. What is the final volume of the gas?

5.38 A gas evolved during the fermentation of glucose (wine making) has a volume of 0.78 L when measured at 20.1°C and 1.00 atm. What was the volume of this gas at the fermentation temperature of 36.5°C and 1.00 atm pressure?

5.39 An ideal gas originally at 0.85 atm and 66°C was allowed to expand until its final volume, pressure, and temperature were 94 mL, 0.60 atm, and 45°C, respectively. What was its initial volume?

5.40 The volume of a gas at STP is 488 mL. Calculate its volume at 22.5 atm and 150°C.

5.41 A gas at 772 mmHg and 35.0°C occupies a volume of 6.85 L. Calculate its volume at STP.

5.42 Dry ice is solid carbon dioxide. A 0.050-g sample of dry ice is placed in an evacuated 4.6-L vessel at 30°C. Calculate the pressure inside the vessel after all the dry ice has been converted to CO_2 gas.

5.43 A volume of 0.280 L of a gas at STP weighs 0.400 g. Calculate the molar mass of the gas.

5.44 A quantity of gas weighing 7.10 g at 741 torr and 44°C occupies a volume of 5.40 L. What is its molar mass?

5.45 The ozone molecules present in the stratosphere absorb much of the harmful radiation from the sun. Typically, the temperature and pressure of ozone in the stratosphere are 250 K and 1.0×10^{-3} atm, respectively. How many ozone molecules are present in 1.0 L of air under these conditions?

5.46 Assuming that air contains 78 percent N_2, 21 percent O_2, and 1 percent Ar, all by volume, how many molecules of each type of gas are present in 1.0 L of air at STP?

5.47 A 2.10-L vessel contains 4.65 g of gas at 1.00 atm and 27.0°C. (a) Calculate the density of the gas in grams per liter. (b) What is the molar mass of the gas?

5.48 Calculate the density of hydrogen bromide (HBr) gas in grams per liter at 733 mmHg and 46°C.

5.49 A certain anesthetic contains 64.9 percent C, 13.5 percent H, and 21.6 percent O by mass. At 120°C and 750 mmHg, 1.00 L of the gaseous compound weighs 2.30 g. What is the molecular formula of the compound?

5.50 A compound has the empirical formula SF_4. At 20°C, 0.100 g of the gaseous compound occupies a volume of 22.1 mL and exerts a pressure of 1.02 atm. What is its molecular formula?

DALTON'S LAW OF PARTIAL PRESSURES

Review Questions

5.51 Define Dalton's law of partial pressures and mole fraction. Does mole fraction have units?

5.52 A sample of air contains only nitrogen and oxygen gases whose partial pressures are 0.80 atm and 0.20 atm, respectively. Calculate the total pressure and the mole fractions of the gases.

Problems

5.53 A mixture of gases contains CH_4, C_2H_6, and C_3H_8. If the total pressure is 1.50 atm and the numbers of moles of the gases present are 0.31 mole for CH_4, 0.25 mole for C_2H_6, and 0.29 mole for C_3H_8, calculate the partial pressures of the gases.

5.54 A 2.5-L flask at 15°C contains a mixture of three gases, N_2, He, and Ne, at partial pressures of 0.32 atm for N_2, 0.15 atm for He, and 0.42 atm for Ne. (a) Calculate the total pressure of the mixture. (b) Calculate the volume in liters at STP occupied by He and Ne if the N_2 is removed selectively.

5.55 Dry air near sea level has the following composition by volume: N_2, 78.08 percent; O_2, 20.94 percent; Ar, 0.93 percent; CO_2, 0.05 percent. The atmospheric pressure is 1.00 atm. Calculate (a) the partial pressure of each gas in atm and (b) the concentration of each gas in moles per liter at 0°C. (*Hint:* Since volume is proportional to the number of moles present, mole fractions of gases can be expressed as ratios of volumes at the same temperature and pressure.)

5.56 A mixture of helium and neon gases is collected over water at 28.0°C and 745 mmHg. If the partial pressure of helium is 368 mmHg, what is the partial pressure of neon? (Vapor pressure of water at 28°C = 28.3 mmHg.)

5.57 A piece of sodium metal undergoes complete reaction with water as follows:

$$2Na(s) + 2H_2O(l) \longrightarrow 2NaOH(aq) + H_2(g)$$

The hydrogen gas generated is collected over water at 25.0°C. The volume of the gas is 246 mL measured at 1.00 atm. Calculate the number of grams of sodium used in the reaction. (Vapor pressure of water at 25°C = 0.0313 atm.)

5.58 A sample of zinc metal is allowed to react completely with an excess of hydrochloric acid:

$$Zn(s) + 2HCl(aq) \longrightarrow ZnCl_2(aq) + H_2(g)$$

The hydrogen gas produced is collected over water at 25.0°C using an arrangement similar to that shown in Figure 5.12. The volume of the gas is 7.80 L, and the pressure is 0.980 atm. Calculate the amount of zinc metal in grams consumed in the reaction. (Vapor pressure of water at 25°C = 23.8 mmHg.)

5.59 Helium is mixed with oxygen gas for deep sea divers. Calculate the percent by volume of oxygen gas in the mixture if the diver has to submerge to a depth where the total pressure is 4.2 atm. The partial pressure of oxygen is maintained at 0.20 atm at this depth.

5.60 A sample of ammonia (NH_3) gas is completely decomposed to nitrogen and hydrogen gases over heated iron wool. If the total pressure is 866 mmHg, calculate the partial pressures of N_2 and H_2.

KINETIC MOLECULAR THEORY OF GASES

Review Questions

5.61 What are the basic assumptions of the kinetic molecular theory of gases?

5.62 What is thermal motion?

5.63 What does the Maxwell speed distribution curve tell us? Does Maxwell's theory work for a sample of 200 molecules? Explain.

5.64 Write the expression for the root-mean-square speed for a gas at temperature T. Define each term in the equation and show the units that are used in the calculations.

5.65 Which of the following two statements is correct? (a) Heat is produced by the collision of gas molecules against one another. (b) When a gas is heated, the molecules collide with one another more often.

5.66 As we know, UF_6 is a much heavier gas than helium. Yet at a given temperature, the average kinetic energies of the samples of the two gases are the same. Explain.

Problems

5.67 Compare the root-mean-square speeds of O_2 and UF_6 at 65°C.

5.68 The temperature in the stratosphere is −23°C. Calculate the root-mean-square speeds of N_2, O_2, and O_3 molecules in this region.

NONIDEAL GAS BEHAVIOR

Review Questions

5.69 Give two pieces of evidence to show that gases do not behave ideally under all conditions.

5.70 Under what set of conditions would a gas be expected to behave most ideally? (a) High temperature and low pressure, (b) high temperature and high pressure, (c) low temperature and high pressure, (d) low temperature and low pressure.

5.71 Write the van der Waals equation for a real gas. Explain clearly the meaning of the corrective terms for pressure and volume.

5.72 The temperature of a real gas that is allowed to expand into a vacuum usually drops. Explain.

Problems

5.73 Using the data shown in Table 5.3, calculate the pressure exerted by 2.50 moles of CO_2 confined in a volume of 5.00 L at 450 K. Compare the pressure with that calculated using the ideal gas equation.

5.74 At 27°C 10.0 moles of a gas in a 1.50-L container exert a pressure of 130 atm. Is this an ideal gas?

5.75 Under the same conditions of temperature and pressure, which of the following gases would behave most ideally: Ne, N_2, or CH_4? Explain.

5.76 From the following data collected at 0°C, comment on whether carbon dioxide behaves as an ideal gas.

P (atm)	0.0500	0.100	0.151	0.202	0.252
V (L)	448.2	223.8	148.8	110.8	89.0

MISCELLANEOUS PROBLEMS

5.77 Discuss the following phenomena in terms of the

gas laws: (a) the pressure in an automobile tire increasing on a hot day, (b) the "popping" of a paper bag, (c) the expansion of a weather balloon as it rises in the air, (d) the loud noise heard when a light bulb shatters.

5.78 Nitroglycerin, an explosive, decomposes according to the equation

$$4C_3H_5(NO_3)_3(s) \longrightarrow 12CO_2(g) + 10H_2O(g) + 6N_2(g) + O_2(g)$$

Calculate the total volume of gases produced when collected at 1.2 atm and 25°C from 2.6×10^2 g of nitroglycerin. What are the partial pressures of the gases under these conditions?

5.79 The empirical formula of a compound is CH. At 200°C, 0.145 g of this compound occupies 97.2 mL at a pressure of 0.74 atm. What is the molecular formula of the compound?

5.80 When ammonium nitrite (NH_4NO_2) is heated, it decomposes to give nitrogen gas. This property is used to inflate some tennis balls. (a) Write a balanced equation for the reaction. (b) Calculate the quantity (in grams) of NH_4NO_2 needed to inflate a tennis ball to a volume of 86.2 mL at 1.20 atm and 22°C.

5.81 The percent by mass of bicarbonate (HCO_3^-) in a certain Alka-Seltzer product is 32.5 percent. Calculate the volume of CO_2 generated (in milliliters) at 37°C and 1.00 atm when a person ingests a 3.29-g tablet. (*Hint:* The reaction is between HCO_3^- and HCl acid in the stomach.)

5.82 The boiling point of liquid nitrogen is −196°C. On the basis of this information alone, do you think nitrogen is an ideal gas?

5.83 In the metallurgical process of refining nickel, the metal is first combined with carbon monoxide to form tetracarbonylnickel, which is a gas at 43°C:

$$Ni(s) + 4CO(g) \longrightarrow Ni(CO)_4(g)$$

This reaction separates nickel from other solid impurities. (a) Starting with 86.4 g of Ni, calculate the pressure of $Ni(CO)_4$ in a container of volume 4.00 L. (Assume the above reaction to go to completion.) (b) On further heating the sample above 43°C, it is observed that the pressure of the gas increases much more rapidly than that predicted based on the ideal gas equation. Explain.

5.84 The partial pressure of carbon dioxide varies with seasons. Would you expect the partial pressure in the Northern Hemisphere to be higher in the summer or winter? Explain.

5.85 A healthy adult exhales about 5.0×10^2 mL of a gaseous mixture with each breath. Calculate the number of molecules present in this volume at 37°C and 1.1 atm. List the major components of this gaseous mixture.

5.86 Sodium bicarbonate ($NaHCO_3$) is called baking soda because when heated, it releases carbon dioxide gas, which is responsible for the rising of cookies, doughnuts, and bread. (a) Calculate the volume (in liters) of CO_2 produced by heating 5.0 g of $NaHCO_3$ at 180°C and 1.3 atm. (b) Ammonium bicarbonate (NH_4HCO_3) has also been used for the same purpose. Suggest one advantage and one disadvantage of using NH_4HCO_3 instead of $NaHCO_3$ for baking.

5.87 A barometer having a cross-sectional area of 1.00 cm^2 at sea level measures a pressure of 76.0 cm of mercury. The pressure exerted by this column of mercury is equal to the pressure exerted by all the air on 1 cm^2 of Earth's surface. Given that the density of mercury is 13.6 g/mL, and the average radius of Earth is 6371 km, calculate the total mass of Earth's atmosphere in kilograms. (*Hint:* The surface area of a sphere is $4\pi r^2$, where r is the radius of the sphere.)

5.88 Some commercial drain cleaners contain two components: sodium hydroxide and aluminum powder. When the mixture is poured down a clogged drain, the following reaction occurs:

$$2NaOH(aq) + 2Al(s) + 6H_2O(l) \longrightarrow 2NaAl(OH)_4(aq) + 3H_2(g)$$

The heat generated in this reaction helps melt away obstructions such as grease, and the hydrogen gas released stirs up the solids clogging the drain. Calculate the volume of H_2 formed at STP if 3.12 g of Al is treated with an excess of NaOH.

5.89 The volume of a sample of pure HCl gas was 189 mL at 25°C and 108 mmHg. It was completely dissolved in about 60 mL of water and titrated with a NaOH solution; 15.7 mL of the NaOH solution were required to neutralize the HCl. Calculate the molarity of the NaOH solution.

5.90 Propane (C_3H_8) burns in oxygen to produce carbon dioxide gas and water vapor (a) Write a balanced equation for this reaction. (b) Calculate the number of liters of carbon dioxide measured at STP that could be produced from 7.45 g of propane.

5.91 Consider the apparatus shown on the next page. When a small amount of water is introduced into the flask by squeezing the bulb of the medicinal dropper, water is squirted upward out of the long glass tubing. Explain this observation. (*Hint:* Hydrogen chloride gas is soluble in water.)

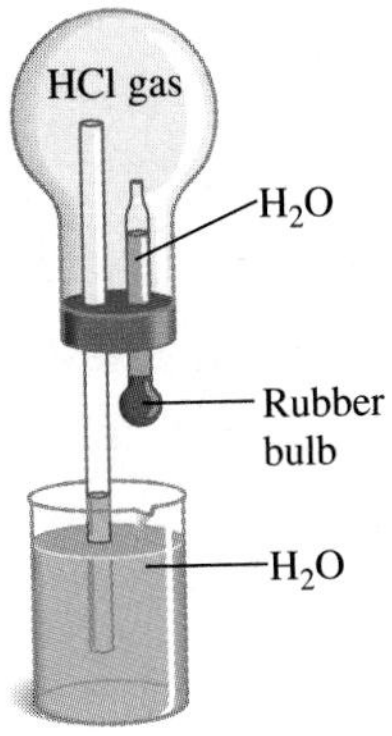

5.92 Nitric oxide (NO) reacts with molecular oxygen as follows:

$$2NO(g) + O_2(g) \longrightarrow 2NO_2(g)$$

Initially NO and O_2 are separated as shown below. When the valve is opened, the reaction quickly goes to completion. Determine what gases remain at the end and calculate their partial pressures. Assume that the temperature remains constant at 25°C.

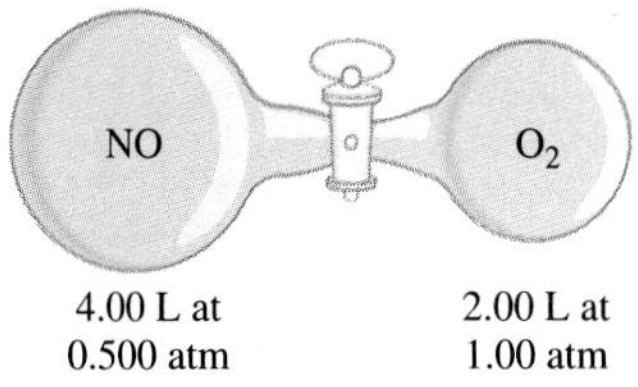

Answers to Practice Exercises: 5.1 0.986 atm. **5.2** 9.29 L. **5.3** 30.6 L. **5.4** 2.6 atm. **5.5** 44.1 g/mol. **5.6** CH_4: 1.29 atm; C_2H_6: 0.0657 atm; C_3H_8: 0.0181 atm. **5.7** 0.0653 g. **5.8** 321 m/s. **5.9** 30.0 atm; 45.5 atm using Equation (5.7).

CHAPTER 6
ENERGY RELATIONSHIPS IN CHEMICAL REACTIONS

The Hindenburg was destroyed in a spectacular fire at Lakehurst, New Jersey, in 1937.

◆ As the giant airship hovered near the landing site, the ground crew scurried about, waiting for the thunderstorms to abate so the dirigible could dock. After two and a half days' journey, the *Hindenburg,* the pride of Germany's trans-Atlantic air fleet, had arrived at Lakehurst, New Jersey, on May 6, 1937. At 7 P.M., as the trial lines were being dropped for descent, the *Hindenburg* suddenly burst into flames. Within minutes, the 800-ft-long craft (about the length of three football fields) was reduced to a tangled mass of charred wreckage. Of the 97 on board, 35 died and many more were injured.

The *Hindenburg* was designed to be used with helium gas, which is chemically unreactive, in order to minimize the threat of an explosion. However, on the eve of World War II, the United States, which has most of the world's supply of helium, would not sell the gas to Germany. And so the *Hindenburg* was filled with hydrogen gas.

What caused the explosion? It is believed that leaking hydrogen, probably from a vent valve, ignited when the airship was struck by lightning. Quickly, the 7 million cubic feet of hydrogen inside the vessel combined with the surrounding air and burned. This chemical reaction violently released the energy stored in molecules of hydrogen. All airships nowadays, such as those you see at football games, are filled with helium, which is nearly as light as hydrogen.

Today hydrogen is used in a different type of air travel—to outer space. Together with oxygen, it makes an excellent rocket propellant. Even though a gallon of gasoline has three times the energy content as a gallon of liquid hydrogen, hydrogen is the fuel of choice for rockets because it weighs ten times less. In space flights, weight consideration is crucial. The large external tank of the space shuttle *Discovery* supplies about 380,000 gallons of liquid hydrogen and 140,000 gallons of liquid oxygen to the three engines that provide the thrust for liftoff. ◆

Liquid hydrogen is the fuel used for space shuttles.

6.1 ENERGY

"Energy" is a much-used term, but it represents a rather abstract concept. For instance, when we feel tired, we might say we haven't any *energy;* and we read about the need to find alternatives to nonrenewable *energy* sources. Unlike matter, energy is known and recognized by its effects. It cannot be seen, touched, smelled, or weighed.

Energy is usually defined as *the capacity to do work.* In Chapter 5 we defined work as "force × distance," but we will soon see that there are other kinds of work. All forms of energy are capable of doing work (that is, of exerting a force over a distance), but not all of them are equally relevant to chemistry. The energy contained in tidal waves, for example, can be harnessed to perform useful work, but the importance of tidal waves to chemistry is minimal. Chemists define ***work*** as *directed energy change resulting from a process.* Kinetic energy—the energy produced by a moving object—is one form of energy that is of particular interest to chemists. Others include radiant energy, thermal energy, chemical energy, and potential energy.

Radiant energy *comes from the sun* (solar energy) and is Earth's primary energy source. Solar energy heats the atmosphere and Earth's surface, stimulates the growth of vegetation through the process known as photosynthesis, and influences global climate patterns.

Thermal energy is *the energy associated with the random motion of atoms and molecules.* In general, thermal energy can be calculated from temperature measurements—the more vigorous the motion of the atoms and molecules in a sample of matter, the hotter the sample is and the greater its thermal energy. However, we need to distinguish carefully between thermal energy and temperature. A cup of coffee at 70°C has a higher temperature than a bathtub filled with warm water at 40°C, but much more thermal energy is stored in the bathtub water because it has a much larger volume and greater mass than the coffee and therefore more water molecules and more molecular motion.

Chemical energy is *stored within the structural units of chemical substances;* its quantity is determined by the type and arrangement of atoms in the substance being considered. When substances participate in chemical reactions, chemical energy is released, stored, or converted to other forms of energy.

Energy is also available by virtue of an object's position. This form of energy is called ***potential energy.*** For instance, because of its altitude, a rock at the top of a cliff has more potential energy and will make a bigger splash in the water below than a similar rock located partway down. Chemical energy can be considered a form of potential energy because it is associated with the relative positions and arrangements of atoms within a substance.

All forms of energy can be interconverted (at least in principle) from one form to another. We feel warm when we stand in sunlight because radiant energy is converted to thermal energy on our skin. When we exercise, chemical energy stored in our bodies is used to produce kinetic energy. When a ball starts to roll downhill, its potential energy is converted to kinetic energy. You can undoubtedly think of many other examples.

Although energy can assume many different forms that are interconvertible, scientists have concluded that energy can be neither destroyed nor created. When one form of energy disappears, some other form of energy (of equal magnitude) must appear, and vice versa. *The total quantity of energy in the universe is thus assumed to remain constant.* This statement is generally known as the ***law of conservation of energy.***

6.2 ENERGY CHANGES IN CHEMICAL REACTIONS

Often the energy changes that take place during chemical reactions are of as much practical interest as the mass relationships we discussed in Chapter 3. For example, combustion reactions involving fuels such as natural gas and oil are carried out in daily life more for the thermal energy they release than for the particular quantities of the products, which are water and carbon dioxide.

Almost all chemical reactions absorb or produce (release) energy, generally in the form of heat. It is important to understand the distinction between thermal energy and heat. ***Heat*** is *the transfer of thermal energy between two bodies that are at different temperatures.* Thus we often speak of the "heat flow" from a hot object to a cold one. Although "heat" itself implies the transfer of energy, we customarily talk of "heat absorbed" or "heat released" when describing the energy changes that occur during a process. *The study of heat changes in chemical reactions* is called ***thermochemistry.***

In order to analyze energy changes associated with chemical reactions we must first define the ***system,*** or *specific part of the universe that is of interest to us.* For chemists, systems usually include substances involved in chemical and physical changes. For example, in an acid-base neutralization experiment, the system may be a beaker containing 50 mL of HCl to which 50 mL of NaOH is added. *The rest of the universe outside the system* is called the ***surroundings.***

There are three types of systems. An ***open system*** *can exchange mass and energy (usually in the form of heat) with its surroundings.* For example, an open system may consist of a quantity of water in an open container, as shown in Figure 6.1(a). If we close the flask, as in Figure 6.1(b), so that no water vapor can escape from or condense into the container, we create a ***closed system,*** which *allows the transfer of energy (heat) but not mass.* By placing the water in a totally insulated container, we construct an ***isolated system,*** which *does not allow the transfer of either mass or energy,* as shown in Figure 6.1(c).

The combustion of acetylene (C_2H_2) gas in oxygen is one of many familiar chemical reactions that release considerable quantities of energy (Figure 6.2):

$$2C_2H_2(g) + 5O_2(g) \longrightarrow 4CO_2(g) + 2H_2O(l) + \text{energy}$$

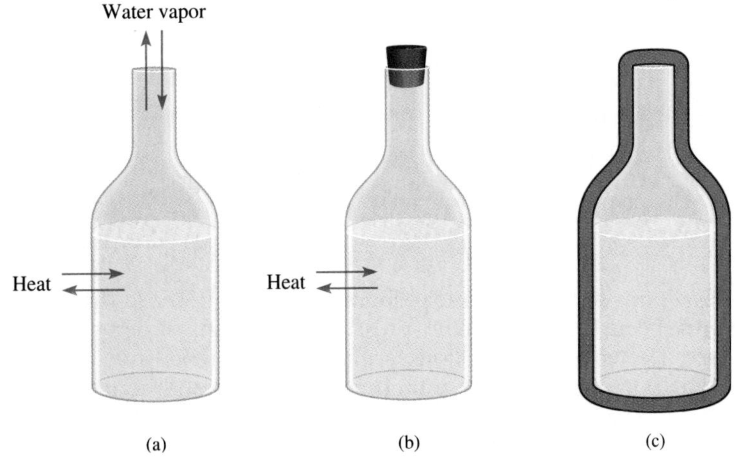

FIGURE 6.1

Three systems represented by water in a flask: (a) an open system, which allows both energy and mass transfer; (b) a closed system, which allows energy but not mass transfer; and (c) an isolated system, which allows neither energy nor mass transfer (here the flask is enclosed by a vacuum jacket).

FIGURE 6.2

An oxyacetylene torch used to weld metals.

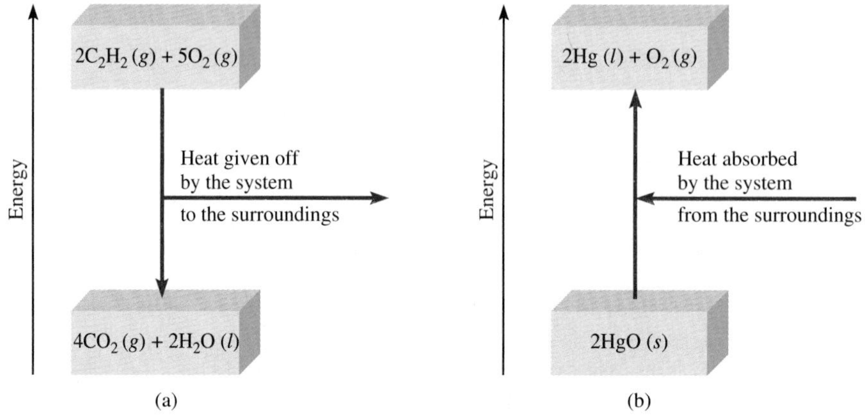

FIGURE 6.3

(a) Energy-level diagram for an exothermic reaction: the combustion of acetylene gas. (b) Energy-level diagram for an endothermic reaction: the decomposition of mercury(II) oxide. The scales in (a) and (b) are not the same. Therefore, the heat released in the combustion of acetylene is not equal to the heat absorbed in the decomposition of mercury(II) oxide.

In this case we label the reacting mixture (acetylene, oxygen, carbon dioxide, and water) the *system* and the rest of the universe the *surroundings.* Since energy cannot be created or destroyed, any energy lost by the system must be gained by the surroundings. Thus the heat generated by the combustion process is transferred from the system to its surroundings. *Any process that gives off heat* (that is, *transfers thermal energy to the surroundings*) is called an ***exothermic process.*** Figure 6.3(a) shows the energy change for the combustion of acetylene gas.

Now consider another reaction, the decomposition of mercury(II) oxide (HgO) at high temperatures:

$$\text{energy} + 2\text{HgO}(s) \longrightarrow 2\text{Hg}(l) + \text{O}_2(g)$$

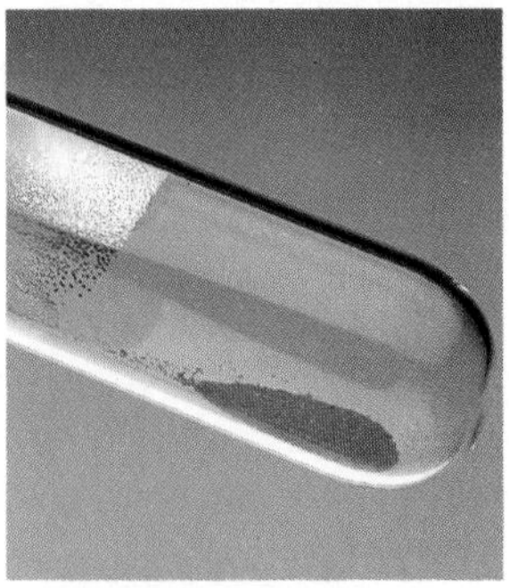

On heating, HgO decomposes to give Hg and O_2.

This is an example of an ***endothermic process,*** *in which heat has to be supplied to the system* (that is, to HgO) *by the surroundings* [Figure 6.3(b)].

From Figure 6.3 you can see that in exothermic reactions the total energy of the products is less than the total energy of the reactants. The difference in the energies is the heat supplied by the system to the surroundings. Just the opposite happens in endothermic reactions. Here, the difference in the energies of the products and reactants is equal to the heat supplied to the system by the surroundings.

6.3 ENTHALPY

Most physical and chemical changes, including those that take place in living systems, occur in the constant-pressure conditions of our atmosphere. In the laboratory, for example, reactions are generally carried out in beakers, flasks, or test tubes that remain open to their surroundings and hence to a pressure of approximately one atmosphere (1 atm). Under these conditions a system absorbs or releases heat during a process. To quantify the heat flow into or out of a system in a constant-pressure process, chemists use a quantity called ***enthalpy,*** represented by the symbol H and defined as $E + PV$. However, it is the *change* in enthalpy, ΔH, that we actually measure. (The Greek letter *delta,* Δ, symbolizes change.) The ***enthalpy of reaction, ΔH,*** is *the difference between the enthalpies of the products and the enthalpies of the reactants:*

$$\Delta H = H(\text{products}) - H(\text{reactants}) \tag{6.1}$$

In other words, ΔH is equal to the heat given off or absorbed.

The enthalpy of reaction can be positive or negative, depending on the process. For an endothermic process (heat absorbed by the system from the surroundings), ΔH is positive (that is, $\Delta H > 0$). For an exothermic process (heat released by the system to the surroundings), ΔH is negative (that is, $\Delta H < 0$). Now let us apply the idea of enthalpy changes to two common processes—the first involving a physical change, the second a chemical change.

Thermochemical Equations

A 0°C and a pressure of 1 atm, ice melts to form liquid water. Measurements show that for every mole of ice converted to liquid water under these conditions, 6.01 kilojoules (kJ) of energy are absorbed by the system (ice). Since ΔH is a positive value, this is an endothermic process, as expected for an energy-absorbing change of melting ice (Figure 6.4). We can write the equation for this physical change as

$$H_2O(s) \longrightarrow H_2O(l) \qquad \Delta H = 6.01 \text{ kJ}$$

As another example, consider the combustion of methane (CH_4), the principal component of natural gas:

$$CH_4(g) + 2O_2(g) \longrightarrow CO_2(g) + 2H_2O(l) \qquad \Delta H = -890.4 \text{ kJ}$$

From experience we know that burning natural gas releases heat to its surroundings, so it is an exothermic process and ΔH must have a negative value.

The equations representing the melting of ice and the combustion of methane not only represent the mass relationships involved but also show the enthalpy changes. *Equations showing both the mass and enthalpy relations* are called ***thermochemical equations.*** The following guidelines are helpful in writing and interpreting thermochemical equations:

Methane gas burning from a Bunsen burner.

- The stoichiometric coefficients always refer to the number of moles of each substance. Thus the equation representing the melting of ice may be "read" as follows: When 1 mole of liquid water is formed from 1 mole of ice at 0°C, the

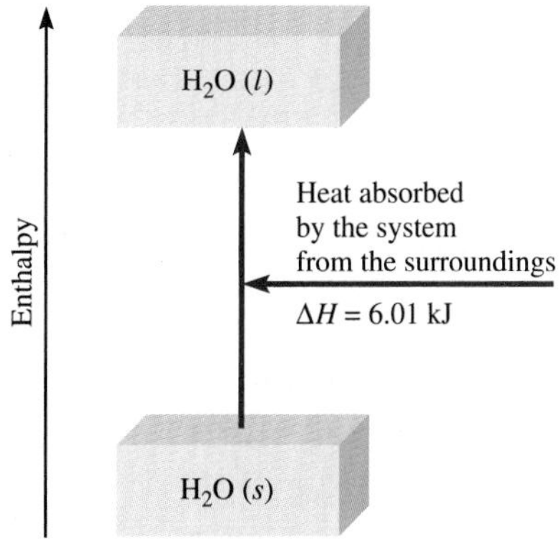

FIGURE 6.4

Melting 1 mole of ice at 0°C (an endothermic process) results in an enthalpy increase in the system of 6.01 kJ.

enthalpy change is 6.01 kJ. For the combustion of methane, we interpret the equation this way: When 1 mole of gaseous methane reacts with 2 moles of gaseous oxygen to form 1 mole of gaseous carbon dioxide and 2 moles of liquid water, the enthalpy change is −890.4 kJ.

- When we reverse an equation, we change the roles of reactants and products. Consequently, the magnitude of ΔH for the equation remains the same, but its sign changes. This is reasonable if you consider the processes involved. For example, if a reaction consumes thermal energy from its surroundings (that is, if it is endothermic), then the reverse of that reaction must release thermal energy back to its surroundings (that is, it must be exothermic). This means that the enthalpy change expression must also change its sign. Thus, when we reverse the melting of ice and the combustion of methane, the thermochemical equations are

$$H_2O(l) \longrightarrow H_2O(s) \qquad \Delta H = -6.01 \text{ kJ}$$

$$CO_2(g) + 2H_2O(l) \longrightarrow CH_4(g) + 2O_2(g) \qquad \Delta H = 890.4 \text{ kJ}$$

and what was an endothermic process now becomes exothermic, and vice versa.

- If we multiply both sides of a thermochemical equation by a factor n, then ΔH must also change by the same factor. Thus, for the melting of ice, if $n = 2$, then

$$2H_2O(s) \longrightarrow 2H_2O(l) \qquad \Delta H = 2(6.01 \text{ kJ}) = 12.0 \text{ kJ}$$

- When writing thermochemical equations, we must always specify the physical states of all reactants and products, because they help determine the actual enthalpy changes. For example, in the equation for the combustion of methane, if we show water vapor rather than liquid water as a product,

$$CH_4(g) + 2O_2(g) \longrightarrow CO_2(g) + 2H_2O(g)$$

the enthalpy change is −802.4 kJ rather than −890.4 kJ because 88.0 kJ of energy are needed to convert 2 moles of liquid water to water vapor; that is,

$$2H_2O(l) \longrightarrow 2H_2O(g) \qquad \Delta H = 88.0 \text{ kJ}$$

6.4 CALORIMETRY

In the laboratory the heat changes in physical and chemical processes are measured with a *calorimeter.* Our discussion of ***calorimetry***—that is, *the measurement of heat changes*—will depend on an understanding of specific heat and heat capacity, so let us consider them first.

SPECIFIC HEAT AND HEAT CAPACITY

The ***specific heat (s)*** of a substance is *the amount of heat required to raise the temperature of one gram of the substance by one degree Celsius.* The **heat capacity (*C*)** of a substance is *the amount of heat required to raise the temperature of a given quantity of the substance by one degree Celsius.* The relationship between the heat capacity and specific heat of a substance is

$$C = ms \tag{6.2}$$

where m is the mass of the substance in grams. For example, the specific heat of water is 4.184 J/g · °C, and the heat capacity of 60.0 g of water is

$$(60.0\ \text{g})(4.184\ \text{J/g}\cdot{}^\circ\text{C}) = 251\ \text{J/}^\circ\text{C}$$

Note that specific heat has the units J/g · °C and heat capacity has the units J/°C. Table 6.1 shows the specific heats of some common substances.

If we know the specific heat and the amount of a substance, then the change in the sample's temperature (Δt) will tell us the amount of heat (q) that has been absorbed or released in a particular process. The equation for calculating the heat change is given by

$$q = ms\,\Delta t \quad (6.3)$$

$$q = C\,\Delta t \quad (6.4)$$

where m is the mass of the sample and Δt is the temperature change:

$$\Delta t = t_{\text{final}} - t_{\text{initial}}$$

The sign convention for q is the same as that for enthalpy change; q is positive for endothermic processes and negative for exothermic processes.

TABLE 6.1
The Specific Heats (s) of Some Common Substances

Substance	s (J/g · °C)
Al	0.900
Au	0.129
C(graphite)	0.720
C(diamond)	0.502
Cu	0.385
Fe	0.444
Hg	0.139
H_2O	4.184
C_2H_5OH (ethanol)	2.46

EXAMPLE 6.1
Calculating Heat Absorbed Using Specific Heat Data

A 466-g sample of water is heated from 8.50°C to 74.60°C. Calculate the amount of heat (in kilojoules) absorbed by the water.

Answer: Using Equation (6.3), we write

$$\begin{aligned} q &= ms\,\Delta t \\ &= (466\ \text{g})(4.184\ \text{J/g}\cdot{}^\circ\text{C})(74.60^\circ\text{C} - 8.50^\circ\text{C}) \\ &= 1.29 \times 10^5\ \text{J} \\ &= 129\ \text{kJ} \end{aligned}$$

PRACTICE EXERCISE

An iron bar of mass 869 g cools from 94°C to 5°C. Calculate the heat released (in kilojoules) by the metal.

Constant-Volume Calorimetry

Heat of combustion is usually measured by placing a known mass of the compound under study in a steel container, called a *constant-volume bomb calorimeter*, which is filled with oxygen at about 30 atm of pressure. The closed bomb is immersed in a known amount of water, as shown in Figure 6.5. The sample is ignited electrically, and the heat produced by the combustion reaction can be calculated accurately by recording the rise in temperature of the water. The heat given off by the sample is absorbed by the water and the calorimeter. The special design of the bomb calorimeter allows us to safely assume that no heat (or mass) is lost to the surroundings during the time it takes to make measurements. Therefore we can

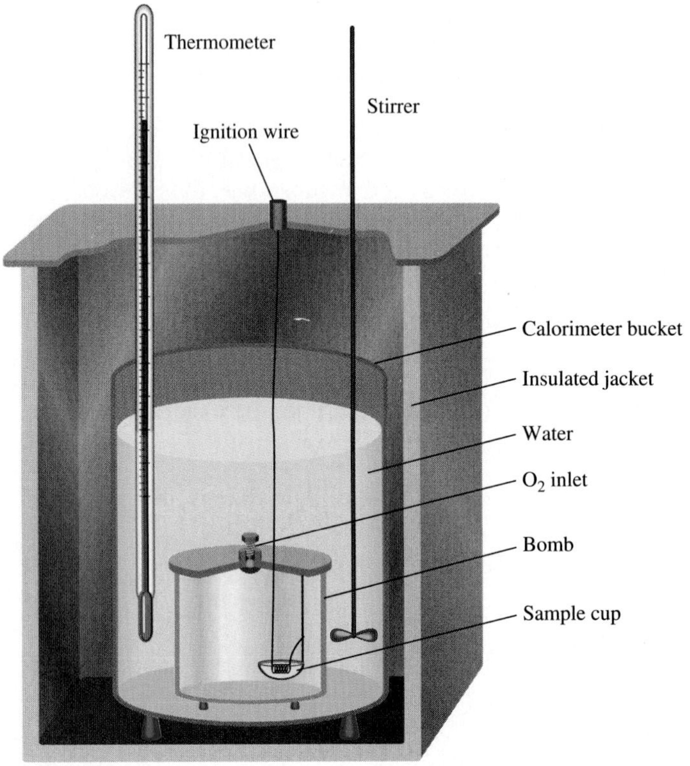

FIGURE 6.5

A constant-volume bomb calorimeter. The calorimeter is filled with oxygen gas before it is placed in the bucket. The sample is ignited electrically, and the heat produced by the reaction can be accurately determined by measuring the temperature increase in the known amount of surrounding water.

call the bomb calorimeter and the water in which it is submerged an isolated system. Because no heat enters or leaves the system throughout the process, we can write

$$q_{\text{system}} = q_{\text{water}} + q_{\text{bomb}} + q_{\text{rxn}} \quad (6.5)$$
$$= 0$$

where q_{water}, q_{bomb}, and q_{rxn} are the heat changes for the water, the bomb, and the reaction, respectively. Thus

$$q_{\text{rxn}} = -(q_{\text{water}} + q_{\text{bomb}}) \quad (6.6)$$

The quantity q_{water} is obtained by

$$q = ms\,\Delta t$$
$$q_{\text{water}} = (m_{\text{water}})(4.184\ \text{J/g}\cdot{}^\circ\text{C})\,\Delta t$$

The product of the mass of the bomb and its specific heat is the heat capacity of the bomb, which remains constant for all experiments carried out in the bomb calorimeter:

$$C_{\text{bomb}} = m_{\text{bomb}} \times s_{\text{bomb}}$$

Hence

$$q_{\text{bomb}} = C_{\text{bomb}}\,\Delta t$$

Note that because reactions in a bomb calorimeter occur under constant-volume rather than constant-pressure conditions, the heat changes do not correspond to the enthalpy change ΔH (see Section 6.3). It is possible to correct the measured heat changes so that they correspond to ΔH values, but the corrections usually are quite small, so we will not concern ourselves with the details of the correction procedure. Finally, it is interesting to note that the energy contents of food and fuel (usually expressed in calories where 1 cal = 4.184 J) are measured with constant-volume calorimeters.

EXAMPLE 6.2
Heat of Combustion from a Constant-Volume Calorimeter

A quantity of 1.435 g of naphthalene ($C_{10}H_8$), a pungent-smelling substance used in moth repellants, was burned in a constant-volume bomb calorimeter. Consequently, the temperature of the water rose from 20.17°C to 25.84°C. If the quantity of water surrounding the calorimeter was exactly 2000 g and the heat capacity of the bomb calorimeter was 1.80 kJ/°C, calculate the heat of combustion of naphthalene on a molar basis; that is, find the molar heat of combustion.

Answer: First we calculate the heat changes for the water and the bomb calorimeter.

$$q = ms\,\Delta t$$

$$q_{\text{water}} = (2000\text{ g})(4.184\text{ J/g}\cdot{}^\circ\text{C})(25.84^\circ\text{C} - 20.17^\circ\text{C})$$

$$= 4.74 \times 10^4\text{ J}$$

$$q_{\text{bomb}} = (1.80 \times 10^3\text{ J/}^\circ\text{C})(25.84^\circ\text{C} - 20.17^\circ\text{C})$$

$$= 1.02 \times 10^4\text{ J}$$

(Note that we changed 1.80 kJ/°C to 1.80×10^3 J/°C.) Next, from Equation (6.6) we write

$$q_{\text{rxn}} = -(4.74 \times 10^4\text{ J} + 1.02 \times 10^4\text{ J})$$

$$= -5.76 \times 10^4\text{ J}$$

The molar mass of naphthalene is 128.2 g, so the heat of combustion of 1 mole of naphthalene is

$$\text{molar heat of combustion} = \frac{-5.76 \times 10^4\text{ J}}{1.435\ \cancel{\text{g C}_{10}\text{H}_8}} \times \frac{128.2\ \cancel{\text{g C}_{10}\text{H}_8}}{1\text{ mol C}_{10}\text{H}_8}$$

$$= -5.15 \times 10^6\text{ J/mol}$$

$$= -5.15 \times 10^3\text{ kJ/mol}$$

PRACTICE EXERCISE

A quantity of 1.922 g of methanol (CH_3OH) was burned in a constant-volume bomb calorimeter. Consequently, the temperature of the water rose by 4.20°C. If the quantity of water surrounding the calorimeter was exactly 2000 g and the heat capacity of the calorimeter was 2.02 kJ/°C, calculate the molar heat of combustion of methanol.

Finally, we note that the heat capacity of a bomb calorimeter is usually determined by burning in it a compound with an accurately known heat of combustion

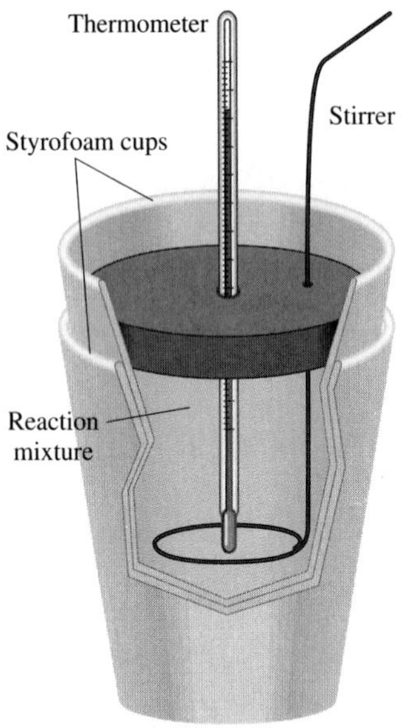

FIGURE 6.6

A constant-pressure calorimeter made of two Styrofoam coffee cups. The outer cup helps to insulate the reacting mixture from the surroundings. Two solutions of known volume containing the reactants at the same temperature are carefully mixed in the calorimeter. The heat produced or absorbed by the reaction can be determined by measuring the temperature change.

TABLE 6.2
Heats of Some Typical Reactions Measured at Constant Pressure

Type of reaction	Example	ΔH (kJ)
Heat of neutralization	$HCl(aq) + NaOH(aq) \rightarrow NaCl(aq) + H_2O(l)$	−56.2
Heat of ionization	$H_2O(l) \rightarrow H^+(aq) + OH^-(aq)$	56.2
Heat of fusion	$H_2O(s) \rightarrow H_2O(l)$	6.01
Heat of vaporization	$H_2O(l) \rightarrow H_2O(g)$	44.0*
Heat of reaction	$MgCl_2(s) + 2Na(l) \rightarrow 2NaCl(s) + Mg(s)$	−180.2

*Measured at 25°C. At 100°C, the value is 40.79 kJ.

value. From the mass of the compound and the temperature rise, we can calculate the heat capacity of the calorimeter (see Problem 6.91).

Constant-Pressure Calorimetry

A simpler device than the constant-volume calorimeter is the constant-pressure calorimeter, which is used to determine the heat changes for noncombustion reactions. An operable constant-pressure calorimeter can be constructed from two Styrofoam coffee cups, as shown in Figure 6.6. This device measures the heat effects of a variety of reactions, such as acid-base neutralization, as well as the heat of solution and heat of dilution. Because the measurements are carried out under constant atmospheric pressure conditions, the heat change for the process (q_{rxn}) is equal to the enthalpy change (ΔH). The measurements are similar to those of a constant-volume calorimeter—we need to know the heat capacity of the calorimeter, as well as the temperature change of the solution. Table 6.2 lists some reactions that have been studied with the constant-pressure calorimeter.

EXAMPLE 6.3
Measuring Heat of Neutralization with a Constant-Pressure Calorimeter

A quantity of 1.00×10^2 mL of 0.500 *M* HCl is mixed with 1.00×10^2 mL of 0.500 *M* NaOH in a constant-pressure calorimeter having a heat capacity of 335 J/°C. The initial temperature of the HCl and NaOH solutions is the same, 22.50°C, and the final temperature of the mixed solution is 24.90°C. Calculate the heat change for the neutralization reaction

$$NaOH(aq) + HCl(aq) \longrightarrow NaCl(aq) + H_2O(l)$$

Assume that the densities and specific heats of the solutions are the same as for water (1.00 g/mL and 4.184 J/g · °C, respectively).

Answer: Assuming that no heat is lost to the surroundings, we write

$$q_{system} = q_{soln} + q_{calorimeter} + q_{rxn} = 0$$

or

$$q_{rxn} = -(q_{soln} + q_{calorimeter})$$

where

$$q_{rxn} = (1.00 \times 10^2 \text{ g} + 1.00 \times 10^2 \text{ g})(4.184 \text{ J/g} \cdot {}^\circ\text{C})(24.90^\circ\text{C} - 22.50^\circ\text{C})$$
$$= 2.01 \times 10^3 \text{ J}$$
$$q_{calorimeter} = (335 \text{ J/}^\circ\text{C})(24.90^\circ\text{C} - 22.50^\circ\text{C})$$
$$= 804 \text{ J}$$

Note that because the density of the solution is 1.00 g/mL, the mass of a 100-mL solution is 100 g. Next we write

$$q_{rxn} = -(2.01 \times 10^3 \text{ J} + 804 \text{ J})$$
$$= -2.81 \times 10^3 \text{ J}$$
$$= -2.81 \text{ kJ}$$

From the molarities given, we know there is 0.0500 mole of HCl in 1.00×10^2 g of the HCl solution and 0.0500 mole of NaOH in 1.00×10^2 g of the NaOH solution. Therefore the heat of neutralization when 1.00 mole of HCl reacts with 1.00 mole of NaOH is

$$\text{heat of neutralization} = \frac{-2.81 \text{ kJ}}{0.0500 \text{ mol}} = -56.2 \text{ kJ/mol}$$

Because the reaction takes place at constant pressure, the heat given off is equal to the enthalpy change.

PRACTICE EXERCISE

A quantity of 4.00×10^2 mL of 0.600 *M* HNO_3 is mixed with 4.00×10^2 mL of 0.300 *M* $Ba(OH)_2$ in a constant-pressure calorimeter having a heat capacity of 387 J/°C. The initial temperature of both solutions is 18.88°C. What is the final temperature of the solution? (Use the result in Example 6.3 for your calculation.)

6.5 STANDARD ENTHALPY OF FORMATION AND REACTION

So far we have learned that we can determine the enthalpy change that accompanies a reaction by measuring the heat absorbed or released (at constant pressure). From Equation (6.1) we see that ΔH can also be calculated if we know the actual enthalpies of all reactants and products. However, there is no way to measure the *absolute* value of the enthalpy of a substance. Only values *relative* to an arbitrary reference can be determined. This problem is similar to the one geographers face in expressing the elevations of specific mountains or valleys. Rather than trying to devise some type of "absolute" elevation scale (perhaps based on distance from the center of Earth?), by common agreement all geographic heights and depths are expressed relative to sea level, an arbitrary reference with a defined elevation of "zero" meters or feet. Similarly, chemists have agreed on an arbitrary reference point with which all enthalpies are compared.

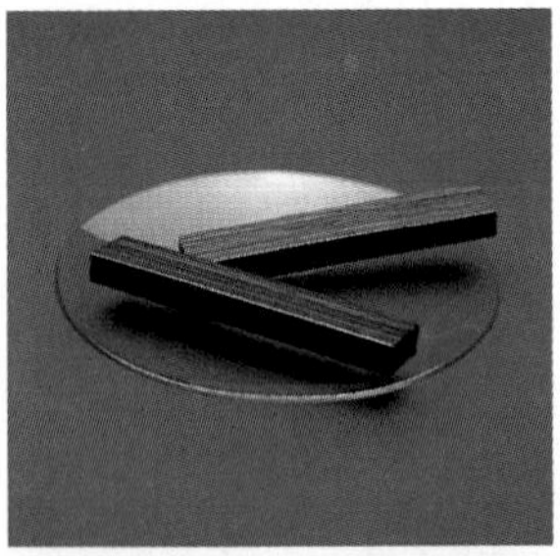

Graphite (top) and diamond (bottom): allotropes of carbon having different values of ΔH_f°.

The "sea level" reference point for all enthalpy expressions is called the ***standard enthalpy of formation (ΔH_f°).*** It is defined as *the heat change that results when one mole of a compound is formed from its elements at a pressure of 1 atm.* At *1 atm,* elements are said to be in the ***standard state,*** hence the term "standard enthalpy." The superscript "o" indicates that the measurement was carried out under standard-state conditions (1 atm), and the subscript "*f*" stands for formation. Although the standard state does not specify a temperature, we will always use ΔH_f° values measured at 25°C.

Table 6.3 lists the standard enthalpies of formation for a number of elements and compounds. (A more complete list of ΔH_f° values is given in Appendix 2.) By convention, *the standard enthalpy of formation of any element in its most stable form is zero.* Take oxygen as an example. Molecular oxygen (O_2) is more stable than the other allotropic form of oxygen, ozone (O_3), at 1 atm and 25°C. Thus we can write $\Delta H_f^\circ (O_2) = 0$, but $\Delta H_f^\circ (O_3) \neq 0$. Similarly, graphite is a more stable allotropic form of carbon than diamond at 1 atm and 25°C, so ΔH_f° (C, graphite) = 0 and ΔH_f° (C, diamond) $\neq$ 0.

The importance of the standard enthalpies of formation is that once we know their values, we can *calculate* the enthalpy of reaction. For example, consider the hypothetical reaction

$$a\text{A} + b\text{B} \longrightarrow c\text{C} + d\text{D}$$

where a, b, c, and d are stoichiometric coefficients. *The enthalpy of a reaction car-*

TABLE 6.3

Standard Enthalpies of Formation of Some Inorganic Substances at 25°C

Substance	ΔH_f° (kJ/mol)	Substance	ΔH_f° (kJ/mol)
$Ag(s)$	0	$H_2O_2(l)$	−187.6
$AgCl(s)$	−127.04	$Hg(l)$	0
$Al(s)$	0	$I_2(s)$	0
$Al_2O_3(s)$	−1669.8	$HI(g)$	25.94
$Br_2(l)$	0	$Mg(s)$	0
$HBr(g)$	−36.2	$MgO(s)$	−601.8
C(graphite)	0	$MgCO_3(s)$	−1112.9
C(diamond)	1.90	$N_2(g)$	0
$CO(g)$	−110.5	$NH_3(g)$	−46.3
$CO_2(g)$	−393.5	$NO(g)$	90.4
$Ca(s)$	0	$NO_2(g)$	33.85
$CaO(s)$	−635.6	$N_2O_4(g)$	9.66
$CaCO_3(s)$	−1206.9	$N_2O(g)$	81.56
$Cl_2(g)$	0	$O(g)$	249.4
$HCl(g)$	−92.3	$O_2(g)$	0
$Cu(s)$	0	$O_3(g)$	142.2
$CuO(s)$	−155.2	S(rhombic)	0
$F_2(g)$	0	S(monoclinic)	0.30
$HF(g)$	−268.61	$SO_2(g)$	−296.1
$H(g)$	218.2	$SO_3(g)$	−395.2
$H_2(g)$	0	$H_2S(g)$	−20.15
$H_2O(g)$	−241.8	$ZnO(s)$	−347.98
$H_2O(l)$	−285.8		

ried out under standard-state conditions, called the ***standard enthalpy of reaction,*** ΔH°_{rxn}, is given by

$$\Delta H^\circ_{rxn} = [c\,\Delta H^\circ_f\,(C) + d\,\Delta H^\circ_f\,(D)] - [a\,\Delta H^\circ_f\,(A) + b\,\Delta H^\circ_f\,(B)] \quad (6.7)$$

where a, b, c, and d all have the unit mole. We can generalize Equation (6.7) as follows:

$$\Delta H^\circ_{rxn} = \Sigma n\,\Delta H^\circ_f\,(\text{products}) - \Sigma m\,\Delta H^\circ_f\,(\text{reactants}) \quad (6.8)$$

where m and n denote the stoichiometric coefficients for the reactants and products, and Σ (sigma) means "the sum of."

Equation (6.7) shows how we can calculate ΔH°_{rxn} from the ΔH°_f values of compounds. To measure ΔH°_f for compounds we can use the direct method or the indirect method.

THE DIRECT METHOD

This method of measuring ΔH°_f applies to compounds that can be readily synthesized from their elements. Suppose we want to know the enthalpy of formation of carbon dioxide. We must measure the enthalpy of the reaction when carbon (graphite) and molecular oxygen in their standard states are converted to carbon dioxide in its standard state:

$$\text{C(graphite)} + O_2(g) \longrightarrow CO_2(g) \qquad \Delta H^\circ_{rxn} = -393.5\text{ kJ}$$

As we know from experience, this combustion reaction easily goes to completion. From Equation (6.7) we can write

$$\Delta H^\circ_{rxn} = (1\text{ mol})\,\Delta H^\circ_f\,(CO_2, g) - [(1\text{ mol})\,\Delta H^\circ_f\,(\text{C, graphite}) + (1\text{ mol})\,\Delta H^\circ_f\,(O_2, g)]$$
$$= -393.5\text{ kJ}$$

Since both graphite and O_2 are stable allotropic forms, it follows that ΔH°_f (C, graphite) and ΔH°_f (O_2, g) are zero. Therefore

$$\Delta H^\circ_{rxn} = (1\text{ mol})\,\Delta H^\circ_f\,(CO_2, g) = -393.5\text{ kJ}$$

or

$$\Delta H^\circ_f\,(CO_2, g) = -393.5\text{ kJ/mol}$$

Note that arbitrarily assigning zero ΔH°_f for each element in its most stable form at the standard state does not affect our calculations in any way. Remember, in thermochemistry we are interested only in enthalpy *changes,* because they can be determined experimentally whereas the absolute enthalpy values cannot. The choice of a zero "reference level" for enthalpy makes calculations easier to handle. Again referring to the terrestrial altitude analogy, we find that Mt. Everest is 8708 ft higher than Mt. McKinley. This difference in altitude is unaffected by the decision to set sea level at 0 ft or at 1000 ft.

Other compounds that can be studied by the direct method are SF_6, P_4O_{10}, and CS_2. The equations representing their syntheses are

$$\text{S(rhombic)} + 3F_2(g) \longrightarrow SF_6(g)$$
$$4\text{P(white)} + 5O_2(g) \longrightarrow P_4O_{10}(s)$$
$$\text{C(graphite)} + 2\text{S(rhombic)} \longrightarrow CS_2(l)$$

Note that S(rhombic) and P(white) are the most stable allotropes of sulfur and phosphorus, respectively, at 1 atm and 25°C, so their ΔH°_f values are zero.

White phosphorus burns in air to form P_4O_{10}.

THE INDIRECT METHOD

Many compounds cannot be directly synthesized from their elements. In some cases, the reaction proceeds too slowly or side reactions produce substances other than the desired compound. In these cases ΔH_f° can be determined by an indirect approach, which is based on the law of heat summation formulated by Henri Hess, a Swiss chemist known as the founder of thermochemistry. ***Hess's law*** can be stated as follows: *When reactants are converted to products, the change in enthalpy is the same whether the reaction takes place in one step or in a series of steps.* In other words, if we can break down the reaction of interest into a series of reactions for which ΔH_{rxn}° can be measured, we can calculate ΔH_{rxn}° for the overall reaction.

Suppose we are interested in the standard enthalpy of formation of methane (CH_4). We might represent the synthesis of CH_4 from its elements as

$$C(\text{graphite}) + 2H_2(g) \longrightarrow CH_4(g)$$

However, this reaction does not take place as written, so we cannot measure the enthalpy change directly. We must employ an indirect route, using Hess's law. To begin with, the following reactions involving C, H_2, and CH_4 with O_2 have all been studied and the ΔH_{rxn}° values accurately determined:

$$\text{(a)} \quad C(\text{graphite}) + O_2(g) \longrightarrow CO_2(g) \qquad \Delta H_{rxn}^\circ = -393.5 \text{ kJ}$$

$$\text{(b)} \quad 2H_2(g) + O_2(g) \longrightarrow 2H_2O(l) \qquad \Delta H_{rxn}^\circ = -571.6 \text{ kJ}$$

$$\text{(c)} \quad CH_4(g) + 2O_2(g) \longrightarrow CO_2(g) + 2H_2O(l) \qquad \Delta H_{rxn}^\circ = -890.4 \text{ kJ}$$

Since we want to obtain one equation containing only C and H_2 as reactants and CH_4 as product, we must reverse (c) to get

$$\text{(d)} \quad CO_2(g) + 2H_2O(l) \longrightarrow CH_4(g) + 2O_2(g) \qquad \Delta H_{rxn}^\circ = 890.4 \text{ kJ}$$

The next step is to add Equations (a), (b), and (d):

$$\begin{array}{lrcll} & C(\text{graphite}) + \cancel{O_2(g)} & \longrightarrow & \cancel{CO_2(g)} & \Delta H_{rxn}^\circ = -393.5 \text{ kJ} \\ + & 2H_2(g) + \cancel{O_2(g)} & \longrightarrow & \cancel{2H_2O(l)} & \Delta H_{rxn}^\circ = -571.6 \text{ kJ} \\ + & \cancel{CO_2(g)} + \cancel{2H_2O(l)} & \longrightarrow & CH_4(g) + \cancel{2O_2(g)} & \Delta H_{rxn}^\circ = 890.4 \text{ kJ} \\ \hline & C(\text{graphite}) + 2H_2(g) & \longrightarrow & CH_4(g) & \Delta H_{rxn}^\circ = -74.7 \text{ kJ} \end{array}$$

As you can see, all the nonessential species (O_2, CO_2, and H_2O) cancel in this operation. Because the above equation represents the synthesis of 1 mole of CH_4 from its elements, we have $\Delta H_f^\circ(CH_4) = -74.7$ kJ/mol.

The general rule in applying Hess's law is to arrange a series of chemical equations (corresponding to a series of steps) in such a way that, when added together, all species will cancel except for the reactants and products that appear in the overall reaction. To achieve this, we often need to multiply some or all of the equations representing the individual steps by the appropriate coefficients.

EXAMPLE 6.4

Applying Hess's Law to Determine the ΔH_f° of a Compound

From the following equations and the enthalpy changes,

$$\text{(a)} \quad C(\text{graphite}) + O_2(g) \longrightarrow CO_2(g) \qquad \Delta H_{rxn}^\circ = -393.5 \text{ kJ}$$

$$\text{(b)} \quad H_2(g) + \tfrac{1}{2}O_2(g) \longrightarrow H_2O(l) \qquad \Delta H_{rxn}^\circ = -285.8 \text{ kJ}$$

TABLE 7.2
Relation between Quantum Numbers and Atomic Orbitals

n	l	m_l	Number of orbitals	Atomic orbital designations
1	0	0	1	$1s$
2	0	0	1	$2s$
	1	$-1, 0, 1$	3	$2p_x, 2p_y, 2p_z$
3	0	0	1	$3s$
	1	$-1, 0, 1$	3	$3p_x, 3p_y, 3p_z$
	2	$-2, -1, 0, 1, 2$	5	$3d_{xy}, 3d_{yz}, 3d_{xz}, 3d_{x^2-y^2}, 3d_{z^2}$
⋮	⋮	⋮	⋮	⋮

s ORBITALS

One of the important questions we ask when studying the properties of atomic orbitals is, What are the shapes of the orbitals? Strictly speaking, an orbital does not have a well-defined shape because the wave function characterizing the orbital extends from the nucleus to infinity. In that sense it is difficult to say what an orbital looks like. On the other hand, it is certainly convenient to think of orbitals in terms of specific shapes, particularly in discussing the formation of chemical bonds between atoms, as we will do in Chapters 9 and 10.

Although in principle an electron can be found anywhere, we know that most of the time it is quite close to the nucleus. Figure 7.15(a) shows a plot of the electron density in a hydrogen atomic 1*s* orbital versus the distance from the nucleus. As you can see, the electron density falls off rapidly as the distance from the nucleus increases. Roughly speaking, there is about a 90 percent probability of finding the electron within a sphere of radius 100 pm (1 pm $= 1 \times 10^{-12}$ m) surrounding the nucleus. Thus we can represent the 1*s* orbital by drawing a ***boundary surface diagram*** that *encloses about 90 percent of the total electron density in an orbital,* as shown in Figure 7.15(b). A 1*s* orbital represented in this manner is merely a sphere.

Figure 7.16 shows boundary surface diagrams for the 1*s*, 2*s*, and 3*s* hydrogen atomic orbitals. All *s* orbitals are spherical in shape but differ in size, which in-

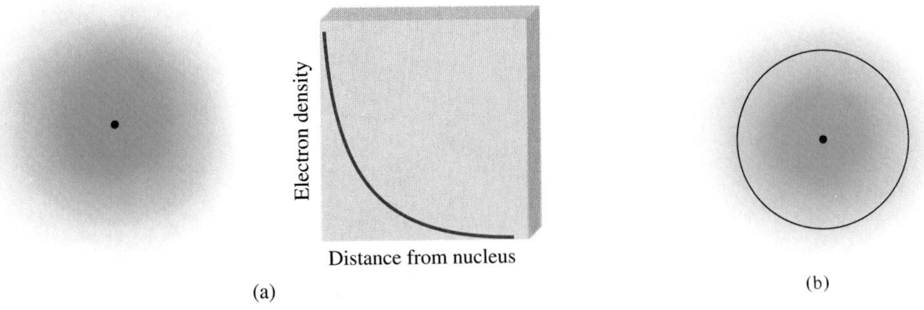

FIGURE 7.15

(a) Plot of electron density in the hydrogen 1s orbital as a function of the distance from the nucleus. The electron density falls off rapidly as the distance from the nucleus increases. (b) Boundary surface diagram of the hydrogen 1s orbital.

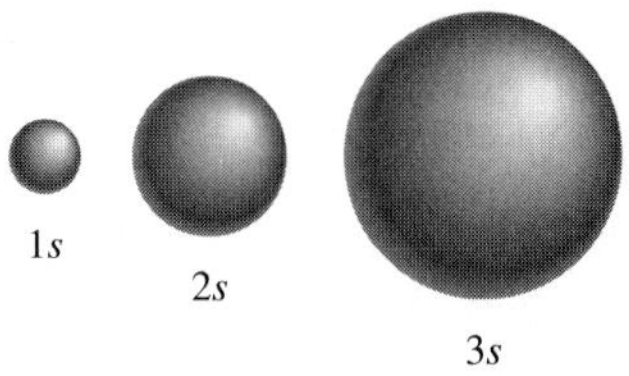

FIGURE 7.16

Boundary surface diagrams of the hydrogen 1s, 2s, and 3s orbitals. Each sphere contains about 90 percent of the total electron density. All s orbitals are spherical. Roughly speaking, the size of an orbital is proportional to n^2, where n is the principal quantum number.

creases as the principal quantum number increases. Although the details of electron density variation within each boundary surface are lost, there is no serious disadvantage. For us the most important features of atomic orbitals are their shapes and *relative* sizes, which are adequately represented by boundary surface diagrams.

p ORBITALS

It should be clear that the *p* orbitals start with the principal quantum number $n = 2$. If $n = 1$, then the angular momentum quantum number l can assume only the value of zero; therefore, there is only a 1*s* orbital. As we saw earlier, when $l = 1$, the magnetic quantum number m_l can have values of $-1, 0, 1$. Starting with $n = 2$ and $l = 1$, we therefore have three 2*p* orbitals: $2p_x$, $2p_y$, and $2p_z$ (Figure 7.17). The letter subscripts indicate the axes along which the orbitals are oriented. These three *p* orbitals are identical in size, shape, and energy; they differ from one another only in orientation. Note, however, that there is no simple relation between the values of m_l and the x, y, and z directions. For our purpose, you need only remember that because there are three possible values of m_l, there are three *p* orbitals with different orientations.

The boundary surface diagrams of *p* orbitals in Figure 7.17 show that each *p* orbital can be thought of as two lobes; the nucleus is at the center of the *p* orbital. Like *s* orbitals, *p* orbitals increase in size from 2*p* to 3*p* to 4*p* orbital and so on.

d ORBITALS AND OTHER HIGHER-ENERGY ORBITALS

When $l = 2$, there are five values of m_l, which correspond to five *d* orbitals. The lowest value of n for a *d* orbital is 3. Because l can never be greater than $n - 1$, when $n = 3$ and $l = 2$, we have five 3*d* orbitals ($3d_{xy}$, $3d_{yz}$, $3d_{xz}$, $3d_{x^2-y^2}$, and $3d_{z^2}$), shown in Figure 7.18. As in the case of the *p* orbitals, the different orientations of the *d* orbitals correspond to the different values of m_l, but again there is no direct correspondence between a given orientation and a particular m_l value. All the 3*d* orbitals in an atom are identical in energy. The *d* orbitals for which n is greater than 3 (4*d*, 5*d*, . . .) have similar shapes.

As we saw in Section 7.7, beyond the *d* orbitals are *f*, *g*, . . . orbitals. The *f* orbitals are important in accounting for the behavior of elements with atomic numbers greater than 57, although their shapes are difficult to represent. We are

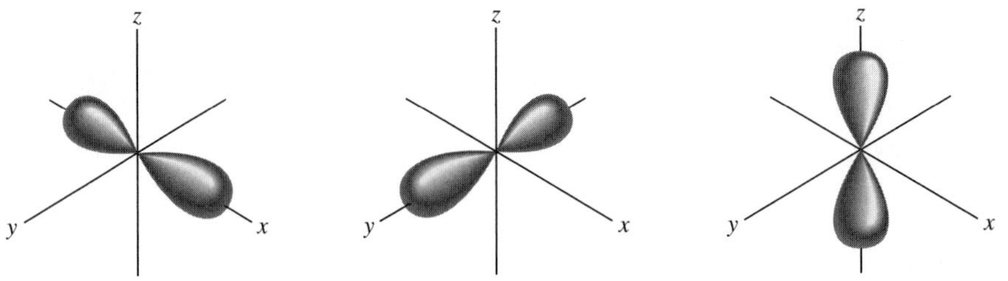

$2p_x$ $2p_y$ $2p_z$

FIGURE 7.17

The boundary surface diagrams of the three 2p orbitals. Except for their differing orientations, these orbitals are identical in shape and energy. The p orbitals of higher principal quantum numbers have similar shapes.

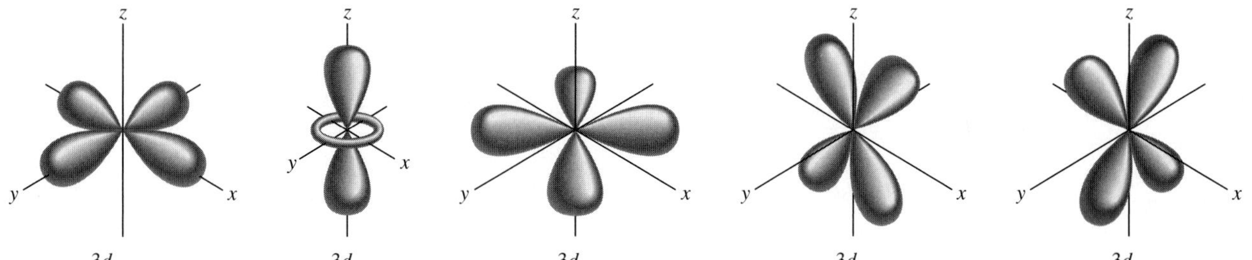

FIGURE 7.18

Boundary surface diagrams of the five 3d orbitals. Although the $3d_{z^2}$ orbital looks different, it is equivalent to the other four orbitals in all other respects. The d orbitals of higher principal quantum numbers have similar shapes.

not concerned here with orbitals having l values greater than 3 (the g orbitals and beyond).

EXAMPLE 7.5
Labeling an Atomic Orbital

List the values of n, l, and m_l for orbitals in the 4d subshell.

Answer: As we saw earlier, the number given in the designation of the subshell is the principal quantum number, so in this case $n = 4$. Since we are dealing with d orbitals, $l = 2$. The values of m_l can vary from $-l$ to l. Therefore, m_l can be $-2, -1, 0, 1, 2$ (which correspond to the five d orbitals).

PRACTICE EXERCISE

Give the values of the quantum numbers associated with the orbitals in the 3p subshell.

EXAMPLE 7.6
Counting the Orbitals Associated with a Principal Quantum Number

What is the total number of orbitals associated with the principal quantum number $n = 3$?

Answer: For $n = 3$, the possible values of l are 0, 1, and 2. Thus there is one 3s orbital ($n = 3, l = 0$, and $m_l = 0$); there are three 3p orbitals ($n = 3, l = 1$, and $m_l = -1, 0, 1$); there are five 3d orbitals ($n = 3, l = 2$, and $m_l = -2, -1, 0, 1, 2$). Therefore the total number of orbitals is $1 + 3 + 5 = 9$.

PRACTICE EXERCISE

What is the total number of orbitals associated with the principal quantum number $n = 4$?

THE ENERGIES OF ORBITALS

Now that we have some understanding of the shapes and sizes of atomic orbitals, we are ready to inquire into their relative energies and how these energy levels help determine the actual arrangement of electrons in atoms.

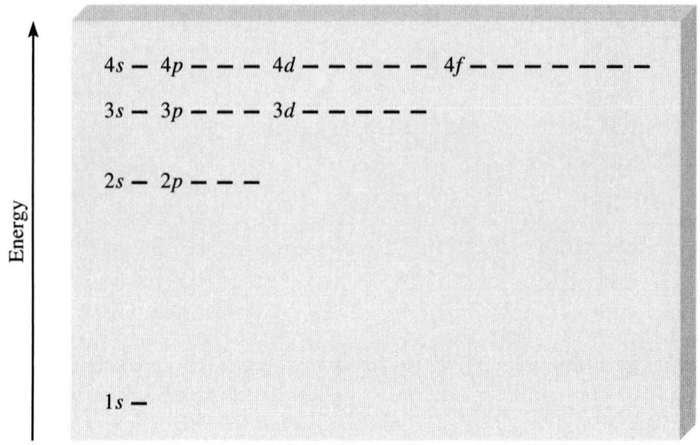

FIGURE 7.19

Orbital energy levels in the hydrogen atom. Each short horizontal line here represents one orbital. Orbitals with the same principal quantum number (n) all have the same energy.

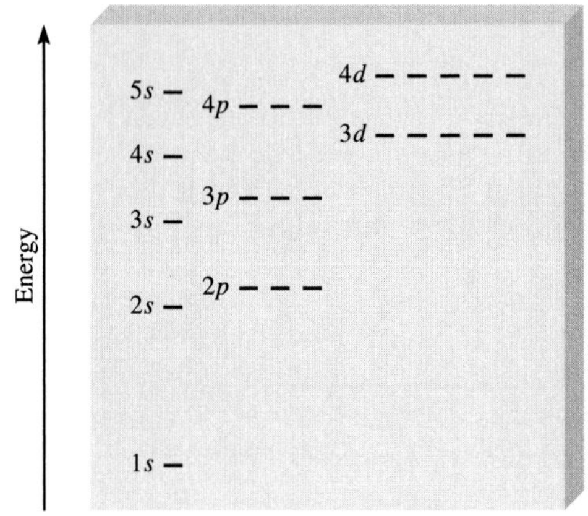

FIGURE 7.20

Orbital energy levels in a many-electron atom. Note that the energy level depends on both n and l values.

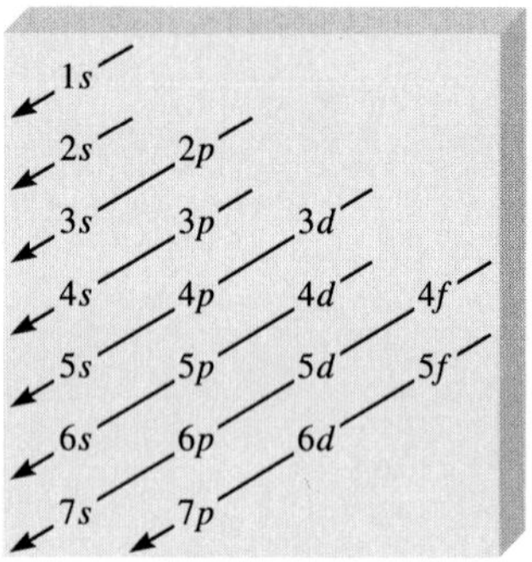

FIGURE 7.21

The order in which atomic subshells are filled in a many-electron atom. Start with the 1s orbital and move downward, following the direction of the arrows. Thus the order goes as follows: $1s \rightarrow 2s \rightarrow 2p \rightarrow 3s \rightarrow 3p \rightarrow 4s \rightarrow 3d \rightarrow \ldots$

According to Equation (7.4), the energy of an electron in a hydrogen atom is determined solely by its principal quantum number. Thus the energies of hydrogen orbitals increase as follows (Figure 7.19):

$$1s < 2s = 2p < 3s = 3p = 3d < 4s = 4p = 4f < \cdots$$

Although the electron density distributions are different in the 2*s* and 2*p* orbitals, hydrogen's electron has the same energy whether it is in the 2*s* orbital or a 2*p* orbital. The 1*s* orbital in a hydrogen atom corresponds to the most stable condition and is called the ground state, and an electron residing in this orbital is most strongly held by the nucleus. An electron in the 2*s*, 2*p*, or higher orbitals in a hydrogen atom is in an excited state.

The energy picture is more complex for many-electron atoms than for hydrogen. The energy of an electron in such an atom depends on its angular momentum quantum number as well as on its principal quantum number (Figure 7.20). For many-electron atoms, the 3*d* energy level is very close to the 4*s* energy level. The total energy of an atom, however, depends not only on the sum of the orbital energies but also on the energy of repulsion between the electrons in these orbitals. It turns out that the total energy of an atom is lower when the 4*s* subshell is filled before a 3*d* subshell. Figure 7.21 depicts the order in which atomic orbitals are filled in a many-electron atom. We will consider specific examples in the next section.

7.9 ELECTRON CONFIGURATION

The four quantum numbers n, l, m_l, and m_s enable us to label completely an electron in any orbital in any atom. For example, the four quantum numbers for a 2*s* orbital electron are $n = 2$, $l = 0$, $m_l = 0$, and either $m_s = +\frac{1}{2}$ or $m_s = -\frac{1}{2}$. In practice it is inconvenient to write out all the individual quantum numbers, and so we use the simplified notation (n, l, m_l, m_s). For the example above, the quantum numbers are either $(2, 0, 0, +\frac{1}{2})$ or $(2, 0, 0, -\frac{1}{2})$. The value of m_s has no effect on the energy, size, shape, or orientation of an orbital, but it plays a profound role in determining how electrons are arranged in an orbital.

EXAMPLE 7.7

Assigning Quantum Numbers to an Electron

List the different ways to write the four quantum numbers that designate an electron in a 3*p* orbital.

Answer: To start with, we know that the principal quantum number n is 3 and the angular momentum quantum number l must be 1 (because we are dealing with a *p* orbital). For $l = 1$, there are three values of m_l given by $-1, 0, 1$. Since the electron spin quantum number m_s can be either $+\frac{1}{2}$ or $-\frac{1}{2}$, we conclude that there are six possible ways to designate the electron:

$$(3, 1, -1, +\tfrac{1}{2}) \qquad (3, 1, -1, -\tfrac{1}{2})$$
$$(3, 1, 0, +\tfrac{1}{2}) \qquad (3, 1, 0, -\tfrac{1}{2})$$
$$(3, 1, 1, +\tfrac{1}{2}) \qquad (3, 1, 1, -\tfrac{1}{2})$$

PRACTICE EXERCISE

List the different ways to write the four quantum numbers that designate the electron in a 5*p* orbital.

The hydrogen atom is a particularly simple system because it contains only one electron. The electron may reside in the 1*s* orbital (the ground state), or it may be found in some higher orbital (an excited state). The situation is different for many-electron atoms. To understand electronic behavior in a many-electron atom, we must know the ***electron configuration*** of the atom. The electron configuration of an atom tells us *how the electrons are distributed among the various atomic orbitals.* We will use the first ten elements (hydrogen to neon) to illustrate the basic rules for writing the electron configurations for the *ground state* of the atoms. (Section 7.10 will describe how these rules can be applied to the remainder of the elements in the periodic table.) Recall that the number of electrons in an atom is equal to its atomic number Z.

Figure 7.19 indicates that the electron in a ground-state hydrogen atom must be in the 1*s* orbital, so its electron configuration is $1s^1$:

$$1s^1$$

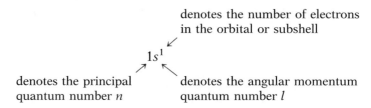

- 1 (the 1): denotes the principal quantum number n
- s: denotes the angular momentum quantum number l
- superscript 1: denotes the number of electrons in the orbital or subshell

The electron configuration can also be represented by an *orbital diagram* that shows the spin of the electron (see Figure 7.14):

H [↑]

where the upward arrow denotes one of the two possible spinning motions of the electron. (We could have represented the electron with the downward arrow.) The box represents an atomic orbital.

The Pauli Exclusion Principle

Wolfgang Pauli (1900–1958)

The hydrogen atom is a straightforward case because there is only one electron present. How do we deal with the electron configurations of atoms that contain more than one electron? In these cases we use the ***Pauli exclusion principle*** as our guide. This principle states that *no two electrons in an atom can have the same four quantum numbers.* If two electrons in an atom should have the same n, l, and m_l values (that is, these two electrons are in the *same* atomic orbital), then they must have different values of m_s. In other words, only two electrons may exist in the same atomic orbital, and these electrons must have opposite spins. Consider the helium atom, which has two electrons. The three possible ways of placing two electrons in the 1s orbital are as follows:

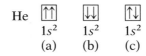

He	[↑↑]	[↓↓]	[↑↓]
	$1s^2$	$1s^2$	$1s^2$
	(a)	(b)	(c)

Diagrams (a) and (b) are ruled out by the Pauli exclusion principle. In (a), both electrons have the same upward spin and would have the quantum numbers (1, 0, 0, $+\frac{1}{2}$); in (b), both electrons have downward spin and would have the quantum numbers (1, 0, 0, $-\frac{1}{2}$). Only the configuration in (c) is physically acceptable, because one electron has the quantum numbers (1, 0, 0, $+\frac{1}{2}$) and the other has (1, 0, 0, $-\frac{1}{2}$). Thus the helium atom has the following configuration:

He [↑↓]
$1s^2$

Note that $1s^2$ is read "one s two," not "one s squared."

Diamagnetism and Paramagnetism

The Pauli exclusion principle is one of the fundamental principles of quantum mechanics. It can be tested by a simple observation. If the two electrons in the 1s orbital of a helium atom had the same, or parallel, spins (↑↑ or ↓↓), their net magnetic fields would reinforce each other. Such an arrangement would make the helium atom ***paramagnetic*** [Figure 7.22(a)]. Paramagnetic substances are those that are *attracted by a magnet.* On the other hand, if the electron spins are paired, or antiparallel to each other (↑↓ or ↓↑), the magnetic effects cancel out and the atom is ***diamagnetic*** [Figure 7.22(b)]. Diamagnetic substances are slightly *repelled by a magnet.*

By experiment we find that the helium atom is diamagnetic in the ground state, in accord with the Pauli exclusion principle. A useful general rule to keep in mind is that any atom with an odd number of electrons *must be* paramagnetic, because we need an even number of electrons for complete pairing. On the other hand, atoms containing an even number of electrons may be either diamagnetic or paramagnetic. We will see the reason for this property shortly.

As another example, consider the lithium atom, which has three electrons. The third electron cannot go into the 1*s* orbital because it would inevitably have the same four quantum numbers as one of the first two electrons. Therefore, this electron "enters" the next (energetically) higher orbital, which is the 2*s* orbital (see Figure 7.20). The electron configuration of lithium is $1s^22s^1$, and its orbital diagram is

Li	↑↓	↑
	$1s^2$	$2s^1$

The lithium atom contains one unpaired electron and is therefore paramagnetic.

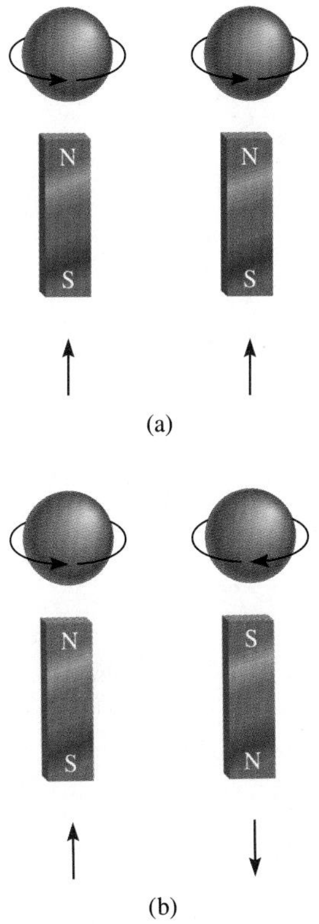

FIGURE 7.22

The (a) parallel and (b) antiparallel spins of two electrons. In (a) the two magnetic fields reinforce each other, and the atom is paramagnetic. In (b) the two magnetic fields cancel each other, and the atom is diamagnetic.

The Shielding Effect in Many-Electron Atoms

But why is the 2*s* orbital lower than the 2*p* orbital on the energy scale in a many-electron atom? In comparing the electron configurations of $1s^22s^1$ and $1s^22p^1$, we note that, in both cases, the 1*s* orbital is filled with two electrons. Because the 2*s* and 2*p* orbitals are larger than the 1*s* orbital, an electron in either of these orbitals will spend more time away from the nucleus (on the average) than an electron in the 1*s* orbital. Thus we can speak of a 2*s* or 2*p* electron being partly "shielded" from the attractive force of the nucleus by the 1*s* electrons. The important consequence of the shielding effect is that it *reduces* the electrostatic attraction between protons in the nucleus and the electron in the 2*s* or 2*p* orbital.

The manner in which the electron density varies as we move outward from the nucleus differs for the 2*s* and 2*p* orbitals. The density for the 2*s* electron near the nucleus is greater than that for the 2*p* electron. In other words, a 2*s* electron spends more time near the nucleus than a 2*p* electron does (on the average). For this reason, the 2*s* orbital is said to be more "penetrating" than the 2*p* orbital, and it is less shielded by the 1*s* electrons. In fact, for the same principal quantum number n, the penetrating power decreases as the angular momentum quantum number l increases, or

$$s > p > d > f > \cdots$$

Since the stability of an electron is determined by how strongly the electron is attracted by the nucleus, it follows that a 2*s* electron will be lower in energy than a 2*p* electron. To put it another way, less energy is required to remove a 2*p* electron than a 2*s* electron because a 2*p* electron is not held quite as strongly by the nucleus. The hydrogen atom has only one electron and, therefore, is without such a shielding effect.

Continuing our discussion of atoms of the first ten elements, the electron configuration of beryllium ($Z = 4$) is $1s^22s^2$, or

Be	↑↓	↑↓
	$1s^2$	$2s^2$

and beryllium atoms are diamagnetic as we would expect.

The electron configuration of boron ($Z = 5$) is $1s^2 2s^2 2p^1$, or

B	[↑↓]	[↑↓]	[↑		]
	$1s^2$	$2s^2$	$2p^1$

Note that the unpaired electron may be in the $2p_x$, $2p_y$, or $2p_z$ orbital. The choice is completely arbitrary because the three p orbitals are equivalent in energy. As the diagram shows, boron atoms are paramagnetic.

Hund's Rule

The electron configuration of carbon ($Z = 6$) is $1s^2 2s^2 2p^2$. The following are different ways of distributing two electrons among three p orbitals:

[↑↓		]	[↑	↓	]	[↑	↑	]
$2p_x\ 2p_y\ 2p_z$	$2p_x\ 2p_y\ 2p_z$	$2p_x\ 2p_y\ 2p_z$
(a)	(b)	(c)

None of the three arrangements violates the Pauli exclusion principle, so we must determine which one will give the greatest stability. The answer is provided by ***Hund's rule,*** which states that *the most stable arrangement of electrons in subshells is the one with the greatest number of parallel spins.* The arrangement shown in (c) satisfies this condition. In both (a) and (b) the two spins cancel each other. Thus the electron configuration of carbon is $1s^2 2s^2 2p^2$, and its orbital diagram is

C	[↑↓]	[↑↓]	[↑	↑	]
	$1s^2$	$2s^2$	$2p_x\ 2p_y\ 2p_z$

Qualitatively, we can understand why (c) is preferred to (a). In (a), the two electrons are in the same $2p_x$ orbital, and their proximity results in a greater mutual repulsion than when they occupy two separate orbitals, say $2p_x$ and $2p_y$. The choice of (c) over (b) is more subtle but can be justified on theoretical grounds.

Measurements of magnetic properties provide the most direct evidence supporting specific electron configurations of elements. Advances in instrument design during the last 20 years or so enable us to determine not only whether an atom is paramagnetic but also how many unpaired electrons are present. The fact that carbon atoms are paramagnetic, each containing two unpaired electrons, is in accord with Hund's rule.

Continuing, the electron configuration of nitrogen ($Z = 7$) is $1s^2 2s^2 2p^3$:

N	[↑↓]	[↑↓]	[↑	↑	↑]
	$1s^2$	$2s^2$	$2p^3$

Again, Hund's rule dictates that all three $2p$ electrons have spins parallel to one another; the nitrogen atom is therefore paramagnetic, containing three unpaired electrons.

The electron configuration of oxygen ($Z = 8$) is $1s^2 2s^2 2p^4$. An oxygen atom is paramagnetic because it has two unpaired electrons:

O	[↑↓]	[↑↓]	[↑↓	↑	↑]
	$1s^2$	$2s^2$	$2p^4$

The electron configuration of fluorine ($Z = 9$) is $1s^2 2s^2 2p^5$. The nine electrons are arranged as follows:

F [↑↓] [↑↓] [↑↓|↑↓|↑]
$1s^2$ $2s^2$ $2p^5$

The fluorine atom is thus paramagnetic, having one unpaired electron.

In neon ($Z = 10$), the $2p$ orbitals are completely filled. The electron configuration of neon is $1s^22s^22p^6$, and *all* the electrons are paired, as follows:

Ne [↑↓] [↑↓] [↑↓|↑↓|↑↓]
$1s^2$ $2s^2$ $2p^6$

The neon atom thus should be diamagnetic, and experimental observation bears out this prediction.

General Rules for Assigning Electrons to Atomic Orbitals

Based on the preceding examples, we can formulate some general rules for determining the maximum number of electrons that can be assigned to the various subshells and orbitals for a given value of n:

- Each shell or principal level of quantum number n contains n subshells. For example, if $n = 2$, then there are two subshells (two values of l) of angular momentum quantum numbers 0 and 1.
- Each subshell of quantum number l contains $2l + 1$ orbitals. For example, if $l = 1$, then there are three p orbitals.
- No more than two electrons can be placed in each orbital. Therefore, the maximum number of electrons is twice the number of orbitals that are employed.

EXAMPLE 7.8
Counting the Number of Electrons in a Principal Level

What is the maximum number of electrons that can be present in the principal level for which $n = 3$?

Answer: When $n = 3$, then $l = 0$, 1, and 2. The number of orbitals for each value of l is given by

Value of l	*Number of orbitals (2l + 1)*
0	1
1	3
2	5

The total number of orbitals is nine. Since each orbital can accommodate two electrons, the maximum number of electrons that can reside in the orbitals is 2×9, or 18. Note that this result can be generalized by the formula $2n^2$. Here we have $n = 3$, so $2(3^2) = 18$.

PRACTICE EXERCISE

Calculate the total number of electrons that can be present in the principal level for which $n = 4$.

At this point let's summarize what our examination has revealed about ground-state electron configurations and the properties of electrons in atoms:

- No two electrons in the same atom can have the same four quantum numbers. This is the Pauli exclusion principle.
- Each orbital can be occupied by a maximum of two electrons. They must have opposite spins, or different electron spin quantum numbers.
- The most stable arrangement of electrons in a subshell is the one that has the greatest number of parallel spins. This is Hund's rule.
- Atoms in which one or more electrons are unpaired are paramagnetic. Atoms in which all the electron spins are paired are diamagnetic.
- In a hydrogen atom, the energy of the electron depends only on its principal quantum number n. In a many-electron atom, the energy of an electron depends on both n and its angular momentum quantum number l.
- In a many-electron atom the subshells are filled in the order shown in Figure 7.21.
- For electrons of the same principal quantum number, their penetrating power, or proximity to the nucleus, decreases in the order $s > p > d > f$. This means that, for example, more energy is required to separate a 3s electron from a many-electron atom than is required to remove a 3p electron.

7.10 THE BUILDING-UP PRINCIPLE

Here we will extend the rules used in writing electron configurations for the first ten elements to the rest of the elements. This process is based on the Aufbau principle. (The German word *Aufbau* means "building up.") The ***Aufbau principle*** is based on the fact that *as protons are added one by one to the nucleus to build up the elements, electrons are similarly added to the atomic orbitals.* Through this process we gain detailed knowledge about the ground-state electron configurations of the elements. As we will see later, knowledge of electron configurations helps us understand and predict the properties of the elements; it also explains why the periodic table works so well.

Table 7.3 gives the ground-state electron configurations of all the known elements from H ($Z = 1$) through Une ($Z = 109$). The electron configurations of all elements except hydrogen and helium are represented by a ***noble gas core,*** which *shows in brackets the noble gas element that most nearly precedes the element being considered,* followed by the symbol for the highest filled subshells in the outermost shells. Notice that the electron configurations of the highest filled subshells in the outermost shells for the elements sodium ($Z = 11$) through argon ($Z = 18$) follow a pattern similar to those of lithium ($Z = 3$) through neon ($Z = 10$).

As mentioned in Section 7.8, the 4s subshell is filled before the 3d subshell in a many-electron atom (see Figure 7.20). Thus the electron configuration of potassium ($Z = 19$) is $1s^22s^22p^63s^23p^64s^1$. Since $1s^22s^22p^63s^23p^6$ is the electron configuration of argon, we can simplify the electron configuration of potassium by writing [Ar]$4s^1$, where [Ar] denotes the "argon core." Similarly, we can write the electron configuration of calcium ($Z = 20$) as [Ar]$4s^2$. The placement of the outermost electron in the 4s orbital (rather than in the 3d orbital) of potassium is strongly supported by experimental evidence. The following comparison also suggests that this

TABLE 7.3

The Ground-State Electron Configurations of the Elements*

Atomic number	Symbol	Electron configuration	Atomic number	Symbol	Electron configuration	Atomic number	Symbol	Electron configuration
1	H	$1s^1$	37	Rb	$[Kr]5s^1$	73	Ta	$[Xe]6s^24f^{14}5d^3$
2	He	$1s^2$	38	Sr	$[Kr]5s^2$	74	W	$[Xe]6s^24f^{14}5d^4$
3	Li	$[He]2s^1$	39	Y	$[Kr]5s^24d^1$	75	Re	$[Xe]6s^24f^{14}5d^5$
4	Be	$[He]2s^2$	40	Zr	$[Kr]5s^24d^2$	76	Os	$[Xe]6s^24f^{14}5d^6$
5	B	$[He]2s^22p^1$	41	Nb	$[Kr]5s^14d^4$	77	Ir	$[Xe]6s^24f^{14}5d^7$
6	C	$[He]2s^22p^2$	42	Mo	$[Kr]5s^14d^5$	78	Pt	$[Xe]6s^14f^{14}5d^9$
7	N	$[He]2s^22p^3$	43	Tc	$[Kr]5s^24d^5$	79	Au	$[Xe]6s^14f^{14}5d^{10}$
8	O	$[He]2s^22p^4$	44	Ru	$[Kr]5s^14d^7$	80	Hg	$[Xe]6s^24f^{14}5d^{10}$
9	F	$[He]2s^22p^5$	45	Rh	$[Kr]5s^14d^8$	81	Tl	$[Xe]6s^24f^{14}5d^{10}6p^1$
10	Ne	$[He]2s^22p^6$	46	Pd	$[Kr]4d^{10}$	82	Pb	$[Xe]6s^24f^{14}5d^{10}6p^2$
11	Na	$[Ne]3s^1$	47	Ag	$[Kr]5s^14d^{10}$	83	Bi	$[Xe]6s^24f^{14}5d^{10}6p^3$
12	Mg	$[Ne]3s^2$	48	Cd	$[Kr]5s^24d^{10}$	84	Po	$[Xe]6s^24f^{14}5d^{10}6p^4$
13	Al	$[Ne]3s^23p^1$	49	In	$[Kr]5s^24d^{10}5p^1$	85	At	$[Xe]6s^24f^{14}5d^{10}6p^5$
14	Si	$[Ne]3s^23p^2$	50	Sn	$[Kr]5s^24d^{10}5p^2$	86	Rn	$[Xe]6s^24f^{14}5d^{10}6p^6$
15	P	$[Ne]3s^23p^3$	51	Sb	$[Kr]5s^24d^{10}5p^3$	87	Fr	$[Rn]7s^1$
16	S	$[Ne]3s^23p^4$	52	Te	$[Kr]5s^24d^{10}5p^4$	88	Ra	$[Rn]7s^2$
17	Cl	$[Ne]3s^23p^5$	53	I	$[Kr]5s^24d^{10}5p^5$	89	Ac	$[Rn]7s^26d^1$
18	Ar	$[Ne]3s^23p^6$	54	Xe	$[Kr]5s^24d^{10}5p^6$	90	Th	$[Rn]7s^26d^2$
19	K	$[Ar]4s^1$	55	Cs	$[Xe]6s^1$	91	Pa	$[Rn]7s^25f^26d^1$
20	Ca	$[Ar]4s^2$	56	Ba	$[Xe]6s^2$	92	U	$[Rn]7s^25f^36d^1$
21	Sc	$[Ar]4s^23d^1$	57	La	$[Xe]6s^25d^1$	93	Np	$[Rn]7s^25f^46d^1$
22	Ti	$[Ar]4s^23d^2$	58	Ce	$[Xe]6s^24f^15d^1$	94	Pu	$[Rn]7s^25f^6$
23	V	$[Ar]4s^23d^3$	59	Pr	$[Xe]6s^24f^3$	95	Am	$[Rn]7s^25f^7$
24	Cr	$[Ar]4s^13d^5$	60	Nd	$[Xe]6s^24f^4$	96	Cm	$[Rn]7s^25f^76d^1$
25	Mn	$[Ar]4s^23d^5$	61	Pm	$[Xe]6s^24f^5$	97	Bk	$[Rn]7s^25f^9$
26	Fe	$[Ar]4s^23d^6$	62	Sm	$[Xe]6s^24f^6$	98	Cf	$[Rn]7s^25f^{10}$
27	Co	$[Ar]4s^23d^7$	63	Eu	$[Xe]6s^24f^7$	99	Es	$[Rn]7s^25f^{11}$
28	Ni	$[Ar]4s^23d^8$	64	Gd	$[Xe]6s^24f^75d^1$	100	Fm	$[Rn]7s^25f^{12}$
29	Cu	$[Ar]4s^13d^{10}$	65	Tb	$[Xe]6s^24f^9$	101	Md	$[Rn]7s^25f^{13}$
30	Zn	$[Ar]4s^23d^{10}$	66	Dy	$[Xe]6s^24f^{10}$	102	No	$[Rn]7s^25f^{14}$
31	Ga	$[Ar]4s^23d^{10}4p^1$	67	Ho	$[Xe]6s^24f^{11}$	103	Lr	$[Rn]7s^25f^{14}6d^1$
32	Ge	$[Ar]4s^23d^{10}4p^2$	68	Er	$[Xe]6s^24f^{12}$	104	Unq	$[Rn]7s^25f^{14}6d^2$
33	As	$[Ar]4s^23d^{10}4p^3$	69	Tm	$[Xe]6s^24f^{13}$	105	Unp	$[Rn]7s^25f^{14}6d^3$
34	Se	$[Ar]4s^23d^{10}4p^4$	70	Yb	$[Xe]6s^24f^{14}$	106	Unh	$[Rn]7s^25f^{14}6d^4$
35	Br	$[Ar]4s^23d^{10}4p^5$	71	Lu	$[Xe]6s^24f^{14}5d^1$	107	Uns	$[Rn]7s^25f^{14}6d^5$
36	Kr	$[Ar]4s^23d^{10}4p^6$	72	Hf	$[Xe]6s^24f^{14}5d^2$	108	Uno	$[Rn]7s^25f^{14}6d^6$
						109	Une	$[Rn]7s^25f^{14}6d^7$

*The symbol [He] is called the *helium core* and represents $1s^2$. [Ne] is called the *neon core* and represents $1s^22s^22p^6$. [Ar] is called the *argon core* and represents $[Ne]3s^23p^6$. [Kr] is called the *krypton core* and represents $[Ar]4s^23d^{10}4p^6$. [Xe] is called the *xenon core* and represents $[Kr]5s^24d^{10}5p^6$. [Rn] is called the *radon core* and represents $[Xe]6s^24f^{14}5d^{10}6p^6$.

is the correct configuration. The chemistry of potassium is very similar to that of lithium and sodium, the first two alkali metals. The outermost electron of both lithium and sodium is in an *s* orbital (there is no ambiguity in assigning their electron configurations); therefore, we expect the last electron in potassium to occupy the 4*s* rather than the 3*d* orbital.

The elements from scandium (Z = 21) to copper (Z = 29) are transition metals. ***Transition metals*** either *have incompletely filled d subshells or readily give rise to cations that have incompletely filled d subshells.* In the series from scandium through copper additional electrons are placed in the 3d orbitals, according to Hund's rule. However, there are two irregularities. The electron configuration of chromium (Z = 24) is [Ar]$4s^1 3d^5$ and not [Ar]$4s^2 3d^4$, as we might expect. A similar break in the pattern is observed for copper, whose electron configuration is [Ar]$4s^1 3d^{10}$ rather than [Ar]$4s^2 3d^9$. The reason for these irregularities is that slightly greater stability is associated with the half-filled ($3d^5$) and completely filled ($3d^{10}$) subshells. Electrons in the same subshells (in this case, the d orbitals) have equal energy but different spatial distributions. Consequently, their shielding of one another is relatively small, and the electrons are more strongly attracted by the nucleus when they have the $3d^5$ configuration. According to Hund's rule, the orbital diagram for Cr is

Cr [Ar] [↑] [↑|↑|↑|↑|↑]
$4s^1$ $3d^5$

Thus, Cr has a total of six unpaired electrons. The orbital diagram for copper is

Cu [Ar] [↑] [↑↓|↑↓|↑↓|↑↓|↑↓]
$4s^1$ $3d^{10}$

Again, extra stability is gained in this case by having completely filled 3d orbitals.

For elements Zn (Z = 30) through Kr (Z = 36), the 4s and 4p subshells fill in a straightforward manner. With rubidium (Z = 37), electrons begin to enter the n = 5 energy level.

The electron configurations in the second transition metal series [yttrium (Z = 39) to silver (Z = 47)] are also irregular, but we will not be concerned with the details here.

The sixth period of the periodic table begins with cesium (Z = 55) and barium (Z = 56), whose electron configurations are [Xe]$6s^1$ and [Xe]$6s^2$, respectively. Next we come to lanthanum (Z = 57). From Figure 7.21 we would expect that after filling the 6s orbital we would place the additional electrons in 4f orbitals. In reality, the energies of the 5d and 4f orbitals are very close; in fact, for lanthanum 4f is slightly higher in energy than 5d. Thus lanthanum's electron configuration is [Xe]$6s^2 5d^1$ and not [Xe]$6s^2 4f^1$.

Following lanthanum are the 14 elements that form the ***lanthanide,*** or ***rare earth, series*** [cerium (Z = 58) to lutetium (Z = 71)]. The rare earth metals *have incompletely filled 4f subshells or readily give rise to cations that have incompletely filled 4f subshells.* In this series, the added electrons are placed in 4f orbitals. After the 4f subshells are completely filled, the next electron enters the 5d subshell of lutetium. Note that the electron configuration of gadolinium (Z = 64) is [Xe]$6s^2 4f^7 5d^1$ rather than [Xe]$6s^2 4f^8$. Again, extra stability is gained by having half-filled subshells ($4f^7$), as is the case for chromium.

The third transition metal series, including lanthanum and hafnium (Z = 72) and extending through gold (Z = 79), is characterized by the filling of the 5d orbitals. The 6s and 6p subshells are filled next, which takes us to radon (Z = 86).

The *last row of elements* belongs to the ***actinide series,*** which starts at thorium (Z = 90). *Most of these elements are not found in nature but have been synthesized.*

With few exceptions, you should be able to write the electron configuration of any element, using Figure 7.21 as a guide. Elements that require particular care are those belonging to the transition metals, the lanthanides, and the actinides. As we noted earlier, at larger values of the principal quantum number n, the order of subshell filling may reverse from one element to the next. Figure 7.23 groups the

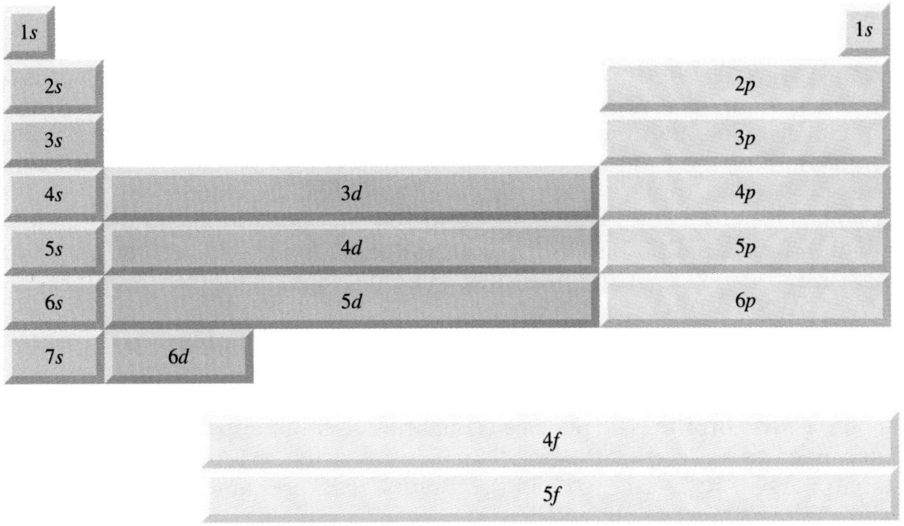

FIGURE 7.23

Classification of groups of elements in the periodic table according to the type of subshell being filled with electrons.

elements according to the type of subshell in which the outermost electrons are placed.

EXAMPLE 7.9

Writing Electron Configurations

Write the electron configurations for sulfur and palladium, which is diamagnetic.

Answer: Sulfur ($Z = 16$)

1. Sulfur has 16 electrons.
2. It takes 10 electrons to complete the first and second periods ($1s^22s^22p^6$). This leaves us 6 electrons to fill the 3*s* orbital and partially fill the 3*p* orbitals. Thus the electron configuration of S is

$$1s^22s^22p^63s^23p^4$$

or [Ne]$3s^23p^4$.

Palladium ($Z = 46$)

1. Palladium has 46 electrons.
2. We need 36 electrons to complete the fourth period, leaving us with 10 more electrons to distribute among the 5*s* and 4*d* orbitals. The three choices are (a) $4d^{10}$, (b) $4d^95s^1$, and (c) $4d^85s^2$. Since atomic palladium is diamagnetic, its electron configuration must be

$$1s^22s^22p^63s^23p^64s^23d^{10}4p^64d^{10}$$

or simply [Kr]$4d^{10}$. The configurations in (b) and (c) both give paramagnetic Pd atoms.

PRACTICE EXERCISE

Write the ground-state electron configuration for phosphorus (P).

SUMMARY

Several major discoveries by physicists in the early twentieth century made a profound impact on our understanding of atoms and molecules and revolutionized physics. Planck's quantum theory explained the way heated solids emit radiant energy. Radiation was already known to have the properties of waves, described in terms of wavelength, frequency, and amplitude. Quantum theory suggested that radiant energy is emitted by atoms and molecules in small discrete amounts (quanta), rather than over a continuous range. This behavior is governed by the relationship $E = h\nu$. Energy is always emitted in whole-number multiples of $h\nu$ ($1\,h\nu$, $2\,h\nu$, $3\,h\nu$, . . .).

Einstein extended the quantum theory to explain the photoelectric effect by proposing that light can behave like a stream of particles (photons).

Bohr developed a model of the hydrogen atom in which the energy of the single electron orbiting the nucleus is quantized—limited to certain energy values determined by an integer, the principal quantum number. An electron in its most stable energy state is said to be in the ground state, and an electron having more energy than its most stable state is said to be in an excited state. In the Bohr model, an electron emits a photon when it drops from a higher-energy orbit (an excited state) to a lower-energy orbit (a less excited state). This model accounts for the specific energies represented by the lines in the hydrogen emission spectrum.

The idea that light has the characteristics of both particles and waves was extended by de Broglie to all matter in motion by the equation $\lambda = h/mu$.

The Schrödinger equation describes the motions and energies of submicroscopic particles in terms of quantum theory. It tells us the possible energy states of the electron in a hydrogen atom and the probability of its location in a particular region surrounding the nucleus. These results can be applied with reasonable accuracy to many-electron atoms.

An atomic orbital is a function that defines the distribution of electron density in space. Orbitals are represented by electron density diagrams or boundary surface diagrams. Four quantum numbers are needed to characterize completely each electron in an atom: the principal quantum number n, which determines the main energy level, or shell, of the orbital; the angular momentum quantum number l, which determines the shape of the orbital; the magnetic quantum number m_l, which determines the orientation of the orbital in space; and the electron spin quantum number m_s, which relates only to the electron's spin on its own axis.

The single s orbital in each energy level is spherical and centered on the nucleus. The three p orbitals each have two lobes, and the pairs of lobes are arranged at right angles to one another. There are five d orbitals, which have more complex shapes and orientations.

The energy of the electron in a hydrogen atom is determined solely by its principal quantum number. In many-electron atoms, both the principal quantum number and the angular momentum quantum number determine the energy of an electron. No two electrons in the same atom can have the same four quantum numbers (the Pauli exclusion principle). The most stable arrangement of electrons in a subshell is the one that has the greatest number of parallel spins (Hund's rule). Atoms with one or more unpaired spins are paramagnetic. Atoms with all electron spins paired are diamagnetic.

Using the Aufbau (building-up) principle we can determine the ground-state electron configurations of all elements based on atomic number and location in the periodic table.

KEY WORDS

QUESTIONS AND PROBLEMS

ELECTROMAGNETIC RADIATION AND THE QUANTUM THEORY

Review Questions

7.1 Define the following terms: wave, wavelength, frequency, amplitude, electromagnetic radiation, quantum.

7.2 What are the units for wavelength and frequency of electromagnetic waves? What is the speed of light, in meters per second and miles per hour?

7.3 List the types of electromagnetic radiation, starting with the radiation having the longest wavelength and ending with the radiation having the shortest wavelength.

7.4 Give the high and low wavelength values that define the visible region.

7.5 Briefly explain Planck's quantum theory. What are the units for the Planck constant?

7.6 Give two familiar, everyday examples to illustrate the concept of quantization.

Problems

7.7 Convert 8.6×10^{13} Hz to nanometers, and 566 nm to hertz.

7.8 (a) What is the frequency of light of wavelength 456 nm? (b) What is the wavelength (in nanometers) of radiation of frequency 2.20×10^9 Hz?

7.9 The average distance between Mars and Earth is about 1.3×10^8 miles. How long would it take TV pictures transmitted from the *Viking* space vehicle on Mars' surface to reach Earth? (1 mile = 1.61 km.)

7.10 How long would it take a radio wave to travel from the planet Venus to Earth? (Average distance from Venus to Earth = 28 million miles.)

7.11 The basic SI unit of time is the second, which is defined as 9,192,631,770 cycles of radiation associated with a certain emission process in the cesium atom. Calculate the wavelength of this radiation (to three significant figures). In which region of the electromagnetic spectrum is this wavelength found?

7.12 The basic SI unit of length is the meter, which is defined as the length equal to 1,650,763.73 wavelengths of the light emitted by a particular energy transition in krypton atoms. Calculate the frequency of the light to three significant figures.

THE PHOTOELECTRIC EFFECT

Review Questions

7.13 Explain what is meant by the photoelectric effect.

7.14 What are photons? What role did Einstein's explanation of the photoelectric effect play in the development of the particle-wave interpretation of the nature of electromagnetic radiation?

Problems

7.15 A photon has a wavelength of 624 nm. Calculate the energy of the photon in joules.

7.16 The blue color of the sky results from the scattering of sunlight by air molecules. The blue light has a frequency of about 7.5×10^{14} Hz. (a) Calculate the wavelength associated with this radiation, and (b) calculate the energy in joules of a single photon associated with this frequency.

7.17 A photon has a frequency of 6.0×10^4 Hz. (a) Convert this frequency into wavelength (in nanometers). Does this frequency fall in the visible region? (b) Calculate the energy (in joules) of this photon.

(c) Calculate the energy (in joules) of 1 mole of photons all with this frequency.

7.18 What is the wavelength in nanometers of radiation that has an energy content of 1.0×10^3 kJ/mol? In which region of the electromagnetic spectrum is this radiation found?

7.19 When copper is bombarded with high-energy electrons, X rays are emitted. Calculate the energy (in joules) associated with the photons if the wavelength of the X rays is 0.154 nm.

7.20 A particular electromagnetic radiation has a frequency of 8.11×10^{14} Hz. (a) What is its wavelength in nanometers? in meters? (b) To what region of the electromagnetic spectrum would you assign it? (c) What is the energy (in joules) of one quantum of this radiation?

BOHR'S THEORY OF THE HYDROGEN ATOM

Review Questions

7.21 What are emission spectra? What are line spectra?

7.22 What is an energy level? Explain the difference between a ground state and an excited state.

7.23 Briefly describe Bohr's theory of the hydrogen atom and how it explains the appearance of an emission spectrum. How does Bohr's theory differ from concepts of classical physics?

7.24 Explain the meaning of the negative sign in Equation (7.4).

Problems

7.25 Explain why elements produce their own characteristic colors when they emit photons.

7.26 When some compounds containing copper are heated in a flame, green light is emitted. How would you determine whether the light is of one wavelength or a mixture of two or more wavelengths?

7.27 Would it be possible for a fluorescent material to emit radiation in the ultraviolet region after absorbing visible light? Explain your answer.

7.28 Explain how astronomers are able to tell which elements are present in distant stars by analyzing the electromagnetic radiation emitted by the stars.

7.29 Consider the following energy levels of a hypothetical atom:

E_4 ———— -1.0×10^{-19} J
E_3 ———— -5.0×10^{-19} J
E_2 ———— -10×10^{-19} J
E_1 ———— -15×10^{-19} J

(a) What is the wavelength of the photon needed to excite an electron from E_1 to E_4? (b) What is the value (in joules) of the quantum of energy of a photon needed to excite an electron from E_2 to E_3? (c) When an electron drops from the E_3 level to the E_1 level, the atom is said to undergo emission. Calculate the wavelength of the photon emitted in this process.

7.30 The first line of the Balmer series occurs at a wavelength of 656.3 nm. What is the energy difference between the two energy levels involved in the emission that results in this spectral line?

7.31 Calculate the wavelength of a photon emitted by a hydrogen atom when its electron drops from the $n = 5$ state to the $n = 3$ state.

7.32 Calculate the frequency and wavelength of the emitted photon when an electron drops from the $n = 4$ to the $n = 2$ level in a hydrogen atom.

7.33 Careful spectral analysis shows that the familiar yellow light of sodium lamps (such as street lamps) is made up of photons of two wavelengths, 589.0 nm and 589.6 nm. What is the difference in energy (in joules) between photons possessing these wavelengths?

7.34 An electron in an orbit of principal quantum number n_i in a hydrogen atom makes a transition to an orbit of principal quantum number 2. The photon emitted has a wavelength of 434 nm. Calculate n_i.

PARTICLE-WAVE DUALITY

Review Questions

7.35 Explain what is meant by the statement that matter and radiation have a "dual nature."

7.36 How does de Broglie's hypothesis account for the fact that the energies of the electron in a hydrogen atom are quantized?

7.37 Why is Equation (7.7) meaningful only to submicroscopic particles such as electrons and atoms and not to macroscopic objects?

7.38 Does a baseball in flight possess wave properties? If so, why can we not determine its wave properties?

Problems

7.39 Thermal neutrons are neutrons that move at speeds comparable to those of air molecules at room temperature. These neutrons are most effective in initiating a nuclear chain reaction among ^{235}U isotopes. Calculate the wavelength (in nanometers) associated with a beam of neutrons moving at 7.00×10^2 m/s. (Mass of a neutron = 1.675×10^{-27} kg.)

7.40 Protons in certain particle accelerators can be accelerated to speeds near that of light. Estimate the wavelength (in nanometers) of such a proton moving at 2.90×10^8 m/s. (Mass of a proton = 1.673×10^{-27} kg.)

7.41 What is the de Broglie wavelength in cm of a 12.4-g hummingbird flying at 1.20×10^2 mph? (1 mile = 1.61 km.)

7.42 What is the de Broglie wavelength associated with a 2.5-g Ping-Pong ball traveling at 35 mph?

QUANTUM MECHANICS AND THE HYDROGEN ATOM

Review Questions

7.43 What are the inadequacies of Bohr's theory?

7.44 What is the Heisenberg uncertainty principle? What is the Schrödinger equation?

7.45 What is the physical significance of the wave function?

7.46 How is the concept of electron density used to describe the position of an electron in the quantum mechanical treatment of an atom?

7.47 Define atomic orbital. How does an atomic orbital differ from an orbit?

7.48 Describe the characteristics of an *s* orbital, a *p* orbital, and a *d* orbital.

7.49 Why is a boundary surface diagram useful in representing an atomic orbital?

7.50 Describe the four quantum numbers used to characterize an electron in an atom.

7.51 Which quantum number defines a shell? Which quantum numbers define a subshell?

7.52 Which of the four quantum numbers (n, l, m_l, m_s) determine the energy of an electron in (a) a hydrogen atom and (b) a many-electron atom?

Problems

7.53 An electron in a certain atom is in the $n = 2$ quantum level. List the possible values of l and m_l that it can have.

7.54 An electron in an atom is in the $n = 3$ quantum level. List the possible values of l and m_l that it can have.

7.55 Give the values of the quantum numbers associated with the following orbitals: (a) 2*p*, (b) 3*s*, (c) 5*d*.

7.56 For the following subshells give the values of the quantum numbers (n, l, and m_l) and the number of orbitals in each subshell: (a) 4*p*, (b) 3*d*, (c) 3*s*, (d) 5*f*.

7.57 Discuss the similarities and differences between a 1*s* and a 2*s* orbital.

7.58 What is the difference between a $2p_x$ and a $2p_y$ orbital?

7.59 List the possible subshells and orbitals associated with the principal quantum number n, if $n = 5$.

7.60 List the possible subshells and orbitals associated with the principal quantum number n, if $n = 6$.

7.61 Calculate the total number of electrons that can occupy (a) one *s* orbital, (b) three *p* orbitals, (c) five *d* orbitals, (d) seven *f* orbitals.

7.62 What is the total number of electrons that can be held in all orbitals having the same principal quantum number n?

7.63 What is the maximum number of electrons that can be found in each of the following subshells? 3*s*, 3*d*, 4*p*, 4*f*, 5*f*.

7.64 Indicate the total number of (a) *p* electrons in N ($Z = 7$); (b) *s* electrons in Si ($Z = 14$); and (c) 3*d* electrons in S ($Z = 16$).

7.65 Make a chart of all allowable orbitals in the first four principal energy levels of the hydrogen atom. Designate each by type (for example, *s*, *p*) and indicate how many orbitals of each type there are.

7.66 Why do the 3*s*, 3*p*, and 3*d* orbitals have the same energy in a hydrogen atom but different energies in a many-electron atom?

7.67 For each of the following pairs of hydrogen orbitals, indicate which is higher in energy: (a) 1*s*, 2*s*; (b) 2*p*, 3*p*; (c) $3d_{xy}$, $3d_{yz}$; (d) 3*s*, 3*d*; (e) 5*s*, 4*f*.

7.68 Which orbital in each of the following pairs is lower in energy in a many-electron atom? (a) 2*s*, 2*p*; (b) 3*p*, 3*d*; (c) 3*s*, 4*s*; (d) 4*d*, 5*f*.

ELECTRON CONFIGURATION AND THE PERIODIC TABLE

Review Questions

7.69 Define the following terms: electron configuration, the Pauli exclusion principle, Hund's rule.

7.70 Explain the meaning of diamagnetic and paramagnetic. Give an example of an atom that is diamagnetic and one that is paramagnetic.

7.71 Explain the meaning of the symbol $4d^6$.

7.72 What is meant by the term "shielding of electrons" in an atom? Using the Li atom as an example, describe the effect of shielding on the energy of electrons in an atom.

7.73 Define and give an example for each of the following terms: transition metals, lanthanides, actinides.

7.74 Explain why the ground-state electron configurations for Cr and Cu are different from what we might expect.

7.75 Explain what is meant by a noble gas core. Write the electron configuration of a "xenon core."

7.76 Comment on the correctness of the following statement: The probability of finding two electrons with the same four quantum numbers in an atom is zero.

Problems

7.77 Which of the following sets of quantum numbers are unacceptable? Explain your answers. (a) $(1, 0, \frac{1}{2}, -\frac{1}{2})$, (b) $(3, 0, 0, +\frac{1}{2})$, (c) $(2, 2, 1, +\frac{1}{2})$, (d) $(4, 3, -2, +\frac{1}{2})$, (e) $(3, 2, 1, 1)$.

7.78 The ground-state electron configurations listed here are incorrect. Explain what mistakes have been made in each and write the correct electron configurations.

Al: $1s^22s^22p^43s^23p^3$
B: $1s^22s^22p^6$
F: $1s^22s^22p^6$

7.79 The atomic number of an element is 73. Are the atoms of this element diamagnetic or paramagnetic?

7.80 Indicate the number of unpaired electrons present in each of the following atoms: B, Ne, P, Sc, Mn, Se, Zr, Ru, Cd, I, W, Pb, Ce, Ho.

7.81 Write the ground-state electron configurations for the following elements: B, V, Ni, As, I, Au.

7.82 Write the ground-state electron configurations for the following elements: Ge, Fe, Zn, Ru, W, Tl.

7.83 The electron configuration of a neutral atom is $1s^22s^22p^63s^2$. Write a complete set of quantum numbers for each of the electrons. Name the element.

7.84 Which of the following species has the most unpaired electrons? S^+, S, or S^-. Explain how you arrived at your answer.

MISCELLANEOUS PROBLEMS

7.85 When a compound containing cesium ion is heated in a Bunsen burner flame, photons with an energy of 4.30×10^{-19} J are emitted. What color is the cesium flame?

7.86 Discuss the current view of the correctness of the following statements. (a) The electron in the hydrogen atom is in an orbit that never brings it closer than 100 pm to the nucleus. (b) Atomic absorption spectra result from transitions of electrons from lower to higher energy levels. (c) A many-electron atom behaves somewhat like a solar system that has a number of planets.

7.87 Distinguish carefully between the following terms: (a) wavelength and frequency, (b) wave properties and particle properties, (c) quantization of energy and continuous variation in energy.

7.88 What is the maximum number of electrons in an atom that can have the following quantum numbers. Label each electron with a particular orbital. (a) $n = 2$, $m_s = +\frac{1}{2}$; (b) $n = 4$, $m_l = +1$; (c) $n = 3$, $l = 2$; (d) $n = 2$, $l = 0$, $m_s = -\frac{1}{2}$; (e) $n = 4$, $l = 3$, $m_l = -2$.

7.89 Identify the following individuals and their contributions to the development of quantum theory: de Broglie, Einstein, Bohr, Planck, Heisenberg, Schrödinger.

7.90 What properties of electrons are used in the operation of an electron microscope?

7.91 In a photoelectric experiment a student uses a light source whose frequency is greater than that needed to eject electrons from a certain metal. However, after continuously shining the light on the same area of the metal for a long period of time the student notices that the maximum kinetic energy of ejected electrons begins to decrease, even though the frequency of the light is held constant. How would you account for this behavior?

7.92 A certain pitcher's fastballs have been clocked at about 120 mph. (a) Calculate the wavelength of a baseball (in nanometers) of mass 0.141 kg at this speed. (b) What is the wavelength of a hydrogen atom at the same speed? (1 mile = 1.61 km.)

7.93 Considering only the ground-state electron configuration, are there more elements that have diamagnetic or paramagnetic atoms? Explain.

Answers to Practice Exercises: **7.1** 8.24 m. **7.2** 3.39×10^3 nm. **7.3** 2.63×10^3 nm. **7.4** 56.6 nm. **7.5** $n = 3, l = 1, m_l = -1, 0, 1$. **7.6** 16. **7.7** $(5, 1, -1, +\frac{1}{2})$, $(5, 1, 0, +\frac{1}{2})$, $(5, 1, 1, +\frac{1}{2})$, $(5, 1, -1, -\frac{1}{2})$, $(5, 1, 0, -\frac{1}{2})$, $(5, 1, 1, -\frac{1}{2})$. **7.8** 32. **7.9** $[Ne]3s^23p^3$.

CHAPTER 8

THE PERIODIC TABLE

◆ In 1848, fourteen-year-old Dmitri Ivanovich Mendeleev took a journey that changed his life. Mendeleev was born in Siberia, the last of 14 children. The family was destitute, but his mother decided that Dmitri Ivanovich should go to Moscow to complete his education. After hitchhiking 14,000 miles across Russia, Mendeleev was refused admission to the University of Moscow because he was a Siberian. Mother and son went on to St. Petersburg, where Mendeleev obtained a grant to train as a secondary school teacher. Eventually he became a professor of inorganic chemistry at the University of St. Petersburg.

Over time Mendeleev noticed a relationship between the molecular masses and physical properties of similar compounds. He wondered whether an analogous relationship existed between atomic masses and the properties of elements. In 1869 Mendeleev drew up a table of elements arranged according to increasing atomic mass. The table revealed a periodic relationship among the elements' properties. Now known as the periodic table of elements, Mendeleev's chart proved to be the key for unlocking the mysteries of atomic structure and chemical bonding.

Mendeleev had discovered a fundamental law of nature. He also realized that there were many more elements to be discovered and predicted their properties using the periodic table. In 1875, when the French chemist Lecoq discovered gallium and gave its density as 4.7 g/cm^3, Mendeleev pointed out that the density should be 5.94 g/cm^3, which is accurate to within 1 percent of the true value. The scientific community was astounded that a theorist knew the properties of a new element better than the chemist who had discovered it!

In addition to formulating the periodic law, Mendeleev wrote textbooks in chemistry and philosophy, helped to develop the oil industry in Russia, and systematized weights and measurements. In his lifetime he was the most famous scientist in Russia. When he died in 1907, students bore huge periodic tables above the funeral procession.

Today Mendeleev's work on the periodic table is regarded as the most important contribution to chemistry in the nineteenth century. Element 101 is named mendelevium (Md) in his honor. ◆

Mendeleev was notorious for cutting his hair only once a year.

8.1 DEVELOPMENT OF THE PERIODIC TABLE

In the nineteenth century, when chemists had only a vague idea of atoms and molecules and did not know of the existence of electrons and protons, they devised the periodic table using their knowledge of atomic masses. Accurate measurements of the atomic masses of many elements had already been made. Arranging elements according to their atomic masses in a periodic table seemed logical to those chemists, who felt that chemical behavior should somehow be related to atomic mass.

In 1864 the English chemist John Newlands noticed that when the known elements were arranged in order of atomic mass, every eighth element had similar properties. Newlands referred to this peculiar relationship as the *law of octaves.* However, this "law" turned out to be inadequate for elements beyond calcium, and Newlands's work was not accepted by the scientific community.

Five years later the Russian chemist Dmitri Mendeleev and the German chemist Lothar Meyer independently proposed a much more extensive tabulation of the elements, based on the regular, periodic recurrence of properties. Mendeleev's classification was a great improvement over Newlands's for two reasons. First, it grouped the elements together more accurately, according to their properties. Equally important, it made possible the prediction of the properties of several elements that had not yet been discovered. For example, Mendeleev proposed the existence of an unknown element that he called eka-aluminum. (*Eka* is a Sanskrit word meaning "first"; thus eka-aluminum would be the first element under aluminum in the same group.) When gallium was discovered four years later, its properties closely matched the predicted properties of eka-aluminum (Table 8.1).

Nevertheless, the early versions of the periodic table had some glaring inconsistencies. For example, the atomic mass of argon (39.95 amu) is greater than that of potassium (39.10 amu). If elements were arranged solely according to increasing atomic mass, argon would appear in the position occupied by potassium in our modern periodic table (see the inside front cover). But no chemist would place argon, an inert gas, in the same group as lithium and sodium, two very reactive metals. This and other discrepancies suggested that some fundamental property other than atomic mass is the basis of the observed periodicity. This property turned out to be associated with atomic number.

Using data from α-scattering experiments (see Section 2.2), Rutherford was able to estimate the number of positive charges in the nucleus of a few elements, but until 1913 there was no general procedure for determining the atomic numbers. In that year a young English physicist, Henry Moseley, discovered a correlation between atomic number and the frequency of X rays generated by the bombardment of the element under study with high-energy electrons. With a few exceptions, Moseley found that the order of increasing atomic number is the same as the order of increasing atomic mass. For example, calcium is the twentieth element in increasing order of atomic mass, and it has an atomic number of 20. The discrepancies that bothered scientists now made sense. The atomic number of argon is 18 and that of potassium is 19, so potassium should follow argon in the periodic table.

A modern periodic table usually shows the atomic number along with the element symbol. As you already know, the atomic number also indicates the number of electrons in the atoms of an element. Electron configurations of elements help explain the recurrence of physical and chemical properties. The importance and usefulness of the periodic table lie in the fact that we can use our understand-

TABLE 8.1
Comparison of the Properties of Eka-Eluminum Predicted by Mendeleev and the Properties of Gallium*

Eka-aluminum (Ea)	Gallium (Ga)
Atomic mass about 68.	Atomic mass 69.9.
Metal of specific gravity 5.9; melting point low; nonvolatile; unaffected by air; should decompose steam at red heat; should dissolve slowly in acids and alkalis.	*Metal* of specific gravity 5.94; melting point 30.15°C; nonvolatile at moderate temperatures; does not change in air; action on steam unknown; dissolves slowly in acids and alkalis.
Oxide: formula Ea_2O_3; specific gravity 5.5; should dissolve in acids to form salts of the type EaX_3. The hydroxide should dissolve in acids and alkalis.	*Oxide:* Ga_2O_3; specific gravity unknown; dissolves in acids, forming salts of the type GaX_3. The hydroxide dissolves in acids and alkalis.
Salts should have tendency to form basic salts; the sulfate should form alums; the sulfide should be precipitated by H_2S or $(NH_4)_2S$. The anhydrous chloride should be more volatile than zinc chloride.	*Salts* hydrolyze readily and form basic salts; alums are known; the sulfide is precipitated by H_2S and by $(NH_4)_2S$ under special conditions. The anhydrous chloride is more volatile than zinc chloride.
The element will probably be discovered by spectroscopic analysis.	Gallium was discovered with the aid of the spectroscope.

*From M. E. Weeks, *Discovery of the Elements,* 6th ed., Chemical Education Publishing Company, Easton, Pa., 1956.

ing of the general properties and trends within a group or a period to predict with considerable accuracy the properties of any element, even though that element may be unfamiliar to us.

8.2 PERIODIC CLASSIFICATION OF THE ELEMENTS

Figure 8.1 shows the periodic table together with the outermost ground-state electron configuration of the elements. (The electron configurations of the elements are also given in Table 7.3.) Starting with hydrogen, we see that the subshells are filled in the order shown in Figure 7.21. According to the type of subshell being filled, the elements can be divided into categories—the representative elements, the noble gases, the transition elements (or transition metals), the lanthanides, and the actinides. Referring to Figure 8.1, the ***representative elements*** (also called *main group elements*) are *the elements in Groups 1A through 7A, all of which have incompletely filled s or p subshells of the highest principal quantum number.* With the exception of helium, the ***noble gases*** *(the Group 8A elements) all have a completely filled p subshell.* (The electron configurations are $1s^2$ for helium and ns^2np^6 for the other noble gases, where n is the principal quantum number for the outermost shell.) The transition metals are the elements in Groups 1B and 3B through 8B, which have incompletely filled d subshells or readily produce cations with incompletely filled d subshells. (These metals are sometimes referred to as the d-block transition elements.) The Group 2B elements are Zn, Cd, and Hg, which

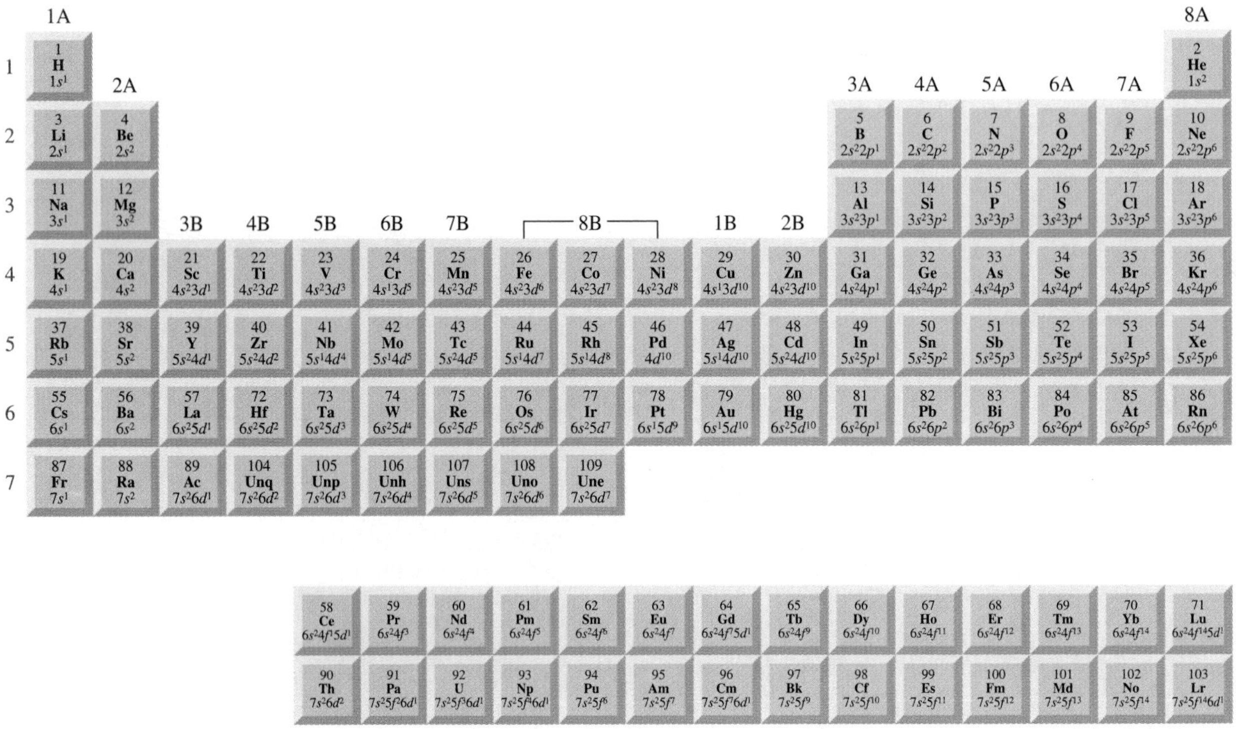
1A
8A
2A
3A 4A 5A 6A 7A
3B 4B 5B 6B 7B 8B 1B 2B
1 2 3 4 5 6 7
1 H $1s^1$
2 He $1s^2$
3 Li $2s^1$
4 Be $2s^2$
5 B $2s^22p^1$
6 C $2s^22p^2$
7 N $2s^22p^3$
8 O $2s^22p^4$
9 F $2s^22p^5$
10 Ne $2s^22p^6$
11 Na $3s^1$
12 Mg $3s^2$
13 Al $3s^23p^1$
14 Si $3s^23p^2$
15 P $3s^23p^3$
16 S $3s^23p^4$
17 Cl $3s^23p^5$
18 Ar $3s^23p^6$
19 K $4s^1$
20 Ca $4s^2$
21 Sc $4s^23d^1$
22 Ti $4s^23d^2$
23 V $4s^23d^3$
24 Cr $4s^13d^5$
25 Mn $4s^23d^5$
26 Fe $4s^23d^6$
27 Co $4s^23d^7$
28 Ni $4s^23d^8$
29 Cu $4s^13d^{10}$
30 Zn $4s^23d^{10}$
31 Ga $4s^24p^1$
32 Ge $4s^24p^2$
33 As $4s^24p^3$
34 Se $4s^24p^4$
35 Br $4s^24p^5$
36 Kr $4s^24p^6$
37 Rb $5s^1$
38 Sr $5s^2$
39 Y $5s^24d^1$
40 Zr $5s^24d^2$
41 Nb $5s^14d^4$
42 Mo $5s^14d^5$
43 Tc $5s^24d^5$
44 Ru $5s^14d^7$
45 Rh $5s^14d^8$
46 Pd $4d^{10}$
47 Ag $5s^14d^{10}$
48 Cd $5s^24d^{10}$
49 In $5s^25p^1$
50 Sn $5s^25p^2$
51 Sb $5s^25p^3$
52 Te $5s^25p^4$
53 I $5s^25p^5$
54 Xe $5s^25p^6$
55 Cs $6s^1$
56 Ba $6s^2$
57 La $6s^25d^1$
72 Hf $6s^25d^2$
73 Ta $6s^25d^3$
74 W $6s^25d^4$
75 Re $6s^25d^5$
76 Os $6s^25d^6$
77 Ir $6s^25d^7$
78 Pt $6s^15d^9$
79 Au $6s^15d^{10}$
80 Hg $6s^25d^{10}$
81 Tl $6s^26p^1$
82 Pb $6s^26p^2$
83 Bi $6s^26p^3$
84 Po $6s^26p^4$
85 At $6s^26p^5$
86 Rn $6s^26p^6$
87 Fr $7s^1$
88 Ra $7s^2$
89 Ac $7s^26d^1$
104 Unq $7s^26d^2$
105 Unp $7s^26d^3$
106 Unh $7s^26d^4$
107 Uns $7s^26d^5$
108 Uno $7s^26d^6$
109 Une $7s^26d^7$
58 Ce $6s^24f^15d^1$
59 Pr $6s^24f^3$
60 Nd $6s^24f^4$
61 Pm $6s^24f^5$
62 Sm $6s^24f^6$
63 Eu $6s^24f^7$
64 Gd $6s^24f^75d^1$
65 Tb $6s^24f^9$
66 Dy $6s^24f^{10}$
67 Ho $6s^24f^{11}$
68 Er $6s^24f^{12}$
69 Tm $6s^24f^{13}$
70 Yb $6s^24f^{14}$
71 Lu $6s^24f^{14}5d^1$
90 Th $7s^26d^2$
91 Pa $7s^25f^26d^1$
92 U $7s^25f^36d^1$
93 Np $7s^25f^46d^1$
94 Pu $7s^25f^6$
95 Am $7s^25f^7$
96 Cm $7s^25f^76d^1$
97 Bk $7s^25f^9$
98 Cf $7s^25f^{10}$
99 Es $7s^25f^{11}$
100 Fm $7s^25f^{12}$
101 Md $7s^25f^{13}$
102 No $7s^25f^{14}$
103 Lr $7s^25f^{14}6d^1$

FIGURE 8.1

The ground-state electron configurations of the elements. For simplicity, only the configurations of the outer electrons are shown.

are neither representative elements nor transition metals. The lanthanides and actinides are sometimes called *f*-block transition elements because they have incompletely filled *f* subshells.

A clear pattern emerges when we examine the electron configurations of the elements in a particular group. The electron configurations for Groups 1A and 2A elements are shown in Table 8.2. We see that all members of the Group 1A alkali metals have similar outer electron configurations; each has a noble gas core and an ns^1 configuration of the outer electron. Similarly, the Group 2A alkaline earth metals have a noble gas core and an ns^2 configuration of the outer electrons. *The outer electrons of an atom, which are those involved in chemical bonding,* are often called the ***valence electrons.*** Having the same number of valence electrons accounts for similarities in chemical behavior among the elements within each of these groups. This observation holds true also for the halogens (the Group 7A elements), which have outer electron configurations of ns^2np^5 and exhibit very similar properties. We must be careful, however, in predicting properties for Groups 3A through 6A. For example, the elements in Group 4A all have the same outer electron configuration, ns^2np^4, but there is much variation in chemical properties among these elements: Carbon is a nonmetal, silicon and germanium are metalloids, and tin and lead are metals.

As a group, the noble gases behave very similarly. With the exception of krypton and xenon, these elements are totally inert chemically. The reason is that these elements all have completely filled outer ns^2np^6 subshells, a condition that represents great stability. Although the outer electron configuration of the transition metals is not always the same within a group and there is no regular pattern in the change of the electron configuration from one metal to the next in the same period, all transition metals share many characteristics that set them apart from other elements. The reason is that these metals all have an incompletely filled *d*

TABLE 8.2

Electron Configurations of Group 1A and Group 2A Elements

Group 1A	Group 2A
Li: [He]$2s^1$	Be: [He]$2s^2$
Na: [Ne]$3s^1$	Mg: [Ne]$3s^2$
K: [Ar]$4s^1$	Ca: [Ar]$4s^2$
Rb: [Kr]$5s^1$	Sr: [Kr]$5s^2$
Cs: [Xe]$6s^1$	Ba: [Xe]$6s^2$
Fr: [Rn]$7s^1$	Ra: [Rn]$7s^2$

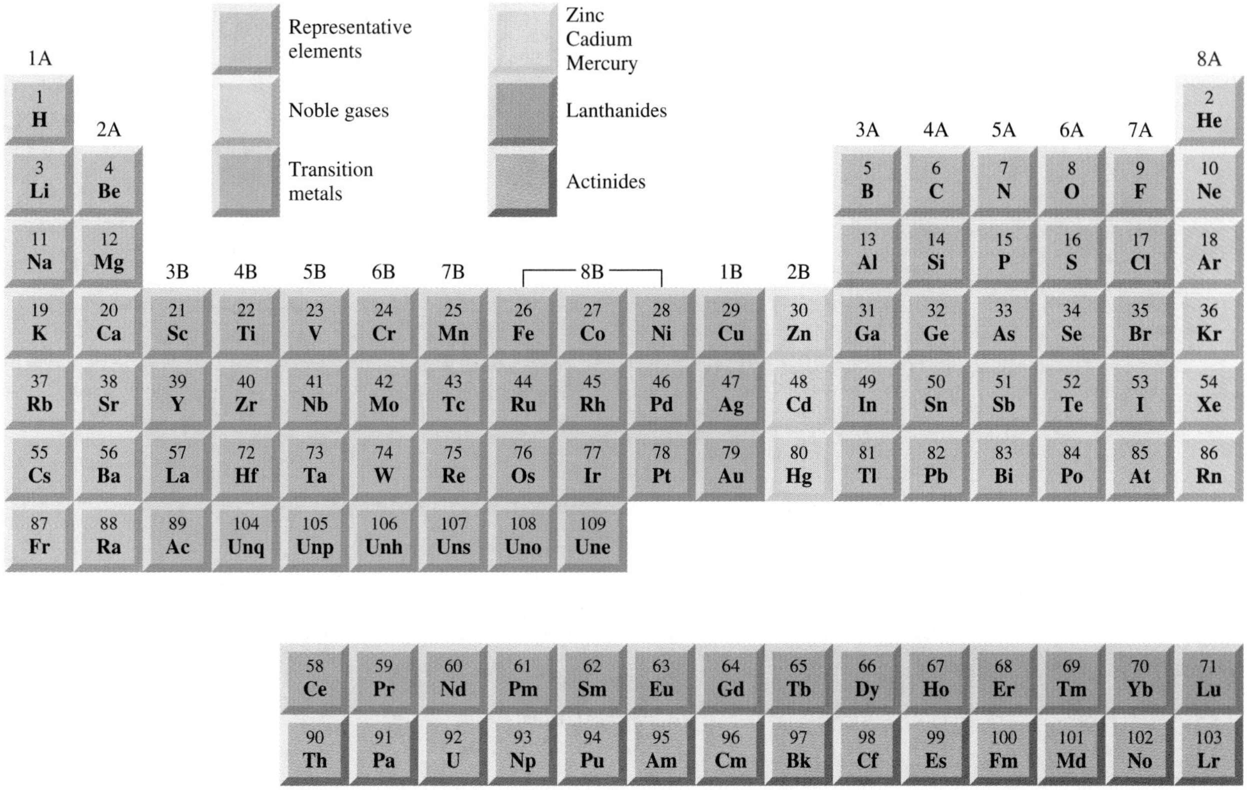

Representative elements
Noble gases
Transition metals
Zinc
Cadium
Mercury
Lanthanides
Actinides
1A
2A
3A
4A
5A
6A
7A
8A
3B
4B
5B
6B
7B
8B
1B
2B
1 H
2 He
3 Li
4 Be
5 B
6 C
7 N
8 O
9 F
10 Ne
11 Na
12 Mg
13 Al
14 Si
15 P
16 S
17 Cl
18 Ar
19 K
20 Ca
21 Sc
22 Ti
23 V
24 Cr
25 Mn
26 Fe
27 Co
28 Ni
29 Cu
30 Zn
31 Ga
32 Ge
33 As
34 Se
35 Br
36 Kr
37 Rb
38 Sr
39 Y
40 Zr
41 Nb
42 Mo
43 Tc
44 Ru
45 Rh
46 Pd
47 Ag
48 Cd
49 In
50 Sn
51 Sb
52 Te
53 I
54 Xe
55 Cs
56 Ba
57 La
72 Hf
73 Ta
74 W
75 Re
76 Os
77 Ir
78 Pt
79 Au
80 Hg
81 Tl
82 Pb
83 Bi
84 Po
85 At
86 Rn
87 Fr
88 Ra
89 Ac
104 Unq
105 Unp
106 Unh
107 Uns
108 Uno
109 Une
58 Ce
59 Pr
60 Nd
61 Pm
62 Sm
63 Eu
64 Gd
65 Tb
66 Dy
67 Ho
68 Er
69 Tm
70 Yb
71 Lu
90 Th
91 Pa
92 U
93 Np
94 Pu
95 Am
96 Cm
97 Bk
98 Cf
99 Es
100 Fm
101 Md
102 No
103 Lr

FIGURE 8.2

Classification of the elements. Note that the Group 2B elements are often classified as transition metals even though they do not exhibit the characteristics of the transition metals.

subshell. Likewise, the lanthanide (and the actinide) elements resemble one another within the series because they have incompletely filled *f* subshells. Figure 8.2 distinguishes the groups of elements discussed here.

EXAMPLE 8.1

Writing an Electron Configuration and Identifying an Element

A neutral atom of a certain element has 15 electrons. Without consulting a periodic table, answer the following questions: (a) What is the electron configuration of the element? (b) How should the element be classified? (c) Are the atoms of this element diamagnetic or paramagnetic?

Answer: (a) Using the building-up principle and knowing the maximum capacity of *s* and *p* subshells, we can write the electron configuration of the element as $1s^2 2s^2 2p^6 3s^2 3p^3$.

(b) Since the 3*p* subshell is not completely filled, this is a representative element. Based on the information given, we cannot say whether it is a metal, a nonmetal, or a metalloid.

(c) According to Hund's rule, the three electrons in the 3*p* orbitals have parallel spins. Therefore, the atoms of this element are paramagnetic, with three unpaired spins. (Remember, we saw in Chapter 7 that any atom that contains an odd number of electrons *must be* paramagnetic.)

PRACTICE EXERCISE

An atom of a certain element has 20 electrons. (a) Write the ground-state configuration of the element; (b) classify the element; (c) determine whether the atoms of the element are diamagnetic or paramagnetic.

ELECTRON CONFIGURATIONS OF CATIONS AND ANIONS

Because many ionic compounds are made up of monatomic anions and/or cations, it is helpful to know how to write the electron configurations of these ionic species. The procedure for writing the electron configurations of ions requires only a slight extension of the method used for neutral atoms. We will group the ions in two categories for discussion.

Ions Derived from Representative Elements

In the formation of a cation from the neutral atom of a representative element, one or more electrons are removed from the highest occupied n shell. Following are the electron configurations of some neutral atoms and their corresponding cations:

Na:	$[Ne]3s^1$	Na^+:	[Ne]
Ca:	$[Ar]4s^2$	Ca^{2+}:	[Ar]
Al:	$[Ne]3s^23p^1$	Al^{3+}:	[Ne]

Note that each ion has a stable noble gas configuration.

In the formation of an anion, one or more electrons are added to the highest partially filled n shell. Consider the following examples:

H:	$1s^1$	H^-:	$1s^2$ or [He]
F:	$1s^22s^22p^5$	F^-:	$1s^22s^22p^6$ or [Ne]
O:	$1s^22s^22p^4$	O^{2-}:	$1s^22s^22p^6$ or [Ne]
N:	$1s^22s^22p^3$	N^{3-}:	$1s^22s^22p^6$ or [Ne]

Again, all the anions have stable noble gas configurations. Thus a characteristic of most representative elements is that ions derived from their neutral atoms have the noble gas outer electron configuration ns^2np^6. *Ions, or atoms and ions, that have the same number of electrons, and hence the same ground-state electron configuration,* are said to be ***isoelectronic.*** Thus H^- and He are isoelectronic; F^-, Na^+, and Ne are isoelectronic; and so on.

Cations Derived from Transition Metals

In Section 7.10 we saw that in the first-row transition metals (Sc to Cu), the 4s orbital is always filled before the 3d orbitals. Consider manganese, whose electron configuration is $[Ar]4s^23d^5$. When the Mn^{2+} ion is formed, we might expect the two electrons to be removed from the 3d orbitals to yield $[Ar]4s^23d^3$. In fact, the electron configuration of Mn^{2+} is $[Ar]3d^5$! The reason is that the electron-electron and electron-nucleus interactions in a neutral atom can be quite different from those in its ion. Thus, whereas the 4s orbital is always filled before the 3d orbital in Mn, electrons are removed from the 4s orbital in forming Mn^{2+} because the 3d orbital is more stable than the 4s orbital in transition metal ions. Therefore, when a cation is formed from an atom of a transition metal, electrons are always removed first from the *ns* orbital and then from the $(n-1)d$ orbitals.

Keep in mind that most transition metals can form more than one cation and that frequently the cations are not isoelectronic with the preceding noble gases.

8.3 PERIODIC VARIATION IN PHYSICAL PROPERTIES

As we have seen, the electron configurations of the elements show a periodic variation with increasing atomic number. Consequently, the elements also display periodic variations in their physical and chemical behavior. In this section and the next two, we will examine some physical properties of elements in a group and across a period and properties that influence the chemical behavior of elements. Before we do so, however, let's look at the concept of effective nuclear charge, which will help us to understand these topics better.

EFFECTIVE NUCLEAR CHARGE

In Chapter 7, we discussed the shielding effect that electrons close to the nucleus have on outer-shell electrons in many-electron atoms. The presence of shielding electrons reduces the electrostatic attraction between the positively charged protons in the nucleus and the outer electrons. Moreover, the repulsive forces between electrons in a many-electron atom further offset the attractive force exerted by the nucleus. The concept of effective nuclear charge allows us to account for the effects of shielding on periodic properties.

Consider, for example, the helium atom, which has the electron configuration $1s^2$. Helium's two protons give the nucleus a charge of $+2$, but the full attractive force of this charge on the two 1*s* electrons is partially offset by electron-electron repulsion. Consequently, we say that each 1*s* electron is shielded from the nucleus by the other 1*s* electron. The *effective nuclear charge,* Z_{eff}, is given by

$$Z_{eff} = Z - \sigma$$

where Z is the actual nuclear charge and σ (sigma) is called the *shielding constant.* σ is greater than zero but smaller than Z.

One example of the effects of electron shielding is the energy required to remove an electron from a many-electron atom. Measurements show that it takes 2373 kJ of energy to remove the first electron from 1 mole of He atoms and 5251 kJ of energy to remove the remaining electron from 1 mole of He^+ ions. The reason it takes so much more energy to remove the second electron is that with only one electron present, there is no shielding and the electron feels the full effect of the $+2$ nuclear charge.

For atoms with three or more electrons, the electrons in a given shell are shielded by electrons in inner shells (that is, shells closer to the nucleus) but not by electrons in outer shells. Thus, for lithium, whose electron configuration is $1s^22s^1$, the 2*s* electron is shielded by the two 1*s* electrons, but the 2*s* electron does not have a shielding effect on the 1*s* electrons. In addition, filled inner shells shield outer electrons more effectively than electrons in the same subshell shield one another. As we will see, electron shielding has important consequences for atomic size and the formation of ions and molecules.

ATOMIC RADIUS

A number of physical properties, including density, melting point, and boiling point, are related to the sizes of atoms, but atomic size is difficult to define. As we saw in Chapter 7, the electron density in an atom extends far beyond the nucleus. In practice, we normally think of atomic size as the volume containing about 90 percent of the total electron density around the nucleus.

Several techniques allow us to estimate the size of an atom. First consider the metallic elements. The structure of metals is quite varied, but all metals share one structural characteristic: Their atoms are linked to one another in an extensive three-dimensional network. Thus the ***atomic radius*** of a metal is *one-half the distance between the two nuclei in two adjacent atoms* [Figure 8.3(a)]. *For elements that exist as diatomic molecules, the atomic radius is one-half the distance between the nuclei of the two atoms in a particular molecule* [Figure 8.3(b)].

Figure 8.4 shows the atomic radii of many elements according to their positions in the periodic table, and Figure 8.5 plots the atomic radii of these elements

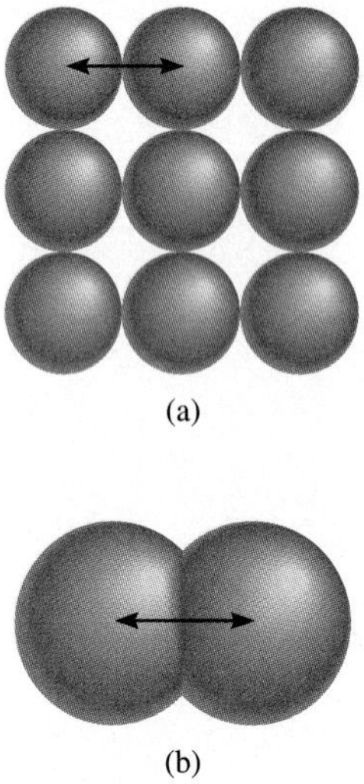

FIGURE 8.3

(a) In a metal, the atomic radius is defined as one-half the distance between the centers of two adjacent atoms. (b) For elements that exist as diatomic molecules, such as iodine, the radius of the atom is defined as one-half the distance between the centers of the atoms in the molecule.

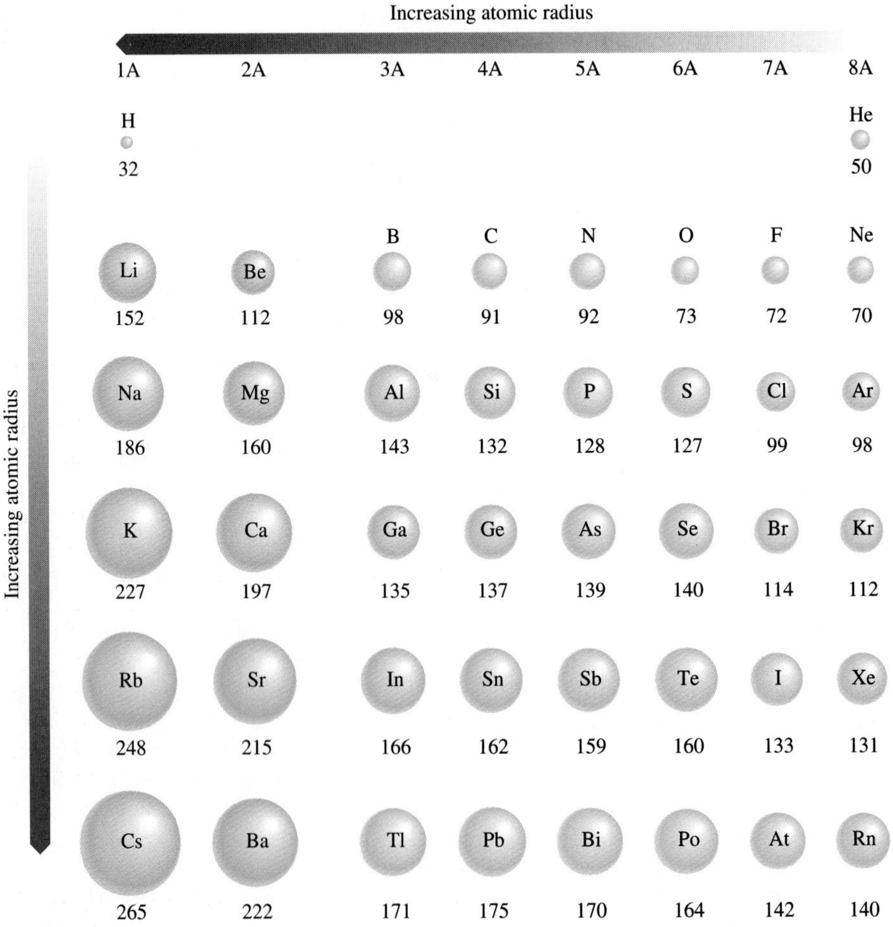

FIGURE 8.4

Atomic radii (in picometers) of representative elements according to their positions in the periodic table.

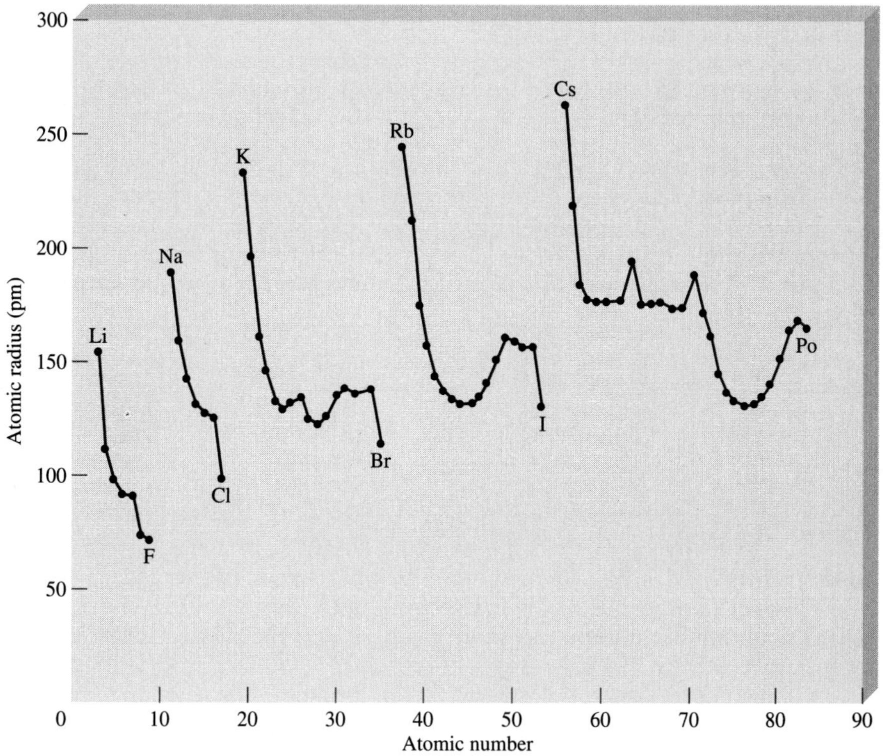

FIGURE 8.5

Plot of atomic radii (in picometers) of elements against their atomic numbers.

against their atomic numbers. Periodic trends are clearly evident. In studying the trends, bear in mind that the atomic radius is determined to a large extent by how strongly the outer-shell electrons are held by the nucleus. The larger the effective nuclear charge, the more strongly held these electrons are and the smaller the atomic radius is. Consider the second-period elements from Li to F. Moving from left to right, we find that the number of electrons in the inner shell ($1s^2$) remains constant while the nuclear charge increases. The electrons that are added to counterbalance the increasing nuclear charge are ineffective in shielding one another. Consequently, the effective nuclear charge increases steadily while the principal quantum number remains constant ($n = 2$). For example, the outer 2*s* electron on lithium is shielded from the nucleus (which has 3 protons) by the two 1*s* electrons. As an approximation, we assume that the shielding effect of the two 1*s* electrons is to cancel two positive charges in the nucleus. Thus the 2*s* electron only feels the attraction due to one proton in the nucleus, or the effective nuclear charge is $+1$. In beryllium ($1s^2 2s^2$) each of the 2*s* electrons is shielded by the inner two 1*s* electrons, which cancel two of the four positive charges in the nucleus. Because the 2*s* electrons do not shield each other as effectively, the net result is that the effective nuclear charge of each 2*s* electron is greater than $+1$. Thus as the effective nuclear charge increases, the atomic radius decreases steadily from lithium to fluorine.

As we go down a group—say, Group 1A—we find that atomic radius increases with increasing atomic number. For alkali metals the outermost electron resides in the *ns* orbital. Since the orbital size increases with the increasing principal quantum number *n*, the size of the metal atoms increases from Li to Cs. The same reasoning can be applied to elements in other groups.

EXAMPLE 8.2
Comparing the Sizes of Atoms

Referring to a periodic table, arrange the following atoms in order of increasing radius: P, Si, N.

Answer: Note that N and P are in the same group (Group 5A). Since atomic radius increases as we go down a group, the radius of N is smaller than that of P. Both Si and P are in the third period, and Si is to the left of P. Because atomic radius decreases as we move from left to right across a period, the radius of P is smaller than that of Si. Thus the order of increasing radius is N < P < Si.

PRACTICE EXERCISE

Arrange the following atoms in order of decreasing radius: C, Li, Be.

Ionic Radius

Ionic radius is *the radius of a cation or an anion.* Ionic radius affects the physical and chemical properties of an ionic compound. For example, the three-dimensional structure of an ionic compound depends on the relative sizes of its cations and anions.

When a neutral atom is converted to an ion, we expect a change in size. If the atom forms an anion, its size (or radius) increases, since the nuclear charge remains the same but the repulsion resulting from the additional electron(s) enlarges the domain of the electron cloud. On the other hand, a cation is smaller than the neutral atom, since removing one or more electrons reduces electron-electron repulsion but the nuclear charge remains the same, so the electron cloud shrinks.

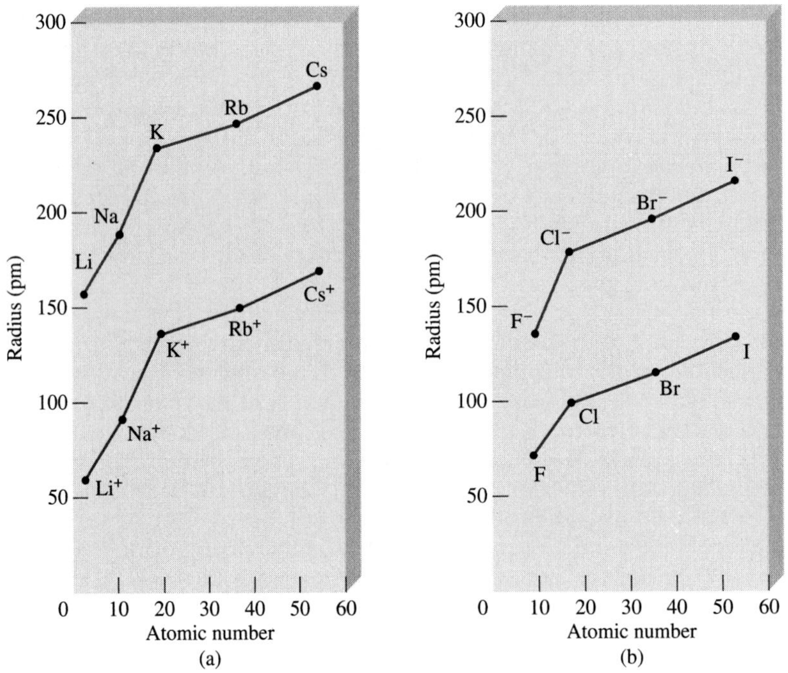

FIGURE 8.6
Comparison of atomic radii with ionic radii. (a) Alkali metals and alkali metal cations. (b) Halogens and halide ions.

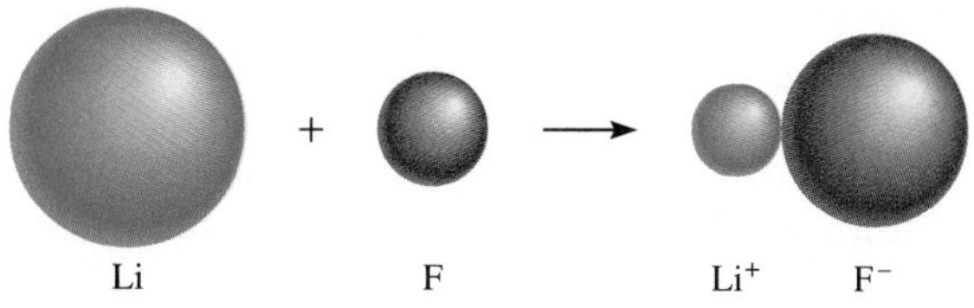

FIGURE 8.7

Changes in size when Li reacts with F to form LiF.

Figure 8.6 shows the changes in size when alkali metals are converted to cations and halogens are converted to anions; Figure 8.7 shows the changes in size when a lithium atom reacts with a fluorine atom to form a LiF unit.

Figure 8.8 shows the radii of ions derived from the more familiar elements, arranged according to their positions in the periodic table. We see that in some places there are parallel trends between atomic radii and ionic radii. For example, from the top to the bottom of the periodic table both the atomic radius and the ionic radius increase. For ions derived from elements in different groups, the comparison in size is meaningful only if the ions are isoelectronic. If we examine isoelectronic ions, we find that cations are smaller than anions. For example, Na^+ is smaller than F^-. Both ions have the same number of electrons, but Na ($Z = 11$) has more protons than F ($Z = 9$). The larger effective nuclear charge of Na^+ results in a smaller radius.

Focusing on isoelectronic cations, we see that the radii of *tripositive ions* (that is, ions that bear three positive charges) are smaller than those of *dipositive ions* (that is, ions that bear two positive charges), which in turn are smaller than *unipositive ions* (that is, ions that bear one positive charge). This trend is nicely illustrated by the sizes of three isoelectronic ions in the third period: Al^{3+}, Mg^{2+}, and Na^+ (see Figure 8.9). The Al^{3+} ion has the same number of electrons as Mg^{2+}, but

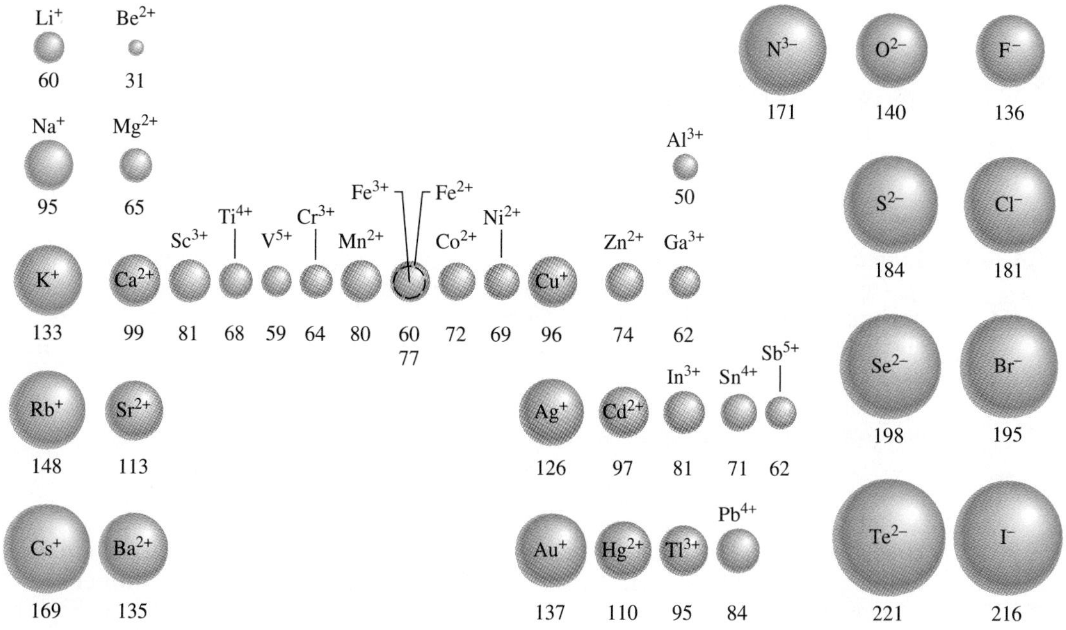

FIGURE 8.8

The radii in picometers of some ions of the more familiar elements shown in their positions in the periodic table.

it has one more proton. Thus the electron cloud in Al^{3+} is pulled inward more than that in Mg^{2+}. The smaller radius of Mg^{2+} compared with that of Na^+ can be similarly explained. Turning to isoelectronic anions, we find that the radius increases as we go from ions with uninegative charge (that is, $-$) to those with dinegative charge (that is, $2-$), and so on. Thus the oxide ion is larger than the fluoride ion because oxygen has one fewer proton than fluorine; the electron cloud is spread out to a greater extent in O^{2-}.

EXAMPLE 8.3

Comparing the Size of Ions

For each of the following pairs, indicate which one of the two species is larger: (a) N^{3-} or F^-; (b) Mg^{2+} or Ca^{2+}; (c) Fe^{2+} or Fe^{3+}?

Answer: (a) N^{3-} and F^- are isoelectronic anions. Since N^{3-} has two fewer protons than F^-, N^{3-} is larger.

(b) Both Mg and Ca belong to Group 2A (the alkaline metals). Because Ca lies below Mg, the Ca^{2+} ion is larger than Mg^{2+}.

(c) Both ions have the same nuclear charge, but Fe^{2+} has one more electron and hence greater electron-electron repulsion. The ionic radius of Fe^{2+} is larger.

PRACTICE EXERCISE

Select the smaller ion in each of the following pairs: (a) K^+, Li^+; (b) Au^+, Au^{3+}; (c) P^{3-}, N^{3-}.

8.4 IONIZATION ENERGY

As we will see throughout this book, the chemical properties of any atom are determined by the configuration of the atom's valence electrons. The stability of these outermost electrons is reflected directly in the atom's ionization energies. ***Ionization energy*** is *the minimum energy required to remove an electron from a gaseous atom in its ground state.* The magnitude of ionization energy is a measure of the effort required to force an atom to give up an electron, or of how "tightly" the electron is held in the atom. The higher the ionization energy, the more difficult it is to remove the electron.

For a many-electron atom, the amount of energy required to remove the first electron from the atom in its ground state,

$$\text{energy} + X(g) \longrightarrow X^+(g) + e^-$$

is called the *first ionization energy* (I_1). In the above equation, X represents a gaseous atom of any element and e^- is an electron. Unlike an atom in the condensed liquid and solid phases, an atom in the gaseous phase is virtually uninfluenced by its neighbors. The *second ionization energy* (I_2) and the *third ionization energy* (I_3) are shown in the following equations:

$$\text{energy} + X^+(g) \longrightarrow X^{2+}(g) + e^- \qquad \text{second ionization}$$

$$\text{energy} + X^{2+}(g) \longrightarrow X^{3+}(g) + e^- \qquad \text{third ionization}$$

The pattern continues for the removal of subsequent electrons.

When an electron is removed from a neutral atom, the repulsion among the remaining electrons decreases. Since the nuclear charge remains constant, more energy is needed to remove another electron from the positively charged ion. Thus, for the same element, ionization energies always increase in the following order:

$$I_1 < I_2 < I_3 < \cdots$$

Table 8.3 lists the ionization energies of the first 20 elements in kilojoules per mole (kJ/mol), that is, the amount of energy in kilojoules needed to remove 1 mole of electrons from 1 mole of gaseous atoms (or ions). By convention, energy absorbed by atoms (or ions) in the ionization process has a positive value. Thus ionization energies are all positive quantities. Figure 8.9 shows the variation of the first ionization energy with atomic number. The plot clearly exhibits the periodicity in the stability of the most loosely held electron. Note that, apart from small irregularities, the ionization energies of elements in a period increase with increasing atomic number. We can explain this trend by referring to the increase in effective nuclear charge from left to right (as in the case of atomic radii variation). A larger effective nuclear charge means a more tightly held outer electron, and hence a higher first ionization energy. A notable feature of Figure 8.9 is the peaks, which correspond to the noble gases. Such high ionization energies are consistent with the fact that most noble gases are chemically unreactive. In fact, helium has the highest first ionization energy of all the elements.

The Group 1A elements (the alkali metals) at the bottom of the graph in Figure 8.9 have the lowest ionization energies. Each of these metals has one valence elec-

TABLE 8.3

The Ionization Energies (kJ/mol) of the First 20 Elements

Z	Element	First	Second	Third	Fourth	Fifth	Sixth
1	H	1312					
2	He	2373	5251				
3	Li	520	7300	11815			
4	Be	899	1757	14850	21005		
5	B	801	2430	3660	25000	32820	
6	C	1086	2350	4620	6220	38000	47261
7	N	1400	2860	4580	7500	9400	53000
8	O	1314	3390	5300	7470	11000	13000
9	F	1680	3370	6050	8400	11000	15200
10	Ne	2080	3950	6120	9370	12200	15000
11	Na	495.9	4560	6900	9540	13400	16600
12	Mg	738.1	1450	7730	10500	13600	18000
13	Al	577.9	1820	2750	11600	14800	18400
14	Si	786.3	1580	3230	4360	16000	20000
15	P	1012	1904	2910	4960	6240	21000
16	S	999.5	2250	3360	4660	6990	8500
17	Cl	1251	2297	3820	5160	6540	9300
18	Ar	1521	2666	3900	5770	7240	8800
19	K	418.7	3052	4410	5900	8000	9600
20	Ca	589.5	1145	4900	6500	8100	11000

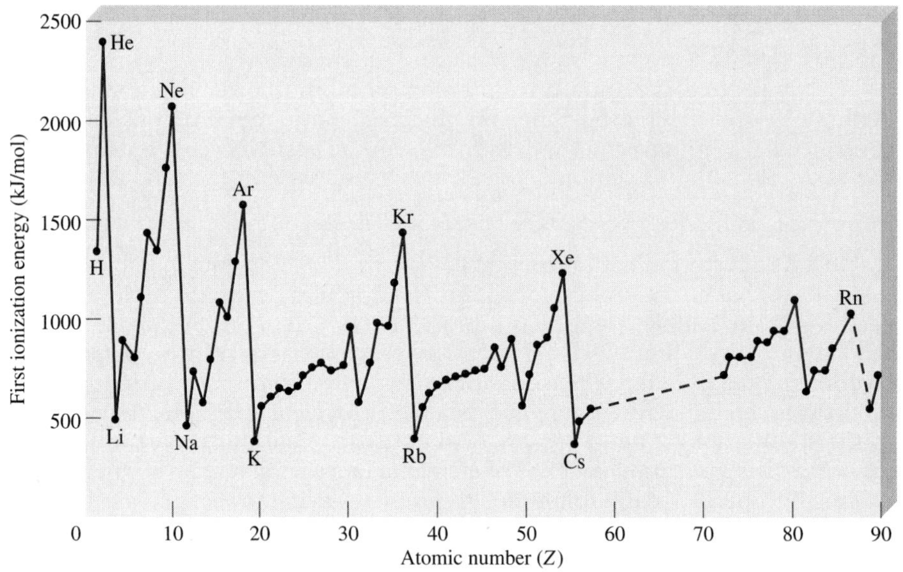

FIGURE 8.9

Variation of the first ionization energy with atomic number. Note that the noble gases have high ionization energies, whereas the alkali metals and alkaline earth metals have low ionization energies.

tron (the outermost electron configuration is ns^1) that is effectively shielded by the completely filled inner shells. Consequently, it is energetically easy to remove an electron from the atom of an alkali metal to form a unipositive ion (Li^+, Na^+, K^+, . . .).

The Group 2A elements (the alkaline earth metals) have higher first ionization energies than the alkali metals do. The alkaline earth metals have two valence electrons (the outermost electron configuration is ns^2). Because these two *s* electrons do not shield each other well, the effective nuclear charge for an alkaline earth metal atom is larger than that for the preceding alkali metal atom. Most alkaline earth compounds contain dipositive ions (Mg^{2+}, Ca^{2+}, Sr^{2+}, Ba^{2+}), which are isoelectronic with the unipositive alkali metal ions preceding them in the same period.

From the preceding discussion, we see that the metals have relatively low ionization energies, whereas nonmetals possess much higher ionization energies. The ionization energies of the metalloids usually fall between those of metals and nonmetals. The difference in ionization energies suggests why metals always form cations and nonmetals form anions in ionic compounds. For a given group, the ionization energy decreases with increasing atomic number (that is, as we move down the group). Elements in the same group have similar outer electron configurations. However, as the principal quantum number n increases, so does the average distance of a valence electron from the nucleus. A greater separation between the electron and the nucleus means a weaker attraction, so that the electron becomes increasingly easier to remove as we go from element to element down a group. Thus the metallic character of the elements within a group increases from top to bottom. This trend is particularly noticeable for elements in Groups 3A to 7A. For example, in Group 4A we note that carbon is a nonmetal, silicon and germanium are metalloids, and tin and lead are metals.

Although the general trend in the periodic table is that of ionization energies increasing from left to right, some irregularities do exist. The first occurs in going from Group 2A to 3A (for example, from Be to B and from Mg to Al). The group 3A elements all have a single electron in the outermost *p* subshell (ns^2np^1), which is well shielded by the inner electrons and the ns^2 electrons. Therefore, less energy is needed to remove a single p electron than to remove a paired s electron from the same principal energy level. This explains the *lower* ionization energies in Group 3A elements compared with those in Group 2A in the same period. The second irregularity occurs between Groups 5A and 6A (for example, from N to O and from P to S). In the Group 5A elements (ns^2np^3) the p electrons are in three separate orbitals according to Hund's rule. In Group 6A (ns^2np^4) the additional electron must be paired with one of the three p electrons. The proximity of two electrons in the same orbital results in greater electrostatic repulsion, which makes it easier to ionize an atom of the Group 6A element, even though the nuclear charge has increased by one unit. Thus the ionization energies in Group 6A are *lower* than those in Group 5A in the same period.

EXAMPLE 8.4

Comparing the Ionization Energies of Elements

(a) Which atom should have a smaller first ionization energy: oxygen or sulfur? (b) Which atom should have a higher second ionization energy: lithium or beryllium?

Answer: (a) Oxygen and sulfur are members of Group 6A. Since the ionization energy of elements decreases as we move down a periodic group, S should have a smaller first ionization energy.

(b) The electron configurations of Li and Be are $1s^22s^1$ and $1s^22s^2$, respectively. For the second ionization process we write

$$\underset{1s^2}{Li^+(g)} \longrightarrow \underset{1s^1}{Li^{2+}(g)}$$

$$\underset{1s^22s^1}{Be^+(g)} \longrightarrow \underset{1s^2}{Be^{2+}(g)}$$

Since $1s$ electrons shield $2s$ electrons much more effectively than they shield each other, we predict that it should be much easier to remove a $2s$ electron from Be^+ than to remove a $1s$ electron from Li^+. Table 8.3 confirms our prediction.

PRACTICE EXERCISE

(a) Which of the following atoms should have a larger first ionization energy: N or P? (b) Which of the following atoms should have a smaller second ionization energy: Na or Mg?

8.5 ELECTRON AFFINITY

Another atomic property that greatly influences chemical behavior is the ability to accept one or more electrons. This ability is measured by ***electron affinity,*** which is *the energy change when an electron is accepted by an atom in the gaseous state.* The equation is

$$X(g) + e^- \longrightarrow X^-(g)$$

TABLE 8.4
Electron Affinities (kJ/mol) of Representative Elements*

1A	2A	3A	4A	5A	6A	7A	8A
H −77							He (21)
Li −58	Be (241)	B −23	C −123	N 0	O −142	F −333	Ne (29)
Na −53	Mg (230)	Al −44	Si −120	P −74	S −200	Cl −348	Ar (35)
K −48	Ca (154)	Ga (−35)	Ge −118	As −77	Se −195	Br −324	Kr (39)
Rb −47	Sr (120)	In −34	Sn −121	Sb −101	Te −190	I −295	Xe (40)
Cs −45	Ba (52)	Tl −48	Pb −101	Bi −100	Po ?	At ?	Rn ?

*The values in parentheses are estimates.

where X is an atom of an element. In accordance with the convention used in thermochemistry (see Chapter 6), we assign a negative value to the electron affinity when energy is released. The more negative the electron affinity, the greater the tendency of the atom to accept an electron. Table 8.4 shows the electron affinity values of some representative elements and the noble gases, and Figure 8.10 plots the values of the first 20 elements versus atomic number. The overall trend is that the tendency to accept electrons increases (that is, electron affinity values become

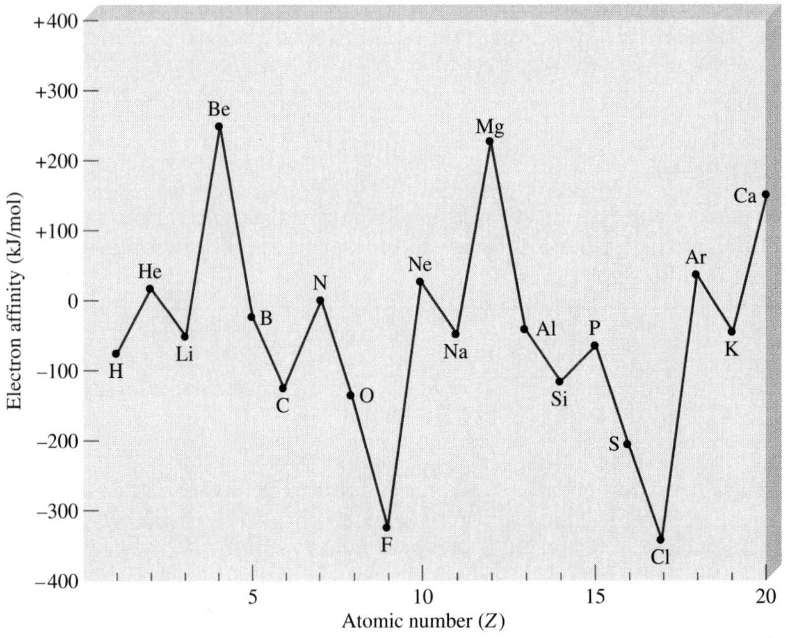

FIGURE 8.10

A plot of electron affinity against atomic number for the first 20 elements.

more negative) as we move from left to right across a period. The electron affinities of metals are generally more positive (or less negative) than those of nonmetals. The values vary little within a given group. The halogens (Group 7A) have the most negative electron affinity values. This is not surprising when we realize that by accepting an electron, each halogen atom assumes the stable electron configuration of the noble gas immediately following it. For example, the electron configuration of F^- is $1s^22s^22p^6$ or [Ne]; for Cl^- it is $[Ne]3s^23p^6$ or [Ar]; and so on. The noble gases, which have filled outer *s* and *p* subshells, have no tendency to accept electrons.

The electron affinity of oxygen has a negative value, which means that the process

$$O(g) + e^- \longrightarrow O^-(g)$$

is favorable. On the other hand, the electron affinity of the O^- ion,

$$O^-(g) + e^- \longrightarrow O^{2-}(g)$$

is positive (780 kJ/mol) even though the O^{2-} ion is isoelectronic with the noble gas Ne. This process is unfavorable in the gas phase because the resulting increase in electron-electron repulsion outweighs the stability gained in achieving a noble gas configuration. However, note that ions such as O^{2-} are common in ionic compounds (for example, Li_2O and MgO); in solids, these ions are stabilized by the neighboring cations.

EXAMPLE 8.5
The Electron Affinities of Alkaline Earth Metals

Explain why the electron affinities of the alkaline earth metals are all positive.

Answer: The outer-shell electron configuration of the alkaline earth metals is ns^2. For the process

$$M(g) + e^- \longrightarrow M^-(g)$$

where M denotes a member of the Group 2A family, the extra electron must enter the *np* subshell, which is effectively shielded by the two *ns* electrons (the *np* electrons are farther away from the nucleus than the *ns* electrons). Consequently, alkaline earth metals have no tendency to pick up an extra electron.

PRACTICE EXERCISE

Is it likely that Ar would form the anion Ar^-?

8.6 CHEMICAL PROPERTIES WITHIN INDIVIDUAL GROUPS

Ionization energy and electron affinity help chemists understand the types of reactions that elements undergo and the nature of the elements' compounds. Utilizing these concepts, we can survey the chemical behavior of the elements systematically, paying particular attention to the relationship between chemical properties and electron configuration.

General Trends in Chemical Properties

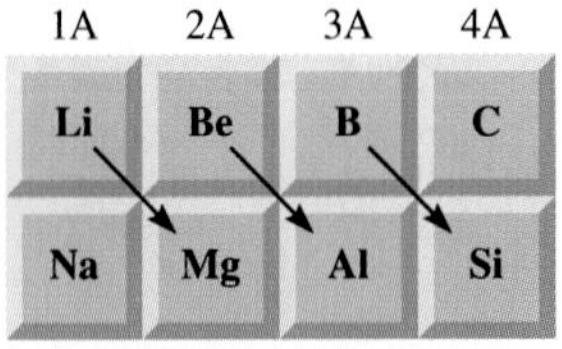

FIGURE 8.11

Diagonal relationships in the periodic table. The relationship holds for lithium and magnesium, beryllium and aluminum, and boron and silicon.

We have said that elements in the same group resemble one another in chemical behavior because they have similar outer electron configurations. This statement, although correct in the general sense, must be applied with caution. Chemists have long known that the first member of each group (that is, the elements in the second period from lithium to fluorine) differs from the rest of the members of the same group. For example, lithium, while exhibiting many of the properties characteristic of the alkali metals, is the only metal in Group 1A that does not form more than one compound with oxygen. Generally, the reason for the differences is the unusually small size of the first member of each group compared with the other members in the same group.

Another trend in chemical behavior of the representative elements is the ***diagonal relationship.*** Diagonal relationship refers to *similarities that exist between pairs of elements in different groups and periods of the periodic table.* Specifically, the first three members of the second period (Li, Be, and B) exhibit many similarities to the elements located diagonally below them in the periodic table (Figure 8.11). The chemistry of lithium resembles that of magnesium in some ways; the same holds for beryllium and aluminum and for boron and silicon. We will see some examples illustrating this relationship later.

In comparing the properties of elements in the same group, bear in mind that the comparison is most valid if we are dealing with elements of the same type. This guideline applies to the elements in Groups 1A and 2A, which are all metals, and to the elements in Group 7A, which are all nonmetals. But Groups 3A through 6A include nonmetals, metalloids, and metals, so it is natural to expect some variations in chemical properties even though the members of the same group have similar outer electron configurations.

Now let us take a closer look at the chemical properties of hydrogen and the representative elements.

Hydrogen ($1s^1$)

There is no totally suitable position for hydrogen in the periodic table. Traditionally hydrogen is shown in group 1A of the periodic table, but you should *not* think of it as a member of that group. Like the alkali metals, it has a single s valence electron and forms a unipositive ion (H^+), which is hydrated in solution. On the other hand, hydrogen also forms the hydride ion (H^-) in ionic compounds such as NaH and CaH_2. In this respect, hydrogen resembles the halogens, all of which form uninegative ions (F^-, Cl^-, Br^-, and I^-). Ionic hydrides react with water to produce hydrogen gas and the corresponding metal hydroxides:

$$2NaH(s) + 2H_2O(l) \longrightarrow 2NaOH(aq) + 2H_2(g)$$

$$CaH_2(s) + 2H_2O(l) \longrightarrow Ca(OH)_2(s) + 2H_2(g)$$

The most important compound of hydrogen is, of course, water, which is formed when hydrogen burns in air:

$$2H_2(g) + O_2(g) \longrightarrow 2H_2O(l)$$

Group 1A Elements (ns^1, $n \geq 2$)

Figure 8.12 shows the Group 1A elements, the alkali metals. All these elements have low ionization energies and therefore a great tendency to lose the single valence electron. In fact, in the vast majority of their compounds they are unipositive

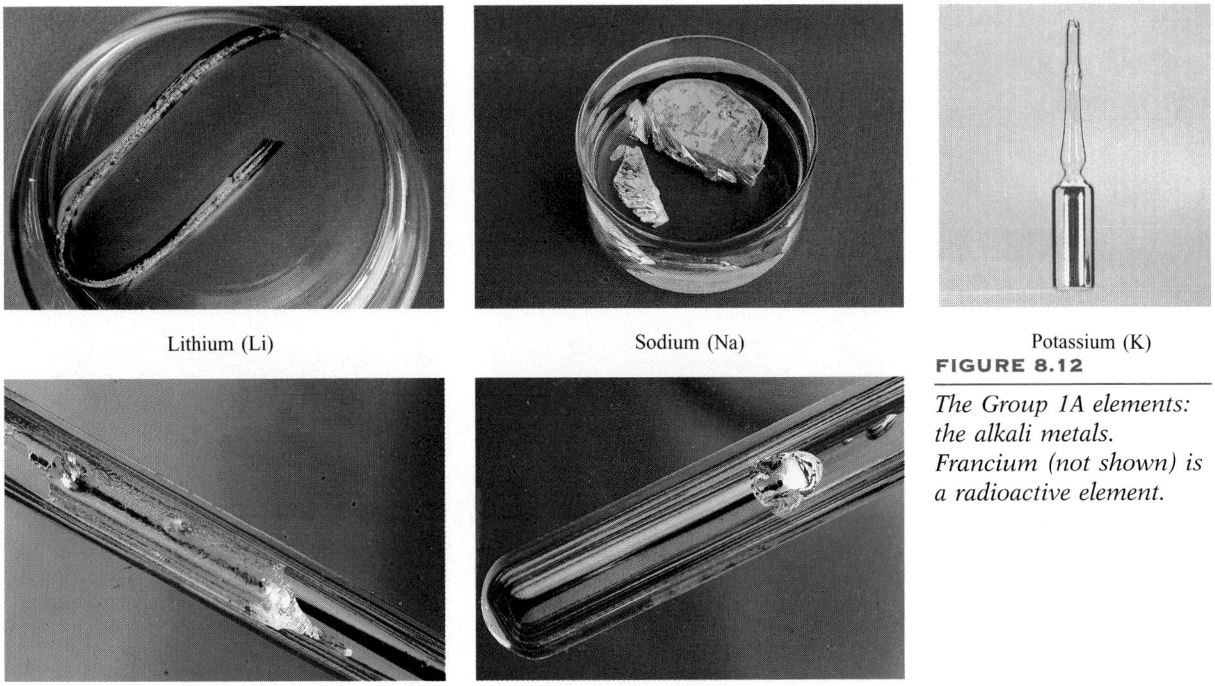
Lithium (Li) Sodium (Na) Potassium (K) Rubidium (Rb) Cesium (Cs)

FIGURE 8.12

The Group 1A elements: the alkali metals. Francium (not shown) is a radioactive element.

ions. These metals are so reactive that they are never found in the free, uncombined state in nature. They react with water to produce hydrogen gas and the corresponding metal hydroxide:

$$2M(s) + 2H_2O(l) \longrightarrow 2MOH(aq) + H_2(g)$$

where M denotes an alkali metal. When exposed to air, they gradually lose their shiny appearance as they combine with oxygen gas to form oxides. Lithium forms lithium oxide (containing the O^{2-} ion):

$$4Li(s) + O_2(g) \longrightarrow 2LI_2O(s)$$

The other alkali metals all form *peroxides* (containing the O_2^{2-} ion) in addition to oxides. For example,

$$2Na(s) + O_2(g) \longrightarrow Na_2O_2(s)$$

Potassium, rubidium, and cesium also form *superoxides* (containing the O_2^- ion):

$$K(s) + O_2(g) \longrightarrow KO_2(s)$$

The reason that different types of oxides are formed when alkali metals react with oxygen has to do with the stability of the oxides in the solid state. Since these oxides are all ionic compounds, their stability depends on how strongly the cations and anions attract one another. Lithium tends to form predominantly lithium oxide because this compound is more stable than lithium peroxide. The formation of other alkali metal oxides can be similarly explained.

Group 2A Elements (ns^2, $n \geq 2$)

Figure 8.13 shows the Group 2A elements. As a group, the alkaline earth metals are

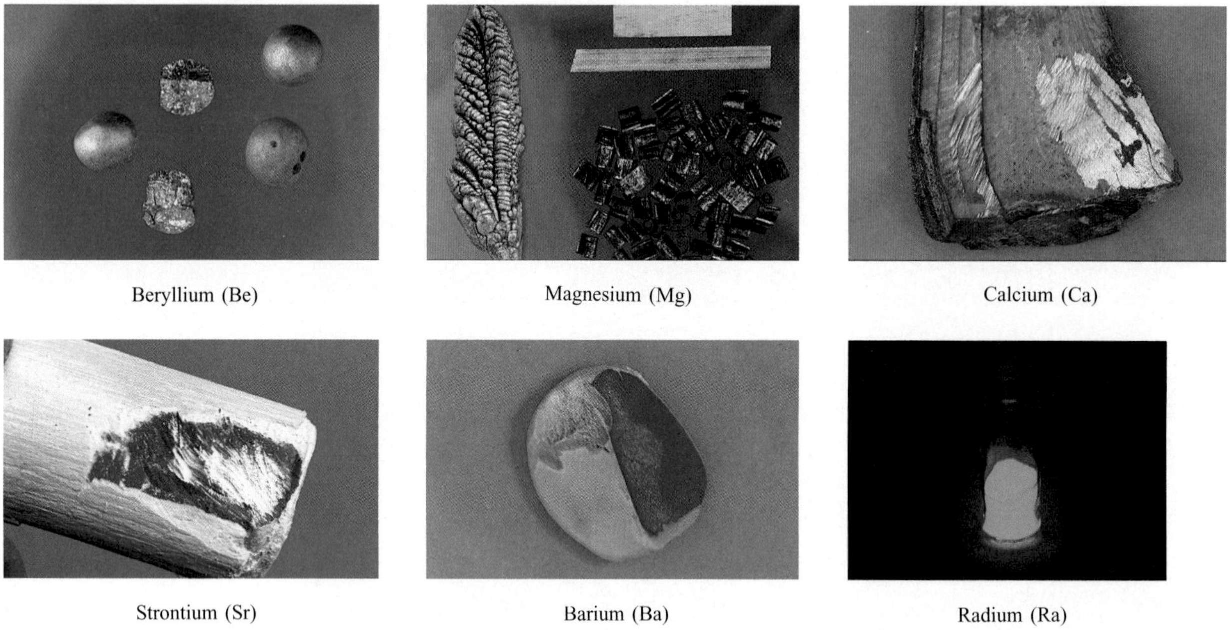

FIGURE 8.13

The Group 2A elements: the alkaline earth metals.

somewhat less reactive than the alkali metals. Both the first and the second ionization energies decrease from beryllium to barium. Thus the tendency is to form M^{2+} ions (where M denotes an alkaline metal atom), and hence the metallic character increases as we go down the group. Most beryllium compounds (BeH_2 and beryllium halides, such as $BeCl_2$) and some magnesium compounds (MgH_2, for example) are molecular rather than ionic in nature. The reactivities of alkaline earth metals with water vary quite markedly. Beryllium does not react with water; magnesium reacts slowly with steam; and calcium, strontium, and barium are reactive enough to attack cold water:

$$Ba(s) + 2H_2O(l) \longrightarrow Ba(OH)_2(aq) + H_2(g)$$

The reactivities of the alkaline earth metals toward oxygen also increase from Be to Ba. Beryllium and magnesium form oxides (BeO and MgO) only at elevated temperatures, whereas CaO, SrO, and BaO are formed at room temperature.

Magnesium reacts with acids to liberate hydrogen gas:

$$Mg(s) + 2H^+(aq) \longrightarrow Mg^{2+}(aq) + H_2(g)$$

Calcium, strontium, and barium also react with acids to produce hydrogen gas. However, because these metals also attack water, two different reactions will occur simultaneously.

The chemical properties of calcium and strontium provide an interesting example of periodic group similarity. Strontium-90, a radioactive isotope, is a major product of an atomic bomb explosion. If an atomic bomb is exploded in the atmosphere, the strontium-90 formed will eventually settle on land and water alike, and reach our bodies via a relatively short food chain. For instance, cows eat contaminated grass and drink contaminated water, and then they pass along strontium-90 in their milk. Because calcium and strontium are chemically similar, Sr^{2+} ions can replace Ca^{2+} ions in our bodies—for example, in bones. Constant exposure of the body to high-energy radiation emitted by the strontium-90 isotopes can lead to anemia, leukemia, and other chronic illnesses.

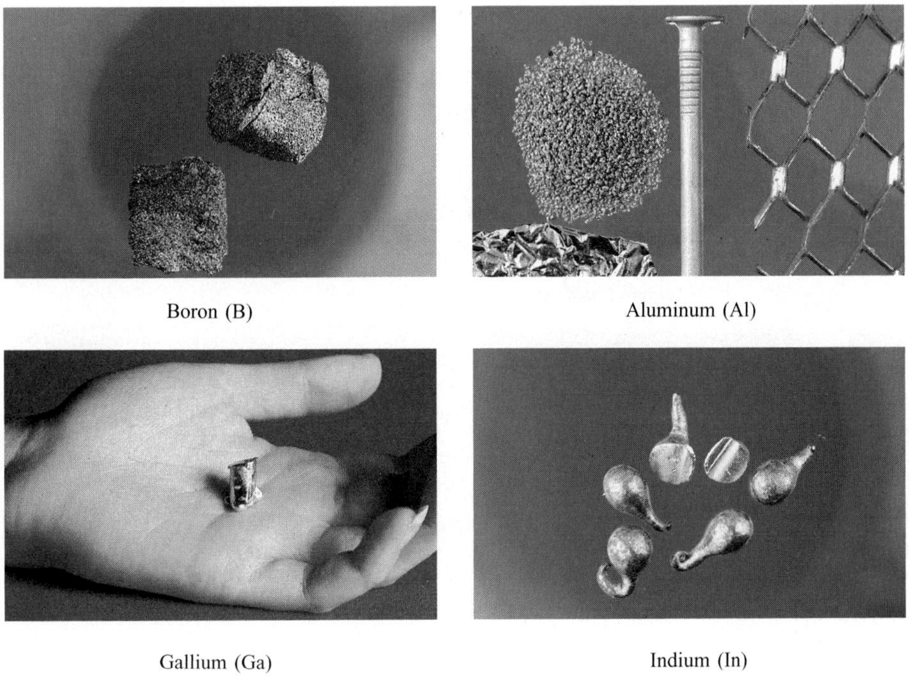

FIGURE 8.14

The Group 3A elements.

Group 3A Elements (ns^2np^1, $n \geq 2$)

B | Al | Ga | In | Tl

The first member of Group 3A, boron, is a metalloid; the rest are metals (Figure 8.14). Boron does not form binary ionic compounds and is unreactive toward oxygen gas and water. The next element, aluminum, readily forms aluminum oxide when exposed to air:

$$4Al(s) + 3O_2(g) \longrightarrow 2Al_2O_3(s)$$

Aluminum that has a protective layer of aluminum oxide is less reactive than elemental aluminum. Aluminum forms only tripositive ions. It reacts with hydrochloric acid as follows:

$$2Al(s) + 6H^+(aq) \longrightarrow 2Al^{3+}(aq) + 3H_2(g)$$

The other Group 3A metallic elements form both unipositive and tripositive ions. Moving down the group, we find that the unipositive ion becomes more stable than the tripositive ion.

The metallic elements in Group 3A also form many molecular compounds. For example, aluminum reacts with hydrogen to form AlH_3, which resembles BeH_2 in properties. (Here is an example of the diagonal relationship.) Thus, as we move from left to right across the periodic table, we begin to see a gradual shift from metallic to nonmetallic character in the representative elements.

Group 4A Elements (ns^2np^2, $n \geq 2$)

C | Si | Ge | Sn | Pb

The first member of Group 4A, carbon, is a nonmetal, and the next two members, silicon and germanium, are metalloids (Figure 8.15). They do not form ionic compounds. The metallic elements of this group, tin and lead, do not react with water but react with acids (hydrochloric acid, for example) to liberate hydrogen gas:

Carbon (graphite)

Carbon (diamond)

Silicon (Si)

Germanium (Ge)

Tin (Sn)

Lead (Pb)

FIGURE 8.15

The Group 4A elements.

$$Sn(s) + 2H^+(aq) \longrightarrow Sn^{2+}(aq) + H_2(g)$$

$$Pb(s) + 2H^+(aq) \longrightarrow Pb^{2+}(aq) + H_2(g)$$

The Group 4A elements form compounds in both the +2 and +4 oxidation states. For carbon and silicon, the +4 oxidation state is the more stable one. For example, CO_2 is more stable than CO, and SiO_2 is a stable compound, but SiO does not exist under normal conditions. As we move down the group, however, the trend in stability is reversed. In tin compounds the +4 oxidation state is only slightly more stable than the +2 oxidation state. In lead compounds the +2 oxidation state is unquestionably the more stable one. The outer electron configuration of lead is $6s^26p^2$, and lead tends to lose only the $6p$ electrons (to form Pb^{2+}) rather than both the $6p$ and $6s$ electrons (to form Pb^{4+}).

Group 5A Elements (ns^2np^3, $n \geq 2$)

N

P

As

Sb

Bi

In Group 5A, nitrogen and phosphorus are nonmetals, arsenic and antimony are metalloids, and bismuth is a metal (Figure 8.16). Thus we expect a larger variation in properties as we move down the group. Elemental nitrogen is a diatomic gas (N_2). It forms a number of oxides (NO, N_2O, NO_2, N_2O_4, and N_2O_5), of which only N_2O_5 is a solid; the others are gases. Nitrogen has a tendency to accept three electrons to form the nitride ion, N^{3-} (thus achieving the electron configuration $2s^22p^6$, which is isoelectronic with neon). Most metallic nitrides (Li_3N and Mg_3N_2, for example) are ionic compounds. Phosphorus exists as P_4 molecules. It forms two solid oxides with the formulas P_4O_6 and P_4O_{10}. The important oxoacids HNO_3 and H_3PO_4 are formed when the following oxides react with water:

$$N_2O_5(g) + H_2O(l) \longrightarrow 2HNO_3(aq)$$

$$P_4O_{10}(s) + 6H_2O(l) \longrightarrow 4H_3PO_4(aq)$$

Arsenic, antimony, and bismuth have extensive three-dimensional structures. Bismuth is a far less reactive metal than those in the preceding groups.

FIGURE 8.16

The Group 5A elements. Molecular nitrogen is a colorless, odorless gas.

Nitrogen (N_2)

White and red phosphorus (P)

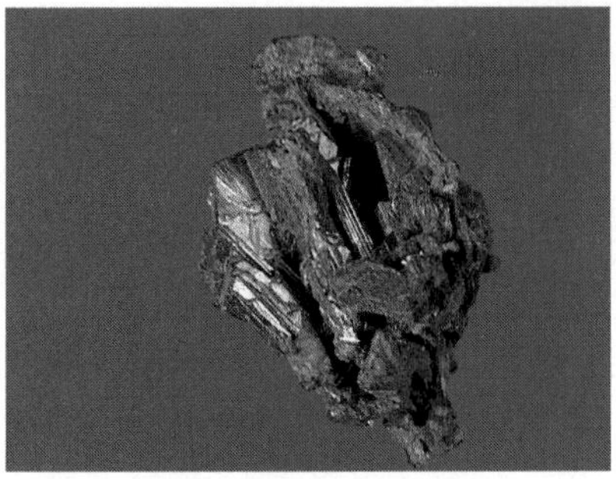

Arsenic (As)

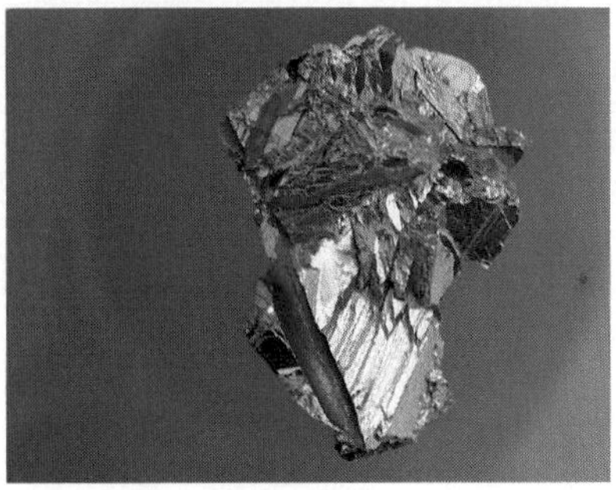

Antimony (Sb)

Bismuth (Bi)

Group 6A Elements (ns^2np^4, $n \geq 2$)

The first three members of Group 6A (oxygen, sulfur, and selenium) are nonmetals, and the last two (tellurium and polonium) are metalloids (Figure 8.17). Oxygen is a diatomic gas; elemental sulfur and selenium have the molecular formulas S_8 and Se_8, respectively; tellurium and polonium have more extensive three-dimensional structures. (Polonium is a radioactive element that is difficult to study in the laboratory.) Oxygen has a tendency to accept two electrons to form the oxide ion (O^{2-}) in many ionic compounds. (The electron configuration of O^{2-} is $1s^22s^22p^6$, which is isoelectronic with Ne.) Sulfur, selenium, and tellurium also form dinegative anions: S^{2-}, Se^{2-}, and Te^{2-}. The elements in this group (especially oxygen) form a large number of molecular compounds with nonmetals. The important compounds of sulfur are SO_2, SO_3, and H_2S. Sulfuric acid is formed when sulfur trioxide dissolves in water:

$$SO_3(g) + H_2O(l) \longrightarrow H_2SO_4(aq)$$

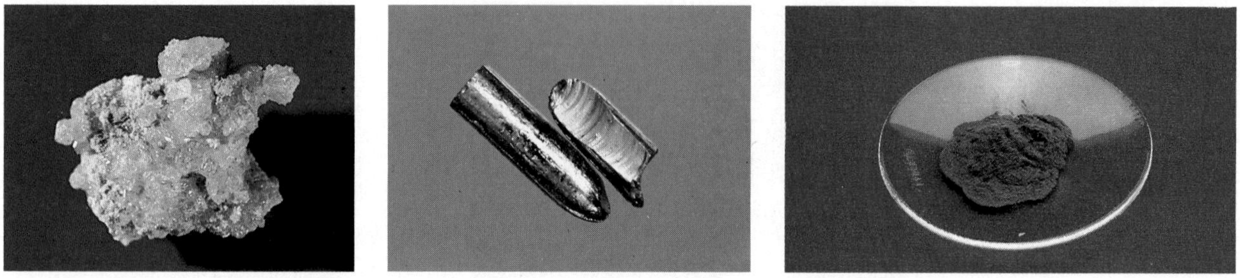

Sulfur (S_8) Selenium (Se_8) Tellurium (Te)

FIGURE 8.17

The Group 6A elements sulfur, selenium, and tellurium. Molecular oxygen is a colorless, odorless gas. Polonium is a radioactive element.

Group 7A Elements (ns^2np^5, $n \geq 2$)

All the halogens are nonmetals with the general formula X_2, where X denotes a halogen element (Figure 8.18). Because of their great reactivity, the halogens are never found in the elemental form in nature. (The last member of Group 7A is astatine, a radioactive element. Little is known about its properties.) Fluorine is so reactive that it attacks water to generate oxygen:

$$2F_2(g) + 2H_2O(l) \longrightarrow 4HF(aq) + O_2(g)$$

Actually the reaction between molecular fluorine and water is quite complex; the products formed depend on reaction conditions. The reaction shown here is one of several possible changes. The halogens have high ionization energies and large negative electron affinities. These facts suggest that they would preferentially form anions of the type X^-. Anions derived from the halogens (F^-, Cl^-, Br^-, and I^-) are called *halides.* They are isoelectronic with the noble gases. For example, F^- is isoelectronic with Ne, Cl^- with Ar, and so on. The vast majority of the alkali metal and alkaline earth metal halides are ionic compounds. The halogens also form many molecular compounds among themselves (such as ICl and BrF_3) and with nonmetallic elements in other groups (such as NF_3, PCl_5, and SF_6). The halogens react with hydrogen to form hydrogen halides:

$$H_2(g) + X_2(g) \longrightarrow 2HX(g)$$

FIGURE 8.18

The Group 7A elements chlorine, bromine, and iodine. Fluorine is a greenish-yellow gas that attacks ordinary glassware. Astatine is a radioactive element.

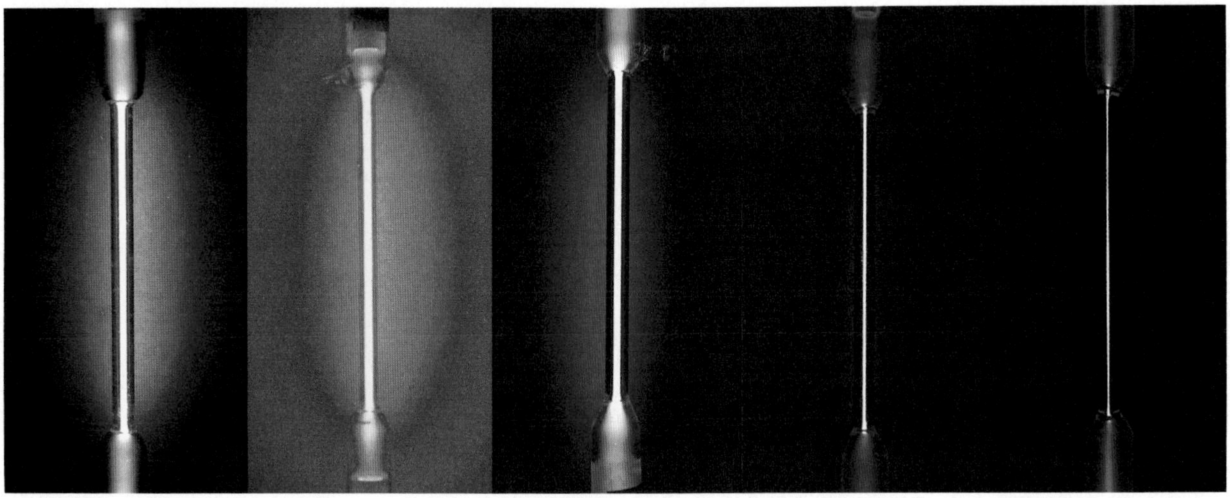

Helium (He) Neon (Ne) Argon (Ar) Krypton (Kr) Xenon (Xe)

FIGURE 8.19

All noble gases are colorless and odorless. These pictures show the colors emitted by the gases from a discharge tube.

When this reaction involves fluorine, it is explosive, but it becomes less and less violent as we substitute chlorine and iodine. The hydrogen halides dissolve in water to form hydrohalic acids. Of this series, hydrofluoric acid (HF) is a weak acid (that is, it is a weak electrolyte), but the other acids (HCl, HBr, and HI) are all strong acids (or strong electrolytes).

Group 8A Elements (ns^2np^6, $n \geq 2$)

All noble gases exist as monatomic species (Figure 8.19). The electron configurations of the noble gases show that their atoms have completely filled outer *ns* and *np* subshells, indicating great stability. (Helium is $1s^2$.) The Group 8A ionization energies are among the highest of all elements, and these gases have no tendency to accept extra electrons. For a number of years these elements were called inert gases, and rightly so. Until 1963 no one had been able to prepare a compound containing any of these elements. But an experiment carried out by the British chemist Neil Bartlett shattered chemists' long-held views of these elements. When he exposed xenon to platinum hexafluoride, a strong oxidizing agent, the following reaction took place (Figure 8.20):

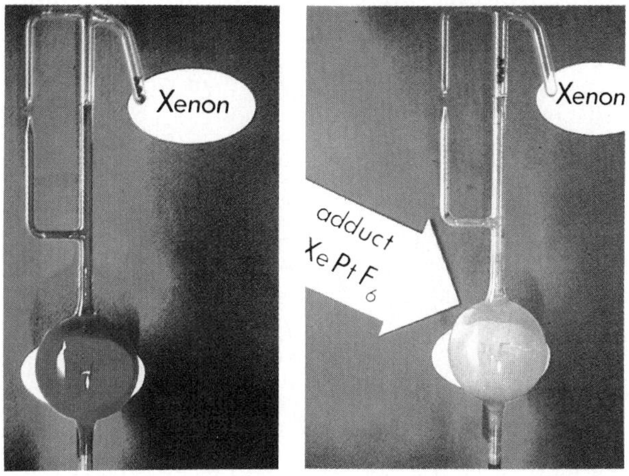

(a) (b)

FIGURE 8.20

(a) Xenon gas (colorless) and PtF_6 (red gas) separated from each other. (b) When the two gases are mixed, a yellow-orange solid compound is formed.

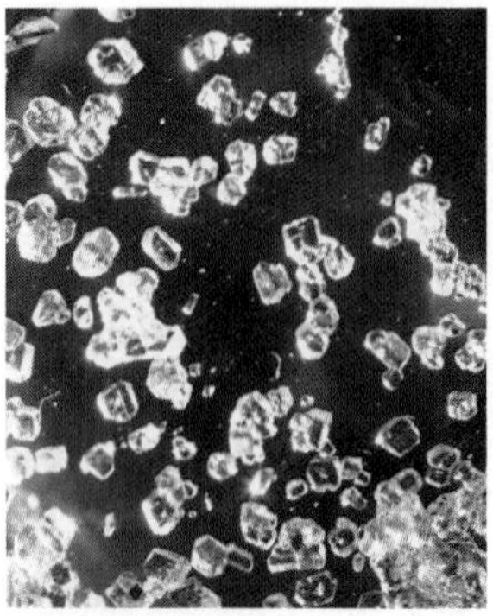

FIGURE 8.21

Crystals of xenon tetrafluoride (XeF_4).

$$Xe(g) + PtF_6(g) \longrightarrow XePtF_6(s)$$

Since then, a number of xenon compounds (XeF_4, XeO_3, XeO_4, $XeOF_4$) and a few krypton compounds (KrF_2, for example) have been prepared (Figure 8.21). Despite the immense interest in the chemistry of the noble gases, however, their compounds do not have any commercial applications, and they are not involved in natural biological processes. No compounds of helium, neon, and argon are known.

Properties of Oxides Across a Period

One way to compare the properties of the representative elements across a period is to examine the properties of a series of similar compounds. Since oxygen combines with almost all elements, we will compare the properties of oxides of the third-period elements to see how metals differ from metalloids and nonmetals. Table 8.5 lists a few general characteristics of the oxides.

We observed earlier that oxygen has a tendency to form the oxide ion. This tendency is greatly favored in reactions with metals that have low ionization energies, namely, those in Groups 1A and 2A and aluminum. Thus Na_2O, MgO, and Al_2O_3 are ionic compounds, as indicated by their high melting points and boiling points. They have extensive three-dimensional structures in which each cation is surrounded by a specific number of anions, and vice versa. As the ionization energies of the elements increase from left to right, so does the molecular nature of the oxides that are formed. Silicon is a metalloid; its oxide (SiO_2) also has a huge three-dimensional network, although no ions are present. The oxides of phosphorus, sulfur, and chlorine are molecular compounds composed of small discrete units. The weak attractions among these molecules result in relatively low melting points and boiling points.

Most oxides can be classified as acidic or basic, depending on whether they produce acids or bases when dissolved in water or react as acids or bases in certain processes. Some oxides are ***amphoteric oxides,*** which means that they *display both acidic and basic properties.* The first two oxides of the third period, Na_2O and MgO, are basic oxides. For example, Na_2O reacts with water to form sodium hydroxide (which is a base):

$$Na_2O(s) + H_2O(l) \longrightarrow 2NaOH(aq)$$

Magnesium oxide is quite insoluble; it does not react with water to any appreciable extent. However, it does react with acids in a manner that resembles an acid-base reaction:

$$MgO(s) + 2HCl(aq) \longrightarrow MgCl_2(aq) + H_2O(l)$$

TABLE 8.5
Some Properties of Oxides of the Third-Period Elements

	Na_2O	MgO	Al_2O_3	SiO_2	P_4O_{10}	SO_3	Cl_2O_7
Type of compound	← Ionic →			← Molecular →			
Structure	← Extensive three-dimensional →				← Discrete molecular units →		
Melting point (°C)	1275	2800	2045	1610	580	16.8	−91.5
Boiling point (°C)	?	3600	2980	2230	?	44.8	82
Acid-base nature	Basic	Basic	Amphoteric	← Acidic →			

Note that the products of this reaction are a salt ($MgCl_2$) and water, the usual products of an acid-base neutralization.

Aluminum oxide is even less soluble than magnesium oxide; it too does not react with water. However, it shows basic properties by reacting with acids:

$$Al_2O_3(s) + 6HCl(aq) \longrightarrow 2AlCl_3(aq) + 3H_2O(l)$$

It also exhibits acidic properties by reacting with bases:

$$Al_2O_3(s) + 2NaOH(aq) + 3H_2O(l) \longrightarrow 2NaAl(OH)_4(aq)$$

Thus Al_2O_3 is classified as an amphoteric oxide because it has properties of both acids and bases. Other amphoteric oxides are ZnO, BeO, and Bi_2O_3.

Silicon dioxide is insoluble and does not react with water. It has acidic properties, however, because it reacts with very concentrated bases:

$$SiO_2(s) + 2NaOH(aq) \longrightarrow Na_2SiO_3(aq) + H_2O(l)$$

For this reason, concentrated bases such as NaOH should not be stored in Pyrex glassware, which is made of SiO_2.

The remaining third-period oxides are acidic, as indicated by their reactions with water to form phosphoric acid (H_3PO_4), sulfuric acid (H_2SO_4), and perchloric acid ($HClO_4$):

$$P_4O_{10}(s) + 6H_2O(l) \longrightarrow 4H_3PO_4(aq)$$

$$SO_3(g) + H_2O(l) \longrightarrow H_2SO_4(aq)$$

$$Cl_2O_7(s) + H_2O(l) \longrightarrow 2HClO_4(aq)$$

This brief examination of oxides of the third-period elements shows that as the metallic character of the elements decreases from left to right across the period, their oxides change from basic to amphoteric to acidic. Normal metallic oxides are usually basic, and most oxides of nonmetals are acidic. The intermediate properties of the oxides (as shown by the amphoteric oxides) are exhibited by elements whose positions are intermediate within the period. Note also that since the metallic character of the elements increases as we move down a particular group of the representative elements, we would expect oxides of elements with higher atomic numbers to be more basic than the lighter elements. This is indeed the case.

SUMMARY

Newlands, Mendeleev, and Meyer devised the periodic table by arranging elements in order of increasing mass. Some discrepancies in the early version of the periodic table were resolved when Moseley showed that elements should be arranged according to increasing atomic number.

Electron configuration directly influences the properties of the elements. The modern periodic table classifies the elements according to their atomic numbers, and thus also by their electron configurations. The configuration of the outermost electrons (called valence electrons) directly affects the properties of the atoms of the representative elements.

Periodic variations in the physical properties of the elements reflect differences in structure. The metallic character of elements decreases across a period from metals through the metalloids to nonmetals and increases from top to bottom within a particular group of representative elements. Atomic size defined in terms of atomic radius also varies periodically, decreasing from left to right and increasing from top to bottom within a group.

Ionization energy is a measure of the tendency of an atom to resist the loss of an electron. The higher the ionization energy, the stronger the nucleus's hold on the electron. Electron affinity is a measure of the tendency of an atom to gain an electron. The more negative the electron affinity, the greater the tendency for the atom to gain an electron. In general, metals have low ionization energies and nonmetals have high (large negative) electron affinities.

Noble gases are very stable because their outer *ns* and *np* subshells are completely filled. The metals of the representative elements (in Groups 1A, 2A, and 3A) tend to lose electrons until their cations become isoelectronic with the noble gases preceding them in the periodic table. The nonmetals of the representative elements (in Groups 5A, 6A, and 7A) tend to accept electrons until their anions become isoelectronic with the noble gases following them in the periodic table.

KEY WORDS

Amphoteric oxide, p. 240
Atomic radius, p. 222
Diagonal relationship, p. 232
Electron affinity, p. 229
Ionic radius, p. 224
Ionization energy, p. 226
Isoelectronic, p. 220
Noble gases, p. 217
Representative elements, p. 217
Valence electrons, p. 218

QUESTIONS AND PROBLEMS

DEVELOPMENT OF THE PERIODIC TABLE

Review Questions

8.1 Briefly describe the significance of Mendeleev's periodic table.

8.2 What is Moseley's contribution to the modern periodic table?

8.3 Describe the general layout of a modern periodic table.

8.4 What is the most important relationship among elements in the same group in the periodic table?

PERIODIC CLASSIFICATION OF THE ELEMENTS

Review Questions

8.5 Which of the following elements are metals, nonmetals, and metalloids? As, Xe, Fe, Li, B, Cl, Ba, P, I, Si.

8.6 Compare the physical and chemical properties of metals and nonmetals.

8.7 Draw a rough sketch of a periodic table (no details are required). Indicate where metals, nonmetals, and metalloids are located.

8.8 What is a representative element? Give names and symbols of four representative elements.

8.9 Without referring to a periodic table, write the name and symbol for an element in each of the following groups: 1A, 2A, 3A, 4A, 5A, 6A, 7A, 8A, transition metals.

8.10 Indicate whether the following elements exist as atomic species, molecular species, or extensive three-dimensional structures in their most stable state at 25°C and 1 atm, and write the molecular or empirical formula for the elements: phosphorus, iodine, magnesium, neon, arsenic, sulfur, boron, selenium, and oxygen.

8.11 You are given a dark shiny solid and asked to determine whether it is iodine or a metallic element. Suggest a nondestructive test that would allow you to arrive at the correct answer.

8.12 Define valence electrons. For representative elements, the number of valence electrons of an element is equal to its group number. Show that this is true for the following elements: Al, Sr, K, Br, P, S, C.

8.13 A neutral atom of a certain element has 17 electrons. Without consulting a periodic table, (a) write the ground-state electron configuration of the element, (b) classify the element, (c) determine whether the atoms of this element are diamagnetic or paramagnetic.

8.14 In the periodic table, the element hydrogen is sometimes grouped with the alkali metals (as in this book) and sometimes with the halogens. Explain why hydrogen can resemble the Group 1A and the Group 7A elements.

8.15 Write the outer electron configurations for (a) the alkali metals, (b) the alkaline earth metals, (c) the halogens, (d) the noble gases.

8.16 Use the first-row transition metals (Sc to Cu) as an example to illustrate the characteristics of the electron configurations of transition metals.

Problems

8.17 Group the following electron configurations in pairs that would represent similar chemical properties of their atoms:
(a) $1s^22s^22p^63s^2$
(b) $1s^22s^22p^3$
(c) $1s^22s^22p^63s^23p^64s^23d^{10}4p^6$
(d) $1s^22s^2$
(e) $1s^22s^22p^6$
(f) $1s^22s^22p^63s^23p^3$

8.18 Group the following electron configurations in pairs that would represent similar chemical properties of their atoms:
(a) $1s^22s^22p^5$
(b) $1s^22s^1$
(c) $1s^22s^22p^6$
(d) $1s^22s^22p^63s^23p^5$
(e) $1s^22s^22p^63s^23p^64s^1$
(f) $1s^22s^22p^63s^23p^64s^23d^{10}4p^6$

8.19 Without referring to a periodic table, write the electron configuration of elements with the following atomic numbers: (a) 9, (b) 20, (c) 26, (d) 33. Classify the elements.

8.20 Specify in what group of the periodic table each of the following elements is found: (a) $[Ne]3s^1$, (b) $[Ne]3s^23p^3$, (c) $[Ne]3s^23p^6$, (d) $[Ar]4s^23d^8$.

8.21 An ion M^{2+} derived from a metal in the first transition metal series has four and only four electrons in the 3*d* subshell. What element might M be?

8.22 A metal ion with a net +3 charge has five electrons in the 3*d* subshell. Identify the metal.

ELECTRON CONFIGURATIONS OF IONS

Review Questions

8.23 What is the characteristic of the electron configuration of stable ions derived from representative elements?

8.24 What do we mean when we say that two ions or an atom and an ion are isoelectronic?

8.25 What is wrong with the statement "The atoms of element X are isoelectronic with the atoms of element Y"?

8.26 Give three examples of first-row transition metal (Sc to Cu) ions whose electron configurations are represented by the argon core.

Problems

8.27 Write ground-state electron configurations for the following ions: (a) Li^+, (b) H^-, (c) N^{3-}, (d) F^-, (e) S^{2-}, (f) Al^{3+}, (g) Se^{2-}, (h) Br^-, (i) Rb^+, (j) Sr^{2+}, (k) Sn^{2+}.

8.28 Write ground-state electron configurations for the following ions, which play important roles in biochemical processes in our bodies: (a) Na^+, (b) Mg^{2+}, (c) Cl^-, (d) K^+, (e) Ca^{2+}, (f) Fe^{2+}, (g) Cu^{2+}, (h) Zn^{2+}.

8.29 Write ground-state electron configurations for the following transition metal ions: (a) Sc^{3+}, (b) Ti^{4+}, (c) V^{5+}, (d) Cr^{3+}, (e) Mn^{2+}, (f) Fe^{2+}, (g) Fe^{3+}, (h) Co^{2+}, (i) Ni^{2+}, (j) Cu^+, (k) Cu^{2+}, (l) Ag^+, (m) Au^+, (n) Au^{3+}, (o) Pt^{2+}.

8.30 Name the ions with +3 charges that have the following electron configurations: (a) $[Ar]3d^3$, (b) [Ar], (c) $[Kr]4d^6$, (d) $[Xe]4f^{14}5d^6$.

8.31 Which of the following species are isoelectronic with each other? C, Cl^-, Mn^{2+}, B^-, Ar, Zn, Fe^{3+}, Ge^{2+}.

8.32 Group the species that are isoelectronic: Be^{2+}, F^-, Fe^{2+}, N^{3-}, He, S^{2-}, Co^{3+}, Ar.

PERIODIC VARIATION IN PHYSICAL PROPERTIES

Review Questions

8.33 Define atomic radius. Does the size of an atom have a precise meaning?

8.34 How does atomic radius change as we move (a) from left to right across the period and (b) from top to bottom in a group?

8.35 Define ionic radius. How does the size change when an atom is converted to (a) an anion and (b) a cation?

8.36 Explain why, for isoelectronic ions, the anions are larger than the cations.

Problems

8.37 On the basis of their positions in the periodic table, select the atom with the larger atomic radius in each of the following pairs: (a) Na, Cs; (b) Be, Ba; (c) N, Sb; (d) F, Br; (e) Ne, Xe.

8.38 Arrange the following atoms in order of decreasing atomic radius: Na, Al, P, Cl, Mg.

8.39 Which is the largest atom in Group 4A?

8.40 Which is the smallest atom in Group 7A?

8.41 Why is the radius of the lithium atom considerably larger than the radius of the hydrogen atom?

8.42 Use the second period of the periodic table as an example to show that the sizes of atoms decrease as we move from left to right. Explain the trend.

8.43 In each of the following pairs, indicate which one of the two species is smaller: (a) Cl or Cl^-, (b) Na or Na^+, (c) O^{2-} or S^{2-}, (d) Mg^{2+} or Al^{3+}, (e) Au^+ or Au^{3+}.

8.44 List the following ions in order of increasing ionic radius: N^{3-}, Na^+, F^-, Mg^{2+}, O^{2-}.

8.45 Explain which of the following ions is larger, and why: Cu^+ or Cu^{2+}.

8.46 Explain which of the following anions is larger, and why: Se^{2-} or Te^{2-}.

8.47 Give the physical states (gas, liquid, or solid) of the representative elements in the fourth period at 1 atm and 25°C: K, Ca, Ga, Ge, As, Se, Br.

8.48 The boiling points of neon and krypton are −245.9°C and −152.9°C, respectively. Using these data, estimate the boiling point of argon. (*Hint:* The properties of argon are intermediate between those of neon and krypton.)

IONIZATION ENERGY

Review Questions

8.49 Define ionization energy. Ionization energy is usually measured in the gaseous state. Why? Why is the second ionization energy always greater than the first ionization energy for any element?

8.50 Sketch an outline of the periodic table and show group and period trends in the first ionization energy of the elements. What types of elements have the highest ionization energies and what types the lowest ionization energies?

Problems

8.51 Use the third period of the periodic table as an example to illustrate the change in first ionization energies of the elements as we move from left to right. Explain the trend.

8.52 Ionization energy usually increases from left to right across a given period. Aluminum, however, has a lower ionization energy than magnesium. Explain.

8.53 The first and second ionization energies of K are 419 kJ/mol and 3052 kJ/mol, and those of Ca are 590 kJ/mol and 1145 kJ/mol, respectively. Compare their values and comment on the differences.

8.54 Two atoms have the electron configurations $1s^22s^22p^6$ and $1s^22s^22p^63s^1$. The first ionization energy of one is 2080 kJ/mol, and that of the other is 496 kJ/mol. Pair each ionization energy with one of the given electron configurations. Justify your choice.

8.55 A hydrogenlike ion is an ion containing only one electron. The energies of the electron in a hydrogenlike ion are given by

$$E_n = -(2.18 \times 10^{-18}\ \mathrm{J})Z^2\left(\frac{1}{n^2}\right)$$

where n is the principal quantum number and Z is the atomic number of the element. Calculate the ionization energy (in kilojoules per mole) of the He^+ ion.

8.56 Plasma is a state of matter in which a gaseous system consists of positive ions and electrons. In the plasma state, a mercury atom would be stripped of its 80 electrons and exist as Hg^{80+}. Use the equation in Problem 8.55 to calculate the energy required for the last step of ionization; that is,

$$Hg^{79+}(g) \longrightarrow Hg^{80+}(g) + e^-$$

ELECTRON AFFINITY

Review Questions

8.57 (a) Define electron affinity. Electron affinity is usually measured with atoms in the gaseous state. Why? (b) Ionization energy is always a positive quantity, whereas electron affinity may be either positive or negative. Explain.

8.58 Explain the trends in electron affinity from aluminum to chlorine (see Table 8.4).

Problems

8.59 Arrange the elements in each of the following groups in order of increasing exothermic electron affinity: (a) Li, Na, K; (b) F, Cl, Br, I.

8.60 Which of the following elements would you expect to have the greatest electron affinity? He, K, Co, S, Cl.

8.61 From the electron-affinity values for the alkali metals, do you think it is possible for these metals to form an anion like M^-, where M represents an alkali metal?

8.62 Explain why alkali metals have a greater affinity for electrons than alkaline earth metals.

VARIATION IN CHEMICAL PROPERTIES

Review Questions

8.63 Explain what is meant by the diagonal relationship. List two pairs of elements that show this relationship.

8.64 Which elements are more likely to form acidic oxides? basic oxides? amphoteric oxides?

Problems

8.65 Use the alkali metals and alkaline earth metals as examples to show how we can predict the chemical properties of elements simply from their electron configurations.

8.66 Based on your knowledge of the chemistry of the alkali metals, predict some of the chemical properties of francium, the last member of the group.

8.67 As a group, the noble gases are very stable chemically (only Kr and Xe are known to form some compounds). Why?

8.68 Why are the Group 1B elements more stable than the Group 1A elements even though they seem to have the same outer electron configuration ns^1 where n is the principal quantum number of the outermost shell?

8.69 How do the chemical properties of oxides change as we move across a period from left to right? as we move down a particular group?

8.70 Predict (and give balanced equations for) the reactions between each of the following oxides and water: (a) Li_2O, (b) CaO, (c) CO_2.

8.71 Write formulas and give names for the binary hydrogen compounds of the second-period elements (Li to F). Describe the changes in physical and chemical properties of these compounds as we move across the period from left to right.

8.72 Which oxide is more basic, MgO or BaO? Why?

MISCELLANEOUS PROBLEMS

8.73 State whether each of the following properties of the representative elements generally increases or decreases (a) from left to right across a period and (b) from top to bottom in a group: metallic character, atomic size, ionization energy, acidity of oxides.

8.74 With reference to the periodic table, name (a) a halogen element in the fourth period, (b) an element similar to phosphorus in chemical properties, (c) the most reactive metal in the fifth period, (d) an element that has an atomic number smaller than 20 and is similar to strontium.

8.75 Why do elements that have high ionization energies usually have more negative electron affinities?

8.76 Arrange the following isoelectronic species in order of (a) increasing ionic radius and (b) increasing ionization energy: O^{2-}, F^-, Na^+, Mg^{2+}.

8.77 Write the empirical (or molecular) formulas of compounds that the elements in the third period (sodium to chlorine) are expected to form with (a) molecular oxygen and (b) molecular chlorine. In each case indicate whether you expect the compound to be ionic or molecular in character.

8.78 Element M is a shiny and highly reactive metal (melting point 63°C), and element X is a highly reactive nonmetal (melting point −7.2°C). They react to form a compound with the empirical formula MX, a colorless, brittle solid that melts at 734°C. When dissolved in water or when in the melt state, the substance conducts electricity. When chlorine gas is bubbled through an aqueous solution containing MX, a reddish-brown liquid appears and Cl^- ions are formed. From these observations, identify M and X. (You may need to consult a handbook of chemistry for the melting-point values.)

8.79 Match each of the elements on the right with its description on the left:

(a) A dark-red liquid	Calcium (Ca)
(b) A colorless gas that burns in oxygen gas	Gold (Au)
	Hydrogen (H_2)
(c) A reactive metal that attacks water	Argon
	Bromine (Br_2)
(d) A shiny metal that is used in jewelry	
(e) A totally inert gas	

8.80 Arrange the following species in isoelectronic pairs: O^+, Ar, S^{2-}, Ne, Zn, Cs^+, N^{3-}, As^{3+}, N, Xe.

8.81 In which of the following are the species written in decreasing radius? (a) Be, Mg, Ba, (b) N^{3-}, O^{2-}, F^-, (c) Tl^{3+}, Tl^{2+}, Tl^+.

8.82 Which of the following properties show a clear periodic variation? (a) first ionization energy, (b) molar mass of the elements, (c) number of isotopes of an element, (d) atomic radius.

8.83 When carbon dioxide is bubbled through a clear calcium hydroxide solution, the solution appears milky. Write an equation for the reaction and explain how this reaction illustrates that CO_2 is an acidic oxide.

8.84 You are given four substances: a fuming red liquid, a dark metallic-looking solid, a pale-yellow gas, and a yellow-green gas that attacks glass. You are told that these substances are the first four members of Group 7A, the halogens. Name each one.

8.85 For each pair of elements listed below, give three properties that show their chemical similarity: (a) sodium and potassium and (b) chlorine and bromine.

8.86 Name the element that forms compounds, under appropriate conditions, with every other element in the periodic table except He, Ne, and Ar.

8.87 Explain why the first electron affinity of sulfur is −200 kJ/mol but the second electron affinity is +649 kJ/mol.

8.88 The H^- ion and the He atom have two 1*s* electrons each. Which of the two species is larger? Explain.

8.89 Acidic oxides are those that react with water to produce acid solutions, while the reactions of basic oxides with water produce basic solutions. Nonmetallic oxides are usually acidic, while those of metals are basic. Predict the products of the following oxides with water: Na_2O, BaO, CO_2, N_2O_5, P_4O_{10}, SO_3. Write an equation for each of the reactions.

8.90 Write the formulas and names of the oxides of the second-period elements (Li to N). Identify the oxides as acidic, basic, or amphoteric.

Answers to Practice Exercises: 8.1 (a) [Ar]4s^2; (b) representative element; (c) diamagnetic. **8.2** Li > Be > C. **8.3** (a) Li^+, (b) Au^{3+}, (c) N^{3-}. **8.4** (a) N, (b) Mg. **8.5** No.

CHAPTER 9

CHEMICAL BONDING I: THE COVALENT BOND

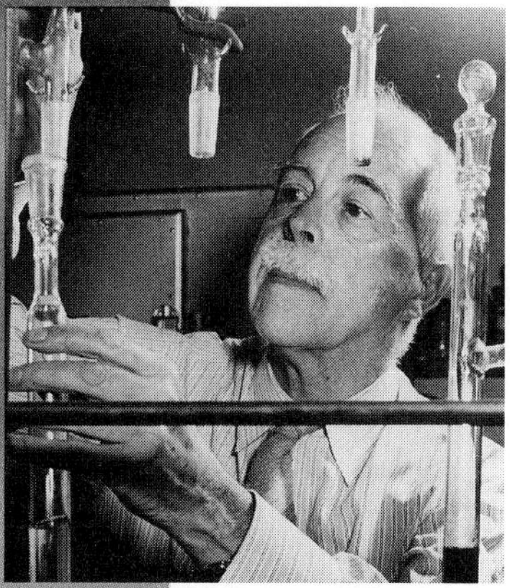

Professor Lewis in his laboratory.

◆ While lecturing to an introductory chemistry class—or so the story goes—Gilbert N. Lewis had an inspiration: Suppose atoms form molecules by sharing electrons. This chemical bonding scheme might account for the fact that some compounds are more stable than their component atoms alone. Based on this insight, Lewis developed the so-called octet rule, which proved to be very useful in explaining the properties and reactions of compounds.

Born in 1875, Lewis grew up in Nebraska. After studying at Harvard University and in Europe, he taught at the Massachusetts Institute of Technology before joining the faculty at the University of California, Berkeley, in 1912. There he molded the chemistry department into one of the world's leading teaching and research facilities.

Although Lewis is best known for his work in chemical bonding, he also made outstanding contributions in acid-base chemistry, thermodynamics, and spectroscopy (the study of interaction between light and matter). In addition, he was the first person to prepare heavy water (D_2O) and to investigate its properties. In spite of his distinguished achievements, Lewis never won a Nobel Prize. He died in 1946. ◆

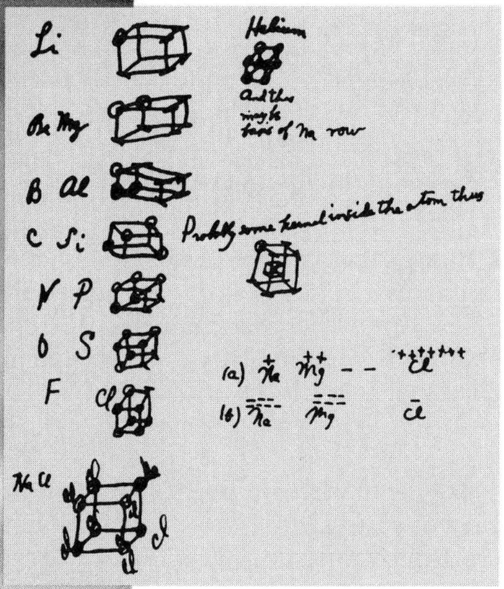

Lewis first sketched his idea about the octet rule on the back of an envelope.

9.1 LEWIS DOT SYMBOLS

The development of the periodic table and concept of electron configuration provided chemists with a rationale for molecule and compound formation. This explanation, formulated by the American chemist Gilbert Lewis, is that atoms react in order to achieve a more stable electron configuration. Maximum stability results when an atom is isoelectronic with a noble gas.

When atoms interact to form a chemical bond, only their outer regions come into contact. For this reason, when we study chemical bonding, we are usually concerned primarily with valence electrons of the atoms. To keep track of valence electrons in a chemical reaction, and to make sure that the total number of electrons does not change, chemists use a system of dots devised by Lewis, called Lewis dot symbols. A ***Lewis dot symbol*** *consists of the symbol of an element and one dot for each valence electron in an atom of the element.* Figure 9.1 shows the Lewis dot symbols of the representative elements and the noble gases. Note that, except for helium, the number of valence electrons in each atom is the same as the group number of the element. For example, Li belongs to Group 1A and has one dot for one valence electron; Be, a Group 2A element, has two valence electrons (two dots); and so on. Elements in the same group have similar outer electron configurations and hence similar Lewis dot symbols. The transition metals, lanthanides, and actinides all have incompletely filled inner shells, and in general we cannot write simple Lewis dot symbols for them.

In this chapter we will learn to use electron configurations and the periodic table to predict the number of bonds an atom of a particular element can form and the stability of the product.

9.2 THE COVALENT BOND

Although the concept of molecules goes back to the seventeenth century, it was not until early in this century that chemists began to understand how and why molecules form. The first major breakthrough was Gilbert Lewis's suggestion that a chemical bond involves electron sharing by atoms. He depicted the formation of a chemical bond in H_2 as

$$\mathrm{H}\cdot + \cdot\mathrm{H} \longrightarrow \mathrm{H}:\mathrm{H}$$

This type of electron pairing is an example of a ***covalent bond,*** *a bond in which two electrons are shared by two atoms.* For the sake of simplicity, the shared pair of electrons is often represented by a single line. Thus the covalent bond in the hydrogen molecule can be written as H—H. In a covalent bond, each electron in a shared pair is attracted to the nuclei of both atoms. This attraction holds the two atoms in H_2 together and is responsible for the formation of covalent bonds in other molecules.

Covalent bonding between many-electron atoms involves only the valence electrons. Consider the fluorine molecule, F_2. The electron configuration of F is $1s^2 2s^2 2p^5$. The $1s$ electrons are low in energy and stay near the nucleus most of the time. For this reason they do not participate in bond formation. Thus each F atom has seven valence electrons (the $2s$ and $2p$ electrons). According to Figure 9.1, there is only one unpaired electron on F, so the formation of the F_2 molecule can be represented as follows:

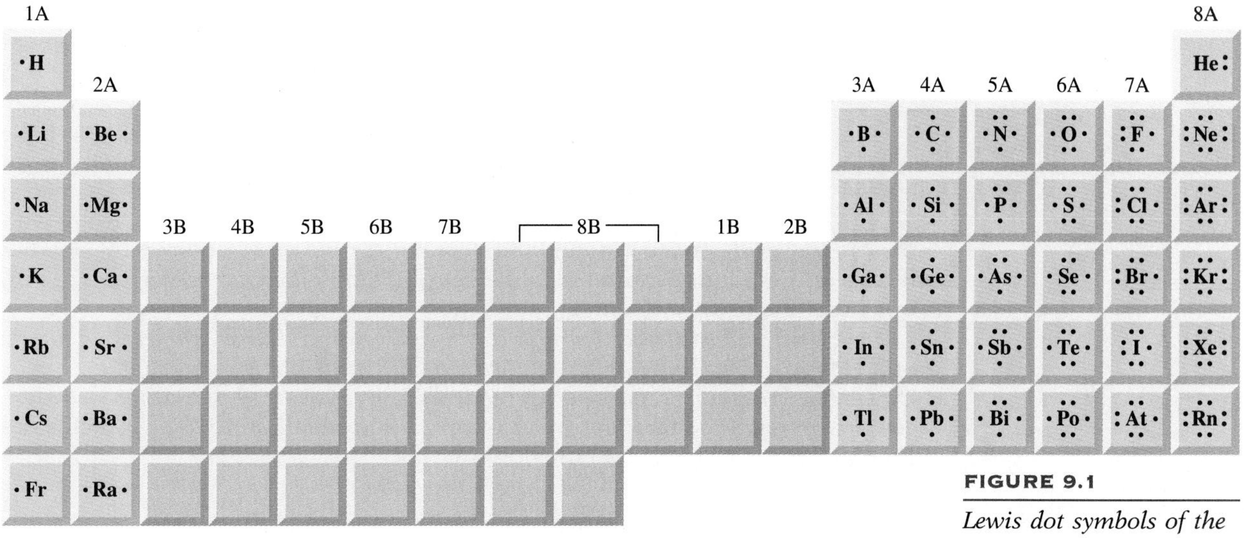

FIGURE 9.1

Lewis dot symbols of the representative elements and the noble gases. The number of unpaired dots corresponds to the number of bonds an atom of the element can form in a compound.

$$:\ddot{F}\cdot + \cdot\ddot{F}: \longrightarrow :\ddot{F}:\ddot{F}: \qquad \text{or} \qquad :\ddot{F}-\ddot{F}:$$

Note that only two valence electrons participate in the formation of F_2. *The valence electrons that are not involved in covalent bond formation* are called ***nonbonding electrons,*** or ***lone pairs.*** Thus each F in F_2 has three lone pairs of electrons:

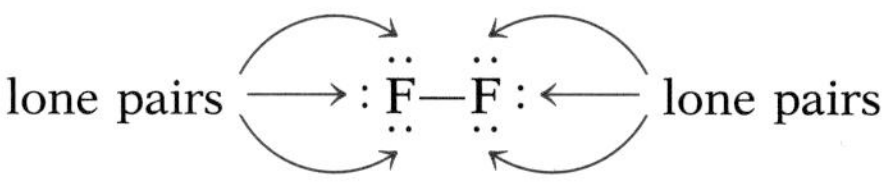

The structures we use to represent H_2 and F_2 are called Lewis structures. A ***Lewis structure*** is *a representation of covalent bonding using Lewis dot symbols in which shared electron pairs are shown either as lines or as pairs of dots between two atoms, and lone pairs are shown as pairs of dots on individual atoms.* Only valence electrons are shown in a Lewis structure.

Next let us consider the Lewis structure of the water molecule. Figure 9.1 shows the Lewis dot symbol for oxygen with two unpaired dots or two unpaired electrons, so we expect that O might form two covalent bonds. Since hydrogen has only one electron, it can form only one covalent bond. Thus the Lewis structure for water is

$$H:\ddot{O}:H \qquad \text{or} \qquad H-\ddot{O}-H$$

In this case, the O atom has two lone pairs. The hydrogen atom has no lone pairs because its only electron is used to form a covalent bond.

In the F_2 and H_2O molecules, the F and O atoms have obtained the stable noble gas configuration by sharing electrons:

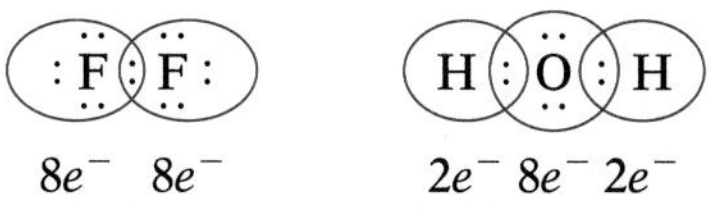

The formation of these molecules illustrates the ***octet rule,*** formulated by Lewis: *An atom other than hydrogen tends to form bonds until it is surrounded by eight valence electrons.* In other words, a covalent bond forms when there are not enough electrons for each atom to have a complete octet. By sharing electrons in a covalent bond, the individual atoms can complete their octets. The requirement for hydrogen is that it attain the electron configuration of helium, or a total of two electrons.

The octet rule works mainly for elements in the second period of the periodic table. These elements have only 2*s* and 2*p* subshells, which can hold a total of eight electrons. When an atom of one of these elements forms a covalent compound, it can attain the noble gas electron configuration [Ne] by sharing electrons with other atoms in the same compound. However, as we will see later, there are a number of important exceptions to the octet rule that give us further insight into the nature of chemical bonding.

Two atoms held together by one electron pair are said to be joined by a ***single bond.*** Many compounds are held together by ***multiple bonds,*** that is, bonds formed when *two atoms share two or more pairs of electrons.* If *two atoms share two pairs of electrons,* the covalent bond is called a ***double bond.*** Double bonds are found in molecules of carbon dioxide (CO_2):

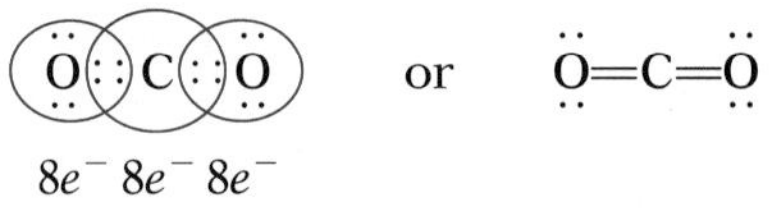

and ethylene (C_2H_4):

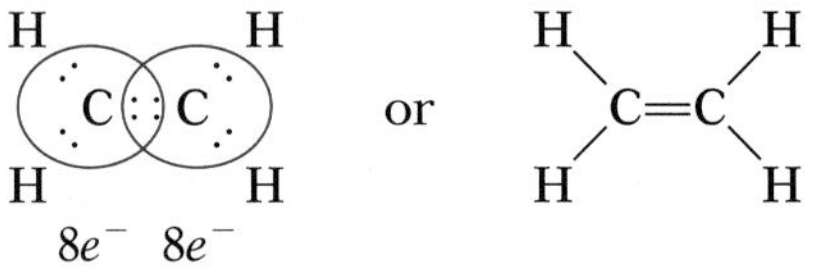

A ***triple bond*** arises when *two atoms share three pairs of electrons,* as in the nitrogen molecule (N_2):

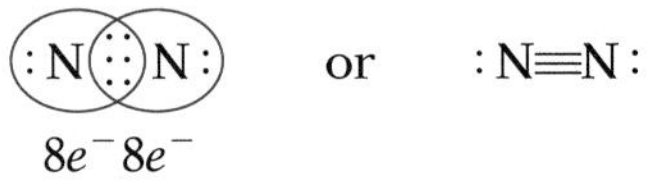

The acetylene molecule (C_2H_2) also contains a triple bond, in this case between two carbon atoms:

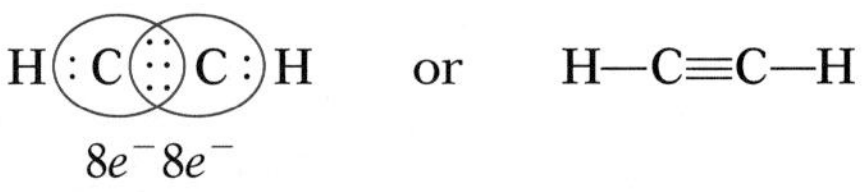

We will discuss the rules for writing Lewis structures containing single, double, and triple bonds shortly.

Multiple bonds are shorter than single covalent bonds. ***Bond length*** is defined as *the distance between the nuclei of two bonded atoms in a molecule* (Figure 9.2). Table 9.1 shows some experimentally determined bond lengths. For a given pair of atoms, such as carbon and nitrogen, triple bonds are shorter than double bonds, which, in turn, are shorter than single bonds. The shorter multiple bonds are also more stable than single bonds, as we will see later.

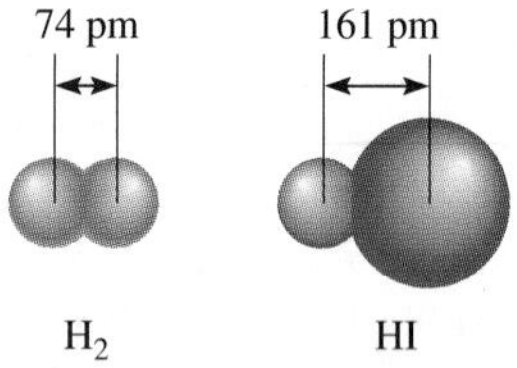

FIGURE 9.2

The bond length is the distance between two bonded nuclei. Bond length measurements (in picometers) are shown for the diatomic molecules H_2 and HI.

TABLE 9.1
Average Bond Lengths of Some Common Single, Double, and Triple Bonds

Bond type	Bond length (pm)
C—H	107
C—O	143
C=O	121
C—C	154
C=C	133
C≡C	120
C—N	143
C=N	138
C≡N	116
N—O	136
N=O	122
O—H	96

9.3 ELECTRONEGATIVITY

A covalent bond, as we have said, is the sharing of an electron pair by two atoms. In a molecule like H_2, in which the atoms are identical, we expect the electrons to be equally shared; that is, the electrons spend the same amount of time in the vicinity of each atom. The situation is different for the HF molecule. Although the H and F atoms are also joined by a covalent bond, the electron pair is not shared equally because H and F are different atoms. Instead, the electrons spend *more* time in the vicinity of one atom than the other. In such a case, the covalent bond is called a *polar covalent bond* or simply a *polar bond.*

Experimental evidence indicates that in the HF molecule the electrons spend more time near the F atom. We can think of this unequal sharing of electrons in terms of a partial transfer of electrons, or a shift in electron density, as it is more commonly described, from H to F (Figure 9.3). This "unequal sharing" of the bonding electron pair results in a relatively greater electron density near the fluorine atom and a correspondingly lower electron density near hydrogen. The HF bond and other polar bonds can be thought of as being intermediate between a (nonpolar) covalent bond, in which the sharing of electrons is exactly equal, and an ***ionic bond,*** in which *the transfer of the electron(s) is nearly complete.* Ionic bonds are found in ionic compounds.

A property that helps us distinguish a nonpolar covalent bond from a polar bond is ***electronegativity,*** *the ability of an atom to attract toward itself the electrons in a chemical bond.* Elements with high electronegativity have a greater tendency to attract electrons than elements with low electronegativity. As we might expect, electronegativity is related to electron affinity and ionization energy. Thus an atom such as fluorine, which has a high electron affinity (tends to pick up electrons easily) and a high ionization energy (does not lose electrons easily), has a high electronegativity. On the other hand, sodium has a low electron affinity, a low ionization energy, and a low electronegativity.

Linus Pauling (1901–1994)

The American chemist Linus Pauling devised a method for calculating *relative* electronegativities of most elements. These values are shown in Figure 9.4. A careful examination of this chart reveals trends and relationships among electronegativity values of different elements. In general, electronegativity increases from left to right across a period in the periodic table, consistent with the decreasing metallic character of the elements. Within each group, electronegativity decreases with increasing atomic number, indicating increasing metallic character. Note that the transition metals do not follow these trends. The most electronegative elements—the halogens, oxygen, nitrogen, and sulfur—are found in the upper right-hand corner of the period table, and the least electronegative elements (the alkali and alkaline earth metals) are clustered near the lower left-hand corner. These trends are readily apparent on a graph, as shown in Figure 9.5.

Atoms of elements with widely different electronegativities tend to form ionic

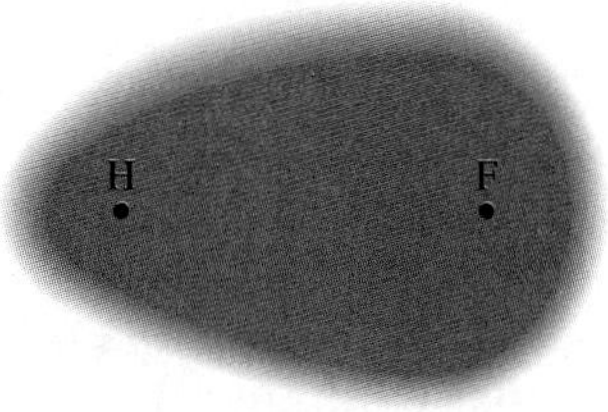

FIGURE 9.3

Electron density distribution in the HF molecule. The dots represent the positions of the nuclei.

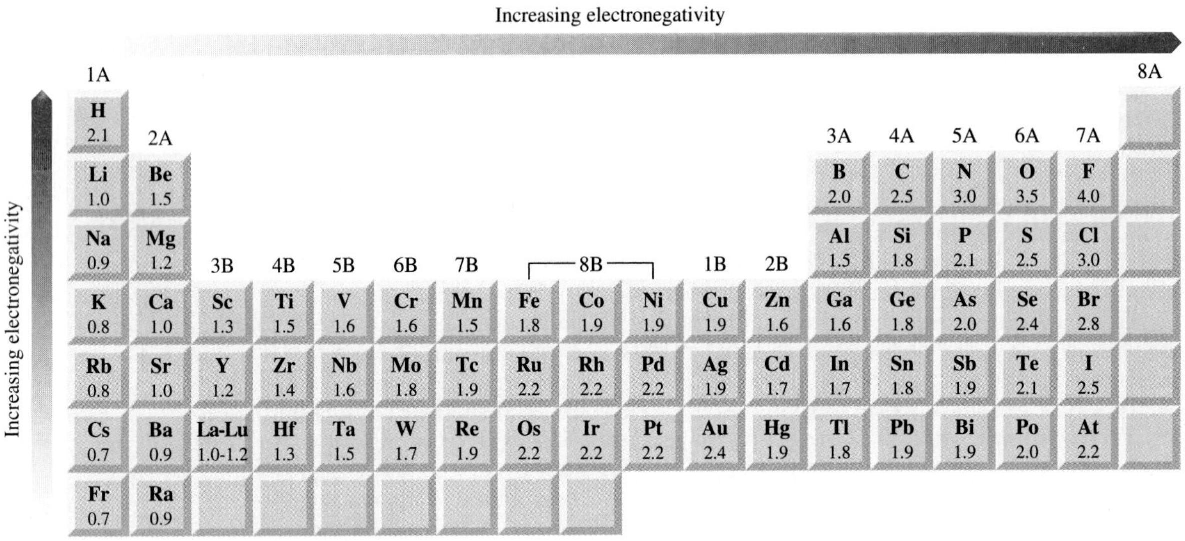

FIGURE 9.4

The electronegativities of common elements.

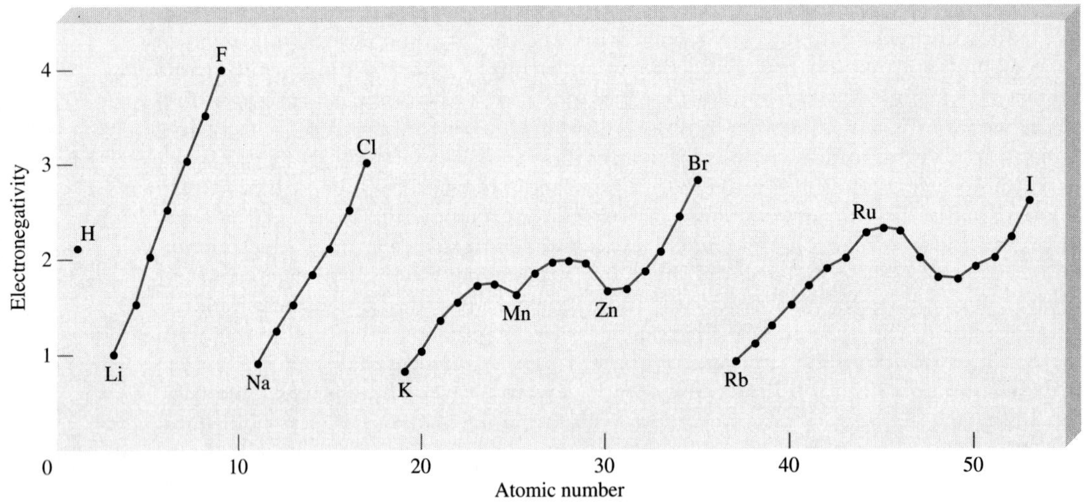

FIGURE 9.5

Variation of electronegativity with atomic number. The halogens have the highest electronegativities, and the alkali metals the lowest.

bonds (such as those that exist in NaCl and CaO compounds) with each other since the atom of the less electronegative element gives up its electron(s) to the atom of the more electronegative element. An ionic bond generally involves an atom of a metallic element and an atom of a nonmetallic element. Atoms of elements with similar electronegativities tend to form polar covalent bonds with each other because the shift in electron density is usually small. Most covalent bonds involve atoms of nonmetallic elements. Only atoms of the same element, which have the same electronegativity, can be joined by a pure covalent bond. These trends and characteristics are what we would expect based on our knowledge of ionization energies and electron affinities.

There is no sharp distinction between a polar bond and an ionic bond, but the following rule is helpful in distinguishing between them. An ionic bond forms when the electronegativity difference between the two bonding atoms is 2.0 or more. This rule applies to most but not all ionic compounds. Sometimes chemists use the quantity *percent ionic character* to describe the nature of a bond. A purely ionic bond has 100 percent ionic character, whereas a nonpolar or purely covalent bond has 0 percent ionic character.

Electronegativity and electron affinity are related but different concepts. Both properties express the tendency of an atom to attract electrons. However, electron affinity refers to an isolated atom's attraction for an additional electron, whereas electronegativity expresses the attraction of an atom in a chemical bond (with another atom) for the shared electrons. Furthermore, electron affinity is an experimentally measurable quantity, whereas electronegativity is a relative number—it is not measurable.

EXAMPLE 9.1

Classifying Chemical Bonds

Classify the following bonds as ionic, polar covalent, or covalent: (a) the bond in HCl, (b) the bond in KF, and (c) the CC bond in H_3CCH_3.

Answer: (a) In Figure 9.4 we see that the electronegativity difference between H and Cl is 0.9, which is appreciable but not large enough (by the 2.0 rule) to qualify HCl as an ionic compound. Therefore, the bond between H and Cl is polar covalent.

(b) The electronegativity difference between K and F is 3.2, well above the 2.0 mark; therefore, the bond between K and F is ionic.

(c) The two C atoms are identical in every respect—they are bonded to each other and each is bonded to three other H atoms. Therefore, the bond between them is purely covalent.

PRACTICE EXERCISE

Specify whether the following bonds are covalent, polar covalent, or ionic: (a) the bond in CsCl, (b) the bond in H_2S, (c) the NN bond in H_2NNH_2.

Electronegativity and Oxidation Number

In Chapter 4 we introduced the rules for assigning oxidation numbers of elements in their compounds. The concept of electronegativity is the basis for these rules. In essence, oxidation number refers to the number of charges an atom would have if electrons were transferred completely to the more electronegative of the bonded atoms in a molecule. Consider the NH_3 molecule, in which the N atom forms three single bonds with the H atoms. Because N is more electronegative than H, electron density will be shifted from H to N. If the transfer were complete, each H would donate an electron to N, which would have a total charge of -3 while each H would have a charge of $+1$. Thus we assign an oxidation number of -3 to N and an oxidation number of +1 to H in NH_3. Oxygen usually has an oxidation number of −2 in its compounds, except in hydrogen peroxide (H_2O_2), whose Lewis structure is

$$\mathrm{H-\ddot{\underset{..}{O}}-\ddot{\underset{..}{O}}-H}$$

A bond between identical atoms makes no contribution to the oxidation number of those atoms because the electron pair of that bond is *equally* shared. Since H has an oxidation number of +1, each O atom has an oxidation number of −1.

Can you see now why fluorine always has an oxidation number of −1? It is the most electronegative element known, and it *always* forms a single bond in its compounds. Therefore, it would bear a −1 charge if electron transfer were complete.

9.4 WRITING LEWIS STRUCTURES

Although the octet rule and Lewis structures do not present a complete picture of covalent bonding, they do help to explain the bonding scheme in many compounds and account for the properties and reactions of molecules. For this reason, you should practice writing Lewis structures of compounds. The basic steps are as follows:

1. Write the skeletal structure of the compound, using chemical symbols and placing bonded atoms next to one another. For simple compounds, this task is fairly easy. For more complex compounds, we must either be given the information or make an intelligent guess about it. In general, the least electronegative atom occupies the central position. Hydrogen and fluorine usually occupy the terminal (end) positions in the Lewis structure.
2. Count the total number of valence electrons present, referring, if necessary, to Figure 9.1. For polyatomic anions, add the number of negative charges to that total. (For example, for the CO_3^{2-} ion we add two electrons because the 2− charge indicates that there are two more electrons than are provided by the neutral atoms.) For polyatomic cations, we subtract the number of positive charges from this total. (Thus for NH_4^+ we subtract one electron because the 1+ charge indicates a loss of one electron from the group of neutral atoms.)
3. Draw a single covalent bond between the central atom and each of the surrounding atoms. Complete the octets of the atoms bonded to the central atom. (Remember that the valence shell of a hydrogen atom is complete with only two electrons.) Electrons belonging to the central or surrounding atoms must be shown as lone pairs if they are not involved in bonding. The total number of electrons to be used is that determined in step 2.
4. If the octet rule is not satisfied for the central atom, try adding double or triple bonds between the surrounding atoms and the central atom, using the lone pairs from the surrounding atoms.

EXAMPLE 9.2
Writing the Lewis Structure of a Molecule

Write the Lewis structure for nitrogen trifluoride (NF_3), in which all three F atoms are bonded to the N atom.

Answer:
Step 1: The skeletal structure of NF_3 is

F N F

F

Step 2: The outer-shell electron configurations of N and F are $2s^22p^3$ and $2s^22p^5$, respectively. Thus there are $5 + (3 \times 7)$, or 26, valence electrons to account for in NF_3.
Step 3: We draw a single covalent bond between N and each F, and complete the octets for the F atoms. We place the remaining two electrons on N:

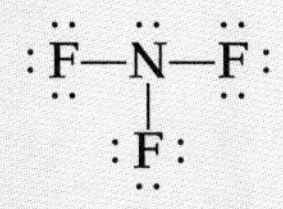

Since this structure satisfies the octet rule for all the atoms, step 4 is not required. To check, we count the valence electrons in NF_3 (in chemical bonds and in lone pairs). The result is 26, the same as the number of valence electrons on three F atoms and one N atom.

PRACTICE EXERCISE

Write the Lewis structure for carbon disulfide (CS_2).

EXAMPLE 9.3
Writing the Lewis Structure of a Polyatomic Anion

Write the Lewis structure for the carbonate ion (CO_3^{2-}).

Answer:
Step 1: We can deduce the skeletal structure of the carbonate ion by realizing that C is less electronegative than O. Therefore, it is most likely to occupy a central position as follows:

O

O C O

Step 2: The outer-shell electron configurations of C and O are $2s^22p^2$ and $2s^22p^4$, respectively, and the ion itself has two negative charges. Thus the total number of electrons is $4 + (3 \times 6) + 2$, or 24.
Step 3: We draw a single covalent bond between C and each O and comply with the octet rule for the O atoms:

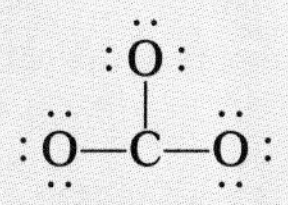

This structure shows all 24 electrons.
Step 4: Although the octet rule is satisfied for the O atoms, it is not for the C atom. Therefore we move a lone pair from one of the O atoms to form another bond with C. Now the octet rule is also satisfied for the C atom:

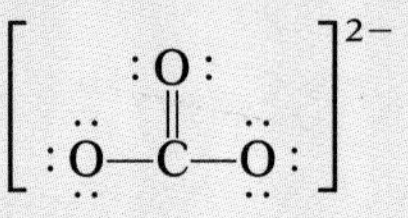

As a final check, we verify that there are 24 valence electrons in the Lewis structure of the carbonate ion.

PRACTICE EXERCISE

Write the Lewis structure for the nitrite ion (NO_2^-).

9.5 FORMAL CHARGE AND LEWIS STRUCTURE

By comparing the number of electrons in an isolated atom with the number of electrons that are associated with the same atom in a Lewis structure, we can determine the distribution of electrons in the molecule and draw the most plausible Lewis structure. The bookkeeping procedure is as follows: In an isolated atom, the number of electrons associated with the atom is simply the number of valence electrons. (As usual, we need not be concerned with the inner electrons.) In a molecule, electrons associated with the atom are the lone pairs on the atom plus the electrons in the bonding pair(s) between the atom and other atom(s). However, because electrons are shared in a bond, we must divide the electrons in a bonding pair equally between the atoms forming the bond. *The difference between the valence electrons in an isolated atom and the number of electrons assigned to that atom in a Lewis structure* is called the atom's ***formal charge.*** We can calculate the formal charge on an atom in a molecule using the equation

$$\begin{array}{l}\text{formal charge on}\\\text{an atom in a}\\\text{Lewis structure}\end{array} = \begin{array}{l}\text{total number of}\\\text{valence electrons}\\\text{in the free atom}\end{array} - \begin{array}{l}\text{total number}\\\text{of nonbonding}\\\text{electrons}\end{array} - \frac{1}{2}\left(\begin{array}{l}\text{total number}\\\text{of bonding}\\\text{electrons}\end{array}\right) \qquad (9.1)$$

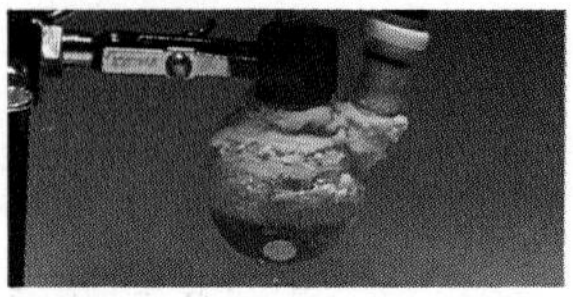

Liquid ozone.

Let us illustrate the concept of formal charge using the ozone molecule (O_3). Proceeding by steps, as we did in Examples 9.2 and 9.3, we draw the skeletal structure of O_3 and then add bonds and electrons to satisfy the octet rule for the two end atoms:

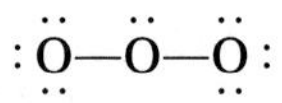

You can see that although all available electrons are used, the octet rule is not satisfied for the central atom. To remedy this, we convert a lone pair on one of the end atoms to a second bond between that end atom and the central atom, as follows:

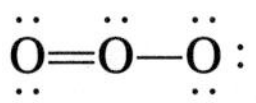

We can now use Equation (9.1) to calculate the formal charges on the O atoms as follows.

- *The central O atom.* In the preceding Lewis structure the central atom has six valence electrons, one lone pair (or two nonbonding electrons), and three bonds (or six bonding electrons). Substituting in Equation (9.1), we write

$$\text{formal charge} = 6 - 2 - \tfrac{1}{2}(6) = +1$$

- *The end O atom in O═O.* This atom has six valence electrons, two lone pairs (or four nonbonding electrons), and two bonds (or four bonding electrons). Thus we write

$$\text{formal charge} = 6 - 4 - \tfrac{1}{2}(4) = 0$$

- *The end O atom in O—O.* This atom has six valence electrons, three lone pairs (or six nonbonding electrons), and one bond (or two bonding electrons). Thus we write

$$\text{formal charge} = 6 - 6 - \tfrac{1}{2}(2) = -1$$

We can now write the Lewis structure for ozone including the formal charges as

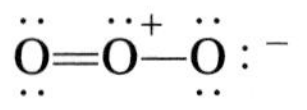

When you write formal charges, the following rules are helpful:

- For neutral molecules, the sum of the formal charges must add up to zero. (This rule applies, for example, to the O_3 molecule.)
- For cations, the sum of the formal charges must equal the positive charge.
- For anions, the sum of the formal charges must equal the negative charge.

Keep in mind that formal charges do not represent actual charge separation within the molecule. In the O_3 molecule, for example, there is no evidence that the central atom bears a net +1 charge or that one of the end atoms bears a −1 charge. Writing these charges on the atoms in the Lewis structure merely helps us keep track of the valence electrons in the molecule.

EXAMPLE 9.4

Writing Formal Charges for a Polyatomic Ion

Write formal charges for the carbonate ion.

Answer: The Lewis structure for the carbonate ion was developed in Example 9.3:

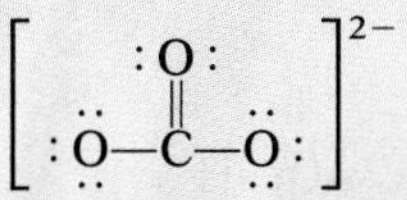

The formal charges on the atoms can be calculated as follows:

The C atom:	formal charge $= 4 - 0 - \tfrac{1}{2}(8) = 0$
The O atom in C═O:	formal charge $= 6 - 4 - \tfrac{1}{2}(4) = 0$
The O atom in C—O:	formal charge $= 6 - 6 - \tfrac{1}{2}(2) = -1$

Thus the Lewis formula for CO_3^{2-} with formal charges is

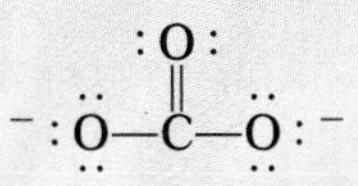

Note that the sum of the formal charges is −2, the same as the charge on the carbonate ion.

PRACTICE EXERCISE

Write formal charges for the nitrite ion (NO_2^-).

Sometimes the rules for writing Lewis structures lead to more than one acceptable representation. In such cases, formal charges often enable us to select a plausible Lewis structure for a given compound. The guidelines we use are summarized as follows:

- For neutral molecules, a Lewis structure in which there are no formal charges is preferable to one in which formal charges are present.
- Lewis structures with large formal charges (+2, +3, and/or −2, −3, and so on) are less plausible than those with small formal charges.
- Among Lewis structures having similar distributions of formal charges, the most plausible structure is the one in which negative formal charges are placed on the more electronegative atoms.

EXAMPLE 9.5

Choosing the Most Plausible Lewis Structure Based on Formal Charges

Formaldehyde, a liquid with a disagreeable odor, traditionally has been used to preserve dead animals. Its molecular formula is CH_2O. Draw the most likely Lewis structure for the compound.

Answer: The two possible skeletal structures are

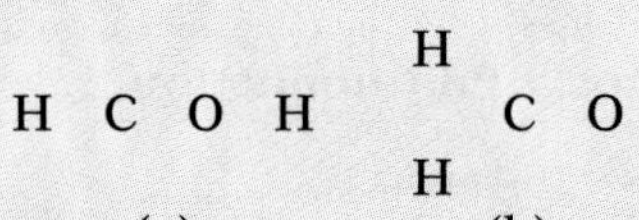

(a) (b)

Following the procedures in previous examples, we can draw a Lewis structure for each of these possibilities:

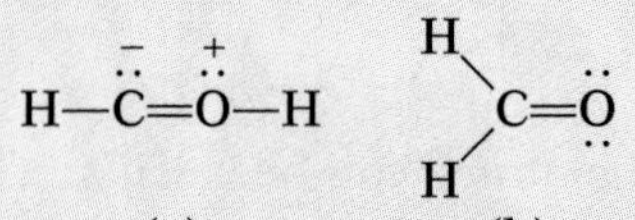

(a) (b)

Since (b) carries no formal charges, it is the more likely structure. This conclusion is confirmed by experimental evidence.

PRACTICE EXERCISE

Draw the most reasonable Lewis structure of a molecule that contains a N atom, a C atom, and a H atom.

9.6 THE CONCEPT OF RESONANCE

In drawing a Lewis structure for ozone (O_3), we satisfied the octet rule for the central atom by drawing a double bond between it and one of the two end O atoms. In fact, we can add the double bond at either end of the molecule, as shown by these two equivalent Lewis structures:

Arbitrarily choosing one of these two Lewis structures to represent ozone presents some problems, however. For one thing, it makes the task of accounting for the known bond lengths in O_3 difficult.

We would expect the O—O bond in O_3 to be longer than the O=O bond because double bonds are known to be shorter than single bonds. Yet experimental evidence shows that both oxygen-to-oxygen bonds are equal in length (128 pm). Therefore neither of the two Lewis structures shown accurately represents the molecule. We resolve this conflict by using *both* Lewis structures to represent O_3:

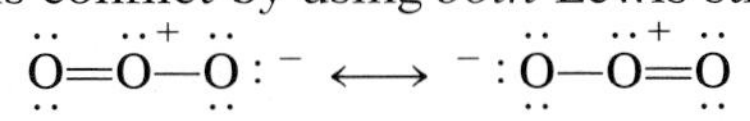

Each of the two structures is called a ***resonance structure.*** A resonance structure, then, is *one of two or more Lewis structures for a single molecule that cannot be described fully with only one Lewis structure.* The symbol ⟷ indicates that the structures shown are resonance structures.

The term ***resonance*** itself means *the use of two or more Lewis structures to represent a particular molecule.* Like the medieval European traveler to Africa who described a rhinoceros as a cross between a griffin and a unicorn, two familiar but imaginary animals, we describe ozone, a real molecule, in terms of two familiar but nonexistent molecules.

A common misconception about resonance is the notion that a molecule such as ozone somehow shifts quickly back and forth from one resonance structure to the other. Keep in mind that *neither* resonance structure adequately represents the actual molecule, which has its own unique, stable structure. "Resonance" is a human invention, designed to address the limitations in these simple bonding models. To extend the rhinoceros analogy, a rhinoceros is a distinct creature, not some oscillation between mythical griffin and unicorn!

The carbonate ion provides another example of resonance:

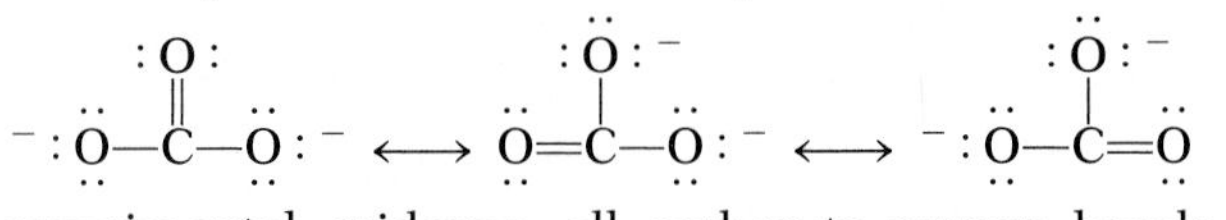

According to experimental evidence, all carbon-to-oxygen bonds in CO_3^{2-} are equivalent. Therefore, the properties of the carbonate ion are best described by considering its resonance structures together.

The concept of resonance applies equally well to organic systems. A well-known example is the benzene molecule (C_6H_6):

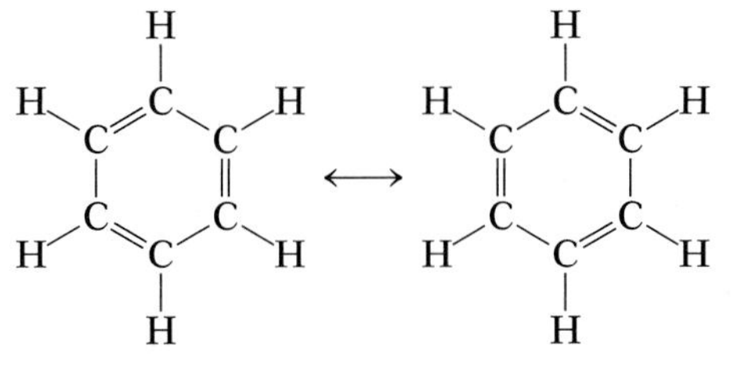

The hexagonal structure of benzene was first proposed by the German chemist August Kekulé (1829–1896).

If one of these resonance structures corresponded to the actual structure of benzene, there would be two different bond lengths between adjacent C atoms, one characteristic of the single bond and the other of the double bond. In fact, the distance between all adjacent C atoms in benzene is 140 pm, which is between the length of a C—C bond (154 pm) and that of a C=C bond (133 pm).

A simpler way of drawing the structure of the benzene molecule and other compounds containing the "benzene ring" is to show only the skeleton and not the carbon and hydrogen atoms. By this convention the resonance structures are

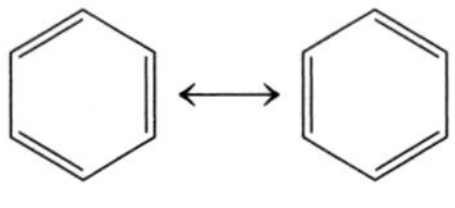

Note that the C atoms at the corners of the hexagon and the H atoms are all omitted, although they are understood to exist. Only the bonds between the C atoms are shown.

Remember this important rule for drawing resonance structures: The positions of electrons, but not those of atoms, can be rearranged in different resonance structures. In other words, the same atoms must be bonded to one another in all the resonance structures for a given species.

EXAMPLE 9.6

Drawing Resonance Structures

Draw resonance structures (including formal charges) for the nitrate ion, NO_3^-, which has the following skeletal arrangement:

O

O N O

Answer: Since nitrogen has five valence electrons and each oxygen has six valence electrons and there is a net negative charge, the total number of valence electrons is $5 + (3 \times 6) + 1 = 24$. The following three resonance structures are all equivalent:

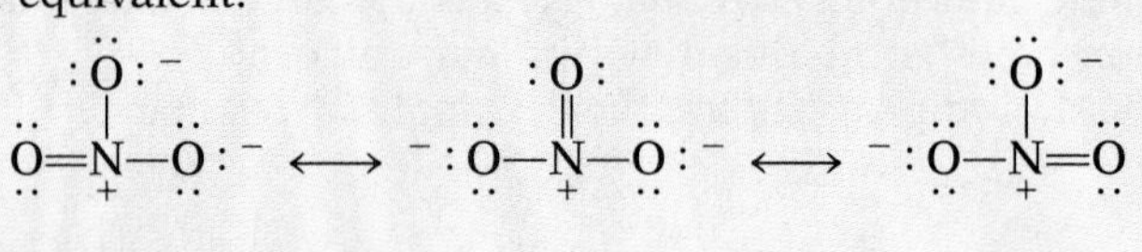

PRACTICE EXERCISE

Draw resonance structures for the nitrite ion (NO_2^-).

A final note on resonance: Although including all the resonance structures is more accurate, for simplicity we often use only one Lewis structure to represent a molecule.

9.7 EXCEPTIONS TO THE OCTET RULE

As mentioned earlier, the octet rule applies mainly to the second-period elements. Exceptions to the octet rule fall into three categories, characterized by an incomplete octet, an odd number of electrons, and more than eight valence electrons around the central atom.

THE INCOMPLETE OCTET

In some compounds the number of electrons surrounding the central atom in a stable molecule is fewer than eight. Consider, for example, beryllium, which is a Group 2A (and a second-period) element. The electron configuration of beryllium is $1s^22s^2$; it has two valence electrons in the 2*s* orbital. In the gas phase, beryllium hydride (BeH_2) exists as discrete molecules. The Lewis structure of BeH_2 is

$$\mathrm{H{-}Be{-}H}$$

As you can see, only four electrons surround the Be atom, and there is no way to satisfy the octet rule for beryllium in this molecule.

Elements in Group 3A, particularly boron and aluminum, also tend to form compounds in which they are surrounded by fewer than eight electrons. Take boron as an example. Since its electron configuration is $1s^22s^22p^1$, it has a total of three valence electrons. Boron forms with the halogens a class of compounds of the general formula BX_3, where X is a halogen atom. Thus, in boron trifluoride there are only six electrons around the boron atom:

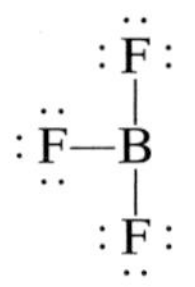

A resonance structure with a double bond between B and F can be drawn that satisfies the octet rule for B. However, the properties of BF_3 are more consistent with a Lewis structure in which there are single bonds between B and each F, as shown above.

Although boron trifluoride is stable, it has a tendency to pick up an unshared electron pair from an atom in another compound, as shown by its reaction with ammonia:

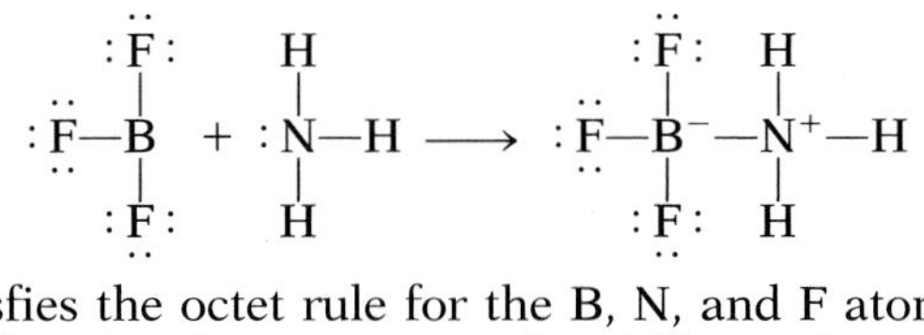

This structure satisfies the octet rule for the B, N, and F atoms.

The B—N bond in the above compound is different from the covalent bonds discussed so far in the sense that both electrons are contributed by the N atom. *A covalent bond in which one of the atoms donates both electrons* is called a ***coordinate covalent bond.*** Although the properties of a coordinate covalent bond do not differ from those of a normal covalent bond (because all electrons are alike no matter what their source), the distinction is useful for keeping track of valence electrons and assigning formal charges.

ODD-ELECTRON MOLECULES

Some molecules contain an *odd* number of electrons. Among them are nitric oxide (NO) and nitrogen dioxide (NO_2):

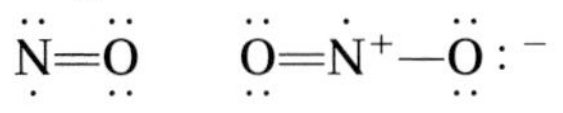

Since we need an even number of electrons for complete pairing (to reach eight), the octet rule clearly cannot be satisfied for all the atoms in any molecule that has an odd number of electrons.

THE EXPANDED OCTET

In a number of compounds there are more than eight valence electrons around an atom. These *expanded octets* are needed only for atoms of elements in and beyond the third period of the periodic table. In addition to the 3*s* and 3*p* orbitals, elements in the third period also have 3*d* orbitals that can be used in bonding. One compound in which there is an expanded octet is sulfur hexafluoride, a very stable compound. The electron configuration of sulfur is $[Ne]3s^23p^4$. In SF_6, each of sulfur's six valence electrons forms a covalent bond with a fluorine atom, so there are twelve electrons around the central sulfur atom:

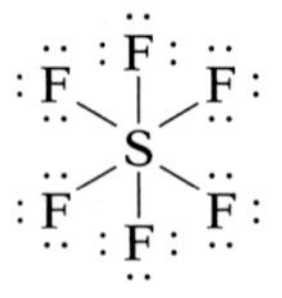

In the next chapter we will see that these twelve electrons, or six bonding pairs, are accommodated in six orbitals that originate from the one 3*s*, the three 3*p*, and two of the five 3*d* orbitals. However, sulfur also forms many compounds in which it does not violate the octet rule. In sulfur dichloride, S is surrounded by only eight electrons and therefore obeys the octet rule:

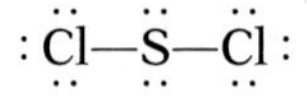

EXAMPLE 9.7

Writing the Lewis Structure of a Molecule That Does Not Satisfy the Octet Rule

Draw the Lewis structure for phosphorus pentafluoride (PF_5), in which all five F atoms are bonded directly to the P atom.

Answer: The outer-shell electron configurations for P and F are $3s^23p^3$ and $2s^22p^5$, respectively, and so the total number of valence electrons is 5 + (5 × 7), or 40. The Lewis structure of PF_5 is

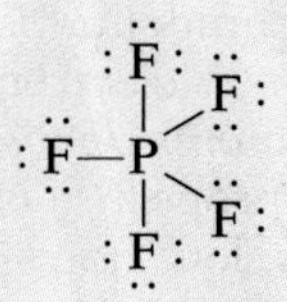

Note that although the octet rule is satisfied for the F atoms, there are ten valence electrons around the P atom, giving it an expanded octet.

PRACTICE EXERCISE

Draw the Lewis structure for aluminum triiodide (AlI_3).

9.8 STRENGTH OF THE COVALENT BOND

The strength of a covalent bond is defined by the amount of energy needed to break it. A quantitative measure of the stability of a molecule is its ***bond dissociation energy*** (or ***bond energy***), which is *the enthalpy change required to break a particular bond in one mole of gaseous molecules.* The experimentally determined bond dissociation energy of the diatomic hydrogen molecule, for example, is

$$H_2(g) \longrightarrow H(g) + H(g) \qquad \Delta H° = 436.4 \text{ kJ}$$

This equation tells us that breaking the covalent bonds in 1 mole of gaseous H_2 molecules requires 436.4 kJ of energy.

Similarly, for the less stable chlorine molecule,

$$Cl_2(g) \longrightarrow Cl(g) + Cl(g) \qquad \Delta H° = 242.7 \text{ kJ}$$

Bond energies can be directly measured also for diatomic molecules containing unlike elements, such as HCl,

$$HCl(g) \longrightarrow H(g) + Cl(g) \qquad \Delta H° = 431.9 \text{ kJ}$$

as well as for molecules containing double and triple bonds:

$$O_2(g) \longrightarrow O(g) + O(g) \qquad \Delta H° = 498.7 \text{ kJ}$$

$$N_2(g) \longrightarrow N(g) + N(g) \qquad \Delta H° = 941.4 \text{ kJ}$$

Measuring the strength of covalent bonds in polyatomic molecules is more complicated. For example, measurements show that more energy is needed to break the first O—H bond in H_2O than is necessary to break the second O—H bond:

$$H_2O(g) \longrightarrow H(g) + OH(g) \qquad \Delta H° = 502 \text{ kJ}$$

$$OH(g) \longrightarrow H(g) + O(g) \qquad \Delta H° = 427 \text{ kJ}$$

In each case, an O—H bond is broken, but the first step is more endothermic than the second. The difference between the two $\Delta H°$ values suggests that the absence of the first O—H bond changes the second one, making it somewhat more easily broken. In fact, the bond energy of the O—H bond varies slightly from one molecule to another, because the chemical environments of different molecules are not the same. This is the case not only for the O—H bond, but also for bonds between any two given atoms in different polyatomic molecules. Thus for polyatomic molecules it is convenient to use the *average bond energy* in calculations. For example, we can measure the energy of the same O—H bond in 10 different polyatomic molecules and obtain the average O—H bond energy by dividing the sum of the bond energies by 10. Table 9.2 lists the average bond energies of a number of bonds found in polyatomic molecules, as well as the bond energies of several diatomic molecules.

USE OF BOND ENERGIES IN THERMOCHEMISTRY

A comparison of the thermochemical changes that take place during a number of reactions (Chapter 6) reveals a strikingly wide variation in the enthalpies of different reactions. For example, the combustion of hydrogen gas in oxygen gas is fairly exothermic:

$$H_2(g) + \tfrac{1}{2}O_2(g) \longrightarrow H_2O(l) \qquad \Delta H° = -285.8 \text{ kJ}$$

TABLE 9.2
Bond Dissociation Energies of Diatomic Molecules* and Average Bond Energies

Bond	Bond energy (kJ/mol)	Bond	Bond energy (kJ/mol)
H—H	436.4	C—S	255
H—N	393	C=S	477
H—O	460	N—N	193
H—S	368	N=N	418
H—P	326	N≡N	941.4
H—F	568.2	N—O	176
H—Cl	431.9	N—P	209
H—Br	366.1	O—O	142
H—I	298.3	O=O	498.7
C—H	414	O—P	502
C—C	347	O=S	469
C=C	620	P—P	197
C≡C	812	P=P	489
C—N	276	S—S	268
C=N	615	S=S	352
C≡N	891	F—F	156.9
C—O	351	Cl—Cl	242.7
C=O†	745	Br—Br	192.5
C—P	263	I—I	151.0

*Bond energies for diatomic molecules (in color) have more significant figures than bond energies for bonds in polyatomic molecules because the bond dissociation energies of diatomic molecules are directly measurable quantities and not averaged over many compounds.
†The C=O bond energy in CO_2 is 799 kJ/mol.

On the other hand, the formation of glucose ($C_6H_{12}O_6$) from water and carbon dioxide, best achieved by photosynthesis, is highly endothermic:

$$6CO_2(g) + 6H_2O(l) \longrightarrow C_6H_{12}O_6(s) + 6O_2(g) \qquad \Delta H^\circ = 2801 \text{ kJ}$$

We can account for such variations by looking at the stability of individual reactant and product molecules. After all, most chemical reactions involve the making and breaking of bonds. Therefore, knowing the bond energies and hence the stability of molecules tells us something about the thermochemical nature of reactions that molecules undergo.

In many cases it is possible to predict the approximate enthalpy of reaction by using the average bond energies. Because energy is always required to break chemical bonds and chemical bond formation is always accompanied by a release of energy, we can estimate the enthalpy of a reaction by counting the total number of bonds broken and formed in the reaction and recording all the corresponding energy changes. The enthalpy of reaction in the *gas phase* is given by

$$\Delta H^\circ = \Sigma\text{BE(reactants)} - \Sigma\text{BE(products)}$$
$$= \text{total energy input} - \text{total energy released} \qquad (9.2)$$

where BE stands for average bond energy and Σ is the summation sign. Equation (9.2) as written takes care of the sign convention for ΔH°. Thus if the total energy input is greater than the total energy released, ΔH° is positive and the reaction is

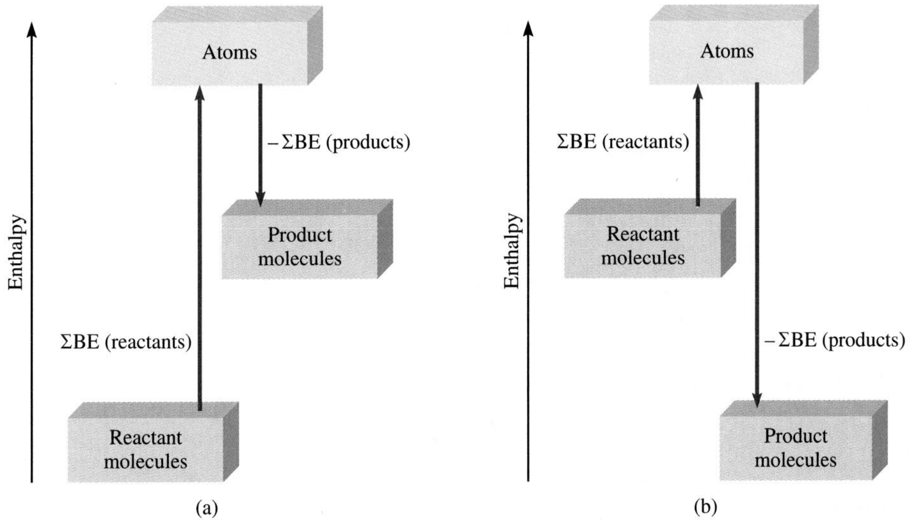

FIGURE 9.6

Bond energy changes in (a) an endothermic reaction and (b) an exothermic reaction.

endothermic. On the other hand, if more energy is released than absorbed, ΔH° is negative and the reaction is exothermic (Figure 9.6). Note that if reactants and products are all diatomic molecules, then Equation (9.2) should yield accurate results since the bond dissociation energies of diatomic molecules are accurately known. If some or all of the reactants and products are polyatomic molecules, Equation (9.2) will yield only approximate results because average bond energies will be used in the calculation.

EXAMPLE 9.8

Using Bond Energies to Estimate the Enthalpy of a Reaction

Estimate the enthalpy change for the combustion of hydrogen gas:

$$2H_2(g) + O_2(g) \longrightarrow 2H_2O(g)$$

Answer: The first step is to count the number of bonds broken and the number of bonds formed. The easiest way to do this is to construct a table:

Type of bonds broken	*Number of bonds broken*	*Bond energy (kJ/mol)*	*Energy change (kJ)*
H—H (H_2)	2	436.4	872.8
O═O (O_2)	1	498.7	498.7

Type of bonds formed	*Number of bonds formed*	*Bond energy (kJ/mol)*	*Energy change (kJ)*
O—H (H_2O)	4	460	1840

Next, we obtain the total energy input and total energy released:

$$\text{total energy input} = 872.8 \text{ kJ} + 498.7 \text{ kJ} = 1372 \text{ kJ}$$

$$\text{total energy released} = 1840 \text{ kJ}$$

Using Equation (9.2) we write

$$\Delta H° = 1372 \text{ kJ} - 1840 \text{ kJ} = -468 \text{ kJ}$$

This result is only an estimate because the bond energy of O—H is an average quantity. Alternatively, we can use Equation (6.8) and the data in Appendix 2 to calculate the enthalpy of reaction:

$$\begin{aligned}\Delta H° &= 2\Delta H_f°(H_2O) - [2\Delta H_f°(H_2) + \Delta H_f°(O_2)] \\ &= (2 \text{ mol})(-241.8 \text{ kJ/mol}) - 0 - 0 \\ &= -483.6 \text{ kJ}\end{aligned}$$

PRACTICE EXERCISE

For the reaction

$$H_2(g) + C_2H_4(g) \longrightarrow C_2H_6(g)$$

(a) estimate the enthalpy of reaction, using the bond energy values in Table 9.2; (b) calculate the enthalpy of reaction, using standard enthalpies of formation. ($\Delta H_f°$ for H_2, C_2H_4, and C_2H_6 are 0, 52.3 kJ/mol, and −84.7 kJ/mol, respectively.)

Note that the estimated value based on average bond energies is quite close to the value calculated using $\Delta H_f°$ data. In general, Equation (9.2) works best for reactions that are quite endothermic or quite exothermic, that is, for $\Delta H° > 100$ kJ or for $\Delta H° < -100$ kJ.

SUMMARY

A Lewis dot symbol shows the number of valence electrons possessed by an atom of a given element. Lewis dot symbols are useful mainly for the representative elements.

In a covalent bond, two electrons (one pair) are shared by two atoms. In multiple covalent bonds, two or three electron pairs are shared by two atoms. Some bonded atoms possess lone pairs, that is, pairs of valence electrons not involved in bonding. The arrangement of bonding electrons and lone pairs around each atom in a molecule is represented by the Lewis structure.

The octet rule predicts that atoms form enough covalent bonds to surround themselves with eight electrons each. When one atom in a covalently bonded pair donates two electrons to the bond, the Lewis structure can include the formal charge on each atom as a means of keeping track of the valence electrons. There are exceptions to the octet rule, particularly for covalent beryllium compounds, elements in Group 3A, and elements in the third period and beyond in the periodic table. Electronegativity is a measure of the ability of an atom to attract electrons in a chemical bond.

For some molecules or polyatomic ions, two or more Lewis structures based on the same skeletal structure satisfy the octet rule and appear chemically reasonable. Such resonance structures taken together represent the molecule or ion.

The strength of a covalent bond is measured in terms of its bond dissociation energy (or bond energy). Bond energies can be used to estimate the enthalpy of reactions.

KEY WORDS

Bond dissociation energy, p. 265
Bond energy, p. 265
Bond length, p. 252
Coordinate covalent bond, p. 263
Covalent bond, p. 250
Double bond, p. 252
Electronegativity, p. 253
Formal charge, p. 258
Ionic bond, p. 253
Lewis dot symbol, p. 250
Lewis structure, p. 251
Lone pair, p. 251
Multiple bond, p. 252
Nonbonding electron, p. 251
Octet rule, p. 251
Resonance, p. 261
Resonance structure, p. 261
Single bond, p. 252
Triple bond, p. 252

QUESTIONS AND PROBLEMS

LEWIS DOT SYMBOLS

Review Questions

9.1 What is a Lewis dot symbol? To what elements does the symbol mainly apply?

9.2 Use the second member of each group from Group 1A to Group 7A to show that the number of valence electrons on an atom of the element is the same as its group number.

9.3 Without referring to Figure 9.1, write Lewis dot symbols for atoms of the following elements: (a) Be, (b) K, (c) Ca, (d) Ga, (e) O, (f) Br, (g) N, (h) I, (i) As, (j) F.

9.4 Write Lewis dot symbols for the following ions: (a) Li^+, (b) Cl^-, (c) S^{2-}, (d) Mg^{2+}, (e) N^{3-}.

9.5 Write Lewis dot symbols for the following atoms and ions: (a) I, (b) I^-, (c) S, (d) S^{2-}, (e) P, (f) P^{3-}, (g) Na, (h) Na^+, (i) Mg, (j) Mg^{2+}, (k) Al, (l) Al^{3+}, (m) Pb, (n) Pb^{2+}.

THE COVALENT BOND

Review Questions

9.6 What is Lewis's contribution to our understanding of the covalent bond?

9.7 Define the following terms: lone pairs, Lewis structure, the octet rule, bond length.

9.8 What is the difference between a Lewis dot symbol and a Lewis structure?

9.9 How many lone pairs are on the underlined atoms in these compounds? H<u>Br</u>, H_2<u>S</u>, <u>C</u>H_4.

9.10 Distinguish among single, double, and triple bonds in a molecule, and give an example of each.

ELECTRONEGATIVITY AND BOND TYPE

Review Questions

9.11 Define electronegativity, and explain the difference between electronegativity and electron affinity. Describe in general how the electronegativities of the elements change according to position in the periodic table.

9.12 What is a polar covalent bond? Name two compounds that contain one or more polar covalent bonds.

Problems

9.13 List the following bonds in order of increasing ionic character: the lithium-to-fluorine bond in LiF, the potassium-to-oxygen bond in K_2O, the nitrogen-to-nitrogen bond in N_2, the sulfur-to-oxygen bond in SO_2, the chlorine-to-fluorine bond in ClF_3.

9.14 Arrange the following bonds in order of increasing ionic character: carbon to hydrogen, fluorine to hydrogen, bromine to hydrogen, sodium to iodine, potassium to fluorine, lithium to chlorine.

9.15 Four atoms are arbitrarily labeled D, E, F, and G. Their electronegativities are as follows: D = 3.8, E = 3.3, F = 2.8, and G = 1.3. If the atoms of these elements form the molecules DE, DG, EG, and DF, how would you arrange these molecules in order of increasing covalent bond character?

9.16 List the following bonds in order of increasing ionic character: cesium to fluorine, chlorine to chlorine, bromine to chlorine, silicon to carbon.

9.17 Classify the following bonds as ionic, polar covalent, or covalent, and give your reasons: (a) the CC bond in H_3CCH_3, (b) the KI bond in KI, (c) the NB bond in H_3NBCl_3, (d) the ClO bond in ClO_2.

9.18 Classify the following bonds as ionic, polar covalent, or covalent, and give your reasons: (a) the SiSi bond in $Cl_3SiSiCl_3$, (b) the SiCl bond in $Cl_3SiSiCl_3$, (c) the CaF bond in CaF_2, (d) the NH bond in NH_3.

LEWIS STRUCTURE AND THE OCTET RULE

Review Questions

9.19 Summarize the essential features of the Lewis

octet rule. The octet rule applies mainly to the second-period elements. Explain.

9.20 Explain the concept of formal charge. Do formal charges on a molecule represent actual separation of charges?

Problems

9.21 Write Lewis structures for the following molecules: (a) ICl, (b) PH_3, (c) P_4 (each P is bonded to three other P atoms), (d) H_2S, (e) N_2H_4, (f) $HClO_3$, (g) $COBr_2$ (C is bonded to O and Br atoms).

9.22 Write Lewis structures for the following ions: (a) O_2^{2-}, (b) C_2^{2-}, (c) NO^+, (d) NH_4^+. Show formal charges.

9.23 The following Lewis structures are incorrect. Explain what is wrong with each one and give a correct Lewis structure for the molecule. (Relative positions of atoms are shown correctly.)

(a) 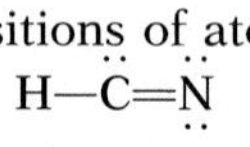

(b) 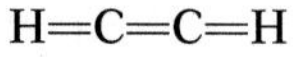

(c) 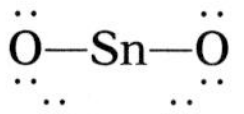

(d) 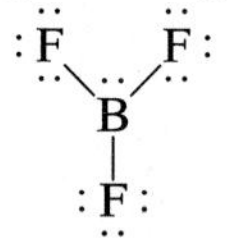

(e) 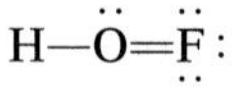

(f) 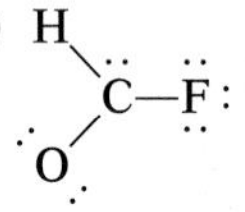

(g) 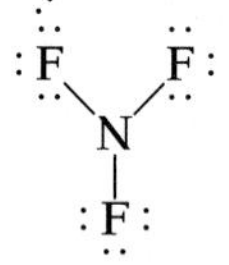

9.24 The skeletal structure of acetic acid in the following structure is correct, but some of the bonds are wrong. (a) Identify the incorrect bonds and explain what is wrong with them. (b) Write the correct Lewis structure for acetic acid.

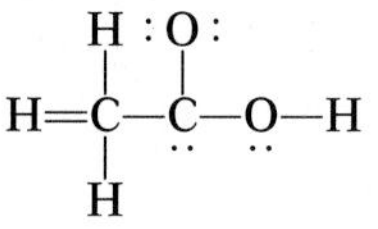

RESONANCE

Review Questions

9.25 Define bond length, resonance, and resonance structure.

9.26 Is it possible to "trap" a resonance structure of a compound for study? Explain.

9.27 The resonance concept is sometimes described by analogy to a mule, which is a cross between a horse and a donkey. Compare this analogy with that used in this chapter, that is, the description of a rhinoceros as a cross between a griffin and a unicorn. Which description is more appropriate? Why?

9.28 Describe the general rules for drawing plausible resonance structures. What are the other two reasons for choosing (b) in Example 9.5?

Problems

9.29 Write Lewis structures for the following species, including all resonance forms, and show formal charges: (a) HCO_2^-, (b) $CH_2NO_2^-$. Relative positions of the atoms are as follows:

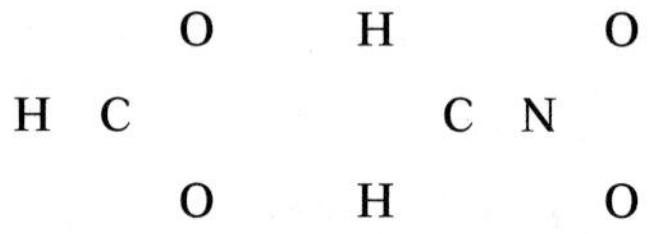

9.30 Draw three resonance structures for the chlorate ion, ClO_3^-. Show formal charges.

9.31 Write three resonance structures for hydrazoic acid, HN_3. The atomic arrangement is HNNN. Show formal charges.

9.32 Draw two resonance structures for diazomethane, CH_2N_2. Show formal charges. The skeletal structure of the molecule is

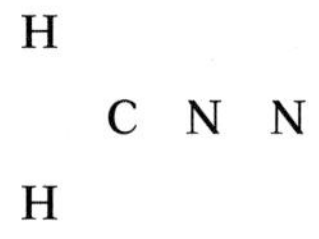

9.33 Draw three reasonable resonance structures for the OCN^- ion. Show formal charges.

9.34 Draw three resonance structures for the molecule N_2O in which the atoms are arranged in the order NNO. Indicate formal charges.

EXCEPTIONS TO THE OCTET RULE

Review Questions

9.35 Why does the octet rule not hold for many compounds containing elements in the third period of the periodic table and beyond?

9.36 Give three examples of compounds that do not satisfy the octet rule. Write a Lewis structure for each.

9.37 Because fluorine has seven valence electrons ($2s^22p^5$), seven covalent bonds in principle could form around the atom. Such a compound might be

FH_7 or FCl_7. These compounds have never been prepared. Why?

9.38 What is a coordinate covalent bond? Is it different from a normal covalent bond?

Problems

9.39 The AlI_3 molecule has an incomplete octet around Al. Draw three resonance structures of the molecule in which the octet rule is satisfied for both the Al and the I atoms. Show formal charges.

9.40 In the vapor phase, beryllium chloride consists of discrete molecular units $BeCl_2$. Is the octet rule satisfied for Be in this compound? If not, can you form an octet around Be by drawing another resonance structure? How plausible is this structure?

9.41 Of the noble gases, only Kr, Xe, and Rn are known to form a few compounds with O and/or F. Write Lewis structures for the following molecules: (a) XeF_2, (b) XeF_4, (c) XeF_6, (d) $XeOF_4$, (e) XeO_2F_2. In each case Xe is the central atom.

9.42 Write a Lewis structure for $SbCl_5$. Is the octet rule obeyed in this molecule?

9.43 Write Lewis structures for SeF_4 and SeF_6. Is the octet rule satisfied for Se?

9.44 Write Lewis structures for the reaction

$$AlCl_3 + Cl^- \longrightarrow AlCl_4^-$$

What kind of bond is between Al and Cl in the product?

BOND ENERGIES

Review Questions

9.45 Define bond dissociation energy. Bond energies of polyatomic molecules are average values. Why?

9.46 Explain why the bond energy of a molecule is usually defined in terms of a gas-phase reaction. Why are bond-breaking processes always endothermic and bond-forming processes always exothermic?

Problems

9.47 From the following data, calculate the average bond energy for the N—H bond:

$$NH_3(g) \longrightarrow NH_2(g) + H(g) \qquad \Delta H° = 435 \text{ kJ}$$

$$NH_2(g) \longrightarrow NH(g) + H(g) \qquad \Delta H° = 381 \text{ kJ}$$

$$NH(g) \longrightarrow N(g) + H(g) \qquad \Delta H° = 360 \text{ kJ}$$

9.48 For the reaction

$$O(g) + O_2(g) \longrightarrow O_3(g) \qquad \Delta H° = -107.2 \text{ kJ}$$

calculate the average bond energy in O_3.

9.49 The bond energy of $F_2(g)$ is 156.9 kJ/mol. Calculate $\Delta H_f°$ for $F(g)$.

9.50 (a) For the reaction

$$2C_2H_6(g) + 7O_2(g) \longrightarrow 4CO_2(g) + 6H_2O(g)$$

predict the enthalpy of reaction from the average bond energies in Table 9.2. (b) Calculate the enthalpy of reaction from the standard enthalpies of formation (see Appendix 2) of the reactant and product molecules, and compare the result with your answer for part (a).

MISCELLANEOUS PROBLEMS

9.51 Match each of the following energy charges with one of the processes given: ionization energy, electron affinity, bond dissociation energy, standard enthalpy of formation.
(a) $F(g) + e^- \rightarrow F^-(g)$
(b) $F_2(g) \rightarrow 2F(g)$
(c) $Na(g) \rightarrow Na^+(g) + e^-$
(d) $Na(s) + \frac{1}{2}F_2(g) \rightarrow NaF(s)$

9.52 The formulas for the fluorides of the third-period elements are NaF, MgF_2, AlF_3, SiF_4, PF_5, SF_6, and ClF_3. Classify these compounds as covalent or ionic.

9.53 Use the ionization energy (see Table 8.3) and electron affinity (see Table 8.4) values to calculate the energy change (in kilojoules) for the following reactions:
(a) $Li(g) + I(g) \rightarrow Li^+(g) + I^-(g)$
(b) $Na(g) + F(g) \rightarrow Na^+(g) + F^-(g)$
(c) $K(g) + Cl(g) \rightarrow K^+(g) + Cl^-(g)$

9.54 Describe some characteristics of an ionic compound such as KF that would distinguish it from a compound such as CO_2.

9.55 Write Lewis structures for BrF_3, ClF_5, and IF_7. Identify those in which the octet rule is not obeyed.

9.56 Write three reasonable resonance structures of the azide ion N_3^- in which the atoms are arranged as NNN. Show formal charges.

9.57 The amide group plays an important role in determining the structure of proteins:

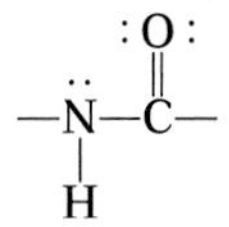

Draw another resonance structure of this group. Show formal charges.

9.58 Give an example of an ion or molecule containing Al that (a) obeys the octet rule, (b) has an expanded octet, and (c) has an incomplete octet.

9.59 Draw four reasonable resonance structures for the PO_3F^{2-} ion. The central P atom is bonded to the three O atoms and to the F atom. Show formal charges.

9.60 Attempts to prepare the following as stable species under atmospheric conditions have failed. Suggest reasons for the failure.

CF_2 CH_5 FH_2^- PI_5

9.61 Draw reasonable resonance structures for the following sulfur-containing ions: (a) HSO_4^-, (b) SO_4^{2-}, (c) HSO_3^-, (d) SO_3^{2-}.

9.62 True or false: (a) Formal charges represent actual separation of charges; (b) ΔH_f° can be estimated from bond energies of reactants and products; (c) all second-period elements obey the octet rule in their compounds; (d) the resonance structures of a molecule can be separated from one another.

9.63 A rule for drawing plausible Lewis structures is that the central atom is invariably less electronegative than the surrounding atoms. Explain why this is so.

9.64 Using the following information:

$$C(s) \longrightarrow C(g) \qquad \Delta H^\circ_{rxn} = 716 \text{ kJ}$$

$$2H_2(g) \longrightarrow 4H(g) \qquad \Delta H^\circ_{rxn} = 872.8 \text{ kJ}$$

and the fact that the average C—H bond energy is 414 kJ/mol, estimate the standard enthalpy of formation of methane (CH_4).

9.65 Based on energy considerations, which of the following two reactions will occur more readily?
(a) $Cl(g) + CH_4(g) \rightarrow CH_3Cl(g) + H(g)$
(b) $Cl(g) + CH_4(g) \rightarrow CH_3(g) + HCl(g)$

(*Hint:* Refer to Table 9.2, and assume that the average bond energy of the C—Cl bond is 338 kJ/mol.)

9.66 Which of the following molecules has the shortest nitrogen-to-nitrogen bond? Explain.

N_2H_4 N_2O N_2 N_2O_4

9.67 Most organic acids can be represented as RCOOH, where COOH is the carboxyl group and R is the rest of the molecule. (For example, R is CH_3 in acetic acid, CH_3COOH). (a) Draw a Lewis structure of the carboxyl group. (b) Upon ionization, the carboxyl group is converted to the carboxylate group, COO^-. Draw resonance structures of the carboxylate group.

9.68 Which of the following molecules or ions are isoelectronic? NH_4^+, C_6H_6, CO, CH_4, N_2, $B_3N_3H_6$.

9.69 The following species have been detected in interstellar space: (a) CH, (b) OH, (c) C_2, (d) HNC, (e) HCO. Draw Lewis structures of these species and indicate whether they are diamagnetic or paramagnetic.

9.70 The amide ion, NH_2^- is a Brønsted base. Represent the reaction between the amide ion and water in terms of Lewis structures.

9.71 Draw Lewis structures of the following organic molecules: (a) tetrafluoroethylene (C_2F_4), (b) propane (C_3H_8), (c) butadiene ($CH_2CHCHCH_2$), (d) propyne (CH_3CCH), (e) benzoic acid (C_6H_5COOH). (To draw C_6H_5COOH, replace a H atom in benzene with a COOH group.)

9.72 The triiodide ion (I_3^-) in which the I atoms are arranged as III is stable, but the corresponding F_3^- ion does not exist. Explain.

9.73 Compare the bond dissociation energy of F_2 with the energy change for the following process:

$$F_2(g) \longrightarrow F^+(g) + F^-(g)$$

Which is the preferred dissociation for F_2, energetically speaking?

9.74 Methyl isocyanate, CH_3NCO, is used to make certain pesticides. In December 1984, water leaked into a tank containing this substance at a chemical plant to produce a toxic cloud that killed thousands of people in Bhopal, India. Draw Lewis structures for this compound, showing formal charges.

9.75 The chlorine nitrate molecule ($ClONO_2$) is believed to be involved in the destruction of ozone in the Antarctic stratosphere. Draw a plausible Lewis structure for the molecule.

9.76 Several resonance structures of the molecule CO_2 are given here. Explain why some of them are likely to be of little importance in describing the bonding in this molecule.

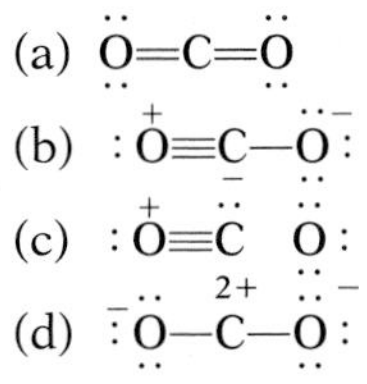

9.77 Draw a Lewis structure for each of the following organic molecules in which the carbon atoms are bonded to each other by single bonds: C_2H_6, C_4H_{10}, C_5H_{12}.

9.78 Draw Lewis structures for the following chlorofluorocarbons (CFCs), which are partly responsible for the depletion of ozone in the stratosphere: $CFCl_3$, CF_2Cl_2, CHF_2Cl, CF_3CHF_2.

9.79 Draw Lewis structures for the following organic molecules, in each of which there is one C=C bond and the rest of the carbon atoms are joined by C—C bonds: C_2H_3F, C_3H_6, C_4H_8.

9.80 Calculate $\Delta H°$ for the reaction

$$H_2(g) + I_2(g) \longrightarrow 2HI(g)$$

using (a) Equation (9.2) and (b) Equation (6.8), given that $\Delta H_f°$ for $I_2(g)$ is 61.0 kJ/mol.

9.81 Draw Lewis structures of the following organic molecules: (a) methanol (CH_3OH); (b) ethanol (CH_3CH_2OH); (c) tetraethyllead [$Pb(CH_2CH_3)_4$], which is used in "leaded" gasoline; (d) methylamine (CH_3NH_2); (e) mustard gas ($ClCH_2CH_2SCH_2CH_2Cl$), a poisonous gas used in World War I; (f) urea [$(NH_2)_2CO$], a fertilizer; (g) glycine (NH_2CH_2COOH), an amino acid.

9.82 Write Lewis structures for the following four isoelectronic species: (a) CO, (b) NO^+, (c) CN^-, (d) N_2. Show formal charges.

9.83 Oxygen forms three types of ionic compounds in which the anions are oxide (O^{2-}), peroxide (O_2^{2-}), and superoxide (O_2^-). Draw Lewis structures of these ions.

9.84 Comment on the correctness of the following statement: All compounds containing a noble gas atom violate the octet rule.

9.85 Vinyl chloride (C_2H_3Cl) differs from ethylene (C_2H_4) in that one of the H atoms is replaced with a Cl atom. It is used to prepare poly(vinyl chloride), which is an important polymer used to make piping. (a) Draw the Lewis structure of vinyl chloride. (b) The repeating unit in poly(vinyl chloride) is —CH_2—CHCl—. Draw a portion of the molecule showing three such repeating units. (c) Calculate the enthalpy change when 1.0×10^3 kg of vinyl chloride react to form poly(vinyl chloride).

9.86 (a) From the following data:

$$F_2(g) \longrightarrow 2F(g) \qquad \Delta H°_{rxn} = 156.9 \text{ kJ}$$

$$F^-(g) \longrightarrow F(g) + e^- \qquad \Delta H°_{rxn} = 333 \text{ kJ}$$

$$F_2^-(g) \longrightarrow F_2(g) + e^- \qquad \Delta H°_{rxn} = 290 \text{ kJ}$$

calculate the bond energy of the F_2^- ion. (b) Explain the difference between the bond energies of F_2 and F_2^-.

9.87 Write three resonance structures for the isocyanate ion (CNO^-). Rank them in importance.

Answers to Practice Exercises: 9.1 (a) Ionic, (b) polar covalent, (c) covalent. **9.2** $\ddot{S}$=C=$\ddot{S}$ **9.3** [$\ddot{O}$=$\ddot{N}$—$\ddot{O}$:]$^-$ **9.4** $\ddot{O}$=$\ddot{N}$—$\ddot{O}$:$^-$ **9.5** H—C≡N: **9.6** $\ddot{O}$=$\ddot{N}$—$\ddot{O}$:$^-$ ⟷ $^-$:$\ddot{O}$—$\ddot{N}$=$\ddot{O}$ **9.7** 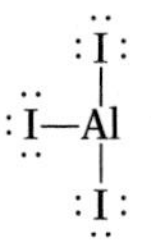 **9.8** (a) −119 kJ, (b) −137.0 kJ.

CHAPTER 10

CHEMICAL BONDING II: MOLECULAR GEOMETRY AND HYBRIDIZATION OF ATOMIC ORBITALS

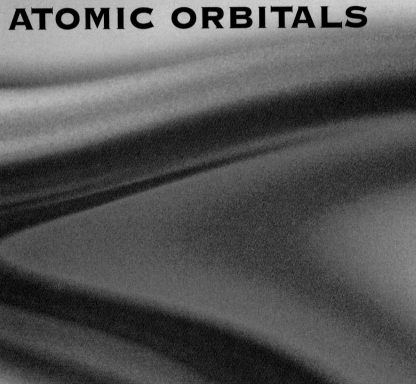

◆ In an effort to create unusual molecules believed to exist in interstellar space, chemists at Rice University used a high-powered laser to vaporize graphite. Among the products formed during this 1985 experiment was a stable species whose molar mass corresponds to the formula C_{60}. Such a molecule would have to have an exotic shape because of its size and the fact that only one type of atom is present. Working with paper, scissors, and Scotch tape, the researchers assembled a closed sphere from 20 paper hexagons and 12 paper pentagons. Significantly, this sphere contains 60 vertices, which correspond to 60 carbon atoms.

Because of its shape, the chemists named the new molecule "buckminsterfullerene," in honor of the late R. Buckminster Fuller, whose engineering and architectural designs often incorporated geodesic domes. But buckminsterfullerene is something of a mouthful, and the new molecule, which also resembles a soccer ball, quickly became known as "buckyball." Subsequent spectroscopic and X-ray measurements have confirmed the predicted structure.

Buckyball has generated tremendous interest in the scientific community for a number of reasons. First, it is a stable allotrope of carbon, as are diamond and graphite. It turns out that buckyball can be prepared under less extreme conditions, and it is a natural component of soot. In addition, buckyball is the most symmetrical molecule known. Furthermore, since buckyball was discovered, chemists have created similar "fullerenes" with more than 60 carbon atoms. In 1992, C_{60} and C_{70} were found in a sample of rock from the Russian town of Shungite, about 250 miles northeast of St. Petersburg.

Buckyball and the other fullerenes represent a whole new concept in molecular architecture. Studies have already shown that these molecules (and compounds derived from them) can act as high-temperature superconductors and lubricants, and they may well find uses as catalysts and antiviral drugs. ◆

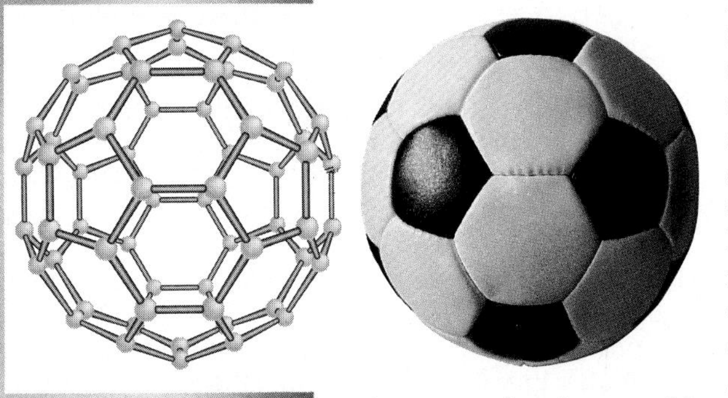

The C_{60} molecule resembles a soccer ball.

10.1 MOLECULAR GEOMETRY

Molecular geometry refers to the three-dimensional arrangement of atoms in a molecule. Many physical and chemical properties, such as melting point, boiling point, density, and the types of reactions that molecules undergo, are affected by a molecule's geometry. In general, bond lengths and the angles between bonds must be determined by experiment. However, there is a simple procedure that allows us to predict with considerable success the overall geometry of a molecule if we know the number of electrons surrounding a central atom in its Lewis structure. The basis of this approach is the assumption that electron pairs in the valence shell of an atom repel one another. The ***valence shell*** is *the outermost electron-occupied shell of an atom; it holds the electrons that are usually involved in bonding.* In a covalent bond, a pair of electrons (often called the *bonding pair*) is responsible for holding two atoms together. However, in a polyatomic molecule, where there are two or more bonds between the central atom and the surrounding atoms, the repulsion between the electrons in different bonding pairs causes them to remain as far apart as possible. The geometry that the molecule ultimately assumes (as defined by the positions of all the atoms) minimizes this repulsion. This approach to the study of molecular geometry is called the ***valence-shell electron-pair repulsion (VSEPR) model,*** because *it accounts for the geometric arrangements of electron pairs around a central atom in terms of the repulsion between electron pairs.*

Two general rules govern the use of the VSEPR model:

- As far as electron-pair repulsion is concerned, double bonds and triple bonds can be treated as though they were single bonds between neighboring atoms. This approximation is good for qualitative purposes. However, you should realize that in reality multiple bonds are "larger" than single bonds; that is, because there are two or three bonds between two atoms, the electron density occupies more space.
- If two or more resonance structures can be drawn for a molecule, we may apply the VSEPR model to any one of them. Furthermore, formal charges are usually not shown.

With the VSEPR model, we can predict the geometry of molecules in a systematic way. To do so, it is convenient to divide molecules into two categories, according to whether or not the central atom has lone pairs.

MOLECULES IN WHICH THE CENTRAL ATOM HAS NO LONE PAIRS

For simplicity we will consider molecules that contain atoms of only two elements, A and B, of which A is the central atom. These molecules have the general formula AB_x, where x is an integer 2, 3, ... (If $x = 1$, we have the diatomic molecule AB, which is linear by definition.) In the vast majority of cases, x is between 2 and 6.

Table 10.1 shows five possible arrangements of electron pairs around the central atom A. As a result of mutual repulsion, the electron pairs stay as far from one another as possible. Note that the table shows arrangements of the electron pairs but not the positions of the surrounding atoms. Molecules in this category (that is, molecules in which the central atom has no lone pairs) have one of these five arrangements of bonding pairs. We will therefore take a close look now at the geometry of molecules with the formulas AB_2, AB_3, AB_4, AB_5, and AB_6.

TABLE 10.1

Arrangement of Electron Pairs about a Central Atom (A) in a Molecule, and Geometry of Some Simple Molecules and Ions in Which the Central Atom Has No Lone Pairs

Number of electron pairs	Arrangement of electron pairs*	Molecular geometry*	Examples
2	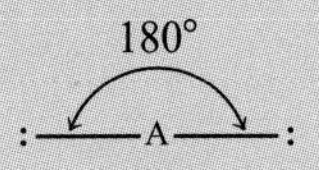 Linear	B—A—B Linear	$BeCl_2$, $HgCl_2$
3	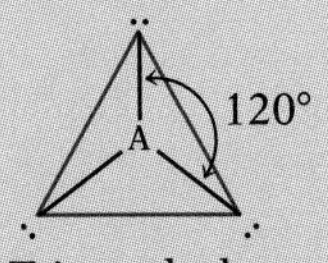 Trigonal planar	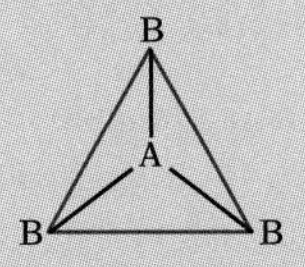 Trigonal planar	BF_3
4	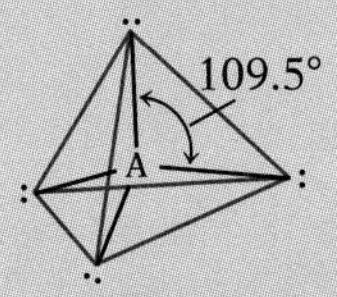 Tetrahedral	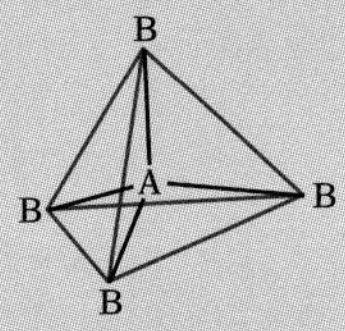 Tetrahedral	CH_4, NH_4^+
5	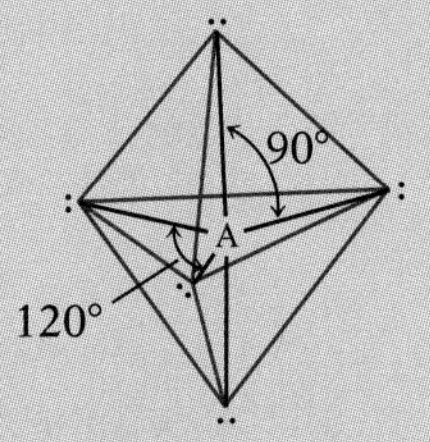 Trigonal bipyramidal	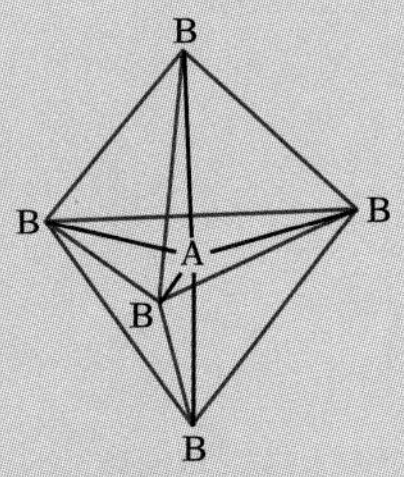 Trigonal bipyramidal	PCl_5
6	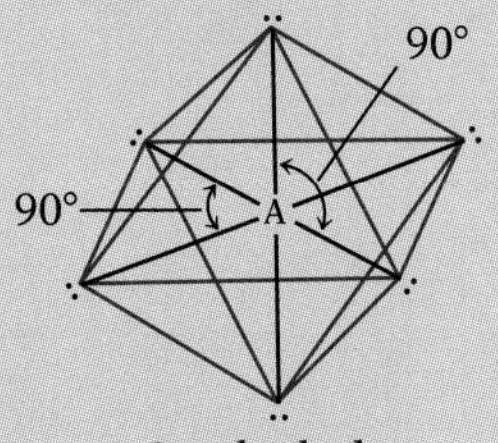 Octahedral	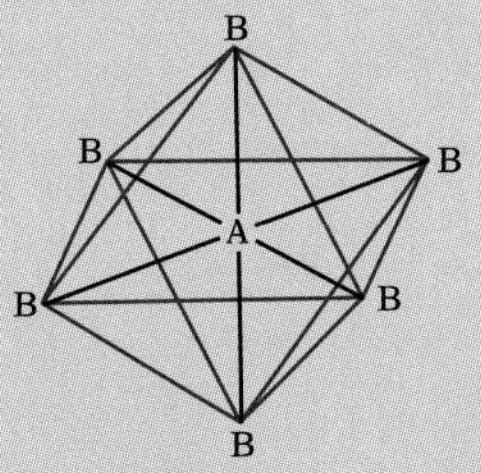 Octahedral	SF_6

*The colored lines are used only to show the overall shapes; they do not represent bonds.

AB_2: Beryllium Chloride ($BeCl_2$)

The Lewis structure of beryllium chloride in the gaseous state is

$$Cl—Be—Cl$$

Because the bonding pairs repel each other, they must be at opposite ends of a straight line in order for them to be as far apart as possible. Thus, the ClBeCl angle is predicted to be 180°, and the molecule is linear (see Table 10.1):

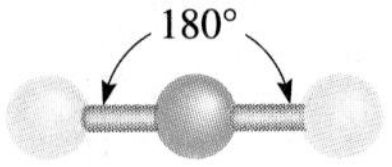

This representation of the $BeCl_2$ molecule is called a *ball-and-stick model.*

AB_3: Boron Trifluoride (BF_3)

Boron trifluoride contains three covalent bonds, or bonding pairs. In the most stable arrangement, the three BF bonds point to the corners of an equilateral triangle with B in the center of the triangle:

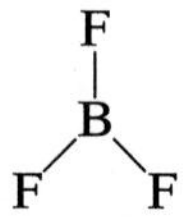

According to Table 10.1, the geometry of BF_3 is *trigonal planar* because the three end atoms are at the corners of an equilateral triangle which is planar:

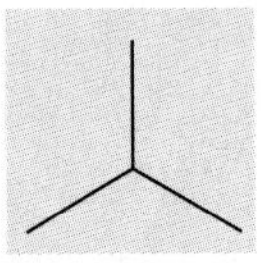
Planar

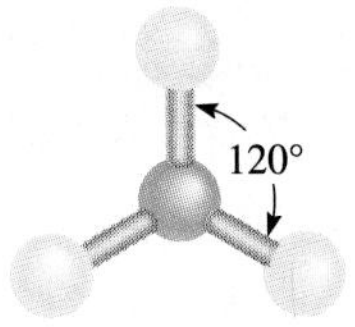

Thus, each of the three FBF angles is 120°, and all four atoms lie in the same plane.

AB_4: Methane (CH_4)

The Lewis structure of methane is

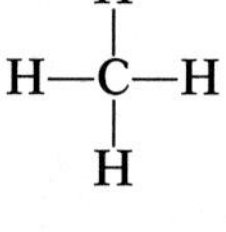

Since there are four bonding pairs, the geometry of CH_4 is tetrahedral (see Table 10.1). A *tetrahedron* has four sides (the prefix *tetra-* means "four"), or four faces, all of which are equilateral triangles. In a tetrahedral molecule, the central atom (C in this case) is located at the center of the tetrahedron and the other four atoms are at the corners. The bond angles are all 109.5°.

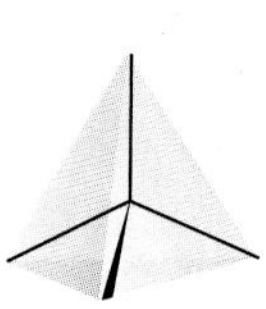
Tetrahedral

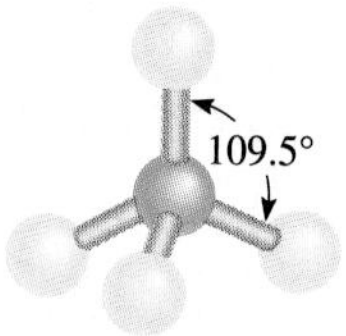

AB_5: Phosphorus Pentachloride (PCl_5)

The Lewis structure of phosphorus pentachloride (in the gas phase) is

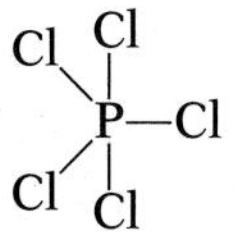

The only way to minimize the repulsive forces among the five bonding pairs is to arrange the PCl bonds in the form of a trigonal bipyramid (see Table 10.1). A trigonal bipyramid can be generated by joining two tetrahedrons along a common base:

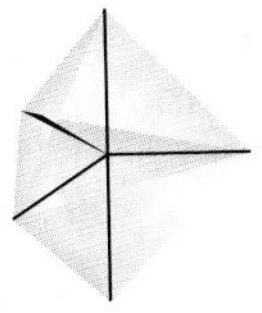

Trigonal bipyramidal

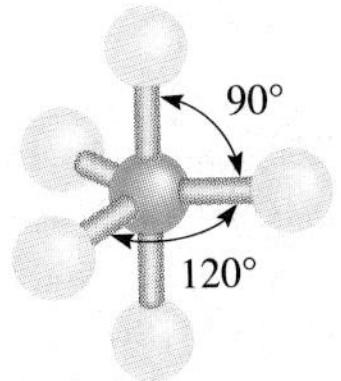

The central atom (P in this case) is located at the center of the common triangle with the other five atoms positioned at the five corners of the trigonal bipyramid. The atoms that are above and below the triangular plane are said to occupy *axial* positions, and those that are in the triangular plane are said to occupy *equatorial* positions. The angle between any two equatorial bonds is 120°; that between an axial bond and an equatorial bond is 90°, and that between the two axial bonds is 180°.

AB_6: Sulfur Hexafluoride (SF_6)

The Lewis structure of sulfur hexafluoride is

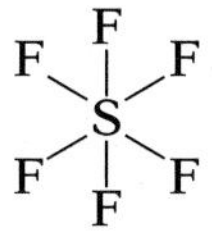

The most stable arrangement of the six SF bonding pairs is that of an octahedron, shown in Table 10.1. An octahedron has eight sides (the prefix *octa-* means "eight"). It can be generated by joining two square pyramids on a common base. The central atom (S in this case) is located at the center of the square base and the other six atoms are at the six corners. All bond angles are 90° except the one made by the bonds between the central atom and the two atoms diametrically opposite each other. That angle is 180°. Since the six bonds are equivalent in an octahedral molecule, we cannot use the terms "axial" and "equatorial" as in a trigonal bipyramidal molecule.

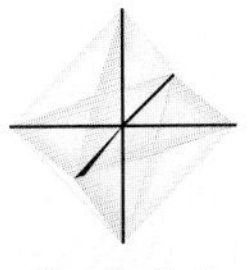

Octahedral

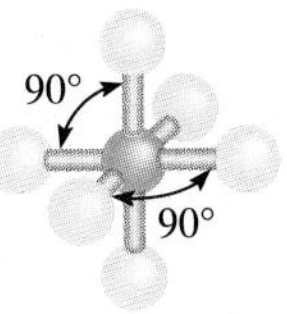

Table 10.1 shows the VSEPR geometries of some simple molecules.

Molecules in Which the Central Atom Has One or More Lone Pairs

Determining the geometry of a molecule is more complicated if the central atom has both lone pairs and bonding pairs. In such molecules there are three types of repulsive forces—those between bonding pairs, those between lone pairs, and those between a bonding pair and a lone pair. In general, according to VSEPR, the repulsive forces decrease in the following order:

lone-pair vs. lone-pair repulsion > lone-pair vs. bonding-pair repulsion > bonding-pair vs. bonding-pair repulsion

Electrons in a bond are held by the attractive forces exerted by the nuclei of the two bonded atoms. These electrons have less "spatial distribution"; that is, they take up less space than lone-pair electrons, which are associated with only one particular atom. Lone-pair electrons in a molecule occupy more space and consequently experience a greater repulsion from neighboring lone pairs and bonding pairs. To keep track of the total number of bonding pairs and lone pairs, we designate molecules with lone pairs as AB_xE_y, where A is the central atom, B is a surrounding atom, and E is a lone pair on A. Both x and y are integers; $x = 2, 3, \ldots$, and $y = 1, 2, \ldots$. Thus the values of x and y indicate the number of surrounding atoms and number of lone pairs on the central atom, respectively. In the scheme here, the simplest molecule would be a triatomic molecule with one lone pair on the central atom; the formula of this molecule is AB_2E.

For molecules in which the central atom has one or more lone pairs, we need to distinguish between the overall arrangement of the electron pairs and the geometry of the molecule. The overall arrangement of the electron pairs refers to the arrangement of *all* electron pairs on the central atom, bonding pairs as well as lone pairs. On the other hand, the geometry of a molecule is described only in terms of the arrangement of its atoms, and hence only the arrangement of bonding pairs is considered. We will see shortly that if lone pairs are present on the central atom, the overall arrangement of the electron pairs is *not* the same as the geometry of the molecule.

This will become clearer as we consider some specific examples.

AB_2E: Sulfur Dioxide (SO_2)

The Lewis structure of sulfur dioxide is

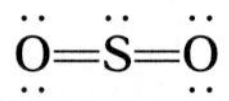

Because in our simplified scheme we treat double bonds as if they were single (p. 276), the SO_2 molecule can be viewed as consisting of three electron pairs on the central S atom. Of these, two are bonding pairs and one is a lone pair. In Table 10.1 we see that the overall arrangement of three electron pairs is trigonal planar. But because one of the electron pairs is a lone pair, the SO_2 molecule has a "bent" shape.

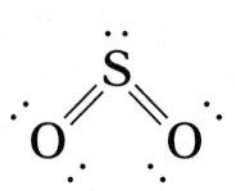

Since the lone-pair versus bonding-pair repulsion is greater than the bonding-pair versus bonding-pair repulsion, the two sulfur-to-oxygen bonds are pushed together slightly and the OSO angle is less than 120°. By experiment the OSO angle is found to be 119.5° (see Problem 10.60).

AB_3E: Ammonia (NH_3)

The ammonia molecule contains three bonding pairs and one lone pair:

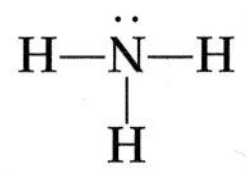

As Table 10.1 shows, the overall arrangement of four electron pairs is tetrahedral. But in NH_3 one of the electron pairs is a lone pair, so the geometry of NH_3 is trigonal pyramidal (so called because it looks like a pyramid, with the N atom at the apex). Because the lone pair repels the bonding pairs more strongly, the three NH bonding pairs are pushed closer together; thus the HNH angle in ammonia is smaller than the ideal tetrahedral angle of 109.5° (Figure 10.1).

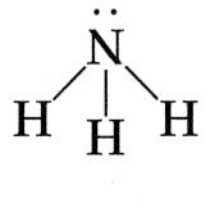

AB_2E_2: Water (H_2O)

A water molecule contains two bonding and two lone pairs:

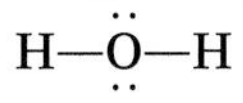

The overall arrangement of the four electron pairs in water is tetrahedral, the same electron-pair arrangement that is found in ammonia. However, unlike ammonia, water has two lone pairs on the central O atom. These lone pairs tend to be as far from each other as possible. Consequently, the two OH bonding pairs are pushed toward each other, so that we can predict an even greater deviation from the tetrahedral angle than in NH_3. As Figure 10.1 shows, the HOH angle is 104.5°. The geometry of H_2O is bent.

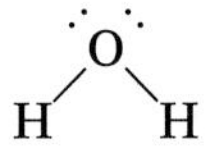

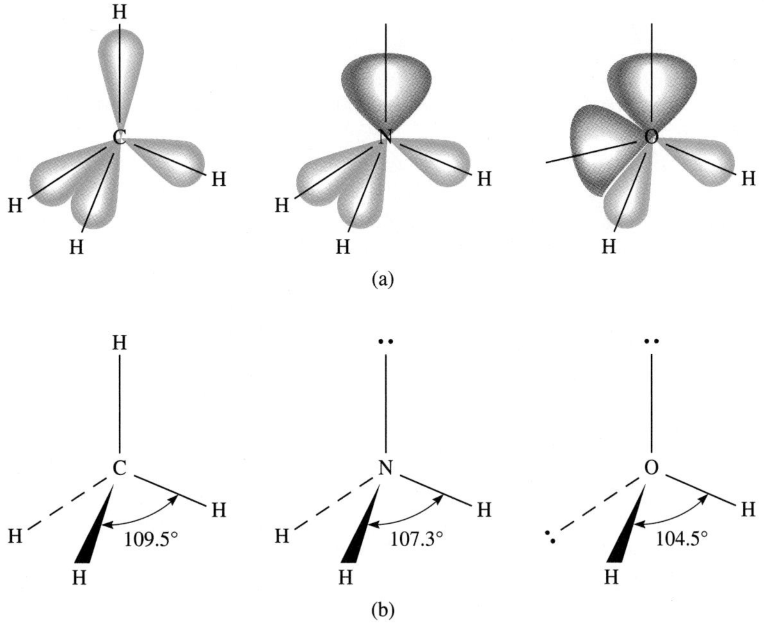

FIGURE 10.1

(a) The relative sizes of the bonding pairs and lone pairs in CH_4, NH_3, and H_2O. (b) The bond angles in CH_4, NH_3, and H_2O. In each diagram the dashed line represents a bond axis behind the plane of the paper, the wedged line represents a bond axis in front of the plane of the paper, and the thin solid lines represent bonds in the plane of the paper.

AB_4E: Sulfur Tetrafluoride (SF_4)

The Lewis structure of SF_4 is

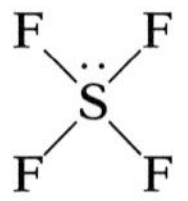

The central sulfur atom has five electron pairs whose arrangement, according to Table 10.1, is trigonal bipyramidal. In the SF_4 molecule, however, one of the electron pairs is a lone pair, so that the molecule must have one of the following geometries:

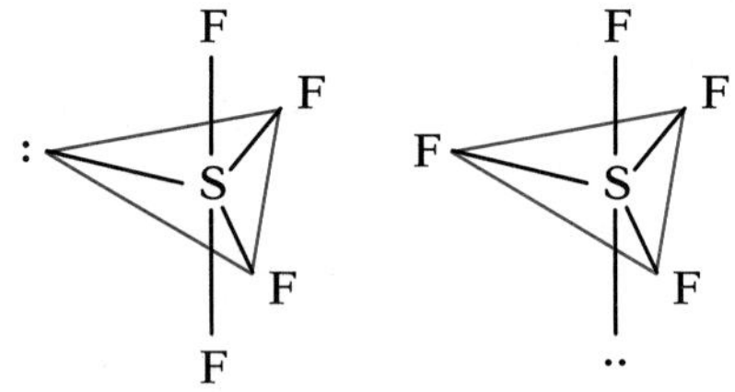

In (a) the lone pair occupies an equatorial position, and in (b) it occupies an axial position. The axial position has three neighboring pairs at 90° and one at 180°, while the equatorial position has two neighboring pairs at 90° and two more at 120°. The repulsion is smaller for (a), and indeed (a) is the structure observed experimentally. The shape shown in (a) is sometimes described as a distorted tetrahedron (or a folded square, or seesaw shape). Experimentally, the angle between the axial F atoms and S is found to be 186°, and that between the equatorial F atoms and S is 116°.

Table 10.2 shows the geometries of simple molecules in which the central atom has one or more lone pairs, including some that we have not discussed.

Geometry of Molecules with More Than One Central Atom

So far we have discussed the geometry of molecules having only one central atom. (The term "central atom" here means an atom that is not a terminal atom in a polyatomic molecule.) The overall geometry of molecules with more than one central atom is difficult to define in most cases. Often we can only describe what the shape is like around each of the central atoms. Consider methanol, CH_3OH, whose Lewis structure is

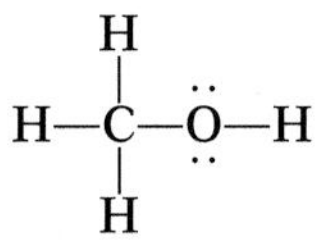

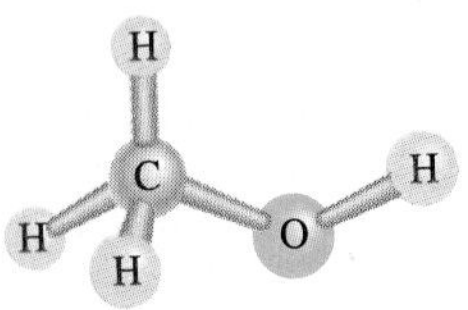

FIGURE 10.2

The geometry of CH_3OH.

The two central (nonterminal) atoms in methanol are C and O. We can say that the three CH and the CO bonding pairs are tetrahedrally arranged about the C atom. The HCH and OCH bond angles are approximately 109°. The O atom here is like the one in water in that it has two lone pairs and two bonding pairs. Therefore, the HOC portion of the molecule is bent, and the angle HOC is approximately equal to 105° (Figure 10.2).

TABLE 10.2

Geometry of Simple Molecules and Ions in Which the Central Atom Has One or More Lone Pairs

Class of molecule	Total number of electron pairs	Number of bonding pairs	Number of lone pairs	Arrangement of electron pairs*	Geometry	Examples
AB_2E	3	2	1	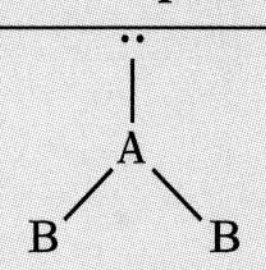 Trigonal planar	Bent	SO_2, O_3
AB_3E	4	3	1	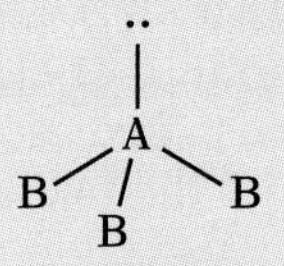 Tetrahedral	Trigonal pyramidal	NH_3
AB_2E_2	4	2	2	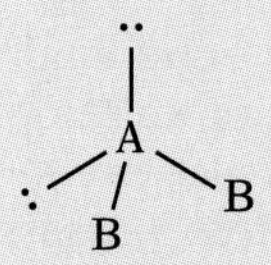 Tetrahedral	Bent	H_2O
AB_4E	5	4	1	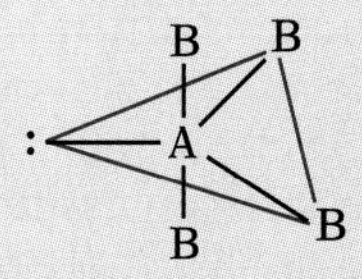 Trigonal bipyramidal	Distorted tetrahedron (or seesaw)	IF_4^+, SF_4, XeO_2F_2
AB_3E_2	5	3	2	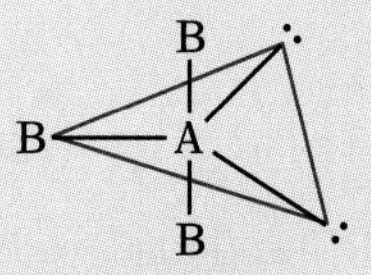 Trigonal bipyramidal	T-shaped	ClF_3
AB_2E_3	5	2	3	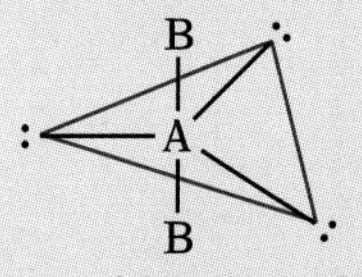 Trigonal bipyramidal	Linear	XeF_2, I_3^-
AB_5E	6	5	1	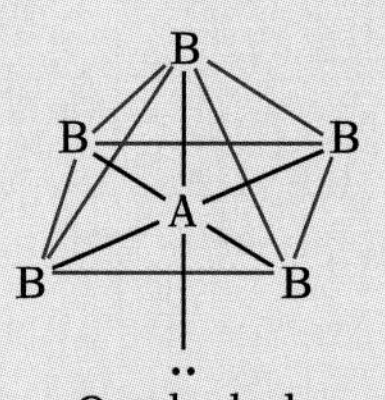 Octahedral	Square pyramidal	BrF_5, $XeOF_4$
AB_4E_2	6	4	2	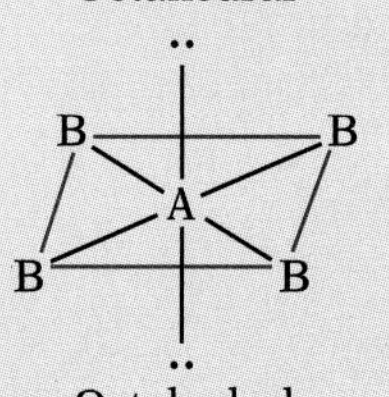 Octahedral	Square planar	XeF_4, ICl_4^-

*The colored lines are used to show the overall shape, not bonds.

Guidelines for Applying the VSEPR Model

Having studied the geometries of molecules in two categories (central atoms with and without lone pairs), let us consider some rules for applying the VSEPR model to all types of molecules:

- Write the Lewis structure of the molecule, considering only the electron pairs around the central atom (that is, the atom that is bonded to more than one other atom).
- Count the number of electron pairs around the central atom (bonding pairs and lone pairs). Treat double and triple bonds as single bonds. Refer to Table 10.1 to predict the overall arrangement of the electron pairs.
- Use Tables 10.1 and 10.2 to predict the geometry of the molecule.
- In predicting bond angles, note that a lone pair repels another lone pair or a bonding pair more strongly than a bonding pair repels another bonding pair. Remember that there is no easy way to predict bond angles accurately when the central atom possesses one or more lone pairs.

The VSEPR model generates reliable predictions of the geometries of a variety of molecular structures. Chemists accept and use the VSEPR approach because of its simplicity and utility. Although some theoretical concerns have been voiced about whether "electron-pair repulsion" actually determines molecular shapes, it is safe to say that such an assumption leads to useful (and generally reliable) predictions. We need not ask more of any model at this stage in the study of chemistry.

EXAMPLE 10.1
Applying the VSEPR Model to Molecules and Ions

Use the VSEPR model to predict the geometry of the following molecules and ions: (a) AsH_3, (b) OF_2, (c) $AlCl_4^-$, (d) I_3^-, (e) C_2H_4.

Answer: (a) The Lewis structure of AsH_3 is

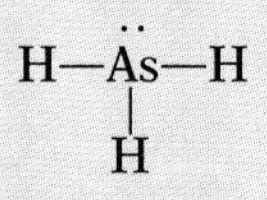

This molecule has three bonding pairs and one lone pair, a combination similar to that in ammonia. Therefore, the geometry of AsH_3 is trigonal pyramidal, like NH_3. We cannot predict the HAsH angle accurately, but we know that it must be less than 109.5° because the repulsion of the bonding electron pairs by the lone pair on As is greater than that between bonding pairs.

(b) The Lewis structure of OF_2 is F—Ö—F. Since there are two lone pairs on the O atom, the OF_2 molecule has a bent shape, like that of H_2O. Again, all we can say about angle size is that the FOF angle should be less than 109.5° due to the stronger repulsive force between the lone pairs and the bonding pairs.

(c) The Lewis structure of $AlCl_4^-$ is

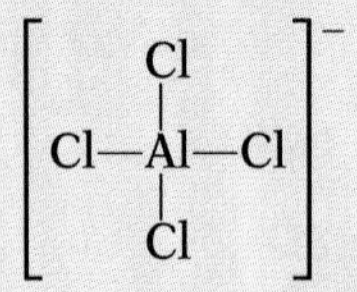

Since the central Al atom has no lone pairs and all four Al—Cl bonds are equivalent, the $AlCl_4^-$ ion should be tetrahedral, and the ClAlCl angles should all be 109.5°.

(d) The Lewis structure of I_3^- is

$$\left[\mathrm{I}-\ddot{\underset{..}{\mathrm{I}}}-\mathrm{I}\right]^-$$

The central I atom has two bonding pairs and three lone pairs. From Table 10.2 we see that the three lone pairs all lie in the triangular plane, and the I_3^- ion should be linear.

(e) The Lewis structure of C_2H_4 is

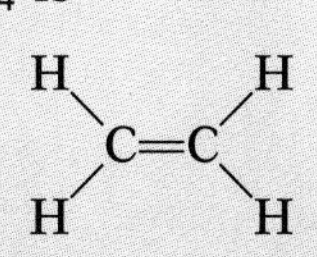

The C═C bond is treated as though it were a single bond. Because there are no lone pairs present, the arrangement around each C atom has a trigonal planar shape like BF_3, discussed earlier. Thus the predicted bond angles in C_2H_4 (HCH and HCC) are all 120°.

PRACTICE EXERCISE

Use the VSEPR model to predict the geometry of (a) $SiBr_4$, (b) CS_2, and (c) NO_3^-.

10.2 DIPOLE MOMENTS

In Section 9.3 we learned that in the hydrogen fluoride molecule, there is a shift in electron density from H to F because the F atom is more electronegative than the H atom. The shift in electron density is symbolized by placing a crossed arrow (⟶ with a cross, +→) above the Lewis structure to indicate the direction of the shift. For example,

$$\overset{+\!\!\longrightarrow}{\mathrm{H}-\ddot{\underset{..}{\mathrm{F}}}:}$$

The consequent charge separation can be represented as

$$\overset{\delta^+}{\mathrm{H}}-\overset{\delta^-}{\ddot{\underset{..}{\mathrm{F}}}}:$$

where δ (delta) denotes a partial charge. This separation of charges is confirmed in an electric field (Figure 10.3). When the field is turned on, HF molecules orient their negative charge centers toward the positive plate and their positive ends toward the negative plate. This alignment of molecules can be detected experimentally. Hydrogen fluoride and other molecules that *have separated positive and negative centers* are called ***polar molecules.***

A quantitative measure of the polarity of a bond is its ***dipole moment,*** μ, which is *the product of the charge Q and the distance r between the charges:*

$$\mu = Q \times r \qquad (10.1)$$

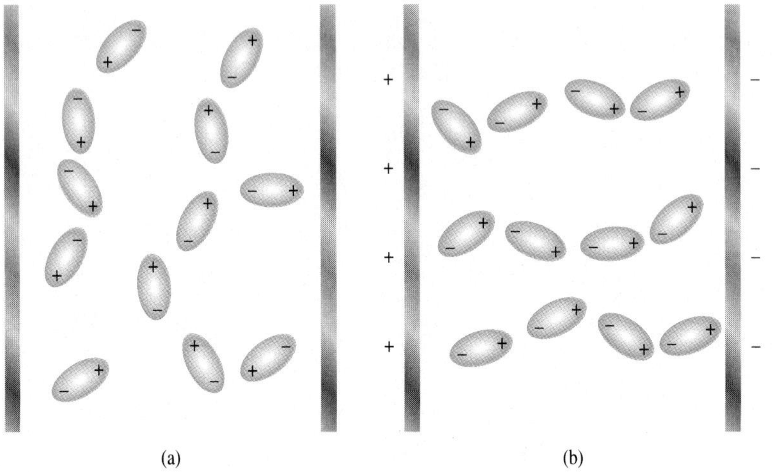

FIGURE 10.3

Behavior of polar molecules (a) in the absence and (b) in the presence of an external electric field. Nonpolar molecules are not affected by an electric field.

Peter Debye (1884–1966)

To maintain electrical neutrality, the charges on both ends of an electrically neutral diatomic molecule must be equal in magnitude and opposite in sign. However, the quantity Q in Equation (10.1) refers only to the magnitude and not its sign, so μ is always positive. Dipole moments are usually expressed in *debye* units (D), named for the Dutch chemist Peter Debye. The conversion factor is

$$1 \text{ D} = 3.33 \times 10^{-30} \text{ C m}$$

where C is coulomb and m is meter.

Diatomic molecules containing atoms of the *same* element (for example, H_2, O_2, and F_2) *do not have dipole moments* and so are ***nonpolar molecules.*** On the other hand, diatomic molecules containing atoms of *different* elements (for example, HCl, CO, and NO) have dipole moments. The dipole moment of a molecule made up of three or more atoms, however, depends on both the polarity of the bonds and molecular geometry. Even if polar bonds are present, the molecule will not necessarily have a dipole moment. Carbon dioxide (CO_2), for example, is a triatomic molecule, so its geometry is either linear or bent:

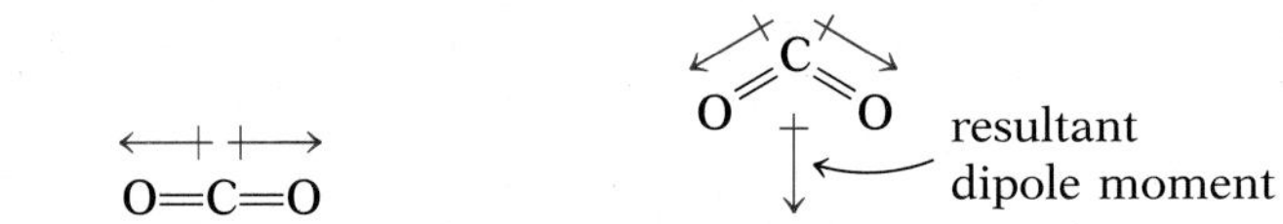

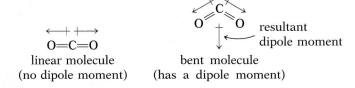

linear molecule (no dipole moment) — bent molecule (has a dipole moment)

The arrows show the shift of electron density from the less electronegative carbon atom to the more electronegative oxygen atom. In each case, the dipole moment of the entire molecule is made up of two *bond moments,* that is, individual dipole moments in the polar C=O bonds. The bond moment is a *vector quantity,* which means that it has both magnitude and direction. The measured dipole moment is equal to the vector sum of the bond moments. The two bond moments in CO_2 are equal in magnitude. Since they point in opposite directions in a linear CO_2 molecule, the sum or resultant dipole moment would be zero. On the other hand, if the CO_2 molecule were bent, the two bond moments would partially reinforce each other, so that the molecule would have a dipole moment. Experimentally it is found that carbon dioxide has no dipole moment. Therefore we conclude that the

TABLE 10.3
Dipole Moments of Some Polar Molecules

Molecule	Geometry	Dipole moment (D)
HF	Linear	1.92
HCl	Linear	1.08
HBr	Linear	0.78
HI	Linear	0.38
H_2O	Bent	1.87
H_2S	Bent	1.10
NH_3	Pyramidal	1.46
SO_2	Bent	1.60

carbon dioxide molecule is linear. The linear nature of carbon dioxide has, in fact, been confirmed through other experimental measurements.

Dipole moment measurements can be used to distinguish between molecules that have the same formula but different structures. For example, the following molecules both exist; they have the same molecular formula ($C_2H_2Cl_2$), the same number and type of bonds, but different molecular structures:

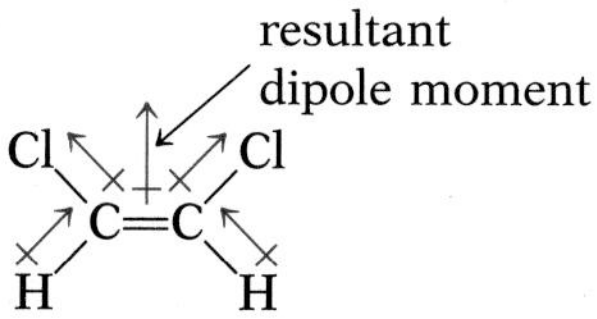

cis-dichloroethylene
$\mu = 1.89$ D

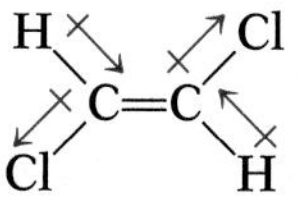

trans-dichloroethylene
$\mu = 0$

Since *cis*-dichloroethylene is a polar molecule but *trans*-dichloroethylene is not, they can readily be distinguished by a dipole moment measurement.

Table 10.3 lists the dipole moments of several polar molecules.

EXAMPLE 10.2
Predicting Dipole Moments

Predict whether each of the following molecules has a dipole moment: (a) IBr, (b) BF_3 (trigonal planar), (c) CH_2Cl_2 (tetrahedral).

Answer: (a) Since IBr (iodine bromide) is diatomic, it has a linear geometry. Bromine is more electronegative than iodine (see Figure 9.4), so IBr is polar with bromine at the negative end.

$$\overset{\longmapsto}{\text{I—Br}}$$

Thus the molecule does have a dipole moment.

(b) Since fluorine is more electronegative than boron, each B—F bond in BF_3 (boron trifluoride) is polar and the three bond moments are equal. However, the symmetry of a trigonal planar shape means that the three bond moments exactly cancel one another:

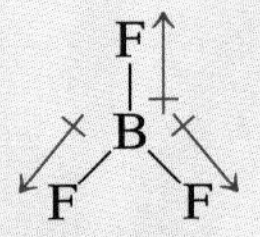

Consequently BF_3 has no dipole moment; it is a nonpolar molecule.

(c) The Lewis structure of CH_2Cl_2 (methylene chloride) is

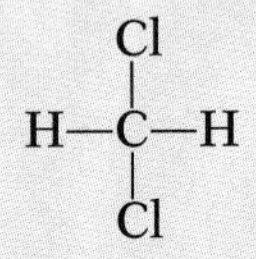

This molecule is similar to CH_4 in that it has an overall tetrahedral shape. However, because not all the bonds are identical, there are three different bond angles: HCH, HCCl, and ClCCl. These bond angles are close to, but not equal to, 109.5°. Since chlorine is more electronegative than carbon, which is more electronegative than hydrogen, the bond moments do not cancel and the molecule possesses a dipole moment:

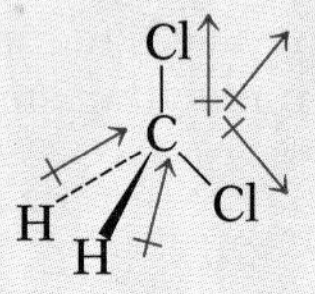

Thus CH_2Cl_2 is a polar molecule.

PRACTICE EXERCISE

Does the SO_2 molecule have a dipole moment?

10.3 HYBRIDIZATION OF ATOMIC ORBITALS

The VSEPR model, based largely on Lewis structures of molecules, provides a relatively simple and straightforward method for predicting the geometry of molecules. But as we noted earlier, the Lewis theory of chemical bonding treats *all* the covalent bonds the same and does not clearly explain why chemical bonds exist. For example, the Lewis theory describes the single bond between the H atoms in H_2 and that between the F atoms in F_2 in essentially the same way—as the pairing of two electrons. Yet these two molecules have quite different bond energies and bond lengths (436.4 kJ/mol and 74 pm for H_2 and 150.6 kJ/mol and 142 pm for F_2). These and many other facts cannot be satisfactorily explained by the Lewis theory. For a more complete explanation of chemical bond formation we must look to quantum mechanics. In fact, the quantum mechanical study of chemical bonding also provides a means for understanding molecular geometry.

In the 1930s *valence bond (VB) theory* was introduced to account for chemical bond formation. VB theory describes covalent bonding as the overlapping of atomic orbitals. This means that the orbitals share a common region in space. Thus the covalent bond in H_2 is formed when the 1*s* orbitals on the two H atoms overlap (Figure 10.4). The region of overlap is a favorable one for electrons to

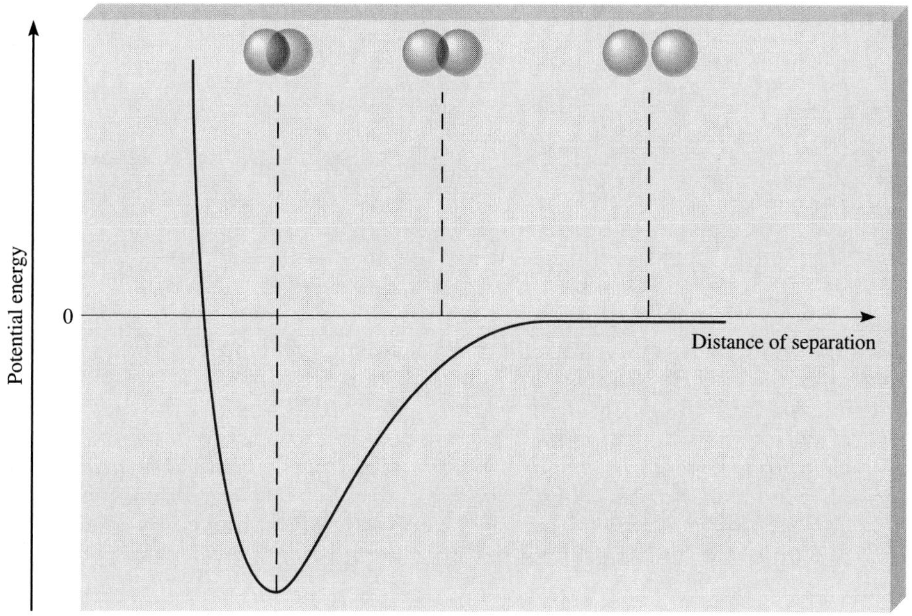

FIGURE 10.4

Change in potential energy of two H atoms with their distance of separation. At the point of minimum potential energy, the H_2 molecule is in its most stable state and the bond length is 74 pm.

reside in, for they are simultaneously attracted to the two positively charged nuclei.

Similarly, a stable F_2 molecule forms when the $2p$ orbitals (containing the unpaired electrons) in the two F atoms overlap to form a covalent bond, and the formation of the HF molecule can be explained by the overlap of the $1s$ orbital in H with the $2p$ orbital in F. Because the orbitals involved are not the same kind in all cases, we can see why the bond energies and bond lengths in H_2, F_2, and HF might be different.

The concept of atomic orbital overlap can also be applied to polyatomic molecules. However, a satisfactory bonding scheme must account for molecular geometry. We will discuss three examples of VB treatment of bonding in polyatomic molecules.

sp HYBRIDIZATION

The $BeCl_2$ (beryllium chloride) molecule is predicted to be linear by VSEPR. The orbital diagram for the valence electrons in Be is

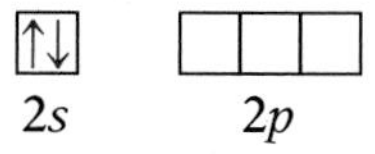

We know that in its ground state Be does not form covalent bonds with Cl because its electrons are paired in the $2s$ orbital. So we turn to hybridization for an explanation of Be's bonding behavior. First a $2s$ electron is promoted (that is, energetically excited) to a $2p$ orbital, resulting in

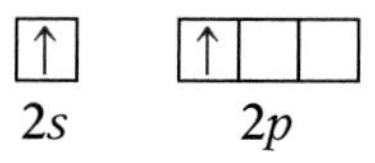

Now two Be orbitals are available for bonding, the $2s$ and $2p$. However, if two Cl atoms were to combine with Be in this excited state, one Cl atom would share a $2s$ electron and the other Cl would share a $2p$ electron, making two nonequivalent

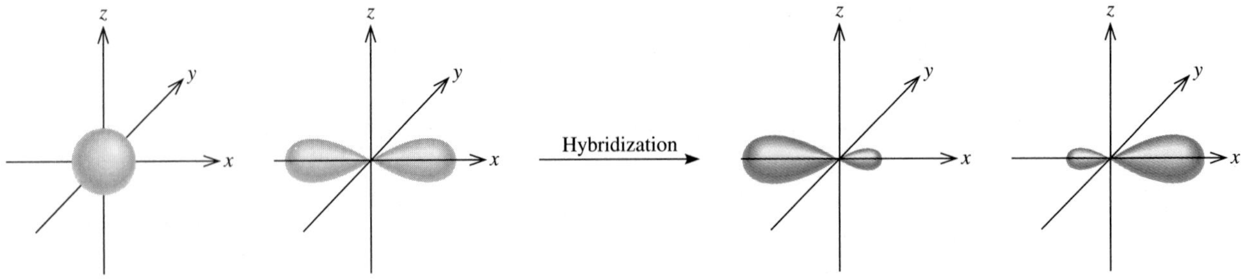

FIGURE 10.5

Formation of sp hybrid orbitals.

BeCl bonds. This scheme contradicts experimental evidence. In the actual $BeCl_2$ molecule, the two BeCl bonds are identical in every respect.

To explain the bonding in beryllium chloride, VB theory uses the concept of ***hybridization.*** Hybridization is *the mixing of atomic orbitals in an atom (usually a central atom) to generate a set of new atomic orbitals,* called ***hybrid orbitals.*** Hybrid orbitals, which are *atomic orbitals obtained when two or more nonequivalent orbitals of the same atom combine,* are used to form covalent bonds. By mixing the 2*s* orbital with one of the 2*p* orbitals we can generate two equivalent *sp* hybrid orbitals:

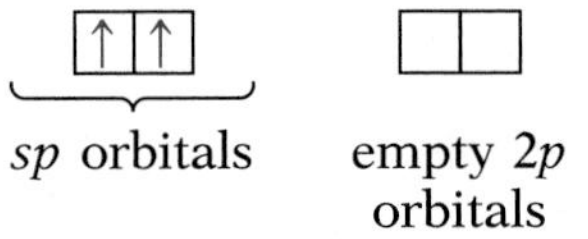

Figure 10.5 shows the shape and orientation of the *sp* orbitals. These two hybrid orbitals lie along the same line, the *x*-axis, so that the angle between them is 180°. Each of the BeCl bonds is then formed by the overlap of a Be *sp* hybrid orbital and a Cl 3*p* orbital, and the resulting $BeCl_2$ molecule has a linear geometry (Figure 10.6).

Note that although an input of energy is required to bring about hybridization, this is more than compensated for by the energy released upon the formation of Be—Cl bonds. (Bond formation is an exothermic process.)

sp^2 HYBRIDIZATION

Next we will look at the BF_3 (boron trifluoride) molecule, known to have planar geometry based on VSEPR. Considering only the valence electrons, we see that the orbital diagram of B is

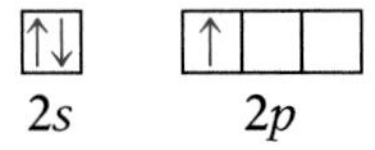

First, we promote a 2*s* electron to an empty 2*p* orbital:

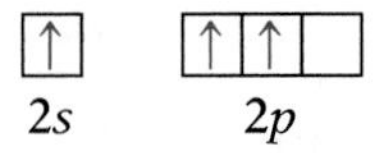

Mixing the 2*s* orbital with the two 2*p* orbitals generates three sp^2 hybrid orbitals:

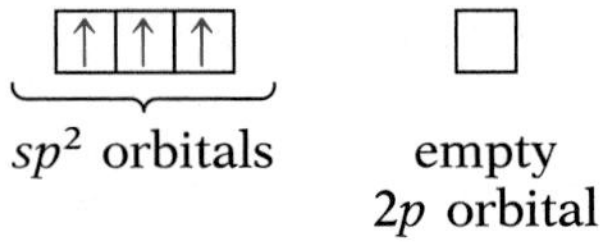

FIGURE 10.6

The linear geometry of $BeCl_2$ can be explained by assuming Be to be sp-hybridized. The two sp hybrid orbitals overlap with the two 3p orbitals of chlorine to form two covalent bonds.

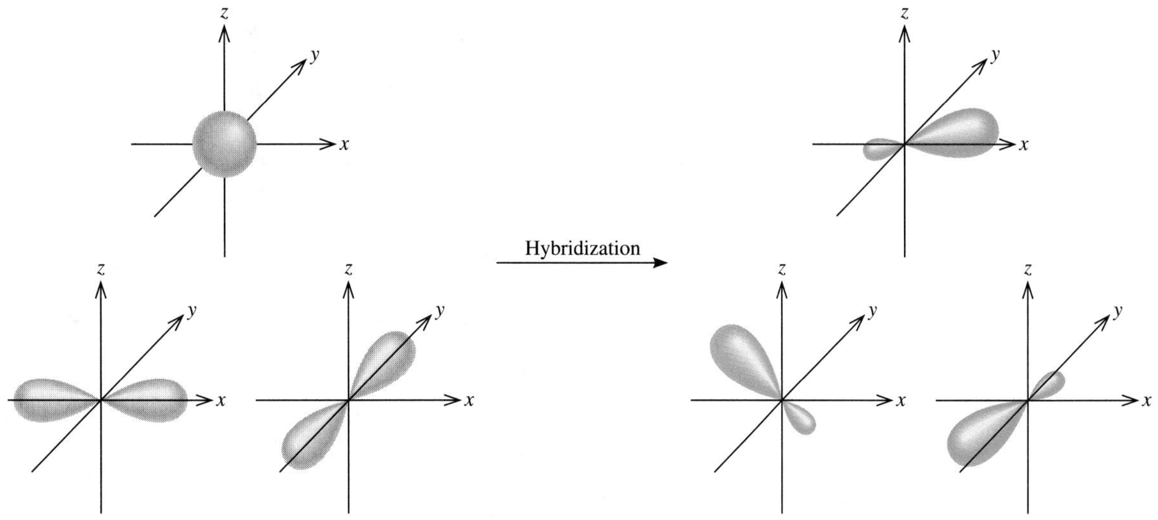

FIGURE 10.7

Formation of sp^2 hybrid orbitals.

These three sp^2 orbitals lie in a plane, and the angle between any two of them is 120° (Figure 10.7). Each of the BF bonds is formed by the overlap of a boron sp^2 hybrid orbital and a fluorine 2*p* orbital (Figure 10.8). The BF_3 molecule is planar with all the FBF angles equal to 120°. This result conforms to experimental findings and also to VSEPR predictions.

sp^3 HYBRIDIZATION

Finally we will consider the CH_4 molecule, which has a tetrahedral geometry. Focusing only on the valence electrons, we can represent the orbital diagram of C as

↑↓		↑	↑	
2*s*		2*p*		

Because the carbon atom has two unpaired electrons (one in each of the two 2*p* orbitals), it can only form two bonds with hydrogen in its ground state. Although the species CH_2 is known, it is very unstable. To account for the four C—H bonds in methane, we promote an electron from the 2*s* orbital to the 2*p* orbital:

↑		↑	↑	↑
2*s*		2*p*		

We can generate four hybrid orbitals by mixing the 2*s* orbital and the three 2*p* orbitals:

↑	↑	↑	↑

sp^3 orbitals

Figure 10.9 shows the shape and orientation of the sp^3 orbitals. These four equivalent hybrid orbitals are directed toward the four corners of a regular tetrahedron. Figure 10.10 shows how the sp^3 hybrid orbitals of carbon and the 1*s* orbitals of hydrogen overlap to form four covalent C—H bonds. Thus CH_4 has a tetrahedral shape, and all the HCH angles are 109.5°.

Another example of sp^3 hybridization is ammonia (NH_3). Table 10.1 shows

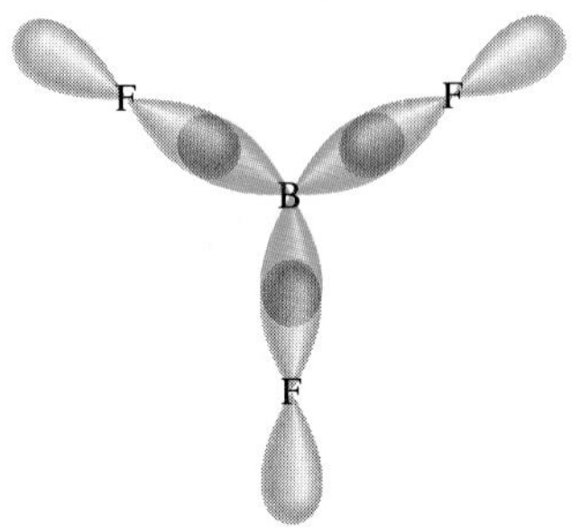

FIGURE 10.8

Overlap of the sp^2 hybrid orbitals of the boron atom with the 2p orbitals of the fluorine atoms. The BF_3 molecule is planar, and all the FBF angles are 120°.

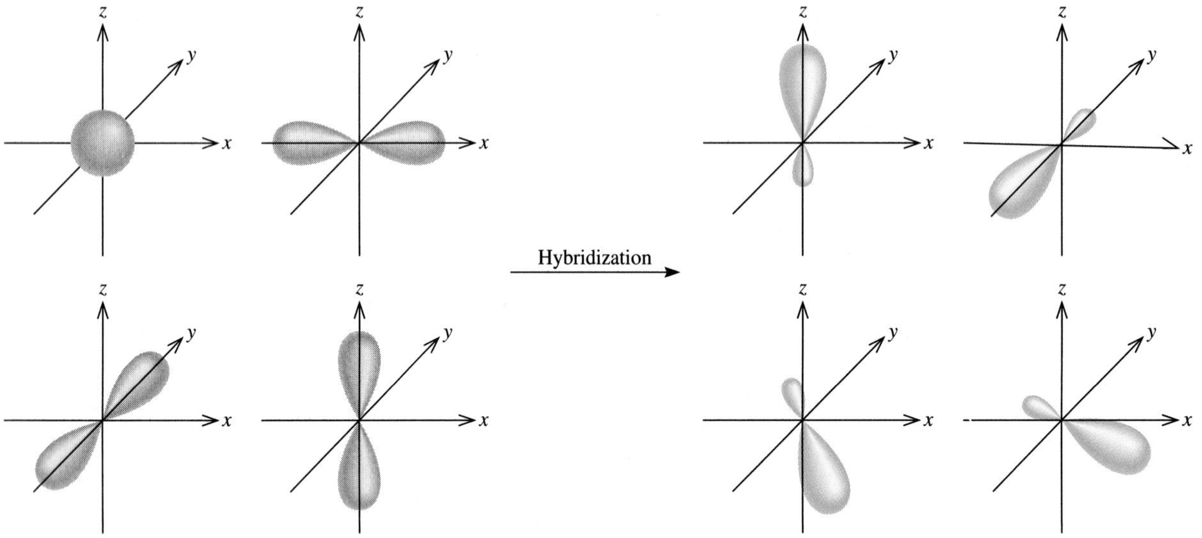

FIGURE 10.9

Formation of sp^3 hybrid orbitals.

that the arrangement of four electron pairs is tetrahedral, so that the bonding in NH_3 can be explained by assuming that N, like C in CH_4, is sp^3-hybridized. The ground-state electron configuration of N is $1s^22s^22p^3$, so that the orbital diagram for the sp^3-hybridized N atom is

↑	↑	↑	↑↓

sp^3 orbitals

Three of the four hybrid orbitals form covalent N—H bonds, and the fourth hybrid orbital accommodates the lone pair on nitrogen (Figure 10.11). Repulsion between the lone-pair electrons and those in the bonding orbitals decreases the HNH bond angles from 109.5° to 107.3°.

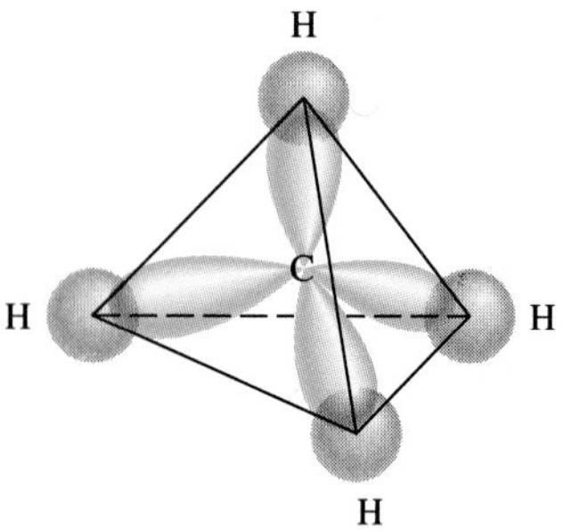

FIGURE 10.10

Formation of four bonds between the carbon sp^3 hybrid orbitals and the hydrogen 1s orbitals in CH_4.

The O atom in H_2O also has four electron pairs. Thus we might expect the O atom to be sp^3-hybridized. However, spectroscopic evidence strongly suggests that the bonding orbitals on O (used to form the O—H bonds) are $2p$ orbitals, rather than sp^3 hybrid orbitals. This means that the O atom in H_2O is *unhybridized.* If this is true, then why isn't the HOH angle 90° since the p orbitals are perpendicular to each other? The reason is that in the H_2O molecule the two O—H bonds need to spread apart to avoid the crowding of the H atoms into the same space (a situation that is energetically unfavorable). Consequently, the bond angle increases from the predicted 90° to 104.5°.

It is important to understand the relationship between hybridization and the VSEPR model. We use hybridization to describe the bonding scheme only when the arrangement of electron pairs has been predicted using VSEPR. If the VSEPR model predicts a tetrahedral arrangement of electron pairs, then we assume that one s and three p orbitals are hybridized to form four sp^3 hybrid orbitals.

You may have noticed an interesting connection between hybridization and the octet rule. Regardless of the type of hybridization, an atom starting with one s and three p orbitals will still possess four orbitals, enough to accommodate a total of eight electrons in a compound. For elements in the second period of the periodic table, eight is the maximum number of electrons that an atom of any of these elements can accommodate in the valence shell. This is the reason that the octet rule is usually obeyed by the second-period elements. The important exceptions are beryllium and boron, which sometimes associate with only four and six valence electrons, respectively, in bonding.

The situation is different for an atom of a third-period element. If we use only the 3*s* and 3*p* orbitals of the atom to form hybrid orbitals in a molecule, then the octet rule applies. However, in some molecules the same atom may use one or more 3*d* orbitals, in addition to the 3*s* and 3*p* orbitals, to form hybrid orbitals. In these cases the octet rule does not hold. We will see specific examples of the participation of the 3*d* orbital in hybridization shortly.

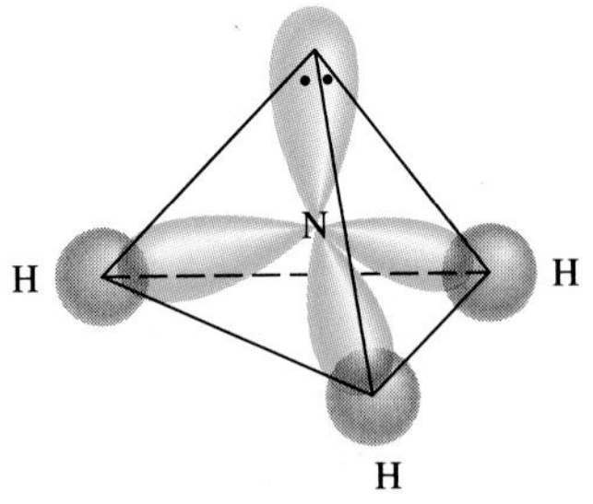

FIGURE 10.11

The N atom is sp^3-hybridized in NH_3. It forms three bonds with the H atoms. One of the sp^3 hybrid orbitals is occupied by the lone pair.

PROCEDURE FOR HYBRIDIZING ATOMIC ORBITALS

Based on the discussion so far, let us summarize what we need to know in order to apply the concept to bonding in polyatomic molecules in general. In essence, hybridization simply extends Lewis theory and the VSEPR model. We must have some idea about the geometry of the molecule to start with. By drawing the Lewis structure of the molecule and predicting the overall arrangement of the electron pairs (both bonding pairs and lone pairs) using the VSEPR model (see Table 10.1), we can deduce the hybridization of the central atom by matching the arrangement of the electron pairs with those of the hybrid orbitals shown in Table 10.4.

Hybridization depends on these assumptions:

- The concept of hybridization is not applied to isolated atoms. It is used only to explain the bonding scheme in a molecule.
- Hybridization is the mixing of at least two nonequivalent atomic orbitals, for example, *s* and *p* orbitals. Therefore, a hybrid orbital is not a pure (that is, native) atomic orbital. Hybrid orbitals and pure atomic orbitals have very different shapes.
- The number of hybrid orbitals generated is equal to the number of pure atomic orbitals that participate in the hybridization process.
- Hybridization requires an input of energy; however, the system more than recovers this energy during bond formation.
- Covalent bonds in polyatomic molecules are formed by the overlap of hybrid orbitals, or of hybrid orbitals with unhybridized ones. Therefore, the hybridization bonding scheme is still within the framework of valence bond theory; electrons in a molecule are assumed to occupy hybrid orbitals of the individual atoms.

EXAMPLE 10.3
Determining the Hybridization State of an Atom

Determine the hybridization state of the central (underlined) atom in each of the following molecules: (a) $\underline{Hg}Cl_2$, (b) $\underline{Al}I_3$, (c) $\underline{P}F_3$. Describe the hybridization process and determine the molecular geometry in each case.

Answer: (a) Referring to Table 7.3, we see that the ground-state electron configuration of Hg is $[Xe]6s^24f^{14}5d^{10}$; therefore, the Hg atom has two valence electrons (the 6*s* electrons). The Lewis structure of $HgCl_2$ is

$$Cl—Hg—Cl$$

There are no lone pairs on the Hg atom, so the arrangement of the two elec-

TABLE 10.4
Important Hybrid Orbitals and Their Shapes

Pure atomic orbitals of the central atom	Hybridization of the central atom	Number of hybrid orbitals	Shape of hybrid orbitals	Examples
s, *p*	*sp*	2	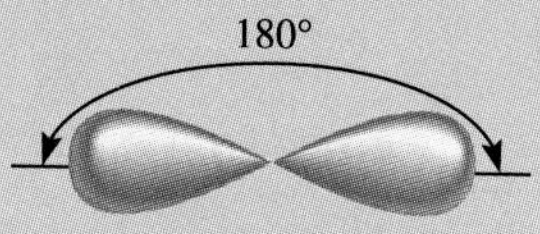 Linear	$BeCl_2$
s, *p*, *p*	sp^2	3	 Planar	BF_3
s, *p*, *p*, *p*	sp^3	4	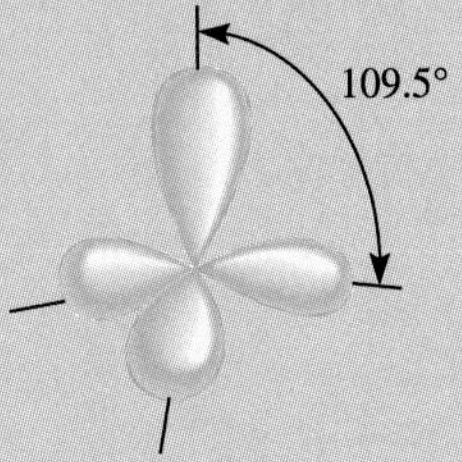 Tetrahedral	CH_4, NH_4^+
s, *p*, *p*, *p*, *d*	sp^3d	5	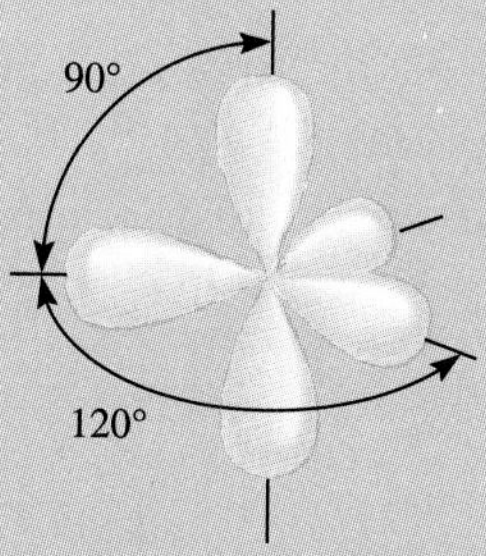 Trigonal bipyramidal	PCl_5
s, *p*, *p*, *p*, *d*, *d*	sp^3d^2	6	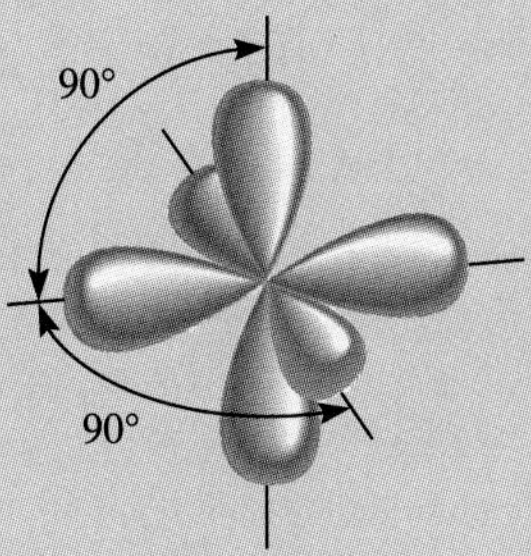 Octahedral	SF_6

tron pairs is linear (see Table 10.4). From Table 10.4 we conclude that Hg is *sp*-hybridized because it has the geometry of the two *sp* hybrid orbitals. The hybridization process can be imagined to take place as follows. First we draw the orbital diagram for the ground state of Hg:

[↑↓] [ ][ ][ ]
6*s* 6*p*

By promoting a 6*s* electron to the 6*p* orbital, we get the excited state:

[↑] [↑][ ][ ]
6*s* 6*p*

The 6*s* and 6*p* orbitals then mix to form two *sp* hybrid orbitals:

[↑][↑] [ ][ ]
sp orbitals — empty 6*p* orbitals

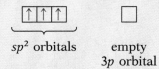

The two Hg—Cl bonds are formed by the overlap of the Hg *sp* hybrid orbitals with the 3*p* orbitals of the Cl atoms. Thus $HgCl_2$ is a linear molecule.

(b) The ground-state electron configuration of Al is $[Ne]3s^23p^1$. Therefore, Al has three valence electrons. The Lewis structure of AlI_3 is

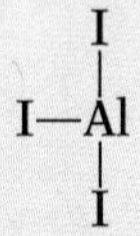

There are three bonding pairs and no lone pair on the Al atom. From Table 10.1 we see that the shape of three electron pairs is trigonal planar, and from Table 10.4 we conclude that Al must be sp^2-hybridized in AlI_3. The orbital diagram of the ground-state Al atom is

[↑↓] [↑][ ][ ]
3*s* 3*p*

By promoting a 3*s* electron into the 3*p* orbital we obtain the following excited state:

[↑] [↑][↑][ ]
3*s* 3*p*

The 3*s* and two 3*p* orbitals then mix to form three sp^2 hybrid orbitals:

[↑][↑][↑] [ ]
sp^2 orbitals — empty 3*p* orbital

The sp^2 hybrid orbitals overlap with the 5*p* orbitals of I to form three covalent Al—I bonds. We predict that the AlI_3 molecule is planar and all the IAlI angles are 120°.

(c) The ground-state electron configuration of P is $[Ne]3s^23p^3$. Therefore, the P atom has five valence electrons. The Lewis structure of PF_3 is

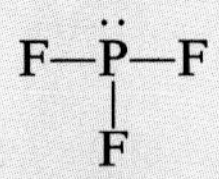

There are three bonding pairs and one lone pair on the P atom. In Table 10.1 we see that the overall arrangement of four electron pairs is tetrahedral, and from Table 10.4 we conclude that P must be sp^3-hybridized. The orbital diagram of the ground-state P atom is

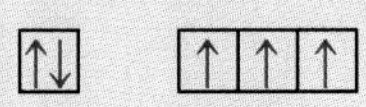

3*s* 3*p*

By mixing the 3*s* and 3*p* orbitals, we obtain four sp^3 hybrid orbitals.

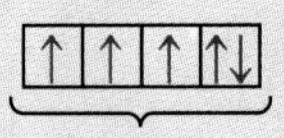

sp^3 orbitals

As in the case of NH_3, one of the sp^3 hybrid orbitals is used to accommodate the lone pair on P. The other three sp^3 hybrid orbitals form covalent P—F bonds with the 2*p* orbitals of F. We predict the geometry of the molecule to be pyramidal; the FPF angle should be somewhat less than 109.5°.

PRACTICE EXERCISE

Determine the hybridization state of the underlined atoms in the following compounds: (a) <u>Si</u>Br_4, (b) <u>B</u>Cl_3.

HYBRIDIZATION OF *s*, *p*, AND *d* ORBITALS

We have seen that hybridization neatly explains bonding that involves *s* and *p* orbitals. For elements in the third period and beyond, however, we cannot always account for molecular geometry by assuming the hybridization of only *s* and *p* orbitals. To understand the formation of molecules with trigonal bipyramidal and octahedral geometries, for instance, we must include *d* orbitals in the hybridization concept.

Consider the SF_6 molecule as an example. In Section 10.1 we saw that this molecule has an octahedral geometry, which is also the arrangement of the six electron pairs. From Table 10.4 we see that the S atom must be sp^3d^2-hybridized in SF_6. The ground-state electron configuration of S is [Ne]$3s^23p^4$:

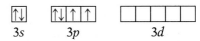

3*s* 3*p* 3*d*

Since the 3*d* level is quite close in energy to the 3*s* and 3*p* levels, we can promote both 3*s* and 3*p* electrons to two of the 3*d* orbitals:

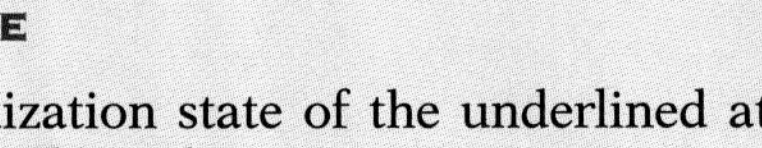

3*s* 3*p* 3*d*

Mixing the 3*s*, three 3*p*, and two 3*d* orbitals generates six sp^3d^2 hybrid orbitals:

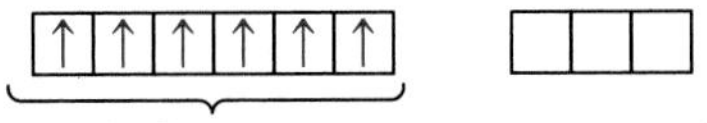

sp^3d^2 orbitals   empty 3*d* orbitals

The six S—F bonds are formed by the overlap of the hybrid orbitals of the S atom and the 2*p* orbitals of the F atoms. Since there are 12 electrons around the S atom, the octet rule is exceeded. *The use of d orbitals in addition to s and p orbitals to form covalent bonds* is an example of ***valence-shell expansion,*** which corresponds to the expanded octet discussed in Section 9.7. The second-period elements, unlike third-period elements, do not have 2*d* levels, so they can never expand their valence shells. Hence atoms of second-period elements can never be surrounded by more than eight electrons in any of their compounds.

EXAMPLE 10.4

Determining the Hybridization State of a Third-Period Element

Describe the hybridization state of phosphorus in phosphorus pentabromide (PBr_5).

Answer: The Lewis structure of PBr_5 is

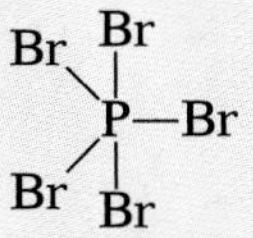

Table 10.1 indicates that the arrangement of five electron pairs is trigonal bipyramidal. Referring to Table 10.4, we find that this is the shape of five sp^3d hybrid orbitals. Thus P must be sp^3d-hybridized in PBr_5. The ground-state electron configuration of P is [Ne]$3s^23p^3$. To describe the hybridization process, we start with the orbital diagram for the ground state of P:

3*s*   3*p*   3*d*

Promoting a 3*s* electron into a 3*d* orbital results in the following excited state:

3*s*   3*p*   3*d*

Mixing the one 3*s*, three 3*p*, and one 3*d* orbitals generates five sp^3d hybrid orbitals:

sp^3d orbitals   empty 3*d* orbitals

These hybrid orbitals overlap with the 4*p* orbitals of Br to form five covalent P—Br bonds. Since there are no lone pairs on the P atom, the geometry of PBr_5 is trigonal bipyramidal.

PRACTICE EXERCISE

Describe the hybridization state of Se in SeF_6.

10.4 HYBRIDIZATION IN MOLECULES CONTAINING DOUBLE AND TRIPLE BONDS

The concept of hybridization is useful also for molecules with double and triple bonds. Consider the ethylene molecule, C_2H_4, as an example. In Example 10.1 we saw that C_2H_4 contains a carbon-carbon double bond and has planar geometry. Both the geometry and the bonding can be understood if we assume that each carbon atom is sp^2-hybridized. Figure 10.12 shows orbital diagrams of this hybridization process. We assume that only the $2p_x$ and $2p_y$ orbitals combine with the 2*s* orbital, and that the $2p_z$ orbital remains unchanged. Figure 10.13 shows that the $2p_z$ orbital is perpendicular to the plane of the hybrid orbitals. Now how do we account for the bonding of the C atoms? As Figure 10.14(a) shows, each carbon atom uses the three sp^2 hybrid orbitals to form two bonds with the two hydrogen 1*s* orbitals and one bond with the sp^2 hybrid orbital of the adjacent C atom. In addition, the two unhybridized $2p_z$ orbitals of the C atoms form another bond by overlapping sideways [Figure 10.14(b)].

A distinction is made between the two types of covalent bonds found in C_2H_4. *In a covalent bond formed by orbitals overlapping end-to-end,* as shown in Figure

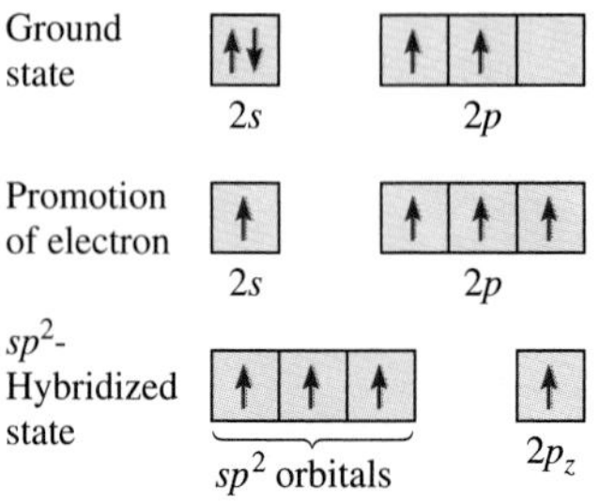

FIGURE 10.12

The sp^2 hybridization of a carbon atom. The 2s orbital is mixed with only two 2p orbitals to form three equivalent sp^2 hybrid orbitals. This process leaves an electron in the unhybridized orbital, the $2p_z$ orbital.

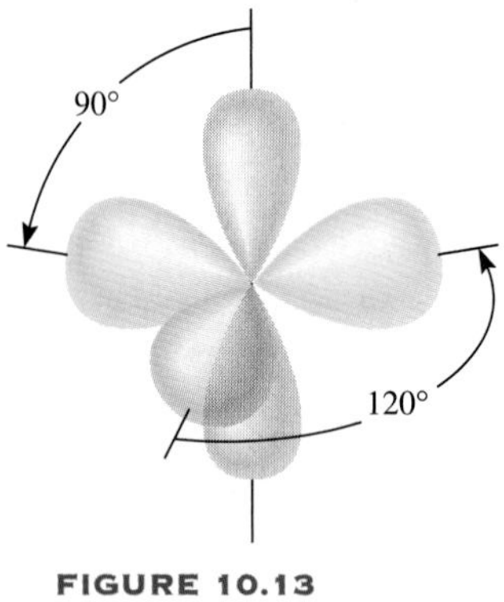

FIGURE 10.13

Each carbon atom in the C_2H_4 molecule has three sp^2 hybrid orbitals and one unhybridized $2p_z$ orbital that is perpendicular to the plane of the hybrid orbitals.

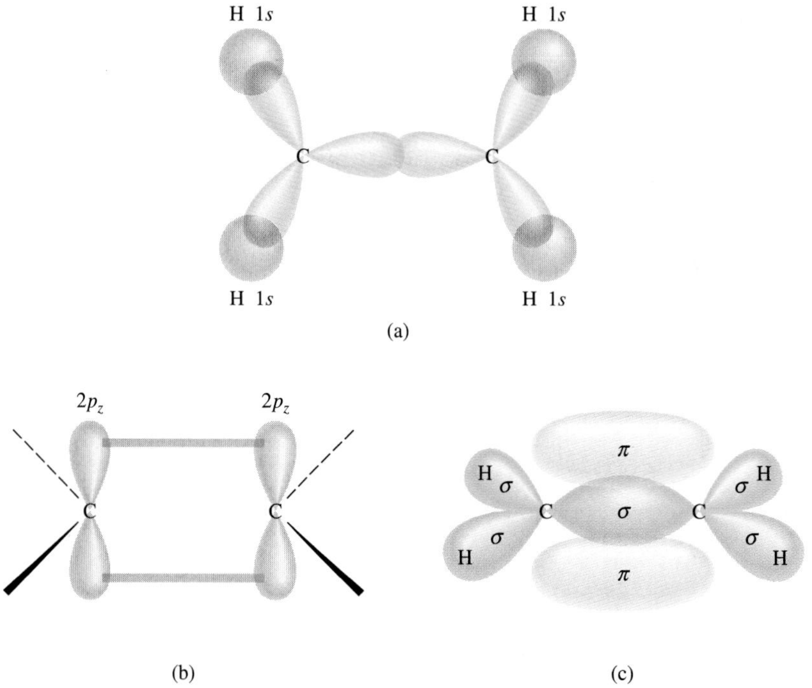

FIGURE 10.14

Bonding in ethylene, C_2H_4. (a) Top view of the sigma bonds between carbon atoms and between carbon and hydrogen atoms. All the atoms lie in the same plane, making C_2H_4 a planar molecule. (b) Side view showing how overlap of the two $2p_z$ orbitals on the two carbon atoms takes place, leading to the formation of a pi bond. (c) The sigma bonds and the pi bond in ethylene. Note that the pi bond lies above and below the plane of the molecule.

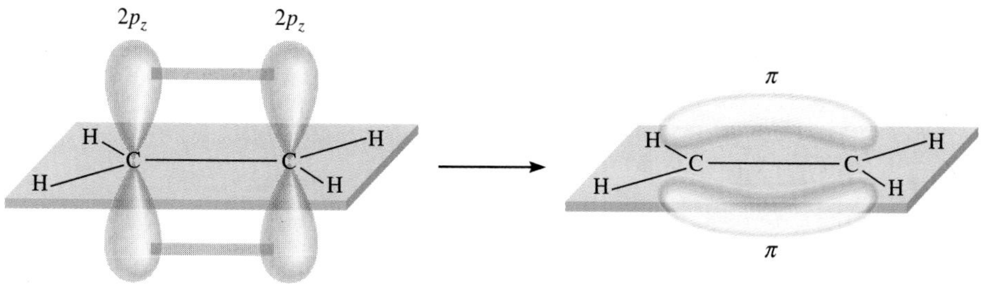

FIGURE 10.15

Another view of pi bond formation in the C_2H_4 molecule. Note that all six atoms lie in the same plane. It is the overlap of the $2p_z$ orbitals that causes the molecule to assume a planar structure.

10.14(a), *the electron density is concentrated between the nuclei of the bonding atoms.* A bond of this kind is called a ***sigma bond (σ bond).*** The three bonds formed by each C atom in Figure 10.14(a) are all sigma bonds. On the other hand, *a covalent bond formed by sideways overlapping orbitals has electron density concentrated above and below the plane of the nuclei of the bonding atoms* and is called a ***pi bond (π bond).*** The two C atoms form a pi bond as shown in Figure 10.14(b). Figure 10.14(c) shows the orientation of both sigma and pi bonds. Figure 10.15 is yet another way of looking at the planar C_2H_4 molecule and the formation of the pi bond. Although we normally represent the carbon-carbon double bond as C=C (as in a Lewis structure), it is important to keep in mind that the two bonds are different types: One is a sigma bond and the other is a pi bond.

The acetylene molecule (C_2H_2) contains a carbon-carbon triple bond. Since the molecule is linear, we can explain its geometry and bonding by assuming that each C atom is *sp*-hybridized (Figure 10.16). As Figure 10.17 shows, the two *sp* hybrid orbitals of each C atom form one sigma bond with a hydrogen 1*s* orbital and another sigma bond with the other C atom. In addition, two pi bonds are formed by the sideways overlap of the unhybridized $2p_y$ and $2p_z$ orbitals. Thus the C≡C bond is made up of one sigma bond and two pi bonds.

The following rule helps us predict hybridization in molecules containing multiple bonds: If the central atom forms a double bond, it is sp^2-hybridized; if it forms two double bonds or a triple bond, it is *sp*-hybridized. Note that this rule applies only to atoms of the second-period elements. Atoms of third-period elements and

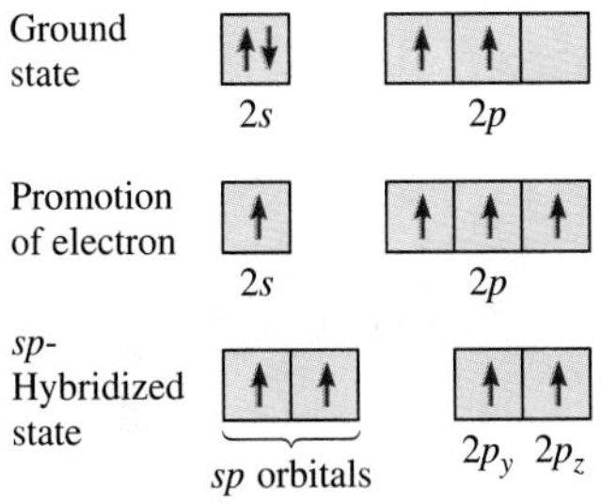

FIGURE 10.16

The sp hybridization of a carbon atom. The 2s orbital is mixed with only one 2p orbital to form two sp hybrid orbitals. This process leaves an electron in each of the two unhybridized 2p orbitals, namely, the $2p_y$ and $2p_z$ orbitals.

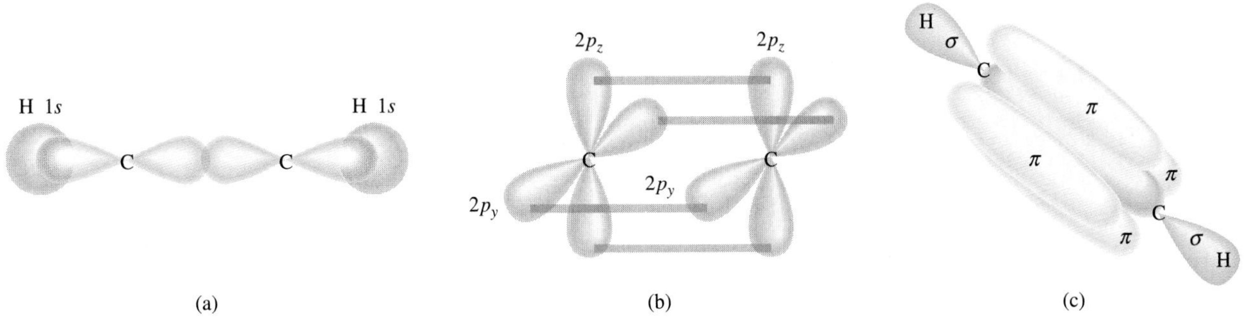

FIGURE 10.17

Bonding in acetylene, C_2H_2. (a) Top view showing the sigma bonds between carbon atoms and between carbon and hydrogen atoms. All the atoms lie along a straight line; therefore acetylene is a linear molecule. (b) Side view showing the overlap of the two $2p_y$ orbitals and of the two $2p_z$ orbitals of the two carbon atoms, which leads to the formation of two pi bonds. (c) Diagram showing both the sigma and the pi bonds.

beyond that form multiple bonds present a more involved problem and will not be dealt with here.

EXAMPLE 10.5

Bonding in a Polyatomic Molecule

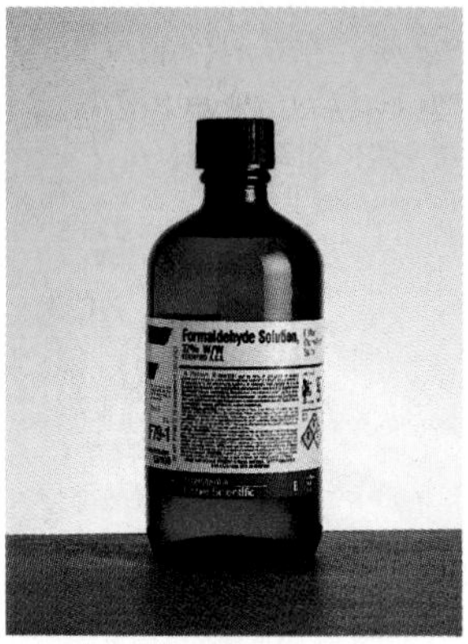

Formaldehyde is a colorless liquid with a disagreeable odor.

Describe the bonding in the formaldehyde molecule whose Lewis structure is

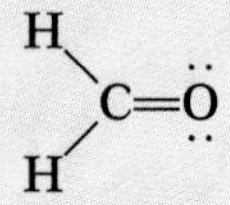

Assume that the O atom is sp^2-hybridized.

Answer: We note that the C atom (a second-period element) has a double bond; therefore, it is sp^2-hybridized. The orbital diagram of the O atom is

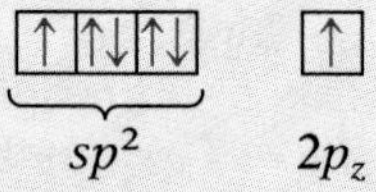

Two of the sp^2 orbitals of the C atom form two sigma bonds with the H atoms, and the third sp^2 orbital forms a sigma bond with an sp^2 orbital of the O atom. The $2p_z$ orbital of the C atom overlaps with the $2p_z$ orbital of the O atom to form a pi bond. The two lone pairs on the O atom are placed in its two remaining sp^2 orbitals (Figure 10.18).

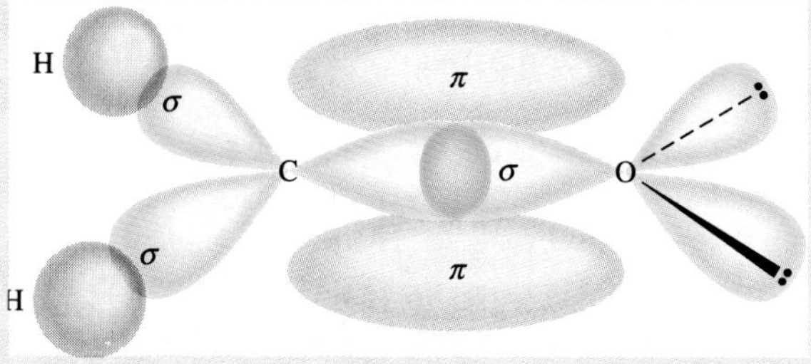

FIGURE 10.18

Bonding in the formaldehyde molecule. A sigma bond is formed by the overlap of the sp^2 hybrid orbital of carbon and the sp^2 hybrid orbital of oxygen; a pi bond is formed by the overlap of the $2p_z$ orbitals of the carbon and oxygen atoms. The two lone pairs on oxygen are placed in the other two sp^2 orbitals of oxygen.

PRACTICE EXERCISE

Describe the bonding in the hydrogen cyanide molecule, HCN. Assume that N is sp-hybridized.

SUMMARY

The VSEPR model for predicting molecular geometry is based on the assumption that valence-shell electron pairs repel one another and tend to stay as far apart as possible. According to the VSEPR model, molecular geometry can be predicted from the number of bonding electron pairs and lone pairs. Lone pairs repel other pairs more strongly than bonding pairs do and thus distort bond angles from those of the ideal geometry.

The dipole moment is a measure of the charge separation in molecules containing atoms of different electronegativities. The dipole moment of a molecule is the resultant of whatever bond moments are present in a molecule. Information about molecular geometry can be obtained from dipole moment measurements.

In valence bond theory, hybridized atomic orbitals are formed by the combination and rearrangement of orbitals of the same atom. The hybridized orbitals are all of equal energy and electron density, and the number of hybridized orbitals is equal to the number of pure atomic orbitals that combine. Valence-shell expansion can be explained by assuming hybridization of *s*, *p*, and *d* orbitals.

In *sp* hybridization, the two hybrid orbitals lie in a straight line; in sp^2 hybridization, the three hybrid orbitals are directed toward the corners of a triangle; in sp^3 hybridization, the four hybrid orbitals are directed toward the corners of a tetrahedron; in sp^3d hybridization, the five hybrid orbitals are directed toward the corners of a trigonal bipyramid; in sp^3d^2 hybridization, the six hybrid orbitals are directed toward the corners of an octahedron.

In an sp^2-hybridized atom (for example, carbon), the one unhybridized *p* orbital can form a pi bond with another *p* orbital. A carbon-carbon double bond consists of a sigma bond and a pi bond. In an *sp*-hybridized carbon atom, the two unhybridized *p* orbitals can form two pi bonds with two *p* orbitals on another atom (or atoms). A carbon-carbon triple bond consists of one sigma bond and two pi bonds.

KEY WORDS

Dipole moment (μ), p. 285
Hybrid orbital, p. 290
Hybridization, p. 290
Nonpolar molecule, p. 286
Pi bond (π bond), p. 299
Polar molecule, p. 285
Sigma bond (σ bond), p. 299
Valence shell, p. 276
Valence-shell electron-pair repulsion (VSEPR) model, p. 276
Valence-shell expansion, p. 297

QUESTIONS AND PROBLEMS

SIMPLE GEOMETRIC SHAPES

Review Questions

10.1 What is molecular geometry? Why is the study of molecular geometry important?

10.2 Sketch the shape of a linear triatomic molecule, a trigonal planar molecule containing four atoms, a tetrahedral molecule, a trigonal bipyramidal molecule, and an octahedral molecule. Give the bond angles in each case.

10.3 How many atoms are directly bonded to the central atom in a tetrahedral molecule, a trigonal bipyramidal molecule, and an octahedral molecule?

VSEPR

Review Questions

10.4 Discuss the basic features of the VSEPR model. Explain why the repulsion decreases in the following order: lone pair–lone pair > lone pair–bonding pair > bonding pair–bonding pair.

10.5 In the trigonal bipyramid arrangement, why does a lone pair occupy an equatorial position rather than an axial position?

10.6 Another possible geometry for CH_4 is square planar, with the four H atoms at the corners of a square and the C atom at the center of the square. Sketch this geometry and compare its stability with that of a tetrahedral CH_4.

Problems

10.7 Predict the geometries of the following species using the VSEPR method: (a) PCl_3, (b) $CHCl_3$, (c) SiH_4, (d) $TeCl_4$.

10.8 What are the geometries of the following species? (a) $AlCl_3$, (b) $ZnCl_2$, (c) $ZnCl_4^{2-}$.

10.9 Predict the geometry of the following molecules using the VSEPR method: (a) $HgBr_2$, (b) N_2O (arrangement of atoms is NNO), (c) SCN^- (arrangement of atoms is SCN).

10.10 What are the geometries of the following ions? (a) NH_4^+, (b) NH_2^-, (c) CO_3^{2-}, (d) ICl_2^-, (e) ICl_4^-, (f) AlH_4^-, (g) $SnCl_5^-$, (h) H_3O^+, (i) BeF_4^{2-}.

10.11 Describe the geometry around each of the three central atoms in the CH_3COOH molecule.

10.12 Which of the following species are tetrahedral? $SiCl_4$, SeF_4, XeF_4, CI_4, $CdCl_4^{2-}$.

DIPOLE MOMENTS AND MOLECULAR GEOMETRY

Review Questions

10.13 Define dipole moment. What are the units and symbol for dipole moment?

10.14 What is the relationship between the dipole moment and bond moment? How is it possible for a molecule to have bond moments and yet be nonpolar?

10.15 Explain why an atom cannot have a permanent dipole moment.

10.16 The bonds in beryllium hydride (BeH_2) molecules are polar, and yet the dipole moment of the molecule is zero. Explain.

Problems

10.17 Arrange the following molecules in order of increasing dipole moment: H_2O, H_2S, H_2Te, H_2Se. (See Table 10.3.)

10.18 The dipole moments of the hydrogen halides decrease from HF to HI (see Table 10.3). Explain this trend.

10.19 List the following molecules in order of increasing dipole moment: H_2O, CBr_4, H_2S, HF, NH_3, CO_2.

10.20 Does the molecule OCS have a higher or lower dipole moment than CS_2?

10.21 Which of the following molecules has a higher dipole moment?

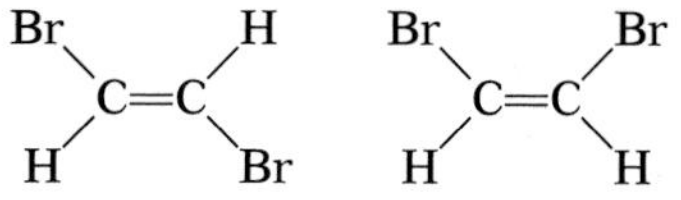

(a) (b)

10.22 Arrange the following compounds in order of increasing dipole moments:

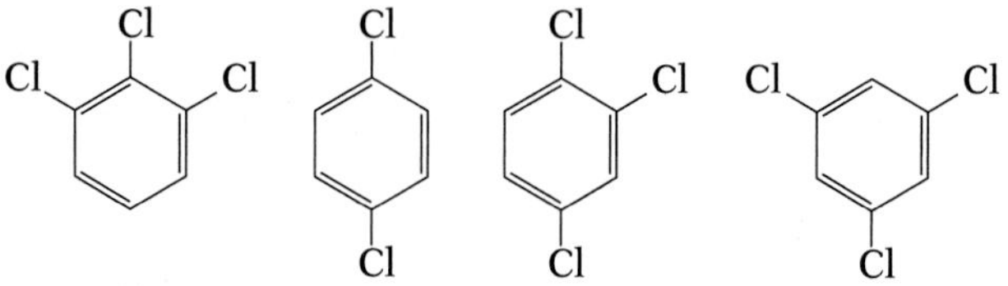

(a) (b) (c) (d)

HYBRIDIZATION

Review Questions

10.23 What is valence bond theory? How does it differ from the Lewis concept of chemical bonding?

10.24 Use valence bond theory to explain the bonding in Cl_2 and HCl. Show the overlap of atomic orbitals in the bonds.

10.25 Draw a potential energy curve for the bond formation in HCl.

10.26 What is the hybridization of atomic orbitals? Why is it impossible for an isolated atom to exist in the hybridized state?

10.27 How does a hybrid orbital differ from a pure atomic orbital? Can two 2*p* orbitals of an atom hybridize to give two hybridized orbitals?

10.28 What is the angle between the following two hybrid orbitals on the same atom? (a) *sp* and *sp* hybrid orbitals, (b) sp^2 and sp^2 hybrid orbitals, (c) sp^3 and sp^3 hybrid orbitals.

10.29 How would you distinguish between a sigma bond and a pi bond?

10.30 Which of the following pairs of atomic orbitals of adjacent nuclei can overlap to form a sigma bond? Which can overlap to form a pi bond? Which cannot overlap (no bond)? Consider the x-axis to be the internuclear axis, that is, the line joining the nuclei of the two atoms. (a) $1s$ and $1s$, (b) $1s$ and $2p_x$, (c) $2p_x$ and $2p_y$, (d) $3p_y$ and $3p_y$, (e) $2p_x$ and $2p_x$, (f) $1s$ and $2s$.

Problems

10.31 Describe the bonding scheme of the AsH_3 molecule in terms of hybridization.

10.32 What is the hybridization of Si in SiH_4 and in $H_3Si{-}SiH_3$?

10.33 Describe the change in hybridization (if any) of the Al atom in the following reaction:

$$AlCl_3 + Cl^- \longrightarrow AlCl_4^-$$

10.34 Consider the reaction

$$BF_3 + NH_3 \longrightarrow F_3B{-}NH_3$$

Describe the changes in hybridization (if any) of the B and N atoms as a result of this reaction.

10.35 What hybrid orbitals are used by nitrogen atoms in the following species? (a) NH_3, (b) $H_2N{-}NH_2$, (c) NO_3^-.

10.36 What are the hybrid orbitals of the carbon atoms in the following molecules?
(a) $H_3C{-}CH_3$
(b) $H_3C{-}CH{=}CH_2$
(c) $CH_3{-}CH_2{-}OH$
(d) $CH_3CH{=}O$
(e) CH_3COOH

10.37 What hybrid orbitals are used by carbon atoms in the following species? (a) CO, (b) CO_2, (c) CN^-.

10.38 What is the state of hybridization of the central N atom in the azide ion, N_3^-? (Arrangement of atoms: NNN.)

10.39 The allene molecule $H_2C{=}C{=}CH_2$ is linear (that is, the three C atoms lie on a straight line). What are the hybridization states of the carbon atoms? Draw diagrams to show the formation of sigma bonds and pi bonds in allene.

10.40 Describe the hybridization of phosphorus in PF_5.

10.41 What is the total number of sigma bonds and pi bonds in each of the following molecules?

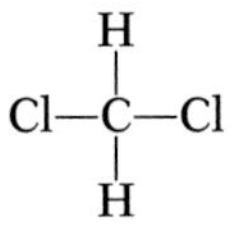
(a)

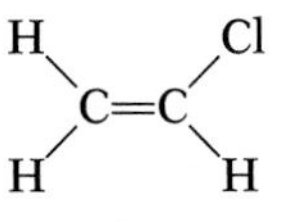
(b)

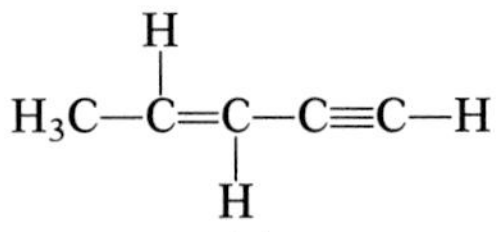
(c)

10.42 How many pi bonds and sigma bonds are there in the tetracyanoethylene molecule?

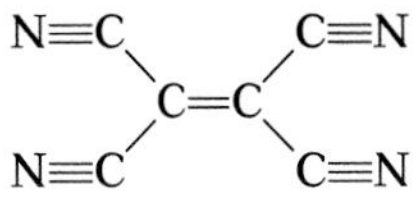

MISCELLANEOUS PROBLEMS

10.43 Which of the following species is not likely to have a tetrahedral shape? (a) $SiBr_4$, (b) NF_4^+, (c) SF_4, (d) $BeCl_4^{2-}$, (e) BF_4^-, (f) $AlCl_4^-$.

10.44 Draw the Lewis structure of mercury(II) bromide. Is this molecule linear or bent? How would you establish its geometry?

10.45 Sketch the bond moments and resultant dipole moments in the following molecules: H_2O, PCl_3, XeF_4, PCl_5, SF_6.

10.46 Although both carbon and silicon are in Group 4A, very few Si=Si bonds are known. Account for the instability of silicon-to-silicon double bonds in general. (*Hint:* Compare the atomic radii of C and Si in Figure 8.4. What effect would the size have on pi bond formation?)

10.47 Predict the geometry of sulfur dichloride (SCl_2) and the hybridization of the sulfur atom.

10.48 Antimony pentafluoride, SbF_5, reacts with XeF_4 and XeF_6 to form the ionic compounds $XeF_3^+SbF_6^-$ and $XeF_5^+SbF_6^-$. Describe the geometries of the cations and anion in these two compounds.

10.49 Write Lewis structures and give the other information requested for the following molecules: (a) BF_3. Shape: planar or nonplanar? (b) ClO_3^-. Shape: planar or nonplanar? (c) H_2O. Show the direction of the resultant dipole moment. (d) OF_2. Polar or nonpolar molecule? (e) NO_2. Estimate the ONO bond angle.

10.50 Predict the bond angles for the following molecules: (a) $BeCl_2$, (b) BCl_3, (c) CCl_4, (d) CH_3Cl, (e) Hg_2Cl_2 (arrangement of atoms: ClHgHgCl), (f) $SnCl_2$, (g) SCl_2, (h) H_2O_2, (i) SnH_4.

10.51 Briefly compare the VSEPR and hybridization approaches to the study of molecular geometry.

10.52 Describe the state of hybridization of arsenic in arsenic pentafluoride (AsF_5).

10.53 Draw Lewis structures and provide the other information requested for the following: (a) SO_3. Polar or nonpolar molecule? (b) PF_3. Polar or nonpolar molecule? (c) F_3SiH. Show the direction of the resultant dipole moment. (d) SiH_3^-. Planar or pyramidal shape? (e) Br_2CH_2. Polar or nonpolar molecule?

10.54 Which of the following molecules are linear? ICl_2^-, IF_2^+, OF_2, SnI_2, $CdBr_2$.

10.55 Draw the Lewis structure for the $BeCl_4^{2-}$ ion. Predict its geometry and describe the state of hybridization of the Be atom.

10.56 The molecule with the formula N_2F_2 exists in both of the following two forms:

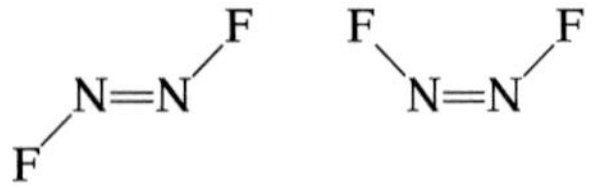

(a) What is the hybridization of N in the molecule? (b) Which of the two forms has a dipole moment?

10.57 Cyclopropane (C_3H_6) has a triangular shape with a C atom bonded to two H atoms and two other C atoms at each corner. Cubane (C_8H_8) is cube-shaped and has a C atom bonded to one H atom and three other C atoms at each corner. (a) Draw the structures of these molecules. (b) Compare the CCC angles in these molecules with those predicted for an sp^3-hybridized C atom. (c) Would you expect these molecules to be easy to prepare in the laboratory?

10.58 The compound 1,2-dichloroethane ($C_2H_4Cl_2$) is nonpolar, while *cis*-dichloroethylene ($C_2H_2Cl_2$) has a dipole moment:

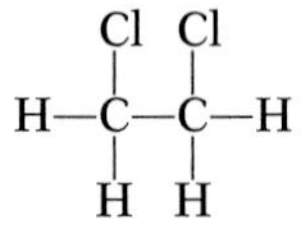

1,2-dichloroethane

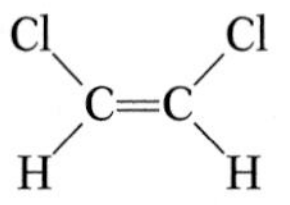

cis-dichloroethylene

The reason for the difference is that groups connected by a single bond can rotate with respect to each other, but no rotation occurs when a double bond connects the groups. On the basis of bonding consideration, explain why rotation occurs in 1,2-dichloroethane but not in *cis*-dichloroethylene.

10.59 Does the following molecule have a dipole moment?

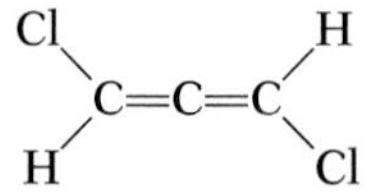

(*Hint:* See the answer to Problem 10.39.)

10.60 The bond angle of SO_2 is very close to 120° (see p. 280), even though there is a lone pair on S. Explain.

10.61 3″-Azido-3″-deoxythymidine, commonly known as AZT, is one of the drugs used to treat acquired immune deficiency syndrome (AIDS). What are the hybridization states of the C and N atoms in this molecule?

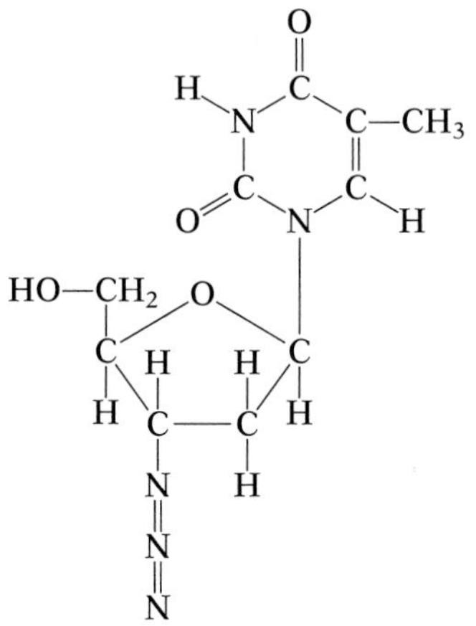

10.62 So-called greenhouse gases, which contribute to global warming, have a dipole moment or can be bent into some shapes that have a dipole moment. Which of the following gases are greenhouse gases? N_2, O_2, O_3, CO, CO_2, NO_2, $CFCl_3$.

10.63 The geometries discussed in this chapter all lend themselves to fairly straightforward elucidation of bond angles. The exception is the tetrahedron, whose bond angles are hard to visualize. Consider the CCl_4 molecule, which has a tetrahedral geometry and is nonpolar. By equating the bond moment along a particular C—Cl bond to the resultant bond moments of the other three C—Cl bonds in the opposite directions, show that the bond angles are all equal to 109.5°.

Answers to Practice Exercises: **10.1** (a) Tetrahedral, (b) linear, (c) trigonal planar; **10.2** Yes; **10.3** (a) sp^3, (b) sp^2; **10.4** sp^3d^2; **10.5** The C atom is *sp*-hybridized. It forms a sigma bond with the H atom and another sigma bond with the N atom. The two unhybridized *p* orbitals on the C atom are used to form two pi bonds with the N atom. The lone pair on the N atom is placed in the *sp* orbital.

CHAPTER 11

INTERMOLECULAR FORCES AND LIQUIDS AND SOLIDS

◆ Copper and aluminum are good conductors of electricity, but they do possess some electrical resistance. In fact, up to 20 percent of electrical energy may be lost (in the form of heat) when these metals are used as electrical transmission cables. Wouldn't it be marvelous if we could produce cables that possessed no electrical resistance?

Actually, it has been known for many years that certain metals and alloys totally lose their resistance when cooled to very low temperatures (around the boiling point of liquid helium, or 4 K). However, it is not practical to use these substances, called superconductors, for transmission of electric power because of the prohibitive cost of maintaining electrical cables at such low temperatures.

In 1986 two physicists in Switzerland discovered a new class of materials that are superconducting at around 30 K. Although 30 K is still a very low temperature, the improvement over the 4 K range was so dramatic that their work generated immense interest and triggered a flurry of research activity. Within months, scientists synthesized compounds containing copper, barium, and a rare earth metal that are superconducting at about 90 K, which is well above the boiling point of liquid nitrogen (77 K). One of the distinct properties of a superconductor is the capacity to levitate magnetized objects.

The levitation of a magnet above a high-temperature superconductor immersed in liquid nitrogen.

There is good reason to be enthusiastic about the new class of superconductors. Because liquid nitrogen is quite cheap (a liter of liquid nitrogen costs *less* than a liter of milk), we can envision electrical power being transmitted over hundreds of miles with no loss of energy. The levitation effect can be used in the operation of trains that ride above (but not in contact with) the tracks. Such trains are fast as well as quiet and smooth-running. Superconductors can be used to build superfast computers, called supercomputers, whose speed is limited by how fast electric current flows. The enormous magnetic fields that can be created in superconductors will result in more powerful particle accelerators, efficient devices for nuclear fusion, and more accurate magnetic resonance imaging techniques for medical use.

An experimental levitation train that operates on superconducting material at liquid helium temperature.

Despite their promise, we are still a long way from seeing the commercial use of high-temperature superconductors. A number of technical problems remain, and so far scientists have not been able to make durable wires out of these materials. Nevertheless, the payoff is so enormous that high-temperature superconductivity is one of the most actively researched areas in physics and chemistry today. ◆

11.1 THE KINETIC MOLECULAR THEORY OF LIQUIDS AND SOLIDS

In Chapter 5 we used the kinetic molecular theory to explain the behavior of gases. That explanation was based on the understanding that a gaseous system is a collection of molecules in constant, random motion. In gases, the distances between molecules are so great (compared with their diameters) that at ordinary temperatures and pressures (say, 25°C and 1 atm), there is no appreciable interaction between the molecules. This rather simple description explains several characteristic properties of gases. Because there is a great deal of empty space in a gas, that is, space not occupied by molecules, gases can be readily compressed. The lack of strong forces between molecules allows a gas to expand to the volume of its container. The large amount of empty space also explains why gases have very low densities under normal conditions.

Liquids and solids are quite a different story. The principal difference between the condensed state (liquids or solids) and the gaseous state lies in the distance between molecules. In a liquid the molecules are so close together that there is very little empty space. Thus liquids are much more difficult to compress than gases and much denser under normal conditions. Molecules in a liquid are held together by one or more types of attractive forces, which will be discussed in the next section. A liquid also has a definite volume, since molecules in a liquid do not break away from the attractive forces. The molecules can, however, move past one another freely, and so a liquid can flow, can be poured, and assumes the shape of its container.

In a solid, molecules are held rigidly in position with virtually no freedom of motion. Many solids are characterized by long-range order; that is, the molecules are arranged in regular configurations in three dimensions. There is even less empty space in a solid than in a liquid. Thus solids are almost incompressible and possess definite shape and volume. With very few exceptions (water being the most important), the density of the solid is higher than that of the liquid for a given substance. Table 11.1 summarizes some of the characteristic properties of the three states of matter.

11.2 INTERMOLECULAR FORCES

Attractive forces between molecules, called ***intermolecular forces,*** are responsible for the nonideal behavior of gases that was described in Chapter 5. They also account for the existence of the condensed states of matter—liquids and solids. As the temperature of a gas drops, the average kinetic energy of its molecules decreases. Eventually, at a sufficiently low temperature, the molecules no longer have enough energy to break away from one another's attraction. At this point, the molecules aggregate to form small drops of liquid. This phenomenon of going from the gaseous to the liquid state is known as *condensation.*

In contrast to intermolecular forces, ***intramolecular forces*** *hold atoms together in a molecule.* (Chemical bonding, discussed in Chapters 9 and 10, involves intramolecular forces.) Intramolecular forces stabilize individual molecules, whereas intermolecular forces are primarily responsible for the bulk properties of matter (for example, melting point and boiling point).

Generally, intermolecular forces are much weaker than intramolecular forces.

TABLE 11.1
Characteristic Properties of Gases, Liquids, and Solids

State of matter	Volume/Shape	Density	Compressibility	Motion of molecules
Gas	Assumes the volume and shape of its container	Low	Very compressible	Very free motion
Liquid	Has a definite volume but assumes the shape of its container	High	Only slightly compressible	Slide past one another freely
Solid	Has a definite volume and shape	High	Virtually incompressible	Vibrate about fixed positions

Thus it usually requires much less energy to evaporate a liquid than to break the bonds in the molecules of the liquid. For example, it takes about 41 kJ of energy to vaporize 1 mole of water at its boiling point; it takes about 930 kJ of energy to break the two O—H bonds in 1 mole of water molecules. The boiling points of substances often reflect the strength of the intermolecular forces operating among the molecules. At the boiling point, enough energy must be supplied to overcome the attraction among molecules so that they can enter the vapor phase. If it takes more energy to separate molecules of substance A than of substance B because A molecules are held together by stronger intermolecular forces than B molecules, then the boiling point of A is higher than that of B. The same principle applies also to the melting points of the substances. In general the melting points of substances increase with the strength of the intermolecular forces.

To understand the properties of condensed matter, we must understand the different types of intermolecular forces. *Dipole-dipole, dipole-induced dipole, and dispersion forces* make up what chemists commonly refer to as ***van der Waals forces.*** Ions and dipoles are attracted to one another by electrostatic forces called *ion-dipole forces;* they are not van der Waals forces. *Hydrogen bonding* is a particularly strong type of dipole-dipole interaction. Because only a few elements can participate in hydrogen bond formation, it is treated as a separate category. Depending on the physical state of a substance (that is, gas, liquid, or solid), the nature of chemical bonds, and the types of elements present, more than one type of interaction may contribute to the total attraction between molecules, as we will see below.

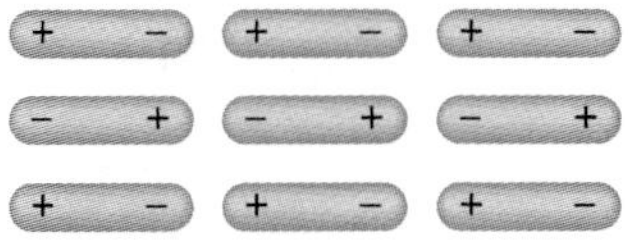

FIGURE 11.1
Molecules that have a permanent dipole moment tend to align in the solid state for maximum attractive interaction.

Dipole-Dipole Forces

Dipole-dipole forces are *forces that act between polar molecules,* that is, between molecules that possess dipole moments (see Section 10.2). Their origin is electrostatic, and they can be understood in terms of Coulomb's law. The larger the dipole moments, the greater the force. Figure 11.1 shows the orientation of polar molecules in a solid. In liquids the molecules are not held as rigidly as in a solid, but they tend to align in such a way that, on the average, the attractive interaction is at a maximum.

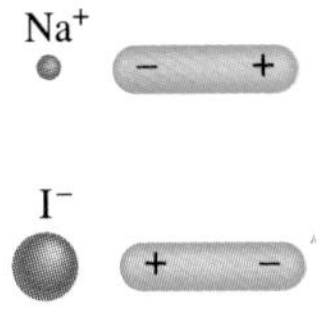

FIGURE 11.2
Two types of ion-dipole interaction.

Ion-Dipole Forces

Coulomb's law also explains ***ion-dipole forces,*** which *occur between an ion (either a cation or an anion) and a polar molecule* (Figure 11.2). The strength of this interaction depends on the charge and the size of the ion and on the magnitude of the dipole moment and the size of the molecule. The charges on cations are generally more concentrated, since cations are usually smaller than anions. Therefore, for equal charges, a cation interacts more strongly with dipoles than does an anion.

Hydration, discussed in Section 4.1, is one example of ion-dipole interaction. In an aqueous NaCl solution, the Na^+ and Cl^- ions are surrounded by water molecules, which have a large dipole moment (1.87 D). When an ionic compound like NaCl dissolves, the water molecules act as an electrical insulator that keeps the ions apart. On the other hand, carbon tetrachloride (CCl_4) is a nonpolar molecule, and therefore it lacks the ability to participate in ion-dipole interaction. What we find in practice is that carbon tetrachloride is a poor solvent for ionic compounds, as are most nonpolar liquids.

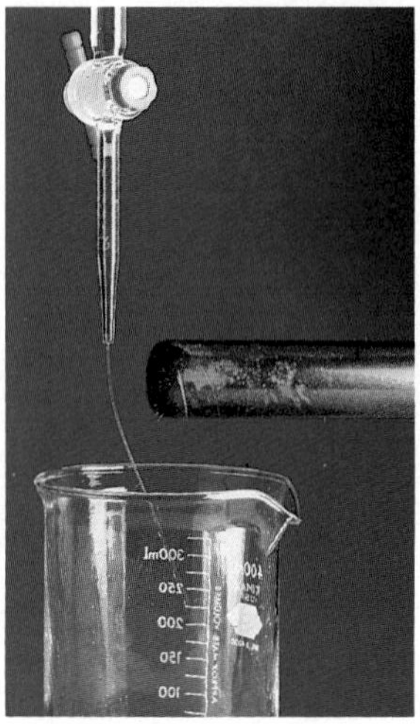

FIGURE 11.3

Deflection of a water stream by a charged ebonite rod.

A convincing demonstration of ion-dipole attraction is that shown in Figure 11.3. Water in a buret is allowed to run into a beaker. If a negatively charged rod (for example, an ebonite rod rubbed with fur) is brought near the stream of water, the water is deflected *toward* the rod. When the ebonite rod is replaced by a positively charged rod (for example, a glass rod rubbed with silk), the water is again attracted to the rod. With the ebonite rod in position, the water molecules orient themselves so that the positive ends of the dipoles lie closer to the rod and they are attracted to the rod's negative charge. When the ebonite rod is replaced with the glass rod, the negative ends of the dipoles are oriented toward the rod. Similar deflection is observed for other polar liquids. No deflection occurs when nonpolar liquids such as hexane (C_6H_{14}) are used instead of water.

Dispersion Forces

So far only ionic species and polar molecules have been discussed. What kind of attractive interaction exists between nonpolar molecules? To learn the answer to this question, consider the arrangement shown in Figure 11.4. If we place an ion or a polar molecule near an atom (or a nonpolar molecule), the electron distribution of the atom (or molecule) is distorted by the force exerted by the ion or the polar molecule. The resulting dipole in the atom (or molecule) is said to be an ***induced dipole*** because *the separation of positive and negative charges in the atom (or a nonpolar molecule) is due to the proximity of an ion or a polar molecule.* The attractive interaction between an ion and the induced dipole is called *ion-induced dipole interaction,* and the attractive interaction between a polar molecule and the induced dipole is called *dipole-induced dipole interaction.*

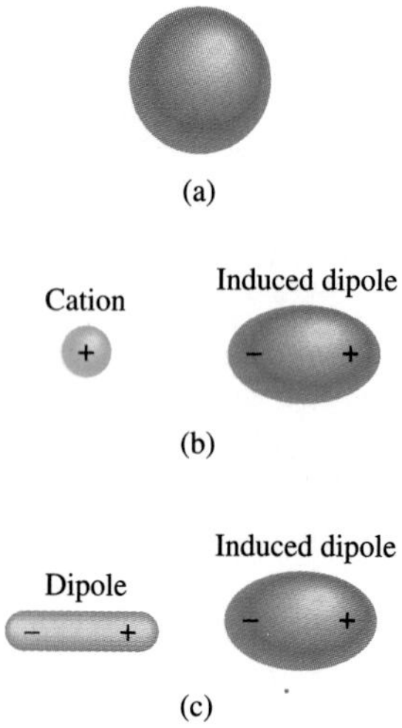

FIGURE 11.4

(a) Spherical charge distribution in a helium atom. Distortion caused by (b) the approach of a cation and (c) the approach of a dipole.

The likelihood of a dipole moment being induced depends not only on the charge on the ion or the strength of the dipole but also on the polarizability of the atom or molecule. ***Polarizability*** is *the ease with which the electron distribution in the atom (or molecule) can be distorted.* Generally, the larger the number of electrons and the more diffuse the electron cloud in the atom or molecule, the greater its polarizability. By *diffuse cloud* we mean an electron cloud that is spread over an appreciable volume, so that the electrons are not held tightly by the nucleus.

Polarizability enables gases containing atoms or nonpolar molecules (for example, He and N_2) to condense. In a helium atom the electrons are moving at some distance from the nucleus. At any instant it is likely that the atom has a dipole moment created by the specific positions of the electrons. This dipole moment is called an *instantaneous dipole* because it lasts for just a tiny fraction of a second. In the next instant the electrons are in different locations and the atom has a new instantaneous dipole and so on. Averaged over time (that is, the time it takes to make a dipole moment measurement), however, the atom has no dipole moment because the instantaneous dipoles all cancel one another. In a collection of He

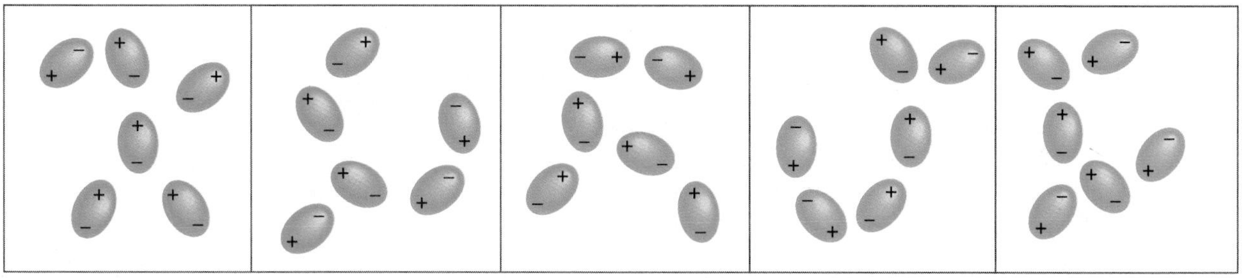

FIGURE 11.5

Induced dipoles interacting with each other. Such patterns exist only momentarily; new arrangements are formed in the next instant. This type of interaction is responsible for the condensation of nonpolar gases.

atoms an instantaneous dipole of one He atom can induce a dipole in each of its nearest neighbors (Figure 11.5). At the next moment, a different instantaneous dipole can create temporary dipoles in the surrounding He atoms. The important point is that this kind of interaction results in an attraction between the He atoms. At very low temperatures (and reduced atomic speeds), this attraction is enough to hold the atoms together, causing the helium gas to condense. The attraction between nonpolar molecules can be explained similarly.

A quantum mechanical interpretation of temporary dipoles was provided by Fritz London, a German physicist, in 1930. London showed that the magnitude of this attractive interaction is directly proportional to the polarizability of the atom or molecule. As we might expect, *the attractive forces that arise as a result of temporary dipoles induced in the atoms or molecules,* called ***dispersion forces,*** may be quite weak. This is certainly true of helium, which has a boiling point of only 4.2 K, or −269°C. (Note that helium has only two electrons, which are tightly held in the 1*s* orbital. Therefore, the helium atom has a small polarizability.)

Dispersion forces usually increase with molar mass because molecules with larger molar mass tend to have more electrons, and dispersion forces increase in strength with the number of electrons. Furthermore, larger molar mass often means a bigger atom whose electron distributions are more easily disturbed because the outer electrons are less tightly held by the nuclei. Table 11.2 compares the melting points of some substances that consist of nonpolar molecules. As expected, the melting point increases as the number of electrons in the molecule increases. Since these are all nonpolar molecules, the only attractive intermolecular forces present are the dispersion forces.

In many cases dispersion forces are comparable to or even greater than the dipole-dipole forces between polar molecules. For a dramatic illustration, let us compare the melting points of CH_3F (−141.8°C) and CCl_4 (−23°C). Although CH_3F has a dipole moment of 1.8 D, it melts at a much lower temperature than CCl_4, a nonpolar molecule. CCl_4 melts at a higher temperature simply because it contains more electrons. As a result, the dispersion forces between CCl_4 molecules are stronger than the dispersion forces plus the dipole-dipole forces between CH_3F molecules. (Keep in mind that dispersion forces exist among species of all types, whether they are neutral or bear a net charge and whether they are polar or nonpolar.)

The following example shows that if we know the kind of species present, we can readily determine the types of intermolecular forces that exist between the species.

TABLE 11.2
Melting Points of Similar Nonpolar Compounds

Compound	Melting point (°C)
CH_4	−182.5
CF_4	−150.0
CCl_4	−23.0
CBr_4	90.0
CI_4	171.0

EXAMPLE 11.1

Identifying Intermolecular Forces

What type(s) of intermolecular forces exist between the following pairs? (a) HBr and H_2S, (b) Cl_2 and CBr_4, (c) I_2 and NO_3^-, (d) NH_3 and C_6H_6.

Answer: (a) Both HBr and H_2S are polar molecules, so the forces between them are dipole-dipole forces. There are also dispersion forces between the molecules.

(b) Both Cl_2 and CBr_4 are nonpolar, so there are only dispersion forces between these molecules.

(c) I_2 is nonpolar, so the forces between it and the ion NO_3^- are ion-induced dipole forces and dispersion forces.

(d) NH_3 is polar, and C_6H_6 is nonpolar. The forces are dipole-induced dipole forces and dispersion forces.

PRACTICE EXERCISE

Name the type(s) of intermolecular forces that exist between molecules (or basic units) in each of the following species: (a) LiF, (b) CH_4, (c) SO_2.

THE HYDROGEN BOND

The ***hydrogen bond*** is *a special type of dipole-dipole interaction between the hydrogen atom in a polar bond, such as N—H, O—H, or F—H, and an electronegative O, N, or F atom.* This interaction is written

$$A—H\cdots B \quad \text{or} \quad A—H\cdots A$$

A and B represent O, N, or F; A—H is one molecule or part of a molecule and B is a part of another molecule; and the dotted line represents the hydrogen bond. The three atoms usually lie along a straight line, but the angle AHB (or AHA) can deviate as much as 30° from linearity.

The average energy of a hydrogen bond is quite large for a dipole-dipole interaction (up to 40 kJ/mol). Thus, hydrogen bonds are a powerful force in determining the structures and properties of many compounds. Figure 11.6 shows several examples of hydrogen bonding.

FIGURE 11.6

Hydrogen bonding in water, ammonia, and hydrogen fluoride. Solid lines represent covalent bonds, and dotted lines represent hydrogen bonds.

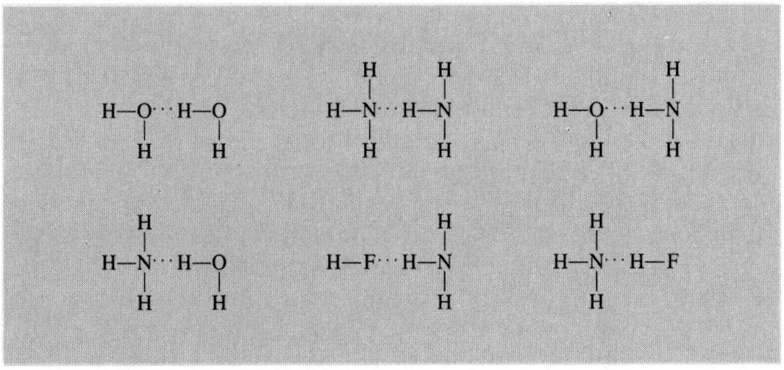

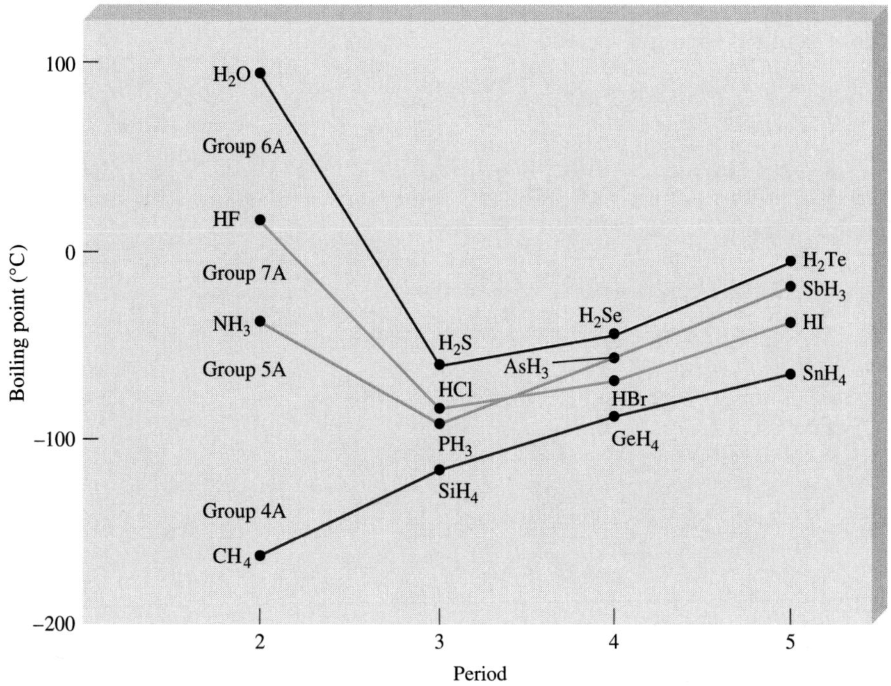

FIGURE 11.7

Boiling points of the hydrogen compounds of Groups 4A, 5A, 6A, and 7A elements. Although normally we expect the boiling point to increase as we move down a group, we see that three compounds (NH_3, H_2O, and HF) behave differently. The anomaly can be explained in terms of intermolecular hydrogen bonding.

Early evidence of hydrogen bonding came from the study of the boiling points of compounds. Normally, the boiling points of a series of similar compounds containing elements in the same group increase with increasing molar mass. But as Figure 11.7 shows, the hydrogen compounds of Groups 5A, 6A, and 7A elements do not follow this trend. In each of these series, the lightest compound (NH_3, H_2O, HF) has the *highest* boiling point, contrary to our expectations based on molar mass. The reason is that there is extensive hydrogen bonding between molecules in these compounds. In solid HF, for example, molecules do not exist as individual units; instead, they form long zigzag chains held together by hydrogen bonds:

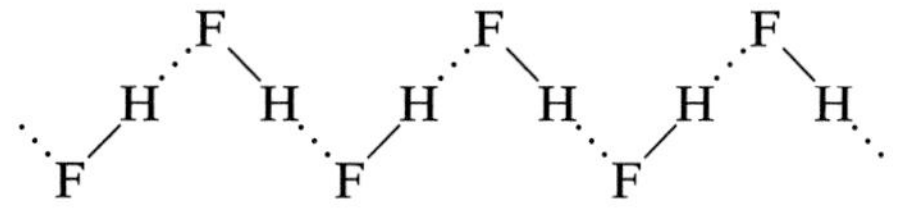

In the liquid state the zigzag chains are broken, but the molecules are still hydrogen-bonded to one another. Molecules held together by hydrogen bonding are more difficult to break apart, so that liquid HF has an unusually high boiling point.

The strength of the hydrogen bond is determined by the coulombic interaction between the lone-pair electrons of the electronegative atom and the hydrogen nucleus. It may seem odd that the boiling point of HF is lower than that of water. Fluorine is more electronegative than oxygen, and so we would expect a stronger hydrogen bond to exist in liquid HF than in H_2O. But H_2O is unique in that each molecule takes part in *four* intermolecular hydrogen bonds, and the H_2O molecules are therefore held together more strongly. We will return to this very important property of water in the next section. The following example shows the type of species that can form hydrogen bonds with water.

EXAMPLE 11.2

Identifying Hydrogen Bonds

Which of the following can form hydrogen bonds with water? CH_3OCH_3, CH_4, F^-, HCOOH, Na^+

Answer: There are no electronegative elements (F, O, or N) in either CH_4 or Na^+. Therefore, only CH_3OCH_3, F^-, and HCOOH can form hydrogen bonds with water.

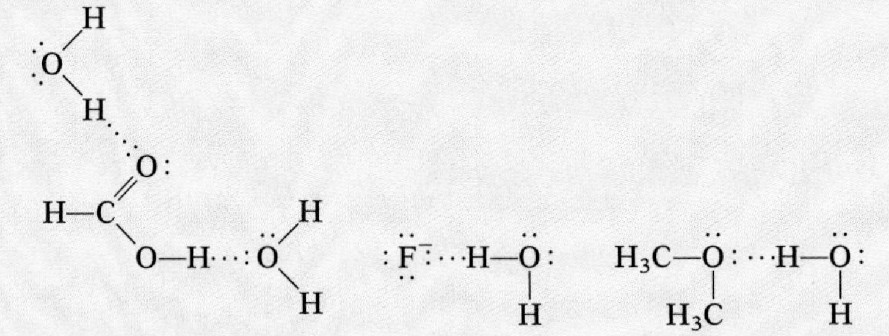

PRACTICE EXERCISE

Which of the following species are capable of hydrogen bonding among themselves? (a) H_2S, (b) C_6H_6, (c) CH_3OH.

The intermolecular forces discussed so far are all attractive in nature. Keep in mind, though, that molecules also exert repulsive forces on one another. Thus when two molecules are brought into contact with each other, the repulsion between the electrons and between the nuclei in the molecules comes into play. The magnitude of the repulsive force rises very steeply as the distance separating the molecules in a condensed state decreases. This is the reason that liquids and solids are so hard to compress. In these states, the molecules are already in close contact with one another, so that they greatly resist being further compressed.

11.3 THE LIQUID STATE

A great many interesting and important chemical reactions occur in water and other liquid solvents, as we will discover in subsequent chapters. In this section we will look at two phenomena associated with liquids: surface tension and viscosity, both of which are attributable to intermolecular forces. We will also discuss the structure and properties of water.

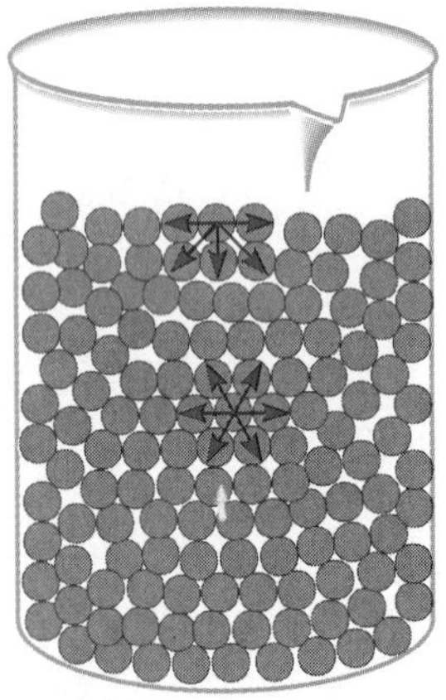

FIGURE 11.8

Intermolecular forces acting on a molecule in the surface layer of a liquid and in the interior region of the liquid.

SURFACE TENSION

We have seen that liquids tend to assume the shapes of their containers. Why, then, does water bead up on a newly waxed car instead of forming a sheet over it? The answer to this question lies in intermolecular forces.

Molecules within a liquid are pulled in all directions by intermolecular forces; there is no tendency for them to be pulled in any one way. However, molecules at the surface are pulled downward and sideways by other molecules, but not upward away from the surface (Figure 11.8). These intermolecular attractions thus tend to pull the molecules into the liquid and cause the surface to tighten like an elastic film. Since there is little or no attraction between polar water molecules and the wax molecules (which are essentially nonpolar) on a freshly waxed car, a drop of water assumes the shape of a small round bead (Figure 11.9).

A measure of the elastic force in the surface of a liquid is surface tension. The ***surface tension*** of a liquid is *the amount of energy required to stretch or increase the surface by unit area.* Liquids in which there are strong intermolecular forces also have high surface tensions. For example, because of hydrogen bonding, water has a considerably greater surface tension than most common liquids.

FIGURE 11.9

Water beads on an apple that has a waxy surface.

Another way in which surface tension manifests itself is in *capillary action.* Figure 11.10(a) shows water rising spontaneously in a capillary tube. A thin film of water adheres to the wall of the glass tube. The surface tension of water causes this film to contract, and as it does, it pulls the water up the tube. Two types of forces bring about capillary action. One is *the intermolecular attraction between like molecules* (in this case, the water molecules), called ***cohesion.*** The other, which is called ***adhesion,*** is *an attraction between unlike molecules,* such as those in water and in the sides of a glass tube. If adhesion is stronger than cohesion, as it is in Figure 11.10(a), the contents of the tube will be pulled upward. This process continues until the adhesive force is balanced by the weight of the water in the tube. This action is by no means universal among liquids, as Figure 11.10(b) shows. In mercury, cohesion is greater than the adhesion between mercury and glass, so that a depression in the liquid level actually occurs when a capillary tube is dipped into mercury.

VISCOSITY

The expression "slow as molasses in January" owes its truth to another physical property of liquids called viscosity. ***Viscosity*** is *a measure of a fluid's resistance to flow.* The greater the viscosity, the more slowly the liquid flows. The viscosity of a liquid usually decreases as temperature increases; thus hot molasses flows much faster than cold molasses.

Liquids that have strong intermolecular forces have higher viscosities than those that have weak intermolecular forces (Table 11.3). Water has a higher viscosity than many other liquids because of its ability to form hydrogen bonds. Interest-

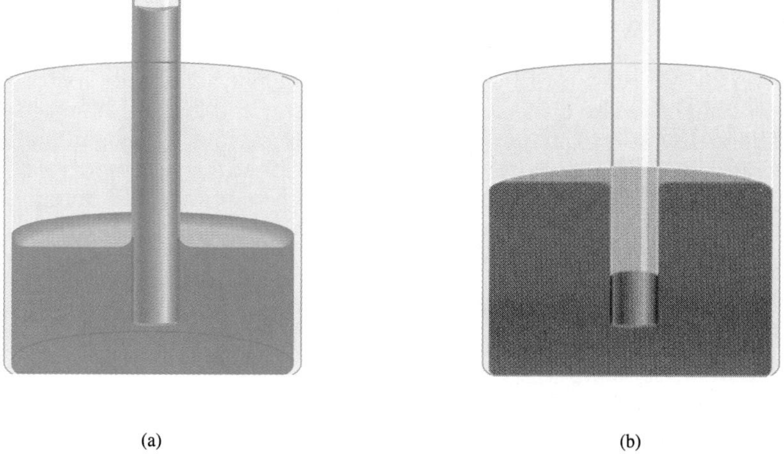

FIGURE 11.10

(a) When adhesion is greater than cohesion, the liquid (for example, water) rises in the capillary tube. (b) When cohesion is greater than adhesion, as it is for mercury, a depression of the liquid in the capillary tube results.

TABLE 11.3
Viscosity of Some Common Liquids at 20°C

Liquid	Viscosity (N s/m^2)*
Acetone (C_3H_6O)	3.16×10^{-4}
Benzene (C_6H_6)	6.25×10^{-4}
Carbon tetrachloride (CCl_4)	9.69×10^{-4}
Ethanol (C_2H_5OH)	1.20×10^{-3}
Ethyl ether ($C_2H_5OC_2H_5$)	2.33×10^{-4}
Glycerol ($C_3H_8O_3$)	1.49
Mercury (Hg)	1.55×10^{-3}
Water (H_2O)	1.01×10^{-3}
Blood	4×10^{-3}

*The SI units of viscosity are newton-second per meter squared.

ingly, the viscosity of glycerol is significantly higher than that of all the other liquids shown. Glycerol has the structure

$$\begin{array}{l} CH_2\text{—}OH \\ | \\ CH\text{—}OH \\ | \\ CH_2\text{—}OH \end{array}$$

Like water, glycerol can form hydrogen bonds. Each glycerol molecule has three —OH groups that can participate in hydrogen bonding with other glycerol molecules. Furthermore, because of their shape, the molecules have a great tendency to become entangled rather than to slip past one another as the molecules in less viscous liquids do. These interactions contribute to its high viscosity.

THE STRUCTURE AND PROPERTIES OF WATER

Water is so common a substance on Earth that we often overlook its unique nature. All life processes involve water. Water is an excellent solvent for many ionic compounds, as well as for other substances capable of forming hydrogen bonds with water.

As Table 6.1 shows, water has a high specific heat. The reason for this is that to raise the temperature of water (that is, to increase the average kinetic energy of water molecules), we must first break the many intermolecular hydrogen bonds. Thus water can absorb a substantial amount of heat while its temperature rises only slightly. The converse is also true: Water can give off much heat with only a slight decrease in its temperature. For this reason, the huge quantities of water that are present in our lakes and oceans can effectively moderate the climate of adjacent land areas by absorbing heat in the summer and giving off heat in the winter, with only small changes in the temperature of the body of water.

The most striking property of water is that its solid form is less dense than its liquid form: An ice cube floats at the surface of water in a glass. This is a virtually unique property. The density of almost all other substances is greater in the solid state than in the liquid state (Figure 11.11).

To understand why water is different, we have to examine the electronic structure of the H_2O molecule. As we saw in Chapter 9, there are two pairs of nonbonding electrons, or two lone pairs, on the oxygen atom:

FIGURE 11.11

(Left) Ice cubes float on water. (Right) Solid benzene sinks to the bottom of liquid benzene.

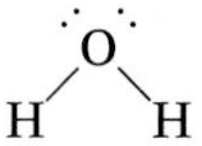

Although many compounds are capable of forming intermolecular hydrogen bonds, the difference between H_2O and other polar molecules, such as NH_3 and HF, is that each oxygen atom can form *two* hydrogen bonds, the same as the number of lone electron pairs on the oxygen atom. Thus water molecules are joined together in an extensive three-dimensional network in which each oxygen atom is approximately tetrahedrally bonded to four hydrogen atoms, two by covalent bonds and two by hydrogen bonds. This equality in the number of hydrogen atoms and lone pairs is not characteristic of NH_3 or HF or, for that matter, of any other molecule capable of forming hydrogen bonds. Consequently, these other molecules can form rings or chains, but not three-dimensional structures.

The highly ordered three-dimensional structure of ice (Figure 11.12) prevents the molecules from getting too close to one another. But consider what happens

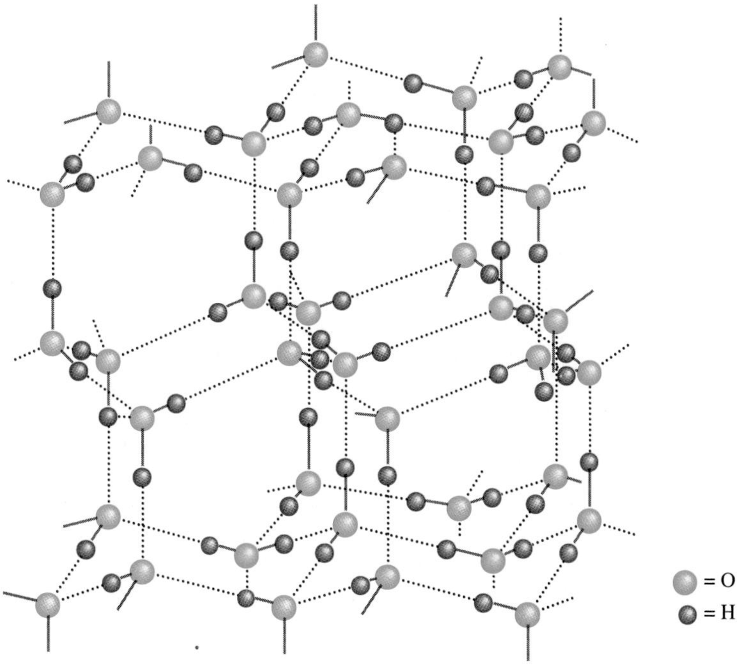

FIGURE 11.12

The three-dimensional structure of ice. Each O atom is bonded to four H atoms. The covalent bonds are shown by short solid lines and the weaker hydrogen bonds by long dotted lines between O and H. The empty space in the structure accounts for the low density of ice.

FIGURE 11.13

Plot of density versus temperature for liquid water. The maximum density of water is reached at 4°C. The density of ice at 0°C is about 0.92 g/cm³.

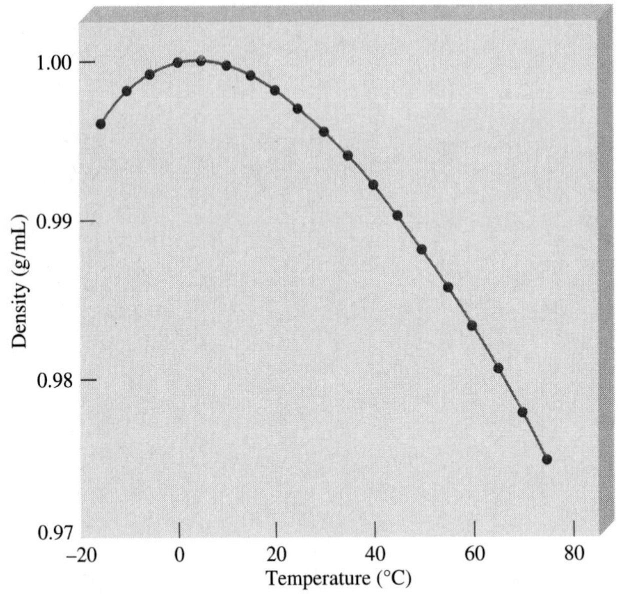

when ice melts. At the melting point, a number of water molecules have enough kinetic energy to pull free of the intermolecular hydrogen bonds. These molecules become trapped in the cavities of the three-dimensional structure, which is broken down into smaller clusters. As a result there are more molecules per unit volume in liquid water than in ice. Thus, since density = mass/volume, the density of water is greater than that of ice. With further heating, more water molecules are released from intermolecular hydrogen bonding, so that just above the melting point water's density tends to rise with temperature. Of course, at the same time, water expands as it is being heated, with the result that its density is decreased. These two processes—the trapping of free water molecules in cavities and thermal expansion—act in opposite directions. From 0°C to 4°C, the trapping prevails and water becomes progressively denser. Beyond 4°C, however, thermal expansion predominates and the density of water decreases with increasing temperature (Figure 11.13).

11.4 CRYSTAL STRUCTURE

Solids can be divided into two categories: crystalline and amorphous. Ice is a ***crystalline solid,*** which *possesses rigid and long-range order; its atoms, molecules, or ions occupy specific positions.* The arrangement of atoms, molecules, or ions in a crystalline solid is such that the net attractive intermolecular forces are at their maximum. The forces responsible for the stability of any crystal can be ionic forces, covalent bonds, van der Waals forces, hydrogen bonds, or a combination of these forces. *Amorphous solids* such as glass lack a well-defined arrangement and long-range molecular order. In this section we will concentrate on the structure of crystalline solids.

The basic repeating structural unit of a crystalline solid is a ***unit cell.*** Figure 11.14 shows a unit cell and its extension in three dimensions. Each sphere *represents an atom, an ion, or a molecule* and is called a ***lattice point.*** In many crystals, the lattice point does not actually contain an atom, ion, or molecule. Rather, there

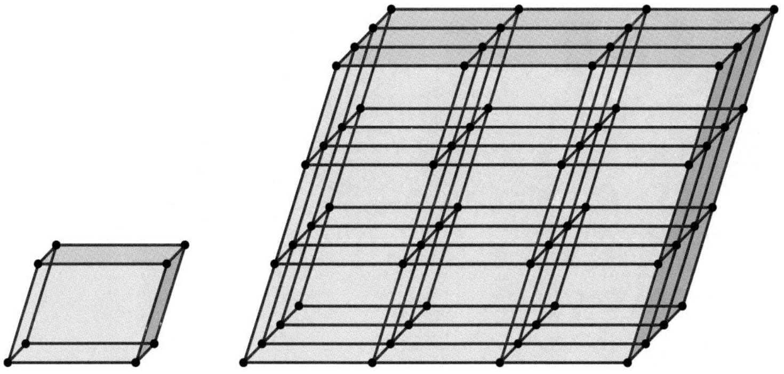

FIGURE 11.14

(a) A unit cell and (b) its extension in three dimensions. The black spheres represent either atoms or molecules.

may be several atoms, ions, or molecules identically arranged about each lattice point. For simplicity, however, we can assume that each lattice point is occupied by an atom. Every crystalline solid can be described in terms of one of the seven types of unit cells shown in Figure 11.15. The geometry of the cubic unit cell is particularly simple because all sides and all angles are equal. Any of the unit cells, when repeated in space in all three dimensions, forms the lattice structure characteristic of a crystallic solid.

PACKING SPHERES

We can understand the general geometric requirements for crystal formation by considering the different ways of packing a number of identical spheres (Ping-Pong balls, for example) to form an ordered three-dimensional structure. The way the spheres are arranged in layers determines what type of unit cell we have.

In the simplest case, a layer of spheres can be arranged as shown in Figure 11.16(a). The three-dimensional structure can be generated by placing a layer

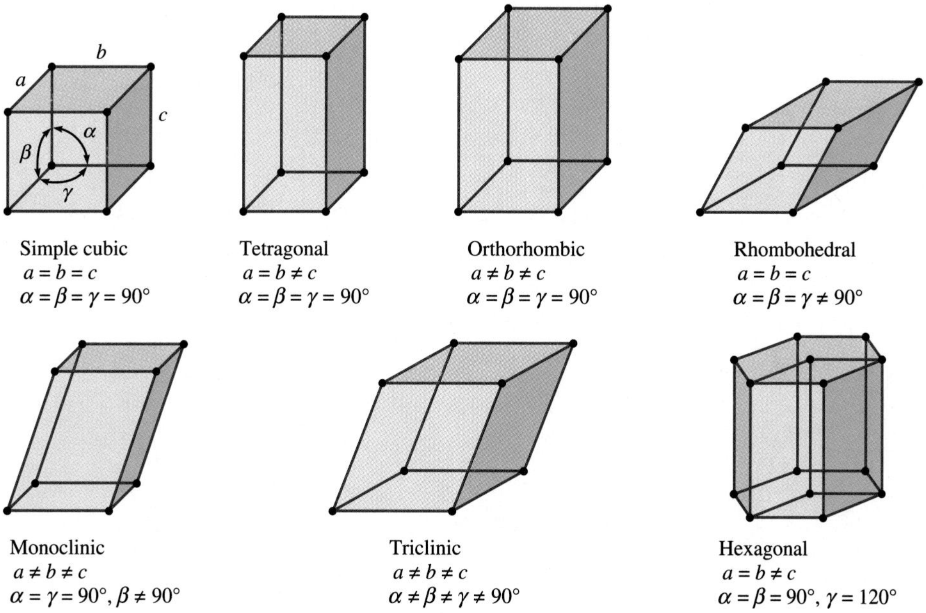

FIGURE 11.15

The seven types of unit cells. Angle α is defined by edges b and c, angle β by edges a and c, and angle γ by edges a and b.

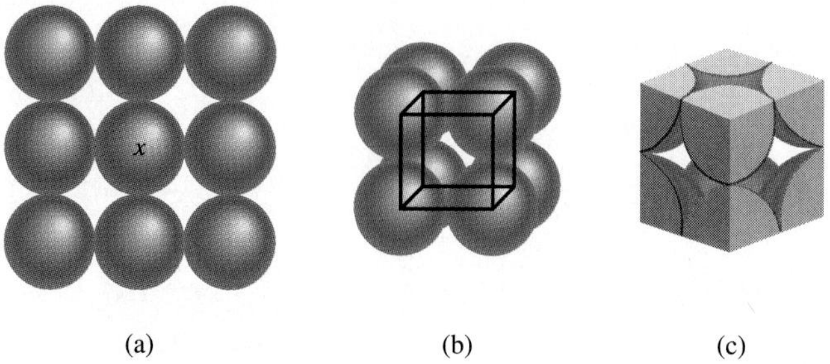

FIGURE 11.16

Arrangement of identical spheres in a simple cubic cell. (a) Top view of one layer of spheres. (b) Definition of a simple cubic cell. (c) Since each sphere is shared by eight unit cells and there are eight corners in a cube, there is the equivalent of one complete sphere inside a simple cubic unit cell.

above and below this layer in such a way that spheres in one layer are directly over the spheres in the layer below it. This procedure can be extended to generate many, many layers, as in the case of a crystal. Focusing on the sphere marked with "x," we see that it is in contact with four spheres in its own layer, one sphere in the layer above, and one sphere in the layer below. Each sphere in this arrangement is said to have a coordination number of 6 because it has six immediate neighbors. The ***coordination number*** is defined as *the number of atoms (or ions) surrounding an atom (or ion) in a crystal lattice.* The basic, repeating unit in this array of spheres is called a *simple cubic cell* (scc) [Figure 11.16(b)].

The other types of cubic cells are the *body-centered cubic cell* (bcc) and the *face-centered cubic cell* (fcc) (Figure 11.17). A body-centered cubic arrangement differs from a simple cube in that the second layer of spheres fits into the depressions of the first layer and the third layer into the depressions of the second layer (Figure 11.18). The coordination number of each sphere in this structure is 8 (each sphere is in contact with four spheres in the layer above and four spheres in the layer below). In the face-centered cubic cell there are spheres at the center of each of the six faces of the cube, in addition to the eight corner spheres.

Because every unit cell in a crystalline solid is adjacent to other unit cells, most of a cell's atoms are shared by neighboring cells. For example, in all types of cubic cells, each corner atom belongs to eight unit cells [Figure 11.19(a)]; a face-centered atom is shared by two unit cells [Figure 11.19(b)]. Since each corner sphere is shared by eight unit cells and there are eight corners in a cube, there will be the

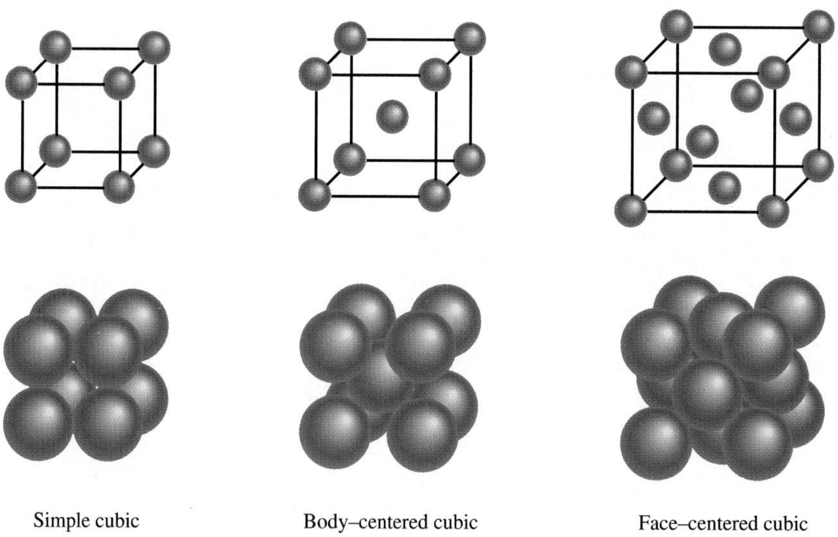

FIGURE 11.17

Three types of cubic cells. In reality, the spheres representing atoms, molecules, or ions are in contact with one another in these cubic cells.

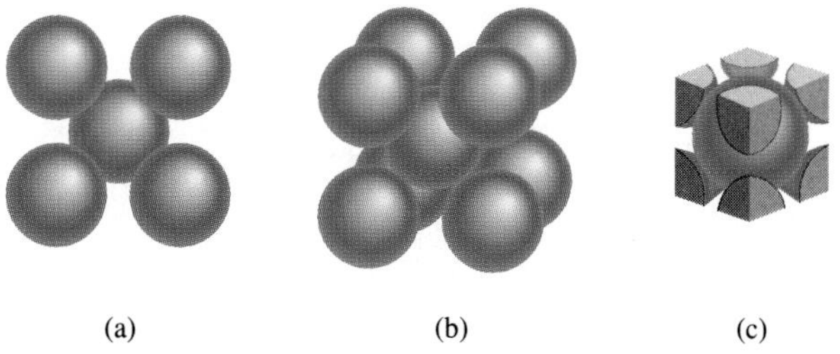

FIGURE 11.18

Arrangement of identical spheres in a body-centered cube. (a) Top view. (b) Definition of a body-centered cubic unit cell. (c) There is the equivalent of two complete spheres inside a body-centered cubic unit cell.

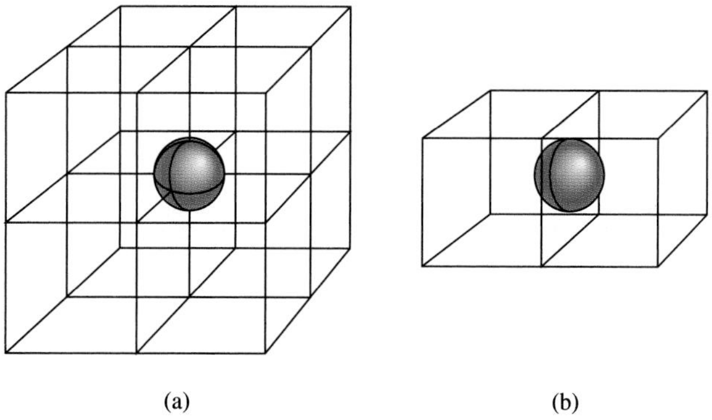

FIGURE 11.19

(a) A corner atom in any cell is shared by eight unit cells. (b) A face-centered atom in a cubic cell is shared by two unit cells.

equivalent of only one complete sphere inside a simple cubic unit cell [see Figure 11.16(c)]. A body-centered cubic cell contains the equivalent of two complete spheres, one in the center and eight shared corner spheres, as shown in Figure 11.18(c). A face-centered cubic cell contains four complete spheres—three from the six face-centered atoms and one from the eight shared corner spheres.

Figure 11.20 summarizes the relationship between the atomic radius r and the edge length a of a simple cubic cell, a body-centered cubic cell, and a face-centered cubic cell. This relationship can be used to determine the density of a crystal, as the following example shows.

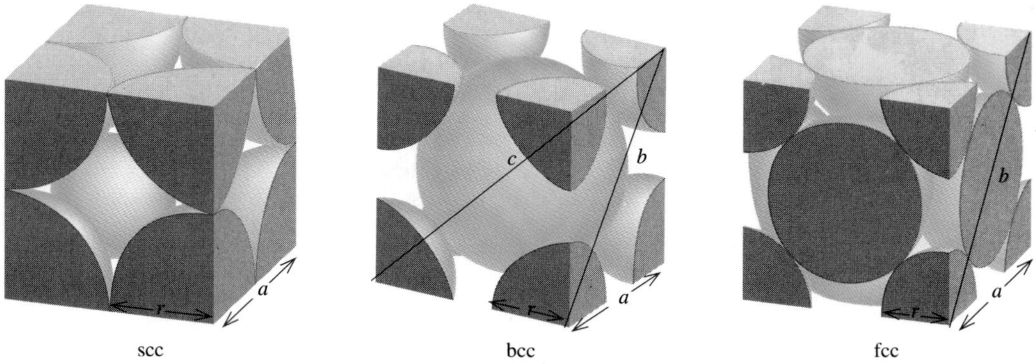

scc: $a = 2r$

bcc:
$$b^2 = a^2 + a^2$$
$$c^2 = a^2 + b^2 = 3a^2$$
$$c = \sqrt{3}a = 4r$$
$$a = \frac{4r}{\sqrt{3}}$$

fcc:
$$b = 4r$$
$$b^2 = a^2 + a^2$$
$$16r^2 = 2a^2$$
$$a = \sqrt{8}r$$

FIGURE 11.20

The relationship between the edge length and radius of atoms for a simple cubic cell, body-centered cubic cell, and face-centered cubic cell.

EXAMPLE 11.3

Calculating Density from Crystal Structure and Atomic Radius

Gold crystals consist of face-centered cubic unit cells. Calculate the density of gold. The atomic radius of gold is 144 pm.

Answer: From Figure 11.20 we see that the relationship between edge length *a* and atomic radius *r* of a face-centered cubic cell is $a = \sqrt{8}r$. Thus

$$a = \sqrt{8}(144 \text{ pm}) = 407 \text{ pm}$$

The volume of the unit cell is

$$V = a^3 = (407 \text{ pm})^3\left(\frac{1 \times 10^{-12} \text{ m}}{1 \text{ pm}}\right)^3\left(\frac{1 \text{ cm}}{1 \times 10^{-2} \text{ m}}\right)^3$$

$$= 6.74 \times 10^{-23} \text{ cm}^3$$

Each unit cell has eight corners and six faces. Therefore, the total number of atoms within such a cell is, according to Figure 11.19, $8 \times \frac{1}{8} + 6 \times \frac{1}{2} = 4$. The mass of a unit cell is

$$m = \frac{4 \text{ atoms}}{1 \text{ unit cell}} \times \frac{197.0 \text{ g}}{1 \text{ mol}} \times \frac{1 \text{ mol}}{6.022 \times 10^{23} \text{ atoms}}$$

$$= 1.31 \times 10^{-21} \text{ g/unit cell}$$

Finally the density of gold is given by

$$d = \frac{m}{V} = \frac{1.31 \times 10^{-21} \text{ g}}{6.74 \times 10^{-23} \text{ cm}^3} = 19.4 \text{ g/cm}^3$$

PRACTICE EXERCISE

When silver crystallizes, it forms face-centered cubic cells. The unit cell edge length is 408.7 pm. Calculate the density of silver.

11.5 BONDING IN SOLIDS

The structure and properties of crystalline solids, such as melting point, density, and hardness, are determined by the attractive forces that hold the particles together. We can classify crystals according to the types of forces between particles: ionic, molecular, covalent, and metallic (Table 11.4).

IONIC CRYSTALS

Ionic crystals consist of ions held together by ionic bonds. The structure of an ionic crystal depends on the charges on the cation and anion and on their radii. We have already discussed the structure of sodium chloride, which has a face-centered cubic lattice (see Figure 2.11). Figure 11.21 shows the structures of three other ionic crystals: CsCl, ZnS, and CaF_2. Because Cs^+ is considerably larger than Na^+, CsCl has the simple cubic lattice structure. ZnS has the *zincblende* structure, which is based on the face-centered cubic lattice. If the S^{2-} ions are located at the lattice points, the Zn^{2+} ions are located one-fourth of the distance along each body diago-

TABLE 11.4
Types of Crystals and General Properties

Type of crystal	Force(s) holding the units together	General properties	Examples
Ionic	Electrostatic attraction	Hard, brittle, high melting point, poor conductor of heat and electricity	NaCl, LiF, MgO, $CaCO_3$
Molecular*	Dispersion forces, dipole-dipole forces, hydrogen bonds	Soft, low melting point, poor conductor of heat and electricity	Ar, CO_2, I_2, H_2O, $C_{12}H_{22}O_{11}$ (sucrose)
Covalent	Covalent bond	Hard, high melting point, poor conductor of heat and electricity	C (diamond),† SiO_2 (quartz)
Metallic	Metallic bond	Soft to hard, low to high melting point, good conductor of heat and electricity	All metallic elements; for example, Na, Mg, Fe, Cu

*Included in this category are crystals made up of individual atoms.
†Diamond is a good thermal conductor.

nal. Other ionic compounds that have the zincblende structure include CuCl, BeS, CdS, and HgS. CaF_2 has the *fluorite* structure. The Ca^{2+} ions are located at the lattice points, and each F^- ion is tetrahedrally surrounded by four Ca^{2+} ions. The compounds SrF_2, BaF_2, $BaCl_2$, and PbF_2 also have the fluorite structure.

Ionic solids have high melting points, an indication of the strong cohesive force holding the ions together. These solids do not conduct electricity because the ions are fixed in position. However, in the molten state (that is, when melted) or dissolved in water, the ions are free to move and the resulting liquid is electrically conducting.

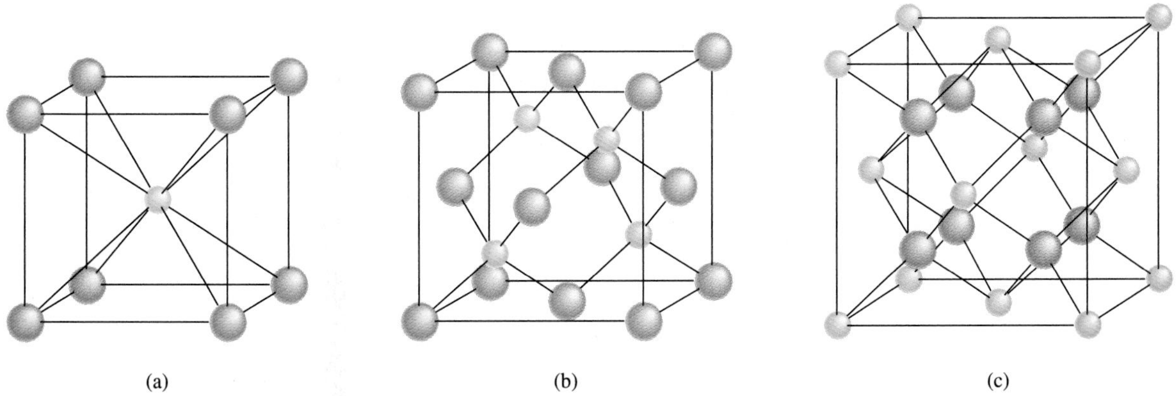

FIGURE 11.21

Crystal structures of (a) CsCl, (b) ZnS, and (c) CaF_2. In each case, the cation is the smaller sphere.

EXAMPLE 11.4

Counting the Number of Ions in a Unit Cell

How many Na^+ and Cl^- ions re in each NaCl unit cell?

Answer: NaCl has a structure based on a face-centered cubic lattice. As Figure 2.10 shows, one whole Na^+ ion is at the center of the unit cell and there are twelve Na^+ ions at the edges. Since each edge Na^+ ion is shared by four unit cells, the total number of Na^+ ions is $1 + (12 \times \frac{1}{4}) = 4$. Similarly, there are six Cl^- ions at the face centers and eight Cl^- ions at the corners. Each face-centered ion is shared by two unit cells, and each corner ion is shared by eight unit cells (see Figure 11.19), so the total number of Cl^- ions is $(6 \times \frac{1}{2}) + (8 \times \frac{1}{8}) = 4$. Thus there are four Na^+ ions and four Cl^- ions in each NaCl unit cell. Figure 11.22 shows the portions of the Na^+ and Cl^- ions *within* a unit cell.

PRACTICE EXERCISE

How many atoms are in a body-centered cube, assuming that all atoms occupy lattice points?

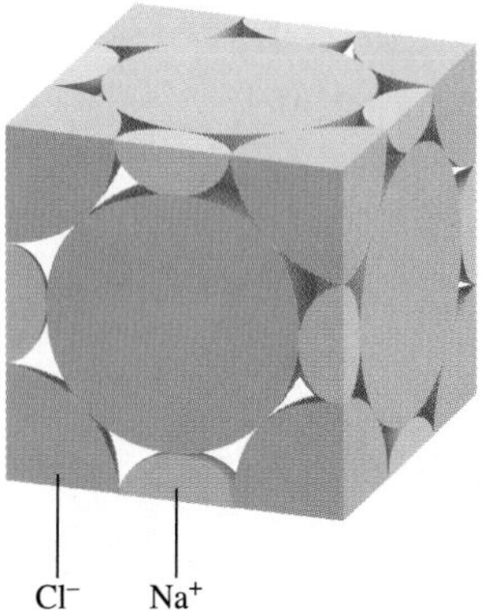

FIGURE 11.22

Portions of Na^+ and Cl^- ions within a face-centered cubic unit cell.

MOLECULAR CRYSTALS

Molecular crystals consist of atoms or molecules held together by van der Waals forces and/or hydrogen bonding. An example of a molecular crystal is solid sulfur dioxide (SO_2), in which the predominant attractive force is dipole-dipole interaction. Intermolecular hydrogen bonding is mainly responsible for the three-dimensional ice lattice (see Figure 11.12). Other examples of molecular crystals are I_2, P_4, and S_8.

In general, except in ice, molecules in molecular crystals are packed together as closely as their size and shape allow. Because van der Waals forces and hydrogen bonding are generally quite weak compared with the covalent and ionic bonds, molecular crystals are more easily broken apart than ionic and covalent crystals. Indeed, most molecular crystals melt below 200°C.

Sulfur.

COVALENT CRYSTALS

In covalent crystals (sometimes called covalent network crystals), atoms are held together entirely by covalent bonds in an extensive three-dimensional network. No discrete molecules are present, as in the case of molecular solids. Well-known examples are the two allotropes of carbon: diamond and graphite (see Figure 8.15). In diamond each carbon atom is tetrahedrally bonded to four other atoms (Figure 11.23). The strong covalent bonds in three dimensions contribute to diamond's unusual hardness (it is the hardest material known) and high melting point (3550°C). In graphite, carbon atoms are arranged in six-membered rings. The atoms are all sp^2-hybridized; each atom is covalently bonded to three other atoms. The unhybridized $2p$ orbital is used in pi bonding. In fact, the electrons in these $2p$ orbitals are free to move around, making graphite a good conductor of electricity in the planes of the bonded carbon atoms. The layers are held together by the weak van der Waals forces. The covalent bonds in graphite account for its hardness; however, because the layers can slide over one another, graphite is slippery to the touch and is effective as a lubricant. It is also used in pencils and in ribbons made for computer printers and typewriters.

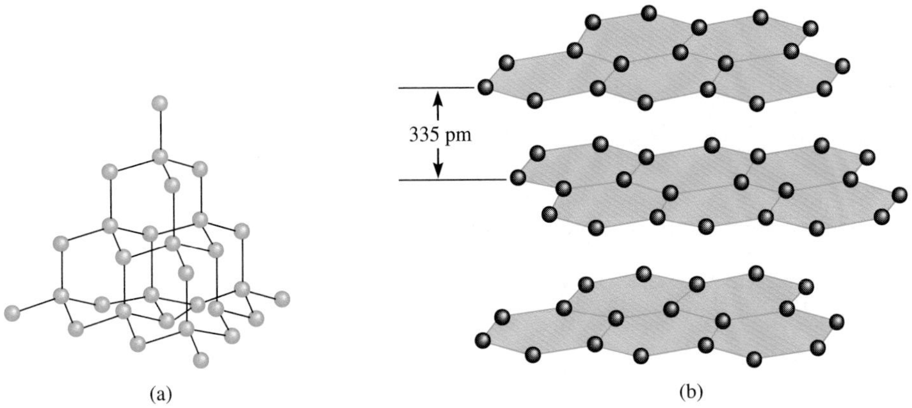

FIGURE 11.23

(a) The structure of diamond. Each carbon is tetrahedrally bonded to four other carbon atoms. (b) The structure of graphite. The distance between successive layers is 335 pm.

Another type of covalent crystal is quartz (SiO_2). The arrangement of silicon atoms in quartz is similar to that of carbon in diamond, but in quartz there is an oxygen atom between each pair of Si atoms. Since Si and O have different electronegativities (see Figure 9.4), the Si—O bond is polar. Nevertheless, SiO_2 is similar to diamond in many respects, such as hardness and high melting point (1610°C).

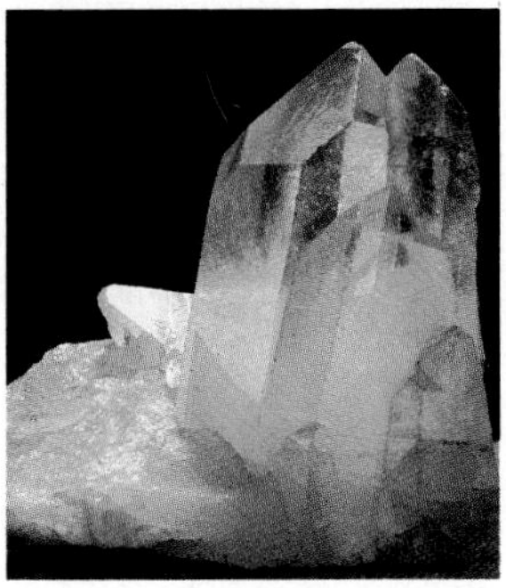

Quartz.

METALLIC CRYSTALS

In a sense, the structure of metallic crystals is the simplest to deal with, since every lattice point in a crystal is occupied by an atom of the same metal. The bonding in metals is quite different from that in other types of crystals. In a metal the bonding electrons are spread (or *delocalized*) over the entire crystal. In fact, metal atoms in a crystal can be imagined as an array of positive ions immersed in a sea of delocalized valence electrons (Figure 11.24). The great cohesive force resulting from delocalization is responsible for a metal's strength, which increases as the number of electrons available for bonding increases. For example, the melting point of sodium, with one valence electron, is 97.6°C, while that of aluminum, with three valence electrons, is 660°C. The mobility of the delocalized electrons makes metals good conductors of heat and electricity.

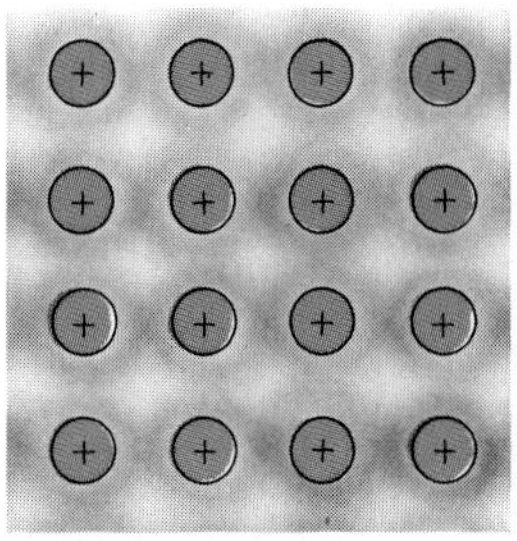

FIGURE 11.24

A cross section of the crystal structure of a metal. Each circled positive charge represents the nucleus and inner electrons of a metal atom. The gray area surrounding the positive metal ions indicates the mobile sea of electrons.

11.6 PHASE CHANGES

The discussions in Chapter 5 and in this chapter have given us an overview of the properties of the three states of matter: gas, liquid, and solid. Each of these states is often referred to as a ***phase,*** which is *a homogeneous part of the system in contact with other parts of the system but separated from them by a well-defined boundary.* An ice cube floating in water makes up two phases of water—the solid phase (ice) and the liquid phase (water). ***Phase changes,*** *transformations from one phase to another,* occur when energy (usually in the form of heat) is added or removed. Phase changes are physical changes that are characterized by changes in molecular order; molecules in the solid state have the most order, and those in the gas phase have the greatest randomness. Keeping in mind the relationship between energy change and the increase or decrease in molecular order will help us understand the nature of phase changes.

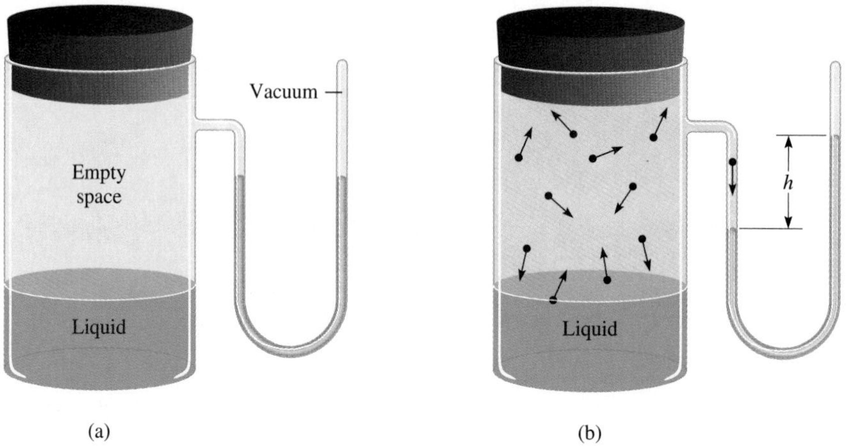

FIGURE 11.25

Apparatus for measuring the vapor pressure of a liquid (a) before the evaporation begins and (b) at equilibrium. In (b) the number of molecules leaving the liquid is equal to the number of molecules returning to the liquid. The difference in the mercury levels (h) gives the equilibrium vapor pressure of the liquid at the specified temperature.

LIQUID-VAPOR EQUILIBRIUM

Vapor Pressure

Molecules in a liquid are not fixed in a rigid lattice. Although they lack the total freedom of gaseous molecules, these molecules are in constant motion. Because liquids are denser than gases, the collision rate among molecules is much higher in the liquid phase than in the gas phase. *At any given temperature, a certain number of the molecules in a liquid possess sufficient kinetic energy to escape from the surface.* This process is called ***evaporation,*** or ***vaporization.***

When a liquid evaporates, its gaseous molecules exert a vapor pressure. Consider the apparatus shown in Figure 11.25. Before the evaporation process starts, the mercury levels in the U-shaped manometer are equal. As soon as a few molecules leave the liquid, a vapor phase is established. The vapor pressure is measurable only when a fair amount of vapor is present. The process of evaporation does not continue indefinitely, however. Eventually, the mercury levels stabilize and no further changes are seen.

What happens at the molecular level during evaporation? In the beginning, the traffic is only one way: Molecules are moving from the liquid to the empty space. Soon the molecules in the space above the liquid establish a vapor phase. *As the concentration of molecules in the vapor phase increases, some molecules return to the liquid phase,* a process called ***condensation.*** Condensation occurs because a molecule striking the liquid surface becomes trapped by intermolecular forces in the liquid.

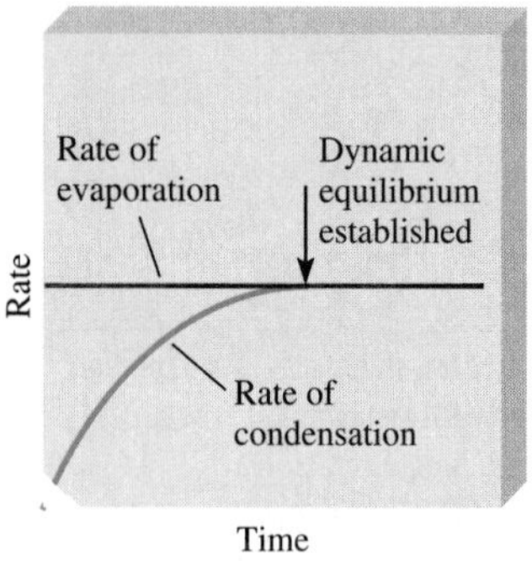

FIGURE 11.26

Comparison of the rates of evaporation and condensation at constant temperature.

The rate of evaporation is constant at any given temperature, and the rate of condensation increases with increasing concentration of molecules in the vapor phase. A state of ***dynamic equilibrium,*** in which *the rate of a forward process is exactly balanced by the rate of the reverse process,* is reached when the rates of condensation and evaporation become equal (Figure 11.26). *The vapor pressure measured under dynamic equilibrium of condensation and evaporation* is called the ***equilibrium vapor pressure.*** We often use the simpler term "vapor pressure" when we talk about the equilibrium vapor pressure of a liquid. This practice is acceptable as long as we know the meaning of the abbreviated term.

It is important to note that the equilibrium vapor pressure is the *maximum* vapor pressure a liquid exerts at a given temperature and that it is constant at constant temperature. Vapor pressure does change with temperature, however.

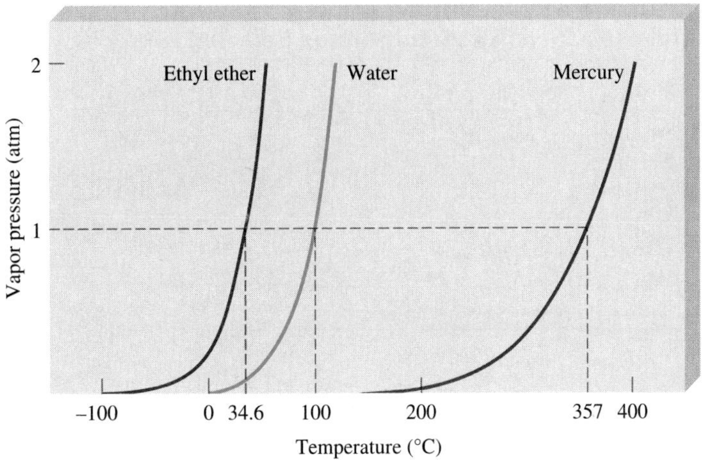

FIGURE 11.27

The increase in vapor pressure with temperature for three liquids. The normal boiling points of the liquids (at 1 atm) are shown on the horizontal axis.

Plots of vapor pressure versus temperature for three different liquids are shown in Figure 11.27. We know that the number of molecules with higher kinetic energies is greater at the higher temperature and therefore so is the evaporation rate. For this reason, the vapor pressure of a liquid always increases with temperature. For example, the vapor pressure of water is 17.5 mmHg at 20°C, but it rises to 760 mmHg at 100°C (see Figure 5.13).

Heat of Vaporization and Boiling Point

A measure of how strongly molecules are held in a liquid is its ***molar heat of vaporization*** (ΔH_{vap}), defined as *the energy* (usually in kilojoules) *required to vaporize one mole of a liquid.* The molar heat of vaporization is directly related to the strength of intermolecular forces that exist in the liquid. If the intermolecular attraction is strong, it takes a lot of energy to free the molecules from the liquid phase. Consequently, the liquid has a relatively low vapor pressure and a high molar heat of vaporization.

A practical way to demonstrate the molar heat of vaporization is by rubbing alcohol on your hands. The heat from your hands increases the kinetic energy of the alcohol molecules. The alcohol evaporates rapidly, extracting heat from your hands and cooling them. The process is similar to perspiration, which is one means by which the human body maintains a constant temperature. Because of the strong intermolecular hydrogen bonding that exists in water, a considerable amount of energy is needed to vaporize the water in perspiration from the body's surface. This energy is supplied by the heat generated in various metabolic processes.

Isopropanol (rubbing alcohol).

You have already seen that the vapor pressure of a liquid increases with temperature. For every liquid there exists a temperature at which the liquid begins to boil. The ***boiling point*** is *the temperature at which the vapor pressure of a liquid is equal to the external pressure.* The *normal* boiling point of a liquid is the boiling point when the external pressure is 1 atm.

At the boiling point, bubbles form within the liquid. When a bubble forms, the liquid originally occupying that space is pushed aside, and the level of the liquid in the container is forced to rise. The pressure exerted *on* the bubble is largely atmospheric pressure, plus some *hydrostatic pressure* (that is, pressure due to the presence of liquid). The pressure *inside* the bubble is due solely to the vapor pressure of

TABLE 11.5
Molar Heats of Vaporization for Selected Liquids

Substance	Boiling point* (°C)	ΔH_{vap} (kJ/mol)
Argon (Ar)	−186	6.3
Methane (CH_4)	−164	9.2
Ethyl ether ($C_2H_5OC_2H_5$)	34.6	26.0
Ethanol (C_2H_5OH)	78.3	39.3
Benzene (C_6H_6)	80.1	31.0
Water (H_2O)	100	40.79
Mercury (Hg)	357	59.0

*Measured at 1 atm.

the liquid. When the vapor pressure equals the external pressure, the bubble rises to the surface of the liquid and bursts. If the vapor pressure in the bubble were lower than the external pressure, the bubble would collapse before it could rise. We can thus conclude that the boiling point of a liquid depends on the external pressure. (We usually ignore the small contribution due to the hydrostatic pressure.) For example, at 1 atm water boils at 100°C, but if the pressure is reduced to 0.5 atm, water boils at only 82°C.

Since the boiling point is defined in terms of the vapor pressure of the liquid, we expect the boiling point to be related to the molar heat of vaporization: The higher ΔH_{vap}, the higher the boiling point. The data in Table 11.5 roughly confirm our prediction. Ultimately, both the boiling point and ΔH_{vap} are determined by the strength of intermolecular forces. For example, argon (Ar) and methane (CH_4), which have weak dispersion forces, have low boiling points and small molar heats of vaporization. Ethyl ether ($C_2H_5OC_2H_5$) has a dipole moment, and the dipole-dipole forces account for its moderately high boiling point and ΔH_{vap}. Both ethanol (C_2H_5OH) and water have strong hydrogen bonding, which accounts for their high boiling points and large ΔH_{vap} values. Strong metallic bonding causes mercury to have the highest boiling point and ΔH_{vap} of this group of liquids. Interestingly, the boiling point of benzene, which is nonpolar, is comparable to that of ethanol. Benzene has a large polarizability, and the dispersion forces among benzene molecules can be as strong as or even stronger than dipole-dipole forces and/or hydrogen bonds.

Critical Temperature and Pressure

The opposite of evaporation is condensation. In principle, a gas can be made to liquefy by either one of two techniques. By cooling a sample of gas we decrease the kinetic energy of its molecules, and eventually molecules aggregate to form small drops of liquid. Alternatively, we may apply pressure to the gas. Under compression, the average distance between molecules is reduced so that they are held together by mutual attraction. Industrial liquefaction processes combine these two methods.

Every substance has a ***critical temperature*** (T_c)*, above which its gas form cannot be made to liquefy, no matter how great the applied pressure.* This is also *the highest temperature at which a substance can exist as a liquid. The minimum pressure that must be applied to bring about liquefaction at the critical temperature* is called the ***critical pressure*** (P_c). The existence of the critical temperature can be qualitatively explained as follows. The intermolecular attraction is a finite quantity

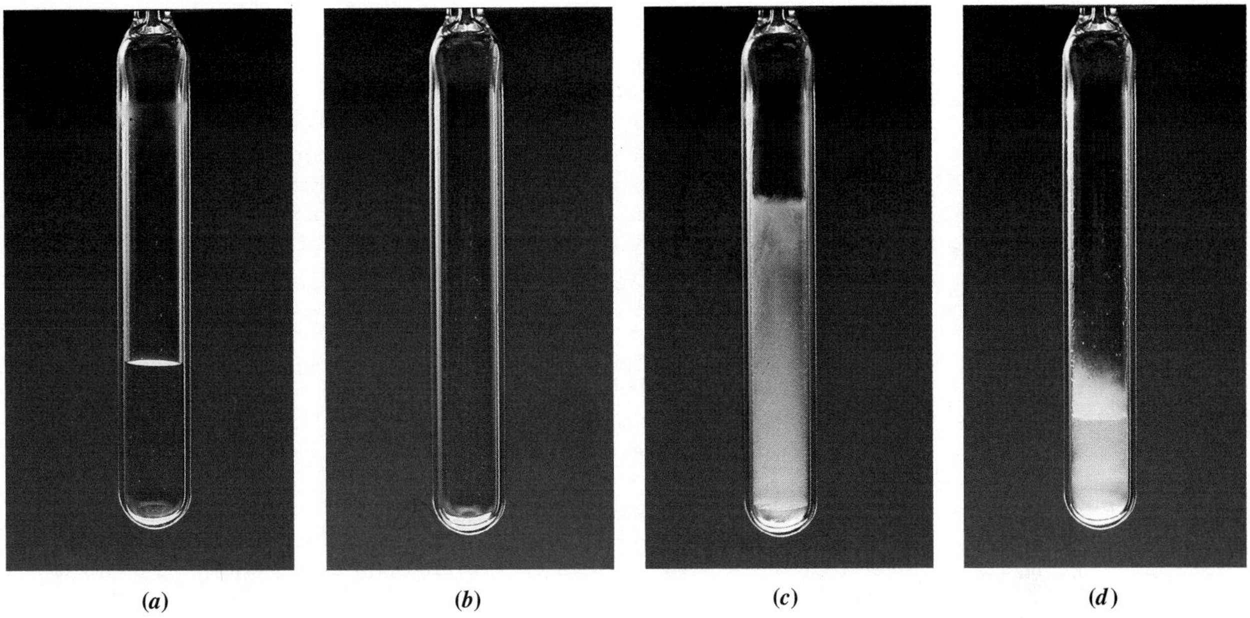

for any given substance. Below T_c, this force is sufficiently strong to hold the molecules together (under some appropriate pressure) in a liquid. Above T_c, molecular motion becomes so energetic that the molecules can always break away from this attraction. Figure 11.28 shows what happens when sulfur hexafluoride is heated above its critical temperature (45.5°C) and then cooled down to below 45.5°C.

Table 11.6 lists the critical temperatures and critical pressures of a number of common substances. Benzene, ethanol, mercury, and water, which have strong intermolecular forces, also have high critical temperatures compared with the other substances listed in the table.

FIGURE 11.28

The critical phenomenon of sulfur hexafluoride. (a) Below the critical temperature. (b) Above the critical temperature. Note that the liquid phase disappears. (c) The substance is cooled just below its critical temperature. The fog represents the condensation of vapor. (d) Finally, the appearance of the liquid phase.

TABLE 11.6
Critical Temperatures and Critical Pressures of Selected Substances

Substance	T_c(°C)	P_c(atm)
Ammonia (NH_3)	132.4	111.5
Argon (Ar)	−186	6.3
Benzene (C_6H_6)	288.9	47.9
Carbon dioxide (CO_2)	31.0	73.0
Ethanol (C_2H_5OH)	243	63.0
Ethyl ether ($C_2H_5OC_2H_5$)	192.6	35.6
Mercury (Hg)	1462	1036
Methane (CH_4)	−83.0	45.6
Molecular hydrogen (H_2)	−239.9	12.8
Molecular nitrogen (N_2)	−147.1	33.5
Molecular oxygen (O_2)	−118.8	49.7
Sulfur hexafluoride (SF_6)	45.5	37.6
Water (H_2O)	374.4	219.5

LIQUID-SOLID EQUILIBRIUM

The transformation of liquid to solid is called *freezing,* and the reverse process is called *melting* or *fusion.* The **melting point** of a solid (or the freezing point of a liquid) is *the temperature at which solid and liquid phases coexist in equilibrium.* The *normal* melting point (or the *normal* freezing point) of a substance is the melting point (or freezing point) measured at 1 atm pressure. We generally omit the word "normal" in referring to the melting point of a substance at 1 atm.

The most familiar liquid-solid equilibrium is that of water and ice. At 0°C and 1 atm, the dynamic equilibrium is represented by

$$\text{ice} \rightleftharpoons \text{water}$$

A practical illustration of this dynamic equilibrium is provided by a glass of ice water. As the ice cubes melt to form water, some of the water between the ice cubes may freeze, thus joining the cubes together. This is not a true dynamic equilibrium; since the glass is not kept at 0°C, all the ice cubes will eventually melt away.

The energy (usually in kilojoules) *required to melt one mole of a solid* is called the ***molar heat of fusion* (ΔH_{fus}).** Table 11.7 shows the molar heats of fusion for the substances listed in Table 11.5. A comparison of the data in the two tables shows that for each substance ΔH_{fus} is smaller than ΔH_{vap}. This is consistent with the fact that molecules in a liquid are fairly closely packed together, so that some energy is needed to bring about the rearrangement from solid to liquid. On the other hand, when a liquid evaporates, its molecules are completely separated from one another and considerably more energy is required to overcome the attractive forces.

SOLID-VAPOR EQUILIBRIUM

Solid iodine in equilibrium with its vapor.

Solids, too, undergo evaporation and therefore possess a vapor pressure. Consider the following dynamic equilibrium:

$$\text{solid} \rightleftharpoons \text{vapor}$$

The process in which molecules go directly from the solid into the vapor phase is called ***sublimation,*** and the reverse process (that is, *from vapor directly to solid*) is called ***deposition.*** Naphthalene (the substance used to make moth balls) has a fairly high vapor pressure for a solid (1 mmHg at 53°C); thus its pungent vapor

TABLE 11.7
Molar Heats of Fusion for Selected Substances

Substance	Melting point* (°C)	ΔH_{fus} (kJ/mol)
Argon (Ar)	−190	1.3
Methane (CH_4)	−183	0.84
Ethyl ether ($C_2H_5OC_2H_5$)	−116.2	6.90
Ethanol (C_2H_5OH)	−117.3	7.61
Benzene (C_6H_6)	5.5	10.9
Water (H_2O)	0	6.01
Mercury (Hg)	−39	23.4

*Measured at 1 atm.

quickly permeates an enclosed space. Generally, because molecules are more tightly held in a solid, the vapor pressure of a solid is much less than that of the corresponding liquid. *The energy* (usually in kilojoules) *required to sublime one mole of a solid,* called the ***molar heat of sublimation (*ΔH_{sub}*),*** is given by the sum of the molar heats of fusion and vaporization:

$$\Delta H_{sub} = \Delta H_{fus} + \Delta H_{vap} \quad (11.1)$$

Strictly speaking, Equation (11.1), which is an illustration of Hess's law, holds if all the phase changes occur at the same temperature. The enthalpy, or heat change, for the overall process is the same whether the substance changes directly from the solid to the vapor form or goes from solid to liquid and then to vapor. Figure 11.29 summarizes the types of phase changes discussed in this section.

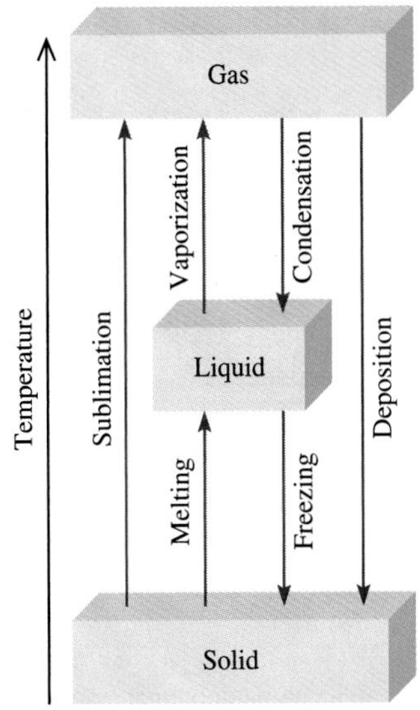

FIGURE 11.29

The various phase changes that a substance can undergo.

11.7 PHASE DIAGRAMS

The overall relationships among the solid, liquid, and vapor phases are best represented in a single graph known as a phase diagram. A ***phase diagram*** summarizes *the conditions at which a substance exists as a solid, liquid, or gas.* In this section we will briefly discuss the phase diagrams of water and carbon dioxide.

WATER

Figure 11.30(a) shows the phase diagram of water. The graph is divided into three regions, each of which represents a pure phase. The line separating any two regions indicates conditions under which these two phases can exist in equilibrium. For example, the curve between the liquid and vapor phases shows the variation of vapor pressure with temperature. (Compare this curve with Figure 11.27.) The other two curves similarly indicate conditions for equilibrium between ice and liquid water and between ice and water vapor. (Note that the solid-liquid boundary line has a negative slope.) The point at which all three curves meet is called the ***triple point.*** For water, this point is at 0.01°C and 0.006 atm. This is the only *temperature and pressure at which all three phases can be in equilibrium with one another.*

Phase diagrams enable us to predict changes in the melting point and boiling point of a substance as a result of changes in the external pressure; we can also

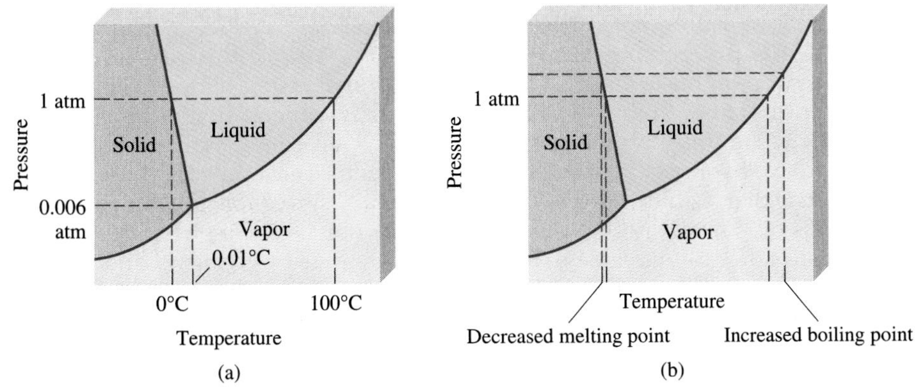

FIGURE 11.30

(a) The phase diagram of water. Note that the liquid-vapor line extends only to the critical temperature of water. (b) This phase diagram tells us that increasing the pressure on ice lowers its melting point and that increasing the pressure of liquid water raises its boiling point.

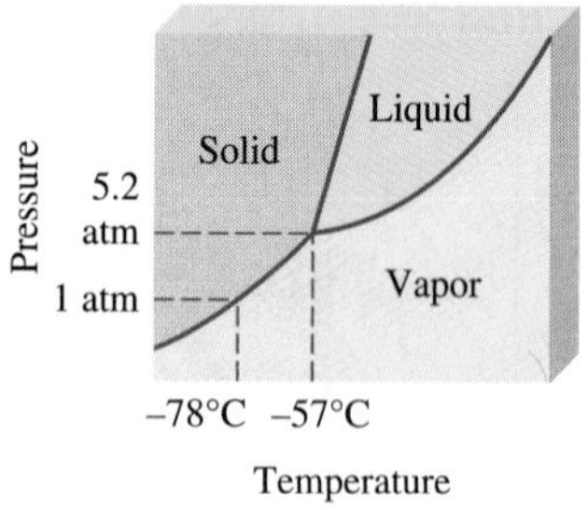

FIGURE 11.31

The phase diagram of carbon dioxide. Note that the solid-liquid boundary line has a positive slope. The liquid phase is not stable below 5.2 atm, so that only the solid and vapor phases can exist under atmospheric conditions.

anticipate directions of phase transitions brought about by changes in temperature and pressure. The normal melting point and boiling point of water, measured at 1 atm, are 0°C and 100°C, respectively. What would happen if melting and boiling were carried out at some other pressure? Figure 11.30(b) shows clearly that increasing the pressure above 1 atm will raise the boiling point and lower the melting point. A decrease in pressure will lower the boiling point and raise the melting point.

CARBON DIOXIDE

The phase diagram of carbon dioxide (Figure 11.31) is similar to that of water, with one important exception—the slope of the curve between solid and liquid is positive. In fact, this holds true for almost all other substances. Water behaves differently because ice is *less* dense than liquid water. The triple point of carbon dioxide is at 5.2 atm and −57°C.

An interesting observation can be made about the phase diagram in Figure 11.31. As you can see, the entire liquid phase lies well above atmospheric pressure; therefore, it is impossible for solid carbon dioxide to melt at 1 atm. Instead, when solid CO_2 is heated to −78°C at 1 atm, it sublimes. In fact, solid carbon dioxide is called dry ice because it looks like ice and *does not melt* (Figure 11.32). Because of this property, dry ice is useful as a refrigerant.

FIGURE 11.32

Under atmospheric conditions, solid carbon dioxide does not melt; it can only sublime. The cold carbon dioxide gas (−78°C) causes nearby water vapor to condense, producing fog.

SUMMARY

All substances exist in one of three states: gas, liquid, or solid. The major difference between the condensed states and the gaseous state is the distance of separation between molecules.

Intermolecular forces act between molecules or between molecules and ions. Generally, these forces are much weaker than bonding forces. Dipole-dipole forces and ion-dipole forces attract molecules with dipole moments to other polar molecules or ions. Dispersion forces are the result of temporary dipole moments induced in ordinarily nonpolar molecules. The extent to which a dipole moment can be induced in a molecule is determined by its polarizability. The term "van der Waals forces" refers to dipole-dipole, dipole-induced dipole, and dispersion forces.

Hydrogen bonding is a relatively strong dipole-dipole force that acts between a polar bond containing a hydrogen atom and the bonded electronegative atoms, O, N, or F. Hydrogen bonds between water molecules are particularly strong.

Liquids tend to assume a geometry that ensures the minimum surface area. Surface tension is the energy needed to expand a liquid surface area; strong intermolecular forces lead to greater surface tension. Viscosity is a measure of the resistance of a liquid to flow; it decreases with increasing temperature.

Water molecules in the solid state form a three-dimensional network in which each oxygen atom is covalently bonded to two hydrogen atoms and is hydrogen-bonded to two hydrogen atoms. This unique structure accounts for the fact that ice is less dense than liquid water. Water is also ideally suited for its ecological role by its high specific heat, another property imparted by its strong hydrogen bonding. Large bodies of water are able to moderate the climate by giving off and absorbing substantial amounts of heat with only small changes in the water temperature.

All solids are either crystalline (with a regular structure of atoms, ions, or molecules) or amorphous (without a regular structure). The basic structural unit of a crystalline solid is the unit cell, which is repeated to form a three-dimensional crystal lattice.

The four types of crystals and the forces that hold their particles together are ionic crystals, held together by ionic bonding; molecular crystals, van der Waals forces and/or hydrogen bonding; covalent crystals, covalent bonding; and metallic crystals, metallic bonding.

A liquid in a closed vessel eventually establishes a dynamic equilibrium between evaporation and condensation. The vapor pressure over the liquid under these conditions is the equilibrium vapor pressure, which is often referred to simply as vapor pressure. At the boiling point, the vapor pressure of a liquid equals the external pressure. The molar heat of vaporization of a liquid is the energy required to vaporize 1 mole of the liquid. The molar heat of fusion of a solid is the energy required to melt 1 mole of the solid.

For every substance there is a temperature, called the critical temperature, above which its gas form cannot be made to liquefy.

The relationships among the phases of a single substance are represented by a phase diagram, in which each region represents a pure phase and the boundaries between the regions show the temperatures and pressures at which the two phases are in equilibrium. At the triple point, all three phases are in equilibrium.

KEY WORDS

QUESTIONS AND PROBLEMS

INTERMOLECULAR FORCES

Review Questions

11.1 Define the following terms and give an example for each category: (a) dipole-dipole interaction, (b) dipole-induced dipole interaction, (c) ion-dipole interaction, (d) dispersion forces, (e) van der Waals forces.

11.2 Explain the term "polarizability." What kind of molecules tend to have high polarizabilities? What is the relationship between polarizability and intermolecular forces?

11.3 Explain the difference between the temporary dipole moment induced in a molecule and the permanent dipole moment in a polar molecule.

11.4 Give some evidence that all molecules exert attractive forces on one another.

11.5 What type of physical properties would you need to consider in comparing the strength of intermolecular forces in solids and in liquids?

11.6 Which elements can take part in hydrogen bonding?

Problems

11.7 The compounds Br_2 and ICl have the same number of electrons, yet Br_2 melts at −7.2°C, whereas ICl melts at 27.2°C. Explain.

11.8 If you lived in Alaska, state which of the following natural gases you would keep in an outdoor storage tank in winter and explain why: methane (CH_4), propane (C_3H_8), or butane (C_4H_{10}).

11.9 The binary hydrogen compounds of the Group 4A elements are CH_4 (−162°C), SiH_4 (−112°C), GeH_4 (−88°C), and SnH_4 (−52°C). The temperatures in parentheses are the corresponding boiling points. Explain the increase in boiling points from CH_4 to SnH_4.

11.10 List the types of intermolecular forces that exist in each of the following species: (a) benzene (C_6H_6), (b) CH_3Cl, (c) PF_3, (d) NaCl, (e) CS_2.

11.11 Ammonia is both a donor and an acceptor of hydrogen in hydrogen bond formation. Draw a diagram to show the hydrogen bonding of an ammonia molecule with two other ammonia molecules.

11.12 Which of the following species are capable of hydrogen bonding among themselves? (a) C_2H_6, (b) HI, (c) KF, (d) BeH_2, (e) CH_3COOH.

11.13 Arrange the following in order of increasing boiling point: RbF, CO_2, CH_3OH, CH_3Br. Explain your arrangement.

11.14 Diethyl ether has a boiling point of 34.5°C, and 1-butanol has a boiling point of 117°C.

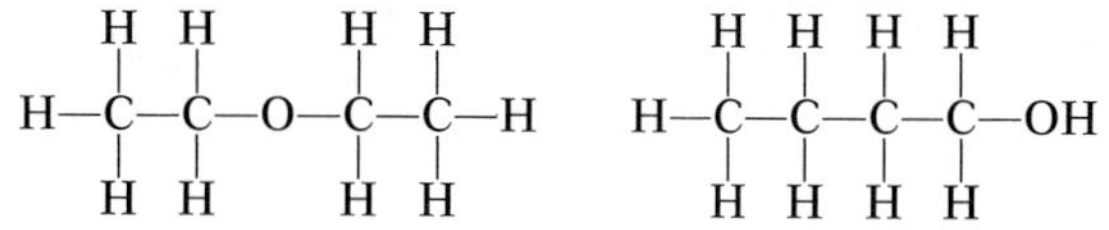

diethyl ether 1-butanol

Both of these compounds have the same numbers and types of atoms. Explain the difference in their boiling points.

11.15 Which member of each of the following pairs of substances would you expect to have a higher boiling point? (a) O_2 or N_2, (b) SO_2 or CO_2, (c) HF or HI.

11.16 State which substance in each of the following pairs you would expect to have the higher boiling point and explain why: (a) Ne or Xe, (b) CO_2 or CS_2, (c) CH_4 or Cl_2, (d) F_2 or LiF, (e) NH_3 or PH_3.

11.17 Explain in terms of intermolecular forces why (a) NH_3 has a higher boiling point than CH_4 and (b) KCl has a higher melting point than I_2.

11.18 What kind of attractive forces must be overcome in order to (a) melt ice, (b) boil molecular bromine, (c) melt solid iodine, and (d) dissociate F_2 into F atoms?

11.19 The following nonpolar molecules have the same number and type of atoms. Which one would you expect to have a higher boiling point?

(*Hint:* Molecules that can be stacked together more easily have greater intermolecular attraction.)

11.20 Explain the difference in the melting points of the following compounds:

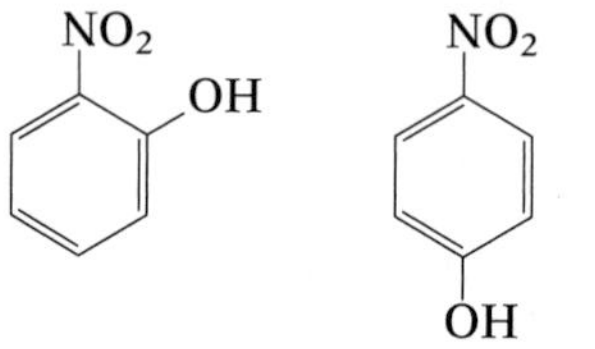

m.p. 45°C m.p. 115°C

(*Hint:* Only one of the two can form intramolecular hydrogen bonds.)

THE LIQUID STATE

Review Questions

11.21 Explain why liquids, unlike gases, are virtually incompressible.

11.22 Define surface tension. What is the relationship between the intermolecular forces that exist in a liquid and its surface tension?

11.23 Despite the fact that stainless steel is much denser than water, a stainless-steel razor blade can be made to float on water. Why?

11.24 Use water and mercury as examples to explain adhesion and cohesion.

11.25 A glass can be filled slightly above the rim with water. Explain why the water does not overflow.

11.26 Draw diagrams showing the capillary action of (a) water and (b) mercury in three tubes of different radii.

11.27 What is viscosity? What is the relationship between the intermolecular forces that exist in a liquid and its viscosity?

11.28 Why does the viscosity of a liquid decrease with increasing temperature?

11.29 Why is ice less dense than water?

11.30 Outdoor water pipes have to be drained or insulated in winter in a cold climate. Why?

Problems

11.31 Predict which of the following liquids has the greater surface tension: ethanol (C_2H_5OH) or dimethyl ether (CH_3OCH_3).

11.32 Predict the viscosity of ethylene glycol

$$\begin{array}{l} CH_2\text{—}OH \\ | \\ CH_2\text{—}OH \end{array}$$

relative to that of ethanol and glycerol (see Table 11.3).

CRYSTALLINE SOLIDS

Review Questions

11.33 Define the following terms: crystalline solid, lattice point, unit cell, coordination number.

11.34 Describe the geometries of the following cubic cells: simple cubic cell, body-centered cubic cell, face-centered cubic cell. Which of these cells would give the highest density for the same type of atoms?

11.35 Describe, with examples, the following types of crystals: (a) ionic crystals, (b) covalent crystals, (c) molecular crystals, (d) metallic crystals.

11.36 A solid is hard, brittle, and electrically nonconducting. Its melt (the liquid form of the substance) and an aqueous solution containing the substance do conduct electricity. Classify the solid.

11.37 A solid is soft and has a low melting point (below 100°C). The solid, its melt, and a solution containing the substance are all nonconductors of electricity. Classify the solid.

11.38 A solid is very hard and has a high melting point. Neither the solid nor its melt conducts electricity. Classify the solid.

11.39 Why are metals good conductors of heat and electricity? Why does the ability of a metal to conduct electricity decrease with increasing temperature?

11.40 Classify the solid states of the elements in the second period of the periodic table.

11.41 The melting points of the oxides of the third-period elements are given in parentheses: Na_2O (1275°C), MgO (2800°C), Al_2O_3 (2045°C), SiO_2 (1610°C), P_4O_{10} (580°C), SO_3 (16.8°C), Cl_2O_7 (−91.5°C). Classify these solids.

11.42 Which of the following are molecular solids and which are covalent solids? Se_8, HBr, Si, CO_2, C, P_4O_6, B, SiH_4.

Problems

11.43 What is the coordination number of each sphere in (a) a simple cubic lattice, (b) a body-centered cubic lattice, and (c) a face-centered cubic lattice? Assume the spheres to be of equal size.

11.44 Calculate the number of spheres in the following unit cells: simple cubic, body-centered cubic, and face-centered cubic cells. Assume that the spheres are of equal size and that they are only at the lattice points.

11.45 Metallic iron crystallizes in a cubic lattice. The unit cell edge length is 287 pm. The density of iron is 7.87 g/cm^3. How many iron atoms are there within a unit cell?

11.46 Barium metal crystallizes in a body-centered cubic lattice (the Ba atoms are at the lattice points only). The unit cell edge length is 502 pm, and the density of Ba is 3.50 g/cm^3. Using this information, calculate Avogadro's number. (*Hint:* First calculate the volume occupied by 1 mole of Ba atoms in the unit cells. Next calculate the volume occupied by one of the Ba atoms in the unit cell.)

11.47 Vanadium crystallizes in a body-centered cubic lattice (the V atoms occupy only the lattice points). How many V atoms are in a unit cell?

11.48 Europium crystallizes in a body-centered cubic lattice (the Eu atoms occupy only the lattice points). The density of Eu is 5.26 g/cm^3. Calculate the unit cell edge length in picometers.

11.49 Crystalline silicon has a cubic structure. The unit cell edge length is 543 pm. The density of the solid is 2.33 g/cm^3. Calculate the number of Si atoms in one unit cell.

11.50 A face-centered cubic cell contains 8 X atoms at the corners of the cell and 6 Y atoms at the faces. What is the empirical formula of the solid?

11.51 Classify the crystalline form of the following substances as ionic crystals, covalent crystals, molecular crystals, or metallic crystals: (a) CO_2, (b) B, (c) S_8, (d) KBr, (e) Mg, (f) SiO_2, (g) LiCl, (h) Cr.

11.52 Explain why diamond is harder than graphite.

PHASE CHANGES

Review Questions

11.53 Define phase change. Name all possible changes that can occur among the vapor, liquid, and solid states of a substance.

11.54 What is the equilibrium vapor pressure of a liquid? How does it change with temperature?

11.55 Use any one of the phase changes to explain what is meant by dynamic equilibrium.

11.56 Define the following terms: (a) molar heat of vaporization, (b) molar heat of fusion, (c) molar heat of sublimation. What are their units?

11.57 How is the molar heat of sublimation related to the molar heats of vaporization and fusion? On what law is this relation based?

11.58 What can we learn about the strength of intermolecular forces in a liquid from its molar heat of vaporization?

11.59 The greater the molar heat of vaporization of a liquid, the greater its vapor pressure. True or false?

11.60 Define boiling point. How does the boiling point of a liquid depend on external pressure? Referring to Table 5.2, what is the boiling point of water when the external pressure is 187.5 mmHg?

11.61 As a liquid is heated at constant pressure, its temperature rises. This trend continues until the boiling point of the liquid is reached. No further rise in the temperature of the liquid can be induced by heating. Explain.

11.62 Define critical temperature. What is the significance of critical temperature in the liquefaction of gases?

11.63 What is the relationship between intermolecular forces in a liquid and the liquid's boiling point and critical temperature? Why is the critical temperature of water greater than that of most other substances?

11.64 How do the boiling points and melting points of water and carbon tetrachloride vary with pressure? Explain any difference in behavior of these two substances.

11.65 Why is solid carbon dioxide called dry ice?

11.66 The vapor pressure of a liquid in a closed container depends on which of the following? (a) The volume above the liquid, (b) the amount of liquid present, (c) temperature.

11.67 Referring to Figure 11.27, estimate the boiling points of ethyl ether, water, and mercury at 0.5 atm.

11.68 Wet clothes dry more quickly on a hot, dry day than on a hot, humid day. Explain.

11.69 Which of the following phase transitions gives off more heat: (a) 1 mole of steam to 1 mole of water at 100°C or (b) 1 mole of water to 1 mole of ice at 0°C?

11.70 A beaker of water is heated to boiling by a Bunsen burner. Would adding another burner raise the boiling point of water? Explain.

Problems

11.71 Calculate the amount of heat (in kilojoules) required to convert 74.6 g of water to steam at 100°C.

11.72 How much heat (in kilojoules) is needed to convert 866 g of ice at −10°C to steam at 126°C? (The specific heats of ice and steam are 2.03 J/g · °C and 1.99 J/g · °C, respectively.)

11.73 How is the rate of evaporation of a liquid affected by (a) temperature, (b) the surface area of a liquid exposed to air, (c) intermolecular forces?

11.74 The molar heats of fusion and sublimation of molecular iodine are 15.27 kJ/mol and 62.30 kJ/mol, respectively. Estimate the molar heat of vaporization of liquid iodine.

11.75 The following compounds are liquid at −10°C; their boiling points are given: butane, −0.5°C; ethanol, 78.3°C; toluene, 110.6°C. At −10°C, which of these liquids would you expect to have the highest vapor pressure? Which the lowest?

11.76 Freeze-dried coffee is prepared by freezing a sample of brewed coffee and then removing the ice component by vacuum-pumping the sample. Describe the phase changes taking place during these processes.

11.77 A student hangs wet clothes outdoors on a winter day when the temperature is −15°C. After a few hours, the clothes are found to be fairly dry. Describe the phase changes in this drying process.

11.78 Steam at 100°C causes more serious burns than water at 100°C. Why?

PHASE DIAGRAMS

Review Questions

11.79 What is a phase diagram? What useful information can be obtained from a phase diagram?

11.80 Explain how water's phase diagram differs from those of most substances. What property of water causes the difference?

Problems

11.81 The blades of ice skates are quite thin, so that the pressure exerted on ice by a skater can be substantial. Explain how this fact enables a person to skate on ice.

11.82 A length of wire is placed on top of a block of ice. The ends of the wire extend over the edges of the ice, and a heavy weight is attached to each end. It is found that the ice under the wire gradually melts, so that the wire slowly moves through the ice block. At the same time, the water above the wire refreezes. Explain the phase changes that accompany this phenomenon.

11.83 The boiling point and freezing point of sulfur dioxide are −10°C and −72.7°C (at 1 atm), respectively. The triple point is −75.5°C and 1.65×10^{-3} atm, and its critical point is at 157°C and 78 atm. On the basis of this information, draw a rough sketch of the phase diagram of SO_2.

11.84 Consider the phase diagram of water shown at the end of this problem. Label the regions. Predict what would happen if we did the following: (a) Starting at A, we raise the temperature at constant pressure. (b) Starting at C, we lower the temperature at constant pressure. (c) Starting at B, we lower the pressure at constant temperature.

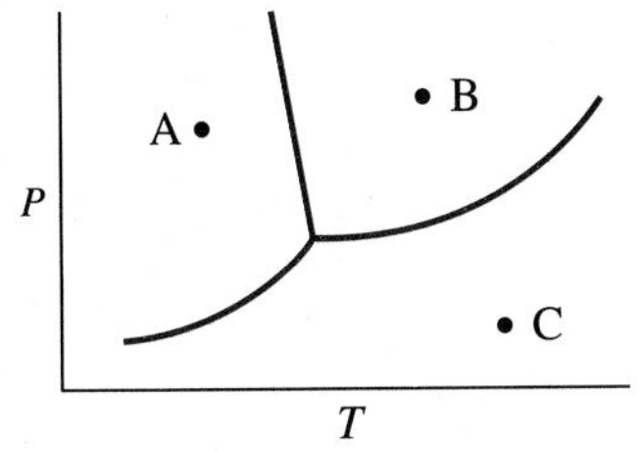

MISCELLANEOUS PROBLEMS

11.85 Name the kinds of attractive forces that must be overcome in order to (a) boil liquid ammonia, (b) melt solid phosphorus (P_4), (c) dissolve CsI in liquid HF, (d) melt potassium metal.

11.86 Which of the following indicates very strong intermolecular forces in a liquid? (a) A very low surface tension, (b) a very low critical temperature, (c) a very low boiling point, (d) a very low vapor pressure.

11.87 At −35°C, liquid HI has a higher vapor pressure than liquid HF. Explain.

11.88 From the following properties of elemental boron, classify it as one of the crystalline solids discussed in Section 11.5: high melting point (2300°C), poor conductor of heat and electricity, insoluble in water, very hard substance.

11.89 Referring to Figure 11.31, determine the stable phase of CO_2 at (a) 4 atm and −60°C and (b) 0.5 atm and −20°C.

11.90 A solid contains X, Y, and Z atoms in a cubic lattice with X atoms in the corners, Y atoms in the body-centered positions, and Z atoms on the faces of the cell. What is the empirical formula of the compound?

11.91 A CO_2 fire extinguisher is located on the outside of a building in Massachusetts. During the winter months, one can hear a sloshing sound when the extinguisher is gently shaken. In the summertime the sound is often absent. Explain. Assume that the extinguisher has no leaks and that it has not been used.

11.92 What is the vapor pressure of mercury at its normal boiling point (357°C)?

11.93 A flask containing water is connected to a powerful vacuum pump. When the pump is turned on, the water begins to boil. After a few minutes, the same water begins to freeze. Eventually, the ice disappears. Explain what happens at each step.

11.94 The liquid-vapor boundary line in the phase diagram of any substance always stops abruptly at a certain point. Why?

11.95 Given the phase diagram of carbon below, answer the following questions: (a) How many triple points are there and what are the phases that can coexist at each triple point? (b) Which has a higher density, graphite or diamond? (c) Synthetic diamond can be made from graphite. Using the phase diagram, how would you go about making diamond?

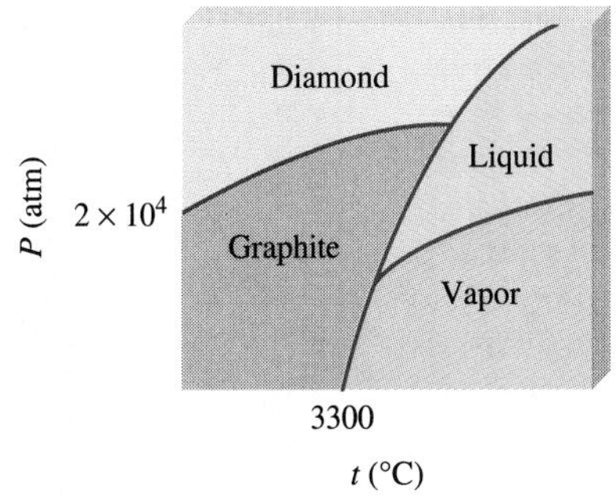

Answers to Practice Exercises: 11.1 (a) Ionic and dispersion forces, (b) dispersion forces, (c) dipole-dipole and dispersion forces; **11.2** CH_3OH; **11.3** 10.50 g/cm^3; **11.4** 2.

CHAPTER 12
PHYSICAL PROPERTIES OF SOLUTIONS

◆ When Svante August Arrhenius (1859–1927) presented his doctoral thesis at the University of Uppsala (Sweden) in May 1884, it was given the lowest possible rating. Consequently he was not awarded an academic appointment at the institution. Arrhenius was bitterly disappointed.

In his thesis, Arrhenius proposed an ionic theory of electrolyte solutions. Experimental work showed that the electrical conductance of certain salt solutions increased with dilution. Arrhenius suggested that these solutions contained a mixture of "active" (electrolyte) and "inactive" (nonelectrolyte) parts. On dilution, the number of active parts increased; in other words, electrolytic dissociation increased with dilution. Incidentally, his theory also explained that it is the ions that carry electric current in solution.

Older chemists and physicists ridiculed the idea of separated electrical charges in solution, but the new generation of scientists readily accepted the notion. The concept of electrolytic dissociation easily accommodated the discovery of electrons and the incorporation of electrons into chemical bonding during the early years of the twentieth century. For his work, Arrhenius was awarded the Nobel Prize in chemistry in 1903.

In addition to his theory of electrolyte solutions, Arrhenius made several valuable contributions to chemistry, the most notable of which was an equation showing the effect of temperature on reaction rates (see Chapter 21).

In his later years Arrhenius began writing about science for laymen and the applications of chemistry to astronomy, biology, and geology. He advanced the idea of *panspermia,* which speculates that life did not begin on Earth but instead came from other planets. ◆

Arrhenius was a precocious child who taught himself to read.

12.1 TYPES OF SOLUTIONS

Most chemical reactions take place, not between pure solids, liquids, or gases, but among ions and molecules dissolved in water or other solvents. In Section 4.1 we noted that a solution is a homogeneous mixture of two or more substances. Since this definition places no restriction on the nature of the substances involved, we can distinguish six types of solutions, depending on the original states (solid, liquid, or gas) of the solution components. Table 12.1 gives examples of each of these types.

Our focus here will be on solutions involving at least one liquid component—that is, gas-liquid, liquid-liquid, and solid-liquid solutions. And, perhaps not too surprisingly, the liquid solvent in most of the solutions we will study is water.

Chemists also characterize solutions by their capacity to dissolve a solute. *A solution that contains the maximum amount of a solute in a given solvent, at a specific temperature,* is called a ***saturated solution.*** Before the saturation point is reached, the solution is said to be ***unsaturated;*** it *contains less solute than it has the capacity to dissolve.* A third type, a ***supersaturated solution,*** *contains more solute than is present in a saturated solution.* Supersaturated solutions are not very stable. In time, some of the solute will come out of a supersaturated solution as crystals. *The process in which dissolved solute comes out of solution and forms crystals* is called ***crystallization.*** Note that both precipitation and crystallization describe the separation of excess solid substance from a supersaturated solution. However, solids formed by the two processes differ in appearance. We normally think of precipitates as being made up of small particles, whereas crystals may be large and well formed (Figure 12.1).

12.2 A MOLECULAR VIEW OF THE SOLUTION PROCESS

In liquids and solids, molecules are held together by intermolecular attractions. These forces also play a central role in the formation of solutions. When one substance (the solute) dissolves in another (the solvent), particles of the solute disperse throughout the solvent. The solute particles occupy positions that are normally taken by solvent molecules. The ease with which a solute particle replaces a solvent molecule depends on the relative strengths of three types of interactions:

TABLE 12.1
Types of Solutions

Solute	Solvent	State of resulting solution	Examples
Gas	Gas	Gas	Air
Gas	Liquid	Liquid	Soda water (CO_2 in water)
Gas	Solid	Solid	H_2 gas in palladium
Liquid	Liquid	Liquid	Ethanol in water
Solid	Liquid	Liquid	NaCl in water
Solid	Solid	Solid	Brass (CuZn), solder (Sn/Pb)

FIGURE 12.1

In a supersaturated sodium acetate solution (left), sodium acetate crystals rapidly form after a small seed crystal is added.

- solvent-solvent interaction
- solute-solute interaction
- solvent-solute interaction

For simplicity, we can imagine the solution process taking place in three distinct steps (Figure 12.2). Step 1 is the separation of solvent molecules, and step 2 entails the separation of solute molecules. These steps require energy input to break attractive intermolecular forces; therefore, they are endothermic. In step 3 the solvent and solute molecules mix. This step may be exothermic or endothermic. The heat of solution ΔH_{soln} is given by

$$\Delta H_{soln} = \Delta H_1 + \Delta H_2 + \Delta H_3$$

If the solute-solvent attraction is stronger than the solvent-solvent attraction and solute-solute attraction, the solution process is favorable; that is, it is exothermic

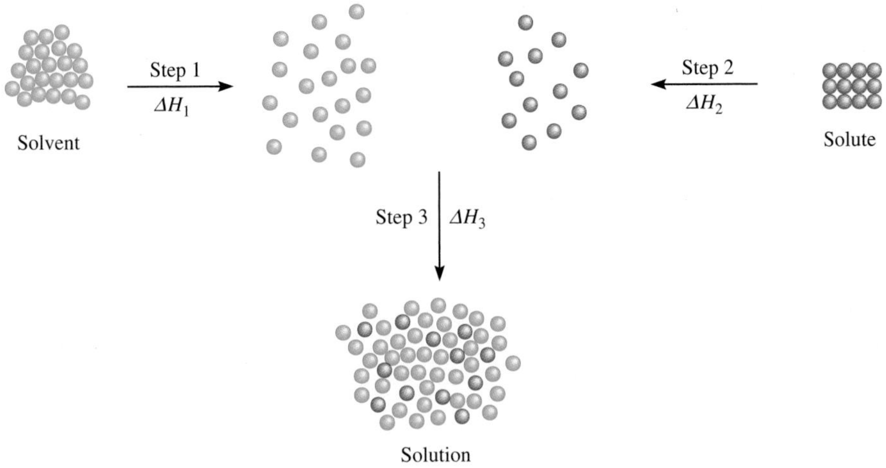

FIGURE 12.2

A molecular view of the solution process: First the solvent and solute molecules are separated (steps 1 and 2). Then the solvent and solute molecules mix (step 3).

($\Delta H_{soln} < 0$). If the solute-solvent interaction is weaker than the solvent-solvent and solute-solute interactions, the solution process is endothermic ($\Delta H_{soln} > 0$).

You may wonder why a solute dissolves in a solvent at all if the attraction among its own molecules is stronger than that between its molecules and the solvent molecules. The solution process, like all physical and chemical processes, is governed by two factors. One is energy, which determines whether a solution process is exothermic or endothermic. The second factor is an inherent tendency toward disorder in all natural events. In much the same way that a deck of new playing cards becomes mixed up after it has been shuffled a few times, when solute and solvent molecules mix to form a solution, there is an increase in randomness or disorder. In the pure state, the solvent and solute possess a fair degree of order, characterized by the more or less regular arrangement of atoms, molecules, or ions in three-dimensional space. Much of this order is destroyed when the solute dissolves in the solvent (see Figure 12.2). Therefore, the solution process is *always* accompanied by an increase in disorder or randomness. It is the increase in disorder of the system that favors the solubility of any substance, even if the solution process is endothermic.

Factors That Affect Solubility

Solubility is a measure of the amount of a solute that will dissolve in a solvent at a specific temperature. The saying "like dissolves like" helps in predicting the solubility of a substance in a solvent. What this expression means is that two substances with intermolecular forces of similar type and magnitude are likely to be soluble in each other. For example, both carbon tetrachloride (CCl_4) and benzene (C_6H_6) are nonpolar liquids. The only intermolecular forces present in these substances are dispersion forces (see Section 11.2). When these two liquids are mixed, they readily dissolve in each other, because the attraction between CCl_4 and C_6H_6 molecules is comparable in magnitude to that between CCl_4 molecules and between C_6H_6 molecules. When *two liquids are completely soluble in each other in all proportions,* as in this case, they are said to be ***miscible.*** Alcohols such as methanol, ethanol, and 1,2-ethylene glycol are miscible with water because of their ability to form hydrogen bonds with water molecules:

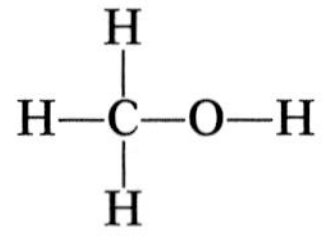
methanol

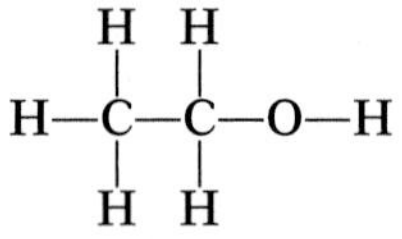
ethanol

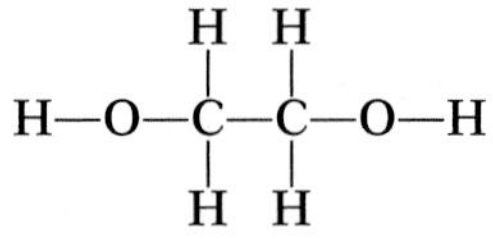
1,2-ethylene glycol

The rules given in Table 4.2 (p. 91) allow us to predict the solubility of a particular ionic compound in water. When sodium chloride dissolves in water, the ions are stabilized in solution by hydration, which involves ion-dipole interaction. In general, we predict that ionic compounds should be much more soluble in polar solvents, such as water, liquid ammonia, and liquid hydrogen fluoride, than in nonpolar solvents, such as benzene and carbon tetrachloride. Since the molecules of nonpolar solvents lack a dipole moment, they cannot effectively solvate the Na^+ and Cl^- ions. (***Solvation*** is *the process in which an ion or a molecule is surrounded by solvent molecules arranged in a specific manner.* When the solvent is water, the process is called *hydration*.) The predominant intermolecular interaction between ions and nonpolar compounds is ion-induced dipole interaction, which is much weaker than ion-dipole interaction. Consequently, ionic compounds usually have extremely low solubility in nonpolar solvents.

EXAMPLE 12.1

Predicting Solubility Based on Intermolecular Forces

Predict the relative solubilities in the following cases: (a) Br_2 in benzene (C_6H_6)($\mu = 0$ D) and in water ($\mu = 1.87$ D), (b) KCl in carbon tetrachloride ($\mu = 0$ D) and in liquid ammonia ($\mu = 1.46$ D.)

Answer: (a) Br_2 is a nonpolar molecule and therefore should be more soluble in C_6H_6, which is also nonpolar, than in water. The only intermolecular forces between Br_2 and C_6H_6 are dispersion forces.

(b) KCl is an ionic compound. For it to dissolve, the individual K^+ and Cl^- ions must be stabilized by ion-dipole interaction. Since carbon tetrachloride has no dipole moment, potassium chloride should be more soluble in liquid ammonia, a polar molecule with a large dipole moment.

PRACTICE EXERCISE

Is iodine (I_2) more soluble in water or in carbon disulfide (CS_2)?

12.3 CONCENTRATION UNITS

Quantitative study of a solution requires that we know its *concentration,* that is, the amount of solute present in a given amount of solution. Chemists use several different concentration units, each of which has advantages as well as limitations. Let us examine the three most common units of concentration: percent by mass, molarity, and molality.

TYPES OF CONCENTRATION UNITS

Percent by Mass

The ***percent by mass*** (also called the *percent by weight* or the *weight percent*) is defined as

$$\text{percent by mass of solute} = \frac{\text{mass of solute}}{\text{mass of solute} + \text{mass of solvent}} \times 100\%$$

$$= \frac{\text{mass of solute}}{\text{mass of soln}} \times 100\%$$

The percent by mass has no units because it is a ratio of two similar quantities.

EXAMPLE 12.2

Calculating the Percent by Mass of a Solution

A sample of 0.892 g of potassium chloride (KCl) is dissolved in 54.6 g of water. What is the percent by mass of KCl in this solution?

Answer:

$$\text{percent by mass of KCl} = \frac{\text{mass of solute}}{\text{mass of soln}} \times 100\%$$

$$= \frac{0.892\text{ g}}{0.892\text{ g} + 54.6\text{ g}} \times 100\%$$

$$= 1.61\%$$

PRACTICE EXERCISE

A sample of 6.44 g of naphthalene ($C_{10}H_8$) is dissolved in 80.1 g of benzene (C_6H_6). Calculate the percent by mass of naphthalene in this solution.

Molarity (*M*)

The molarity unit was defined in Section 4.5 as the number of moles of solute in 1 liter of solution; that is,

$$\text{molarity} = \frac{\text{moles of solute}}{\text{liters of soln}}$$

Thus, molarity has the units of mole per liter (mol/L).

Molality (*m*)

Molality is *the number of moles of solute dissolved in 1 kg (1000 g) of solvent*—that is,

$$\text{molality} = \frac{\text{moles of solute}}{\text{mass of solvent (kg)}}$$

For example, to prepare a 1 *molal,* or 1 *m,* sodium sulfate (Na_2SO_4) aqueous solution, we need to dissolve 1 mole (142.0 g) of the substance in 1000 g (1 kg) of water. Depending on the nature of the solute-solvent interaction, the final volume of the solution will be either greater or less than 1000 mL. It is also possible, though very unlikely, that the final volume could be equal to 1000 mL.

EXAMPLE 12.3

Calculating the Molality of a Solution

Calculate the molality of a sulfuric acid solution containing 24.4 g of sulfuric acid in 198 g of water. The molar mass of sulfuric acid is 98.08 g.

Answer: From the known molar mass of sulfuric acid, we can calculate the molality in two steps. First we need to find the number of grams of sulfuric acid dissolved in 1000 g (1 kg) of water. Next we must convert the number of grams into the number of moles. Combining these two steps, we write

$$\text{molality} = \frac{\text{moles of solute}}{\text{mass of solvent (kg)}}$$

$$\text{molality} = \frac{24.4 \text{ g } H_2SO_4}{198 \text{ g } H_2O} \times \frac{1000 \text{ g } H_2O}{1 \text{ kg } H_2O} \times \frac{1 \text{ mol } H_2SO_4}{98.08 \text{ g } H_2SO_4}$$

$$= 1.26 \text{ mol } H_2SO_4/\text{kg } H_2O$$

$$= 1.26\ m$$

PRACTICE EXERCISE

What is the molality of a solution containing 7.78 g of urea [$(NH_2)_2CO$] in 203 g of water?

Comparison of Concentration Units

The choice of a concentration unit is based on the purpose of the measurement. The advantage of molarity is that it is generally easier to measure the volume of a solution, using precisely calibrated volumetric flasks, than to weigh the solvent, as we saw in Section 4.5. For this reason, molarity is often preferred over molality. On the other hand, molality is independent of temperature, since the concentration is expressed in number of moles of solute and mass of solvent, whereas the volume of a solution typically increases with increasing temperature. A solution that is 1.0 *M* at 25°C may become 0.97 *M* at 45°C because of the increase in volume. This concentration dependence on temperature can significantly affect the accuracy of an experiment.

Percent by mass is similar to molality in that it is independent of temperature. Furthermore, since it is defined in terms of ratio of mass of solute to mass of solution, we do not need to know the molar mass of the solute in order to calculate the percent by mass.

EXAMPLE 12.4

Converting Molality to Molarity

Calculate the molarity of a 0.396 *m* glucose ($C_6H_{12}O_6$) solution. The molar mass of glucose is 180.2 g, and the density of the solution is 1.16 g/mL.

Answer: The mass of the solution must be converted to volume in working problems of this type. The density of the solution is used as a conversion factor. Since a 0.396 *m* glucose solution contains 0.396 mole of glucose in 1 kg of water, the total mass of the solution is

$$\left(0.396 \text{ mol } C_6H_{12}O_6 \times \frac{180.2 \text{ g}}{1 \text{ mol } C_6H_{12}O_6}\right) + 1000 \text{ g } H_2O \text{ soln} = 1071 \text{ g}$$

From the known density of the solution (1.16 g/mL), we can calculate the molarity as follows:

$$\text{molarity} = \frac{0.396 \text{ mol } C_6H_{12}O_6}{1071 \text{ g soln}} \times \frac{1.16 \text{ g soln}}{1 \text{ mL soln}} \times \frac{1000 \text{ mL soln}}{1 \text{ L soln}}$$

$$= \frac{0.429 \text{ mol } C_6H_{12}O_6}{1 \text{ L soln}}$$

$$= 0.429\ M$$

PRACTICE EXERCISE

Calculate the molarity of a 1.74 *m* sucrose ($C_{12}H_{22}O_{11}$) solution whose density is 1.12 g/mL.

EXAMPLE 12.5

Converting Molarity to Molality

The density of a 2.45 *M* aqueous methanol (CH_3OH) solution is 0.976 g/mL. What is the molality of the solution? The molar mass of methanol is 32.04 g.

Answer: The total mass of 1 L of a 2.45 *M* solution of methanol is

$$1 \text{ L soln} \times \frac{1000 \text{ mL soln}}{1 \text{ L soln}} \times \frac{0.976 \text{ g soln}}{1 \text{ mL soln}} = 976 \text{ g soln}$$

Since this solution contains 2.45 moles of methanol, the amount of water in the solution is

$$976 \text{ g soln} - \left(2.45 \text{ mol } CH_3OH \times \frac{32.04 \text{ g } CH_3OH}{1 \text{ mol } CH_3OH}\right) = 898 \text{ g } H_2O$$

Now the molality of the solution can be calculated:

$$\begin{aligned} \text{molality} &= \frac{2.45 \text{ mol } CH_3OH}{898 \text{ g } H_2O} \times \frac{1000 \text{ g } H_2O}{1 \text{ kg } H_2O} \\ &= \frac{2.73 \text{ mol } CH_3OH}{1 \text{ kg } H_2O} \\ &= 2.73\ m \end{aligned}$$

PRACTICE EXERCISE

Calculate the molality of a 5.86 *M* ethanol (C_2H_5OH) solution whose density is 0.927 g/mL.

EXAMPLE 12.6

Converting Percent by Mass to Molality

Calculate the molality of a 35.4 percent (by mass) aqueous solution of phosphoric acid (H_3PO_4). The molar mass of phosphoric acid is 98.00 g.

Answer: In solving this type of problem it is convenient to assume that we start with 100.0 g of the solution. If the mass of phosphoric acid is 35.4 percent or 35.4 g, the percent and mass of water must be 100.0% − 35.4% = 64.6% and 64.6 g. From the known molar mass of phosphoric acid, we can calculate the molality in two steps. First we need to find the number of grams of phosphoric acid dissolved in 1000 g (1 kg) of water. Next, we must convert the number of grams into the number of moles. Combining these two steps we write

$$\text{molality} = \frac{35.4 \text{ g } H_3PO_4}{64.6 \text{ g } H_2O} \times \frac{1000 \text{ g } H_2O}{1 \text{ kg } H_2O} \times \frac{1 \text{ mol } H_3PO_4}{98.00 \text{ g } H_3PO_4}$$

$$= 5.59 \text{ mol } H_3PO_4/\text{kg } H_2O$$

$$= 5.59\ m$$

PRACTICE EXERCISE

Calculate the molality of a 44.6 percent by mass aqueous solution of sodium chloride.

12.4 EFFECT OF TEMPERATURE ON SOLUBILITY

Recall that solubility is defined as the maximum amount of a solute that will dissolve in a given quantity of solvent *at a specific temperature.* For most substances, temperature affects solubility. In this section we will consider the effects of temperature on the solubility of solids and gases.

SOLID SOLUBILITY AND TEMPERATURE

Figure 12.3 shows the temperature dependence of the solubility of some ionic compounds in water. In most but certainly not all cases, the solubility of a solid substance increases with temperature. However, there is no clear correlation between the sign of ΔH_{soln} and the variation of solubility with temperature. For ex-

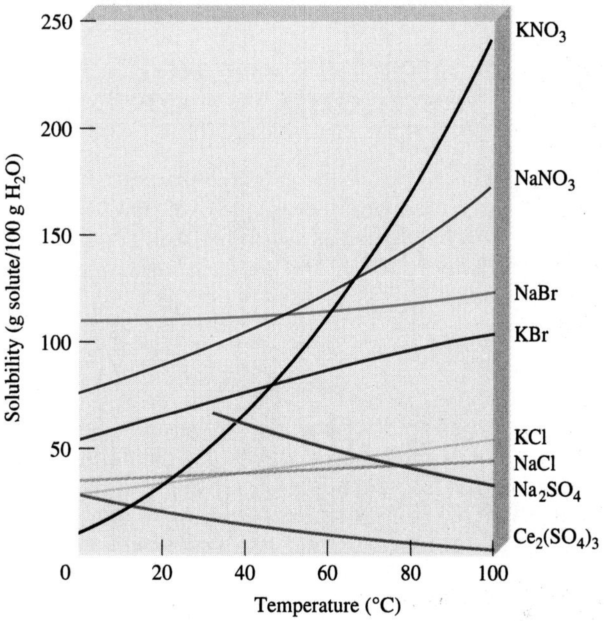

FIGURE 12.3

Temperature dependence of the solubility of some ionic compounds in water.

ample, the solution process of $CaCl_2$ is exothermic and that of NH_4NO_3 is endothermic. But the solubility of both compounds increases with increasing temperature. In general, the effect of temperature on solubility is best determined experimentally.

Gas Solubility and Temperature

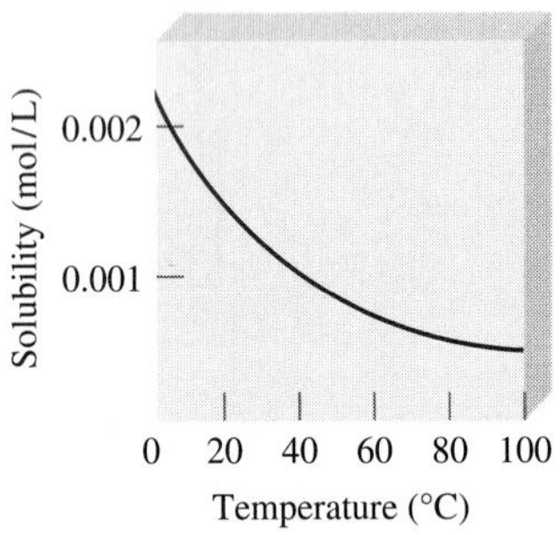

FIGURE 12.4

Dependence on temperature of the solubility of O_2 gas in water. Note that the solubility decreases as temperature increases. The pressure of the gas over the solution is 1 atm.

The solubility of gases in water usually decreases with increasing temperature (Figure 12.4). When water is heated in a beaker, you can see bubbles of air forming on the side of the glass before the water boils. As the temperature rises, the dissolved air molecules begin to "boil out" of the solution long before the water itself boils.

The reduced solubility of molecular oxygen in hot water has a direct bearing on ***thermal pollution,*** that is, *the heating of the environment*—usually waterways—*to temperatures that are harmful to its living inhabitants.* It is estimated that every year in the United States some 100,000 billion gallons of water are used for industrial cooling, mostly in electric power and nuclear power production. This process heats up the water, which is then returned to the rivers and lakes from which it was taken. Ecologists have become increasingly concerned about the effect of thermal pollution on aquatic life. Fish, like all other cold-blooded animals, have much more difficulty coping with rapid temperature fluctuation in the environment than humans do. An increase in water temperature accelerates their rate of metabolism, which generally doubles with each 10°C rise. The speedup of metabolism increases the fish's need for oxygen at the same time that the supply of oxygen decreases because of its lower solubility in heated water. Effective ways to cool power plants while doing only minimal damage to the biological environment are being sought.

On the lighter side, a knowledge of the variation of gas solubility with temperature can improve one's performance in a popular recreational sport—fishing. On a hot summer day, an experienced fisherman usually picks a deep spot in the river or lake to cast the bait. Because the oxygen content is greater in the deeper, cooler region, most fish will be found there.

12.5 Effect of Pressure on the Solubility of Gases

For all practical purposes, external pressure has no influence on the solubilities of liquids and solids, but it does greatly affect the solubility of gases. The quantitative relationship between gas solubility and pressure is given by ***Henry's law,*** which states that *the solubility of a gas in a liquid is proportional to the pressure of the gas over the solution:*

$$c \propto P$$

$$c = kP \tag{12.1}$$

Here c is the molar concentration (moles per liter) of the dissolved gas; P is the pressure (in atmospheres) of the gas over the solution; and, for a given gas, k is a constant that depends only on temperature. The constant k has the units mol/L · atm. You can see that when the pressure of the gas is 1 atm, c is *numerically* equal to k.

Henry's law can be understood qualitatively in terms of the kinetic molecular theory. The amount of gas that will dissolve in a solvent depends on how frequently the molecules in the gas phase collide with the liquid surface and become

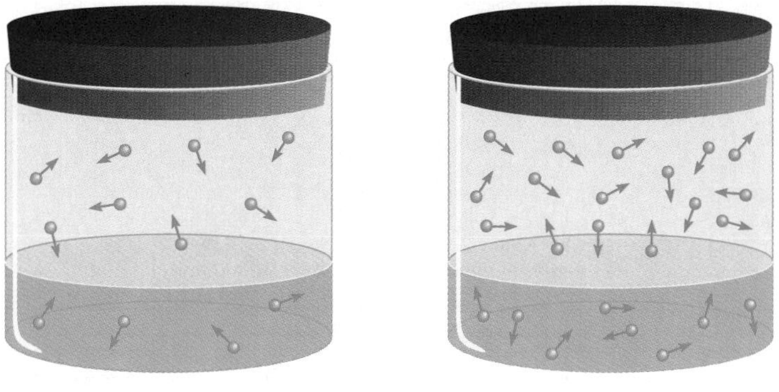

FIGURE 12.5

A molecular interpretation of Henry's law. When the partial pressure of the gas over the solution increases from (a) to (b), the concentration of the dissolved gas also increases according to Equation (12.1).

trapped by the condensed phase. Suppose we have a gas in dynamic equilibrium with a solution [Figure 12.5(a)]. At every instant, the number of gas molecules entering the solution is equal to the number of dissolved molecules moving into the gas phase. When the partial pressure is increased, more molecules dissolve in the liquid because more molecules are striking the surface of the liquid. This process continues until the concentration of the solution is again such that the number of molecules leaving the solution per second equals the number entering the solution [Figure 12.5(b)]. Because of the increased concentration of molecules in both the gas and solution phases, this number is greater in (b) than in (a), where the partial pressure is lower.

EXAMPLE 12.7

Applying Henry's Law

The solubility of pure nitrogen gas at 25°C and 1 atm is 6.8×10^{-4} mol/L. What is the concentration of nitrogen dissolved in water under atmospheric conditions? The partial pressure of nitrogen gas in the atmosphere is 0.78 atm.

Answer: The first step is to calculate the quantity k in Equation (12.1):

$$c = kP$$

$$6.8 \times 10^{-4}\ \text{mol/L} = k\ (1\ \text{atm})$$

$$k = 6.8 \times 10^{-4}\ \text{mol/L} \cdot \text{atm}$$

Therefore, the solubility of nitrogen gas in water is

$$\begin{aligned} c &= (6.8 \times 10^{-4}\ \text{mol/L} \cdot \text{atm})(0.78\ \text{atm}) \\ &= 5.3 \times 10^{-4}\ \text{mol/L} \\ &= 5.3 \times 10^{-4}\ M \end{aligned}$$

The decrease in solubility is the result of lowering the pressure from 1 atm to 0.78 atm.

PRACTICE EXERCISE

Calculate the concentration (in M) of oxygen in water at 25°C for a partial pressure of 0.22 atm. The Henry's law constant for oxygen is 3.5×10^{-4} mol/L · atm.

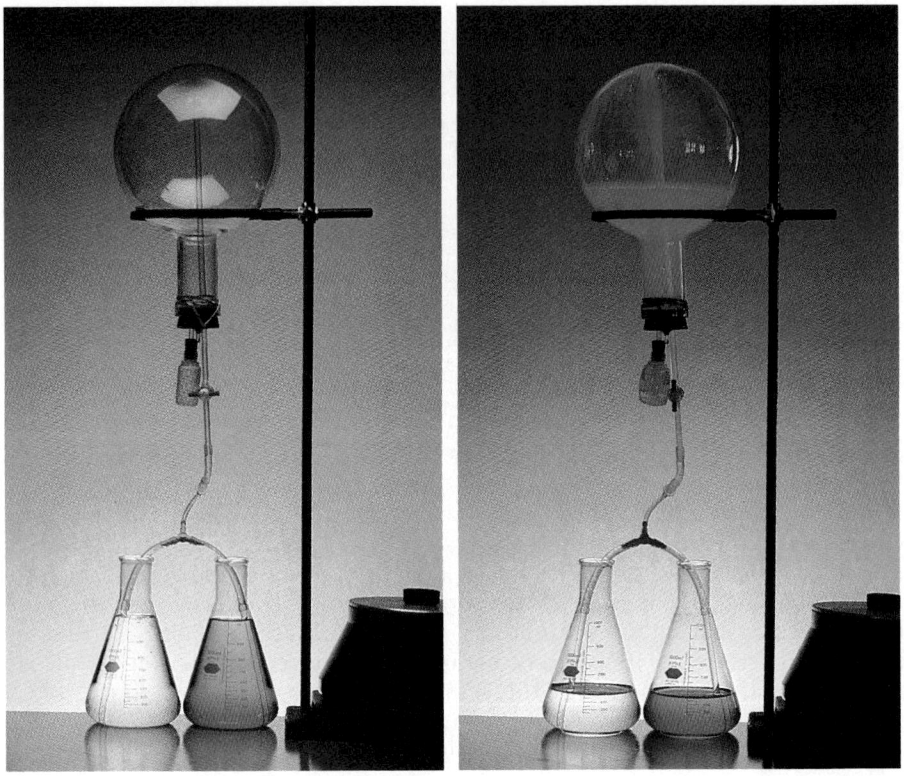

FIGURE 12.6

The ammonia fountain display. (Left) The inverted round-bottomed flask is filled with ammonia gas. (Right) When a small amount of water is introduced into the flask by squeezing the polyethylene bottle, most of the ammonia gas is dissolved in the water, creating a partial vacuum. The pressure from the atmosphere forces the liquids from the two Erlenmeyer flasks into the round-bottomed flask. The mixing of the two liquids initiates a chemical reaction, accompanied by the emission of blue light.

Most gases obey Henry's law, but there are some important exceptions. For example, if the dissolved gas *reacts* with water, higher solubilities can result. The solubility of ammonia is much higher than expected because of the reaction (Figure 12.6)

$$NH_3 + H_2O \rightleftharpoons NH_4^+ + OH^-$$

Carbon dioxide also reacts with water, as follows:

$$CO_2 + H_2O \rightleftharpoons H_2CO_3$$

12.6 COLLIGATIVE PROPERTIES

Several important properties of solutions depend on the number of solute particles in solution and not on the nature of the solute particles. These properties are called ***colligative properties*** (or collective properties) because they are bound together by a common origin; that is, they all depend on the number of solute particles present, whether these particles are atoms, ions, or molecules. The colligative properties are vapor-pressure lowering, boiling-point elevation, freezing-point depression, and osmotic pressure. For our discussion of colligative properties of nonelectrolyte solutions it is important to keep in mind that we are talking about relatively dilute solutions, that is, solutions whose concentrations are $\leq 0.2\ M$.

VAPOR-PRESSURE LOWERING

If a solute is ***nonvolatile*** (that is, it *does not have a measurable vapor pressure*), the vapor pressure of its solution is always less than that of the pure solvent. Thus the relationship between solution vapor pressure and solvent vapor pressure depends on the concentration of the solute in the solution. This relationship is given by ***Raoult's law*** (after the French chemist Francois Raoult), which states that *the partial pressure of a solvent over a solution, P_1, is given by the vapor pressure of the pure solvent, P_1°, times the mole fraction of the solvent in the solution, X_1:*

$$P_1 = X_1 P_1^\circ \tag{12.2}$$

In a solution containing only one solute, $X_1 = 1 - X_2$, where X_2 is the mole fraction of the solute (see Section 5.4). Equation (12.2) can therefore be rewritten as

$$P_1 = (1 - X_2)P_1^\circ$$

$$P_1^\circ - P_1 = \Delta P = X_2 P_1^\circ \tag{12.3}$$

We see that the *decrease* in vapor pressure, ΔP, is directly proportional to the concentration (measured in mole fraction) of the solute present.

EXAMPLE 12.8
Calculating Molality from Vapor-Pressure Lowering

At 25°C, the vapor pressure of pure water is 23.76 mmHg and that of an aqueous urea solution is 22.98 mmHg. Estimate the molality of the solution.

Answer: From Equation (12.3) we write

$$\Delta P = (23.76 - 22.98)\ \text{mmHg} = X_2(23.76\ \text{mmHg})$$

$$X_2 = 0.033$$

By definition

$$X_2 = \frac{n_2}{n_1 + n_2}$$

where n_1 and n_2 are the numbers of moles of solvent and solute, respectively. Since the solution is dilute, we can assume that n_1 is much larger than n_2, and we can write

$$X_2 = \frac{n_2}{n_1 + n_2} \simeq \frac{n_2}{n_1} \qquad (n_1 \gg n_2)$$

$$n_2 = n_1 X_2$$

The number of moles of water in 1 kg of water is

$$1000\ \text{g H}_2\text{O} \times \frac{1\ \text{mol H}_2\text{O}}{18.02\ \text{g H}_2\text{O}} = 55.49\ \text{mol H}_2\text{O}$$

and the number of moles of urea present in 1 kg of water is

$$n_2 = n_1 X_2 = (55.49\ \text{mol})(0.033)$$

$$= 1.8\ \text{mol}$$

Thus the concentration of the urea solution is 1.8 *m*.

PRACTICE EXERCISE

The vapor pressure of a glucose ($C_6H_{12}O_6$) solution is 17.01 mmHg at 20°C, while that of pure water is 17.25 mmHg at the same temperature. Calculate the molality of the solution.

Why is the vapor pressure of a solution less than that of its pure solvent? As was mentioned in Section 12.2, one driving force in physical and chemical processes is the increase in disorder—the greater the disorder created, the more favorable the process. Vaporization increases the disorder of a system because molecules in a vapor are not as closely packed as and therefore have less order than those in a liquid. Because a solution is more disordered than a pure solvent, the difference in disorder between a solution and a vapor is less than that between a pure solvent and a vapor. Thus solvent molecules have less of a tendency to leave a solution than to leave the pure solvent to become vapor, and the vapor pressure of a solution is less than that of the solvent.

If both components of a solution are ***volatile*** (that is, *have measurable vapor pressure*), the vapor pressure of the solution is the sum of the individual partial pressures. Raoult's law holds equally well in this case:

$$P_A = X_A P_A^\circ$$

$$P_B = X_B P_B^\circ$$

where P_A and P_B are the partial pressures over the solution for components A and B; P_A° and P_B° are the vapor pressures of the pure substances; and X_A and X_B are their mole fractions. The total pressure is given by Dalton's law of partial pressure [Equation (5.11)]:

$$P_T = P_A + P_B$$

Benzene and toluene have similar structures and therefore similar intermolecular forces:

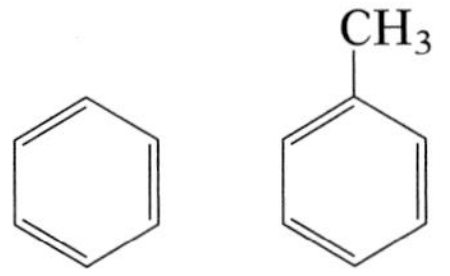

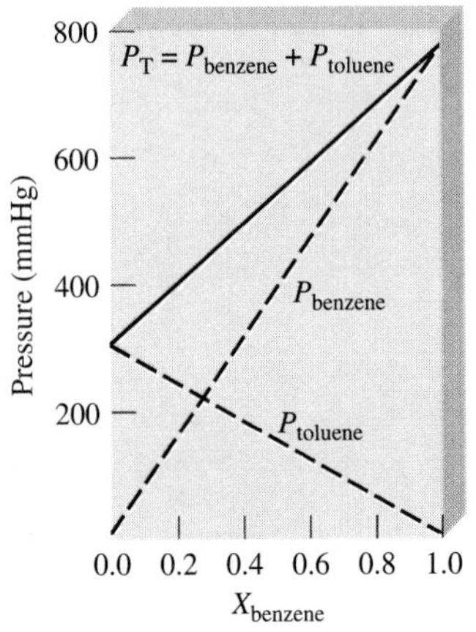

FIGURE 12.7

The dependence of the partial pressures of benzene and toluene on their mole fractions in a benzene-toluene solution ($X_{toluene} = 1 - X_{benzene}$) at 80°C. This solution is said to be ideal because the vapor pressures obey Raoult's law.

In a solution of benzene and toluene, the vapor pressure of each component obeys Raoult's law. Figure 12.7 shows the dependence of the total vapor pressure (P_T) in a benzene-toluene solution on the composition of the solution. Note that we need only express the composition of the solution in terms of the mole fraction of one component. For every value of $X_{benzene}$, the mole fraction of toluene, $X_{toluene}$, is given by $(1 - X_{benzene})$. The benzene-toluene solution is one of the few examples of an ***ideal solution,*** which is *any solution that obeys Raoult's law.* One characteristic of an ideal solution is that the heat of solution, ΔH_{soln}, is always zero.

Boiling-Point Elevation

Because the presence of a *nonvolatile* solute lowers the vapor pressure of a solution, it must also affect the boiling point of the solution. The boiling point of a

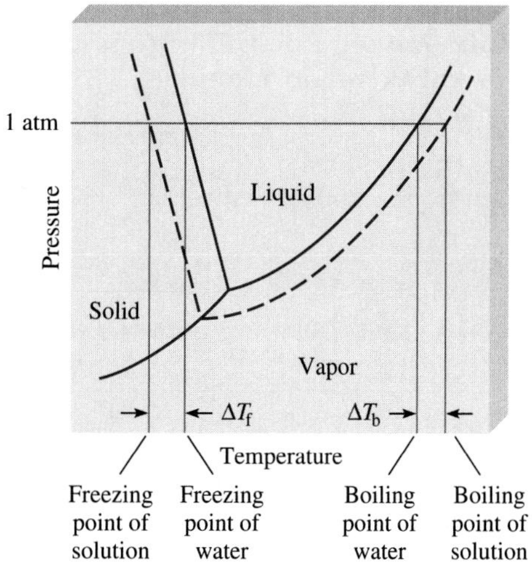

FIGURE 12.8

Phase diagram illustrating the boiling-point elevation and freezing-point depression of aqueous solutions. The dashed curves pertain to the solution, and the solid curves to the pure solvent. As you can see, the boiling point of the solution is higher than that of water, and the freezing point of the solution is lower than that of water.

solution is the temperature at which its vapor pressure equals the external atmospheric pressure (see Section 11.6). Figure 12.8 shows the phase diagram of water and the changes that occur in an aqueous solution. Because at any temperature the vapor pressure of the solution is lower than that of the pure solvent, the liquid-vapor curve for the solution lies below that for the pure solvent. Consequently, the solution curve dotted line intersects the horizontal line that marks $P = 1$ atm at a *higher* temperature than the normal boiling point of the pure solvent. This graphical analysis shows that the boiling point of the solution is higher than that of water. The *boiling-point elevation,* ΔT_b, is defined as

$$\Delta T_b = T_b - T_b^\circ$$

where T_b is the boiling point of the solution and T_b° the boiling point of the pure solvent. Because ΔT_b is proportional to the vapor-pressure lowering, it is also proportional to the concentration (molality) of the solution. That is,

$$\Delta T_b \propto m$$

$$\Delta T_b = K_b m \qquad (12.4)$$

where m is the molality of the solution and K_b is the *molal boiling-point elevation constant.* The units of K_b are °C/m.

It is important to understand the choice of concentration unit here. We are dealing with a system (the solution) whose temperature is *not* kept constant, so we cannot express the concentration units in molarity because molarity changes with temperature.

Table 12.2 lists the value of K_b for several common solvents. Using the boiling-point elevation constant for water and Equation (12.4), you can see that if the molality m of an aqueous solution is 1.00, the boiling point will be 100.52°C.

De-icing of airplanes is based on freezing-point depression.

FREEZING-POINT DEPRESSION

A nonscientist may remain forever unaware of the boiling-point elevation phenomenon, but a careful observer living in a cold climate is familiar with freezing-point depression. Ice on frozen roads and sidewalks melts when sprinkled with salts such as NaCl or $CaCl_2$. This method of thawing succeeds because it depresses the freezing point of water.

TABLE 12.2
Molal Boiling-Point Elevation and Freezing-Point Depression Constants of Several Common Liquids

Solvent	Normal freezing point (°C)*	K_f (°C/*m*)	Normal boiling point (°C)*	K_b (°C/*m*)
Water	0	1.86	100	0.52
Benzene	5.5	5.12	80.1	2.53
Ethanol	−117.3	1.99	78.4	1.22
Acetic acid	16.6	3.90	117.9	2.93
Cyclohexane	6.6	20.0	80.7	2.79

*Measured at 1 atm.

Figure 12.8 shows that lowering the vapor pressure of the solution shifts the solid-liquid curve to the left. Consequently, this line intersects the horizontal line at a temperature *lower* than the freezing point of water. The *freezing-point depression*, ΔT_f, is defined as

$$\Delta T_f = T_f^\circ - T_f$$

where T_f° is the freezing point of the pure solvent and T_f the freezing point of the solution. Again, ΔT_f is proportional to the concentration of the solution:

$$\Delta T_f \propto m$$

$$\Delta T_f = K_f m \qquad (12.5)$$

where m is the concentration of the solute in molality units, and K_f is the *molal freezing-point depression constant* (see Table 12.2). Like K_b, K_f has the units °C/m.

Note that the solute must be nonvolatile in the case of boiling-point elevation, but no such restriction applies to freezing-point depression. For example, methanol (CH_3OH), a fairly volatile liquid that boils at only 65°C, has sometimes been used as an antifreeze in automobile radiators.

In cold climates, antifreeze must be used in car radiators in winter.

EXAMPLE 12.9
Calculating Freezing-Point Depression

Ethylene glycol (EG), $CH_2(OH)CH_2(OH)$, is a common automobile antifreeze. It is water soluble and fairly nonvolatile (b.p. 197°C). Calculate the freezing point of a solution containing 651 g of this substance in 2505 g of water. Would you keep this substance in your car radiator during the summer? The molar mass of ethylene glycol is 62.01 g.

Answer: The number of moles of ethylene glycol in 1000 g, or 1 kg, of water is

$$651 \text{ g EG} \times \frac{1 \text{ mol EG}}{62.01 \text{ g EG}} \times \frac{1 \text{ kg}}{2.505 \text{ kg}} = 4.19 \text{ mol EG}$$

Thus the molality of the solution is 4.19 m. From Equation (12.5) and Table 12.2 we have

$$\Delta T_f = (1.86°C/m)(4.19\ m)$$
$$= 7.79°C$$

Since pure water freezes at 0°C, the solution will freeze at −7.79°C. We can calculate boiling-point elevation in the same way as follows:

$$\Delta T_b = (0.52°C/m)(4.19\ m)$$
$$= 2.2°C$$

Because the solution will boil at 102.2°C, it would be preferable to leave the antifreeze in your car radiator in summer to prevent the solution from boiling.

PRACTICE EXERCISE

Calculate the boiling point and freezing point of a solution containing 478 g of ethylene glycol in 3202 g of water.

Osmotic Pressure

Many chemical and biological processes depend on the selective passage of solvent molecules through a porous membrane from a dilute solution to a more concentrated one. Figure 12.9 illustrates this phenomenon. The left compartment of the apparatus contains pure solvent; the right compartment contains a solution. The two compartments are separated by a ***semipermeable membrane,*** which *allows solvent molecules to pass through but blocks the passage of solute molecules.* At the start, the water levels in the two tubes are equal [see Figure 12.9(a)]. After some time, the level in the right tube begins to rise; this continues until equilibrium is reached. *The net movement of solvent molecules through a semipermeable membrane from a pure solvent or from a dilute solution to a more concentrated solution* is called ***osmosis.*** The ***osmotic pressure*** (π) of a solution is *the pressure required to stop osmosis.* As shown in Figure 12.9(b), this pressure can be measured directly from the difference in the final fluid levels.

What causes water to move spontaneously from left to right in this case? Compare the vapor pressure of pure water and that of water from a solution (Figure

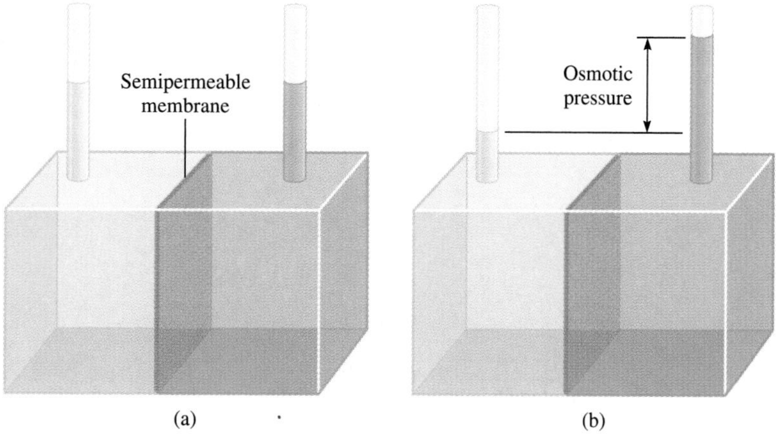

FIGURE 12.9

Osmotic pressure. (a) The levels of the pure solvent (left) and of the solution (right) are equal at the start. (b) During osmosis, the level on the solution side rises as a result of the net flow of solvent from left to right. The osmotic pressure is equal to the hydrostatic pressure exerted by the column of fluid in the right tube at equilibrium.

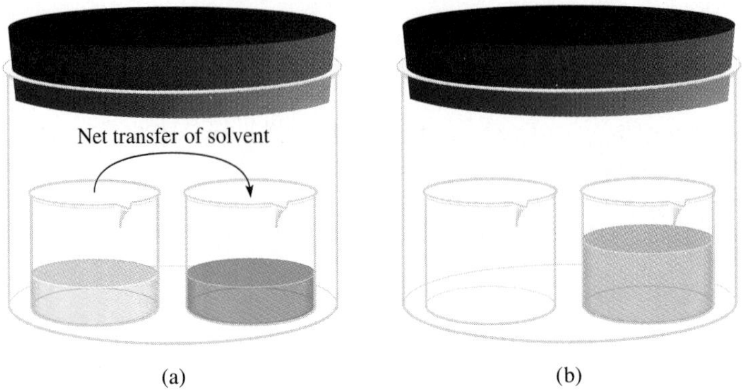

FIGURE 12.10

(a) Unequal vapor pressures inside the container leads to a net transfer of water from the left beaker (which contains pure water) to the right one (which contains a solution). (b) At equilibrium, all the water in the left beaker has been transferred to the right beaker. The driving force for this solvent transfer is analogous to the osmosis phenomenon shown in Figure 12.9.

12.10). Because the vapor pressure of pure water is higher, there is a net transfer of water from the left beaker to the right one. Given enough time, the transfer will continue to completion. A similar force causes water to move into the solution during osmosis.

Although osmosis is a common and well-studied phenomenon, relatively little is known about how the semipermeable membrane stops some molecules yet allows others to pass. In some cases, it is simply a matter of size. A semipermeable membrane may have pores small enough to let only the solvent molecules through. In other cases, a different mechanism may be responsible for the membrane's selectivity—for example, the solvent's greater "solubility" in the membrane.

The osmotic pressure of a solution is given by

$$\pi = MRT \tag{12.6}$$

where M is the molarity of solution, R the gas constant (0.0821 L · atm/K · mol), and T the absolute temperature. The osmotic pressure, π, is expressed in atmospheres. Since osmotic pressure measurements are carried out at constant temperature, we express the concentration here in terms of the more convenient units of molarity rather than molality.

Like boiling-point elevation and freezing-point depression, osmotic pressure is directly proportional to the concentration of solution. This is what we would expect, bearing in mind that all colligative properties depend only on the number of solute particles in solution. If two solutions are of equal concentration and, hence, of the same osmotic pressure, they are said to be *isotonic*. If two solutions are of unequal osmotic pressures, the more concentrated solution is said to be *hypertonic* and the more dilute solution is described as *hypotonic* (Figure 12.11).

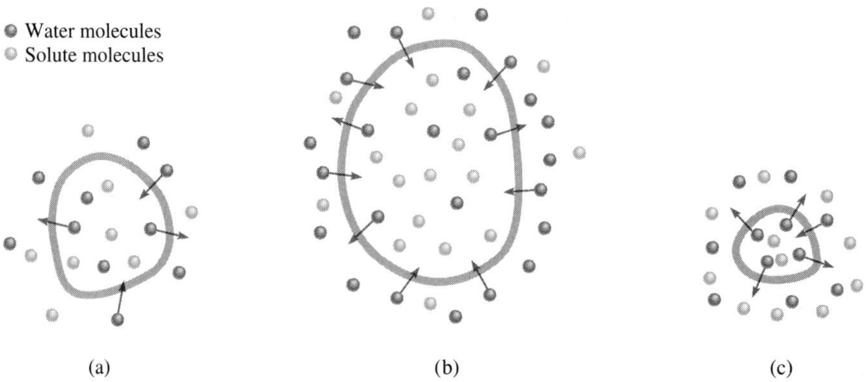

FIGURE 12.11

A cell in (a) an isotonic solution, (b) a hypotonic solution, and (c) a hypertonic solution. The cell remains unchanged in (a), swells in (b), and shrinks in (c).

The osmotic pressure phenomenon manifests itself in many interesting applications. To study the contents of red blood cells, which are protected from the external environment by a semipermeable membrane, biochemists use a technique called hemolysis. The red blood cells are placed in a hypotonic solution. Because the hypotonic solution is less concentrated than the interior of the cell, water moves into the cells, as shown in Figure 12.11(b). The cells swell and eventually burst, releasing hemoglobin and other molecules.

Home preserving of jam and jelly provides another example of the use of osmotic pressure. A large quantity of sugar is actually essential to the preservation process because the sugar helps to kill bacteria that may cause botulism. As Figure 12.11(c) shows, when a bacterial cell is in a hypertonic (high-concentration) sugar solution, the intracellular water tends to move out of the bacterial cell to the more concentrated solution by osmosis. This process, known as *crenation,* causes the cell to shrink and, eventually, to cease functioning. The natural acidity of fruits also inhibits bacteria growth.

Osmotic pressure also is the major mechanism for transporting water upward in plants. Because leaves constantly lose water to the air, in a process called *transpiration,* the solute concentrations in leaf fluids increase. Water is pushed up through the trunk, branches, and stems of trees by osmotic pressure. Up to 10 to 15 atm pressure is necessary to transport water to the leaves at the tops of California's redwoods, which reach about 120 m in height. (The capillary action discussed in Section 11.3 is responsible for the rise of water only up to a few centimeters.)

California redwoods.

Using Colligative Properties to Determine Molar Mass

The colligative properties of nonelectrolyte solutions provide a means of determining the molar mass of a solute. Theoretically, any of the four colligative properties are suitable for this purpose. In practice, however, only freezing-point depression and osmotic pressure are used because they show the most pronounced changes.

EXAMPLE 12.10
Finding Molar Mass from Freezing-Point Depression

A 7.85-g sample of a compound with the empirical formula C_5H_4 is dissolved in 301 g of benzene. The freezing point of the solution is 1.05°C below that of pure benzene. What are the molar mass and molecular formula of this compound?

Answer: From Equation (12.5) and Table 12.2 we can write

$$\text{molality} = \frac{\Delta T_f}{K_f} = \frac{1.05°\text{C}}{5.12°\text{C}/m} = 0.205\ m$$

The number of moles of solute in 301 g, or 0.301 kg, of solvent is given by

$$\frac{0.205\ \text{mol}}{1\ \text{kg solvent}} \times 0.301\ \text{kg solvent} = 0.0617\ \text{mol}$$

Finally, we calculate the molar mass of the solute as follows:

$$\frac{7.85\ \text{g}}{0.0617\ \text{mol}} = 127\ \text{g/mol}$$

Since the formula mass of C_5H_4 is 64 g and the molar mass is 127 g, the molecular formula of the compound is $C_{10}H_8$ (naphthalene).

PRACTICE EXERCISE

A solution of 0.85 g of the organic compound mesitol in 100.0 g of benzene has a freezing point of 5.16°C. What are the molality of the solution and the molar mass of mesitol?

EXAMPLE 12.11

Finding Molar Mass from Osmotic Pressure

A solution is prepared by dissolving 35.0 g of hemoglobin (Hb) in enough water to make up 1 L in volume. If the osmotic pressure of the solution is found to be 10.0 mmHg at 25°C, calculate the molar mass of hemoglobin.

Answer: First we calculate the concentration of the solution:

$$\pi = MRT$$

$$M = \frac{\pi}{RT}$$

$$= \frac{10.0 \text{ mmHg} \times \dfrac{1 \text{ atm}}{760 \text{ mmHg}}}{(0.0821 \text{ L} \cdot \text{atm/K} \cdot \text{mol})(298 \text{ K})}$$

$$= 5.38 \times 10^{-4}\, M$$

The volume of the solution is 1 L, so it must contain 5.38×10^{-4} mol of Hb. We use this quantity to calculate the molar mass:

$$\text{moles of Hb} = \frac{\text{mass of Hb}}{\text{molar mass of Hb}}$$

$$\text{molar mass of Hb} = \frac{\text{mass of Hb}}{\text{moles of Hb}}$$

$$= \frac{35.0 \text{ g}}{5.38 \times 10^{-4} \text{ mol}}$$

$$= 6.51 \times 10^{4} \text{ g/mol}$$

PRACTICE EXERCISE

A 202-mL benzene solution containing 2.47 g of an organic polymer has an osmotic pressure of 8.63 mmHg at 21°C. Calculate the molar mass of the polymer.

A pressure such as 10.0 mmHg in Example 12.11 can be measured easily and accurately. For this reason, osmotic pressure measurements are very useful for determining the molar masses of large molecules, such as proteins. To see how much more practical the osmotic pressure technique is than freezing-point depression would be, let us estimate the change in freezing point of the same hemoglobin solution. If a solution is quite dilute, we can assume that molarity is roughly equal to molality. (Molarity would be equal to molality if the density of the solution were 1 g/mL.) Hence, from Equation (12.5) we write

$$\Delta T_f = (1.86°C/m)(5.38 \times 10^{-4}\ m)$$
$$= 1.00 \times 10^{-3}°C$$

The thousandth-of-a-degree freezing-point depression is too small a temperature change to measure accurately. For this reason, the freezing-point depression technique is more suitable for determining the molar mass of smaller and more soluble molecules, those having molar masses of 500 g or less, since the freezing-point depressions of their solutions are much greater.

Colligative Properties of Electrolytes

The colligative properties of electrolytes require a slightly different approach than the one used for the colligative properties of nonelectrolytes. The reason is that electrolytes dissociate into ions in solution, and so one unit of an electrolyte compound separates into two or more particles when it dissolves. (Remember, it is the number of solute particles that determines the colligative properties of a solution.) For example, each unit of NaCl dissociates into two ions—Na^+ and Cl^-. Thus the colligative properties of a 0.1 *m* NaCl solution should be twice as great as those of a 0.1 *m* solution containing a nonelectrolyte, such as sucrose. Similarly, we would expect a 0.1 *m* $CaCl_2$ solution to depress the freezing point by three times as much as a 0.1 *m* sucrose solution. To account for this effect we must modify the equations for colligative properties as follows:

$$\Delta T_b = iK_b m$$
$$\Delta T_f = iK_f m$$
$$\pi = iMRT$$

The variable i is the *van't Hoff factor,* which is defined as

$$i = \frac{\text{actual number of particles in soln after dissociation}}{\text{number of formula units initially dissolved in soln}}$$

Thus i should be 1 for all nonelectrolytes. For strong electrolytes such as NaCl and KNO_3, i should be 2, and for strong electrolytes such as Na_2SO_4 and $MgCl_2$, i should be 3.

In reality, the colligative properties of electrolyte solutions are usually smaller than anticipated because at higher concentrations, electrostatic forces come into play, drawing cations and anions together. *A cation and an anion held together by electrostatic forces* are called an ***ion pair.*** The formation of an ion pair reduces the number of particles in solution by one, causing a reduction in the colligative properties (Figure 12.12). Table 12.3 shows the experimentally measured values of i and

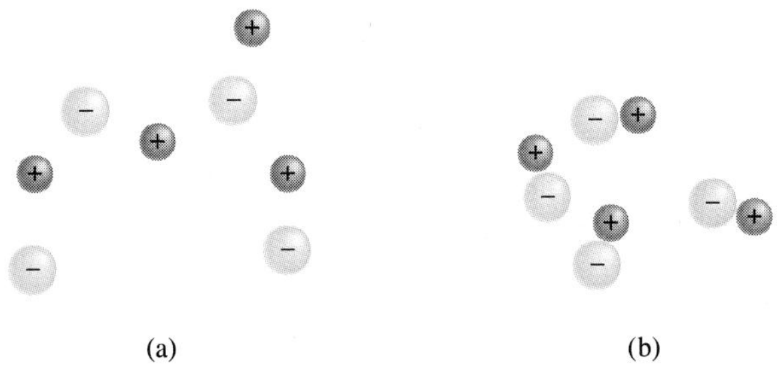

FIGURE 12.12

(a) Free ions and (b) ion pairs in solution. Such an ion pair bears no net charge and therefore cannot conduct electricity in solution.

TABLE 12.3
The van't Hoff Factor of 0.0500 *M* Electrolyte Solutions at 25°C

Electrolyte	*i* (measured)	*i* (calculated)
Sucrose*	1.0	1.0
HCl	1.9	2.0
NaCl	1.9	2.0
$MgSO_4$	1.3	2.0
$MgCl_2$	2.7	3.0
$FeCl_3$	3.4	4.0

*Sucrose is a nonelectrolyte. It is listed here for comparison only.

those calculated assuming complete dissociation. As you can see, the agreement is close but not perfect, indicating that the extent of ion-pair formation in these solutions is appreciable.

SUMMARY

Solutions are homogeneous mixtures of two or more substances, which may be solids, liquids, or gases. The ease of dissolution of a solute in a solvent is governed by intermolecular forces. Energy and the increase in disorder that result when molecules of the solute and solvent mix to form a solution are the forces driving the solution process.

The concentration of a solution can be expressed as percent by mass, mole fraction, molarity, and molality. The circumstances dictate which units are appropriate.

A rise in temperature usually increases the solubility of solid and liquid substances and decreases the solubility of gases. According to Henry's law, the solubility of a gas in a liquid is directly proportional to the partial pressure of the gas over the solution.

Raoult's law states that the partial pressure of a substance A over a solution is related to the mole fraction (X_A) of A and to the vapor pressure (P_A°) of pure A as follows: $P_A = X_A P_A^\circ$. An ideal solution obeys Raoult's law over the entire range of concentration. In practice, very few solutions exhibit ideal behavior.

Vapor-pressure lowering, boiling-point elevation, freezing-point depression, and osmotic pressure are colligative properties of solutions; that is, they are properties that depend only on the number of solute particles that are present and not on their nature. In electrolyte solutions, the interaction between ions leads to the formation of ion pairs. The van't Hoff factor provides a measure of the extent of dissociation of electrolytes in solution.

KEY WORDS

Percent by mass, p. 343
Raoult's law, p. 351
Saturated solution, p. 340
Semipermeable membrane, p. 355
Solvation, p. 342
Supersaturated solution, p. 340
Thermal pollution, p. 348
Unsaturated solution, p. 340
Volatile, p. 352

QUESTIONS AND PROBLEMS

THE SOLUTION PROCESS

Review Questions

12.1 Briefly describe the solution process at the molecular level. Use the dissolution of a solid in a liquid as an example.

12.2 Basing your answer on intermolecular force considerations, explain what "like dissolves like" means.

12.3 What is solvation? What are the factors that influence the extent to which solvation occurs? Give two examples of solvation, including one that involves ion-dipole interaction and another in which dispersion forces come into play.

12.4 As you know, some solution processes are endothermic and others are exothermic. Provide a molecular interpretation for the difference.

12.5 Explain why the solution process invariably leads to an increase in disorder.

12.6 Describe the factors that affect the solubility of a solid in a liquid. What does it mean to say that two liquids are miscible?

Problems

12.7 Why is naphthalene ($C_{10}H_8$) more soluble than CsF in benzene?

12.8 Explain why ethanol (C_2H_5OH) is not soluble in cyclohexane (C_6H_{12}).

12.9 Arrange the following compounds in order of increasing solubility in water: O_2, LiCl, Br_2, methanol (CH_3OH).

12.10 Explain the variations in solubility in water of the alcohols listed below:

Compound	*Solubility in water g/100 g, 20°C*
CH_3OH	∞
CH_3CH_2OH	∞
$CH_3CH_2CH_2OH$	∞
$CH_3CH_2CH_2CH_2OH$	9
$CH_3CH_2CH_2CH_2CH_2OH$	2.7

Note: ∞ means the alcohol and water are completely miscible in all proportions.

CONCENTRATION UNITS

Review Questions

12.11 Define the following concentration terms and give their units: percent by mass, molarity, molality. Compare their advantages and disadvantages.

12.12 Outline the steps required for conversion among molarity, molality, and percent by mass.

Problems

12.13 Calculate the percent by mass of the solute in each of the following aqueous solutions: (a) 5.50 g of NaBr in 78.2 g of solution, (b) 31.0 g of KCl in 152 g of water, (c) 4.5 g of toluene in 29 g of benzene.

12.14 Calculate the amount of water (in grams) that must be added to (a) 5.00 g of urea [$(NH_2)_2CO$] in the preparation of a 16.2 percent by mass solution and (b) 26.2 g of $MgCl_2$ in the preparation of a 1.5 percent by mass solution.

12.15 Calculate the molality of each of the following solutions: (a) 14.3 g of sucrose ($C_{12}H_{22}O_{11}$) in 676 g of water, (b) 7.20 moles of ethylene glycol ($C_2H_6O_2$) in 3546 g of water.

12.16 Calculate the molality of each of the following aqueous solutions: (a) 2.50 *M* NaCl solution (density of solution = 1.08 g/mL), (b) 48.2 percent by mass KBr solution.

12.17 Calculate the molalities of the following aqueous solutions: (a) 1.22 *M* sugar ($C_{12}H_{22}O_{11}$) solution (density of solution = 1.12 g/mL), (b) 0.87 *M* NaOH solution (density of solution = 1.04 g/mL), (c) 5.24 *M* $NaHCO_3$ solution (density of solution = 1.19 g/mL).

12.18 For dilute solutions in which the density of the solution is roughly equal to that of the pure solvent, the molarity of the solution is equal to its molality. Show that this statement is correct for a 0.010 *M* aqueous urea [$(NH_2)_2CO$] solution.

12.19 The alcohol content of hard liquor is normally given in terms of the "proof," which is defined as twice the percentage by volume of ethanol

(C_2H_5OH) present. Calculate the number of grams of alcohol present in 1.00 L of 75 proof gin. The density of ethanol is 0.798 g/mL.

12.20 The concentrated sulfuric acid we use in the laboratory is 98.0 percent H_2SO_4 by mass. Calculate the molality and molarity of the acid solution. The density of the solution is 1.83 g/mL.

12.21 Calculate the molarity and the molality of NH_3 for a solution of 30.0 g of NH_3 in 70.0 g of water. The density of the solution is 0.982 g/mL.

12.22 The density of an aqueous solution containing 10.0 percent of ethanol (C_2H_5OH) by mass is 0.984 g/mL. (a) Calculate the molality of this solution. (b) Calculate its molarity. (c) What volume of the solution would contain 0.125 mole of ethanol?

EFFECT OF TEMPERATURE AND PRESSURE ON SOLUBILITY

Review Questions

12.23 How do the solubilities of most ionic compounds in water change with temperature?

12.24 What is the effect of pressure on the solubility of a liquid in liquid and of a solid in liquid?

Problems

12.25 A 3.20-g sample of a salt dissolves in 9.10 g of water to give a saturated solution at 25°C. What is the solubility (in g salt/100 g of H_2O) of the salt?

12.26 The solubility of KNO_3 is 155 g per 100 g of water at 75°C and 38.0 g at 25°C. What mass (in grams) of KNO_3 will crystallize out of solution if exactly 100 g of its saturated solution at 75°C are cooled at 25°C?

GAS SOLUBILITY

Review Questions

12.27 Discuss the factors that influence the solubility of a gas in a liquid. Explain why the solubility of a gas in a liquid usually decreases with increasing temperature.

12.28 What is thermal pollution? Why is it harmful to aquatic life?

12.29 What is Henry's law? Define each term in the equation, and give its units. Explain the law in terms of the kinetic molecular theory of gases.

12.30 Give two exceptions to Henry's law.

12.31 A student is observing two beakers of water. One beaker is heated to 30°C, and the other is heated to 100°C. In each case, bubbles form in the water. Are these bubbles of the same origin? Explain.

12.32 A man bought a goldfish in a pet shop. Upon returning home, he put the goldfish in a bowl of recently boiled water that had been cooled quickly. A few minutes later the fish was dead. Explain what happened to the fish.

Problems

12.33 A beaker of water is initially saturated with dissolved air. Explain what happens when He gas at 1 atm is bubbled through the solution for a long time.

12.34 A miner working 260 m below sea level opened a carbonated soft drink during a lunch break. To his surprise, the soft drink tasted rather "flat." Shortly afterward, the miner took an elevator to the surface. During the trip up, he could not stop belching. Why?

12.35 The solubility of CO_2 in water at 25°C and 1 atm is 0.034 mol/L. What is its solubility under atmospheric conditions? (The partial pressure of CO_2 in air is 0.0003 atm.) Assume that CO_2 obeys Henry's law.

12.36 The solubility of N_2 in blood at 37°C and at a partial pressure of 0.80 atm is 5.6×10^{-4} mol/L. A deep-sea diver breathes compressed air with the partial pressure of N_2 equal to 4.0 atm. Assume that the total volume of blood in the body is 5.0 L. Calculate the amount of N_2 gas released (in liters) when the diver returns to the surface of the water, where the partial pressure of N_2 is 0.80 atm.

COLLIGATIVE PROPERTIES OF NONELECTROLYTES

Review Questions

12.37 What are colligative properties? What is the meaning of the word "colligative" in this context?

12.38 Give two examples of a volatile liquid and two examples of a nonvolatile liquid.

12.39 Define Raoult's law. Define each term in the equation representing Raoult's law, and give its units. What is an ideal solution?

12.40 Define boiling-point elevation and freezing-point depression. Write the equations relating boiling-point elevation and freezing-point depression to the concentration of the solution. Define all the terms, and give their units.

12.41 How is the lowering in vapor pressure related to a rise in the boiling point of a solution?

12.42 Use a phase diagram to show the difference in freezing point and boiling point between an aqueous urea solution and pure water.

12.43 What is osmosis? What is a semipermeable membrane?

12.44 Write the equation relating osmotic pressure to the concentration of a solution. Define all the terms and give their units.

12.45 What does it mean when we say that the osmotic pressure of a sample of seawater is 25 atm at a certain temperature?

12.46 Explain why molality is used for boiling-point elevation and freezing-point depression calculations and molarity is used in osmotic pressure calculations.

12.47 Describe how you would use the freezing-point depression and osmotic pressure measurements to determine the molar mass of a compound. Why is the boiling-point elevation phenomenon normally not used for this purpose?

12.48 Explain why it is essential that fluids used in intravenous injections have approximately the same osmotic pressure as blood.

Problems

12.49 A solution is prepared by dissolving 396 g of sucrose ($C_{12}H_{22}O_{11}$) in 624 g of water. What is the vapor pressure of this solution at 30°C? (The vapor pressure of water is 31.8 mmHg at 30°C.)

12.50 How many grams of sucrose ($C_{12}H_{22}O_{11}$) must be added to 552 g of water to give a solution with a vapor pressure 2.0 mmHg less than that of pure water at 20°C? (The vapor pressure of water at 20°C is 17.5 mmHg.)

12.51 The vapor pressure of benzene is 100.0 mmHg at 26.1°C. Calculate the vapor pressure of a solution containing 24.6 g of camphor ($C_{10}H_{16}O$) dissolved in 98.5 g of benzene. (Camphor is a low-volatility solid.)

12.52 The vapor pressures of ethanol (C_2H_5OH) and 1-propanol (C_3H_7OH) at 35°C are 100 mmHg and 37.6 mmHg, respectively. Assume ideal behavior and calculate the partial pressures of ethanol and 1-propanol at 35°C over a solution of ethanol in 1-propanol, in which the mole fraction of ethanol is 0.300.

12.53 The vapor pressure of ethanol (C_2H_5OH) at 20°C is 44 mmHg, and the vapor pressure of methanol (CH_3OH) at the same temperature is 94 mmHg. A mixture of 30.0 g of methanol and 45.0 g of ethanol is prepared (and may be assumed to behave as an ideal solution). (a) Calculate the vapor pressure of methanol and ethanol above this solution at 20°C. (b) Calculate the mole fraction of methanol and ethanol in the vapor above this solution at 20°C.

12.54 How many grams of urea [$(NH_2)_2CO$] must be added to 450 g of water to give a solution with a vapor pressure 2.50 mmHg less than that of pure water at 30°C? (The vapor pressure of water at 30°C is 31.8 mmHg.)

12.55 What are the boiling point and freezing point of a 2.47 *m* solution of naphthalene in benzene? (The boiling point and freezing point of benzene are 80.1°C and 5.5°C, respectively.)

12.56 An aqueous solution contains the amino acid glycine (NH_2CH_2COOH). Assuming no ionization of the acid, calculate the molality of the solution if it freezes at −1.1°C.

12.57 Pheromones are compounds secreted by the females of many insect species to attract males. One of these compounds contains 80.78% C, 13.56% H, and 5.66% O. A solution of 1.00 g of this pheromone in 8.50 g of benzene freezes at 3.37°C. What are the molecular formula and molar mass of the compound? (The normal freezing point of pure benzene is 5.50°C.)

12.58 The elemental analysis of an organic solid extracted from gum arabic showed that it contained 40.0% C, 6.7% H, and 53.3% O. A solution of 0.650 g of the solid in 27.8 g of the solvent diphenyl gave a freezing-point depression of 1.56°C. Calculate the molar mass and molecular formula of the solid. (K_f for diphenyl is 8.00°C/m.)

12.59 How many liters of the antifreeze ethylene glycol [$CH_2(OH)CH_2(OH)$] would you add to a car radiator containing 6.50 L of water if the coldest winter temperature in your area is −20°C? Calculate the boiling point of this water–ethylene glycol mixture. The density of ethylene glycol is 1.11 g/mL.

12.60 A solution is prepared by condensing 4.00 L of a gas, measured at 27°C and 748 mmHg pressure, into 58.0 g of benzene. Calculate the freezing point of this solution.

12.61 The molar mass of benzoic acid (C_6H_5COOH) determined by measuring the freezing-point depression in benzene is twice that expected for the molecular formula, $C_7H_6O_2$. Explain this apparent anomaly.

12.62 A solution of 2.50 g of a compound of empirical formula C_6H_5P in 25.0 g of benzene is observed to freeze at 4.3°C. Calculate the molar mass of the solute and its molecular formula.

12.63 What is the osmotic pressure (in atmospheres) of a 12.36 *M* aqueous urea solution at 22.0°C?

12.64 A solution containing 0.8330 g of a protein of unknown structure in 170.0 mL of aqueous solution was found to have an osmotic pressure of 5.20 mmHg at 25°C. Determine the molar mass of the protein.

12.65 A quantity of 7.480 g of an organic compound is dissolved in water to make 300.0 mL of solution. The solution has an osmotic pressure of 1.43 atm at 27°C. The analysis of this compound shows it to contain 41.8% C, 4.7% H, 37.3% O, and 16.3% N. Calculate the molecular formula of the organic compound.

12.66 A solution of 6.85 g of a carbohydrate in 100.0 g of water has a density of 1.024 g/mL and an osmotic pressure of 4.61 atm at 20.0°C. Calculate the molar mass of the carbohydrate.

COLLIGATIVE PROPERTIES OF ELECTROLYTE SOLUTIONS

Review Questions

12.67 Why is the discussion of the colligative properties of electrolyte solutions more involved than that of nonelectrolyte solutions?

12.68 Define ion pairs. What effect does ion-pair formation have on the colligative properties of a solution? How does the ease of ion-pair formation depend on (a) charges on the ions, (b) size of the ions, (c) nature of the solvent (polar versus nonpolar), (d) concentration?

12.69 In each case, indicate which of the following pairs of compounds is more likely to form ion pairs in water: (a) NaCl or Na_2SO_4, (b) $MgCl_2$ or $MgSO_4$, (c) LiBr or KBr.

12.70 Define the van't Hoff factor. What information does this quantity provide?

Problems

12.71 Which of the following two aqueous solutions has (a) the higher boiling point, (b) the higher freezing point, and (c) the lower vapor pressure: 0.35 *m* $CaCl_2$ or 0.90 *m* urea? State your reasons.

12.72 Consider two aqueous solutions, one of sucrose ($C_{12}H_{22}O_{11}$) and the other of nitric acid (HNO_3), both of which freeze at −1.5°C. What other properties do these solutions have in common?

12.73 Arrange the following solutions in order of decreasing freezing point: 0.10 *m* Na_3PO_4, 0.35 *m* NaCl, 0.20 *m* $MgCl_2$, 0.15 *m* $C_6H_{12}O_6$, 0.15 *m* CH_3COOH.

12.74 Arrange the following aqueous solutions in order of decreasing freezing point and explain your reasons: 0.50 *m* HCl, 0.50 *m* glucose, 0.50 *m* acetic acid.

12.75 What are the normal freezing points and boiling points of the following solutions? (a) 21.2 g NaCl in 135 mL of water, (b) 15.4 g of urea in 66.7 mL of water.

12.76 At 25°C the vapor pressure of pure water is 23.76 mmHg and that of seawater is 22.98 mmHg. Assuming that seawater contains only NaCl, estimate its concentration in molality units.

12.77 Both NaCl and $CaCl_2$ are used to melt ice on roads in winter. What advantages do these substances have over sucrose or urea in lowering the freezing point of water?

12.78 A 0.86 percent by mass solution of NaCl is called "physiological saline" because its osmotic pressure is equal to that of the solution in blood cells. Calculate the osmotic pressure of this solution at normal body temperature (37°C). Note that the density of the saline solution is 1.005 g/mL.

12.79 The osmotic pressure of 0.010 *M* solutions of $CaCl_2$ and urea at 25°C are 0.605 atm and 0.245 atm, respectively. Calculate the van't Hoff factor for the $CaCl_2$ solution.

12.80 Calculate the osmotic pressure of a 0.0500 *M* $MgSO_4$ solution at 22°C. (*Hint:* See Table 12.3.)

MISCELLANEOUS PROBLEMS

12.81 Lysozyme is an enzyme that cleaves bacterial cell walls. A sample of lysozyme extracted from chicken egg white has a molar mass of 13,930 g. A quantity of 0.100 g of this enzyme is dissolved in 150 g of water at 25°C. Calculate the vapor-pressure lowering, the depression in freezing point, the elevation in boiling point, and the osmotic pressure of this solution. (The vapor pressure of water at 25°C = 23.76 mmHg.)

12.82 Solutions A and B have osmotic pressures of 2.4

atm and 4.6 atm, respectively, at a certain temperature. What is the osmotic pressure of a solution prepared by mixing equal volumes of A and B at the same temperature?

12.83 A cucumber placed in concentrated brine (salt water) shrivels into a pickle. Explain.

12.84 Two liquids A and B have vapor pressures of 76 mmHg and 132 mmHg, respectively, at 25°C. What is the total vapor pressure of the ideal solution made up of (a) 1.00 mole of A and 1.00 mole of B and (b) 2.00 moles of A and 5.00 moles of B?

12.85 Calculate the van't Hoff factor of Na_3PO_4 in a 0.40 *m* solution whose boiling point is 100.78°C.

12.86 A 262-mL sample of a sugar solution containing 1.22 g of the sugar has an osmotic pressure of 30.3 mmHg at 35°C. What is the molar mass of the sugar?

12.87 Consider the following three mercury manometers. One of them has 1 mL of water placed on top of the mercury, another has 1 mL of a 1 *m* urea solution placed on top of the mercury, and the third one has 1 mL of a 1 *m* NaCl solution placed on top of the mercury. Identify X, Y, and Z with these solutions.

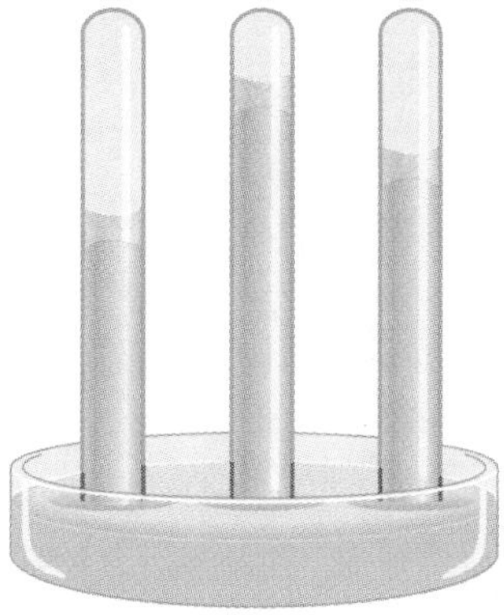

12.88 A forensic chemist is given a white powder for analysis. She dissolves 0.50 g of the substance in 8.0 g of benzene. The solution freezes at 3.9°C. Can the chemist conclude that the compound is cocaine ($C_{17}H_{21}NO_4$)? What assumptions are made in the analysis?

12.89 "Time-release" drugs have the advantage of releasing the drug to the body at a constant rate so that the drug concentration at any time is not too high as to have harmful side effects or too low as to be ineffective. A schematic diagram of a pill that works on this basis is shown below. Explain how it works.

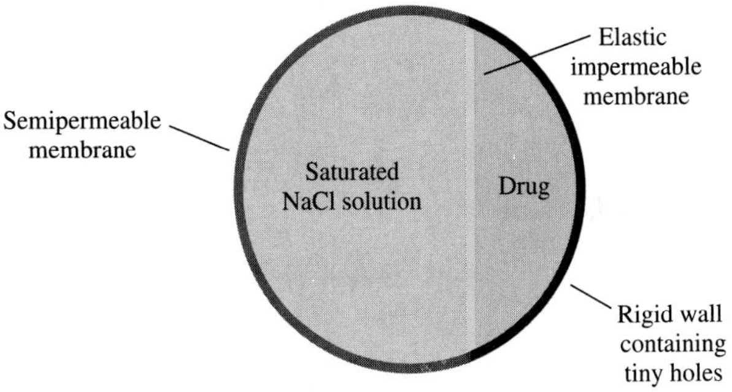

12.90 Concentrated hydrochloric acid is usually available at 37.7 percent by mass. What is its concentration in molarity? (The density of the solution is 1.19 g/mL.)

Answers to Practice Exercises: 12.1 Carbon disulfide; **12.2** 7.44%; **12.3** 0.638 *m*; **12.4** 1.22 *M*; **12.5** 8.92 *m*; **12.6** 13.8 *m*; **12.7** 7.7×10^{-5} *M*; **12.8** 0.78 *m*; **12.9** 101.3°C; −4.48°C; **12.10** 0.066 *m*, 1.3×10^2 g/mol; **12.11** 2.60×10^4 g/mol.

CHAPTER 13

INTRODUCTION TO ORGANIC CHEMISTRY

◆ On a hot summer day the beaches are crowded with sun worshippers trying to get a tan. While in Western cultures a suntan is considered a badge of good health and vitality, it is actually the body's reaction to harmful radiation.

The damaging radiation from the sun is mainly in the ultraviolet range, which is divided into three regions called UV-C, UV-B, and UV-A. The most detrimental to the skin is UV-C, which has the shortest wavelength (200–290 nm) and highest energy. Fortunately, most of UV-C radiation is absorbed by the ozone layer in the stratosphere. UV-B (290–320 nm) reaches Earth's surface in small amounts and is responsible for the redness and blistered and peeling skin associated with sunburn. (The redness is due to the increased flow of blood to blood vessels beneath the skin, which widen in response to the radiation.) UV-B rays are believed to cause skin cancer. The least energetic radiation, UV-A (320–400 nm), is sometimes called the "sun's tanning rays."

When UV-A or UV-B strikes the cells beneath the skin, or melanocytes, they produce a UV-absorbing dark pigment called *melanin*. This substance screens out part of the radiation and helps to minimize damage to the underlying layers of skin. In addition, the melanocytes start dividing more rapidly than usual in order to replace damaged cells in the outer layers. Normally it takes about four weeks for new cells to reach the surface, where they are shed as part of the skin's renewal cycle. Prolonged exposure to the sun speeds up this process, so that the large number of melanin-containing cells arrive at the surface in a few days, causing what we called a "suntan." Unless exposure to the sun continues, the dark-pigmented cells are sloughed off, and the tan fades. Dark-skinned people have more melanocytes and a wider distribution of melanin in the upper layers of the skin and therefore have greater built-in protection against UV radiation.

Suntan lotions and sunscreens block UV rays selectively. Most contain synthetic compounds derived from benzene, the parent of literally millions of organic compounds:

Pure benzene absorbs UV-C radiation, but adding appropriate groups of atoms shifts absorption to the longer wavelengths. Among the most widely used sunscreens agents are para-aminobenzoic acid (PABA), benzophenone, and other structurally related compounds:

Commercial sunscreens.

para-aminobenzoic acid

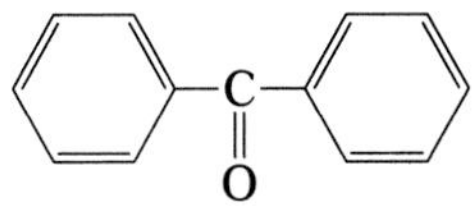

benzophenone

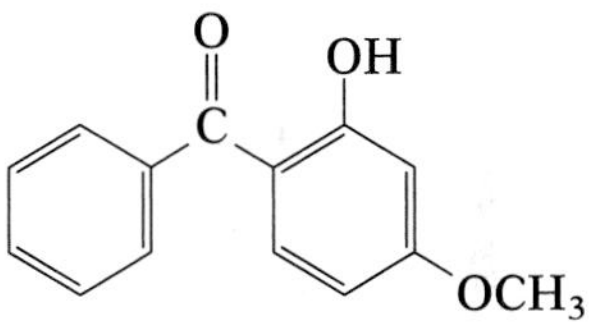

2-hydroxy-4-methoxybenzophenone

These compounds absorb the more harmful UV-B rays. The sunscreen protection factor (SPF) that is prominently displayed on sunscreen labels indicates the degree of protection given by the product. A sunscreen with an SPF of 7, for example, will allow only one-seventh of the sun's radiation to reach treated skin, compared with untreated skin, for the same length of exposure.

For people who desire a tan but cannot spend hours in the sun, there are "sunless" tanning agents, some of which contain dihydroxyacetone

$$HOCH_2—\overset{\overset{\displaystyle O}{\|}}{C}—CH_2OH$$

and react with amino acids on the surface of the skin to produce a brown pigment similar to melanin. ◆

13.1 THE STUDY OF ORGANIC CHEMISTRY

Organic chemistry is *the study of carbon compounds.* (The important exceptions are carbon monoxide, carbon dioxide, carbonates, and bicarbonates, which are regarded as inorganic compounds.) Carbon can form more compounds than any other element because carbon atoms are able not only to form single, double, and triple carbon-carbon bonds, but also to link up with each other in straight chains, branched chains, and rings. To date, over 8 million organic compounds are known. This number is far greater than the hundred thousand or so inorganic compounds that have been characterized. The food we eat, the clothing we wear, and most household items and industrial products are made of organic compounds. With the exception of water, calcium phosphate, and metal cations and inorganic anions like chloride and phosphate, our bodies consist mostly of organic materials.

For easier study, organic chemists have divided organic compounds into a relatively few classes, according to the *functional groups* they contain. A ***functional group*** is *a group of atoms that is largely responsible for the chemical behavior of the parent molecule.* Different molecules containing the same kind of functional group or groups undergo similar reactions. Thus, by learning the characteristic properties of a few functional groups, we can study and predict the properties of many organic compounds.

13.2 HYDROCARBONS

Before discussing functional groups we will examine a class of compounds that forms the framework of all organic compounds—the hydrocarbons. ***Hydrocarbons*** are *made up of only two elements, hydrogen and carbon.* Hydrocarbons are divided into two main classes—aliphatic and aromatic. ***Aliphatic hydrocarbons*** *do not contain the benzene group or the benzene ring,* whereas ***aromatic hydrocarbons*** *contain one or more benzene rings.* Aliphatic hydrocarbons are further divided into alkanes, cycloalkanes, alkenes, and alkynes (Figure 13.1).

ALKANES

Alkanes *have the general formula C_nH_{2n+2}, where $n = 1,2, \ldots$.* The essential characteristic of alkane hydrocarbon molecules is that *only single covalent bonds are*

FIGURE 13.1

Classification of hydrocarbons.

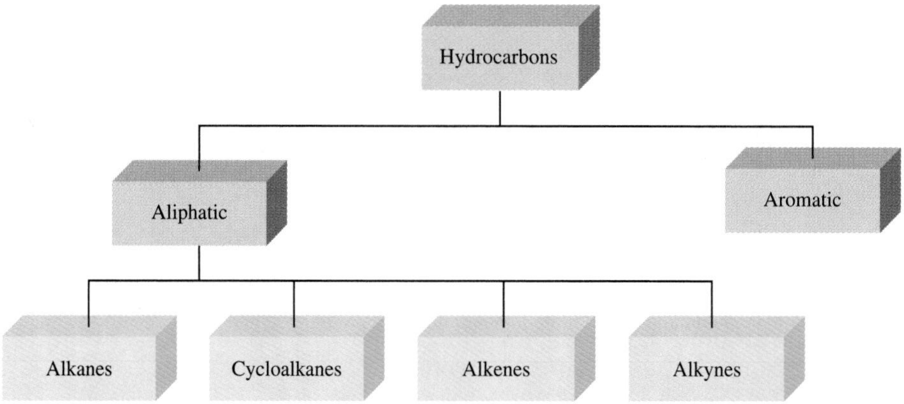

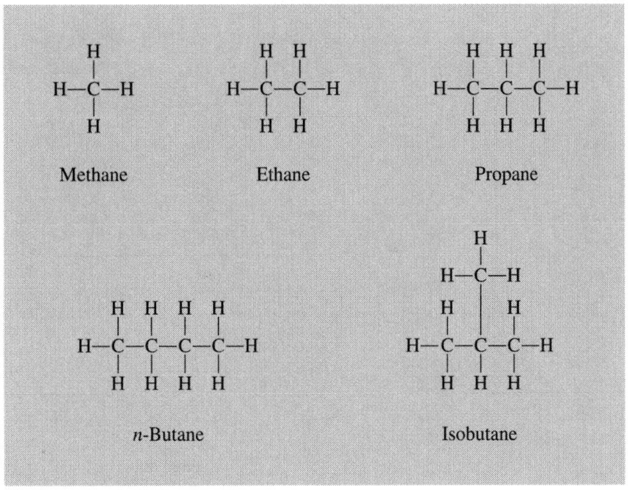

FIGURE 13.2

Structures of the first four alkanes. Note that butane can exist in two structurally different forms, called structural isomers.

present. In these compounds the bonds are said to be saturated, meaning that *no more hydrogen can be added to the carbon atoms.* Thus the alkanes are known as ***saturated hydrocarbons.***

The simplest member of the family (that is, with $n = 1$) is CH_4 (methane), which occurs naturally by the anaerobic bacterial decomposition of vegetable matter under water. Because it was first collected in marshes, methane became known as "marsh gas." A rather improbable but proven source of methane is termites. When these voracious insects consume wood, microorganisms in their digestive systems break down cellulose (the major component of wood) into methane, carbon dioxide, and other compounds. An estimated 170 million tons of methane are produced annually by termites! It is also produced in some sewage treatment processes. Commercially, methane is obtained from natural gas.

Termites are a natural source of methane.

Figure 13.2 shows the structures of the first four alkanes (that is, from $n = 1$ to $n = 4$). Natural gas contains mostly methane, plus small portions of ethane, propane, and butane. In Chapter 10 we discussed the bonding scheme of methane. Indeed the carbon atoms in all the alkanes can be assumed to be sp^3-hybridized. The structures of ethane and propane are straightforward, for there is only one way to join the carbon atoms in these molecules. Butane, however, has two possible bonding schemes and can exist as either *n*-butane (*n* stands for normal) or isobutane. The two structures of butane are ***structural isomers,*** *molecules that have the same molecular formula but a different order of linking the atoms.*

In the alkane series, as the number of carbon atoms increases, the number of structural isomers increases rapidly. For example, butane, C_4H_{10}, has 2 isomers; decane, $C_{10}H_{22}$, has 75 isomers; and the alkane $C_{30}H_{62}$ has over 400 million, or 4×10^8, isomers! Obviously, most of these isomers do not exist in nature and have not been synthesized. Nevertheless, the numbers help to explain why carbon is found in so many more compounds than any other element.

EXAMPLE 13.1

Determining the Number of Structural Isomers

How many structural isomers can be identified for pentane, C_5H_{12}?

Answer: For small hydrocarbon molecules (eight or fewer carbon atoms), it is relatively easy to determine the number of structural isomers by trial and error. The first step is to write down the straight-chain structure:

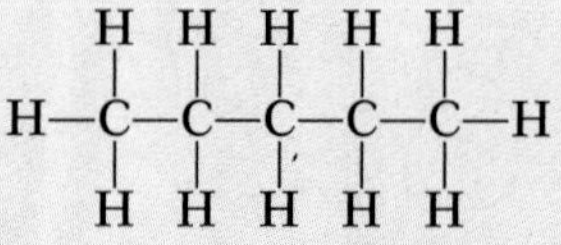

n-pentane
(b.p. 36.1°C)

The second structure, by necessity, must be a branched chain:

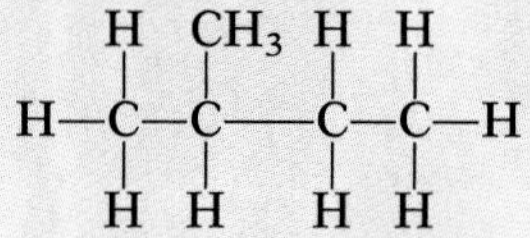

2-methylbutane
(b.p. 27.9°C)

Yet another branched-chain structure is possible:

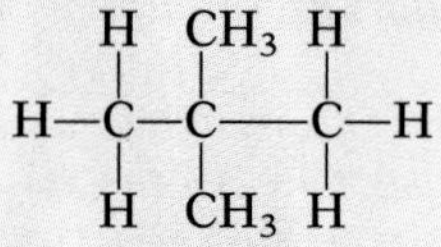

2,3-dimethylpropane
(b.p. 9.5°C)

We can draw no other structure for an alkane having the molecular formula C_5H_{12}. Thus pentane has three structural isomers, in which the numbers of carbon and hydrogen atoms remain unchanged despite the differences in structure.

PRACTICE EXERCISE

How many structural isomers are there in the alkane C_6H_{14}?

Table 13.1 shows the melting and boiling points of the straight-chain isomers of the first ten alkanes. The first four are gases at room temperature; and pentane (C_5H_{12}) through decane ($C_{10}H_{22}$) are liquids. As molecular size increases, so does the boiling point, because of increased dispersion forces (see Section 11.2). The higher alkanes as well as other hydrocarbons are obtained from petroleum by a process called fractional distillation.

Crude oil is a major source of hydrocarbons.

Alkane Nomenclature

The names of the straight-chain alkanes are fairly easy to remember. As Table 13.1 shows, except for the first four members, the number of carbon atoms in each alkane is reflected in the Greek prefix (see Table 2.4). When one or more hydrogen atoms are replaced by other groups, the name of the compound must indicate the locations of carbon atoms where replacements are made. Let us first consider groups that are themselves derived from alkanes. For example, when a hydrogen atom is removed from methane, we are left with the CH_3 fragment, which is called

TABLE 13.1
The First Ten Straight-Chain Alkanes*

Name of hydrocarbon	Molecular formula	Number of carbon atoms	Melting point (°C)	Boiling point (°C)
Methane	CH_4	1	−182.5	−161.6
Ethane	$CH_3—CH_3$	2	−183.3	−88.6
Propane	$CH_3—CH_2—CH_3$	3	−189.7	−42.1
Butane	$CH_3—(CH_2)_2—CH_3$	4	−138.3	−0.5
Pentane	$CH_3—(CH_2)_3—CH_3$	5	−129.8	36.1
Hexane	$CH_3—(CH_2)_4—CH_3$	6	−95.3	68.7
Heptane	$CH_3—(CH_2)_5—CH_3$	7	−90.6	98.4
Octane	$CH_3—(CH_2)_6—CH_3$	8	−56.8	125.7
Nonane	$CH_3—(CH_2)_7—CH_3$	9	−53.5	150.8
Decane	$CH_3—(CH_2)_8—CH_3$	10	−29.7	174.0

*By "straight chain," we mean that the carbon atoms are joined along one line. This does not mean, however, that these molecules are linear. Each carbon is sp^3- (not sp-) hybridized.

a *methyl* group. Similarly, removing a hydrogen atom from the ethane molecule gives an *ethyl* group, or C_2H_5. Because they are derived from alkanes, they are also called *alkyl groups.* Table 13.2 lists the names of several common alkyl groups.

With the aid of Tables 13.1 and 13.2 and the rule that the locations of other groups are denoted by numbering the atoms along the longest carbon chain, we are now ready to name some substituted alkanes. Note that we always start numbering from the end of the longest chain that is closer to the carbon atom bearing the substituted group. Consider the following example:

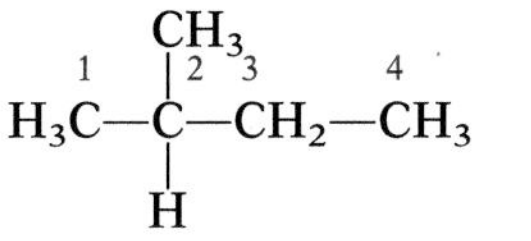

Since the longest carbon chain contains four C atoms, the parent name of the compound is butane. We note that a methyl group is attached to carbon number 2; therefore, the name of the compound is 2-methylbutane. Here is another example:

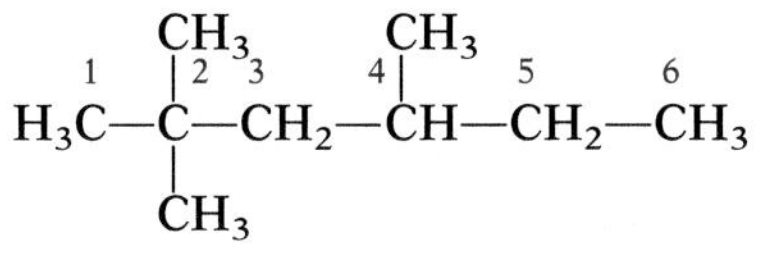

The parent name of this compound is hexane. Since two methyl groups are attached to carbon number 2 and one methyl group is attached to carbon number 4, we call the compound 2,2,4-trimethylhexane. Note that the total number of methyl groups is indicated with the prefix "tri."

Of course, we can have many different types of substituents on alkanes. Table 13.3 lists the names of some common functional groups. Thus the compound

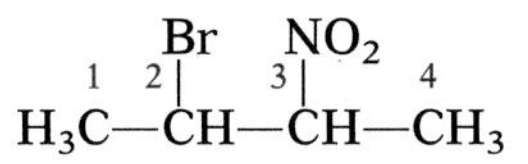

is called 2-bromo-3-nitrobutane. If the Br is attached to the first carbon atom

TABLE 13.2
Common Alkyl Groups

Name	Formula
Methyl	$—CH_3$
Ethyl	$—CH_2—CH_3$
n-Propyl	$—(CH_2)_2—CH_3$
n-Butyl	$—(CH_2)_3—CH_3$
Isopropyl	CH_3 —C—H CH_3
t-Butyl*	CH_3 $—C—CH_3$ CH_3

*The letter *t* stands for tertiary.

TABLE 13.3
Names of Common Functional Groups

Functional group	Name
$—NH_2$	Amino
—F	Fluoro
—Cl	Chloro
—Br	Bromo
—I	Iodo
$—NO_2$	Nitro
$—CH{=}CH_2$	Vinyl

$$\overset{1}{Br—CH_2}—\overset{2}{CH_2}—\overset{3}{\underset{}{CH}}(NO_2)—\overset{4}{CH_3}$$

the compound becomes 1-bromo-3-nitrobutane.

Reactions of Alkanes

Alkanes are not reactive substances. However, under suitable conditions they do undergo several kinds of reactions—including combustion. The burning of natural gas, gasoline, and fuel oil involves the combustion of alkanes. These reactions are all highly exothermic:

$$CH_4(g) + 2O_2(g) \longrightarrow CO_2(g) + 2H_2O(l) \qquad \Delta H° = -890.4 \text{ kJ}$$

$$2C_2H_6(g) + 7O_2(g) \longrightarrow 4CO_2(g) + 6H_2O(l) \qquad \Delta H° = -3119 \text{ kJ}$$

These and similar reactions have long been utilized in industrial processes and in domestic heating and cooking.

The *halogenation* of alkanes—that is, the replacement of one or more hydrogen atoms by halogen atoms—is another well-studied kind of reaction. When a mixture of methane and chlorine is heated above 100°C or irradiated with light of a suitable wavelength, methyl chloride is produced:

$$CH_4(g) + Cl_2(g) \longrightarrow \underset{\text{methyl chloride}}{CH_3Cl(g)} + HCl(g)$$

If chlorine gas is present in a sufficient amount, the reaction can proceed further:

$$CH_3Cl(g) + Cl_2(g) \longrightarrow \underset{\text{methylene chloride}}{CH_2Cl_2(l)} + HCl(g)$$

$$CH_2Cl_2(l) + Cl_2(g) \longrightarrow \underset{\text{chloroform}}{CHCl_3(l)} + HCl(g)$$

$$CHCl_3(l) + Cl_2(g) \longrightarrow \underset{\text{carbon tetrachloride}}{CCl_4(l)} + HCl(g)$$

A great deal of experimental evidence suggests that in the initial step of the first halogenation reaction the covalent bond in Cl_2 breaks and two chlorine atoms form:

$$Cl_2 + \text{energy} \longrightarrow Cl\cdot + Cl\cdot$$

We know it is the Cl—Cl bond that breaks first when the mixture is heated or irradiated because the bond energy of Cl_2 is 242.7 kJ/mol, whereas about 414 kJ/mol are needed to break the C—H bonds in CH_4.

A chlorine atom is called a ***free radical*** (or *radical*), *a species containing an unpaired electron* (shown by a single dot). Chlorine radicals are highly reactive and attack methane molecules according to the equation

$$CH_4 + Cl\cdot \longrightarrow \cdot CH_3 + HCl$$

This reaction produces hydrogen chloride and the methyl radical $\cdot CH_3$. The methyl radical is another reactive species; it combines with molecular chlorine to give methyl chloride and a chlorine radical:

$$\cdot CH_3 + Cl_2 \longrightarrow CH_3Cl + Cl\cdot$$

The production of methylene chloride from methyl chloride and further reactions can be explained in the same way.

Alkanes in which one or more hydrogen atoms have been replaced by a halo-

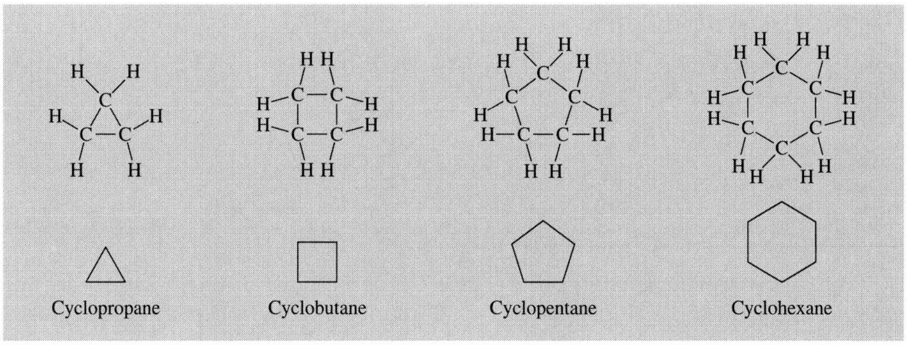

FIGURE 13.3

Structures of the first four cycloalkanes and their simplified forms.

gen atom are called *alkyl halides*. Among the large number of alkyl halides, the best known are chloroform ($CHCl_3$), carbon tetrachloride (CCl_4), and methylene chloride (CH_2Cl_2, also called dichloromethane). Chloroform is a volatile, sweet-tasting liquid used for many years as an anesthetic. However, because of its toxicity—it can severely damage the liver, kidneys, and heart—it is no longer used as an anesthetic. Carbon tetrachloride, another toxic liquid, serves as a dry-cleaning agent, for it removes grease stains from clothing. Methylene chloride is used as a solvent to decaffeinate coffee and as a paint remover.

Cycloalkanes

Alkanes whose carbon atoms are joined in rings are known as ***cycloalkanes.*** They have the general formula C_nH_{2n}, where $n = 3, 4, \ldots$. The structures of some simple cycloalkanes are given in Figure 13.3. Many biologically significant substances contain one or more such ring systems. Theoretical analysis shows that cyclohexane can assume two different geometries that are relatively free of strain. By "strain" we mean that bonds are compressed, stretched, or twisted out of their normal orientations as predicted by sp^3 hybridization. The most stable geometry for cyclohexane is the *chair form* (Figure 13.4).

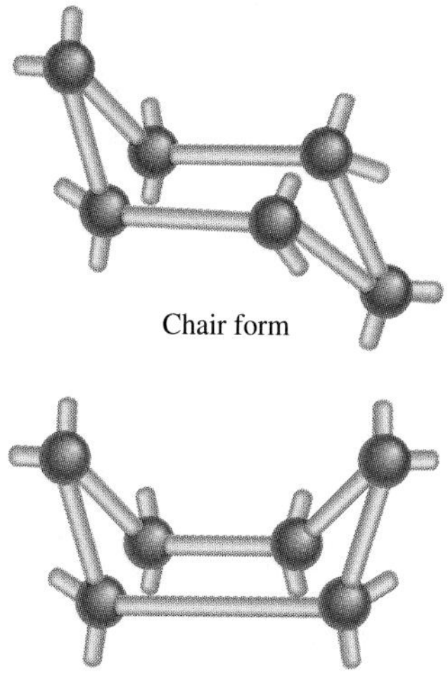

FIGURE 13.4

Alternative cyclohexane structures. The most stable shape is the chair form, while the boat form is less stable. Hydrogen atoms are not shown in these diagrams.

ALKENES

The ***alkenes*** (also called *olefins*) *contain at least one carbon-carbon double bond. Alkenes have the general formula C_nH_{2n}, where $n = 2, 3, \ldots$.* The simplest alkene is C_2H_4, ethene (commonly called ethylene), in which both carbon atoms are sp^2-hybridized and the double bond is made up of a sigma bond and a pi bond (see Section 10.4).

Alkene Nomenclature

In naming alkenes it is necessary to indicate the positions of the carbon-carbon double bonds. The names of compounds containing C=C bonds end with "-ene." As with the alkanes, the name of the parent compound is determined by the number of carbon atoms in the longest chain (see Table 13.1), as shown here:

$$\underset{\text{1-butene}}{CH_2{=}CH{-}CH_2{-}CH_3} \qquad \underset{\text{2-butene}}{H_3C{-}CH{=}CH{-}CH_3}$$

The numbers in 1-butene and 2-butene indicate the lowest-numbered carbon atom in the chain that is part of the C=C bond of the alkene. The name "butene" means that this is a compound with four carbon atoms in the longest chain.

Geometric Isomers

2-Butene is interesting because it exists in two different forms known as geometric isomers:

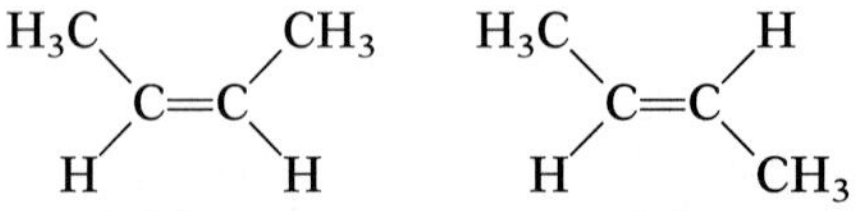

cis-2-butene *trans*-2-butene

Geometric isomers are *compounds with the same type, number, and order of attachment of atoms and the same chemical bonds, but different spatial arrangements of the atoms; such isomers cannot be interconverted without breaking a chemical bond.* Geometric isomers come in pairs. We use the terms "cis" and "trans" to distinguish one geometric isomer of a compound from the other. Cis means that two particular atoms (or groups of atoms) are adjacent to each other, and trans means that the atoms (or groups of atoms) are on opposite sides in the structural formula. Thus in *cis*-2-butene the two H atoms (and the two methyl groups) are adjacent to each other, whereas in *trans*-2-butene these atoms (and the methyl groups) are on opposite side of the C=C bond. Cis and trans isomers of the same compound generally have quite different physical properties such as density, melting point, boiling point, and dipole moment.

The existence of cis and trans alkene isomers can be understood by considering the characteristics of the double bond. In a compound such as ethane, C_2H_6, the rotation of the two methyl groups about the carbon-carbon single bond (which is a sigma bond) is quite unrestricted. The situation is different for molecules like ethylene, C_2H_4, which contains a carbon-carbon double bond, because of the pi bond between the two carbon atoms. Rotation about the carbon-carbon linkage does not affect the C—C sigma bond, but it does move the two $2p_z$ orbitals out of alignment for overlap and, hence, partially or totally destroys the pi bond (see Figure 10.14). Complete disruption of the C—C pi bond requires an input of energy on the order of 270 kJ/mol. For this reason, the rotation of a carbon-carbon double bond is considerably restricted, but not impossible. While ethylene itself does not have geometric isomers because all four of the groups attached to the C atoms are the same, other alkenes may have geometric isomers, which cannot be interconverted without breaking a chemical bond. Cis-trans conversion (also called geometric isomerization) is usually accomplished by treating alkenes with heat or light.

EXAMPLE 13.2

Distinguishing between Structural and Geometric Isomers

Draw Lewis structures for all compounds with the molecular formula $C_2H_2Cl_2$ and determine which are geometric isomers.

Answer: Following the procedure outlined in Chapter 9, we can draw three Lewis structures as follows:

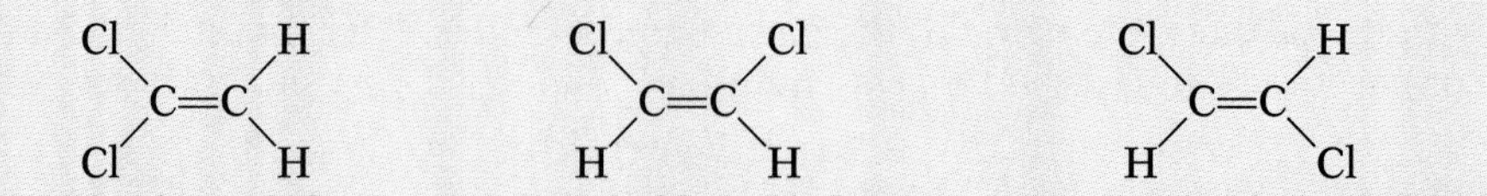

1,1-dichloroethylene *cis*-1,2-dichloroethylene *trans*-1,2-dichloroethylene

(For simplicity we have omitted the lone pairs on the Cl atoms.) The geometric isomers are *cis*- and *trans*-1,2-dichloroethylene. The first molecule, 1,1-dichloroethylene, is a structural isomer to the geometric isomers because its atoms are attached in a different order.

PRACTICE EXERCISE

Which of the following molecules can exist as cis and trans isomers? (a) $(CH_3)_2C{=}CH_2$, (b) $(CH_3)HC{=}C(CH_3)H$, (c) $FHC{=}CFCl$.

Properties and Reactions of Alkenes

Ethylene is an extremely important substance because it is used in large quantities in manufacturing organic polymers (very large molecules, which are discussed in the next chapter) and in preparing many other organic chemicals. Ethylene and other alkenes are prepared industrially by the *cracking* process, that is, the thermal decomposition of a large hydrocarbon into smaller molecules. When ethane is heated to about 800°C in the presence of platinum, it undergoes the following reaction:

$$C_2H_6(g) \xrightarrow[\text{catalyst}]{\text{Pt}} CH_2{=}CH_2(g) + H_2(g)$$

The platinum acts as a *catalyst,* a substance that speeds up a reaction without being used up in the process and therefore does not appear on either side of the equation. Other alkenes can be prepared by cracking the higher members of the alkane family.

Alkenes are classified as ***unsaturated hydrocarbons,*** *compounds with double or triple carbon-carbon bonds.* Unsaturated hydrocarbons commonly undergo ***addition reactions*** in which *one molecule adds to another to form a single product.* An example of an addition reaction is ***hydrogenation,*** which is *the addition of molecular hydrogen to compounds containing C=C and C≡C bonds*

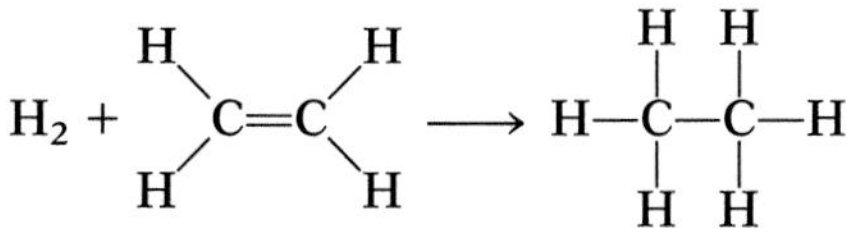

Hydrogenation is an important process in the food industry. Vegetable oils have considerable nutritional value, but some oils must be hydrogenated to eliminate some of the C=C bonds before they can be used to prepare food. Upon exposure to air, *polyunsaturated* molecules—molecules with many C=C bonds—undergo oxidation to yield unpleasant-tasting products (vegetable oil that has oxidized is said to be rancid). In the hydrogenation process, a small amount of nickel catalyst is added to the oil and the mixture is exposed to hydrogen gas at high temperature and pressure. Afterward, the nickel is removed by filtration. Hydrogenation reduces the number of double bonds in the molecule but does not completely eliminate them. If all the double bonds are eliminated, the oil becomes hard and brittle. Under controlled conditions, suitable cooking oils and margarine may be prepared by hydrogenation from vegetable oils extracted from cottonseed, corn, and soybeans.

Other addition reactions to the C=C bond include

$$C_2H_4(g) + HX(g) \longrightarrow CH_3{-}CH_2X(g)$$

$$C_2H_4(g) + X_2(g) \longrightarrow CH_2X{-}CH_2X(g)$$

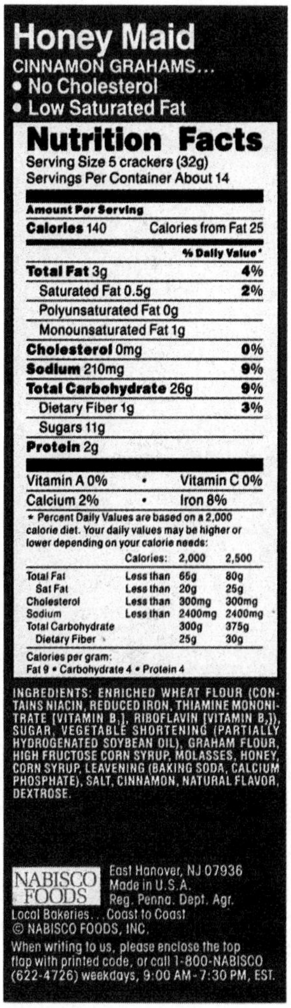

Label showing hydrogenated oil.

FIGURE 13.5

When ethylene gas is bubbled through an aqueous bromine solution, the brownish-red color gradually disappears due to the formation of 1,2-dibromoethane, which is colorless.

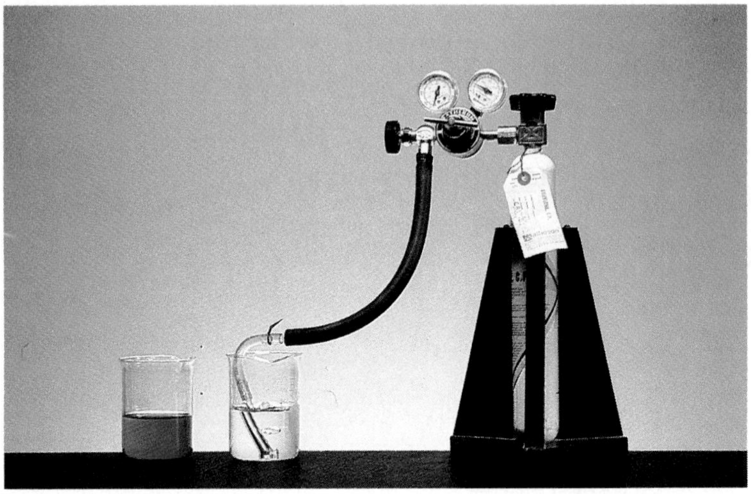

where X represents a halogen (Cl, Br, or I). Figure 13.5 shows the reaction between ethylene and aqueous bromine to form 1,2-dibromoethane (CH_2BrCH_2Br).

ALKYNES

Alkynes *contain at least one carbon-carbon triple bond.* They have the *general formula* C_nH_{2n-2}*, where* $n = 2, 3, \ldots$.

Alkyne Nomenclature

Names of compounds containing C≡C bonds end with "-yne." Again, the name of the parent compound is determined by the number of carbon atoms in the longest chain (see Table 13.1 for names of alkane counterparts). As in the case of alkenes, the names of alkynes indicate the position of the carbon-carbon triple bond, as, for example, in

$$\underset{\text{1-butyne}}{HC{\equiv}C{-}CH_2{-}CH_3} \qquad \underset{\text{2-butyne}}{H_3C{-}C{\equiv}C{-}CH_3}$$

The reaction between calcium carbide and water produces acetylene, a flammable gas.

Properties and Reactions of Alkynes

The simplest alkyne is ethyne, better known as acetylene (C_2H_2). The structure and bonding of C_2H_2 were discussed in Section 10.4. Acetylene is a colorless gas (b.p. −84°C) prepared by the reaction between calcium carbide and water:

$$CaC_2(s) + 2H_2O(l) \longrightarrow C_2H_2(g) + Ca(OH)_2(aq)$$

Acetylene has many important uses in industry. Because of its high heat of combustion

$$2C_2H_2(g) + 5O_2(g) \longrightarrow 4CO_2(g) + 2H_2O(l) \qquad \Delta H° = -2599.2 \text{ kJ}$$

acetylene burned in an "oxyacetylene torch" gives an extremely hot flame (about 3000°C). Thus oxyacetylene torches are used to weld metals (see p. 103). Acetylene is unstable (relative to its elements) and has a tendency to decompose:

$$C_2H_2(g) \longrightarrow 2C(s) + H_2(g)$$

In the presence of a suitable catalyst or when the gas is kept under pressure, this reaction can occur with explosive violence. For the gas to be transported safely, it must be dissolved in an inert organic solvent such as acetone at moderate pressure. Liquid acetylene is very sensitive to shock and is highly explosive.

Acetylene, an unsaturated hydrocarbon, can be hydrogenated to yield ethylene:

$$C_2H_2(g) + H_2(g) \longrightarrow C_2H_4(g)$$

It undergoes the following addition reactions with hydrogen halides and halogens:

$$C_2H_2(g) + HX(g) \longrightarrow CH_2{=}CHX(g)$$
$$C_2H_2(g) + X_2(g) \longrightarrow CHX{=}CHX(g)$$
$$C_2H_2(g) + 2X_2(g) \longrightarrow CHX_2{-}CHX_2(g)$$

Methylacetylene (propyne), $CH_3{-}C{\equiv}C{-}H$, is the next member in the alkyne family. It undergoes reactions similar to those of acetylene.

Aromatic Hydrocarbons

Benzene is the parent compound of this large family of organic substances. As we saw in Section 9.6, the properties of benzene are best represented by both of the following resonance structures:

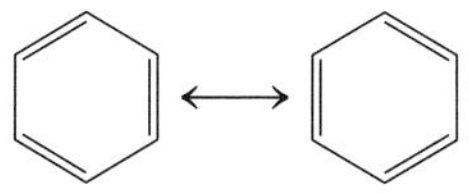

An electron micrograph of benzene molecules clearly shows the ring structure.

For simplicity, however, we will use only one resonance structure to represent compounds containing the benzene ring.

The naming of monosubstituted benzenes, that is, benzenes in which one H atom has been replaced by another atom or a group of atoms, is quite straightforward, as shown below:

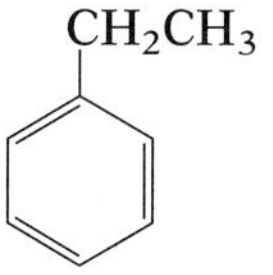

ethylbenzene

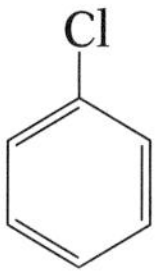

chlorobenzene

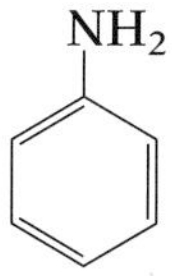

aminobenzene (aniline)

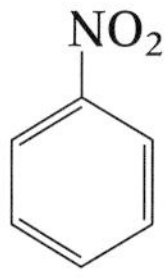

nitrobenzene

If more than one substituent is present, we must indicate the location of the second group relative to the first. The systematic way to accomplish this is to number the carbon atoms as follows:

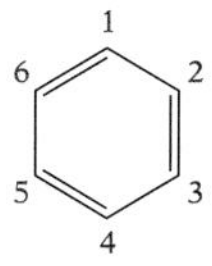

Three different dibromobenzenes are possible:

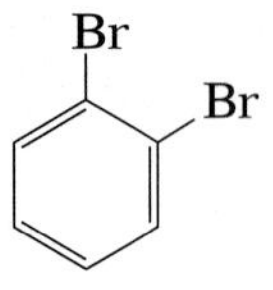

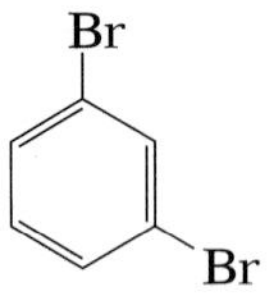

1,2-dibromobenzene (*o*-dibromobenzene) | 1,3-dibromobenzene (*m*-dibromobenzene) | 1,4-dibromobenzene (*p*-dibromobenzene)

The prefixes *o-* (*ortho-*), *m-* (*meta-*), and *p-* (*para-*) are also used to denote the relative positions of the two substituted groups, as shown above for the dibromobenzenes. Compounds in which the two substituted groups are different are named accordingly. Thus,

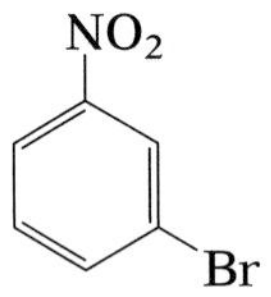

is named 3-bromonitrobenzene, or *m*-bromonitrobenzene.

Finally we note that the group containing benzene minus a hydrogen atom (C_6H_5) is called the *phenyl group*. Thus the following molecule is called 2-phenylpropane:

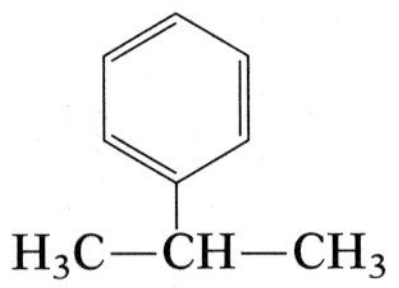

Properties and Reactions of Aromatic Compounds

Benzene is a colorless, flammable liquid (b.p. 80.1°C) obtained chiefly from petroleum and to a lesser extent coal tar. Perhaps the most remarkable chemical property of benzene is its relative inertness. Although it has the same empirical formula as acetylene (CH) and a high degree of unsaturation (three double bonds), it is much less reactive than either ethylene or acetylene. Benzene can be hydrogenated, but only with difficulty. The following reaction is carried out at significantly higher temperatures and pressures than are similar reactions for the alkenes:

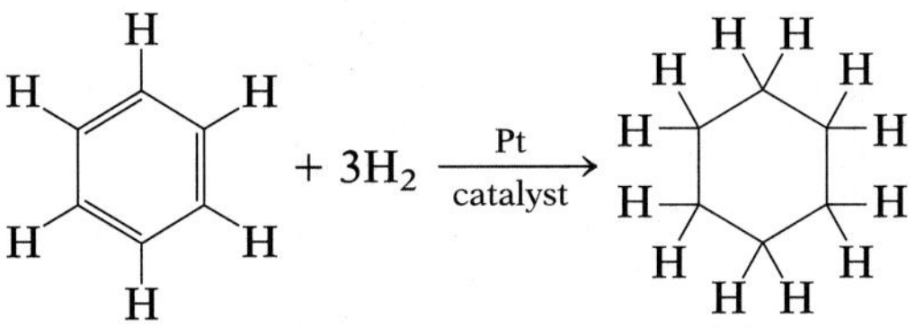

cyclohexane

We saw earlier that alkenes react readily with halogens to form addition products. The most common reaction of halogens with benzene is the ***substitution***

reaction, in which *an atom or a group of atoms in a molecule is replaced by another atom or group of atoms:*

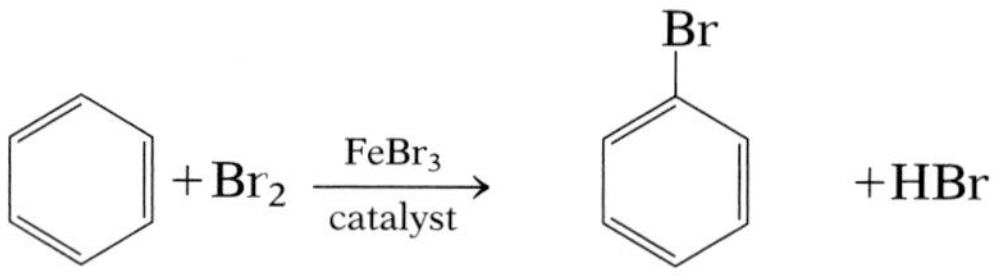

bromobenzene

Alkyl groups can be substituted into the ring system by allowing benzene to react with an alkyl halide using $AlCl_3$ as the catalyst:

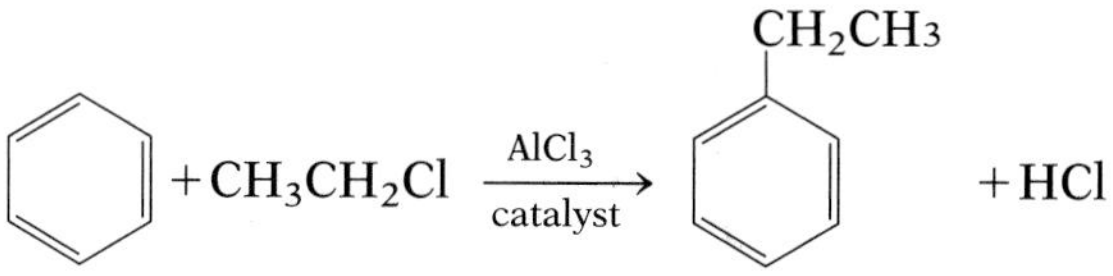

ethylbenzene

An enormously large number of compounds can be generated from substances in which benzene rings are fused together. Some of these *polycyclic* aromatic hydrocarbons are shown in Figure 13.6. The best known of these compounds is naphthalene, an important component of mothballs. These and many other similar compounds are present in coal tar. Some of the compounds with larger numbers of rings are powerful carcinogens—that is, they can cause cancer in humans and other animals. Benz(*a*)anthracene occurs, for example, in cigarette smoke and in burned meat.

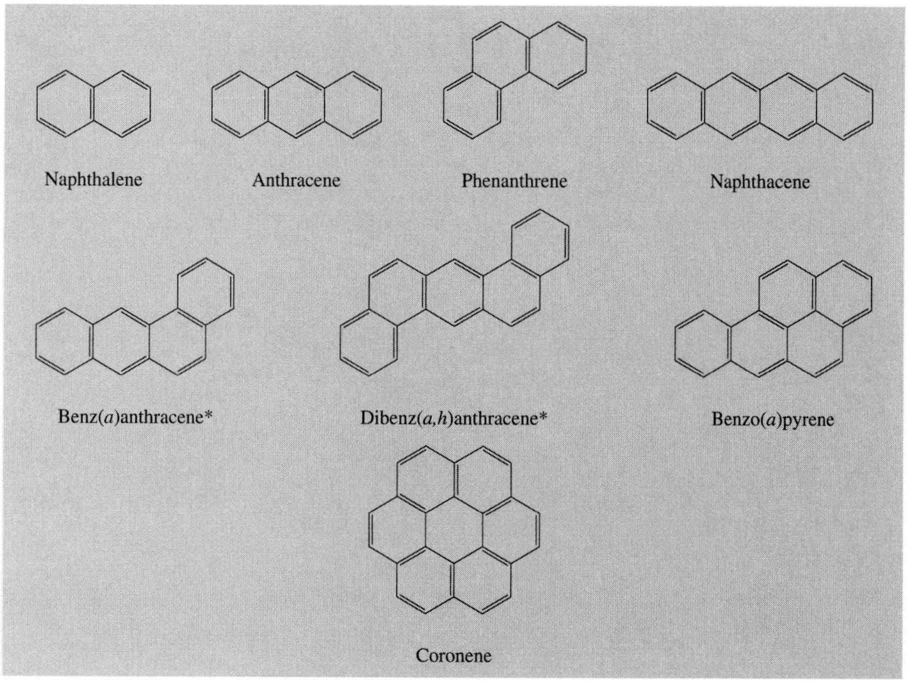

FIGURE 13.6

Polycyclic aromatic hydrocarbons. Compounds marked with an asterisk are potent carcinogens. The letters a and h denote the ways the benzene rings are fused together. An enormous number of such compounds exist in nature.

13.3 FUNCTIONAL GROUPS

We will now examine the most common organic functional groups and the kinds of reactions they undergo. In particular, we will focus on classes of compounds in which the functional groups include oxygen or nitrogen.

ALCOHOLS

All ***alcohols*** *contain the hydroxyl group, —OH.* Some common alcohols are shown in Figure 13.7. Ethyl alcohol (CH_3CH_2OH), or ethanol, is by far the best known. It is produced biologically by the fermentation of sugar or starch. In the absence of oxygen the enzymes (which are biological catalysts) present in bacteria and yeast catalyze the reaction

$$C_6H_{12}O_6(aq) \xrightarrow{\text{enzyme}} 2CH_3CH_2OH(aq) + 2CO_2(g)$$

This process gives off energy, which microorganisms, in turn, use for growth and other functions.

Ethanol is prepared commercially by the addition of water to ethylene at about 280°C and 300 atm, with sulfuric acid as the catalyst:

$$CH_2{=}CH_2(g) + H_2O(g) \xrightarrow[\text{catalyst}]{H_2SO_4} CH_3CH_2OH(g)$$

Ethanol has countless applications as a solvent for organic chemicals and as a starting compound for the manufacture of dyes, synthetic drugs, cosmetics, explosives, and so on. It is also a familiar constituent of beverages. Ethanol is the only nontoxic (more properly, the least toxic) of the straight-chain alcohols; our bodies produce an enzyme, called *alcohol dehydrogenase,* which helps metabolize ethanol by oxidizing it to acetaldehyde:

$$CH_3CH_2OH \longrightarrow \underset{\text{acetaldehyde}}{CH_3CHO} + H_2$$

This equation is a simplified version of what actually takes place; the H atoms are taken up by other molecules, so that no H_2 gas is evolved.

Ethanol can be oxidized also by inorganic oxidizing agents, such as acidified potassium dichromate ($K_2Cr_2O_7$), to acetic acid:

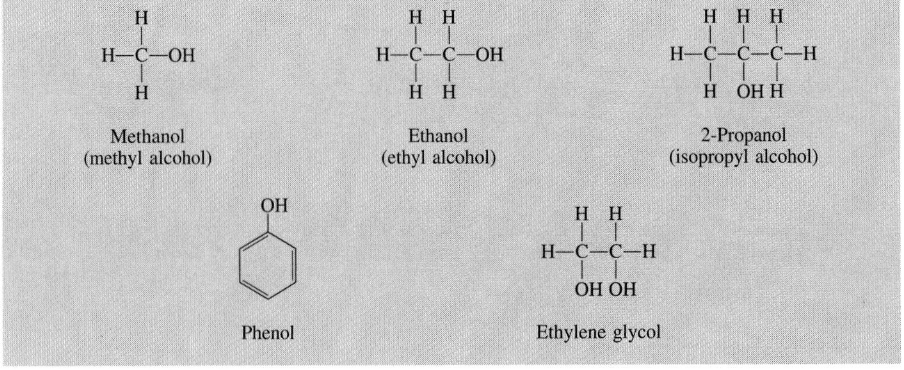

FIGURE 13.7

Common alcohols. Note that all the compounds contain the OH group. The properties of phenol are quite different from those of the aliphatic alcohols.

$$\underset{\text{}}{3CH_3CH_2OH} + \underset{\text{orange-yellow}}{2K_2Cr_2O_7} + 8H_2SO_4 \longrightarrow 3CH_3COOH + \underset{\text{green}}{2Cr_2(SO_4)_3} + 2K_2SO_4 + 11H_2O$$

Left: A potassium dichromate solution. Right: A chromium(III) sulfate solution.

This reaction has been employed by law enforcement agencies to test drivers suspected of being drunk. A sample of the driver's breath is drawn into a device called a breath analyzer, where it is reacted with an acidic potassium dichromate solution. From the color change (orange-yellow to green) it is possible to determine the alcohol content in the driver's blood.

The simplest aliphatic alcohol is methanol, CH_3OH. Called *wood alcohol,* it was prepared at one time by the dry distillation of wood. It is now synthesized industrially by the reaction of carbon monoxide and molecular hydrogen at high temperatures and pressures, with Fe_2O_3 as a catalyst:

$$CO(g) + 2H_2(g) \xrightarrow[\text{catalyst}]{Fe_2O_3} CH_3OH(g)$$

Methanol is highly toxic. Ingestion of only a few milliliters can cause nausea and blindness. Ethanol intended for industrial use is often mixed with methanol to prevent people from drinking it. Ethanol containing methanol or other toxic substances is called *denatured alcohol.*

The alcohols are very weakly acidic; they do not react with strong bases, such as NaOH. The alkali metals react with alcohols to produce hydrogen:

$$2CH_3OH + 2Na \longrightarrow \underset{\text{sodium methoxide}}{2NaOCH_3} + H_2$$

However, the reaction is much less violent than that between Na and water (see p. 103) because water is a stronger acid than alcohols are:

$$2H_2O + 2Na \longrightarrow 2NaOH + H_2$$

Two other familiar aliphatic alcohols are isopropyl alcohol, commonly known as rubbing alcohol, and ethylene glycol, which is used as an antifreeze. Note that ethylene glycol has two —OH groups and so can form hydrogen bonds with water molecules more effectively than compounds that have only one —OH group (see Figure 13.7). Most alcohols—especially those with low molar masses—are highly flammable.

Alcohols react more slowly with sodium metal than does water.

ETHERS

Ethers *contain the R—O—R' linkage, where R and R' are either an alkyl group or a group derived from an aromatic hydrocarbon.* They are formed by condensation reactions with alcohols. A ***condensation reaction*** is characterized by *the joining of two molecules and the elimination of a small molecule, usually water:*

$$CH_3OH + HOCH_3 \xrightarrow{H_2SO_4} \underset{\text{dimethyl ether}}{CH_3OCH_3} + H_2O$$

Like alcohols, ethers are extremely flammable. When left standing in air, they have a tendency to slowly form explosive peroxides, which contain the —O—O— linkage:

$$\underset{\text{diethyl ether}}{C_2H_5OC_2H_5} + O_2 \longrightarrow C_2H_5O—\overset{\displaystyle CH_3}{\underset{\displaystyle H}{\overset{|}{\underset{|}{C}}}}—O—O—H$$

Diethyl ether, commonly known as "ether," was used as an anesthetic for many years. It produces unconsciousness by depressing the activity of the central nervous system. The major disadvantages of diethyl ether are its irritating effects on the respiratory system and the occurrence of postanesthetic nausea and vomiting. "Neothyl," or methyl propyl ether, $CH_3OCH_2CH_2CH_3$, is currently favored as an anesthetic because it is relatively free of side effects.

Aldehydes and Ketones

Under mild oxidation conditions, it is possible to convert alcohols to aldehydes and ketones:

$$CH_3OH + \tfrac{1}{2}O_2 \longrightarrow \underset{\text{formaldehyde}}{H_2C{=}O} + H_2O$$

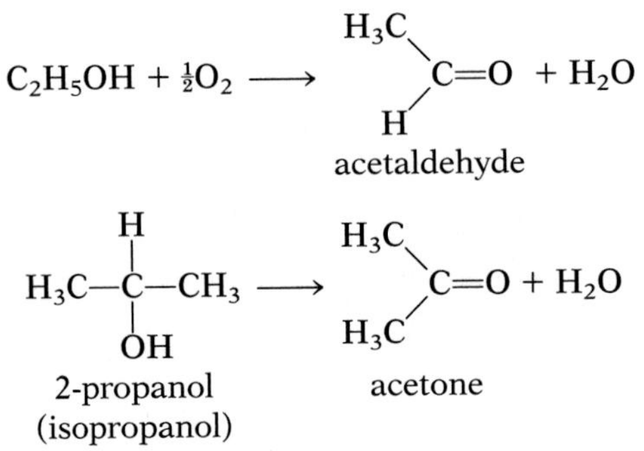

The functional group in these compounds is the carbonyl group, $>C{=}O$. The difference between ***aldehydes*** and ***ketones*** is that *in aldehydes at least one hydrogen atom is bonded to the carbon atom in the carbonyl group, whereas in ketones no hydrogen atoms are bonded to this carbon atom.*

Formaldehyde ($H_2C{=}O$), the simplest aldehyde, has a tendency to *polymerize;* that is, the individual molecules join together to form a compound of high molar mass. This action gives off much heat and is often explosive, so formaldehyde is usually prepared and stored in aqueous solution (to reduce the concentration). This rather disagreeable-smelling liquid is used as a starting material in the polymer industry (see Chapter 14) and in the laboratory as a preservative for animal specimens. Interestingly, the higher-molar-mass aldehydes, such as cinnamic aldehyde

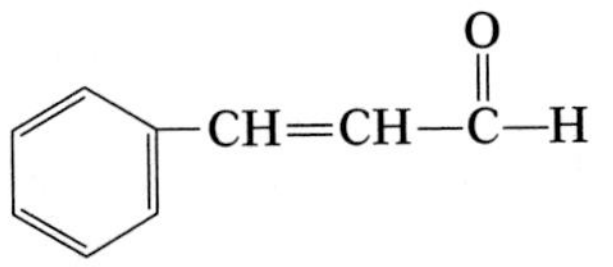

often have a pleasant odor and can be used in the manufacture of perfumes. Cinnamic aldehyde also gives cinnamon its characteristic aroma.

Ketones generally are less reactive than aldehydes. The simplest ketone is acetone, a pleasant-smelling liquid that is used mainly as a solvent for organic compounds. A good example of this is its use as nail polish remover.

Carboxylic Acids

Under appropriate conditions both alcohols and aldehydes can be oxidized to ***carboxylic acids,*** *acids that contain the carboxyl group, —COOH:*

$$CH_3CH_2OH + O_2 \longrightarrow CH_3COOH + H_2O$$

$$CH_3CHO + \tfrac{1}{2}O_2 \longrightarrow CH_3COOH$$

These reactions occur so readily that wine must be protected from atmospheric oxygen while in storage. Otherwise, it would soon taste like vinegar due to the formation of acetic acid. Figure 13.8 shows the structure of some of the common carboxylic acids.

Carboxylic acids are widely distributed in nature; they are found in both the plant and animal kingdoms. All protein molecules are made of ***amino acids,*** which are *a special kind of carboxylic acid that contains at least one carboxyl group (—COOH) and at least one amino group (—NH_2, a nitrogen-containing functional group):*

$$\underset{\text{amino group}}{-NH_2} \qquad \underset{\text{carboxyl group}}{-\overset{\overset{\displaystyle O}{\|}}{C}-OH}$$

Unlike the inorganic acids HCl, HNO_3, and H_2SO_4, carboxylic acids are generally weak. They react with alcohols, usually with catalysis by an inorganic acid, to form pleasant-smelling derivatives of carboxylic acids called esters:

$$CH_3COOH + HOCH_2CH_3 \xrightarrow[\text{catalyst}]{H_2SO_4} \underset{\text{ethyl acetate}}{CH_3COOCH_2CH_3} + H_2O$$

Other common reactions of carboxylic acids are neutralization

$$CH_3COOH + NaOH \longrightarrow CH_3COONa + H_2O$$

and the formation of acid halides, such as acetyl chloride

$$CH_3COOH + PCl_5 \longrightarrow \underset{\substack{\text{acetyl}\\ \text{chloride}}}{CH_3COCl} + HCl + \underset{\substack{\text{phosphoryl}\\ \text{chloride}}}{POCl_3}$$

Acid halides are reactive compounds used as intermediates in the preparation of many other organic compounds, most notably esters.

Esters

Esters *have the general formula R′COOR, where R′ can be H, an alkyl, or an aromatic hydrocarbon group and R is an alkyl or an aromatic hydrocarbon group.* Esters

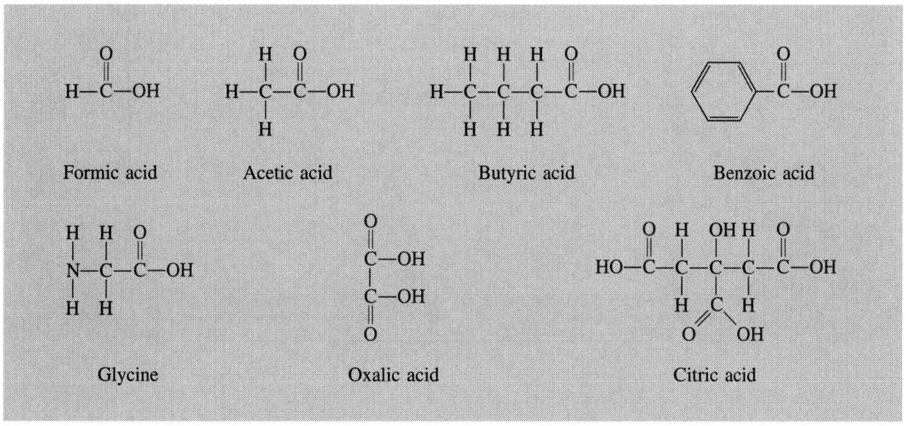

FIGURE 13.8

Some common carboxylic acids. Note that they all contain the COOH group. (Glycine is one of the amino acids found in proteins.)

The odor of fruits is mainly due to the ester compounds in them.

are used in the manufacture of perfumes and as flavoring agents in the confectionery and soft-drink industries. Many fruits owe their characteristic smell and flavor to the presence of esters. For example, bananas contain isopentyl acetate [$CH_3COOCH_2CH_2CH(CH_3)_2$], oranges contain octyl acetate ($CH_3COOC_8H_{17}$), and apples contain methyl butyrate ($CH_3CH_2CH_2COOCH_3$).

The functional group in esters is —COOR. In the presence of an acid catalyst, such as HCl, esters undergo a reaction with water (a *hydrolysis* reaction) to regenerate a carboxylic acid and an alcohol. For example, in acid solution, ethyl acetate is converted to acetic acid:

$$\underset{\text{ethyl acetate}}{CH_3COOC_2H_5} + H_2O \rightleftharpoons \underset{\text{acetic acid}}{CH_3COOH} + \underset{\text{ethanol}}{C_2H_5OH}$$

However, this reaction does not go to completion because the reverse reaction, that is, the formation of an ester from an alcohol and an acid, also occurs to an appreciable extent. On the other hand, when the hydrolysis reaction is run in aqueous NaOH solution, ethyl acetate is converted to sodium acetate, which does not react with ethanol, so this reaction goes to completion from left to right:

$$\underset{\text{ethyl acetate}}{CH_3COOC_2H_5} + NaOH \longrightarrow \underset{\text{sodium acetate}}{CH_3COO^-Na^+} + \underset{\text{ethanol}}{C_2H_5OH}$$

The term ***saponification*** (meaning *soapmaking*) was originally used to describe the reaction between an ester and sodium hydroxide to yield soap molecules (sodium stearate):

$$\underset{\text{ethyl stearate}}{C_{17}H_{35}COOC_2H_5} + NaOH \longrightarrow \underset{\text{sodium stearate}}{C_{17}H_{35}COO^-Na^+} + C_2H_5OH$$

Saponification is now *a general term for alkaline hydrolysis of any type of ester.* Soap molecules are characterized by a long nonpolar hydrocarbon chain and a polar head (the $—COO^-$ group). The hydrocarbon chain is readily soluble in oily substances, while the ionic carboxylate group ($—COO^-$) remains outside the oily nonpolar surface. Figure 13.9 shows the action of soap.

AMINES

Amines are *organic bases. They have the general formula R_3N, where one of the R groups must be an alkyl group or an aromatic hydrocarbon group.* Like ammonia, amines react with water as follows:

$$RNH_2 + H_2O \longrightarrow RNH_3^+ + OH^-$$

Like all bases, the amines form salts when allowed to react with acids:

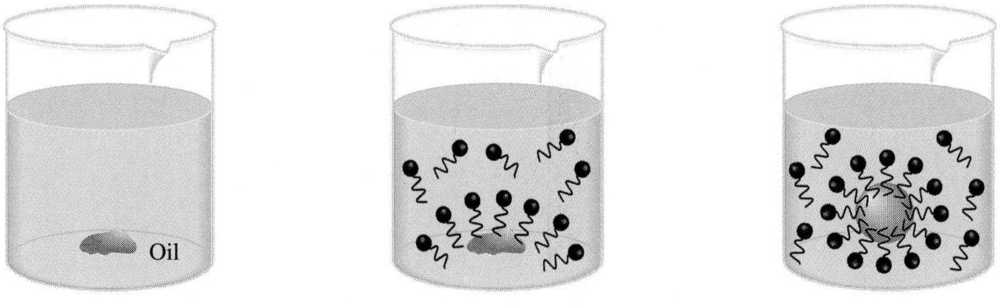

(a) (b) (c)

FIGURE 13.9

The cleansing action of soap. The soap molecule is represented by a polar head and zigzag hydrocarbon tail. An oily spot (a) can be removed by soap (b) because the nonpolar tail dissolves in the oil (like dissolves like), and the entire system becomes soluble in water because the exterior portion is now ionic.

$$\underset{\text{ethylamine}}{CH_3CH_2NH_2} + HCl \longrightarrow \underset{\text{ethylammonium chloride}}{CH_3CH_2NH_3^+Cl^-}$$

These salts are usually colorless, odorless solids. Many of the aromatic amines are carcinogenic.

Summary of Functional Groups

Table 13.4 summarizes the common functional groups. A compound can and often does contain more than one functional group. Generally, the reactivity of a compound is determined by the number and types of functional groups in its makeup.

TABLE 13.4
Important Functional Groups and Their Reactions

Functional group	Name	Typical reactions
$>C=C<$	Carbon-carbon double bond	Addition reactions with halogens, hydrogen halides, and water; hydrogenation to yield alkanes
$-C\equiv C-$	Carbon-carbon triple bond	Addition reactions with halogens, hydrogen halides; hydrogenation to yield alkenes and alkanes
$-\ddot{\underset{..}{O}}-H$	Hydroxyl	Esterification (formation of an ester) with carboxylic acids; oxidation to aldehydes, ketones, and carboxylic acids
$>C=\ddot{\underset{..}{O}}$	Carbonyl	Reduction to yield alcohols; oxidation of aldehydes to yield carboxylic acids
$-\overset{\overset{:O:}{\parallel}}{C}-\ddot{\underset{..}{O}}-H$	Carboxyl	Esterification with alcohols; reaction with phosphorus pentachloride to yield acid chlorides
$-\overset{\overset{:O:}{\parallel}}{C}-\ddot{\underset{..}{O}}-R$	Ester	Hydrolysis to yield acids and alcohols
$-\ddot{N}(R)R$	Amine	Formation of ammonium salts with acids

EXAMPLE 13.3
Reactions of Functional Groups

Cholesterol is a major component of gallstones. It has received much attention because of the suspected correlation between the cholesterol level in the blood and certain types of heart disease, for example, atherosclerosis. Predict the products of the reaction of cholesterol with (a) Br_2, (b) H_2 (in the presence of Pt catalyst), and (c) CH_3COOH, based on cholesterol's structure:

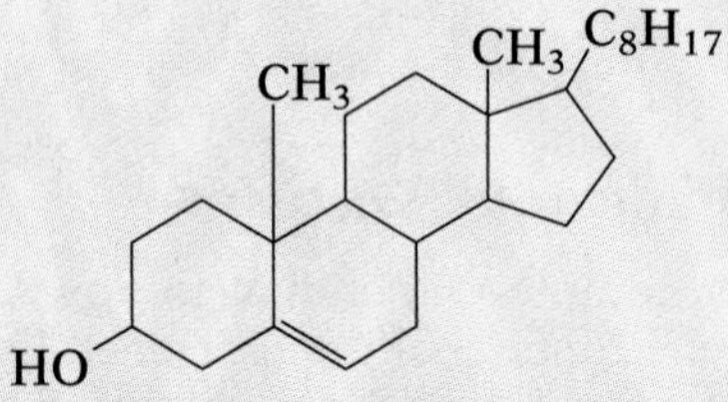

Answer: The first step is to identify the functional groups in cholesterol. There are two: the hydroxyl group and the carbon-carbon double bond.
(a) The reaction with bromine results in the addition of bromine to the double-bonded carbons, which become single-bonded.
(b) This is a hydrogenation reaction. Again, the carbon-carbon double bond is converted to a carbon-carbon single bond.
(c) The acid reacts with the hydroxyl group to form an ester and water. Figure 13.10 shows the products of these reactions.

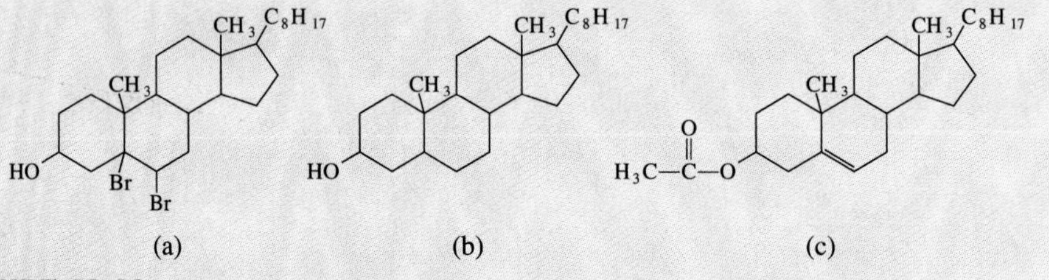

FIGURE 13.10

The products formed by the reaction of cholesterol with (a) molecular bromine, (b) molecular hydrogen, and (c) acetic acid.

PRACTICE EXERCISE

Predict the products of the following reaction:

$CH_3OH + CH_3CH_2COOH \longrightarrow ?$

13.4 CHIRALITY—THE HANDEDNESS OF MOLECULES

Many organic compounds can exist as mirror-image twins, in which one partner may cure disease, quell headache, or smell good, while its mirror-reversed counterpart may be poisonous, smell repugnant, or simply be inert. Compounds that come as mirror-image pairs are sometimes compared with left-handed and right-handed gloves and are referred to as ***chiral,*** or *handed,* molecules. While every molecule can have a mirror image, the difference between chiral and *achiral* (meaning nonchiral) molecules is that only the twins of the former are nonsuperimposable.

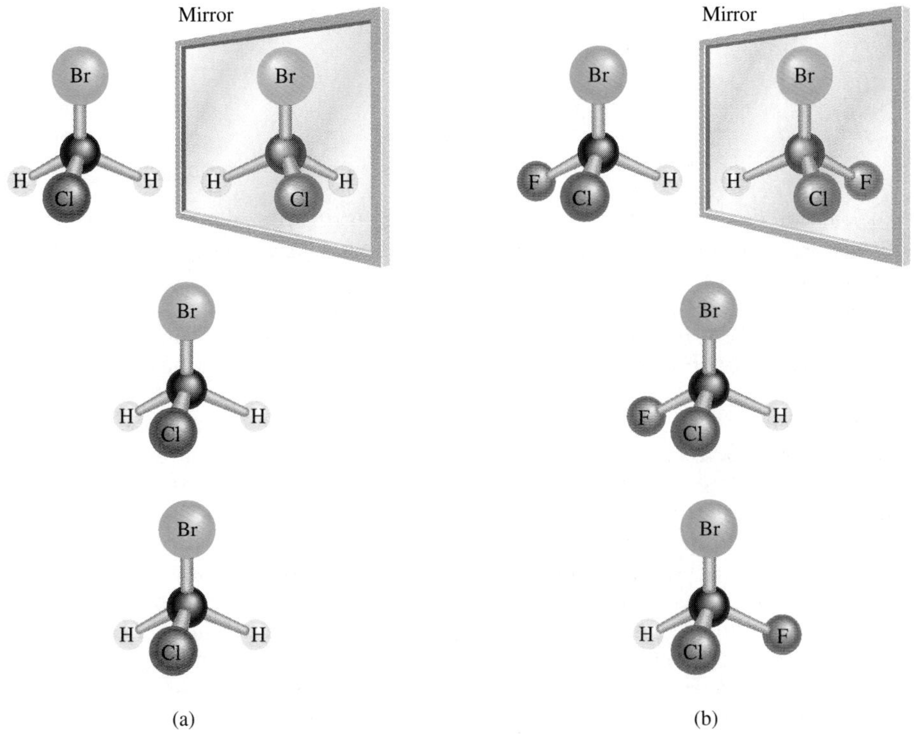

FIGURE 13.11

(a) The CH_2ClBr molecule and its mirror image. Since the molecule and its mirror image are superimposable, the molecule is said to be achiral. (b) The CHFClBr molecule and its mirror image. Since the molecule and its mirror image are not superimposable, no matter how we rotate one with respect to the other, the molecule is said to be chiral.

Consider the substituted methanes CH_2ClBr and CHFClBr. Figure 13.11 shows perspective drawings of these two molecules and their mirror images. The two mirror images of (a) are superimposable, but those of (b) are not, no matter how we rotate the molecules. Thus the CHFClBr molecule is chiral. Careful observation shows that most simple chiral molecules contain at least one *asymmetric* carbon atom—that is, a carbon atom bonded to four different atoms or groups of atoms.

EXAMPLE 13.4

Determining the Chirality of a Molecule

Is the following molecule chiral?

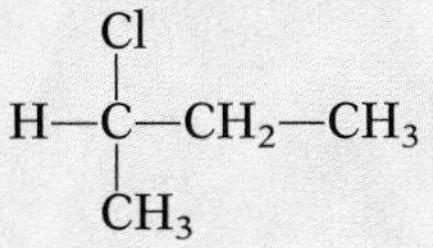

Answer: We note that the central carbon atoms is bonded to a hydrogen atom, a chlorine atom, a $—CH_3$ group, and a $—CH_2—CH_3$ group. Therefore the central carbon atom is asymmetric and the molecule is chiral.

PRACTICE EXERCISE:

Is the following molecule chiral?

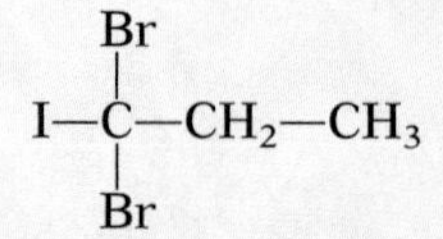

The nonsuperimposable mirror images of a chiral compound are called ***optical isomers*** or ***enantiomers.*** Like geometric isomers, optical isomers come in pairs. However, the optical isomers of a compound have identical physical and chemical properties, such as melting point, boiling point, and chemical reactivity toward molecules that are not chiral themselves. Chiral molecules are said to be optically active because of their ability to rotate the plane of polarization of polarized light as it passes through them. Unlike ordinary light, which vibrates in all directions, plane-polarized light vibrates only in a single plane. To *study the interaction between plane-polarized light and optical isomers* we use a ***polarimeter,*** shown schematically in Figure 13.12. A beam of unpolarized light first passes through a Polaroid sheet, called the polarizer, and then through a sample tube containing a solution of a chiral compound. As the polarized light passes through the sample tube, its plane of polarization is rotated either to the right or to the left. This rotation can be measured directly by turning the analyzer in the appropriate direction until minimal light transmission is achieved (Figure 13.13). If the plane of polarization is rotated to the right, the isomer is said to be dextrorotatory (D); it is levarotatory (L) if the rotation is to the left. The D and L isomers of a chiral substance always rotate the light by the same amount, but in opposite directions. Thus, in *an equimolar mixture of two enantiomers,* called a ***racemic mixture,*** the net rotation is zero.

Chirality plays an important role in biological systems. As we will see in Chapter 14, protein molecules have many asymmetric carbon atoms and their functions are often influenced by their chirality. Because the enantiomers of a chiral compound usually behave very differently from each other in the body, chiral twins are

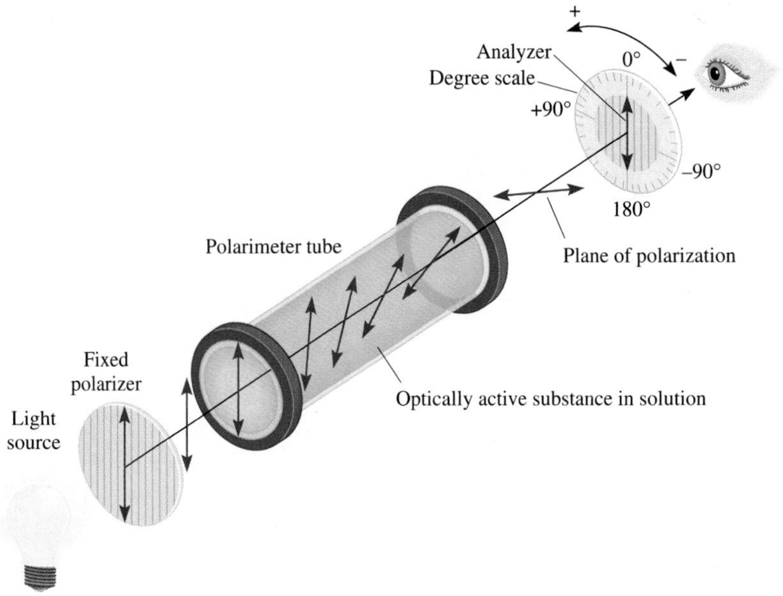

FIGURE 13.12

Operation of a polarimeter. Initially, the tube is filled with an achiral compound. The analyzer is rotated so that its plane of polarization is perpendicular to that of the polarizer. Under this condition, no light reaches the observer. Next, a chiral compound is placed in the tube as shown. The plane of polarization of the polarized light is rotated as it travels through the tube so that some light reaches the observer. Rotating the analyzer (either to the left or to the right) until no light reaches the observer again allows the angle of optical rotation to be measured.

FIGURE 13.13

With one Polaroid sheet over a picture, light passes through. With a second sheet of Polaroid placed over the first so that the axes of polarization of the sheets are perpendicular, little or no light passes through. If the axes of polarization of the two sheets were parallel, light would pass through.

coming under increasing scrutiny among pharmaceutical manufacturers. More than half of the most prescribed drugs in 1995 are chiral. In most of these cases only one enantiomer of the drug works as a medicine, while the other form is useless or less effective or may even cause serious side effects. The best-known case in which the use of a racemic mixture of a drug had tragic consequences occurred in Europe in the late 1950s. The drug thalidomide was prescribed for pregnant women there as an antidote to morning sickness. But by 1962, the drug had to be withdrawn from the market after thousands of deformed children had been born to mothers who had taken it. Only later did researchers discover that the sedative properties of thalidomide belong to D-thalidomide and that L-thalidomide is a potent mutagen. (A *mutagen* is a substance that causes gene mutation, usually leading to deformed offspring.)

Figure 13.14 shows the two enantiomeric forms of another drug, ibuprofen. This popular pain reliever is sold as a racemic mixture, but only the one on the left is potent. The other form is ineffective but also harmless. Organic chemists today are actively researching ways to synthesize enantiomerically pure drugs, or "chiral drugs." Chiral drugs contain only one enantiomeric form both for efficiency and for protection against possible side effects from its mirror-image twin.

SUMMARY

Because carbon atoms can link up with other carbon atoms in straight and branched chains, carbon can form more compounds than any other element.

Methane, CH_4, is the simplest of the alkanes, a family of hydrocarbons with the general formula C_nH_{2n+2}. The cycloalkanes are a subfamily of alkanes whose carbon atoms are joined in a ring. Alkanes and cycloalkanes are saturated hydrocarbons. Ethylene, $CH_2{=}CH_2$, is the simplest of alkenes, a class of hydrocarbons

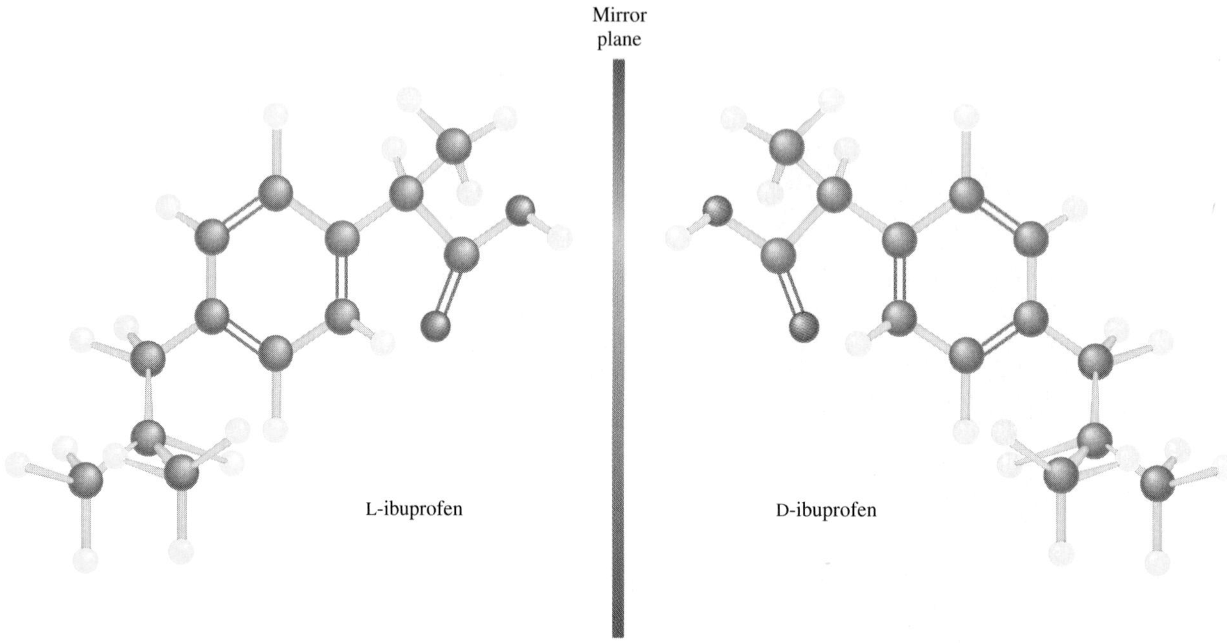

FIGURE 13.14

The L *and* D *forms of ibuprofen are mirror images of each other. The gray spheres are C atoms, the yellow spheres H atoms, and the red spheres O atoms. There is only one asymmetric carbon atom in the molecule. Can you spot it?*

containing carbon-carbon double bonds and having the general formula C_nH_{2n}. Acetylene, CH≡CH, is the simplest of the alkynes, which are compounds that have the general formula C_nH_{2n-2} and contain carbon-carbon triple bonds. Compounds that contain one or more benzene rings are called aromatic hydrocarbons.

Functional groups determine the chemical reactivity of molecules in which they are found. Classes of compounds characterized by their functional groups include alcohols, ethers, aldehydes and ketones, carboxylic acids, esters, and amines.

Chirality refers to molecules that have nonsuperimposable mirror images. Most chiral molecules contain one or more asymmetric carbon atoms. Chiral molecules are widespread in biological systems and are important in drug design.

KEY WORDS

QUESTIONS AND PROBLEMS

HYDROCARBONS

Review Questions

13.1 Define the following terms: hydrocarbon, aliphatic hydrocarbon, aromatic hydrocarbon, alkane, cycloalkane, alkene, alkyne, free radical.

13.2 What do "saturated" and "unsaturated" mean when applied to hydrocarbons? Give examples of a saturated hydrocarbon and an unsaturated hydrocarbon.

13.3 Alkenes exhibit geometric isomerism because rotation about the C=C bond is restricted. Explain what this means.

13.4 Why is it that alkanes and alkynes, unlike alkenes, have no geometric isomers?

13.5 Explain why carbon is able to form so many more compounds than any other element.

13.6 Compare the typical reactions undergone by ethylene and benzene.

Problems

13.7 Draw all possible structural isomers for the following alkane: C_7H_{16}.

13.8 How many distinct chloropentanes, $C_5H_{11}Cl$, could be produced in the direct chlorination of *n*-pentane, $CH_3(CH_2)_3CH_3$? Draw the structure of each molecule.

13.9 Draw all possible isomers for the molecule C_4H_8.

13.10 Draw all possible isomers for the molecule C_3H_5Br.

13.11 The structural isomers of pentane, C_5H_{12}, have quite different boiling points (see Example 13.1). Explain the observed variation in boiling point, in terms of structure.

13.12 Discuss how you can determine which of the following compounds might be alkanes, cycloalkanes, alkenes, or alkynes, without drawing their formulas: (a) C_6H_{12}, (b) C_4H_6, (c) C_5H_{12}, (d) C_7H_{14}, (e) C_3H_4.

13.13 Would you expect cyclobutadiene to be a stable molecule? Explain.

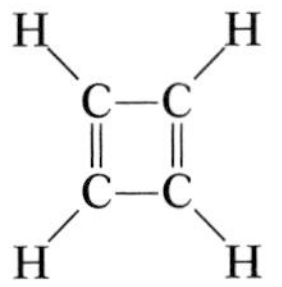

13.14 Suggest two chemical tests that would help you distinguish between these two compounds:

$$CH_3CH_2CH_2CH_2CH_3 \qquad CH_3CH_2CH_2CH{=}CH_2$$

The benzene and cyclohexane molecules both contain six-membered rings. Benzene is a planar molecule, and cyclohexane is nonplanar. Explain.

13.15 Draw the structures of *cis*-2-butene and *trans*-2-butene. Which of the two compounds would have the higher heat of hydrogenation? Explain.

13.16 Acetylene is an unstable compound. It has a tendency to form benzene as follows:

$$3C_2H_2(g) \longrightarrow C_6H_6(l)$$

Calculate the standard enthalpy change in kilojoules for this reaction at 25°C.

13.17 How many different isomers can be derived from ethylene by replacement of two hydrogen atoms by a fluorine atom and a chlorine atom? Draw their structures and name them. Indicate which are structural isomers and which are geometric isomers.

13.18 Geometric isomers are not restricted to compounds containing the C=C bond. For example, certain disubstituted cycloalkanes can exist in the cis and the trans forms. Label the following molecules as the cis and trans isomer of the same compound:

(a)
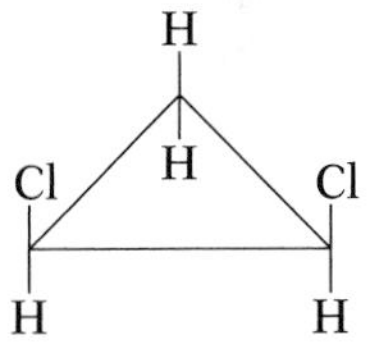

(b)
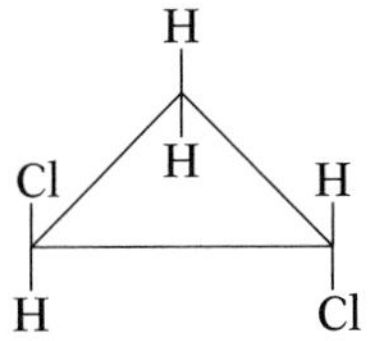

FUNCTIONAL GROUPS

Review Questions

13.19 Define functional group. Why is it logical and useful to classify organic compounds according to their functional groups?

13.20 Draw the Lewis structure for each of the following functional groups: alcohol, ether, aldehyde, ketone, carboxylic acid, ester, amine.

Problems

13.21 Draw structures for molecules with the following formulas: (a) CH_4O, (b) C_2H_6O, (c) $C_3H_6O_2$, (d) C_3H_8O.

13.22 Classify each of the following molecules as alcohol, aldehyde, ketone, carboxylic acid, amine, or ether:

(a) $H_3C—O—CH_2—CH_3$

(b) $H_3C—CH_2—NH_2$

(c) 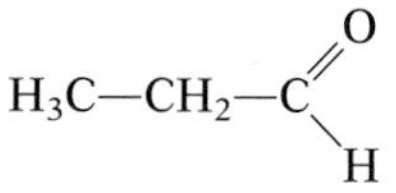

(d) 

(e) 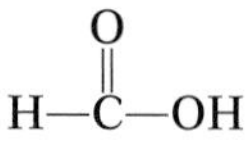

(f) $H_3C—CH_2CH_2—OH$

(g) 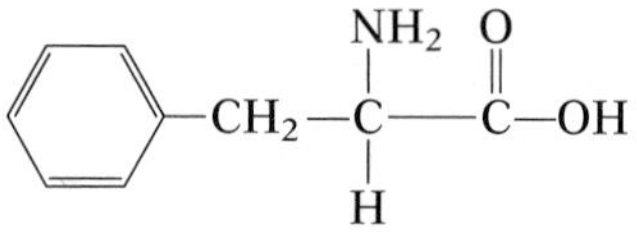

13.23 Generally aldehydes are more susceptible to oxidation in air than are ketones. Use acetaldehyde and acetone as examples and show why ketones such as acetone are more stable than aldehydes in this respect.

13.24 Complete the following equation and identify the products:

$$HCOOH + CH_3OH \longrightarrow$$

13.25 A compound having the molecular formula $C_4H_{10}O$ does not react with sodium metal. In the presence of light, the compound reacts with Cl_2 to form three compounds having the formula C_4H_9OCl. Draw a structure for the original compound that is consistent with this information.

13.26 A compound has the empirical formula $C_5H_{12}O$. Upon controlled oxidation, it is converted into a compound of empirical formula $C_5H_{10}O$, which behaves as a ketone. Draw possible structures of the original compound and the final compound.

13.27 Predict the product or products of each of the following reactions:

(a) $C_2H_5OH + HCOOH \longrightarrow$

(b) $H—C\equiv C—CH_3 + H_2 \longrightarrow$

(c) 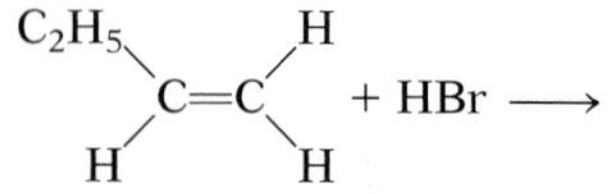

13.28 Identify the functional groups in each of the following molecules:

(a) $CH_3CH_2COCH_2CH_2CH_3$

(b) $CH_3COOC_2H_5$

(c) $CH_3CH_2OCH_2CH_2CH_2CH_3$

CHIRALITY

Review Questions

13.29 What factor determines that a compound is chiral?

13.30 Give examples of a chiral substituted alkane and an achiral substituted alkane.

Problems

13.31 Identify the chiral molecules in Problem 13.8.

13.32 Which of the following amino acids are chiral: (a) $CH_3CH(NH_2)COOH$, (b) $CH_2(NH_2)COOH$, (c) $CH_2(OH)CH(NH_2)COOH$?

NOMENCLATURE

Problems

13.33 Name the following compounds:

(a) 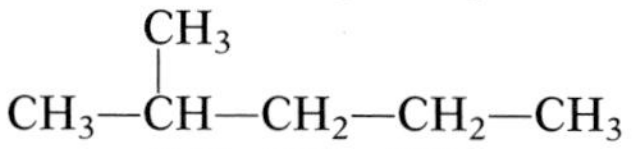

(b) 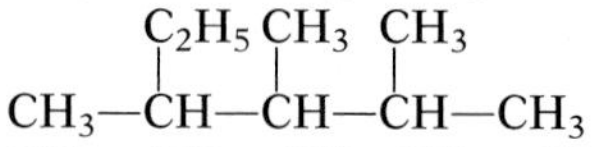

(c) 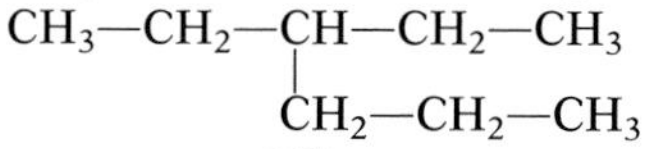

(d) 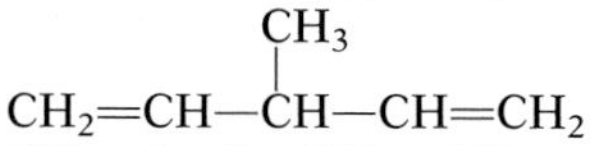

(e) $CH_3—C\equiv C—CH_2—CH_3$

(f) 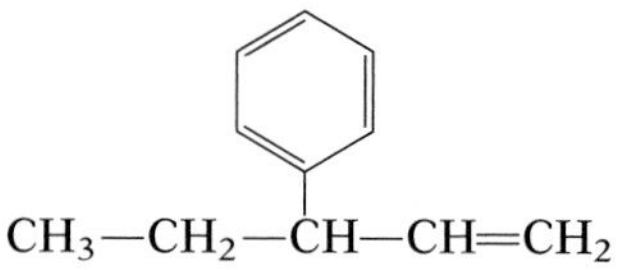

13.34 Write the structural formulas for the following organic compounds: (a) 3-methylhexane, (b) 1,3,5-trichlorocyclohexane, (c) 2,3-dimethylpentane, (d) 2-phenyl-4-bromopentane, (e) 3,4,5-trimethyloctane.

13.35 Write the structural formulas for the following compounds: (a) *trans*-2-pentene, (b) 2-ethyl-1-butene, (c) 4-ethyl-*trans*-2-heptene, (d) 3-phenylbutyne, (e) 2-nitro-4-bromotoluene.

13.36 Name the following compounds:

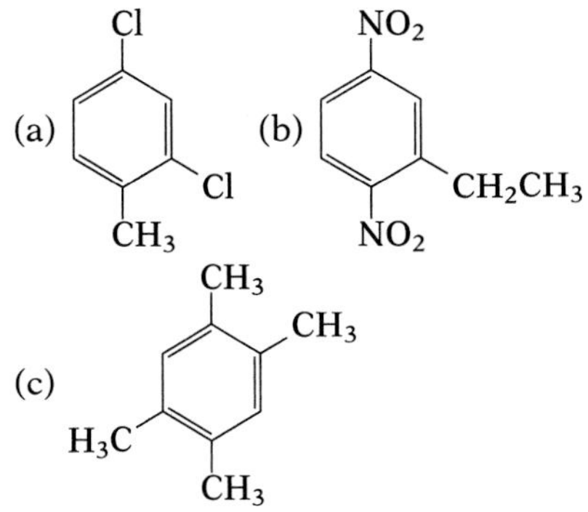

MISCELLANEOUS PROBLEMS

13.37 Draw all the possible structural isomers for the molecule having the formula C_7H_7Cl. The molecule contains one benzene ring.

13.38 Given these data

$$C_2H_4(g) + 3O_2(g) \longrightarrow 2CO_2(g) + 2H_2O(l) \quad \Delta H^\circ = -1411 \text{ kJ}$$

$$2C_2H_2(g) + 5O_2(g) \longrightarrow 4CO_2(g) + 2H_2O(l) \quad \Delta H^\circ = -2599 \text{ kJ}$$

$$H_2(g) + \tfrac{1}{2}O_2(g) \longrightarrow H_2O(l) \quad \Delta H^\circ = -285.8 \text{ kJ}$$

calculate the heat of hydrogenation for acetylene:

$$C_2H_2(g) + H_2(g) \longrightarrow C_2H_4(g)$$

13.39 An organic compound is found to contain 37.5 percent carbon, 3.2 percent hydrogen, and 59.3 percent fluorine by mass. The following pressure and volume data were obtained for 1.00 g of this substance at 90°C:

P (atm)	V (L)
2.00	0.332
1.50	0.409
1.00	0.564
0.50	1.028

The molecule is known to have no dipole moment. (a) What is the empirical formula of this substance? (b) Does this substance behave as an ideal gas? (c) What is its molecular formula? (d) Draw the Lewis structure of this molecule and describe its geometry. (e) What is the systematic name of this compound?

13.40 Which of the following compounds can form hydrogen bonds with water molecules? (a) Carboxylic acids, (b) alkenes, (c) ethers, (d) aldehydes, (e) alkanes, (f) amines.

13.41 Name at least one commercial use for each of the following compounds: (a) 2-propanol (isopropyl alcohol), (b) acetic acid, (c) naphthalene, (d) methanol, (e) ethanol, (f) ethylene glycol, (g) methane, (h) ethylene.

13.42 How many liters of air (78% N_2, 22% O_2 by volume) at 20°C and 1.00 atm are needed for the complete combustion of 1.0 L of octane, C_8H_{18}, a typical gasoline component, of density 0.70 g/mL?

13.43 How many carbon-carbon sigma bonds are present in each of the following molecules: (a) 2-butyne, (b) anthracene (see Figure 13.6), (c) 2,3-dimethylpentane.

13.44 How many carbon-carbon sigma bonds are present in each of the following molecules: (a) benzene, (b) cyclobutane, (c) 2-methyl-3-ethylpentane.

13.45 Draw all the structural isomers of the formula $C_4H_8Cl_2$. Indicate those that are chiral and give them systematic names.

13.46 Combustion of 20.63 mg of compound Y, which contains only C, H, and O, with excess oxygen gave 57.94 mg of CO_2 and 11.85 mg of H_2O. (a) Calculate how many milligrams of C, H, and O were present in the original sample of Y. (b) Derive the empirical formula of Y. (c) Suggest a plausible structure for Y if the empirical formula is the same as the molecular formula.

13.47 Combustion of 3.795 mg of liquid B, which contains only C, H, and O, with excess oxygen gave 9.708 mg of CO_2 and 3.969 mg of H_2O. In a molar mass determination, 0.205 g of B vaporized at 1.00 atm and 200.0°C occupied a volume of 89.8 mL. Derive the empirical formula, molar mass, and molecular formula of B and draw three plausible structures.

13.48 Draw the structures of the geometric isomers of the molecule $CH_3—CH{=}CH—CH{=}CH—CH_3$.

13.49 Suppose that benzene contained three distinct single bonds and three distinct double bonds. How many different isomers would there be for dichlorobenzene ($C_6H_4Cl_2$)? Draw all your proposed structures.

13.50 Beginning with 3-methyl-1-butyne, show how you would prepare the following compounds:

```
       Br CH3
       |  |
(a) CH2=C—CH—CH3

             CH3
             |
(b) CH2Br—CBr2—CH—CH3
```

(c) 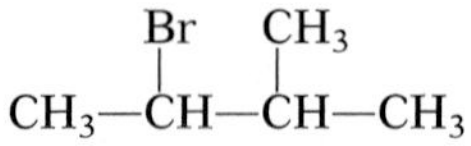

13.51 Draw structures for the following compounds: (a) cyclopentane, (b) *cis*-2-butene, (c) 2-hexanol, (d) 1,4-dibromobenzene, (e) 2-butyne.

13.52 Indicate the classes of compounds having the formulas
(a) $H—C{\equiv}C—CH_3$
(b) C_4H_9OH
(c) $CH_3OC_2H_5$
(d) C_2H_5CHO
(e) C_6H_5COOH
(f) CH_3NH_2

13.53 Ethanol, C_2H_5OH, and dimethyl ether, CH_3OCH_3, are structural isomers. Compare their melting points, boiling points, and solubilities in water.

13.54 Amines are Brønsted bases. The unpleasant smell of fish is due to certain amines. Explain why cooks often add lemon juice to suppress the smell (in addition to enhancing the flavor).

13.55 You are given two bottles, each of which contains a colorless liquid. You are told that one liquid is cyclohexane and the other is benzene. Suggest one chemical test that would allow you to distinguish between these two liquids.

13.56 Write the structural formula of an aldehyde that is an isomer of acetone.

13.57 State which member of each of the following pairs of compounds is the more reactive and explain why: (a) propane and cyclopropane, (b) ethylene and methane.

13.58 The alcohol content in a 60.0-g sample of blood from a driver required 28.6 mL of 0.0765 M $K_2Cr_2O_7$ for titration. Should the police prosecute the individual for drunken driving? (The current legal limit of blood alcohol content in the United States is 0.1 percent by mass.)

13.59 Octane number is assigned to gasoline to indicate the tendency toward knocking in an automobile's engine. The higher the octane number, the more smoothly the fuel will burn without knocking. Branched-chain aliphatic hydrocarbons have higher octane numbers than straight-chain aliphatic hydrocarbons, and aromatic hydrocarbons have the highest octane numbers. (a) Arrange the following compounds in order of decreasing octane numbers: 2,2,4-trimethylpentane, toluene, n-heptane, and 2-methylhexane. (b) Oil refineries carry out *catalytic reforming* in which a straight-chain hydrocarbon, in the presence of a catalyst, is converted to an aromatic molecule and a useful by-product. Write an equation for the conversion from n-heptane to toluene. (c) In recent years, methyl tert-butyl ether has been widely used as an antiknocking agent to enhance the octane number of gasoline. Write the structural formula of the compound.

13.60 How many asymmetric carbon atoms are present in each of the following compounds? [The structure in (d) is thalidomide.]

(a) 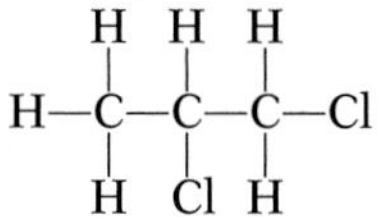

(b) 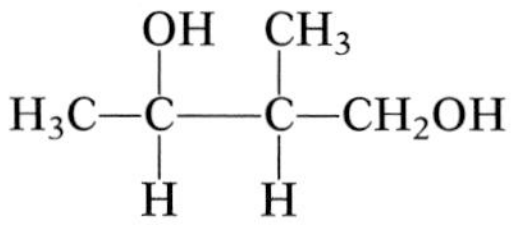

(c) 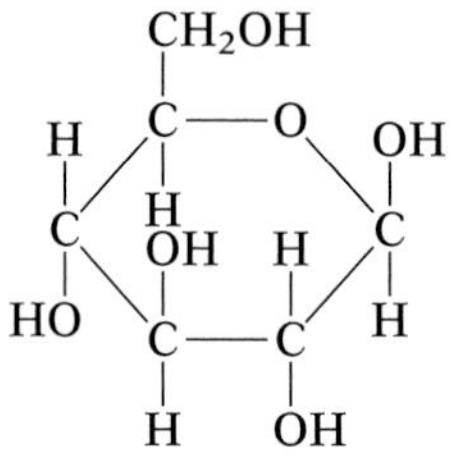

(d) 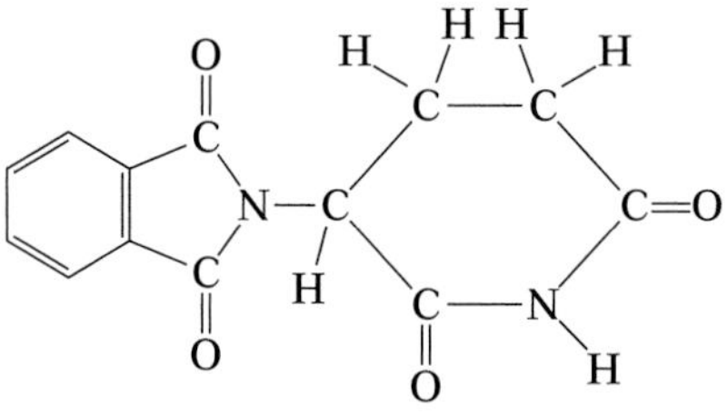

Answers to Practice Exercises: 13.1 5; **13.2** (b) and (c); **13.3** $CH_3CH_2COOCH_3$ and H_2O; **13.4** No.

CHAPTER 14

ORGANIC POLYMERS—SYNTHETIC AND NATURAL

◆ Art museums sometimes call in chemists, or art conservation scientists, to help identify the materials and methods used to produce a work of art. Scientists have participated in art conservation for nearly a century, but only in the last forty years, since the introduction of sensitive analytical techniques, have chemists made a major impact in this field. The sensitivity of these techniques helps to overcome the major problem in conservation science—small sample size. Samples removed from works of art must be essentially microscopic in order to minimize damage.

Conservation scientists perform analyses on paintings, frames, works of art on paper, sculpture, and textiles. Their work can be crucial to the historical accuracy and durability of restoration efforts. In addition, these scientists are occasionally called upon to authenticate pieces that may be worth millions of dollars—or a great deal less.

In one case, scientists at the National Gallery of Art in Washington, D.C., analyzed paint samples from *The Annunciation with Saint Francis and Saint Louis of Toulouse,* a fifteenth-century work by the Italian artist Cosima Tura, to assist in its restoration. The four panels that make up the piece needed cleaning and restoration. After the initial cleaning, it could be seen that the red areas of the painting were in good condition, but the blue pigments were flaking off. Determining what medium, or matrix, was used to bind the pigment to the canvas would help conservators with the delicate task of repairing the painting.

At the time Tura executed his *Annunciation,* the popular media for artists' paints were organic substances containing *proteins*—high-molar-mass compounds consisting primarily of carbon, hydrogen, nitrogen, and oxygen. The paints Tura used could have contained any of four proteins: collagen (from animal-skin glue), milk protein (casein), egg yolk, and egg white. Using hydrolysis and sophisticated chromatographic techniques, which separate compounds according to their molar mass, scientists were able to analyze minuscule paint samples and identify the protein components in the *Annunciation.* They found that the painting was worked in mixed media. The blues and greens were bound in collagen from animal skin, whereas the reds and browns were painted in tempera, an egg-yolk medium that was preferred by artists during the Middle Ages, before the introduction of oil-based paint early in the fifteenth century. Tempera not only dries quickly and adheres well (think of the challenge of washing dishes after you've had eggs for breakfast), but also does not distort the color of pigments. Paintings rendered in tempera have retained their brilliance and clarity over hundreds of years. Collagen-based paints are less oily and more brittle than tempera. Their use in the predominantly blue garments in Tura's painting accounts for the flaking in those areas.

Part of* The Annunciation *before (left) and after (right) conservation treatment.

In addition to protein-based and oil paints, many twentieth-century painters have worked in synthetic polymers. (Polymers are molecules with very high molar masses and can be natural, like proteins, or synthetic, like acrylic.) With advanced techniques similar to the one used to identify the media in Tura's *Annunciation,* chemists at the National Gallery and in museum laboratories around the world can analyze these paints as well as other protein-based media, such as ivory, bone, wool, and leather. ◆

14.1 PROPERTIES OF POLYMERS

A ***polymer*** is *a molecular compound with a high molar mass, ranging into thousands and millions of grams.* The physical properties of these so-called *macromolecules* differ greatly from those of small, ordinary molecules, and special techniques are required to study them.

Polymers are divided into two classes: natural and synthetic. Natural polymers include proteins, nucleic acids, cellulose (polysaccharides), and rubber (polyisoprene). Most synthetic polymers are organic compounds. Familiar examples are nylon, poly(hexamethylene adipamide); Dacron, poly(ethylene terephthalate); and Lucite or Plexiglas, poly(methyl methacrylate).

The development of polymer chemistry began in the 1920s. Chemists were making great progress in clarifying the chemical structures of various substances, but they were generally puzzled by the behavior of certain materials, including wood, gelatin, cotton, and rubber. For example, when rubber, with the known empirical formula of C_5H_8, was dissolved in an organic solvent, the solution displayed several unusual properties—high viscosity, low osmotic pressure, and negligible depression in the freezing point. These observations strongly suggested the presence of very high molar mass solutes. Convincing evidence was provided by the German chemist Hermann Staudinger, who clearly showed that substances such as rubber and gelatin are enormously large molecules, each of which contains many thousands of atoms held together by covalent bonds.

Once the structures of macromolecules were understood, the way was open for manufacturing polymers, which now pervade almost every aspect of our daily lives. About 90 percent of today's chemists, including biochemists, work with polymers.

14.2 SYNTHETIC ORGANIC POLYMERS

Because of their size, we might expect molecules containing thousands of carbon and hydrogen atoms to form an enormous number of structural and geometric isomers (if C=C bonds are present). However, these molecules all contain rather *simple repeating units,* called ***monomers,*** which severely restrict the number of isomers that can exist. Synthetic polymers are made by joining monomers together, one at a time. Here we will discuss two types of reactions that lead to polymerization: addition reactions and condensation reactions.

ADDITION REACTIONS

In Chapter 13 we saw that addition reactions occur with unsaturated compounds containing C=C and C≡C bonds. Hydrogenation and reactions between hydrogen halides and alkenes and alkynes are examples of addition reactions.

Consider the addition reaction by which a polymer is formed from ethylene monomers. The first step involves the generation of free radicals from an initiator molecule R_2:

$$R_2 \longrightarrow 2R\cdot$$

where R_2 contains a weak covalent bond that can be readily broken with heat. The

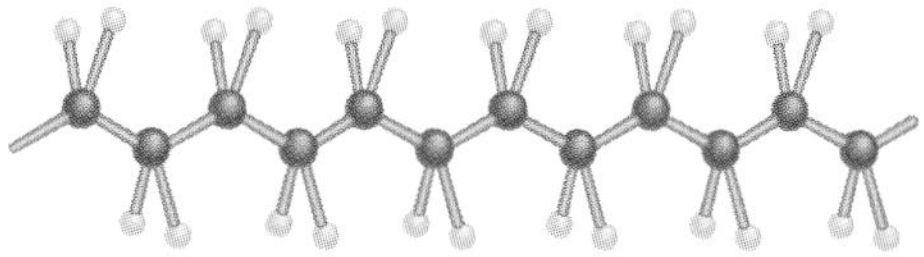

FIGURE 14.1

Structure of polyethylene. Each carbon atom is sp^3-hybridized.

R· radical attacks or adds onto an ethylene molecule by breaking the pi bond between the C atoms to form a new radical as follows:

$$R\cdot + H_2C{=}CH_2 \longrightarrow R{-}CH_2{-}CH_2\cdot$$

which further reacts with another ethylene molecule, and so on:

$$R{-}CH_2{-}CH_2\cdot + CH_2{=}CH_2 \longrightarrow R{-}CH_2{-}CH_2{-}CH_2{-}CH_2\cdot$$

Very quickly a long chain of CH_2 groups is built. Eventually, this process ends with the combination of two long-chain radicals to give the polymer called polyethylene (Figure 14.1):

$$R{+}CH_2{-}CH_2{+}_nCH_2CH_2\cdot + R{+}CH_2{-}CH_2{+}_nCH_2CH_2\cdot \longrightarrow R{+}CH_2{-}CH_2{+}_nCH_2CH_2{-}CH_2CH_2{+}CH_2{-}CH_2{+}_nR$$

where ${+}CH_2{-}CH_2{+}_n$ is a convenient shorthand convention for representing the repeating unit in the polymer. The value of n is on the order of hundreds.

Polyethylene is one of the most common and important polymers. About 10 million tons of the substance are produced in the United States each year for use in pipes, bottles, electrical insulation, toys, building materials, and mailing envelopes. The properties of polyethylene depend on the structure and treatment of the polymer. Polyethylene with a straight chain, that is, a polymer in which propagation of the growing chain proceeds along the same direction, can be packed closely together and is called high-density polyethylene (HDPE). An important property of HDPE is that it possesses a fair amount of crystallinity, which means that in the folded geometry of the polymer segments, the $-CH_2-CH_2-$ chains line up in an orderly fashion (Figure 14.2). The whole molecule, however, does not possess the long-range order we find in a solid crystalline substance. "Crystalline" polymers are harder than rubber, yet due to the lack of long-range order, they are flexible and tough, not brittle like glass. Milk jugs are made of HDPE.

Milk jugs made of high-density polyethylene (HDPE).

Low-density polyethylene (LPDE) is synthesized under different conditions and has a branched-chain structure, which cannot pack closely together, so it

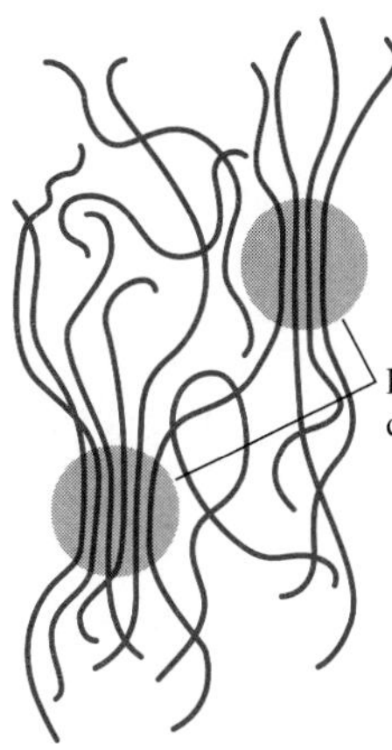

FIGURE 14.2

Regions of crystallinity in high-density polyethylene.

FIGURE 14.3

The bottle on the left is made of rigid, high-density polyethylene. The one on the right is made of flexible, low-density polyethylene.

FIGURE 14.4

Common mailing envelopes made of Tyvek.

A giant bubble of tough, transparent PVC film.

forms a meshlike structure:

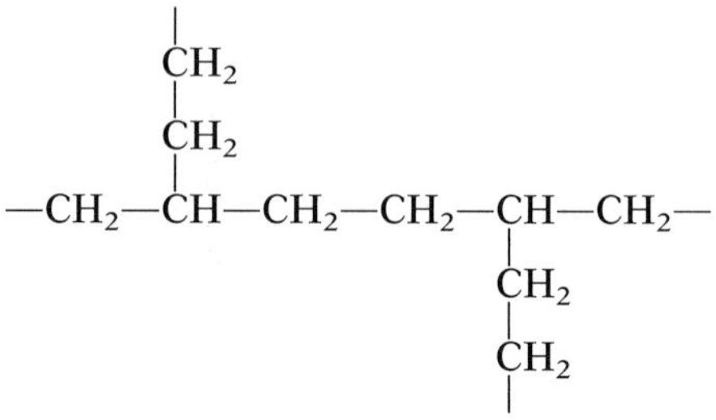

LDPE does not possess any crystallinity; its softness and flexibility make it suitable for plastic bags, refrigerator dishes, and electrical insulation. Figure 14.3 shows examples of HDPE and LDPE.

A special type of polyethylene, called Tyvek, is formed by compressing intertwined HDPE fibers into sheets. Tyvek has great mechanical strength and is used to make mailing envelopes and to insulate houses (Figure 14.4).

Polymers that are made from only one type of monomer, as polyethylene is, are called ***homopolymers.*** Other important homopolymers synthesized by addition reactions are Teflon, polytetrafluoroethylene (Figure 14.5), and poly(vinyl chloride) (PVC). For polytetrafluoroethylene

$$n\mathrm{CF_2{=}CF_2} \longrightarrow -\!(\mathrm{CF_2{-}CF_2})\!-_n$$

the monomer is tetrafluoroethylene ($\mathrm{CF_2{=}CF_2}$). For PVC the monomer is vinyl chloride ($\mathrm{CHCl{=}CH_2}$):

$$n\mathrm{CHCl{=}CH_2} \longrightarrow -\!(\mathrm{CH_2{-}\underset{\displaystyle Cl}{\underset{|}{C}H}})\!-_n$$

PVC is a hard, rigid material of high molar mass that is used in garden hoses and plastic plumbing pipe, among other things. With the addition of a *plasticizer,* the polymer becomes more flexible. The effect of the plasticizer is to reduce the interaction between chains and thus to make the polymer more pliable. Over a period of time, the plasticizer may leach out and an object made of this polymer-plasticizer blend, such as garden hose, will stiffen and possibly crack.

The chemistry becomes more complex if the starting units in an addition reaction are asymmetric:

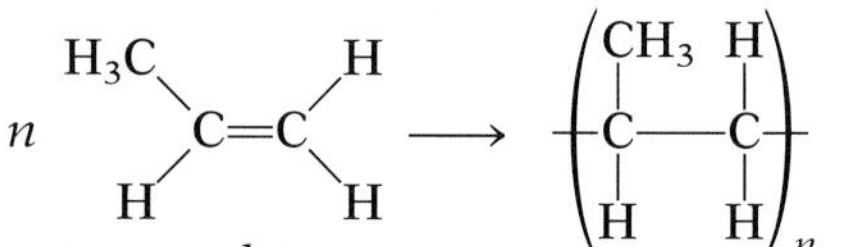

propylene polypropylene

Several structurally different polymers can result from an addition reaction involving propylenes (Figure 14.6). If the additions occur randomly, we obtain *atactic* polypropylenes, which do not pack together well. These polymers are rubbery, amorphous (noncrystalline), and relatively weak. Two other possibilities are an *isotactic* structure, in which substituent groups (R groups) are all on the same side of the carbon atoms, and a *syndiotactic* form, in which the R groups alternate on the left and right sides of the carbon backbone of the polymer. Of these, the isotactic isomer has a higher melting point and greater crystallinity and is endowed with superior mechanical properties.

A major problem that the polymer industry faced in the beginning was how to

FIGURE 14.5

A cooking utensil coated with Silverstone, which contains polytetrafluoroethylene.

selectively synthesize either the isotactic or syndiotactic polymer without having it contaminated by other products. The answer came from the Italian chemist Giulio Natta and the German chemist Karl Ziegler, who demonstrated that certain catalysts, including triethylaluminum [$Al(C_2H_5)_3$] and titanium trichloride ($TiCl_3$), promote the formation only of specific isomers. Their work opened up almost limitless possibilities for chemists to design polymers to suit any purpose.

Rubber is probably the best-known organic polymer and the only true hydrocarbon polymer found in nature. It is formed by the radical addition of the monomer isoprene:

$$n CH_2{=}\overset{\displaystyle CH_3}{\underset{\displaystyle |}{C}}{-}CH{=}CH_2 \longrightarrow \left(\begin{array}{c} CH_3 \quad\quad H \\ C{=}C \\ {-}CH_2 \quad CH_2{-} \end{array}\right)_n \quad \text{and/or} \quad \left(\begin{array}{c} CH_2 \quad\quad H \\ C{=}C \\ CH_3 \quad CH_2{-} \end{array}\right)_n$$

isoprene   poly-*cis*-iosprene   poly-*trans*-isoprene

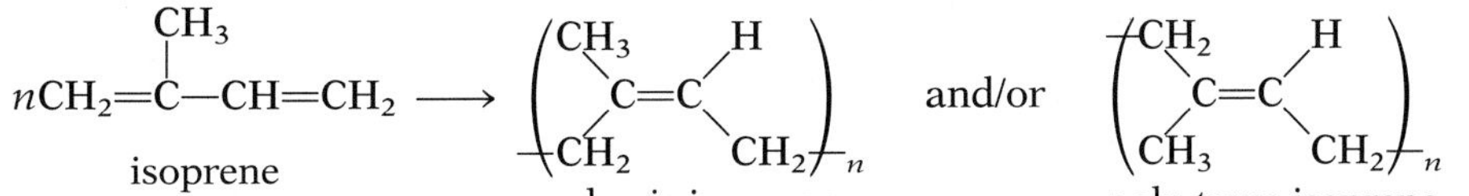

Giulio Natta (1903–1979)

Actually, polymerization can result in either poly-*cis*-isoprene or poly-*trans*-isoprene—or a mixture of both, depending on reaction conditions. Note that in the

(a)

(b)

(c)

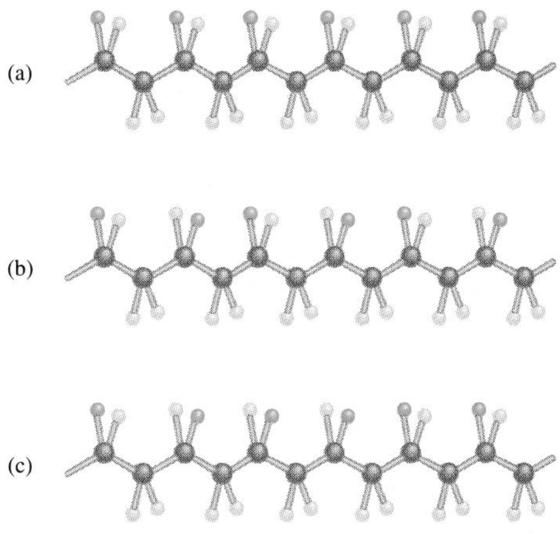

FIGURE 14.6

Stereoisomers of polymers. When the R group (green sphere) is CH_3, the polymer is polypropylene. (a) When the R groups are all on one side of the chain, the polymer is said to be isotactic. (b) When the R groups alternate from side to side, the polymer is said to be syndiotactic. (c) When the R groups are disposed at random, the polymer is atactic.

FIGURE 14.7

Latex (aqueous suspension of rubber particles) being collected from a rubber tree.

cis isomer the two CH_2 groups are on the same sides of the C=C bond, whereas the same groups are across from each other in the trans isomer. Natural rubber is poly-*cis*-isoprene, which is extracted from the tree *Hevea brasiliensis* (Figure 14.7).

An unusual and very useful property of rubber is its elasticity. Rubber can be extended up to ten times its length and it will return to its original size. In contrast, a piece of copper wire can be stretched only a small percentage of its length and still return to its original size. Unstretched rubber is amorphous. Stretched rubber, however, possesses a fair amount of crystallinity and order. The elastic property of rubber is due to the flexibility of the long-chain molecules. In the bulk state, however, rubber is a tangle of polymeric chains; and if the external force applied to stretch it is strong enough, individual chains slip past one another, thereby causing the rubber to lose most of its elasticity. In 1839, the American chemist Charles Goodyear discovered that natural rubber could be cross-linked with sulfur (using zinc oxide as the catalyst) to prevent chain slippage (Figure 14.8). This process, known as *vulcanization,* paved the way for many practical and commercial uses of rubber—for instance, in automobile tires and dentures.

Bubble gums contain synthetic styrene-butadiene rubber.

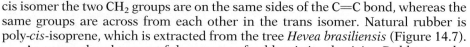

During World War II a shortage of natural rubber in the United States prompted an intensive program to produce synthetic rubber. Most synthetic rubbers (called *elastomers*) are formed from products of the petroleum industry such as ethylene, propylene, and butadiene. For example, chloroprene molecules polymerize readily to form polychloroprene, commonly known as *neoprene,* which has properties that are comparable or even superior to those of natural rubber:

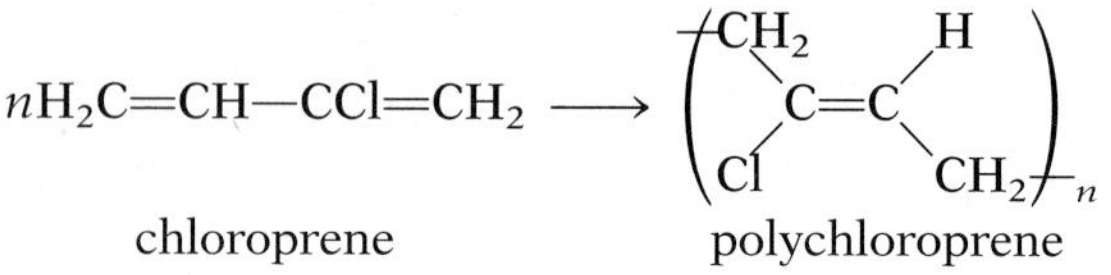

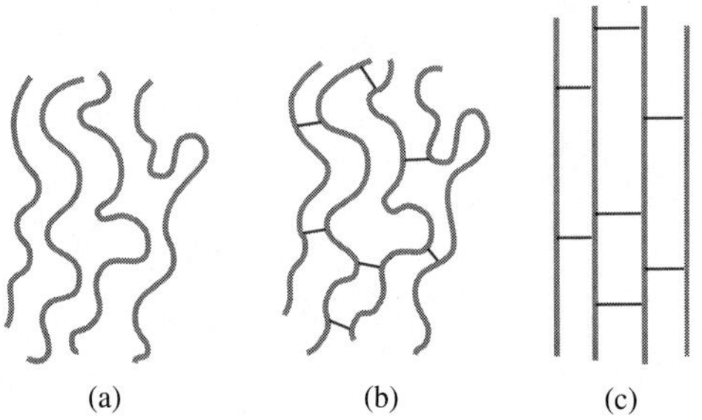

FIGURE 14.8

Rubber molecules ordinarily are bent and convoluted. Parts (a) and (b) represent the long chains before and after vulcanization, respectively; part (c) shows the alignment of molecules when stretched. Without vulcanization, these molecules would slip past one another.

Another important synthetic rubber is formed by the addition of butadiene to styrene in a 3:1 ratio to give styrene-butadiene rubber (SBR) (Table 14.1). Because

TABLE 14.1
Some Monomers and Their Common Synthetic Polymers

Monomer		Polymer	
Formula	**Name**	**Name and formula**	**Uses**
$H_2C{=}CH_2$	Ethylene	Polyethylene $-(CH_2-CH_2)_n-$	Plastic piping, bottles, electrical insulation, toys
$H_2C{=}\overset{H}{\underset{CH_3}{C}}$	Propylene	Polypropylene $-(\underset{CH_3}{CH}-CH_2-\underset{CH_3}{CH}-CH_2)_n-$	Packaging film, carpets, crates for soft-drink bottles, lab wares, toys
$H_2C{=}\overset{H}{\underset{Cl}{C}}$	Vinyl chloride	Poly(vinyl chloride) (PVC) $-(CH_2-\underset{Cl}{CH})_n-$	Piping, siding, gutters, floor tile, clothing, toys
$H_2C{=}\overset{H}{\underset{CN}{C}}$	Acrylonitrile	Polyacrylonitrile (PAN) $-(CH_2-\underset{CN}{CH})_n-$	Carpets, knitwear
$F_2C{=}CF_2$	Tetrafluoroethylene	Polytetrafluoroethylene (Teflon) $-(CF_2-CF_2)_n-$	Coating on cooking utensils, electrical insulation, bearings
$H_2C{=}\overset{COOCH_3}{\underset{CH_3}{C}}$	Methyl methacrylate	Poly(methyl methacrylate) (Plexiglas) $-(CH_2-\overset{COOCH_3}{\underset{CH_3}{C}})_n-$	Optical equipment, home furnishing
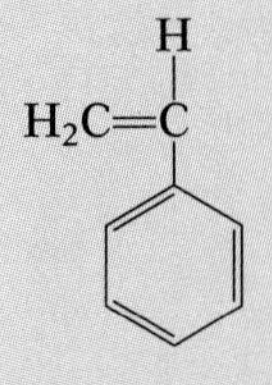	Styrene	Polystyrene 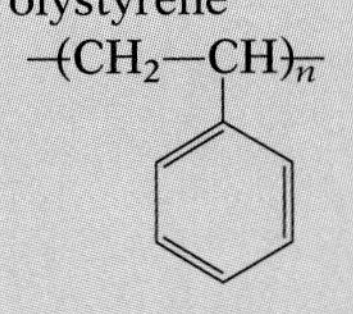	Containers, thermal insulation (ice buckets, water coolers), toys
$H_2C{=}\overset{H}{C}-\overset{H}{C}{=}CH_2$	Butadiene	Polybutadiene $-(CH_2CH{=}CHCH_2)_n-$	Tire tread, coating resin
See above structures	Butadiene and styrene	Styrene-butadiene rubber (SBR) 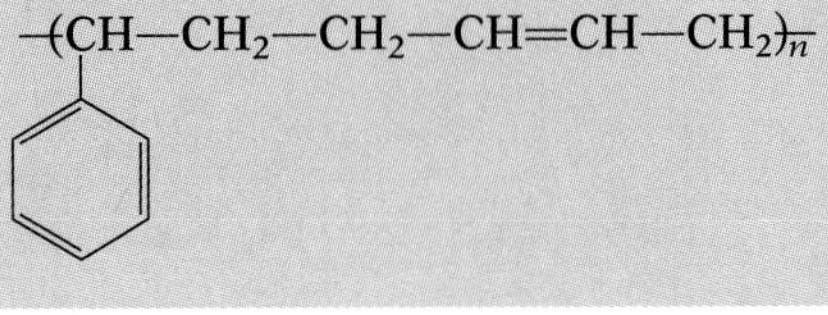	Synthetic rubber

two different types of monomer are needed to synthesize SBR, this process is an example of copolymerization. ***Copolymerization*** is *the formation of a polymer containing two or more different monomers.*

Wallace Carothers (1896–1937)

Condensation Reactions

The condensation reaction was introduced in Chapter 13 (see p. 381). One of the best-known polycondensation processes is the reaction between hexamethylenediamine and adipic acid, shown in Figure 14.9. The final product, called nylon 66, was first made by the American chemist Wallace Carothers in 1931. (The nylons are described by a numbering system that indicates the number of carbon atoms in the monomer units. Because there are six carbon atoms each in hexamethylenediamine and adipic acid, it was logical to name the product nylon 66.) Figure 14.10 shows how nylon 66 is prepared in the laboratory. Nylon can be processed into fibers that are stronger, tougher, and more elastic than natural fibers such as silk and wool (Figure 14.11). It is used in clothing, carpets, ropes, and many other

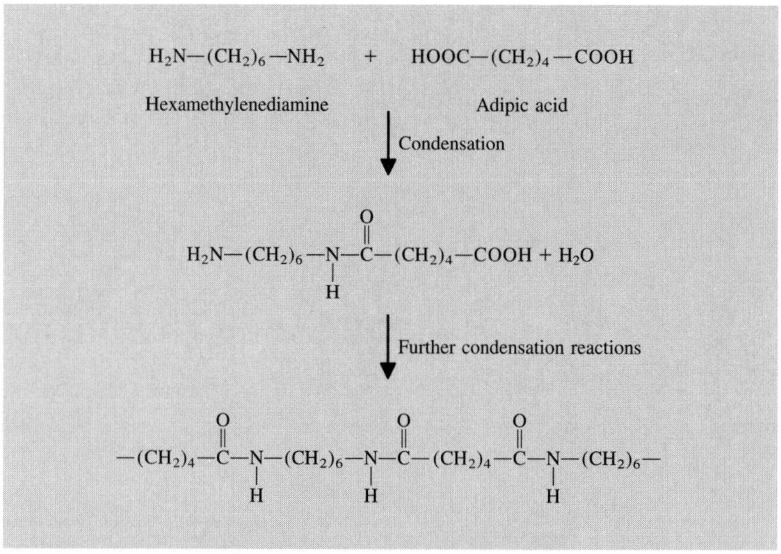

FIGURE 14.9

The formation of nylon by the condensation reaction between hexamethylenediamine and adipic acid.

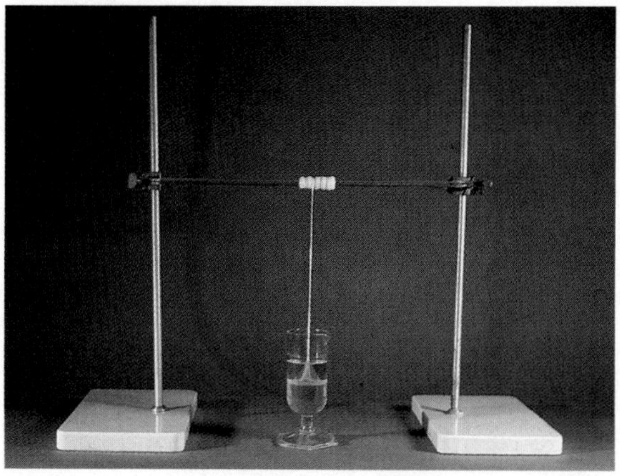

FIGURE 14.10

The nylon rope trick. Adding a solution of adipoyl chloride (an adipic acid derivative in which the OH groups have been replaced by Cl groups) in cyclohexane to an aqueous solution of hexamethylenediamine causes nylon to form at the interfacial layer. It then can be drawn off.

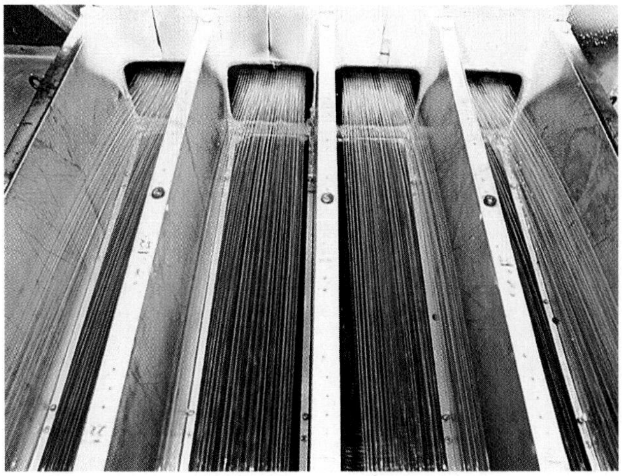

FIGURE 14.11

Molten nylon is forced through the tiny holes of a device called a spinneret. The fiber filaments solidify as they are cooled.

household and industrial items. The versatility of nylons is so great that the annual production of nylons and related substances in the United States now amounts to several billion pounds.

Condensation reactions are also used to produce polymers such as Dacron (polyester):

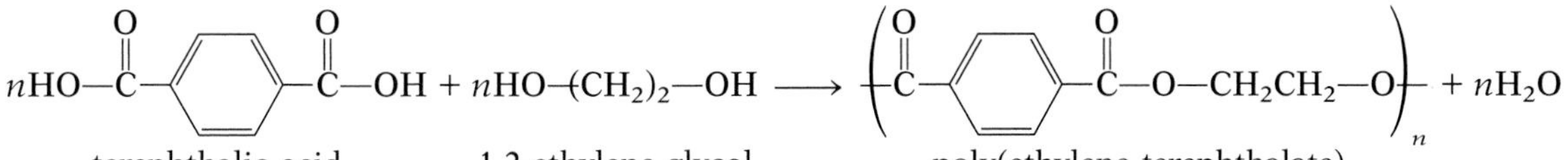

terephthalic acid 1,2-ethylene glycol poly(ethylene terephthalate)

Polyesters are used in fibers, films, and plastic bottles.

14.3 PROTEINS

Having discussed the synthetic organic polymers and natural rubber, we will now look at some other natural polymers. We will study the structure and properties of proteins in this section and then go on to discuss the properties of nucleic acids in Section 14.4.

Proteins are *polymers of amino acids* (see p. 383). They play a key role in nearly all biological processes. Most enzymes, the catalysts of biochemical reactions, are proteins. Proteins facilitate a wide range of other functions, such as transport and storage of vital substances, coordinated motion, mechanical support, and protection against diseases.

The simplest amino acid is glycine:

$$H_2N{-}CH_2{-}\overset{\overset{\displaystyle O}{\|}}{C}{-}OH$$

which is usually written as NH_2CH_2COOH. Twenty different amino acids are the building blocks for the proteins in the human body. With the exception of glycine, all of the amino acids are chiral.

Proteins have high molar masses, ranging from about 5000 g to 1×10^7 g. The

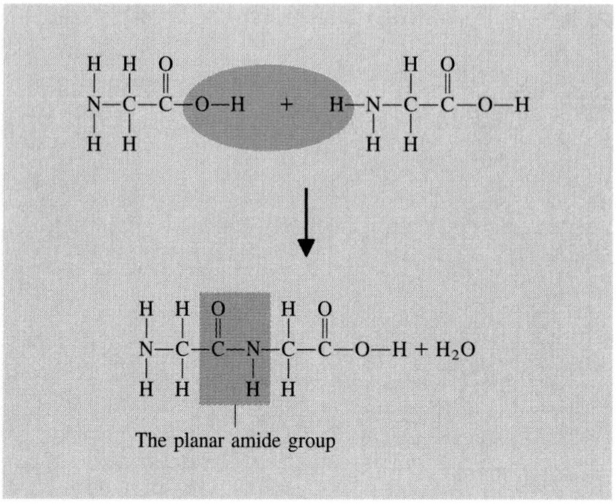

FIGURE 14.12

Condensation reaction between two amino acid molecules (glycine). An amino acid contains at least one amino group (NH_2) and one carboxyl group (COOH). The reaction joins the two molecules together, with the elimination of one water molecule. The resultant carbon-to-nitrogen bond is called a peptide bond.

mass composition of proteins by elements is remarkably constant: carbon, 50 to 55 percent; hydrogen, 7 percent; oxygen, 23 percent; nitrogen, 16 percent; and sulfur, 1 percent.

The first step in the synthesis of a protein molecule is a condensation reaction between two amino acids, as shown in Figure 14.12. The resulting carbon-nitrogen bond is called a *peptide bond,* and the —CO—NH— group is called the *amide* group. (Note that nylon 66 also contains a peptide bond.) The molecule formed from two amino acids is called a *dipeptide.* Either end of the dipeptide can engage in a condensation reaction with another amino acid to form a *tripeptide,* and so on. The final product, the protein molecule, is a *polypeptide;* it can also be thought of as a polymer of amino acids.

An amino acid unit in a polypeptide chain is called a *residue.* Typically, a polypeptide chain contains 100 or more amino acid residues. The sequence of amino acids in a polypeptide chain is written conventionally from left to right, starting with the amino-terminal residue (that is, the end with an unreacted $—NH_2$ group) and ending with the carboxyl-terminal residue (that is, the end with an unreacted —COOH group). For a simple dipeptide, like one formed from glycine and alanine, the two residues may be linked in two different sequences. Figure 14.13 shows that alanylglycine and glycylalanine are different molecules. With 20 different amino

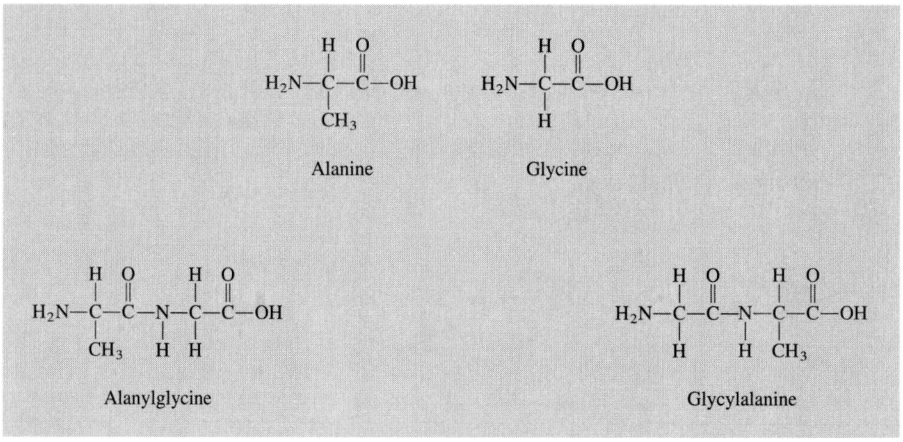

FIGURE 14.13

The formation of two dipeptides from two different amino acids. Alanylglycine is different from glycylalanine in that in alanylglycine the amino and methyl groups are bonded to the same carbon atom.

acids to choose from, 20^2, or 400, different dipeptides can be generated. Even for a very small protein such as insulin, which contains only 50 amino acid residues, the number of chemically different structures that is possible is of the order of 20^{50}, or 10^{65}! This is an incredibly large number when you consider that the total number of atoms in our galaxy is about 10^{68}.

It is estimated that the human body contains about 100,000 different kinds of protein molecules, a large number by ordinary standards, but very small compared with 10^{65}. With so many possibilities for protein synthesis, it is remarkable that generation after generation of cells can produce identical proteins for specific physiological functions.

To understand the properties of a protein molecule, we must know its structure. This prerequisite leads to two questions: What is the amino acid sequence in the polypeptide chain? What is the overall shape of the molecule? Under suitable conditions, proteins can be degraded to yield their amino acid components. The type and number of amino acids present in the original protein can be determined readily by a number of techniques. But the task of determining the sequence or order in which these amino acids are joined together is more involved and requires an instrument called an amino acid "sequencer," which is capable of cleaving amino acid residues from a polypeptide chain one at a time for analysis.

Once the amino acid sequence is known, we can move on to the question about the shape of the protein molecule. Is the polypeptide chain stretched out like a raw noodle, or is it folded in a particular pattern? In the 1930s Linus Pauling and his coworkers conducted a systematic investigation of protein structure. First they studied the geometry of the basic repeating group, that is, the amide group. The carbon-to-nitrogen bond in the amide group actually has considerable "double-bond character," which means that both of the following resonance structures contribute to the overall bonding properties:

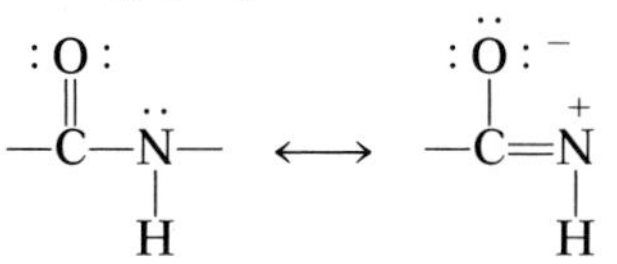

Because it is more difficult (that is, it would take more energy) to twist a double bond than a single bond, the four atoms in the amide group become locked in the same plane, and this restriction has a profound influence on the geometry of the polypeptide chain (Figure 14.14). Figure 14.15 depicts the repeating amide group in a polypeptide chain.

Pauling suggested that there are two common structures for protein molecules, called the *α-helix* and the *β-pleated sheet.* The α-helical structure of a polypeptide chain is shown in Figure 14.16. The helix is stabilized by *intramolecular* hydrogen bonds between NH and CO groups in the main chain, giving rise to an overall rodlike shape. The CO group of each amino acid is hydrogen-bonded to the NH group of the amino acid that is four residues away in the sequence. In this manner all the main-chain CO and NH groups take part in hydrogen bonding. It is known that the structure of a number of proteins, including myoglobin and hemo-

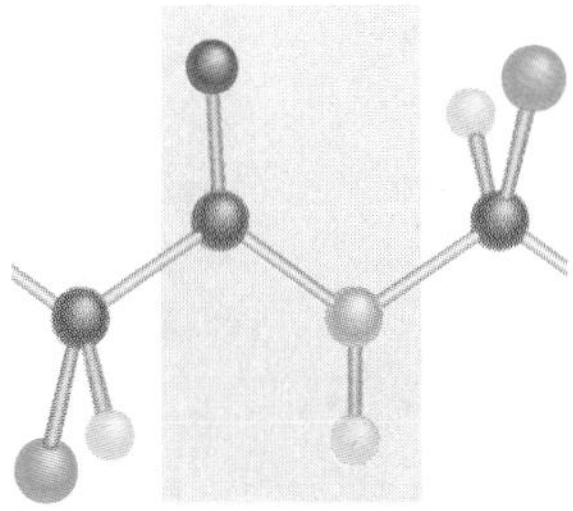

FIGURE 14.14

The planar amide group in protein. Rotation about the carbon-to-nitrogen bond in the amide group is hindered because it has considerable double-bond character. The gray atoms represent carbon; blue, nitrogen; red, oxygen; green, R group; and yellow, hydrogen.

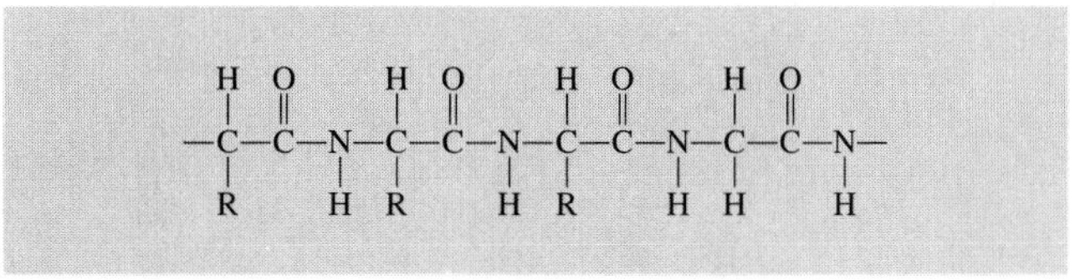

FIGURE 14.15

A polypeptide chain. Note the repeating units of the amide group. The symbol R represents part of the structure characteristic of the individual amino acids. For glycine, R is simply a H atom.

globin, is to a great extent α-helical in nature (Figure 14.17). Other proteins, such as the digestive enzyme chymotrypsin and the electron carrier cytochrome *c*, are almost devoid of α-helical structure.

The β-pleated structure is markedly different from the α-helix in that it is like a sheet rather than a rod. The polypeptide chain is almost fully extended, and each chain forms many *intermolecular* hydrogen bonds with adjacent chains. Figure 14.18 shows the two different types of β-pleated structures, called *parallel* and *antiparallel.* Silk molecules possess the β structure. Because its polypeptide chains are already in extended form, silk lacks elasticity and extensibility, but it is quite strong due to the many intermolecular hydrogen bonds.

The structure of a protein is ultimately responsible for the protein's activities: catalytic action, binding of oxygen and other molecules, and so on. Consider the enzyme chymotrypsin. Chymotrypsin belongs to a class of enzymes called *digestive enzymes,* which catalyze the degradation of food molecules. Without these enzymes it would take about 50 years to digest a meal! We owe our lives to their efficiency. If the polypeptide chain of chymotrypsin were to fold in a different way, the enzyme would not be able to catalyze the degradation of food molecules.

Pauling showed that it is possible to predict a protein structure purely from a knowledge of the geometry of its fundamental building blocks—amino acids. It turns out, however, that many proteins have structures that do not correspond to the α-helical or β structure. Chemists now realize that the three-dimensional structures of these biopolymers are maintained by several types of intermolecular forces in addition to hydrogen bonding. These forces include van der Waals forces (see Chapter 11) and other intermolecular forces (Figure 14.19). The delicate balance of the various interactions can be appreciated by considering an example: When glutamic acid, one of the amino acid residues in two of the four polypeptide chains in hemoglobin, is replaced by valine, another amino acid, the protein molecules aggregate to form insoluble polymers, causing the disease known as sickle cell anemia.

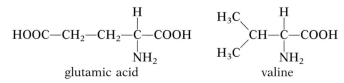

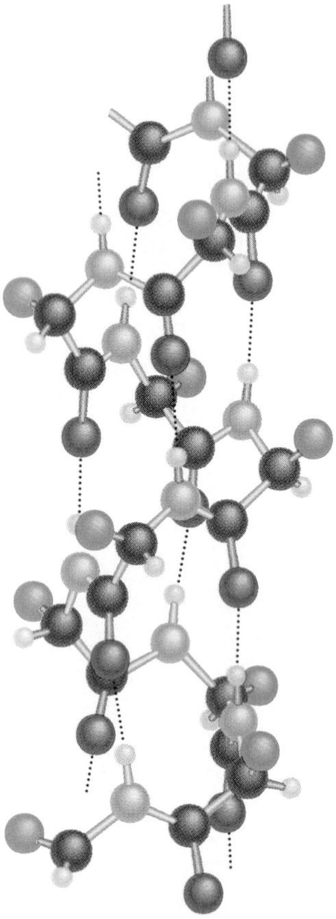

FIGURE 14.16

The α-helical structure of a polypeptide chain. The structure is held in position by intramolecular hydrogen bonds, shown as dotted lines. For the color key, see Figure 14.14.

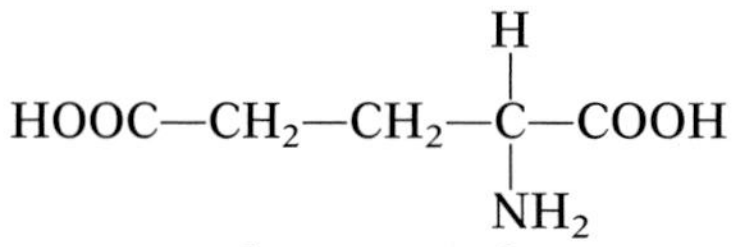

glutamic acid

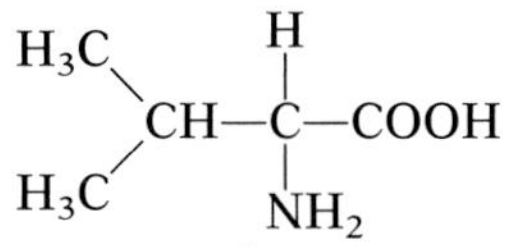

valine

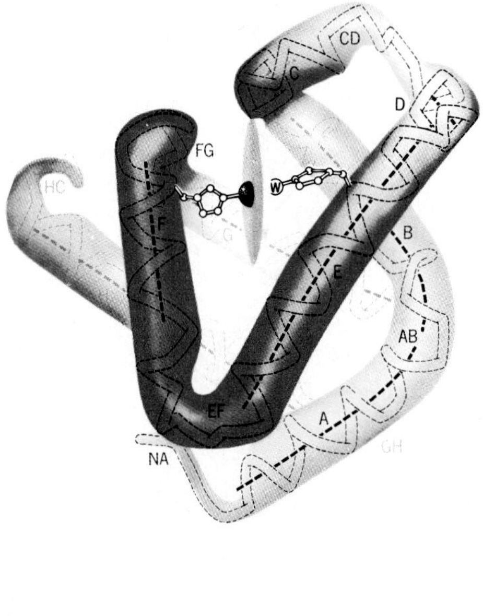

FIGURE 14.17

The myoglobin molecule. The α-helical structure in myoglobin is quite evident. (The disk suspended between different regions of the folded molecule is the iron-containing heme group.)

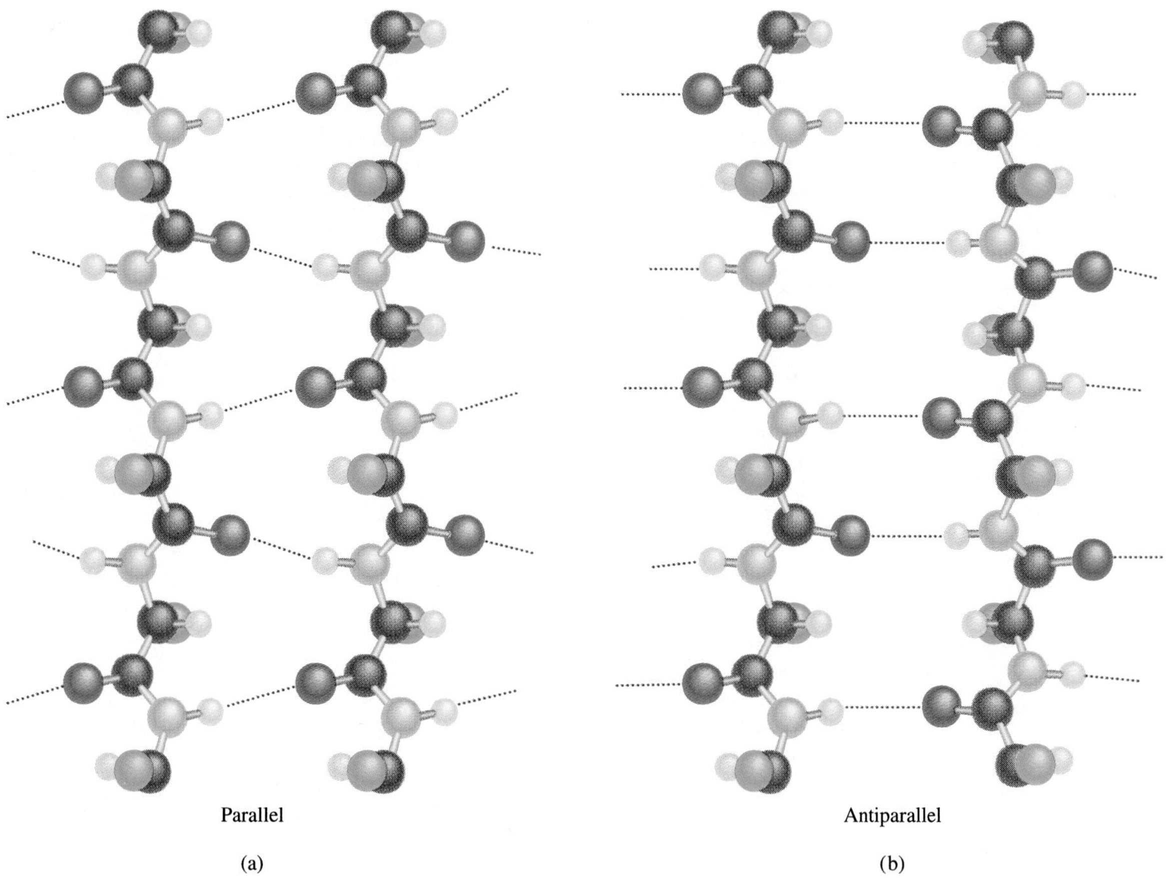

FIGURE 14.18

Hydrogen bonds (a) in a parallel β-pleated sheet, in which all the polypeptide chains run in the same direction, and (b) in an antiparallel β-pleated sheet, in which adjacent polypeptide chains run in opposite directions. For the color key, see Figure 14.14.

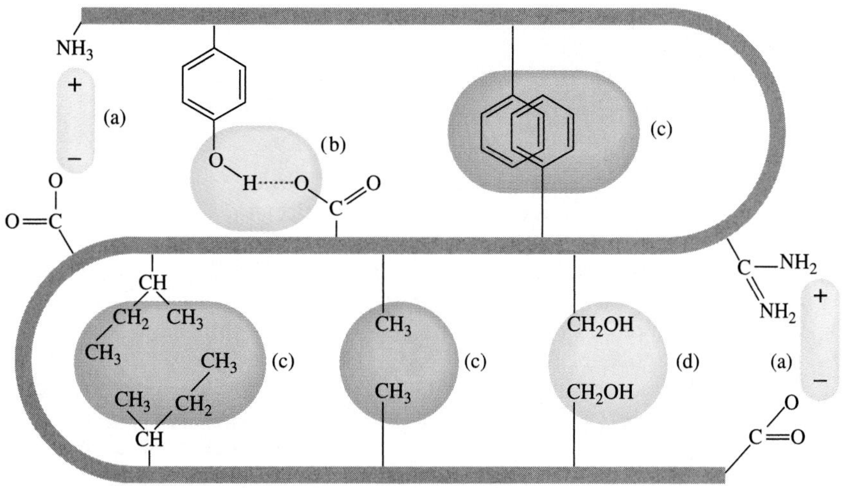

FIGURE 14.19

Intermolecular forces in a protein molecule:
(a) ionic forces,
(b) hydrogen bonding,
(c) dispersion forces, and
(d) dipole-dipole forces.

When proteins are heated above body temperature or when they are subjected to unusual acid or base conditions or treated with special reagents called *denaturants,* they lose some or all of their three-dimensional structure. *Proteins in this state no longer exhibit normal biological activities* and are said to be ***denatured proteins.*** If a protein is denatured under mild conditions, the original structure can often be regenerated by removing the denaturant or by restoring the temperature to its normal physiological value. This process is called *reversible denaturation.*

14.4 NUCLEIC ACIDS

Nucleic acids are *high-molar-mass polymers that store genetic information and control protein synthesis.* There are two types of nucleic acids: deoxyribonucleic acid (DNA) and ribonucleic acid (RNA). ***DNA** carries all the genetic information necessary to carry on reproduction and to sustain an organism throughout its lifetime.* ***RNA** transcribes these instructions and controls the synthesis of proteins.* In this section we discuss the composition and structure of these vital molecules.

DNA molecules are among the largest molecules known; they have molar masses of up to tens of billions of grams. On the other hand, RNA molecules vary greatly in size, some having a molar mass of no more than about 25,000 g. Compared with proteins, which are made of up to 20 different amino acids, nucleic acids are fairly simple in composition. DNA and RNA molecules contain only four types of building blocks: purines, pyrimidines, a sugar, and a phosphate group (Figure 14.20). The two purines, adenine and guanine, are found in both DNA and RNA. In addition, DNA contains the pyrimidines thymine and cytosine, along with the sugar deoxyribose. RNA also contains thymine and another pyrimidine, uracil, plus the sugar ribose. The purines and pyrimidines are collectively called *bases.*

Despite their comparatively simple composition, the structures of nucleic acids are fairly complex and it took many years to analyze them. In the 1940s, the American biochemist Erwin Chargaff studied DNA molecules obtained from various organisms and noticed the following regularities in base composition:

1. The amount of adenine (a purine) is equal to that of thymine (a pyrmidine); that is, A = T, or A/T = 1.
2. The amount of cytosine (a pyrimidine) is equal to that of guanine (a purine); that is, C = G, or C/G = 1.
3. The total number of purine bases is equal to the total number of pyrimidine bases; that is, A + G = C + T.

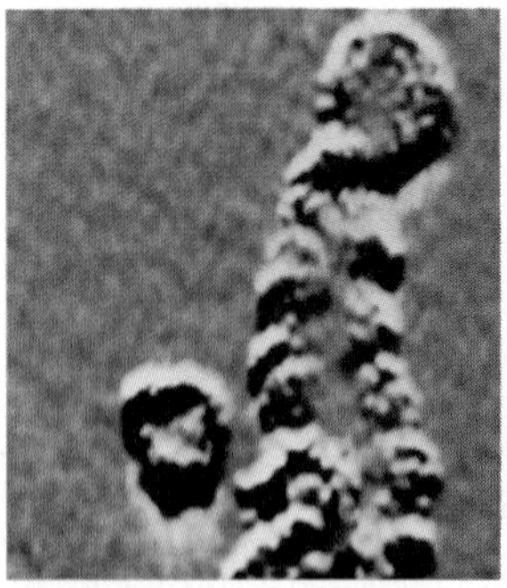

An electron micrograph of a DNA molecule. The double-helical structure is evident.

These observations are now known as *Chargaff's rules.* Based on chemical analyses and information obtained from X-ray diffraction measurements, the American biologist James Watson and the British biologist Francis Crick came up with the double-helical structure for the DNA molecule in 1953. Watson and Crick determined that the DNA molecule is made up of two helical strands. *The repeating unit in each strand* is called a ***nucleotide,*** *which consists of a base-deoxyribose-phosphate linkage* (Figure 14.21).

The key to the double-helical structure is the formation of hydrogen bonds between bases in the two strands. Although hydrogen bonds can form between any two bases, or *base pairs,* Watson and Crick found that the most favorable couplings are between adenine and thymine and between cytosine and guanine (Figure

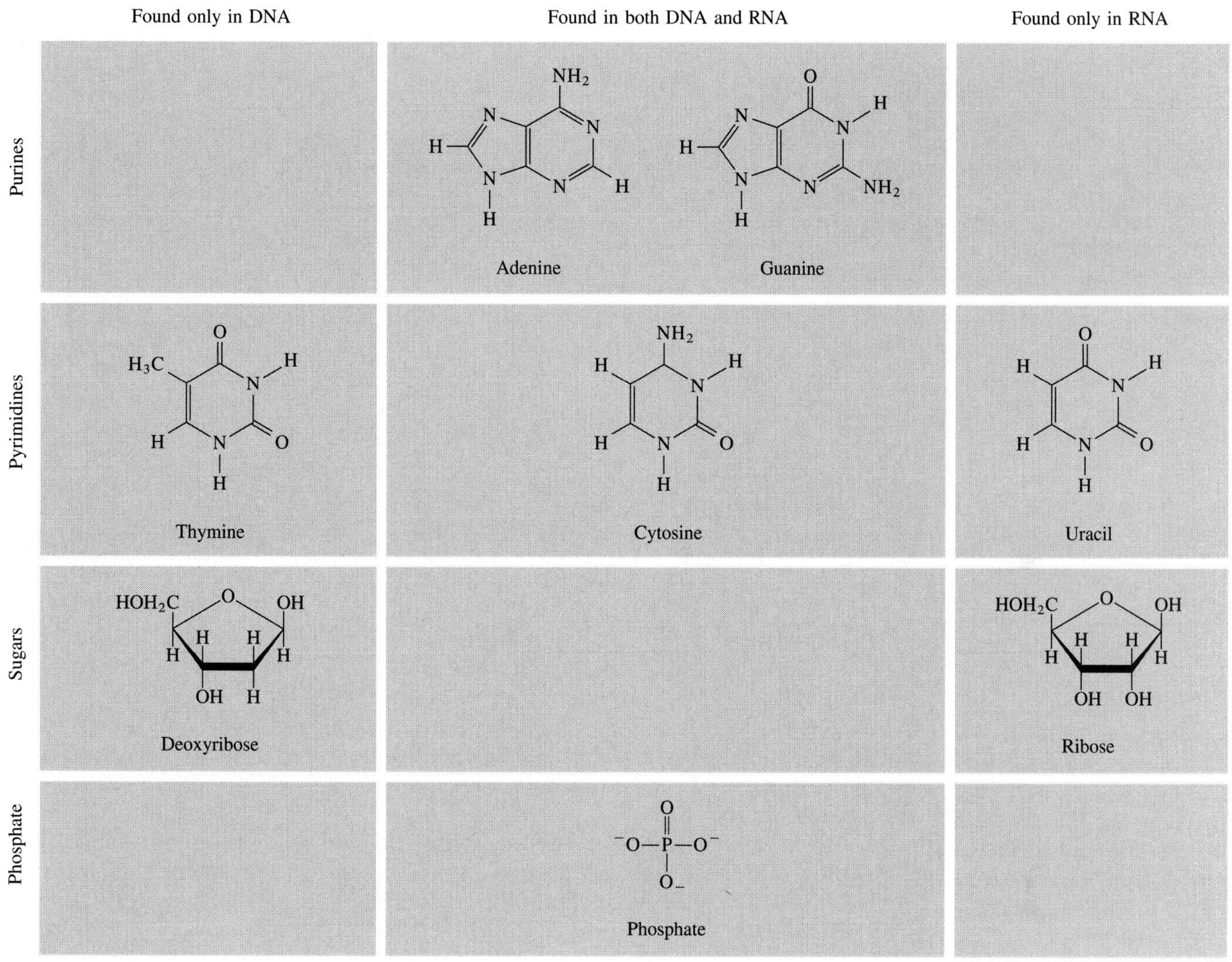

FIGURE 14.20

The components of the nucleic acids DNA and RNA.

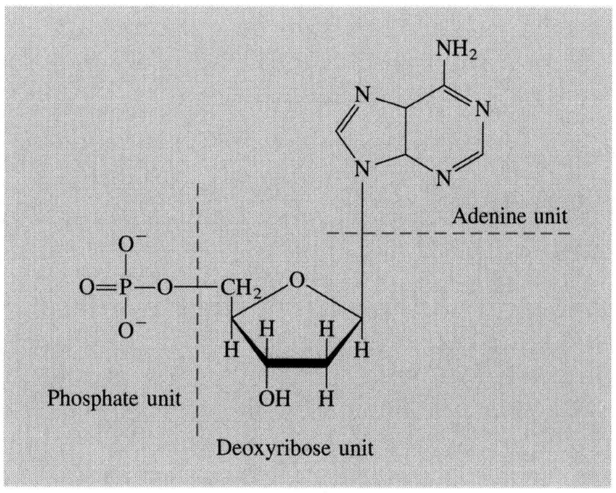

FIGURE 14.21

Structure of a nucleotide, one of the repeating units in DNA.

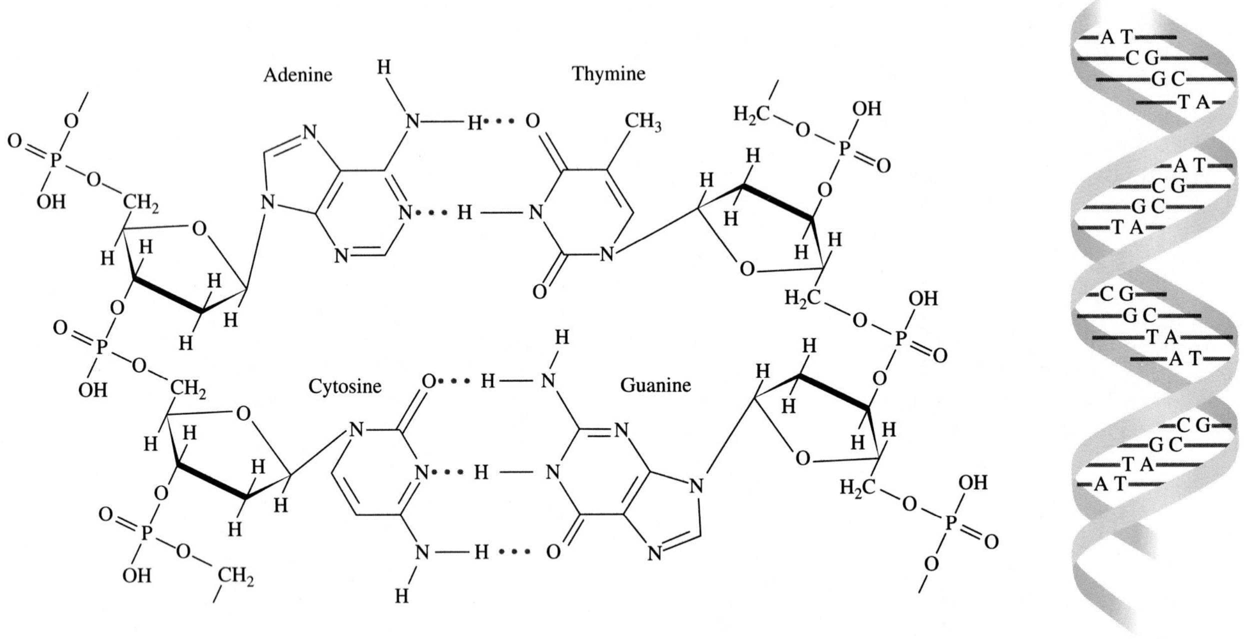

(a) (b)

FIGURE 14.22

(a) Base-pair formation by adenine and thymine and by cytosine and guanine. (b) The double-helical strand of a DNA molecule is held together by hydrogen bonds (and other intermolecular forces) between base pairs A-T and C-G.

14.22). Note that this scheme is consistent with Chargaff's rules, because every purine base is hydrogen-bonded to a pyrimidine base. Other attractive forces such as dipole-dipole and van der Waals forces between the base pairs also help to stabilize the double helix.

Chemical analysis shows that the composition of RNA does not obey Chargaff's rules. In other words, the purine-to-pyrimidine ratio is not equal to 1 as it is in DNA. This and other evidence rules out a double-helical RNA structure. In fact, the RNA molecule is a single-strand polynucleotide. There are actually three types of RNA molecules, called *messenger* RNA (*m*RNA), *ribosomal* RNA (*r*RNA), and *transfer* RNA (*t*RNA). These RNAs have similar composition but differ from one another in molar mass, overall structure, and biological function. Certain RNAs can function as enzymes.

DNA and RNA, like so many other bio-organic polymers, can be only partially understood in terms of their structure and composition. Their intricate chemical interactions and their orchestration of the processes of life set them apart from other substances and are the basis of much research in biochemistry and molecular biology.

SUMMARY

Polymers are large molecules made up of small, repeating units called monomers. Proteins, nucleic acids, cellulose, and rubber are natural polymers. Nylon, Dacron, and Lucite are examples of synthetic polymers. The properties of a polymer vary, depending on whether its structure is atactic, isotactic, or syndiotactic.

Organic polymers can be synthesized via addition reactions or condensation reactions. Synthetic rubbers include polychloroprene and styrene-butadiene rubber, which is made by the copolymerization of styrene and butadiene.

Proteins are polymers of amino acids. Twenty different amino acids make up all the proteins found in the human body; all except glycine are chiral. Most enzymes are proteins. Structure is very closely related to the function and properties of proteins. Hydrogen bonding and other intermolecular forces largely determine the three-dimensional structure of proteins.

Nucleotides are the building blocks of DNA and RNA. The DNA nucleotides each contain a purine or pyrmidine base, a deoxyribose molecule, and a phosphate group. The nucleotides of RNA are similar but contain different bases and ribose instead of deoxyribose.

KEY WORDS

Copolymerization, p. 404	Homopolymer, p. 400	Nucleotide, p. 410	Protein, p. 405
Denatured protein, p. 410	Monomer, p. 398	Polymer, p. 398	RNA, p. 410
DNA, p. 410	Nucleic acids, p. 410		

QUESTIONS AND PROBLEMS

SYNTHETIC ORGANIC POLYMERS

Review Questions

14.1 Define the following terms: copolymerization, homopolymer, monomer, polymer.

14.2 Name ten objects that are either totally or partly made of synthetic organic polymers.

14.3 Calculate the molar mass of a particular polyethylene sample, $\text{+}CH_2\text{—}CH_2\text{+}_n$, where $n = 4600$.

14.4 Describe the two major mechanisms in organic polymer synthesis.

14.5 What are Natta-Ziegler catalysts? What are their roles in polymer synthesis?

14.6 In Chapter 12 you learned about the colligative properties of solutions. Which of the colligative properties is suitable for determining the molar mass of a polymer? Why?

Problems

14.7 Teflon is formed by a radical addition reaction involving the monomer tetrafluoroethylene. Show the mechanism for this reaction.

14.8 Vinyl chloride, $H_2C\text{=}CHCl$, undergoes copolymerization with 1,1-dichloroethylene, $H_2C\text{=}CCl_2$, to form a polymer commercially known as Saran. Draw the structure of the polymer, showing the repeating monomer units.

14.9 Kevlar is a polymer used to make bullet-proof vests, among other things. It is formed by the copolymerization of the following two monomers:

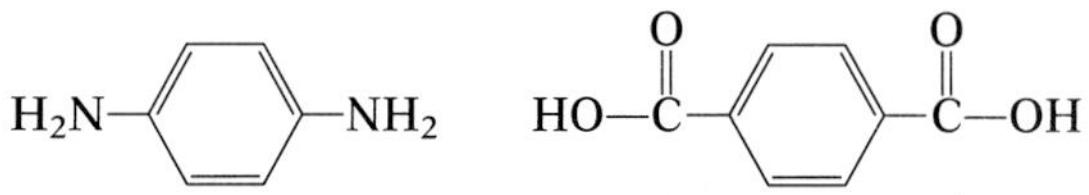

Sketch a portion of the polymer chain showing several monomer units. Write an overall equation for the condensation reaction.

14.10 Describe the formation of polystyrene.

14.11 Deduce plausible monomers for polymers with the following repeating units:

(a) $\text{+}CH_2\text{—}CF_2\text{+}_n$

(b) 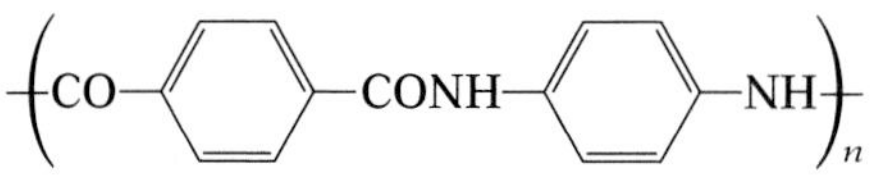

14.12 Deduce plausible monomers for polymers with the following repeating units:

(a) $\text{+}CH_2\text{—}CH\text{=}CH\text{—}CH_2\text{+}_n$

(b) $\text{+}CO\text{+}CH_2\text{+}_6NH\text{+}_n$

PROTEINS

Review Questions

14.13 Discuss the characteristics of an amide group and its importance in protein structure.

14.14 What is the α-helical structure in proteins?

14.15 Describe the β-pleated structure present in some proteins.

14.16 Discuss the main functions of proteins in living systems.

Problems

14.17 Draw structures of the dipeptides that can be formed from the reaction between the amino acids glycine and alanine.

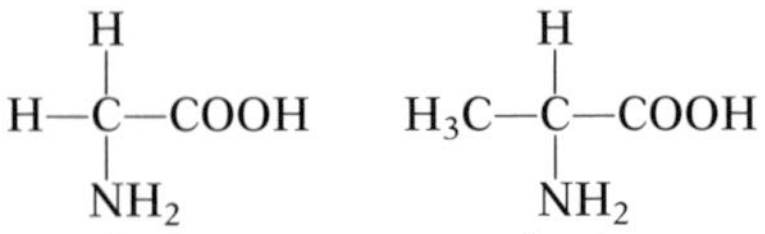

glycine alanine

14.18 Draw structures of the dipeptides that can be formed from the reaction between the amino acids glycine and valine. (For structures see Problem 14.17 and page 408.)

14.19 The amino acid glycine can be condensed to form a polymer called polyglycine. Draw the repeating monomer unit of this polymer.

14.20 The following are data obtained on the rate of product formation of an enzyme-catalyzed reaction:

Temperature (°C)	*Rate of product formation (M/s)*
10	0.0025
20	0.0048
30	0.0090
35	0.0086
45	0.0012

Comment on the dependence of rate on temperature. (No calculations are required.)

NUCLEIC ACIDS

Review Questions

14.21 Describe the structure of a nucleotide.

14.22 What is the difference between ribose and deoxyribose?

14.23 What are Chargaff's rules?

14.24 Describe the role of hydrogen bonding in maintaining the double-helical structure of DNA.

MISCELLANEOUS PROBLEMS

14.25 Discuss the importance of hydrogen bonding in biological systems. Use proteins and nucleic acids as examples.

14.26 Proteins vary widely in structure, whereas nucleic acids have rather uniform structures. How do you account for this major difference?

14.27 If untreated, fevers of 104°F or higher may lead to brain damage. Why?

14.28 The "melting point" of a DNA molecule is the temperature at which the double-helical strand breaks apart. Suppose you are given two DNA samples. One sample contains 45 percent C-G base pairs while the other contains 64 percent C-G base pairs. The total number of bases is the same in each sample. Which of the two samples has a higher melting point? Why?

14.29 When fruits such as apples and pears are cut, the exposed parts begin to turn brown. This is the result of an oxidation reaction catalyzed by enzymes present in the fruits. Often the browning action can be prevented or slowed by adding a few drops of lemon juice to the exposed areas. What is the chemical basis for this treatment?

14.30 When an egg is hard-boiled, the egg white, which contains mostly proteins, undergoes thermal denaturation and forms a gel. Occasionally the eggshell cracks during the heating process. As a result, some of the egg white leaks into the water and forms white strands. One remedy is to add salt or vinegar to the water before boiling. How does this procedure work?

14.31 Nylon can be destroyed easily by strong acids. Explain the chemical basis for the destruction. (*Hint:* The products are the starting materials of the polymerization reaction.)

14.32 Despite what you may have read in science fiction novels or seen in horror movies, it is extremely unlikely that insects can ever grow to human size. Why? (*Hint:* Insects do not have hemoglobin, the oxygen-carrying molecules, in their blood.)

14.33 How many different tripeptides can be formed by lysine and alanine?

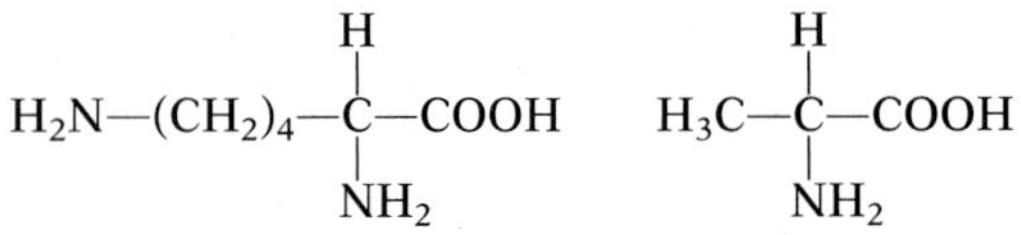

lysine alanine

14.34 Chemical analysis shows that hemoglobin contains 0.34% of Fe by mass. What is the minimum possible molar mass of hemoglobin? The actual molar mass of hemoglobin is four times this minimum value. What conclusion can you draw from these data?

14.35 The folding of a polypeptide chain depends not only on its amino acid sequence but also on the nature of the solvent. Discuss the types of interactions that might occur between water molecules and the amino acid residues of the polypeptide chain. Which groups would be exposed on the exterior of the protein in contact with water and which groups would be buried in the interior of the protein?

14.36 What kind of intermolecular forces are responsible for the aggregation of hemoglobin molecules that leads to sickle cell anemia? (*Hint:* The amino acid valine has a nonpolar side chain, that is, the methyl groups.)

14.37 Draw structures of the nucleotides containing the following components: (a) deoxyribose and cytosine, (b) ribose and uracil.

CHAPTER 15
CHEMICAL EQUILIBRIUM

Haber is considered the greatest authority on the relations between scientific research and industry.

◆ In the early 1900s there was a shortage of nitrogen compounds, which were used as agricultural fertilizers and as explosives. Chemists at that time were interested in converting atmospheric nitrogen into usable compounds (a process called *nitrogen fixation*), such as ammonia. In 1912 Fritz Haber, a German chemist, developed a method, which now bears his name, for synthesizing ammonia directly from nitrogen and hydrogen:

$$N_2(g) + 3H_2(g) \rightleftharpoons 2NH_3(g)$$

This process is sometimes called the Haber-Bosch process to give credit also to Karl Bosch, the engineer who developed the equipment for the industrial production of ammonia, which has to be carried out at 500°C and 500 atm. Haber's success was based on his knowledge of factors affecting gaseous equilibrium and the choice of proper catalysts. His effort is believed to have prolonged World War I by a few years because it allowed Germany to continue manufacturing explosives after its supply of sodium nitrate from Chile was cut off by the Allied naval blockade. In 1995 about 35 million pounds of ammonia were produced in the United States by the Haber process; most of it was used for fertilizer.

Fritz Haber was born in Prussia in 1868. Besides the synthesis of ammonia, he did important work in electrochemistry and combustion chemistry, and he was the director of the prestigious Kaiser Wilhelm Institute. Haber's notable scientific failure was his attempt to extract gold from the oceans—he had overestimated the gold concentration in seawater by a thousand times! The activity that most tarnished his name, however, was his involvement in the introduction of poison gas (chlorine) on the battlefield. The decision to award Haber the Nobel Prize in chemistry in 1918 generated considerable controversy and criticism, something that rarely happens with Nobel awards in the physical sciences.

When the Nazis came to power in 1933, Haber was expelled from Germany because he was Jewish. He died of a heart attack in Switzerland the next year. ◆

Liquid ammonia being applied to the soil before planting.

15.1 THE CONCEPT OF EQUILIBRIUM

Few chemical reactions proceed in only one direction. Most are, at least to some extent, reversible. At the start of a reversible process, the reaction proceeds toward the formation of products. As soon as some product molecules are formed, the reversible process—that is, the formation of reactant molecules from product molecules—begins to take place. *When the rates of the forward and reverse reactions are equal and the concentrations of the reactions and products no longer change with time,* ***chemical equilibrium*** is reached.

Chemical equilibrium is a dynamic process. As such, it can be likened to the movement of skiers at a busy ski resort, where the number of skiers carried up the mountain on the chair lift is equal to the number coming down the slopes. Thus, while there is a constant transfer of skiers, the number of people at the top and the number at the bottom of the slope do not change.

Liquid water in equilibrium with its vapor at room temperature.

Note that a chemical equilibrium reaction involves different substances as reactants and products. Equilibrium between two phases of the same substance is called ***physical equilibrium*** because *the changes that occur are physical processes.* The vaporization of water in a closed container at a given temperature is an example of physical equilibrium. In this instance, the number of H_2O molecules leaving and the number returning to the liquid phase are equal:

$$H_2O(l) \rightleftharpoons H_2O(g)$$

(Recall from Chapter 4 that the double arrow means that the reaction is reversible.) The study of physical equilibrium yields useful information, such as the equilibrium vapor pressure (see Section 11.6). However, chemists are particularly interested in chemical equilibrium processes, such as the reversible reaction involving nitrogen dioxide (NO_2) and dinitrogen tetroxide (N_2O_4). The progress of the reaction

$$N_2O_4(g) \rightleftharpoons 2NO_2(g)$$

can be monitored easily because N_2O_4 is a colorless gas, whereas NO_2 has a dark-brown color that makes it sometimes visible in polluted air. Suppose that a known amount of N_2O_4 is injected into an evacuated flask. Some brown color appears immediately, indicating the formation of NO_2 molecules. The color intensifies as the dissociation of N_2O_4 continues until eventually equilibrium is reached. Beyond that point, no further change in color is observed. By experiment we find that we can also reach the equilibrium state by starting with pure NO_2 or with a mixture of NO_2 and N_2O_4. In each case, we observe an initial change in color, due either to the formation of NO_2 (if the color intensifies) or to the depletion of NO_2 (if the color fades), and then the final state in which the color of NO_2 no longer changes. Depending on the temperature of the reacting system and on the initial amounts of NO_2 and N_2O_4, the concentrations of NO_2 and N_2O_4 at equilibrium differ from system to system (Figure 15.1).

Table 15.1 shows some experimental data for this reaction at 25°C. The gas concentrations are expressed in molarity, which can be calculated from the number of moles of the gases present initially and at equilibrium and the volume of the flask in liters. Analysis of the data at equilibrium shows that although the ratio $[NO_2]/[N_2O_4]$ gives scattered values, the ratio $[NO_2]^2/[N_2O_4]$ gives a nearly constant value that averages 4.63×10^{-3}. This value is called the *equilibrium constant, K,* for the reaction at 25°C. Expressed mathematically, the equilibrium constant for the NO_2–N_2O_4 equilibrium is

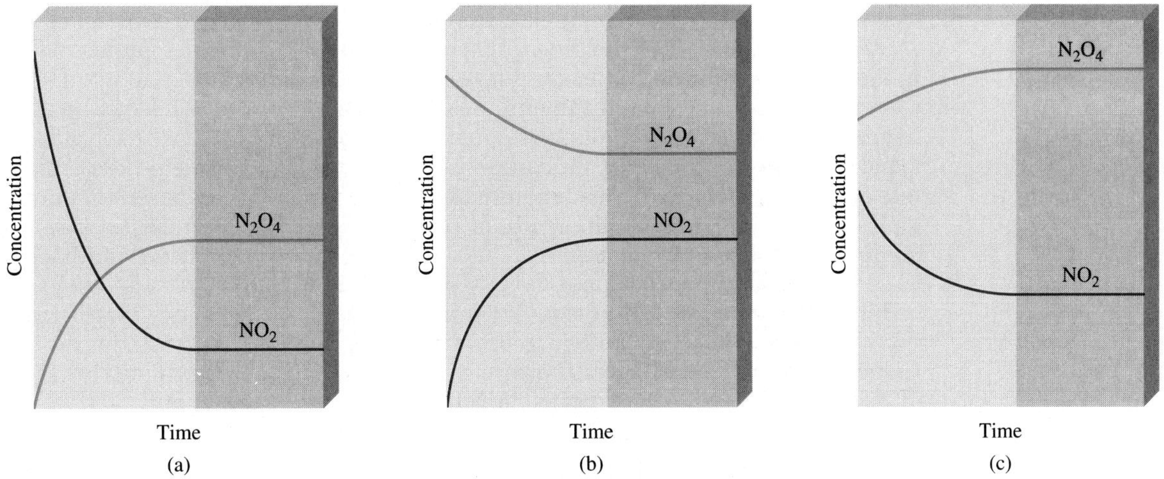

FIGURE 15.1

Change in the concentrations of NO_2 and N_2O_4 with time. (a) Initially only NO_2 is present. (b) Initially only N_2O_4 is present. (c) Initially a mixture of NO_2 and N_2O_4 is present.

$$K = \frac{[NO_2]^2}{[N_2O_4]} = 4.63 \times 10^{-3} \tag{15.1}$$

Note that the exponent 2 for $[NO_2]$ in this expression is the same as the stoichiometric coefficient for NO_2 in the reversible reaction.

We can generalize this discussion by considering the following reversible reaction:

$$a\text{A} + b\text{B} \rightleftharpoons c\text{C} + d\text{D}$$

where a, b, c, and d are the stoichiometric coefficients for the reacting species A, B, C, and D. The equilibrium constant for the reaction at a particular temperature is

$$K = \frac{[\text{C}]^c[\text{D}]^d}{[\text{A}]^a[\text{B}]^b} \tag{15.2}$$

TABLE 15.1

The NO_2–N_2O_4 System at 25°C

Initial Concentrations (M)		Equilibrium Concentrations (M)		Ratio of Concentrations at Equilibrium	
$[NO_2]$	$[N_2O_4]$	$[NO_2]$	$[N_2O_4]$	$\frac{[NO_2]}{[N_2O_4]}$	$\frac{[NO_2]^2}{[N_2O_4]}$
0.000	0.670	0.0547	0.643	0.0851	4.65×10^{-3}
0.0500	0.446	0.0457	0.448	0.102	4.66×10^{-3}
0.0300	0.500	0.0475	0.491	0.0967	4.60×10^{-3}
0.0400	0.600	0.0523	0.594	0.0880	4.60×10^{-3}
0.200	0.000	0.0204	0.0898	0.227	4.63×10^{-3}

Equation (15.2) is the mathematical form of the *law of mass action.* It *relates the concentrations of reactants and products at equilibrium* in terms of a quantity called the ***equilibrium constant.*** The equilibrium constant is defined by a quotient. The numerator is obtained by multiplying together the equilibrium concentrations of the *products,* each raised to a power equal to its stoichiometric coefficient in the balanced equation. The same procedure is applied to the equilibrium concentrations of *reactants* to obtain the denominator. This formulation is based on purely empirical evidence, such as the study of reactions like NO_2–N_2O_4. However, the equilibrium constant has its origin in thermodynamics, to be discussed in Chapter 19.

Finally, we note that if the equilibrium constant is much greater than 1 (that is, $K \gg 1$), the equilibrium will lie to the right of the reaction arrows and favor the products. Conversely, if the equilibrium constant is much smaller than 1 (that is, $K \ll 1$), the equilibrium will lie to the left and favor the reactants.

15.2 WAYS OF EXPRESSING EQUILIBRIUM CONSTANTS

To use equilibrium constants, we must express them in terms of the reactant and product concentrations. Our only guidance is the law of mass action [Equation (15.2)]. However, because the concentrations of the reactants and products can be expressed in different units and because the reacting species are not always in the same phase, there may be more than one way to express the equilibrium constant for the *same* reaction. To begin with, we will consider reactions in which the reactants and products are in the same phase.

HOMOGENEOUS EQUILIBRIA

The term ***homogeneous equilibrium*** applies to reactions in which *all reacting species are in the same phase.* An example of homogeneous gas-phase equilibrium is the dissociation of N_2O_4. The equilibrium constant, as given in Equation (15.1), is

$$K_c = \frac{[NO_2]^2}{[N_2O_4]}$$

Note that the subscript in K_c denotes that the concentrations of the reacting species are expressed in moles per liter. The concentrations of reactants and products in gaseous reactions can also be expressed in terms of their partial pressures. From Equation (5.7) we see that at constant temperature the pressure P of a gas is directly related to the concentration in moles per liter of the gas; that is, $P = (n/V)RT$. Thus, for the equilibrium process

$$N_2O_4(g) \rightleftharpoons 2NO_2(g)$$

we can write

$$K_P = \frac{P^2_{NO_2}}{P_{N_2O_4}} \tag{15.3}$$

where P_{NO_2} and $P_{N_2O_4}$ are the equilibrium partial pressures (in atmospheres) of NO_2 and N_2O_4, respectively. The subscript in K_P tells us that equilibrium concentrations are expressed in terms of pressure.

In general, K_c is not equal to K_P, since the partial pressures of reactants and products are not equal to their concentrations expressed in moles per liter. A simple relationship between K_P and K_c can be derived as follows. Let us consider the following equilibrium in the gas phase:

$$a\mathrm{A}(g) \rightleftharpoons b\mathrm{B}(g)$$

where a and b are stoichiometric coefficients. The equilibrium constant K_c is

$$K_c = \frac{[\mathrm{B}]^b}{[\mathrm{A}]^a}$$

and the expression for K_P is

$$K_P = \frac{P_\mathrm{B}^b}{P_\mathrm{A}^a}$$

where P_A and P_B are the partial pressures of A and B. Assuming ideal gas behavior,

$$P_\mathrm{A}V = n_\mathrm{A}RT$$

$$P_\mathrm{A} = \frac{n_\mathrm{A}RT}{V}$$

where V is the volume of the container in liters. Also

$$P_\mathrm{B}V = n_\mathrm{B}RT$$

$$P_\mathrm{B} = \frac{n_\mathrm{B}RT}{V}$$

Substituting these relations into the expression for K_P, we obtain

$$K_P = \frac{\left(\dfrac{n_\mathrm{B}RT}{V}\right)^b}{\left(\dfrac{n_\mathrm{A}RT}{V}\right)^a} = \frac{\left(\dfrac{n_\mathrm{B}}{V}\right)^b}{\left(\dfrac{n_\mathrm{A}}{V}\right)^a}(RT)^{b-a}$$

Now both n_A/V and n_B/V have the units of moles per liter and can be replaced by [A] and [B], so that

$$K_P = \frac{[\mathrm{B}]^b}{[\mathrm{A}]^a}(RT)^{\Delta n}$$

$$= K_c(RT)^{\Delta n} \qquad (15.4)$$

where

$$\Delta n = b - a$$

$$= \text{moles of gaseous products} - \text{mole of gaseous reactants}$$

Since pressure is usually expressed in atmospheres, the gas constant R is given by 0.0821 L · atm/K · mol, and we can write the relationship between K_P and K_c as

$$K_P = K_c(0.0821\ T)^{\Delta n} \qquad (15.5)$$

In general $K_P \neq K_c$ except in the special case when $\Delta n = 0$. In that case Equation (15.5) can be written as

$$K_P = K_c(0.0821T)^0$$

$$= K_c$$

As another example of homogeneous equilibrium, let us consider the ionization of acetic acid (CH_3COOH) in water:

$$CH_3COOH(aq) + H_2O(l) \rightleftharpoons CH_3COO^-(aq) + H_3O^+(aq)$$

The equilibrium constant is

$$K_c' = \frac{[CH_3COO^-][H_3O^+]}{[CH_3COOH][H_2O]}$$

(We use the prime for K_c here to distinguish it from the final form of equilibrium constant to be derived below.) However, in 1 L, or 1000 g, of water, there are 1000 g/(18.02 g/mol), or 55.5 moles, of water. Therefore, the "concentration" of water, or $[H_2O]$, is 55.5 mol/L, or 55.5 *M*. This is a large quantity compared with the concentrations of other species in solution (usually 1 *M* or smaller), and we can assume that it does not change appreciably during the course of a reaction. Thus we may treat $[H_2O]$ as a constant and rewrite the equilibrium constant as

$$K_c = \frac{[CH_3COO^-][H_3O^+]}{[CH_3COOH]}$$

where

$$K_c = K_c'[H_2O]$$

Note that it is general practice not to include units for the equilibrium constant. In thermodynamics, *K* is defined to have no units since every concentration (molarity) or pressure (atmospheres) term is actually a ratio to a standard value, which is 1 *M* or 1 atm. This procedure eliminates all units but does not alter the numerical parts of the concentration or pressure. Consequently, *K* has no units. We will extend this practice to acid-base equilibria and solubility equilibria (to be discussed in Chapter 17).

EXAMPLE 15.1

Writing Expressions for K_c and K_P

Write expressions for K_c, and K_P, if applicable, for the following reversible reactions at equilibrium:

(a) $2NO(g) + O_2(g) \rightleftharpoons 2NO_2(g)$
(b) $CH_3COOH(aq) + C_2H_5OH(aq) \rightleftharpoons CH_3COOC_2H_5(aq) + H_2O(l)$

Answer: (a)

$$K_c = \frac{[NO_2]^2}{[NO]^2[O_2]} \qquad K_P = \frac{P_{NO_2}^2}{P_{NO}^2 P_{O_2}}$$

(b) The equilibrium constant K_c' is given by

$$K_c' = \frac{[CH_3COOC_2H_5][H_2O]}{[CH_3COOH][C_2H_5OH]}$$

Because the water produced in the reaction is negligible compared with the water solvent, the concentration of water does not change. Thus we can write the new equilibrium constant as

$$K_c = \frac{[CH_3COOC_2H_5]}{[CH_3COOH][C_2H_5OH]}$$

PRACTICE EXERCISE

Write K_c and K_P for the following reaction:

$$2N_2O_5(g) \rightleftharpoons 4NO_2(g) + O_2(g)$$

EXAMPLE 15.2
Calculating Equilibrium Partial Pressure

The equilibrium constant K_P for the reaction

$$PCl_5(g) \rightleftharpoons PCl_3(g) + Cl_2(g)$$

is found to be 1.05 at 250°C. If the equilibrium partial pressures of PCl_5 and PCl_3 are 0.875 atm and 0.463 atm, respectively, what is the equilibrium partial pressure of Cl_2 at 250°C?

Answer: First, we write K_P in terms of the partial pressures of the reacting species.

$$K_P = \frac{P_{PCl_3}P_{Cl_2}}{P_{PCl_5}}$$

Knowing the partial pressures, we write

$$1.05 = \frac{(0.463)(P_{Cl_2})}{(0.875)}$$

or

$$P_{Cl_2} = \frac{(1.05)(0.875)}{(0.463)} = 1.98 \text{ atm}$$

Note that we have added atm as the unit for P_{Cl_2}.

PRACTICE EXERCISE

The equilibrium constant K_P for the reaction

$$2NO_2(g) \rightleftharpoons 2NO(g) + O_2(g)$$

is 158 at 1000 K. Calculate P_{O_2} if $P_{NO_2} = 0.400$ atm and $P_{NO} = 0.270$ atm.

EXAMPLE 15.3
Converting K_c to K_P

For the reaction

$$N_2(g) + 3H_2(g) \rightleftharpoons 2NH_3(g)$$

K_P is 4.3×10^{-4} at 375°C. Calculate K_c for the reaction.

Answer: From Equation (15.5) we write

$$K_c = \frac{K_P}{(0.0821T)^{\Delta n}}$$

Since $T = 648$ K and $\Delta n = 2 - 4 = -2$, we have

$$K_c = \frac{4.3 \times 10^{-4}}{(0.0821 \times 648)^{-2}}$$
$$= 1.2$$

PRACTICE EXERCISE

The equilibrium constant (K_c) for the reaction

$$N_2O_4(g) \rightleftharpoons 2NO_2(g)$$

is 4.63×10^{-3} at 25°C. What is the value of K_P at this temperature?

HETEROGENEOUS EQUILIBRIA

The mineral calcite is made of calcium carbonate, as are chalk and marble.

A reversible reaction involving reactants and products that are in different phases leads to a ***heterogeneous equilibrium.*** For example, when calcium carbonate is heated in a closed vessel, the following equilibrium is attained:

$$CaCO_3(s) \rightleftharpoons CaO(s) + CO_2(g)$$

The two solids and one gas constitute three separate phases. At equilibrium, we might write the equilibrium constant as

$$K'_c = \frac{[CaO][CO_2]}{[CaCO_3]} \qquad (15.6)$$

However, the "concentration" of a solid, like its density, is an intensive property and does not depend on how much of the substance is present. [Note that the units of concentration (moles per liter) can be converted to units of density (grams per cubic centimeters) and vice versa.] For this reason, the terms $[CaCO_3]$ and $[CaO]$ are themselves constants and can be combined with the equilibrium constant. We can simplify the above equilibrium expression by writing

$$\frac{[CaCO_3]}{[CaO]}K'_c = K_c = [CO_2] \qquad (15.7)$$

where K_c, the "new" equilibrium constant, is now conveniently expressed in terms of a single concentration, that of CO_2. Keep in mind that the value of K_c does not depend on how much $CaCO_3$ and CaO are present, as long as some of each is present in equilibrium (Figure 15.2).

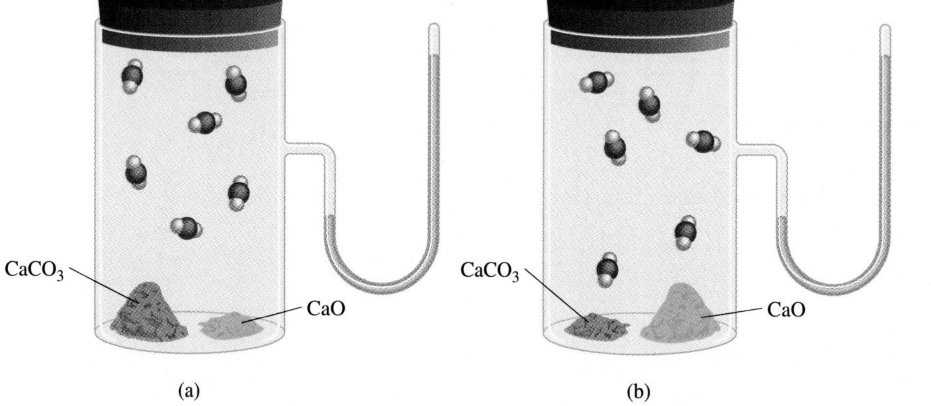

FIGURE 15.2

Regardless of the amount of $CaCO_3$ and CaO present, the equilibrium pressure of CO_2 is the same in (a) and (b) at the same temperature.

Alternatively, we can express the equilibrium constant as

$$K_P = P_{CO_2} \tag{15.8}$$

The equilibrium constant in this case is numerically equal to the pressure of CO_2 gas, an easily measurable quantity.

What has been said about solids also applies to liquids. Thus if a liquid is a reactant or a product, we can treat its concentration as constant and omit it from the equilibrium constant expression.

EXAMPLE 15.4

Calculating the K_P and K_c of a Heterogeneous Equilibrium

Consider the following heterogeneous equilibrium:

$$CaCO_3(s) \rightleftharpoons CaO(s) + CO_2(g)$$

At 800°C, the pressure of CO_2 is 0.236 atm. Calculate (a) K_P and (b) K_c for the reaction at this temperature.

Answer: (a) Using Equation (15.8), we write

$$K_P = P_{CO_2}$$
$$= 0.236$$

(b) From Equation (15.5), we know

$$K_P = K_c(0.0821T)^{\Delta n}$$

$T = 800 + 273 = 1073\ K$ and $\Delta n = 1$, so we substitute these in the equation and obtain

$$0.236 = K_c(0.0821 \times 1073)$$
$$K_c = 2.68 \times 10^{-3}$$

PRACTICE EXERCISE

Consider the following equilibrium at 295 K:

$$NH_4HS(s) \rightleftharpoons NH_3(g) + H_2S(g)$$

The partial pressures of the gases are the same and equal to 0.265 atm. Calculate K_P and K_c for the reaction.

THE FORM OF *K* AND THE EQUILIBRIUM EQUATION

Before closing this section, we should note the two following important rules about writing equilibrium constants:

- When the equation for a reversible reaction is written in the opposite direction, the equilibrium constant becomes the reciprocal of the original equilibrium constant. Thus if we write the NO_2–N_2O_4 equilibrium at 25°C as

$$N_2O_4(g) \rightleftharpoons 2NO_2(g)$$

then

$$K_c = \frac{[NO_2]^2}{[N_2O_4]} = 4.63 \times 10^{-3}$$

However, we can represent the equilibrium equally well as

$$2NO_2(g) \rightleftharpoons N_2O_4(g)$$

and the equilibrium constant is now given by

$$K'_c = \frac{[N_2O_4]}{[NO_2]^2} = \frac{1}{K_c} = \frac{1}{4.63 \times 10^{-3}} = 216$$

You can see that $K_c = 1/K'_c$ or $K_cK'_c = 1.00$. Either K_c or K'_c is a valid equilibrium constant, but it is meaningless to say that the equilibrium constant for the NO_2–N_2O_4 system is 4.63×10^{-3}, or 216, unless we also specify how the equilibrium equation is written.

- The value of K also depends on how the equilibrium equation is balanced. Consider the following two ways of describing the same equilibrium:

$$\tfrac{1}{2}N_2O_4(g) \rightleftharpoons NO_2(g) \qquad K'_c = \frac{[NO_2]}{[N_2O_4]^{1/2}}$$

$$N_2O_4(g) \rightleftharpoons 2NO_2(g) \qquad K_c = \frac{[NO_2]^2}{[N_2O_4]}$$

Looking at the exponents we see that $K'_c = \sqrt{K_c}$. In Table 15.1 we find $K_c = 4.63 \times 10^{-3}$; therefore $K'_c = 0.0680$.

Thus if you double a chemical equation throughout, the corresponding equilibrium constant will be the square of the original value; if you triple the equation, the equilibrium constant will be the cube of the original value, and so on. The NO_2–N_2O_4 example illustrates once again the need to write the particular chemical equation when quoting the numerical value of an equilibrium constant.

Summary of the Rules for Writing Equilibrium Constant Expressions

- The concentrations of the reacting species in the condensed phase are expressed in moles per liter; in the gaseous phase, the concentrations can be expressed in moles per liter or in atmospheres. K_c is related to K_P by a simple equation [Equation (15.5)].
- The concentrations of pure solids, pure liquids (in heterogeneous equilibria), and solvents (in homogeneous equilibria) do not appear in the equilibrium constant expressions.
- The equilibrium constant (K_c or K_P) is dimensionless.
- In quoting a value for the equilibrium constant, we must specify the balanced equation and the temperature.

15.3 What Does the Equilibrium Constant Tell Us?

We have seen that the equilibrium constant for a given reaction can be calculated from known equilibrium concentrations. Once we know the value of the equilibrium constant, we can use Equation (15.2) to calculate unknown equilibrium concentrations—remembering, of course, that the equilibrium constant has a constant value only if the temperature does not change. In general, the equilibrium constant helps us to predict the direction in which a reaction mixture will proceed to achieve equilibrium and to calculate the concentrations of reactants and products once equilibrium has been reached. These uses of the equilibrium constant will be explored in this section.

PREDICTING THE DIRECTION OF A REACTION

The equilibrium constant K_c for the reaction

$$H_2(g) + I_2(g) \rightleftharpoons 2HI(g)$$

is 54.3 at 430°C. Suppose that in a certain experiment we place 0.243 mole of H_2, 0.146 mole of I_2, and 1.98 moles of HI all in a 1.00-L container at 430°C. Will there be a net reaction to form more H_2 and I_2 or more HI? Inserting the starting concentrations in the equilibrium constant expression, we write

$$\frac{[HI]_0^2}{[H_2]_0[I_2]_0} = \frac{(1.98)^2}{(0.243)(0.146)} = 111$$

where the subscript 0 indicates initial concentrations. Because the quotient $[HI]_0^2/[H_2]_0[I_2]_0$ is greater than K_c, this system is not at equilibrium. Consequently, some of the HI will react to form more H_2 and I_2 (decreasing the value of the quotient). Thus the net reaction proceeds from right to left to reach equilibrium.

The quantity obtained by substituting the initial concentrations into the equilibrium constant expression is called the ***reaction quotient*** (Q_c). To determine in which direction the net reaction will proceed to achieve equilibrium, we compare the values of Q_c and K_c. The three possible cases are as follows:

- $Q_c > K_c$ The ratio of initial concentrations of products to reactants is too large. To reach equilibrium, products must be converted to reactants. The system proceeds from right to left (consuming products, forming reactants) to reach equilibrium.
- $Q_c = K_c$ The initial concentrations are equilibrium concentrations. The system is at equilibrium.
- $Q_c < K_c$ The ratio of initial concentrations of products to reactants is too small. To reach equilibrium, reactants must be converted to products. The system proceeds from left to right (consuming reactants, forming products) to reach equilibrium.

EXAMPLE 15.5
Using Q_c to Predict the Direction of a Reaction

At the start of a reaction, there are 0.249 mol N_2, 3.21×10^{-2} mol H_2, and 6.42×10^{-4} mol NH_3 in a 3.50-L reaction vessel at 375°C. If the equilibrium constant (K_c) for the reaction

$$N_2(g) + 3H_2(g) \rightleftharpoons 2NH_3(g)$$

is 1.2 at this temperature, decide whether the system is at equilibrium. If it is not, predict which way the net reaction will proceed.

Answer: The initial concentrations of the reacting species are

$$[N_2]_0 = \frac{0.249 \text{ mol}}{3.50 \text{ L}} = 0.0711\ M$$

$$[H_2]_0 = \frac{3.21 \times 10^{-2}\ \text{mol}}{3.50\ \text{L}} = 9.17 \times 10^{-3}\ M$$

$$[NH_3]_0 = \frac{6.42 \times 10^{-4}\ \text{mol}}{3.50\ \text{L}} = 1.83 \times 10^{-4}\ M$$

Next we write

$$\frac{[NH_3]_0^2}{[N_2]_0[H_2]_0^3} = \frac{(1.83 \times 10^{-4})^2}{(0.0711)(9.17 \times 10^{-3})^3} = 0.611 = Q_c$$

Since Q_c is smaller than K_c (1.2), the system is not at equilibrium. The net result will be an increase in the concentration of NH_3 and a decrease in the concentrations of N_2 and H_2. That is, the net reaction will proceed from left to right until equilibrium is reached.

PRACTICE EXERCISE

The equilibrium constant (K_c) for the reaction

$$2NO(g) + Cl_2(g) \rightleftharpoons 2NOCl(g)$$

is 6.5×10^4 at 35°C. In a certain experiment 2.0×10^{-2} mol of NO, 8.3×10^{-3} mol of Cl_2, and 6.8 mol of NOCl are mixed in a 2.0-L flask. In which direction will the system proceed to reach equilibrium?

Calculating Equilibrium Concentrations

If we know the equilibrium constant for a particular reaction, we can calculate the concentrations in the equilibrium mixture from a knowledge of the initial concentrations. Depending on the information given, the calculation may be straightforward or complex. In the most common situation, only the initial reactant concentrations are given. Let us consider the following system involving a pair of geometric isomers in an organic solvent (Figure 15.3), which has an equilibrium constant (K_c) of 24.0 at 200°C:

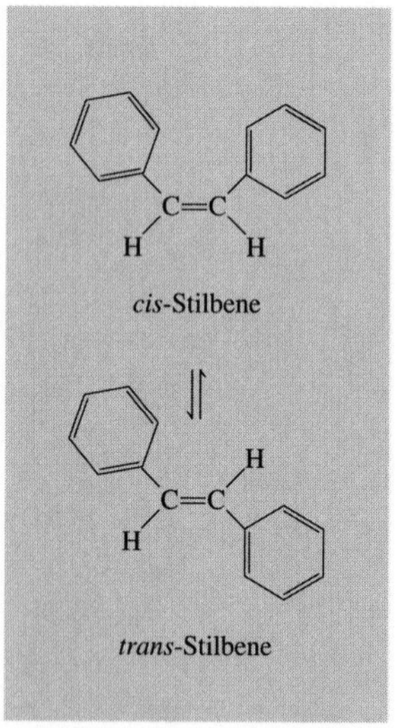

FIGURE 15.3

Structures of cis-stilbene and trans-stilbene.

$$cis\text{-stilbene} \rightleftharpoons trans\text{-stilbene}$$

Suppose that only *cis*-stilbene is initially present at a concentration of 0.850 mol/L. How do we calculate the concentrations of *cis*- and *trans*-stilbene at equilibrium? From the stoichiometry of the reaction we see that for every mole of *cis*-stilbene converted, 1 mole of *trans*-stilbene is formed. Let x be the equilibrium concentration of *trans*-stilbene in moles per liter; therefore, the equilibrium concentration of *cis*-stilbene must be $(0.850 - x)$ mol/L. It is useful to summarize the changes in concentration as follows:

	cis-stilbene ⇌	*trans*-stilbene
Initial (M):	0.850	0
Change (M):	$-x$	$+x$
Equilibrium (M):	$(0.850 - x)$	x

A positive (+) change represents an increase and a negative (−) change indicates a decrease in concentration at equilibrium. Next we set up the equilibrium constant expression

$$K_c = \frac{[\textit{trans}\text{-stilbene}]}{[\textit{cis}\text{-stilbene}]}$$

$$24.0 = \frac{x}{0.850 - x}$$

$$x = 0.816\ M$$

Having solved for x, we calculate the equilibrium concentrations of *cis*-stilbene and *trans*-stilbene as follows:

$$[\textit{cis}\text{-stilbene}] = (0.850 - 0.816)\ M = 0.034\ M$$

$$[\textit{trans}\text{-stilbene}] = 0.816\ M$$

We summarize our approach to solving equilibrium constant problems as follows:

1. Express the equilibrium concentrations of all species in terms of the initial concentrations and a single unknown x, which represents the change in concentration.
2. Write the equilibrium constant expression in terms of the equilibrium concentrations. Knowing the value of the equilibrium constant, solve for x.
3. Having solved for x, calculate the equilibrium concentrations of all species.

EXAMPLE 15.6

Calculating Equilibrium Concentrations

A mixture of 0.500 mol H_2 and 0.500 mol I_2 was placed in a 1.00-L stainless-steel flask at 430°C. Calculate the concentrations of H_2, I_2, and HI at equilibrium. The equilibrium constant K_c for the reaction $H_2(g) + I_2(g) \rightleftharpoons 2HI(g)$ is 54.3 at this temperature.

Answer:

Step 1: The stoichiometry of the reaction is 1 mol H_2 reacting with 1 mol I_2 to yield 2 mol HI. Let x be the depletion in concentration (moles per liter) of either H_2 or I_2 at equilibrium. It follows that the equilibrium concentration of HI must be $2x$. We summarize the changes in concentrations as follows:

	H_2	+	I_2	$\rightleftharpoons$	2HI
Initial (M):	0.500		0.500		0.000
Change (M):	$-x$		$-x$		$+2x$
Equilibrium (M):	$(0.500 - x)$		$(0.500 - x)$		$2x$

Step 2: The equilibrium constant is given by

$$K_c = \frac{[HI]^2}{[H_2][I_2]}$$

Substituting, we get

$$54.3 = \frac{(2x)^2}{(0.500 - x)(0.500 - x)}$$

Taking the square root of both sides, we get

$$7.37 = \frac{2x}{0.500 - x}$$

$$x = 0.393\ M$$

Step 3: At equilibrium, the concentrations are

$$[H_2] = (0.500 - 0.393)\ M = 0.107\ M$$

$$[I_2] = (0.500 - 0.393)\ M = 0.107\ M$$

$$[HI] = 2 \times 0.393\ M = 0.786\ M$$

You can check you answers by calculating K_c using the equilibrium concentrations.

PRACTICE EXERCISE

Consider the reaction in Example 15.6. Starting with a concentration of 0.040 *M* for HI, calculate the concentrations of HI, H_2, and I_2 at equilibrium.

EXAMPLE 15.7

Calculating Equilibrium Concentrations

For the same reaction and temperature as in Example 15.6, suppose that the initial concentrations of H_2, I_2, and HI are 0.00623 *M*, 0.00414 *M*, and 0.0224 *M*, respectively. Calculate the concentrations of these species at equilibrium.

Answer:

Step 1: Let *x* be the depletion in concentration (moles per liter) for H_2 and I_2 at equilibrium. From the stoichiometry of the reaction it follows that the increase in concentration for HI must be 2*x*. Next we write

	H_2	+	I_2	$\rightleftharpoons$	2HI
Initial (*M*):	0.00623		0.00414		0.0224
Change (*M*):	$-x$		$-x$		$+2x$
Equilibrium (*M*):	$(0.00623 - x)$		$(0.00414 - x)$		$(0.0224 + 2x)$

Step 2: The equilibrium constant is

$$K_c = \frac{[HI]^2}{[H_2][I_2]}$$

Substituting, we get

$$54.3 = \frac{(0.0224 + 2x)^2}{(0.00623 - x)(0.00414 - x)}$$

It is not possible to solve this equation by the square root shortcut, as the starting concentrations $[H_2]$ and $[I_2]$ are unequal. Instead, we must first carry out the multiplications

$$54.3(2.58 \times 10^{-5} - 0.0104x + x^2) = 5.02 \times 10^{-4} + 0.0896x + 4x^2$$

Collecting terms, we get

$$50.3x^2 - 0.654x + 8.98 \times 10^{-4} = 0$$

This is a quadratic equation of the form $ax^2 + bx + c = 0$. The solution for a quadratic equation (see Appendix 3) is

$$x = \frac{-b \pm \sqrt{b^2 - 4ac}}{2a}$$

Here we have $a = 50.3$, $b = -0.654$, and $c = 8.98 \times 10^{-4}$, so that

$$x = \frac{0.654 \pm \sqrt{(-0.654)^2 - 4(50.3)(8.98 \times 10^{-4})}}{2 \times 50.3}$$

$$x = 0.0114\ M \qquad \text{or} \qquad x = 0.00156\ M$$

The first solution is physically impossible since the amounts of H_2 and I_2 reacted would be more than those originally present. The second solution gives the correct answer. Note that in solving quadratic equations of this type, one answer is always physically impossible, so the choice of which value to use for x is easy to make.

Step 3: At equilibrium, the concentrations are

$$[H_2] = (0.00623 - 0.00156)\ M = 0.00467\ M$$

$$[I_2] = (0.00414 - 0.00156)\ M = 0.00258\ M$$

$$[HI] = (0.0224 + 2 \times 0.00156)\ M = 0.0255\ M$$

PRACTICE EXERCISE

At 1280°C the equilibrium constant (K_c) for the reaction

$$Br_2(g) \rightleftharpoons 2Br(g)$$

is 1.1×10^{-3}. If the initial concentrations are $[Br_2] = 6.3 \times 10^{-2}\ M$ and $[Br] = 1.2 \times 10^{-2}\ M$, calculate the concentrations of these species at equilibrium.

15.4 FACTORS THAT AFFECT CHEMICAL EQUILIBRIUM

Chemical equilibrium represents a balance between forward and reverse reactions. In most cases, this balance is quite delicate. Changes in experimental conditions may disturb the balance and shift the equilibrium position so that more or less of the desired product is formed. When we say that an equilibrium position shifts to the right, for example, we mean that the net reaction is now from left to right. At our disposal are the following experimentally controllable variables: concentration, pressure, volume, and temperature. Here we will examine how each of these variables affects a reacting system at equilibrium. In addition, we will examine the effect of a catalyst on equilibrium.

Le Chatelier's Principle

Henry Le Chatelier (1850–1936). French chemist and industrial manager.

There is a general rule that helps us to predict the direction in which an equilibrium reaction will move when a change in concentration, pressure, volume, or temperature occurs. The rule, known as ***Le Chatelier's principle,*** states that *if an external stress is applied to a system at equilibrium, the system adjusts in such a way that the stress is partially offset.* The word "stress" here means a change in concentration, pressure, volume, or temperature that removes a system from the equilibrium state. We will use Le Chatelier's principle to assess the effects of such changes.

Changes in Concentrations

Iron(III) thiocyanate [$Fe(SCN)_3$] dissolves readily in water to give a red solution. The red color is due to the presence of hydrated $FeSCN^{2+}$ ion. The equilibrium between undissociated $FeSCN^{2+}$ and the Fe^{3+} and SCN^- ions is given by

$$\underset{\text{red}}{FeSCN^{2+}(aq)} \rightleftharpoons \underset{\text{pale yellow}}{Fe^{3+}(aq)} + \underset{\text{colorless}}{SCN^-(aq)}$$

What happens if we add some sodium thiocyanate (NaSCN) to this solution? In this case, the stress applied to the equilibrium system is an increase in the concentration of SCN^- (from the dissociation of NaSCN). To offset this stress, some Fe^{3+} ions react with the added SCN^- ions, and the equilibrium shifts from right to left:

$$FeSCN^{2+}(aq) \longleftarrow Fe^{3+}(aq) + SCN^-(aq)$$

Consequently, the red color of the solution deepens (Figure 15.4). Similarly, if we added iron(III) nitrate [$Fe(NO_3)_3$] to the original solution, the red color would also deepen because the additional Fe^{3+} ions [from $Fe(NO_3)_3$] would shift the equilibrium from right to left. Both Na^+ and NO_3^- are colorless spectator ions.

Now suppose we add some oxalic acid ($H_2C_2O_4$) to the original solution. Oxalic acid ionizes in water to form the oxalate ion, $C_2O_4^{2-}$, which binds strongly to the Fe^{3+} ions. The formation of the stable yellow ion $Fe(C_2O_4)_3^{3-}$ removes free Fe^{3+} ions in solution. Consequently, more $FeSCN^{2+}$ units dissociate and the equilibrium shifts from left to right:

$$FeSCN^{2+}(aq) \longrightarrow Fe^{3+}(aq) + SCN^-(aq)$$

The red solution will turn yellow due to the formation of $Fe(C_2O_4)_3^{3-}$ ions.

This experiment demonstrates that at equilibrium all reactants and products are present in the reacting system. Second, increasing the concentrations of the

(a) (b) (c) (d)

FIGURE 15.4

Effect of concentration change on the position of equilibrium. (a) An aqueous $Fe(SCN)_3$ solution. The color of the solution is due to both the red $FeSCN^{2+}$ and the yellow Fe^{3+} species. (b) After the addition of some NaSCN to the solution in (a), the equilibrium shifts to the left. (c) After the addition of some $Fe(NO_3)_3$ to the solution in (a), the equilibrium shifts to the left. (d) After the addition of some $H_2C_2O_4$ to the solution in (a), the equilibrium shifts to the right. The yellow color is due to the $Fe(C_2O_4)_3^{3-}$ ions.

products (Fe^{3+} or SCN^-) shifts the equilibrium to the left, and decreasing the concentration of the product Fe^{3+} shifts the equilibrium to the right. These results are just as predicted by Le Chatelier's principle.

EXAMPLE 15.8

Effect of a Change in Concentration on the Equilibrium Position

At 720°C, the equilibrium constant K_c for the reaction

$$N_2(g) + 3H_2(g) \rightleftharpoons 2NH_3(g)$$

is 2.37×10^{-3}. In a certain experiment, the equilibrium concentrations are $[N_2] = 0.683\ M$, $[H_2] = 8.80\ M$, and $[NH_3] = 1.05\ M$. Suppose some NH_3 is added to the mixture so that its concentration is increased to 3.65 *M*. (a) Use Le Chatelier's principle to predict the direction that the net reaction will shift to reach a new equilibrium. (b) Confirm your prediction by calculating the reaction quotient Q_c and comparing its value with K_c.

Answer: (a) The stress applied to the system is the addition of NH_3. To offset this stress, some NH_3 reacts to produce N_2 and H_2 until a new equilibrium is established. The net reaction therefore shifts from right to left; that is,

$$N_2(g) + 3H_2(g) \longleftarrow 2NH_3(g)$$

(b) At the instant when some of the NH_3 is added, the system is no longer at equilibrium. The reaction quotient is given by

$$Q_c = \frac{[NH_3]_0^2}{[N_2]_0[H_2]_0^3}$$

$$= \frac{(3.65)^2}{(0.683)(8.80)^3}$$

$$= 2.86 \times 10^{-2}$$

Since this value is greater than 2.37×10^{-3}, the net reaction shifts from right to left until Q_c equals K_c.

Figure 15.5 shows qualitatively the changes in concentrations of the reacting species.

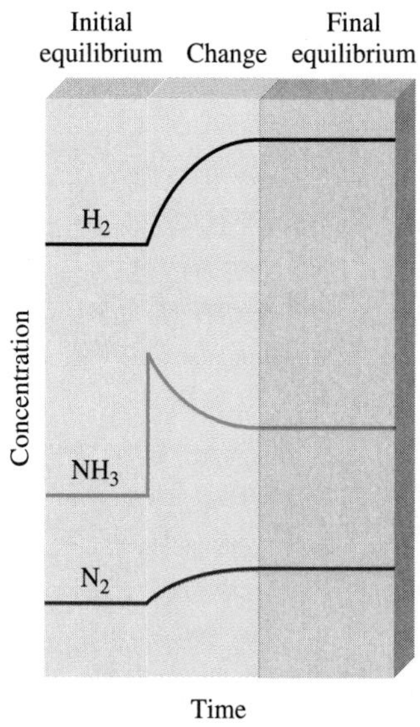

FIGURE 15.5

Changes in concentration of H_2, N_2, and NH_3 after the addition of NH_3 to the equilibrium mixture.

PRACTICE EXERCISE

At 430°C, the equilibrium constant (K_P) for the reaction

$$2NO(g) + O_2(g) \rightleftharpoons 2NO_2(g)$$

is 1.5×10^5. In one experiment, the initial pressures for NO, O_2, and NO_2 are 2.1×10^{-3} atm, 1.1×10^{-2} atm, and 0.14 atm, respectively. Calculate Q_P and predict the direction that the net reaction will shift to reach equilibrium at the same temperature.

CHANGES IN PRESSURE AND VOLUME

Changes in pressure ordinarily do not affect the concentrations of reacting species in condensed phases (say, in an aqueous solution) because liquids and solids are virtually incompressible. On the other hand, concentrations of gases are greatly affected by changes in pressure. Let us look again at Equation (5.7):

$$PV = nRT$$

$$P = \left(\frac{n}{V}\right)RT$$

Thus P and V are related to each other inversely: The greater the pressure, the smaller the volume, and vice versa. Note, too, that the term (n/V) is the concentration of the gas in moles per liter, and it varies directly with pressure.

Suppose that the equilibrium system

$$N_2O_4(g) \rightleftharpoons 2NO_2(g)$$

is in a cylinder fitted with a movable piston. What happens if we increase the pressure on the gases by pushing down on the piston at constant temperature? Since the volume decreases, the concentration (n/V) of both NO_2 and N_2O_4 increases. Because the concentration of NO_2 is squared, the increase in pressure increases the numerator more than the denominator. The system is no longer at equilibrium, so we write

$$Q_c = \frac{[NO_2]_0^2}{[N_2O_4]_0}$$

Thus $Q_c > K_c$, and the net reaction will shift to the left until $Q_c = K_c$. Conversely, a decrease in pressure (increase in volume) would result in $Q_c < K_c$; the net reaction would shift to the right until $Q_c = K_c$.

In general, an increase in pressure (decrease in volume) favors the net reaction that decreases the total number of moles of gases (the reverse reaction, in this case), and a decrease in pressure (increase in volume) favors the net reaction that increases the total number of moles of gases (here, the forward reaction). For reactions in which there is no change in the number of moles of gases, a pressure (or volume) change has no effect on the position of equilibrium.

It is possible to change the pressure of a system without changing its volume. Suppose the NO_2–N_2O_4 system is contained in a stainless-steel vessel whose volume is constant. We can increase the total pressure in the vessel by adding an inert gas (helium, for example) to the equilibrium system. Adding helium to the equilibrium mixture at constant volume increases the total gas pressure and decreases the mole fractions of both NO_2 and N_2O_4; but the partial pressure of each gas, given by the product of its mole fraction and total pressure (see Section 5.5), does not change. Thus the presence of an inert gas in such a case does not affect the equilibrium.

EXAMPLE 15.9

Affecting the Equilibrium Position with a Change in Pressure and Volume

Consider the following equilibrium systems:

(a) $2PbS(s) + 3O_2(g) \rightleftharpoons 2PbO(s) + 2SO_2(g)$
(b) $PCl_5(g) \rightleftharpoons PCl_3(g) + Cl_2(g)$
(c) $H_2(g) + CO_2(g) \rightleftharpoons H_2O(g) + CO(g)$

Predict the direction of the net reaction in each case as a result of increasing the pressure (decreasing the volume) on the system at constant temperature.

Answer: (a) Consider only the gaseous molecules. In the balanced equation there are 3 moles of gaseous reactants and 2 moles of gaseous products. Therefore the net reaction will shift toward the products (to the right) when the pressure is increased.

(b) The number of moles of products is 2 and that of reactants is 1; therefore the net reaction will shift to the left, toward the reactants.

(c) The number of moles of products is equal to the number of moles of reactants, so a change in pressure has no effect on the equilibrium.

PRACTICE EXERCISE

Consider the reaction

$$2NOCl(g) \rightleftharpoons 2NO(g) + Cl_2(g)$$

Predict the direction of the net reaction as a result of decreasing the pressure (increasing the volume) on the system at constant temperature.

Changes in Temperature

A change in concentration, pressure, or volume may alter the equilibrium position, but it does not change the value of the equilibrium constant. Only a change in temperature can alter the equilibrium constant.

The formation of NO_2 from N_2O_4 is an endothermic process:

$$N_2O_4(g) \longrightarrow 2NO_2(g) \qquad \Delta H^\circ = 58.0 \text{ kJ}$$

and the reverse reaction is exothermic:

$$2NO_2(g) \longrightarrow N_2O_4(g) \qquad \Delta H^\circ = -58.0 \text{ kJ}$$

At equilibrium the net heat effect is zero because there is no net reaction. What happens if the following equilibrium system

$$N_2O_4(g) \rightleftharpoons 2NO_2(g)$$

is heated at constant volume? Since endothermic processes absorb heat from the surroundings, heating favors the dissociation of N_2O_4 into NO_2 molecules. Consequently, the equilibrium constant, given by

$$K_c = \frac{[NO_2]^2}{[N_2O_4]}$$

increases with temperature (Figure 15.6).

As another example, consider the equilibrium between the following ions:

$$\underset{\text{blue}}{CoCl_4^{2-}} + 6H_2O \rightleftharpoons \underset{\text{pink}}{Co(H_2O)_6^{2+}} + 4Cl^-$$

The formation of $CoCl_4^{2-}$ is endothermic. On heating, the equilibrium shifts to the left and the solution turns blue. Cooling favors the exothermic reaction [the formation of $Co(H_2O)_6^{2+}$] and the solution turns pink (Figure 15.7).

We summarize the results with the statement: *A temperature increase favors an endothermic reaction, and a temperature decrease favors an exothermic reaction.*

FIGURE 15.6

(a) Two bulbs containing a mixture of NO_2 and N_2O_4 gases at equilibrium. (b) When one bulb is immersed in ice water (left), its color becomes lighter, indicating the formation of the colorless gas N_2O_4. When the other bulb is immersed in hot water, its color darkens, indicating the formation of NO_2.

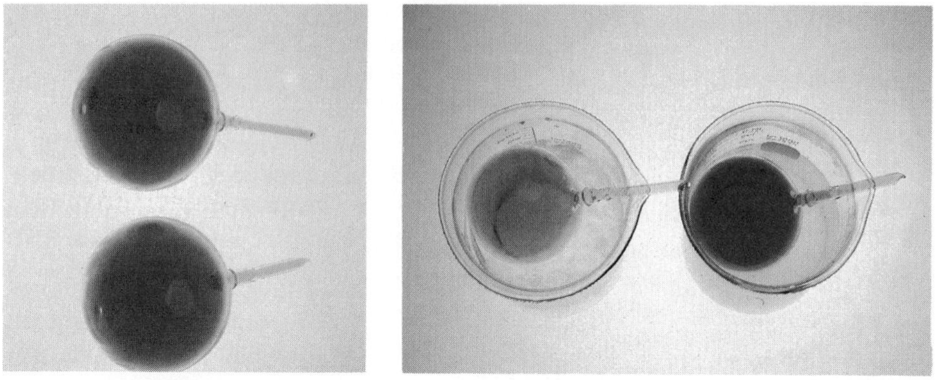

(a) **(b)**

FIGURE 15.7

(Left) Heating favors the formation of the blue $CoCl_4^{2-}$ ion. (Right) Cooling favors the formation of the red $Co(H_2O)_6^{2+}$ ion.

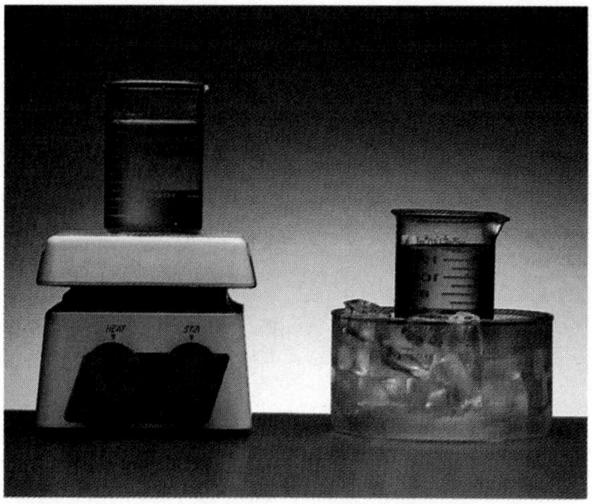

The Effect of a Catalyst

A catalyst increases the rate at which a reaction occurs (see Chapter 21). For a reversible reaction, a catalyst affects the rate in the forward and reverse direction to the same extent. Therefore, the presence of a catalyst does not alter the equilibrium constant, nor does it shift the position of an equilibrium system. Adding a catalyst to a reaction mixture that is not at equilibrium will speed up both the forward and reverse rates to achieve an equilibrium mixture much faster. The same equilibrium mixture could be obtained without the catalyst, but we might have to wait much longer for the results.

Le Chatelier's principle finds many applications in industrial processes. Consider the Haber process of ammonia synthesis on page 417:

$$N_2(g) + 3H_2(g) \rightleftharpoons 2NH_3(g) \qquad \Delta H^\circ = -92.6 \text{ kJ}$$

Because the forward reaction results in a decrease of the number of moles of gases, the reaction is carried out at very high pressures—between 500 atm and 1000 atm—to maximize the yield. The exothermic nature of ammonia formation suggests that a low-temperature operation would be desirable. However, the process is actually run at about 500°C, with catalysts, to increase the rate of the reaction even though the equilibrium constant is smaller at the high temperature. In practice, the reaction is not allowed to reach equilibrium. Instead, ammonia is constantly removed from the reacting mixture so as to shift the net reaction from left to right.

Summary of Factors That May Affect the Equilibrium Position

We have considered four ways to affect a reacting system at equilibrium. It is important to remember that, of the four, *only a change in temperature changes the value of the equilibrium constant.* Changes in concentration, pressure, and volume can alter the equilibrium concentrations of the reacting mixture, but they cannot change the equilibrium constant as long as the temperature does not change. A catalyst can help establish equilibrium faster, but it has no effect on the equilibrium constant or on the equilibrium concentrations of the reacting species.

EXAMPLE 15.10
Factors That May Affect the Equilibrium Position or Equilibrium Constant

Consider the following equilibrium process:

$$N_2F_4(g) \rightleftharpoons 2NF_2(g) \qquad \Delta H^\circ = 38.5 \text{ kJ}$$

Predict the changes in the equilibrium if (a) the reacting mixture is heated at constant volume; (b) NF_2 gas is removed from the reacting mixture at constant temperature and volume; (c) the pressure on the reacting mixture is decreased at constant temperature; and (d) an inert gas, such as helium, is added to the reacting mixture at constant volume and temperature.

Answer: (a) Since the forward reaction is endothermic, an increase in temperature favors the formation of NF_2. The equilibrium constant

$$K_c = \frac{[NF_2]^2}{[N_2F_4]}$$

will therefore increase with increasing temperature.

(b) The stress here is the removal of NF_2 gas. To offset it, more N_2F_4 will decompose to form NF_2. The equilibrium constant K_c remains unchanged, however.

(c) A decrease in pressure (which is accompanied by an increase in gas volume) favors the formation of more gas molecules, that is, the forward reaction. Thus, more NF_2 gas will be formed. The equilibrium constant will remain unchanged.

(d) Adding helium to the equilibrium mixture at constant volume will not shift the equilibrium.

PRACTICE EXERCISE

Consider the equilibrium

$$3O_2(g) \rightleftharpoons 2O_3(g) \qquad \Delta H^\circ = 284 \text{ kJ}$$

What would be the effect of (a) increasing the pressure on the system by decreasing the volume, (b) increasing pressure by adding O_2 to the system, (c) decreasing the temperature, (d) adding a catalyst?

SUMMARY

Chemists are concerned with dynamic equilibria between phases (physical equilibria) and between reacting substances (chemical equilibria). For the general chemical reaction

$$aA + bB \longrightarrow cC + dD$$

the concentrations of reactants and products at equilibrium (in moles per liter) are related by the equilibrium constant expression

$$K_c = \frac{[C]^c[D]^d}{[A]^a[B]^b}$$

The equilibrium constant may also be expressed in terms of the equilibrium partial pressures (in atmospheres) of gases, as K_P.

An equilibrium in which all reactants and products are in the same phase is a homogeneous equilibrium. If the reactants and products are not all in the same phase, the equilibrium is called heterogeneous. The concentrations of pure solids, pure liquids, and solvents are constant and do not appear in the equilibrium constant expression of a reaction. The value of K depends on how the chemical equation is balanced, and the equilibrium constant for the reverse of a particular reaction is the reciprocal of the equilibrium constant of that reaction.

The reaction quotient Q has the same form as the equilibrium constant expression, but it applies to a reaction that may not be at equilibrium. If $Q > K$, the reaction will proceed from right to left to achieve equilibrium. If $Q < K$, the reaction will proceed from left to right to achieve equilibrium.

Le Chatelier's principle states that if an external stress is applied to a system at chemical equilibrium, the system will adjust to partially offset the stress. Only a change in temperature changes the value of the equilibrium constant for a particular reaction. Changes in concentration, pressure, or volume may change the equilibrium concentrations of reactants and products. The addition of a catalyst hastens the attainment of equilibrium but does not affect the equilibrium concentrations of reactants and products.

KEY WORDS

Chemical equilibrium, p. 418
Equilibrium constant, p. 420
Heterogeneous equilibrium, p. 424
Homogeneous equilibrium, p. 420
Le Chatelier's principle, p. 432
Physical equilibrium, p. 418
Reaction quotient (Q_c), p. 427

QUESTIONS AND PROBLEMS

CONCEPT OF EQUILIBRIUM

Review Questions

15.1 Define equilibrium. Give two examples of a dynamic equilibrium.

15.2 Explain the difference between physical equilibrium and chemical equilibrium. Give two examples of each.

15.3 Briefly describe the importance of equilibrium in the study of chemical reactions.

15.4 Consider the equilibrium system $3A \rightleftharpoons B$. Sketch the change in concentrations of A and B with time for the following situations: (a) Initially only A is present; (b) initially only B is present; (c) initially both A and B are present (with A in higher concentration). In each case, assume that the concentration of B is higher than that of A at equilibrium.

EQUILIBRIUM CONSTANT EXPRESSIONS

Review Questions

15.5 Define homogeneous equilibrium and heterogeneous equilibrium. Give two examples of each.

15.6 What do the symbols K_c and K_P represent?

15.7 Write the expressions for the equilibrium constants K_P of the following thermal decompositions:
(a) $2NaHCO_3(s) \rightleftharpoons Na_2CO_3(s) + CO_2(g) + H_2O(g)$
(b) $2CaSO_4(s) \rightleftharpoons 2CaO(s) + 2SO_2(g) + O_2(g)$

15.8 Write equilibrium constant expressions for K_c and for K_P, if applicable, for the following processes:
(a) $2CO_2(g) \rightleftharpoons 2CO(g) + O_2(g)$
(b) $3O_2(g) \rightleftharpoons 2O_3(g)$
(c) $CO(g) + Cl_2(g) \rightleftharpoons COCl_2(g)$
(d) $H_2O(g) + C(s) \rightleftharpoons CO(g) + H_2(g)$
(e) $HCOOH(aq) \rightleftharpoons H^+(aq) + HCOO^-(aq)$
(f) $2HgO(s) \rightleftharpoons 2Hg(l) + O_2(g)$

15.9 Write the equilibrium constant expressions for K_c and K_P, if applicable, for the following reactions:
(a) $2NO_2(g) + 7H_2(g) \rightleftharpoons 2NH_3(g) + 4H_2O(l)$
(b) $2ZnS(s) + 3O_2(g) \rightleftharpoons 2ZnO(s) + 2SO_2(g)$
(c) $C(s) + CO_2(g) \rightleftharpoons 2CO(g)$
(d) $C_6H_5COOH(aq) \rightleftharpoons C_6H_5COO^-(aq) + H^+(aq)$

CALCULATING EQUILIBRIUM CONSTANTS

Review Questions

15.10 Write the equation relating K_c and K_P and define all the terms.

Problems

15.11 The equilibrium constant (K_c) for the reaction

$$2HCl(g) \rightleftharpoons H_2(g) + Cl_2(g)$$

is 4.17×10^{-34} at 25°C. What is the equilibrium constant for the reaction

$$H_2(g) + Cl_2(g) \rightleftharpoons 2HCl(g)$$

at the same temperature?

15.12 Consider the following equilibrium process at 700°C:

$$2H_2(g) + S_2(g) \rightleftharpoons 2H_2S(g)$$

Analysis shows that there are 2.50 moles of H_2, 1.35×10^{-5} mole of S_2, and 8.70 moles of H_2S present in a 12.0-L flask. Calculate the equilibrium constant K_c for the reaction.

15.13 What is the K_P at 1273°C for the reaction

$$2CO(g) + O_2(g) \rightleftharpoons 2CO_2(g)$$

if K_c is 2.24×10^{22} at the same temperature?

15.14 The equilibrium constant K_P for the reaction

$$2SO_3(g) \rightleftharpoons 2SO_2(g) + O_2(g)$$

is 5.0×10^{-4} at 302°C. What is K_c for this reaction?

15.15 Consider the following reaction:

$$N_2(g) + O_2(g) \rightleftharpoons 2NO(g)$$

If the equilibrium partial pressures of N_2, O_2, and NO are 0.15 atm, 0.33 atm, and 0.050 atm, respectively, at 2200°C, what is K_P?

15.16 A reaction vessel contains NH_3, N_2, and H_2 at equilibrium at a certain temperature. The equilibrium concentrations are $[NH_3] = 0.25\ M$, $[N_2] = 0.11\ M$, and $[H_2] = 1.91\ M$. Calculate the equilibrium constant K_c for the synthesis of ammonia if the reaction is represented as
(a) $N_2(g) + 3H_2(g) \rightleftharpoons 2NH_3(g)$
(b) $\frac{1}{2}N_2(g) + \frac{3}{2}H_2(g) \rightleftharpoons NH_3(g)$

15.17 The equilibrium constant K_c for the reaction

$$I_2(g) \rightleftharpoons 2I(g)$$

is 3.8×10^{-5} at 727°C. Calculate K_c and K_P for the equilibrium

$$2I(g) \rightleftharpoons I_2(g)$$

at the same temperature.

15.18 The pressure of the reacting mixture

$$CaCO_3(s) \rightleftharpoons CaO(s) + CO_2(g)$$

at equilibrium is 0.105 atm at 350°C. Calculate K_P and K_c for this reaction.

15.19 The equilibrium constant K_P for the reaction

$$PCl_5(g) \rightleftharpoons PCl_3(g) + Cl_2(g)$$

is 1.05 at 250°C. The reaction starts with a mixture of PCl_5, PCl_3, and Cl_2 at pressures 0.177 atm, 0.223 atm, and 0.111 atm, respectively, at 250°C. When the mixture comes to equilibrium at that temperature, which pressures will have decreased and which will have increased? Explain why.

15.20 Ammonium carbamate, $NH_4CO_2NH_2$, decomposes as follows:

$$NH_4CO_2NH_2(s) \rightleftharpoons 2NH_3(g) + CO_2(g)$$

Starting with only the solid, it is found that at 40°C the total gas pressure (NH_3 and CO_2) is 0.363 atm. Calculate the equilibrium constant K_P.

15.21 Consider the following reaction at 1600°C:

$$Br_2(g) \rightleftharpoons 2Br(g)$$

When 1.05 moles of Br_2 are put in a 0.980-L flask, 1.20 percent of the Br_2 undergoes dissociation. Calculate the equilibrium constant K_c for the reaction.

15.22 Pure phosgene gas ($COCl_2$), 3.00×10^{-2} mol, was placed in a 1.50-L container. It was heated at 800 K, and at equilibrium the pressure of CO was found to be 0.497 atm. Calculate the equilibrium constant K_P for the reaction

$$CO(g) + Cl_2(g) \rightleftharpoons COCl_2(g)$$

15.23 Consider the equilibrium

$$2NOBr(g) \rightleftharpoons 2NO(g) + Br_2(g)$$

If nitrosyl bromide, NOBr, is 34 percent dissociated at 25°C and the total pressure is 0.25 atm, calculate K_P and K_c for the dissociation at this temperature.

15.24 A 2.50-mole quantity of NOCl was initially placed in a 1.50-L reaction chamber at 400°C. After equilibrium was established, it was found that 28.0 percent of the NOCl had dissociated:

$$2NOCl(g) \rightleftharpoons 2NO(g) + Cl_2(g)$$

Calculate the equilibrium constant K_c for the reaction.

CALCULATING EQUILIBRIUM CONCENTRATIONS

Review Questions

15.25 Define reaction quotient. How does it differ from equilibrium constant?

15.26 Outline the steps for calculating the concentrations of reacting species in an equilibrium reaction.

Problems

15.27 The equilibrium constant K_P for the reaction

$$2SO_2(g) + O_2(g) \rightleftharpoons 2SO_3(g)$$

is 5.60×10^4 at 350°C. SO_2 and O_2 are mixed initially at 0.350 atm and 0.762 atm, respectively, at 350°C. When the mixture equilibrates, is the total pressure less than or greater than the sum of the initial pressures, 1.112 atm?

15.28 For the synthesis of ammonia

$$N_2(g) + 3H_2(g) \rightleftharpoons 2NH_3(g)$$

the equilibrium constant K_c at 375°C is 0.65. Starting with $[H_2]_0 = 0.76\ M$, $[N_2]_0 = 0.60\ M$, and $[NH_3]_0 = 0.48\ M$, when this mixture comes to equilibrium, which gases will have increased in concentration and which will have decreased in concentration?

15.29 For the reaction

$$H_2(g) + CO_2(g) \rightleftharpoons H_2O(g) + CO(g)$$

at 700°C, $K_c = 0.534$. Calculate the number of moles of H_2 formed at equilibrium if a mixture of 0.300 mole of CO and 0.300 mole of H_2O is heated to 700°C in a 10.0-L container.

15.30 A sample of pure NO_2 gas heated to 1000 K decomposes:

$$2NO_2(g) \rightleftharpoons 2NO(g) + O_2(g)$$

The equilibrium constant K_P is 158. Analysis shows that the partial pressure of O_2 is 0.25 atm at equilibrium. Calculate the pressure of NO and NO_2 in the mixture.

15.31 The equilibrium constant K_c for the reaction

$$H_2(g) + Br_2(g) \rightleftharpoons 2HBr(g)$$

is 2.18×10^6 at 730°C. Starting with 3.20 moles HBr in a 12.0-L reaction vessel, calculate the concentrations of H_2, Br_2, and HBr at equilibrium.

15.32 The dissociation of molecular iodine into iodine atoms is represented as

$$I_2(g) \rightleftharpoons 2I(g)$$

At 1000 K, the equilibrium constant K_c for the reaction is 3.80×10^{-5}. Suppose you start with 0.0456 mole of I_2 in a 2.30-L flask at 1000 K. What are the concentrations of the gases at equilibrium?

15.33 The equilibrium constant K_c for the decomposition of phosgene, $COCl_2$, is 4.63×10^{-3} at 527°C:

$$COCl_2(g) \rightleftharpoons CO(g) + Cl_2(g)$$

Calculate the equilibrium partial pressure of all the components, starting with pure phosgene at 0.760 atm.

15.34 Consider the following equilibrium process at 686°C:

$$CO_2(g) + H_2(g) \rightleftharpoons CO(g) + H_2O(g)$$

The equilibrium concentrations of the reacting species are [CO] = 0.050 *M*, $[H_2]$ = 0.045 *M*, $[CO_2]$ = 0.086 *M*, and $[H_2O]$ = 0.040 *M*. (a) Calculate K_c for the reaction at 686°C. (b) If the concentration of CO_2 were raised to 0.50 mol/L by the addition of CO_2, what would be the concentrations of all the gases when equilibrium is reestablished?

15.35 Consider the heterogeneous equilibrium process:

$$C(s) + CO_2(g) \rightleftharpoons 2CO(g)$$

At 700°C, the total pressure of the system is found to be 4.50 atm. If the equilibrium constant K_P is 1.52, calculate the equilibrium partial pressures of CO_2 and CO.

15.36 The equilibrium constant K_c for the reaction

$$H_2(g) + CO_2(g) \rightleftharpoons H_2O(g) + CO(g)$$

is 4.2 at 1650°C. Initially 0.80 mole H_2 and 0.80 mole CO_2 are injected into a 5.0-L flask. Calculate the concentration of each species at equilibrium.

LE CHATELIER'S PRINCIPLE

Review Questions

15.37 Explain Le Chatelier's principle. How can this principle help us maximize the yields of reactions?

15.38 Use Le Chatelier's principle to explain why the equilibrium vapor pressure of a liquid increases with increasing temperature.

15.39 List four factors that can shift the position of an equilibrium. Which one can alter the value of the equilibrium constant?

15.40 What is meant by "the position of an equilibrium"? Does the addition of a catalyst have any effects on the position of an equilibrium?

Problems

15.41 Consider the following equilibrium system:

$$SO_2(g) + Cl_2(g) \rightleftharpoons SO_2Cl_2(g)$$

Predict how the equilibrium position would change if (a) Cl_2 gas were added to the system, (b) SO_2Cl_2 were removed from the system, (c) SO_2 were removed from the system. The temperature remains constant.

15.42 Heating solid sodium bicarbonate in a closed vessel established the following equilibrium:

$$2NaHCO_3(s) \rightleftharpoons Na_2CO_3(s) + H_2O(g) + CO_2(g)$$

What would happen to the equilibrium position if (a) some of the CO_2 were removed from the system, (b) some solid Na_2CO_3 were added to the system, (c) some of the solid $NaHCO_3$ were removed from the system? The temperature remains constant.

15.43 Consider the following equilibrium systems:
(a) $A \rightleftharpoons 2B \quad \Delta H° = 20.0$ kJ
(b) $A + B \rightleftharpoons C \quad \Delta H° = -5.4$ kJ
(c) $A \rightleftharpoons B \quad \Delta H° = 0.0$ kJ
Predict the change in the equilibrium constant K_c that would occur in each case if the temperature of the reacting system were raised.

15.44 What effect does an increase in pressure have on each of the following systems at equilibrium?
(a) $A(s) \rightleftharpoons 2B(s)$
(b) $2A(l) \rightleftharpoons B(l)$
(c) $A(s) \rightleftharpoons B(g)$
(d) $A(g) \rightleftharpoons B(g)$
(e) $A(g) \rightleftharpoons 2B(g)$
The temperature is kept constant. In each case, the reactants are in a cylinder fitted with a movable piston.

15.45 Consider the equilibrium

$$2I(g) \rightleftharpoons I_2(g)$$

What would be the effect on the position of equilibrium of (a) increasing the total pressure on the system by decreasing its volume, (b) adding I_2 to the reaction mixture, (c) decreasing the temperature?

15.46 Consider the following equilibrium process:

$$PCl_5(g) \rightleftharpoons PCl_3(g) + Cl_2(g) \qquad \Delta H° = 92.5 \text{ kJ}$$

Predict the direction of the shift in equilibrium when (a) the temperature is raised, (b) more chlorine gas is added to the reaction mixture, (c) some PCl_3 is removed from the mixture, (d) the pressure on the gases is increased, (e) a catalyst is added to the reaction mixture.

15.47 Consider the reaction

$$2SO_2(g) + O_2(g) \rightleftharpoons 2SO_3(g) \qquad \Delta H° = -198.2 \text{ kJ}$$

Comment on the changes in the concentrations of SO_2, O_2, and SO_3 at equilibrium if we were to (a) increase the temperature, (b) increase the pressure, (c) increase SO_2, (d) add a catalyst, (e) add helium at constant volume.

15.48 In the uncatalyzed reaction

$$N_2O_4(g) \rightleftharpoons 2NO_2(g)$$

at 100°C the pressures of the gases at equilibrium are $P_{N_2O_4} = 0.377$ atm and $P_{NO_2} = 1.56$ atm. What would happen to these pressures if a catalyst were present?

15.49 Consider the gas-phase reaction

$$2CO(g) + O_2(g) \rightleftharpoons 2CO_2(g)$$

Predict the shift in the equilibrium position when helium gas is added to the equilibrium mixture (a) at constant pressure and (b) at constant volume.

15.50 Consider the following reaction at equilibrium in a closed container:

$$CaCO_3(s) \rightleftharpoons CaO(s) + CO_2(g)$$

What would happen if (a) the volume is increased, (b) some CaO is added to the mixture, (c) some $CaCO_3$ is removed, (d) some CO_2 is added to the mixture, (e) a few drops of a NaOH solution are added to the mixture, (f) a few drops of a HCl solution are added to the mixture (ignore the reaction between CO_2 and water), (g) temperature is increased?

MISCELLANEOUS PROBLEMS

15.51 Consider the statement: The equilibrium constant of a reacting mixture of solid NH_4Cl and gaseous NH_3 and HCl is 0.316. List three important pieces of information that are missing from this statement.

15.52 Pure NOCl gas was heated at 240°C in a 1.00-L container. At equilibrium the total pressure was 1.00 atm and the NOCl pressure was 0.64 atm.

$$2NOCl(g) \rightleftharpoons 2NO(g) + Cl_2(g)$$

(a) Calculate the partial pressures of NO and Cl_2 in the system. (b) Calculate the equilibrium constant K_P.

15.53 Consider the following reaction:

$$N_2(g) + O_2(g) \rightleftharpoons 2NO(g)$$

The equilibrium constant K_P for the reaction is 1.0×10^{-15} at 25°C and 0.050 at 2200°C. Is the formation of nitric oxide endothermic or exothermic? Explain your answer.

15.54 Baking soda (sodium bicarbonate) undergoes thermal decomposition as follows:

$$2NaHCO_3(s) \rightleftharpoons Na_2CO_3(s) + CO_2(g) + H_2O(g)$$

Would we obtain more CO_2 and H_2O by adding extra baking soda to the reaction mixture in (a) a closed vessel or (b) an open vessel?

15.55 Consider the following reaction at equilibrium:

$$A(g) \rightleftharpoons 2B(g)$$

From the following data, calculate the equilibrium constant (both K_P and K_c) at each temperature. Is the reaction endothermic or exothermic?

Temperature (°C)	*[A]*	*[B]*
200	0.0125	0.843
300	0.171	0.764
400	0.250	0.724

15.56 The equilibrium constant K_P for the reaction

$$2H_2O(g) \rightleftharpoons 2H_2(g) + O_2(g)$$

is found to be 2×10^{-42} at 25°C. (a) What is K_c for the reaction at the same temperature? (b) The very small value of K_P (and K_c) indicates that the reaction overwhelmingly favors the formation of water molecules. Explain why, despite this fact, a mixture of hydrogen and oxygen gases can be kept at room temperature without any change.

15.57 Consider the following reacting system:

$$2NO(g) + Cl_2(g) \rightleftharpoons 2NOCl(g)$$

What combination of temperature and pressure would maximize the yield of NOCl? [*Hint:* $\Delta H_f^\circ(NOCl) = 51.7$ kJ/mol. You will also need to consult Appendix 2.]

15.58 At a certain temperature and a total pressure of 1.2 atm, the partial pressures of an equilibrium mixture

$$2A(g) \rightleftharpoons B(g)$$

are $P_A = 0.60$ atm and $P_B = 0.60$ atm. (a) Calculate the K_P for the reaction at this temperature. (b) If the total pressure were increased to 1.5 atm, what would be the partial pressures of A and B at equilibrium?

15.59 The decomposition of ammonium hydrogen sulfide

$$NH_4HS(s) \rightleftharpoons NH_3(g) + H_2S(g)$$

is an endothermic process. A 6.1589-g sample of the solid is placed in an evacuated 4.000-L vessel at exactly 24°C. After equilibrium has been established, the total pressure inside is 0.709 atm. Some solid NH_4HS remains in the vessel. (a) What is the K_P for the reaction? (b) What percentage of the solid has decomposed? (c) If the volume of the vessel were doubled at constant temperature, what would happen to the amount of solid in the vessel?

15.60 Consider the reaction

$$2NO(g) + O_2(g) \rightleftharpoons 2NO_2(g)$$

At 430°C, an equilibrium mixture consists of 0.020 mole of O_2, 0.040 mole of NO, and 0.96 mole of NO_2. Calculate K_P for the reaction, given that the total pressure is 0.20 atm.

15.61 When heated, ammonium carbamate decomposes as follows:

$$NH_4CO_2NH_2(s) \rightleftharpoons 2NH_3(g) + CO_2(g)$$

At a certain temperature the equilibrium pressure of the system is 0.318 atm. Calculate K_P for the reaction.

15.62 A mixture of 0.47 mole of H_2 and 3.59 moles of HCl is heated to 2800°C. Calculate the equilibrium partial pressures of H_2, Cl_2, and HCl if the total pressure is 2.00 atm. The K_P for the reaction $H_2(g) + Cl_2(g) \rightleftharpoons 2HCl(g)$ is 193 at 2800°C.

15.63 Consider the reaction in a closed container:

$$N_2O_4(g) \rightleftharpoons 2NO_2(g)$$

Initially, 1 mole of N_2O_4 is present. At equilibrium, α mole of N_2O_4 has dissociated to form NO_2. (a) Derive an expression for K_P in terms of α and P, the total pressure. (b) How does the expression in (a) help you predict the shift in equilibrium due to an increase in P? Does your prediction agree with Le Chatelier's principle?

15.64 One mole of N_2 and three moles of H_2 are placed in a flask at 397°C. Calculate the total pressure of the system at equilibrium if the mole fraction of NH_3 is found to be 0.21. The K_P for the reaction is 4.31×10^{-4}.

15.65 At 1130°C the equilibrium constant (K_c) for the reaction

$$2H_2S(g) \rightleftharpoons 2H_2(g) + S_2(g)$$

is 2.25×10^{-4}. If $[H_2S] = 4.84 \times 10^{-3}\,M$ and $[H_2] = 1.50 \times 10^{-3}\,M$, calculate $[S_2]$.

15.66 A quantity of 6.75 g of SO_2Cl_2 was placed in a 2.00-L flask. At 648 K, there is 0.0345 mole of SO_2 present. Calculate K_c for the reaction

$$SO_2Cl_2(g) \rightleftharpoons SO_2(g) + Cl_2(g)$$

15.67 The formation of SO_3 from SO_2 and O_2 is an intermediate step in the manufacture of sulfuric acid, and it is also responsible for the acid rain phenomenon. The equilibrium constant (K_P) for the reaction

$$2SO_2(g) + O_2(g) \rightleftharpoons 2SO_3(g)$$

is 0.13 at 830°C. In one experiment 2.00 moles of SO_2 and 2.00 moles of O_2 were initially present in a flask. What must the total pressure be at equilibrium in order to have an 80.0 percent yield of SO_3?

15.68 Consider the dissociation of iodine:

$$I_2(g) \rightleftharpoons 2I(g)$$

A 1.00-g sample of I_2 is heated at 1200°C in a 500-mL flask. At equilibrium the total pressure is 1.51 atm. Calculate K_P for the reaction. [*Hint:* Use the result in 15.63(a). The degree of dissociation α is given by the ratio of observed pressure over calculated pressure, assuming no dissociation.]

15.69 Eggshells are composed mostly of calcium carbonate ($CaCO_3$) formed by the reaction

$$Ca^{2+}(aq) + CO_3^{2-}(aq) \rightleftharpoons CaCO_3(s)$$

The carbonate ions are supplied by carbon dioxide produced as a result of metabolism. Explain why eggshells are thinner in the summer, when the rate of chicken panting is greater. Suggest a remedy for this situation.

15.70 The equilibrium constant K_P for the following reaction is found to be 4.31×10^{-4} at 397°C:

$$N_2(g) + 3H_2(g) \rightleftharpoons 2NH_3(g)$$

In a certain experiment a student starts with 0.862 atm of N_2 and 0.373 atm of H_2 in a constant-volume vessel at 200°C. Calculate the partial pressures of all species when equilibrium is reached.

15.71 A quantity of 0.20 mole of carbon dioxide was heated at a certain temperature with an excess of graphite in a closed container until the following equilibrium was reached:

$$C(s) + CO_2(g) \rightleftharpoons 2CO(g)$$

Under this condition, the average molar mass of the gases was found to be 35 g/mol. (a) Calculate the mole fractions of CO and CO_2. (b) What is the K_P for the equilibrium if the total pressure was 11 atm? (*Hint:* The average molar mass is the sum of the products of the mole fraction of each gas and its molar mass.)

15.72 When dissolved in water, glucose (corn sugar) and fructose (fruit sugar) exist in equilibrium as follows:

$$\text{fructose} \rightleftharpoons \text{glucose}$$

A chemist prepared a 0.244 *M* fructose solution at 25°C. At equilibrium, it was found that its concentration had decreased to 0.113 *M*. (a) Calculate the equilibrium constant for the reaction. (b) At equilibrium, what percentage of fructose was converted to glucose?

15.73 At room temperature, solid iodine is in equilibrium with its vapor through sublimation and deposition (see Figure 8.18). Describe how you would use radioactive iodine, in either solid or vapor form, to show that there is a dynamic equilibrium between these two phases.

15.74 At 1024°C, the pressure of oxygen gas from the decomposition of copper(II) oxide (CuO) is 0.49 atm:

$$4CuO(s) \rightleftharpoons 2Cu_2O(s) + O_2(g)$$

(a) What is the K_P for the reaction? (b) Calculate the fraction of CuO decomposed if 0.16 mole of it is placed in a 2.0-L flask at 1024°C. (c) What would be the fraction if a 1.0-mole sample of CuO were used? (d) What is the smallest amount of CuO (in moles) that would establish the equilibrium?

15.75 A mixture containing 3.9 moles of NO and 0.88 mole of CO_2 was allowed to react in a flask at a certain temperature according to the equation

$$NO(g) + CO_2(g) \rightleftharpoons NO_2(g) + CO(g)$$

At equilibrium, 0.11 mole of CO_2 was present. Calculate the equilibrium constant K_c of this reaction.

15.76 The equilibrium constant K_c for the reaction

$$H_2(g) + I_2(g) \rightleftharpoons 2HI(g)$$

is 54.3 at 430°C. At the start of the reaction there are 0.714 mole of H_2, 0.984 mole of I_2, and 0.886 mole of HI in a 2.40-L reaction chamber. Calculate the concentrations of the gases at equilibrium.

15.77 On heating, a gaseous compound A dissociates as follows:

$$A(g) \rightleftharpoons B(g) + C(g)$$

In an experiment A was heated at a certain temperature until its equilibrium pressure reached 0.14P, where P is the total pressure. Calculate the equilibrium constant (K_P) of this reaction.

15.78 When a gas was heated at atmospheric pressure and 25°C, its color was found to deepen. Heating above 150°C caused the color to fade, and at 550°C the color was barely detectable. However, at 550°C, the color was partially restored by increasing the pressure of the system. Which of the following best fits the above description? Justify your choice. (a) A mixture of hydrogen and bromine, (b) pure bromine, (c) a mixture of nitrogen dioxide and dinitrogen tetroxide. (*Hint:* Bromine has a reddish color and nitrogen dioxide is a brown gas. The other gases are colorless.)

15.79 In this chapter we learned that a catalyst has no effect on the position of an equilibrium because it speeds up both the forward and reverse rates to the same extent. To test this statement, consider a situation in which an equilibrium of the type

$$2A(g) \rightleftharpoons B(g)$$

is established inside a cylinder fitted with a weightless piston. The piston is attached by a string to the cover of a box containing a catalyst. When the piston moves upward (expanding against atmospheric pressure), the cover is lifted and the catalyst is exposed to the gases. When the piston moves downward, the box is closed. Assume that the catalyst only speeds up the forward reaction ($2A \rightarrow B$) but does not affect the reverse process ($B \rightarrow 2A$). Supposing the catalyst is suddenly exposed to the equilibrium system as shown below, describe what would happen subsequently. How does this "thought" experiment convince you that no such catalyst can exist?

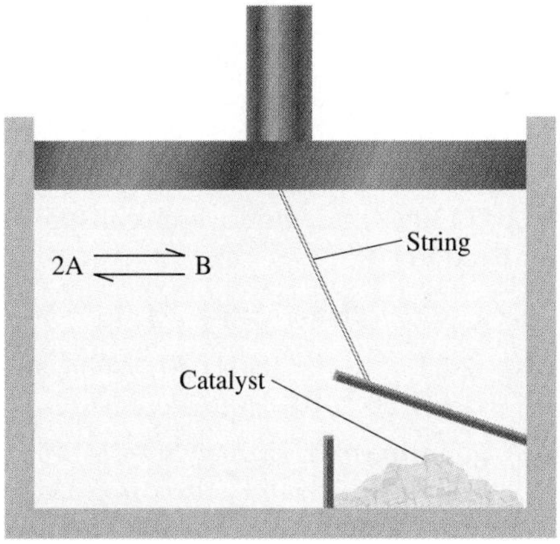

15.80 The equilibrium constant K_c for the following reaction is 0.65 at 375°C.

$$N_2(g) + 3H_2(g) \rightleftharpoons 2NH_3(g)$$

(a) What is the value of K_P for this reaction?
(b) What is the value of the equilibrium constant K_c for $2NH_3(g) \rightleftharpoons N_2(g) + 3H_2(g)$?
(c) What is K_c for $\frac{1}{2}N_2(g) + \frac{3}{2}H_2(g) \rightleftharpoons NH_3(g)$?
(d) What are the values of K_P for the reactions described in (b) and (c)?

15.81 A sealed glass bulb contains a mixture of NO_2 and N_2O_4 gases. When the bulb is heated from 20°C to 40°C, what happens to the following properties of the gases? (a) Color, (b) pressure, (c) average molar mass, (d) degree of dissociation (from N_2O_4 to NO_2), (e) density. Assume that volume remains constant. (*Hint:* NO_2 is a brown gas; N_2O_4 is colorless.)

15.82 At 20°C, the vapor pressure of water is 0.0231 atm. Calculate K_P and K_c for the process

$$H_2O(l) \rightleftharpoons H_2O(g)$$

15.83 Industrially, sodium metal is obtained by electrolyzing molten sodium chloride. The reaction at the cathode is $Na^+ + e^- \rightarrow Na$. We might expect that the potassium metal would also be prepared by electrolyzing molten potassium chloride. However, potassium metal is soluble in molten potassium chloride and therefore is hard to recover. Furthermore, potassium vaporizes readily at the operating temperature, creating hazardous conditions. Instead, potassium is prepared by the distillation of molten potassium chloride in the presence of sodium vapor at 892°C:

$$Na(g) + KCl(l) \rightleftharpoons NaCl(l) + K(g)$$

In view of the fact that potassium is a stronger reducing agent than sodium, explain why this approach works. (The boiling points of sodium and potassium are 892°C and 770°C, respectively.)

15.84 In the gas phase, nitrogen dioxide is actually a mixture of nitrogen dioxide (NO_2) and dinitrogen tetroxide (N_2O_4). If the density of such a mixture at 74°C and 1.3 atm is 2.9 g/L, calculate the partial pressures of the gases and K_P.

Answers to Practice Exercises: 15.1 $K_c = \dfrac{[NO_2]^4[O_2]}{[N_2O_5]^2}$, $K_P = \dfrac{P_{NO_2}^4 P_{O_2}}{P_{N_2O_5}^2}$; **15.2** 347 atm; **15.3** 0.113; **15.4** $K_P = 0.0702$; $K_c = 1.20 \times 10^{-4}$; **15.5** From right to left; **15.6** [HI] = 0.031 *M*, $[H_2] = 4.3 \times 10^{-3}$ *M*, $[I_2] = 4.3 \times 10^{-3}$ *M*; **15.7** $[Br_2] = 0.065$ *M*, $[Br] = 8.4 \times 10^{-3}$ *M*; **15.8** $Q_P = 4.0 \times 10^5$, the net reaction will shift from right to left; **15.9** Left to right; **15.10** The equilibrium will shift from (a) left to right, (b) left to right, and (c) right to left. (d) A catalyst has no effect on the equilibrium.

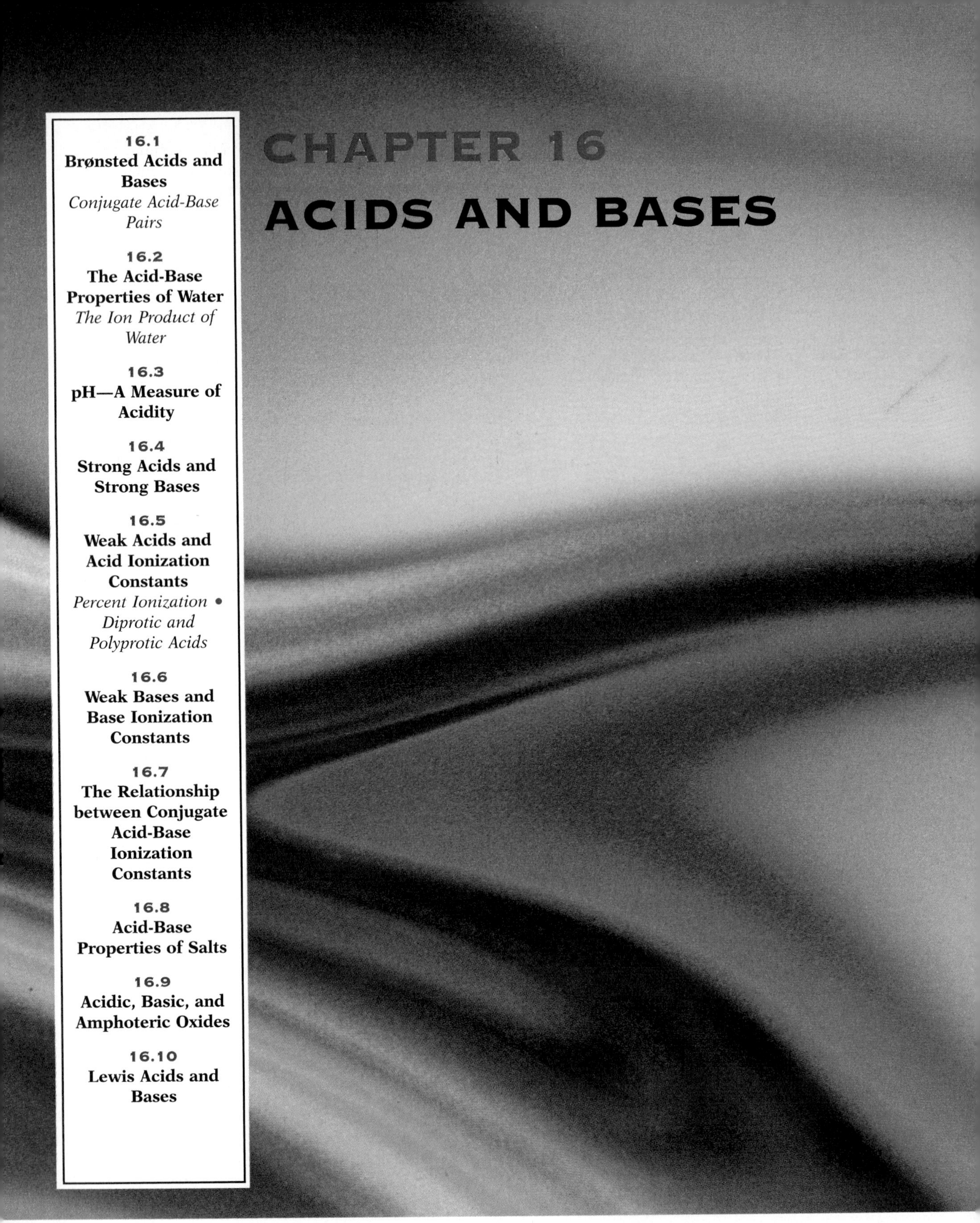

CHAPTER 16
ACIDS AND BASES

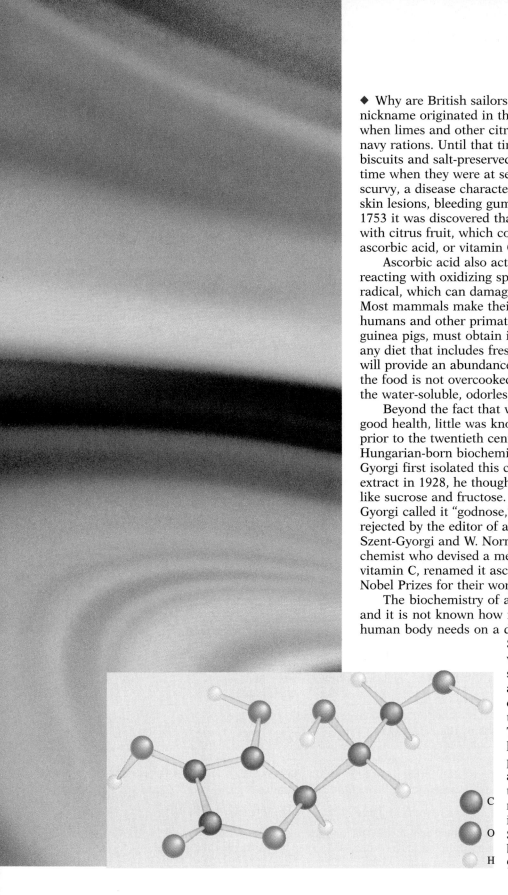

◆ Why are British sailors called "limeys"? The nickname originated in the late eighteenth century when limes and other citrus fruits became part of navy rations. Until that time sailors often ate only biscuits and salt-preserved meats for months at a time when they were at sea. Many developed scurvy, a disease characterized by aching joints, skin lesions, bleeding gums, and loose teeth. In 1753 it was discovered that scurvy can be cured with citrus fruit, which contains large amounts of ascorbic acid, or vitamin C ($C_6H_8O_6$).

Ascorbic acid also acts as an antioxidant, reacting with oxidizing species such as the $\cdot OH$ radical, which can damage an organism's DNA. Most mammals make their own vitamin C, but humans and other primates, as well as bats and guinea pigs, must obtain it from their food. Almost any diet that includes fresh fruit and vegetables will provide an abundance of vitamin C, as long as the food is not overcooked; heat readily destroys the water-soluble, odorless white solid.

Beyond the fact that vitamin C is essential for good health, little was known about its chemistry prior to the twentieth century. Indeed, when the Hungarian-born biochemist Albert von Szent-Gyorgi first isolated this compound from a cabbage extract in 1928, he thought it might be a sugar, like sucrose and fructose. Consequently, Szent-Gyorgi called it "godnose," a name that was firmly rejected by the editor of a scientific journal. Later, Szent-Gyorgi and W. Norman Haworth, a British chemist who devised a method for synthesizing vitamin C, renamed it ascorbic acid. Both won Nobel Prizes for their work on vitamin C in 1937.

The biochemistry of ascorbic acid is complex, and it is not known how much vitamin C the human body needs on a daily basis. Nevertheless, Szent-Gyorgi, together with Linus Pauling and some other scientists, advocated taking large doses of vitamin C tablets to maintain good health. They believed that a high level of vitamin C helps to prevent the common cold and protect against certain cancers. These claims remain controversial, but it is worth noting that Szent-Gyorgi and Pauling both lived to the age of 93. ◆

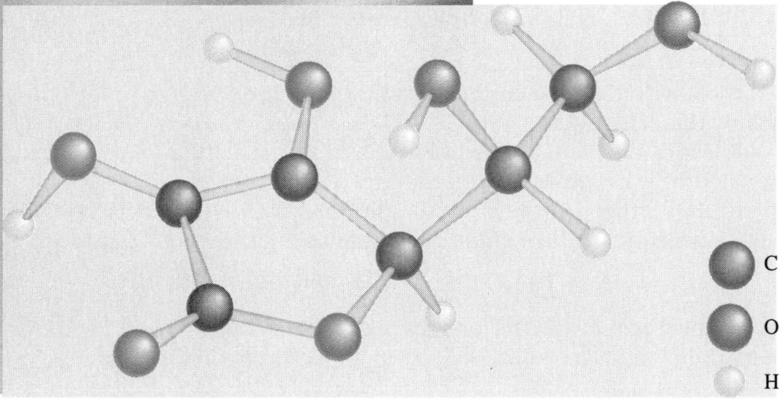

16.1 BRØNSTED ACIDS AND BASES

In Chapter 4 we defined a Brønsted acid as a substance capable of donating a proton, and a Brønsted base as a substance capable of accepting a proton. These definitions are generally suitable for discussion of the properties and reactions of acids and bases.

CONJUGATE ACID-BASE PAIRS

An extension of the Brønsted definition of acids and bases is the concept of the ***conjugate acid-base pair,*** which can be defined as *an acid and its conjugate base or a base and its conjugate acid.* The conjugate base of a Brønsted acid is the species that remains when one proton has been removed from the acid. Conversely, a conjugate acid results from the addition of a proton to a Brønsted base.

Every Brønsted acid has a conjugate base, and every Brønsted base has a conjugate acid. For example, the chloride ion (Cl^-) is the conjugate base formed from the acid HCl, and H_2O is the conjugate base of the acid H_3O^+. Similarly, the ionization of acetic acid can be represented as

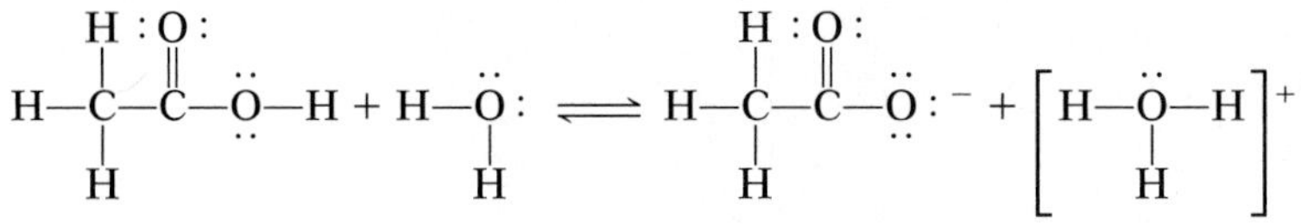

$$\underset{\text{acid}_1}{CH_3COOH(aq)} + \underset{\text{base}_2}{H_2O(l)} \rightleftharpoons \underset{\text{base}_1}{CH_3COO^-(aq)} + \underset{\text{acid}_2}{H_3O^+(aq)}$$

The subscripts 1 and 2 designate the two conjugate acid-base pairs. Thus the acetate ion (CH_3COO^-) is the conjugate base of CH_3COOH. Both the ionization of HCl (see Section 4.3) and the ionization of CH_3COOH are examples of Brønsted acid-base reactions.

The Brønsted definition also allows us to classify ammonia as a base because of its ability to accept a proton:

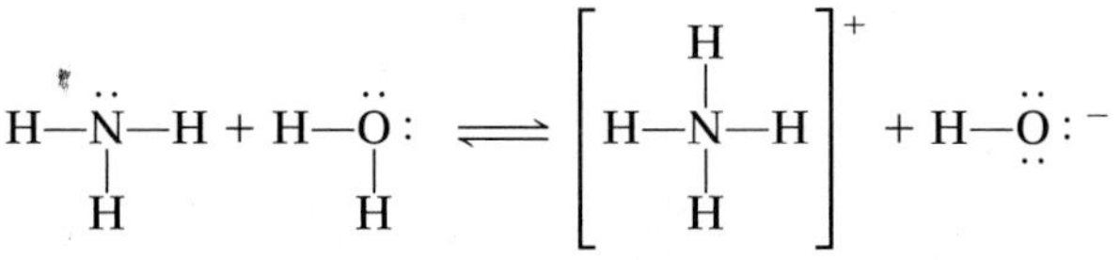

$$\underset{\text{base}_1}{NH_3(aq)} + \underset{\text{acid}_2}{H_2O(l)} \rightleftharpoons \underset{\text{acid}_1}{NH_4^+(aq)} + \underset{\text{base}_2}{OH^-(aq)}$$

In this case, NH_4^+ is the conjugate acid of the base NH_3, and OH^- is the conjugate base of the acid H_2O.

A less clear-cut case is that of NaOH, which, strictly speaking, is not a Brønsted base because it cannot accept a proton. However, NaOH is a strong electrolyte that ionizes completely in solution. The hydroxide ion (OH^-) formed by that ionization *is* a Brønsted base because it can accept a proton:

$$H_3O^+(aq) + OH^-(aq) \rightleftharpoons 2H_2O(l)$$

Thus, when we call NaOH or any other metal hydroxide a base, we are actually referring to the OH^- species derived from the hydroxide.

EXAMPLE 16.1
Identifying Conjugate Acid-Base Pairs

Identify the conjugate acid-base pairs in the following reaction:

$$NH_3(aq) + HF(aq) \rightleftharpoons NH_4^+(aq) + F^-(aq)$$

Answer: The conjugate acid-base pairs are (1) HF (acid) and F^- (base) and (2) NH_4^+ (acid) and NH_3 (base).

PRACTICE EXERCISE

Identify the conjugate acid-base pairs for the reaction

$$CN^- + H_2O \rightleftharpoons HCN + OH^-$$

Finally we note that it is equally acceptable to represent the proton in aqueous solutions as H^+ or as H_3O^+. The formula H^+ is less cumbersome in calculations involving hydrogen ion concentrations and in calculations involving equilibrium constants, whereas H_3O^+ is more useful when discussing Brønsted acid-base properties.

16.2 THE ACID-BASE PROPERTIES OF WATER

Water, as we know, is a unique solvent. One of its special properties is its ability to act both as an acid and as a base. Water functions as a base in reactions with acids such as HCl and CH_3COOH, and it functions as an acid in reactions with bases such as NH_3. Water is a very weak electrolyte and therefore a poor conductor of electricity, but it does undergo ionization to a small extent:

$$H_2O(l) \rightleftharpoons H^+(aq) + OH^-(aq) \qquad (16.1)$$

This reaction is sometimes called the *autoionization* of water. To describe the acid-base properties of water in the Brønsted framework, we express its autoionization as

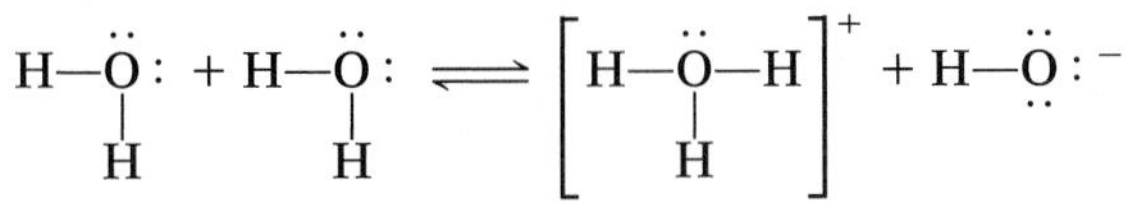

or

$$\underset{\text{acid}_1}{H_2O} + \underset{\text{base}_2}{H_2O} \rightleftharpoons \underset{\text{acid}_2}{H_3O^+} + \underset{\text{base}_1}{OH^-}$$

The acid-base conjugate pairs are (1) H_2O (acid) and OH^- (base) and (2) H_3O^+ (acid) and H_2O (base).

THE ION PRODUCT OF WATER

In the study of acid-base reactions in aqueous solutions, the important quantity is the hydrogen ion concentration. Expressing the proton as H^+ rather than H_3O^+, we can write the equilibrium constant for the autoionization of water, Equation (16.1), as

$$K_c = \frac{[H^+][OH^-]}{[H_2O]}$$

Since a very small fraction of water molecules are ionized, the concentration of water, that is, $[H_2O]$, remains virtually unchanged. Therefore

$$K_c[H_2O] = K_w = [H^+][OH^-] \tag{16.2}$$

The equilibrium constant K_w is called the ***ion-product constant,*** which is *the product of the molar concentrations of H^+ and OH^- ions at a particular temperature.*

In pure water at 25°C, the concentrations of H^+ and OH^- ions are equal and found to be $[H^+] = 1.0 \times 10^{-7}\ M$ and $[OH^-] = 1.0 \times 10^{-7}\ M$. Thus, from Equation (16.2), at 25°C

$$K_w = (1.0 \times 10^{-7})(1.0 \times 10^{-7}) = 1.0 \times 10^{-14}$$

Note that whether we have pure water or a solution of dissolved species, the following relation *always* holds at 25°C:

$$\begin{aligned} K_w &= [H^+][OH^-] \\ &= 1.0 \times 10^{-14} \end{aligned}$$

Whenever $[H^+] = [OH^-]$, the aqueous solution is said to be neutral. In an acidic solution there is an excess of H^+ ions and $[H^+] > [OH^-]$. In a basic solution there is an excess of hydroxide ions, so $[H^+] < [OH^-]$. In practice we can change the concentration of either H^+ or OH^- ions in solution, but we cannot vary them independently. If we adjust the solution so that $[H^+] = 1.0 \times 10^{-6}\ M$, the OH^- concentration *must* change to

$$\begin{aligned} [OH^-] &= \frac{K_w}{[H^+]} \\ &= \frac{1.0 \times 10^{-14}}{1.0 \times 10^{-6}} = 1.0 \times 10^{-8}\ M \end{aligned}$$

EXAMPLE 16.2

Calculating $[H^+]$ from $[OH^-]$

The concentration of OH^- ions in a certain household ammonia cleaning solution is 0.0025 *M*. Calculate the concentration of the H^+ ions.

Answer: Rearranging Equation (16.2), we write

$$\begin{aligned} [H^+] &= \frac{K_w}{[OH^-]} \\ &= \frac{1.0 \times 10^{-14}}{0.0025} = 4.0 \times 10^{-12}\ M \end{aligned}$$

Since $[H^+] < [OH^-]$, the solution is basic, as we would expect from the earlier discussion of the reaction of ammonia with water.

PRACTICE EXERCISE

Calculate the concentration of OH^- ions in a HCl solution whose hydrogen ion concentration is 1.3 *M*.

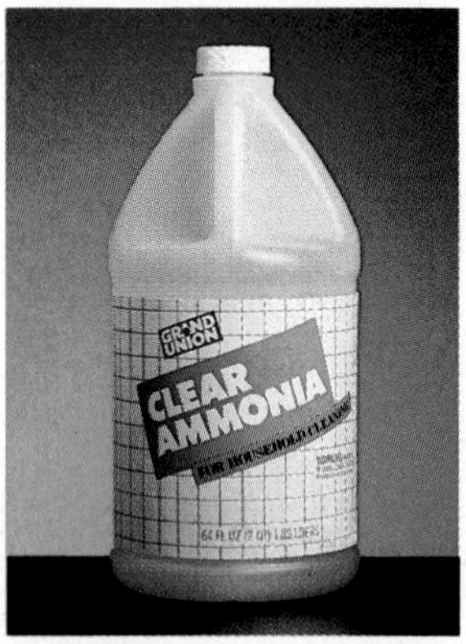

Many household cleaning fluids contain ammonia.

16.3 PH—A MEASURE OF ACIDITY

Because the concentrations of H^+ and OH^- ions in aqueous solutions are frequently very small numbers and therefore inconvenient to work with, the Danish biochemist Soren Sorensen in 1909 proposed a more practical measure called pH. The ***pH*** of a solution is defined as *the negative logarithm of the hydrogen ion concentration (in moles per liter):*

$$\text{pH} = -\log\,[\text{H}^+] \tag{16.3}$$

Keep in mind that Equation (16.3) is simply a definition designed to give us convenient numbers to work with. The negative logarithm gives us a positive number for pH, which otherwise would be negative due to the small value of $[H^+]$. (See Appendix 3 for a discussion of logarithms and significant figures.) Furthermore, the term $[H^+]$ in Equation (16.3) pertains only to the *numerical part* of the expression for hydrogen ion concentration, for we cannot take the logarithm of units. Thus, like the equilibrium constant, the pH of a solution is a dimensionless quantity.

Since pH is simply a way to express hydrogen ion concentration, acidic and basic solutions at 25°C can be identified by their pH values, as follows:

Acidic solutions:	$[H^+] > 1.0 \times 10^{-7}\,M$, pH < 7.00
Basic solutions:	$[H^+] < 1.0 \times 10^{-7}\,M$, pH > 7.00
Neutral solutions:	$[H^+] = 1.0 \times 10^{-7}\,M$, pH = 7.00

Notice that pH increases as $[H^+]$ decreases.

In the laboratory, the pH of a solution is measured with a pH meter (Figure 16.1). Table 16.1 lists the pHs for a number of common fluids.

A pOH scale analogous to the pH scale can be devised using the negative logarithm of the hydroxide ion concentration. Thus we define pOH as

$$\text{pOH} = -\log\,[\text{OH}^-] \tag{16.4}$$

Now consider again the ion-product constant for water:

$$[\text{H}^+][\text{OH}^-] = K_w = 1.0 \times 10^{-14}$$

Taking the negative logarithm of both sides, we obtain

$$-(\log\,[\text{H}^+] + \log\,[\text{OH}^-]) = -\log\,(1.0 \times 10^{-14})$$

$$-\log\,[\text{H}^+] - \log\,[\text{OH}^-] = 14.00$$

TABLE 16.1
The pHs of Some Common Fluids

Sample	pH value
Gastric juice in the stomach	1.0–2.0
Lemon juice	2.4
Vinegar	3.0
Grapefruit juice	3.2
Orange juice	3.5
Urine	4.8–7.5
Water exposed to air*	5.5
Saliva	6.4–6.9
Milk	6.5
Pure water	7.0
Blood	7.35–7.45
Tears	7.4
Milk of magnesia	10.6
Household ammonia	11.5

*Water exposed to air for a long period of time absorbs atmospheric CO_2 to form carbonic acid, H_2CO_3.

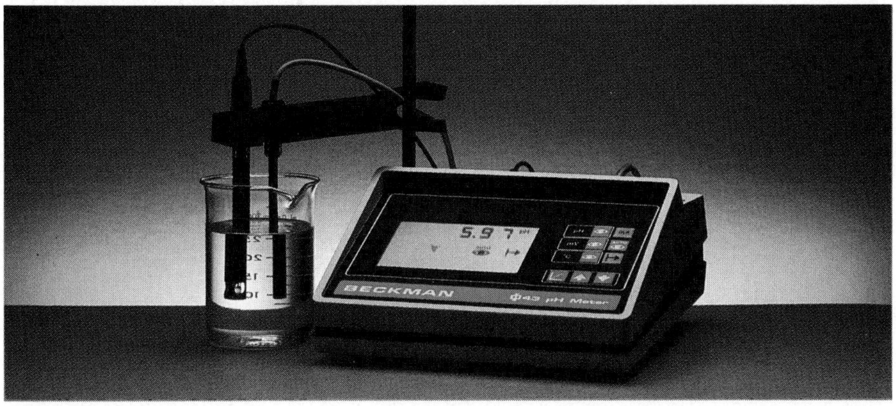

FIGURE 16.1

A pH meter is commonly used in the laboratory to determine the pH of a solution. Although many pH meters have scales marked with values from 1 to 14, pH values can, in fact, be less than 1 and greater than 14.

From the definitions of pH and pOH we obtain

$$pH + pOH = 14.00 \tag{16.5}$$

Equation (16.5) provides us with another way to express the relationship between the H^+ ion concentration and the OH^- ion concentration.

EXAMPLE 16.3
Calculating pH from $[H^+]$

The concentration of H^+ ions in a bottle of table wine was 3.2×10^{-4} *M* right after the cork was removed. Only half of the wine was consumed. The other half, after it had been standing open to the air for a month, was found to have a hydrogen ion concentration equal to 1.0×10^{-3} *M*. Calculate the pH of the wine on these two occasions.

Answer: When the bottle was first opened, $[H^+] = 3.2 \times 10^{-4}$ *M*, which we substitute in Equation (16.3).

$$\begin{aligned} pH &= -\log [H^+] \\ &= -\log (3.2 \times 10^{-4}) = 3.49 \end{aligned}$$

On the second occasion, $[H^+] = 1.0 \times 10^{-3}$ *M*, so that

$$pH = -\log (1.0 \times 10^{-3}) = 3.00$$

Note that the increase in hydrogen ion concentration (or decrease in pH) is largely the result of the conversion of some of the alcohol (ethanol) to acetic acid, a reaction that takes place in the presence of molecular oxygen.

PRACTICE EXERCISE

Calculate the pH of a HNO_3 solution whose hydrogen ion concentration is 0.76 *M*.

EXAMPLE 16.4
Calculating $[H^+]$ from pH

The pH of rainwater collected in a certain region of the northeastern United States on a particular day was 4.82. Calculate the H^+ ion concentration of the rainwater.

Answer: From Equation (16.3)

$$4.82 = -\log [H^+]$$

Taking the antilog of both sides of this equation (see Appendix 3) gives

$$1.5 \times 10^{-5}\ M = [H^+]$$

Because the pH is between 4 and 5, we can expect $[H^+]$ to be between 1×10^{-4} *M* and 1×10^{-5} *M*. Therefore, the answer is reasonable.

PRACTICE EXERCISE

The pH of a certain fruit juice is 3.33. Calculate the H^+ ion concentration.

16.4 STRONG ACIDS AND STRONG BASES

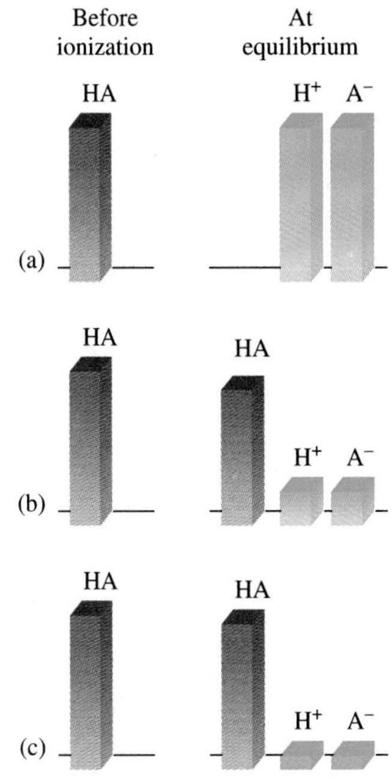

FIGURE 16.2

The extent of ionization of (a) a strong acid that undergoes 100 percent ionization, (b) a weak acid, and (c) a very weak acid.

Strong acids *are strong electrolytes, which,* for most practical purposes, *are assumed to ionize completely in water* (Figure 16.2). Most of the strong acids are inorganic acids: hydrochloric acid (HCl), nitric acid (HNO_3), perchloric acid ($HClO_4$), and sulfuric acid (H_2SO_4):

$$HCl(aq) + H_2O(l) \longrightarrow H_3O^+(aq) + Cl^-(aq)$$

$$HNO_3(aq) + H_2O(l) \longrightarrow H_3O^+(aq) + NO_3^-(aq)$$

$$HClO_4(aq) + H_2O(l) \longrightarrow H_3O^+(aq) + ClO_4^-(aq)$$

$$H_2SO_4(aq) + H_2O(l) \longrightarrow H_3O^+(aq) + HSO_4^-(aq)$$

Note that H_2SO_4 is a diprotic acid; we show only the first stage of ionization here. At equilibrium, solutions of strong acids will not contain any nonionized acid molecules.

Most acids *ionize only to a limited extent in water.* Such acids are classified as ***weak acids.*** At equilibrium, aqueous solutions of weak acids contain a mixture of nonionized acid molecules, H_3O^+ ions, and the conjugate base. Examples of weak acids are hydrofluoric acid (HF), acetic acid (CH_3COOH), and the ammonium ion (NH_4^+). Within this category, the strength of acids can vary greatly due to differences in extent to ionization. The limited ionization of weak acids is related to the equilibrium constant for ionization, which we will study in the next section.

What has been said about strong acids also applies to strong bases, which include hydroxides of alkali metals and certain alkaline earth metals, such as NaOH, KOH, and $Ba(OH)_2$. ***Strong bases*** *are all strong electrolytes that ionize completely in water:*

$$NaOH(s) \xrightarrow{H_2O} Na^+(aq) + OH^-(aq)$$

$$KOH(s) \xrightarrow{H_2O} K^+(aq) + OH^-(aq)$$

$$Ba(OH)_2(s) \xrightarrow{H_2O} Ba^{2+}(aq) + 2OH^-(aq)$$

Weak bases, like weak acids, *are weak electrolytes.* Ammonia is a weak base. It ionizes to a very limited extent in water:

$$NH_3(aq) + H_2O(l) \rightleftharpoons NH_4^+(aq) + OH^-(aq)$$

Table 16.2 lists some important conjugate acid-base pairs, in order of their relative strengths. It is useful to keep the following points in mind:

- If an acid is strong, its conjugate base has no measurable strength.
- H_3O^+ is the strongest acid that can exist in aqueous solution. Acids stronger than H_3O^+ react with water to produce H_3O^+ and their conjugate bases. Thus HCl, which is a stronger acid than H_3O^+, reacts with water completely to form H_3O^+ and Cl^-:

  $$HCl(aq) + H_2O(l) \longrightarrow H_3O^+(aq) + Cl^-(aq)$$

  Acids weaker than H_3O^+ react with water to a much smaller extent, producing H_3O^+ and their conjugate bases. For example, the following equilibrium lies primarily to the left:

  $$HF(aq) + H_2O(l) \rightleftharpoons H_3O^+(aq) + F^-(aq)$$

- The OH^- ion is the strongest base that can exist in aqueous solution. Bases

TABLE 16.2
Relative Strengths of Conjugate Acid-Base Pairs

(Acid strength increases upward; base strength increases downward.)

	Acid	Conjugate base
Strong acids	$HClO_4$ (perchloric acid)	ClO_4^- (perchlorate ion)
	HI (hydroiodic acid)	I^- (iodide ion)
	HBr (hydrobromic acid)	Br^- (bromide ion)
	HCl (hydrochloric acid)	Cl^- (chloride ion)
	H_2SO_4 (sulfuric acid)	HSO_4^- (hydrogen sulfate ion)
	HNO_3 (nitric acid)	NO_3^- (nitrate ion)
	H_3O^+ (hydronium ion)	H_2O (water)
Weak acids	HSO_4^- (hydrogen sulfate ion)	SO_4^{2-} (sulfate ion)
	HF (hydrofluoric acid)	F^- (fluoride ion)
	HNO_2 (nitrous acid)	NO_2^- (nitrite ion)
	HCOOH (formic acid)	$HCOO^-$ (formate ion)
	CH_3COOH (acetic acid)	CH_3COO^- (acetate ion)
	NH_4^+ (ammonium ion)	NH_3 (ammonia)
	HCN (hydrocyanic acid)	CN^- (cyanide ion)
	H_2O (water)	OH^- (hydroxide ion)
	NH_3 (ammonia)	NH_2^- (amide ion)

Acid strength increases | Base strength increases

stronger than OH^- react with water to produce OH^- and their conjugate acids. For example, the oxide ion (O^{2-}) is a stronger base than OH^-, so it reacts with water completely as follows:

$$O^{2-}(aq) + H_2O \longrightarrow 2OH^-(aq)$$

For this reason the oxide ion does not exist in aqueous solutions. (Note that one of the two OH^- ions produced is identified as the conjugate acid of the O^{2-} ion.)

EXAMPLE 16.5
Calculating the pH of a Strong Acid and a Strong Base

Calculate the pH of (a) a 1.0×10^{-3} *M* HCl solution and (b) a 0.020 *M* $Ba(OH)_2$ solution.

Answer: (a) Since HCl is a strong acid, it is completely ionized in solution:

$$HCl(aq) \longrightarrow H^+(aq) + Cl^-(aq)$$

The concentration of all the species (HCl, H^+, and Cl^-) before and after ionization can be represented as follows:

	$HCl(aq)$	$\longrightarrow$	$H^+(aq)$	+	$Cl^-(aq)$
Initial (*M*):	1.0×10^{-3}		0.0		0.0
Change (*M*):	-1.0×10^{-3}		$+1.0 \times 10^{-3}$		$+1.0 \times 10^{-3}$
Final (*M*):	0.0		1.0×10^{-3}		1.0×10^{-3}

A positive (+) change represents an increase and a negative (−) change indicates a decrease in concentration. Thus

$$[H^+] = 1.0 \times 10^{-3}\ M$$

$$pH = -\log (1.0 \times 10^{-3})$$
$$= 3.00$$

(b) $Ba(OH)_2$ is a strong base; each $Ba(OH)_2$ unit produces two OH^- ions:

$$Ba(OH)_2(aq) \longrightarrow Ba^{2+}(aq) + 2OH^-(aq)$$

The changes in the concentrations of all the species can be represented as follows:

	$Ba(OH)_2(aq)$ $\longrightarrow$	$Ba^{2+}(aq)$ +	$2OH^-(aq)$
Initial (M):	0.020	0.00	0.00
Change (M):	−0.020	+0.020	+2(0.020)
Final (M):	0.00	0.020	0.040

Thus

$$[OH^-] = 0.040\ M$$
$$pOH = -\log 0.040 = 1.40$$

Therefore

$$pH = 14.00 - pOH$$
$$= 14.00 - 1.40$$
$$= 12.60$$

Note that in both (a) and (b) we have neglected the contribution of the autoionization of water to $[H^+]$ and $[OH^-]$ because $1.0 \times 10^{-7}\ M$ is so small compared with $1.0 \times 10^{-3}\ M$ and $0.040\ M$.

PRACTICE EXERCISE

Calculate the pH of a $1.5 \times 10^{-2}\ M$ $Ba(OH)_2$ solution.

16.5 WEAK ACIDS AND ACID IONIZATION CONSTANTS

As we have seen, the vast majority of acids are weak acids. Consider a weak monoprotic acid HA. Its ionization in water is represented by

$$HA(aq) + H_2O(l) \rightleftharpoons H_3O^+(aq) + A^-(aq)$$

or simply

$$HA(aq) \rightleftharpoons H^+(aq) + A^-(aq)$$

The *equilibrium constant for* this *acid ionization,* which we call the ***acid ionization constant,*** K_a, is given by

$$K_a = \frac{[H^+][A^-]}{[HA]}$$

At a given temperature, the strength of the acid HA is measured quantitatively by

TABLE 16.3
Ionization Constants of Some Weak Acids at 25°C

Name of acid	Formula	Structure	K_a	Conjugate base	K_b
Hydrofluoric acid	HF	H—F	7.1×10^{-4}	F^-	1.4×10^{-11}
Nitrous acid	HNO_2	O=N—O—H	4.5×10^{-4}	NO_2^-	2.2×10^{-11}
Acetylsalicylic acid (aspirin)	$C_9H_8O_4$	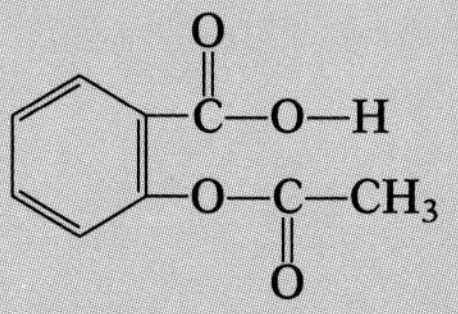	3.0×10^{-4}	$C_9H_7O_4^-$	3.3×10^{-11}
Formic acid	$HCOOH$	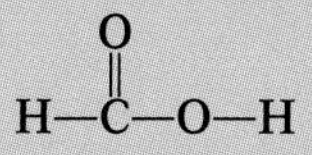	1.7×10^{-4}	$HCOO^-$	5.9×10^{-11}
Ascorbic acid*	$C_6H_8O_6$	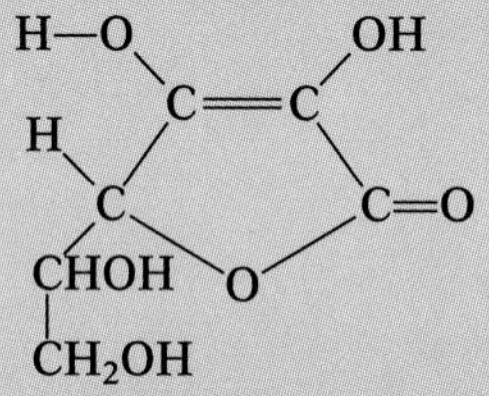	8.0×10^{-5}	$C_6H_7O_6^-$	1.3×10^{-10}
Benzoic acid	C_6H_5COOH	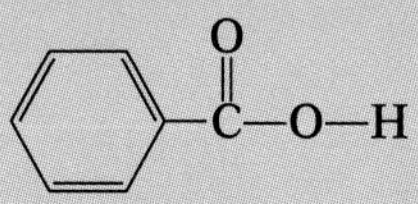	6.5×10^{-5}	$C_6H_5COO^-$	1.5×10^{-10}
Acetic acid	CH_3COOH	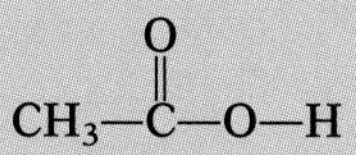	1.8×10^{-5}	CH_3COO^-	5.6×10^{-10}
Hydrocyanic acid	HCN	H—C≡N	4.9×10^{-10}	CN^-	2.0×10^{-5}
Phenol	C_6H_5OH	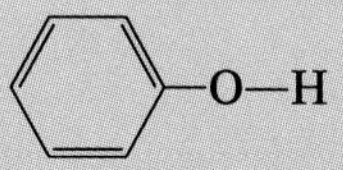	1.3×10^{-10}	$C_6H_5O^-$	7.7×10^{-5}

*For ascorbic acid it is the upper left hydroxyl group that is associated with this ionization constant.

the magnitude of K_a. The larger K_a, the stronger the acid—that is, the greater the concentration of H^+ ions at equilibrium due to its ionization.

Because the ionization of weak acids is never complete, all species (the nonionized acid, the H^+ ions, and the A^- ions) are present at equilibrium. Table 16.3 lists a number of weak acids and their K_a values in order of decreasing acid strength. Note that at equilibrium the predominant species in solution (other than the solvent) is the nonionized acid. Although the compounds listed in Table 16.3 are all weak acids, within the group there is great variation in acid strength. For example, K_a for HF (7.1×10^{-4}) is about 1.5 million times that for HCN (4.9×10^{-10}).

We can calculate K_a from the initial concentration of the acid and the pH of the solution, and we can use K_a and the initial concentration of the acid to calculate equilibrium concentrations of all the species and pH of solution. In calculating the equilibrium concentrations in a weak acid, we follow essentially the same procedure outlined in Section 15.3; the systems may be different, but the calculations are based on the same principle, the law of mass action [Equation 15.2)]. In this case, the three basic steps are:

1. Express the equilibrium concentrations of all species in terms of the initial concentrations and a single unknown x, which represents the change in concentration.
2. Write the acid ionization constant in terms of the equilibrium concentrations. Knowing the value of K_a, we can solve for x.
3. Having solved for x, calculate the equilibrium concentrations of all species and/or the pH of the solution.

Unless otherwise stated, we will assume that the temperature is 25°C for all such calculations.

EXAMPLE 16.6

Ionization of a Weak Monoprotic Acid

Calculate the concentrations of the nonionized acid and of the ions of a 0.100 *M* formic acid (HCOOH) solution at equilibrium.

Answer:

Step 1: Table 16.3 shows that HCOOH is a weak acid. Since it is monoprotic, one HCOOH molecule ionizes to give one H^+ ion and one $HCOO^-$ ion. Let x be the equilibrium concentration of H^+ and $HCOO^-$ ions in moles per liter. Then the equilibrium concentration of HCOOH must be $(0.100 - x)$ mol/L or $(0.100 - x)$ *M*. We can now summarize the changes in concentrations as follows:

	$HCOOH(aq)$ $\rightleftharpoons$	$H^+(aq)$ +	$HCOO^-(aq)$
Initial (*M*):	0.100	0.000	0.000
Change (*M*):	$-x$	$+x$	$+x$
Equilibrium (*M*):	$(0.100 - x)$	x	x

Step 2: According to Table 16.3

$$K_a = \frac{[H^+][HCOO^-]}{[HCOOH]} = 1.7 \times 10^{-4}$$

$$\frac{x^2}{0.100 - x} = 1.7 \times 10^{-4}$$

This equation can be rewritten as

$$x^2 + 1.7 \times 10^{-4}x - 1.7 \times 10^{-5} = 0$$

which fits the quadratic equation $ax^2 + bx + c = 0$.

We can often apply a simplifying approximation to this type of problem. Since HCOOH is a weak acid, the extent of its ionization must be small. Therefore x is small compared with 0.100. As a general rule, if the quantity x that is subtracted from the original concentration of the acid (0.100 *M* in this case) is equal to or less than 5 percent of the original concentration, we can assume $0.100 - x \simeq 0.100$. This removes x from the denominator of the equilibrium constant expression. If x is more than 5 percent of the original value, then we must solve the quadratic equation. In case of doubt, we can always solve for x by the approximate method and then check the validity of our approximation. Assuming that $0.100 - x \simeq 0.100$, then

$$\frac{x^2}{0.100 - x} \simeq \frac{x^2}{0.100} = 1.7 \times 10^{-4}$$

$$x^2 = 1.7 \times 10^{-5}$$

Taking the square root of both sides, we obtain

$$x = 4.1 \times 10^{-3}\ M$$

Step 3: At equilibrium, therefore,

$$[H^+] = 4.1 \times 10^{-3}\ M$$

$$[HCOO^-] = 4.1 \times 10^{-3}\ M$$

$$[HCOOH] = (0.100 - 0.0041)\ M$$

$$= 0.096\ M$$

To check the validity of our approximation

$$\frac{0.0041\ M}{0.100\ M} \times 100\% = 4.1\%$$

This shows that the quantity x is less than 5 percent of the original concentration of the acid. Thus our approximation was justified.

Note that we omitted the contribution to $[H^+]$ by water. Except in very dilute acid solutions, this omission is always valid, because the hydrogen ion concentration due to water is negligibly small compared with that due to the acid.

PRACTICE EXERCISE

Calculate the concentration of H^+, A^-, and nonionized HA in a 0.20 M HA solution. The K_a of HA is 2.7×10^{-4}.

EXAMPLE 16.7

Ionization of a Weak Monoprotic Acid

Calculate the pH of a 0.052 M nitrous acid (HNO_2) solution.

Answer:

Step 1: From Table 16.3 we see that HNO_2 is a weak acid. Letting x be the equilibrium concentration of H^+ and NO_2^- ions in moles per liter, we summarize:

	$HNO_2(aq)$	$\rightleftharpoons$	$H^+(aq)$	+	$NO_2^-(aq)$
Initial (M):	0.052		0.00		0.00
Change (M):	$-x$		$+x$		$+x$
Equilibrium (M):	$(0.052 - x)$		x		x

Step 2: From Table 16.3,

$$K_a = \frac{[H^+][NO_2^-]}{[HNO_2]} = 4.5 \times 10^{-4}$$

$$\frac{x^2}{0.052 - x} = 4.5 \times 10^{-4}$$

Applying the approximation $0.052 - x \simeq 0.052$, we obtain

$$\frac{x^2}{0.052 - x} \simeq \frac{x^2}{0.052} = 4.5 \times 10^{-4}$$

$$x^2 = 2.3 \times 10^{-5}$$

Taking the square root of both sides gives

$$x = 4.8 \times 10^{-3}\ M$$

To test the approximation

$$\frac{0.0048\ M}{0.052\ M} \times 100\% = 9.2\%$$

This shows that x is more than 5 percent of the original concentration, so the approximation is not valid. There are two ways to get the correct answer.

1. *The quadratic equation.* We can solve the following quadratic equation:

$$x^2 + 4.5 \times 10^{-4}x - 2.3 \times 10^{-5} = 0$$

Using the quadratic formula,

$$\begin{aligned} x &= \frac{-b \pm \sqrt{b^2 - 4ac}}{2a} \\ &= \frac{-4.5 \times 10^{-4} \pm \sqrt{(4.5 \times 10^{-4})^2 - 4(1)(-2.3 \times 10^{-5})}}{2(1)} \end{aligned}$$

Thus

$$x = 4.6 \times 10^{-3}\ M \qquad \text{or} \qquad x = -5.0 \times 10^{-3}\ M$$

The second solution is physically impossible, since the concentrations of ions produced as a result of ionization cannot be negative. Therefore, the solution is given by the positive root—that is,

$$x = 4.6 \times 10^{-3}\ M$$

2. *The method of successive approximation.* In this method, we first solve for x by assuming $0.052 - x \simeq 0.052$ as before. Next, we use the approximate value of x ($4.8 \times 10^{-3}\ M$) to find a more exact value for the concentration of HNO_2:

$$[HNO_2] = 0.052\ M - 4.8 \times 10^{-3}\ M = 0.047\ M$$

Substituting the value of $[HNO_2]$ in the expression for K_a, we write

$$\frac{x^2}{0.047} = 4.5 \times 10^{-4}$$

$$x^2 = 2.1 \times 10^{-5}$$

$$x = 4.6 \times 10^{-3}\ M$$

Using 4.6×10^{-3} *M* for *x*, we can recalculate $[HNO_2]$ and solve for *x* again. This time we find that the answer is 4.6×10^{-3} *M*. In general, we apply the method of successive approximation until the value of *x* obtained for the last step does not differ from the previous step. (In the majority of cases you will need to apply this method only twice to get the correct answer.)

Step 3: At equilibrium

$$[H^+] = 4.6 \times 10^{-3}\ M$$

and

$$\begin{aligned} pH &= -\log (4.6 \times 10^{-3}) \\ &= 2.34 \end{aligned}$$

PRACTICE EXERCISE

What is the pH of a 0.122 *M* monoprotic acid whose K_a is 5.7×10^{-4}?

EXAMPLE 16.8

Determining K_a from pH Measurement

The pH of a 0.100 *M* solution of a weak monoprotic acid HA is 2.85. What is the K_a of the acid?

Answer: In this case we are given the pH, from which we can obtain the equilibrium concentrations, and we are asked to calculate the acid ionization constant. We can follow the same basic steps.

Step 1: First we need to calculate the hydrogen ion concentration from the pH.

$$\begin{aligned} pH &= -\log [H^+] \\ 2.85 &= -\log [H^+] \end{aligned}$$

Taking the antilog of both sides, we get

$$[H^+] = 1.4 \times 10^{-3}\ M$$

Next we summarize the changes:

	$HA(aq)$	$\rightleftharpoons$	$H^+(aq)$	+	$A^-(aq)$
Initial (*M*):	0.100		0.000		0.000
Change (*M*):	−0.0014		+0.0014		+0.0014
Equilibrium (*M*):	(0.100 − 0.0014)		0.0014		0.0014

Step 2: The acid ionization constant is given by

$$\begin{aligned} K_a = \frac{[H^+][A^-]}{[HA]} &= \frac{(0.0014)(0.0014)}{(0.100 - 0.0014)} \\ &= 2.0 \times 10^{-5} \end{aligned}$$

PRACTICE EXERCISE

The pH of a 0.060 *M* weak monoprotic acid is 3.44. Calculate the K_a of the acid.

Percent Ionization

We have seen that the magnitude of K_a indicates the strength of an acid. Another way to study the strength of an acid is to measure its ***percent ionization,*** which is

$$\text{percent ionization} = \frac{\text{ionized acid concentration at equilibrium}}{\text{initial concentration of acid}} \times 100\%$$

The stronger the acid, the greater the percent ionization. For a monoprotic acid HA, the concentration of the acid that undergoes ionization is equal to the concentration of the H^+ ions or the concentration of the A^- ions at equilibrium. Therefore, we can write the percent ionization as

$$\text{percent ionization} = \frac{[H^+]}{[HA]_0} \times 100\%$$

where $[H^+]$ is the concentration at equilibrium and $[HA]_0$ is the initial concentration. Note that this expression is the same as that used to check the approximation in Example 16.6. Thus the percent ionization of a 0.100 *M* formic acid is 4.1%.

The extent to which a weak acid ionizes depends on the initial concentration of the acid. The more dilute the solution, the greater the percent ionization (Figure 16.3). In qualitative terms, when an acid is diluted, initially the number of particles (nonionized acid molecules plus ions) per unit volume is reduced. According to Le Chatelier's principle (see Section 15.4), to counteract this "stress" (that is, the dilution), the equilibrium shifts from nonionized acid to H^+ and its conjugate base to produce more particles (ions).

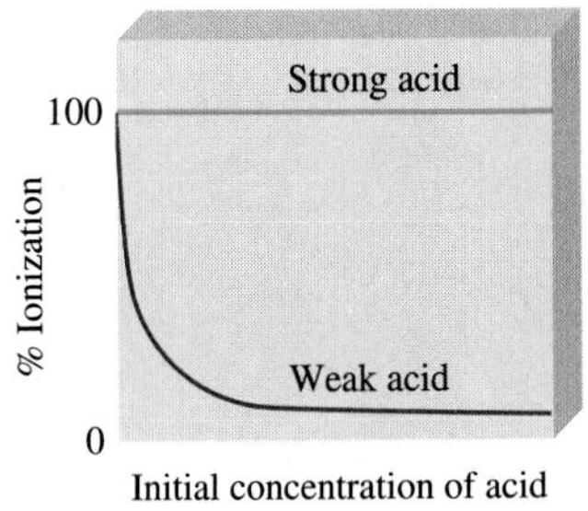

FIGURE 16.3

Dependence of percent ionization on initial concentration of acid. Note that at very low concentrations all acids (weak and strong) are almost completely ionized.

Diprotic and Polyprotic Acids

Diprotic and polyprotic acids may yield more than one hydrogen ion per molecule. These acids ionize in a stepwise manner, that is, they lose one proton at a time. An ionization constant expression can be written for each ionization stage. Consequently, two or more equilibrium constant expressions must often be used to calculate the concentrations of species in the acid solution. For example, for H_2CO_3 we write

$$H_2CO_3(aq) \rightleftharpoons H^+(aq) + HCO_3^-(aq) \qquad K_{a_1} = \frac{[H^+][HCO_3^-]}{[H_2CO_3]}$$

$$HCO_3^-(aq) \rightleftharpoons H^+(aq) + CO_3^{2-}(aq) \qquad K_{a_2} = \frac{[H^+][CO_3^{2-}]}{[HCO_3^-]}$$

Note that the conjugate base in the first ionization stage becomes the acid in the second ionization stage.

Table 16.4 shows the ionization constants of several diprotic acids and a polyprotic acid. For a given acid, the first ionization constant is much larger than the second ionization constant, and so on. This trend is reasonable because it is easier to remove a H^+ ion from a neutral molecule than to remove another H^+ from a negatively charged ion derived from the molecule.

EXAMPLE 16.9

Calculating the pH of a Diprotic Acid

Oxalic acid is a poisonous substance used chiefly as a bleaching and cleansing agent (for example, to remove bathtub rings). Calculate the concentrations of all the species present at equilibrium in a 0.10 *M* solution.

TABLE 16.4

Ionization Constants of Some Common Diprotic and Polyprotic Acids in Water at 25°C

Name of acid	Formula	Structure	K_a	Conjugate base	K_b
Sulfuric acid	H_2SO_4	H—O—S(=O)(=O)—O—H	very large	HSO_4^-	Very small
Hydrogen sulfate ion	HSO_4^-	H—O—S(=O)(=O)—O$^-$	1.3×10^{-2}	SO_4^{2-}	7.7×10^{-13}
Oxalic acid	$C_2H_2O_4$	H—O—C(=O)—C(=O)—O—H	6.5×10^{-2}	$C_2HO_4^-$	1.5×10^{-13}
Hydrogen oxalate ion	$C_2HO_4^-$	H—O—C(=O)—C(=O)—O$^-$	6.1×10^{-5}	$C_2O_4^{2-}$	1.6×10^{-10}
Sulfurous acid*	H_2SO_3	H—O—S(=O)—O—H	1.3×10^{-2}	HSO_3^-	7.7×10^{-13}
Hydrogen sulfite ion	HSO_3^-	H—O—S(=O)—O$^-$	6.3×10^{-8}	SO_3^{2-}	1.6×10^{-7}
Carbonic acid	H_2CO_3	H—O—C(=O)—O—H	4.2×10^{-7}	HCO_3^-	2.4×10^{-8}
Hydrogen carbonate ion	HCO_3^-	H—O—C(=O)—O$^-$	4.8×10^{-11}	CO_3^{2-}	2.1×10^{-4}
Hydrosulfuric acid	H_2S	H—S—H	9.5×10^{-8}	HS^-	1.1×10^{-7}
Hydrogen sulfide ion†	HS^-	H—S$^-$	1×10^{-19}	S^{2-}	1×10^{5}
Phosphoric acid	H_3PO_4	H—O—P(=O)(—O—H)—O—H	7.5×10^{-3}	$H_2PO_4^-$	1.3×10^{-12}
Dihydrogen phosphate ion	$H_2PO_4^-$	H—O—P(=O)(—O—H)—O$^-$	6.2×10^{-8}	HPO_4^{2-}	1.6×10^{-7}
Hydrogen phosphate ion	HPO_4^{2-}	H—O—P(=O)(—O$^-$)—O$^-$	4.8×10^{-13}	PO_4^{3-}	2.1×10^{-2}

*H_2SO_3 has never been isolated and exists in only minute concentration in aqueous solution of SO_2. The K_a value here refers to the process $SO_2(g) + H_2O(l) \rightleftharpoons H^+(aq) + HSO_3^-(aq)$.

†The ionization constant of HS^- is very low and difficult to measure. The value listed here is only an estimate.

Answer: Oxalic acid is a diprotic acid (see Table 16.4). We begin with the first stage of ionization.

Step 1:

	$C_2H_2O_4(aq)$ $\rightleftharpoons$	$H^+(aq)$ +	$C_2HO_4^-(aq)$
Initial (*M*):	0.10	0.00	0.00
Change (*M*):	$-x$	$+x$	$+x$
Equilibrium (*M*):	$(0.10 - x)$	x	x

Step 2: With the ionization constant from Table 16.4, we write

$$K_{a_1} = \frac{[H^+][C_2HO_4^-]}{[C_2H_2O_4]} = 6.5 \times 10^{-2}$$

$$\frac{x^2}{0.10 - x} = 6.5 \times 10^{-2}$$

Since the ionization constant is fairly large, we cannot apply the approximation $0.10 - x \simeq 0.10$. Instead, we must solve the quadratic equation

$$x^2 + 6.5 \times 10^{-2}x - 6.5 \times 10^{-3} = 0$$

The solution gives $x = 0.054\ M$.

Step 3: When the equilibrium for the first stage of ionization is reached, the concentrations are

$$[H^+] = 0.054\ M$$

$$[C_2HO_4^-] = 0.054\ M$$

$$[C_2H_2O_4] = (0.10 - 0.054)\ M = 0.046\ M$$

Next we consider the second stage of ionization.

Step 1: Let y be the equilibrium concentration of $C_2O_4^{2-}$ in moles per liter. Thus the equilibrium concentration of $C_2HO_4^-$ must be $(0.054 - y)M$. We have

	$C_2HO_4^-(aq)$ $\rightleftharpoons$	$H^+(aq)$	+ $C_2O_4^{2-}(aq)$
Initial (*M*):	0.054	0.054	0.00
Change (*M*):	$-y$	$+y$	$+y$
Equilibrium (*M*):	$(0.054 - y)$	$(0.054 + y)$	y

Step 2: Using the ionization constant in Table 16.4, we write

$$K_{a_2} = \frac{[H^+][C_2O_4^{2-}]}{[C_2HO_4^-]} = 6.1 \times 10^{-5}$$

$$\frac{(0.054 + y)y}{(0.054 - y)} = 6.1 \times 10^{-5}$$

Because the ionization constant is small, we can make the following approximations:

$$0.054 + y \simeq 0.054$$

$$0.054 - y \simeq 0.054$$

Next we substitute them in the equation, which reduces to

$$y = 6.1 \times 10^{-5}\ M$$

Step 3: At equilibrium, therefore

$$[C_2H_2O_4] = 0.046\ M$$

$$[C_2HO_4^-] = (0.054 - 6.1 \times 10^{-5})\ M \simeq 0.054\ M$$
$$[H^+] = (0.054 + 6.1 \times 10^{-5})\ M \simeq 0.054\ M$$
$$[C_2O_4^{2-}] = 6.1 \times 10^{-5}\ M$$
$$[OH^-] = 1.0 \times 10^{-14}/0.054 = 1.9 \times 10^{-13}\ M$$

PRACTICE EXERCISE

Calculate the concentrations of $C_2H_2O_4$, $C_2HO_4^-$, $C_2O_4^{2-}$, and H^+ in a 0.20 *M* oxalic acid solution.

Example 16.9 shows that for diprotic acids if $K_{a_1} \gg K_{a_2}$, then the concentration of the H^+ ions at equilibrium may be assumed to result only from the first stage of ionization. Furthermore, the concentration of the conjugate base for the second-stage ionization is numerically equal to K_{a_2}.

Phosphoric acid (H_3PO_4) is an important polyprotic acid with three ionizable hydrogen atoms:

$$H_3PO_4(aq) \rightleftharpoons H^+(aq) + H_2PO_4^-(aq) \qquad K_{a_1} = \frac{[H^+][H_2PO_4^-]}{[H_3PO_4]} = 7.5 \times 10^{-3}$$

$$H_2PO_4^-(aq) \rightleftharpoons H^+(aq) + HPO_4^{2-}(aq) \qquad K_{a_2} = \frac{[H^+][HPO_4^{2-}]}{[H_2PO_4^-]} = 6.2 \times 10^{-8}$$

$$HPO_4^{2-}(aq) \rightleftharpoons H^+(aq) + PO_4^{3-}(aq) \qquad K_{a_3} = \frac{[H^+][PO_4^{3-}]}{[HPO_4^{2-}]} = 4.8 \times 10^{-13}$$

We see that phosphoric acid is a weak polyprotic acid and that its ionization constants decrease markedly for the second and third stages. Thus we can predict that, in a solution containing phosphoric acid, the concentration of the nonionized acid is the highest, and the only other species present in significant concentrations are H^+ and $H_2PO_4^-$ ions.

16.6 WEAK BASES AND BASE IONIZATION CONSTANTS

Weak bases are treated like weak acids. When ammonia dissolves in water, it undergoes the reaction

$$NH_3(aq) + H_2O(l) \rightleftharpoons NH_4^+(aq) + OH^-(aq)$$

The production of hydroxide ions in this *base ionization* reaction means that, in this solution at 25°C, $[OH^-] > [H^+]$, and therefore pH > 7.

Compare with the total concentration of water, very few water molecules are consumed by the reaction, so we can treat $[H_2O]$ as a constant. Thus we can write the equilibrium constant as

$$K[H_2O] = K_b = \frac{[NH_4^+][OH^-]}{[NH_3]}$$
$$= 1.8 \times 10^{-5}$$

where K_b, *the equilibrium constant for base ionization,* is called the ***base ionization***

TABLE 16.5
Ionization Constants of Some Common Weak Bases at 25°C

Name of base	Formula	Structure	K_b*	Conjugate acid	K_a
Ethylamine	$C_2H_5NH_2$	$CH_3—CH_2—\ddot{N}H—H$	5.6×10^{-4}	$C_2H_5\overset{+}{N}H_3$	1.8×10^{-11}
Methylamine	CH_3NH_2	$CH_3—\ddot{N}H—H$	4.4×10^{-4}	$CH_3\overset{+}{N}H_3$	2.3×10^{-11}
Caffeine	$C_8H_{10}N_4O_2$	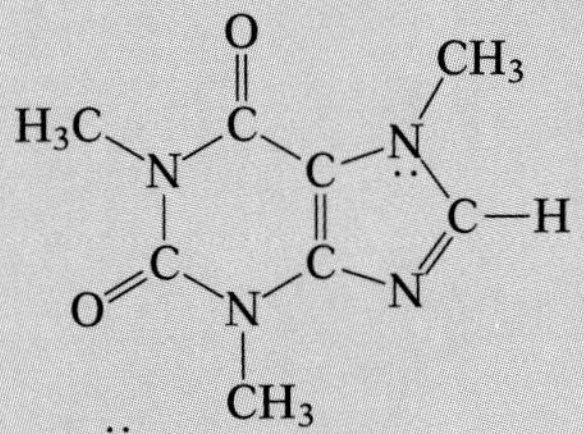	4.1×10^{-4}	$C_8H_{11}\overset{+}{N}_4O_2$	2.4×10^{-11}
Ammonia	NH_3	$H—\ddot{N}H—H$	1.8×10^{-5}	NH_4^+	5.6×10^{-10}
Pyridine	C_5H_5N	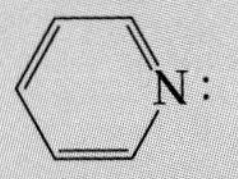	1.7×10^{-9}	$C_5H_5\overset{+}{N}H$	5.9×10^{-6}
Aniline	$C_6H_5NH_2$	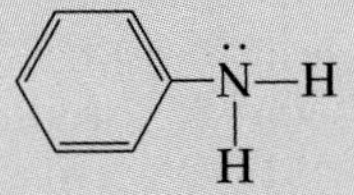	3.8×10^{-10}	$C_6H_5\overset{+}{N}H_3$	2.6×10^{-5}
Urea	N_2H_4CO	$H—\ddot{N}H—C(=O)—\ddot{N}H—H$	1.5×10^{-14}	$H_2NCO\overset{+}{N}H_3$	0.67

*The nitrogen atom with the lone pair accounts for each compound's basicity. In the case of urea, K_b can be associated with either nitrogen atom.

constant. Table 16.5 lists a number of common weak bases and their ionization constants. Note that the basicity of all these compounds is attributable to the lone pair of electrons on the nitrogen atom.

EXAMPLE 16.10
Calculating the pH of a Weak Base Solution

What is the pH of a 0.400 M ammonia solution?

Answer: The procedure is essentially the same as the one we use for weak acids.

Step 1: Let x be the concentration in moles per liter of NH_4^+ and OH^- ions at equilibrium. Next we summarize:

	$NH_3(aq)$	$+\ H_2O(l)$	$\rightleftharpoons\ NH_4^+(aq)$	$+\ OH^-(aq)$
Initial (M):	0.400		0.000	0.000
Change (M):	$-x$		$+x$	$+x$
Equilibrium (M):	$(0.400 - x)$		x	x

Step 2: Using the base ionization constant listed in Table 16.5, we write

$$K_b = \frac{[NH_4^+][OH^-]}{[NH_3]} = 1.8 \times 10^{-5}$$

$$\frac{x^2}{0.400 - x} = 1.8 \times 10^{-5}$$

Applying the approximation $0.400 - x \simeq 0.400$ gives

$$\frac{x^2}{0.400 - x} \approx \frac{x^2}{0.400} = 1.8 \times 10^{-5}$$

$$x^2 = 7.2 \times 10^{-6}$$

$$x = 2.7 \times 10^{-3}\ M$$

Step 3: At equilibrium, $[OH^-] = 2.7 \times 10^{-3}\ M$. Thus

$$\text{pOH} = -\log(2.7 \times 10^{-3})$$
$$= 2.57$$
$$\text{pH} = 14.00 - 2.57$$
$$= 11.43$$

Note that we omitted the contribution to $[OH^-]$ from water. This condition is similar to that for ignoring the contribution to $[H^+]$ from water (see p. 458).

PRACTICE EXERCISE

Calculate the pH of a 0.26 *M* methylamine solution (see Table 16.4).

16.7 THE RELATIONSHIP BETWEEN CONJUGATE ACID-BASE IONIZATION CONSTANTS

An important relationship between the acid ionization constant and the ionization constant of its conjugate base can be derived as follows, using acetic acid as an example:

$$CH_3COOH(aq) \rightleftharpoons H^+(aq) + CH_3COO^-(aq)$$

$$K_a = \frac{[H^+][CH_3COO^-]}{[CH_3COOH]}$$

The conjugate base, CH_3COO^-, reacts with water according to the equation

$$CH_3COO^-(aq) + H_2O(l) \rightleftharpoons CH_3COOH(aq) + OH^-(aq)$$

and we can write the base ionization constant as

$$K_b = \frac{[CH_3COOH][OH^-]}{[CH_3COO^-]}$$

The product of these two ionization constants is given by

$$K_aK_b = \frac{[H^+][CH_3COO^-]}{[CH_3COOH]} \times \frac{[CH_3COOH][OH^-]}{[CH_3COO^-]}$$
$$= [H^+][OH^-]$$
$$= K_w$$

This result may seem strange at first, but it can be understood by realizing that the sum of reaction (1) and (2) below is simply the autoionization of water.

$$\begin{array}{llr} (1) & CH_3COOH(aq) \rightleftharpoons H^+(aq) + CH_3COO^-(aq) & K_a \\ (2) & CH_3COO^-(aq) + H_2O(l) \rightleftharpoons CH_3COOH(aq) + OH^-(aq) & K_b \\ \hline (3) & H_2O(l) \rightleftharpoons H^+(aq) + OH^-(aq) & K_w \end{array}$$

This example illustrates one of the rules for chemical equilibria: When two reactions are added to give a third reaction, the equilibrium constant for the third reaction is the product of the equilibrium constants for the two added reactions. Thus for any conjugate acid-base pair it is always true that

$$K_aK_b = K_w \tag{16.6}$$

Expressing Equation (16.6) as

$$K_a = \frac{K_w}{K_b} \qquad K_b = \frac{K_w}{K_a}$$

enables us to draw an important conclusion: The stronger the acid (the larger K_a), the weaker its conjugate base (the smaller K_b), and vice versa (see Tables 16.3, 16.4, and 16.5).

We can use Equation (16.6) to calculate the K_b of the conjugate base (CH_3COO^-) of CH_3COOH as follows. We find the K_a value of CH_3COOH in Table 16.3 and write

$$K_b = \frac{K_w}{K_a}$$
$$= \frac{1.0 \times 10^{-14}}{1.8 \times 10^{-5}}$$
$$= 5.6 \times 10^{-10}$$

16.8 ACID-BASE PROPERTIES OF SALTS

As defined in Section 4.3, a salt is an ionic compound formed by the reaction between an acid and a base. Salts are strong electrolytes that completely dissociate in water and in some cases react with water. The term ***salt hydrolysis*** describes *the reaction of an anion or a cation of a salt, or both, with water.* Salt hydrolysis usually affects the pH of a solution.

SALTS THAT PRODUCE NEUTRAL SOLUTIONS

When $NaNO_3$, formed by the reaction between NaOH and HNO_3, dissolves in water, it dissociates completely as follows:

$$NaNO_3(s) \xrightarrow{H_2O} Na^+(aq) + NO_3^-(aq)$$

The hydrated Na^+ ion neither donates nor accepts H^+ ions. The NO_3^- ion is the conjugate base of the strong acid HNO_3, and it has no affinity for H^+ ions. Consequently, a solution containing Na^+ and NO_3^- ions is neutral, with a pH of 7. It is generally true that salts containing an alkali metal ion or alkaline earth metal ion (except Be^{2+}) and the conjugate base of a strong acid (for example, Cl^-, Br^-, and NO_3^-) do not undergo hydrolysis, and their solutions are also neutral.

Salts That Produce Basic Solutions

The dissociation of sodium acetate (CH_3COONa) in water is given by

$$CH_3COONa(s) \xrightarrow{H_2O} Na^+(aq) + CH_3COO^-(aq)$$

The hydrated Na^+ ion has no acidic or basic properties. The acetate ion CH_3COO^-, however, is the conjugate base of the weak acid CH_3COOH and therefore has an affinity for H^+ ions. The hydrolysis reaction is given by

$$CH_3COO^-(aq) + H_2O(l) \rightleftharpoons CH_3COOH(aq) + OH^-(aq)$$

Because this reaction produces OH^- ions, the sodium acetate solution will be basic. The equilibrium constant for this hydrolysis reaction is the base ionization constant expression for CH_3COO^-, so we write (see p. 466)

$$K_b = \frac{[CH_3COOH][OH^-]}{[CH_3COO^-]} = 5.6 \times 10^{-10}$$

Salts That Produce Acidic Solutions

When a salt derived from a strong acid and a weak base dissolves in water, the solution becomes acidic. For example, consider the process

$$NH_4Cl(s) \xrightarrow{H_2O} NH_4^+(aq) + Cl^-(aq)$$

The Cl^- ion has no affinity for H^+. The ammonium ion NH_4^+ is the weak conjugate acid of the weak base NH_3 and ionizes as follows:

$$NH_4^+(aq) + H_2O(l) \rightleftharpoons NH_3(aq) + H_3O^+(aq)$$

or simply

$$NH_4^+(aq) \rightleftharpoons NH_3(aq) + H^+(aq)$$

Since this reaction produces H^+ ions, the pH of the solution decreases. As you can see, the hydrolysis of the NH_4^+ ion is the same as the ionization of the NH_4^+ acid. The equilibrium constant (or ionization constant) for this process is given by

$$K_a = \frac{[NH_3][H^+]}{[NH_4^+]} = \frac{K_w}{K_b} = \frac{1.0 \times 10^{-14}}{1.8 \times 10^{-5}} = 5.6 \times 10^{-10}$$

Salts that contain small, highly charged metal cations (for example, Al^{3+}, Cr^{3+}, Fe^{3+}, Bi^{3+}, and Be^{2+}) and the conjugate bases of strong acids also produce an acidic solution. For example, when aluminum chloride ($AlCl_3$) dissolves in water, the Al^{3+} ions take the hydrated form $Al(H_2O)_6^{3+}$ (Figure 16.4). Let us consider one bond between the metal ion and the oxygen atom of one of the six water molecules in $Al(H_2O)_6^{3+}$:

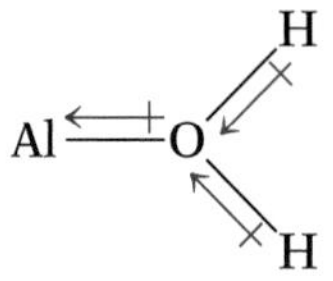

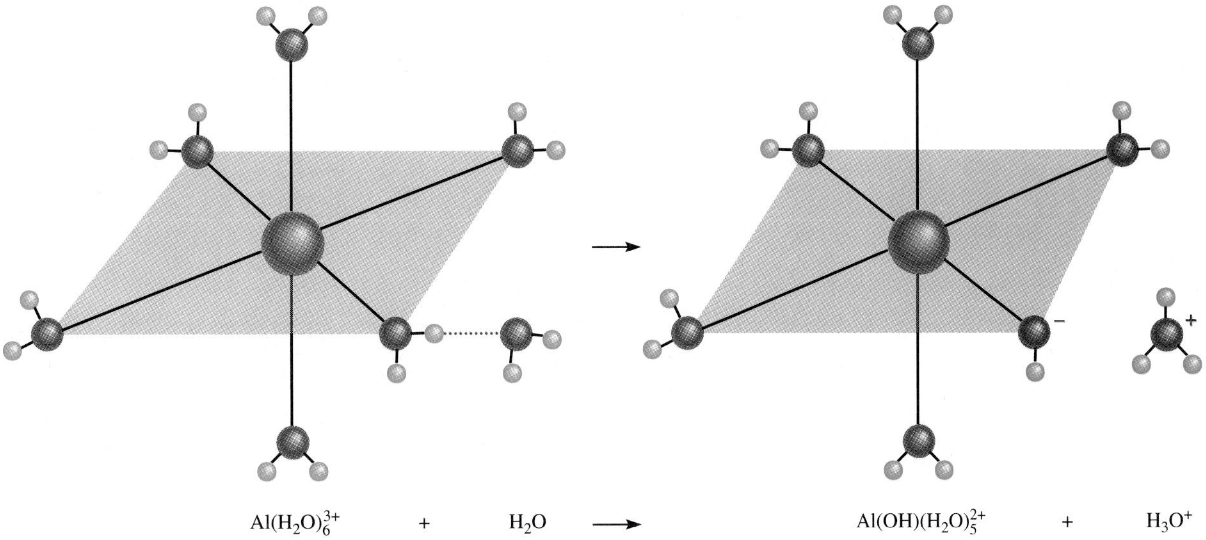

FIGURE 16.4

The six H_2O molecules surround the Al^{3+} ion octahedrally. The attraction of the small Al^{3+} ion for the lone pairs on the oxygen atoms is so great that the O—H bonds in a H_2O molecule attached to the metal cation are weakened, allowing the loss of a proton (H^+) to an incoming H_2O molecule. This metal cation hydrolysis makes the solution acidic.

The positively charged Al^{3+} ion draws electron density toward itself, making the O—H bond more polar. Consequently, the H atoms have a greater tendency to ionize than those in water molecules not involved in hydration. The resulting ionization process can be written as

$$Al(H_2O)_6^{3+}(aq) + H_2O(l) \rightleftharpoons Al(OH)(H_2O)_5^{2+}(aq) + H_3O^+(aq)$$

or simply

$$Al(H_2O)_6^{3+}(aq) \rightleftharpoons Al(OH)(H_2O)_5^{2+}(aq) + H^+(aq)$$

The equilibrium constant for the metal cation hydrolysis is given by

$$K_a = \frac{[Al(OH)(H_2O)_5^{2+}][H^+]}{[Al(H_2O)_6^{3+}]} = 1.3 \times 10^{-5}$$

Note that the $Al(OH)(H_2O)_5^{2+}$ species can undergo further ionization as

$$Al(OH)(H_2O)_5^{2+}(aq) \rightleftharpoons Al(OH)_2(H_2O)_4^{+}(aq) + H^+(aq)$$

and so on. However, generally it is sufficient to consider only the first step of hydrolysis.

The extent of hydrolysis is greatest for the smallest and most highly charged ions because a "compact" highly charged ion is more effective in polarizing the O—H bond and facilitating ionization. This is the reason that relatively large ions of low charge such as Na^+ and K^+ do not undergo hydrolysis.

Salts in Which Both the Cation and the Anion Hydrolyze

So far we have considered salts in which only one ion undergoes hydrolysis. For salts derived from a weak acid and a weak base, both the cation and the anion

TABLE 16.6
Acid-Base Properties of Salts

Type of salt	Examples	Ions that undergo hydrolysis	pH of solution
Cation from strong base; anion from strong acid	NaCl, KI, KNO_3, RbBr, $BaCl_2$	None	$\simeq 7$
Cation from strong base; anion from weak acid	CH_3COONa, KNO_2	Anion	>7
Cation from weak base; anion from strong acid	NH_4Cl, NH_4NO_3	Cation	<7
Cation from weak base; anion from weak acid	NH_4NO_2, CH_3COONH_4, NH_4CN	Anion and cation	<7 if $K_b < K_a$ $\simeq 7$ if $K_b \simeq K_a$ >7 if $K_b > K_a$
Small, highly charged cation; anion from strong acid	$AlCl_3$, $Fe(NO_3)_3$	Hydrated cation	<7

hydrolyze. However, whether a solution containing such a salt is acidic, basic, or neutral depends on the relative strengths of the weak acid and the weak base. Since the mathematics associated with this type of system is rather involved, we limit ourselves to making qualitative predictions about these solutions. We consider three situations:

- $K_b > K_a$. If K_b for the anion is greater than K_a for the cation, then the solution must be basic because the anion will hydrolyze to a greater extent than the cation. At equilibrium, there will be more OH^- ions than H^+ ions.
- $K_b < K_a$. Conversely, if K_b of the anion is smaller than K_a of the cation, the solution will be acidic because cation hydrolysis will be more extensive than anion hydrolysis.
- $K_a \simeq K_b$. If K_a is approximately equal to K_b, the solution will be nearly neutral.

Table 16.6 summarizes the behavior in aqueous solution of the salts discussed in this section.

EXAMPLE 16.11
Predicting the Acid-Base Properties of Salt Solutions

Predict whether the following solutions will be acidic, basic, or near neutral: (a) NH_4I, (b) $CaCl_2$, (c) KCN, (d) $Fe(NO_3)_3$.

Answer: (a) The cation is NH_4^+, which will hydrolyze to produce NH_3 and H^+. The I^- anion is the conjugate base of the strong acid HI. Therefore, I^- will not hydrolyze and the solution is acidic.

(b) The Ca^{2+} cation does not hydrolyze. The Cl^- anion also does not hydrolyze. The solution will be near neutral.

(c) The K^+ cation does not hydrolyze. The CN^- is the conjugate base of the weak acid HCN and will hydrolyze to give OH^- and HCN. The solution will be basic.

(d) Fe^{3+} is a small metal ion with a high charge and will hydrolyze to produce H^+ ions. The NO_3^- ion does not hydrolyze. The solution will be acidic.

PRACTICE EXERCISE

Predict the pH of the following salt solutions: (a) $LiClO_4$, (b) Na_3PO_4, (c) $Al_2(SO_4)_3$, (d) NH_4CN.

Finally we note that some anions can act either as an acid or as a base. For example, the bicarbonate ion (HCO_3^-) can ionize or undergo hydrolysis as follows:

$$HCO_3^-(aq) + H_2O(l) \rightleftharpoons H_3O^+(aq) + CO_3^{2-}(aq) \qquad K_a = 4.8 \times 10^{-11}$$

$$HCO_3^-(aq) + H_2O(l) \rightleftharpoons H_2CO_3(aq) + OH^-(aq) \qquad K_b = 2.4 \times 10^{-8}$$

Since $K_b > K_a$, we predict that the hydrolysis reaction will outweigh the ionization process. Thus a solution of sodium bicarbonate, say, will be basic.

16.9 ACIDIC, BASIC, AND AMPHOTERIC OXIDES

As we saw in Chapter 8, oxides can be classified as acidic, basic, or amphoteric. Our discussion of acid-base reactions would be incomplete without examining the properties of these compounds.

Figure 16.5 shows the formulas of a number of oxides of the representa-

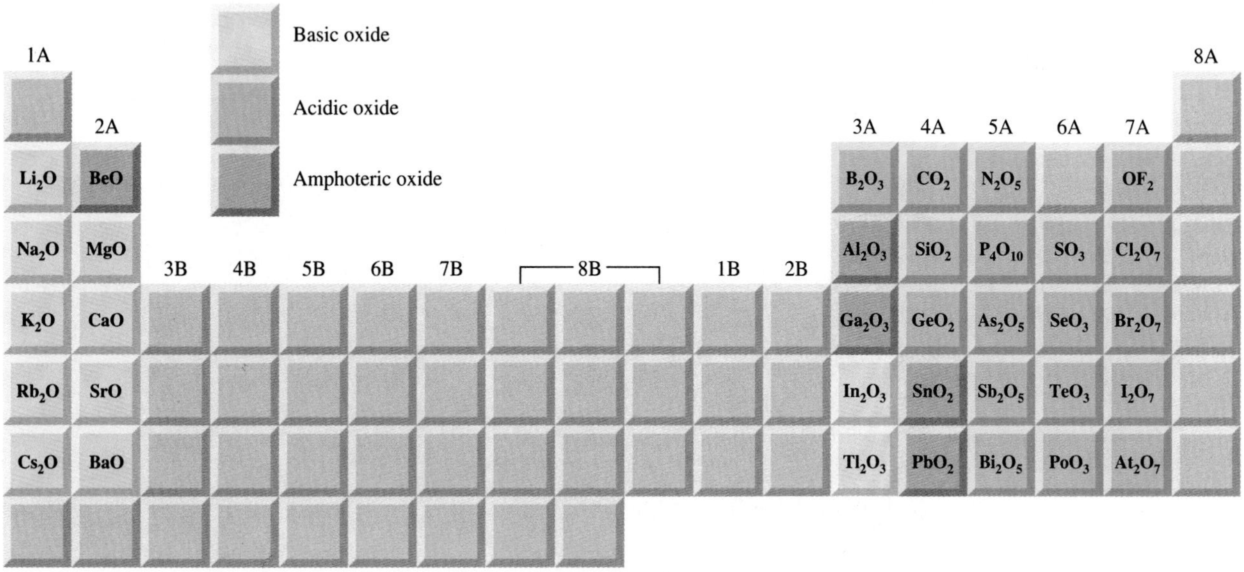

FIGURE 16.5

The formulas of a number of oxides of the representative elements in their highest oxidation states.

tive elements in their highest oxidation states. Note that all alkali metal oxides and all alkaline earth metal oxides except BeO are basic. Beryllium oxide and several metallic oxides in Groups 3A and 4A are amphoteric. Nonmetallic oxides in which the oxidation number of the representative element is high are acidic (for example, N_2O_5, SO_3, and Cl_2O_7). Those in which the oxidation number of the representative element is low (for example, CO and NO) show no measurable acidic properties. No nonmetallic oxides are known to have basic properties.

The basic metallic oxides react with water to form metal hydroxides:

$$Na_2O(s) + H_2O(l) \longrightarrow 2NaOH(aq)$$

$$BaO(s) + H_2O(l) \longrightarrow Ba(OH)_2(aq)$$

The reactions between acidic oxides and water are as follows:

$$CO_2(g) + H_2O(l) \rightleftharpoons H_2CO_3(aq)$$

$$SO_3(g) + H_2O(l) \rightleftharpoons H_2SO_4(aq)$$

$$N_2O_5(g) + H_2O(l) \rightleftharpoons 2HNO_3(aq)$$

$$P_4O_{10}(s) + 6H_2O(l) \rightleftharpoons 4H_3PO_4(aq)$$

$$Cl_2O_7(g) + H_2O(l) \rightleftharpoons 2HClO_4(aq)$$

A forest damaged by acid rain.

The reaction between CO_2 and H_2O explains why when pure water is exposed to air (which contains CO_2) it gradually reaches a pH of about 5.5 (Figure 16.6). The reaction between SO_3 and H_2O is largely responsible for acid rain.

Reactions between acidic oxides and bases and those between basic oxides and acids resemble normal acid-base reactions in that the products are a salt and water:

$$\underset{\text{acidic oxide}}{CO_2(g)} + \underset{\text{base}}{2NaOH(aq)} \longrightarrow \underset{\text{salt}}{Na_2CO_3(aq)} + \underset{\text{water}}{H_2O(l)}$$

$$\underset{\text{basic oxide}}{BaO(s)} + \underset{\text{acid}}{2HNO_3(aq)} \longrightarrow \underset{\text{salt}}{Ba(NO_3)_2(aq)} + \underset{\text{water}}{H_2O(l)}$$

As Figure 16.5 shows, aluminum oxide (Al_2O_3) is amphoteric. Depending on the reaction conditions, it can behave either as an acidic oxide or as a basic oxide.

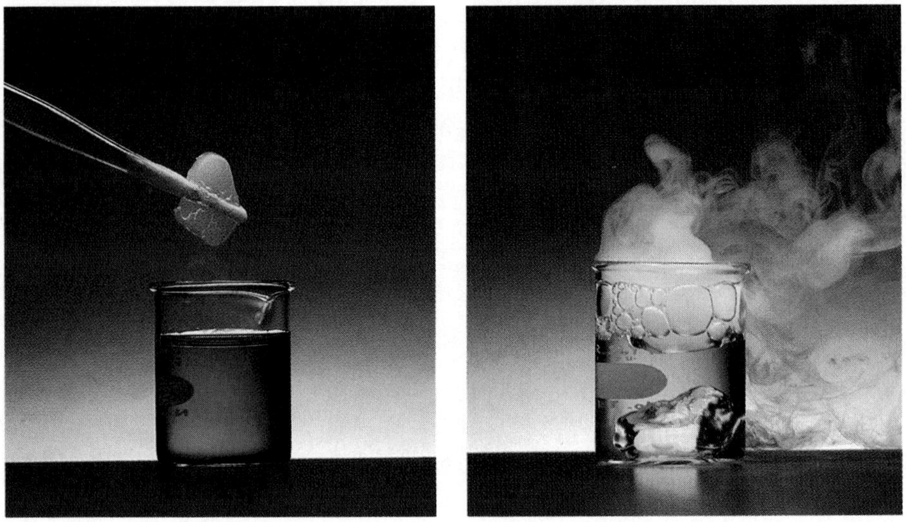

FIGURE 16.6

(Left) A beaker of water to which a few drops of bromothymol blue indicator have been added. (Right) When dry ice is added to the water, the CO_2 reacts to form carbonic acid, which turns the solution acidic and the color from blue to yellow.

For example, Al_2O_3 acts as a base with hydrochloric acid to produce a salt ($AlCl_3$) and water:

$$Al_2O_3(s) + 6HCl(aq) \longrightarrow 2AlCl_3(aq) + 3H_2O(l)$$

and acts as an acid with sodium hydroxide:

$$Al_2O_3(s) + 2NaOH(aq) + 3H_2O(l) \longrightarrow 2NaAl(OH)_4(aq)$$

Note that only a salt, $NaAl(OH)_4$ [containing the Na^+ and $Al(OH)_4^-$ ions], is formed in the latter reaction; no water is produced. Nevertheless, this reaction can still be classified as an acid-base reaction because Al_2O_3 neutralizes NaOH.

Some transition metal oxides in which the metal has a high oxidation number act as acidic oxides. Two examples are manganese(VII) oxide (Mn_2O_7) and chromium(VI) oxide (CrO_3), both of which react with water to produce acids:

$$Mn_2O_7(l) + H_2O(l) \longrightarrow \underset{\text{permanganic acid}}{2HMnO_4(aq)}$$

$$CrO_3(s) + H_2O(l) \longrightarrow \underset{\text{chromic acid}}{H_2CrO_4(aq)}$$

16.10 LEWIS ACIDS AND BASES

Acid-base properties so far have been discussed in terms of the Brønsted theory. To behave as a Brønsted base, for example, a substance must be able to accept protons. By this definition both the hydroxide ion and ammonia are bases:

$$H^+ + {}^-:\ddot{\underset{..}{O}}\text{—}H \longrightarrow H\text{—}\ddot{\underset{..}{O}}\text{—}H$$

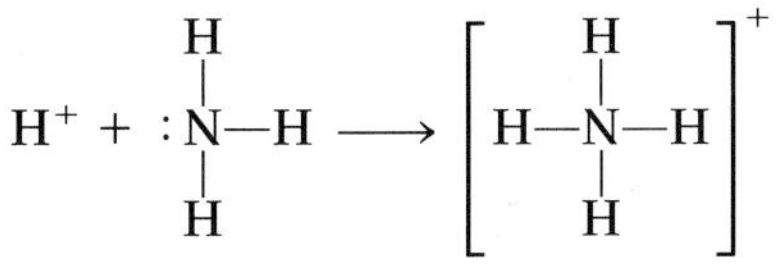

In each case, the atom to which the proton becomes attached possesses at least one unshared pair of electrons. This characteristic property of the OH^- ion, of NH_3, and of other Brønsted bases suggests a more general definition of acids and bases.

The American chemist G. N. Lewis formulated such a definition. According to Lewis's definition, a base is *a substance that can donate a pair of electrons,* and an acid is *a substance that can accept a pair of electrons.* For example, in the protonation of ammonia, NH_3 acts as a ***Lewis base*** because it donates a pair of electrons to the proton H^+, which acts as a ***Lewis acid*** by accepting the pair of electrons. A Lewis acid-base reaction, therefore, is one that involves the donation of a pair of electrons from one species to another. Such a reaction does not produce a salt and water.

The significance of the Lewis concept is that it is much more general than other definitions; it includes as acid-base reactions many reactions that do not involve Brønsted acids. Consider, for example, the reaction between boron trifluoride (BF_3) and ammonia (Figure 16.7):

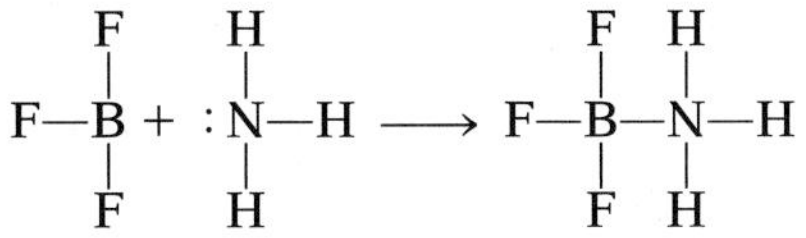

acid base

FIGURE 16.7

A Lewis acid-base reaction involving BF_3 and NH_3.

In Section 10.3 we saw that the B atom in BF_3 is sp^2-hybridized. The vacant, unhybridized $2p$ orbital accepts the pair of electrons from NH_3. So BF_3 functions as an acid according to the Lewis definition, even though it does not contain an ionizable proton. Note that a coordinate covalent bond is formed between the B and N atoms.

Another Lewis acid containing boron is boric acid. Boric acid (a weak acid used in eyewash) is an oxoacid with the following structure:

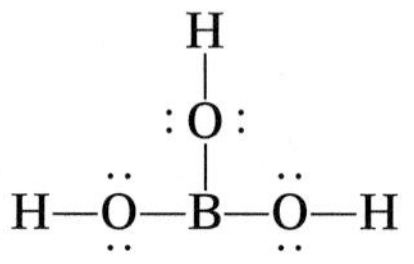

boric acid

Note that boric acid does not ionize in water to produce a H^+ ion. Instead, its reaction with water is

$$B(OH)_3(aq) + H_2O(l) \rightleftharpoons B(OH)_4^-(aq) + H^+(aq)$$

In this Lewis acid-base reaction, boric acid accepts a pair of electrons from the hydroxide ion that is derived from the H_2O molecule.

The hydration of carbon dioxide to produce carbonic acid

$$CO_2(g) + H_2O(l) \rightleftharpoons H_2CO_3(aq)$$

can be understood in the Lewis framework as follows: The first step involves donation of a lone pair on the oxygen atom in H_2O to the carbon atom in CO_2. An orbital is vacated on the C atom to accommodate the lone pair by removal of the electron pair in the C—O pi bond. These shifts of electrons are indicated by the curved arrows.

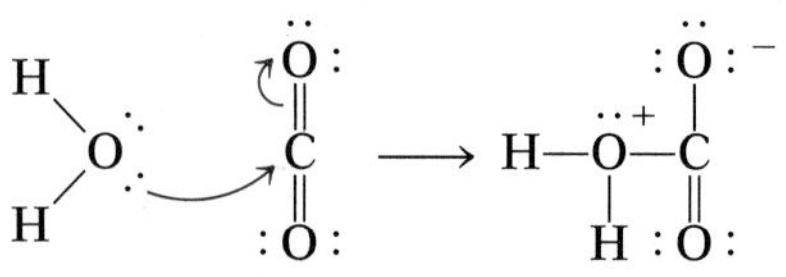

Therefore, H_2O is a Lewis base and CO_2 is a Lewis acid. Next, a proton is transferred onto the O atom bearing a negative charge to form H_2CO_3.

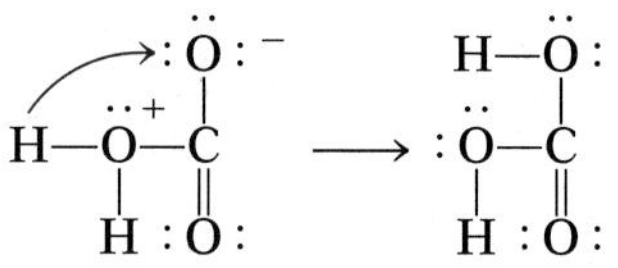

SUMMARY

Brønsted acids donate protons and Brønsted bases accept protons. Every Brønsted acid has a conjugate Brønsted base and vice versa.

The acidity of an aqueous solution is expressed as its pH, which is defined as the negative logarithm of the hydrogen ion concentration (in moles per liter). At 25°C, an acidic solution has pH < 7, a basic solution has pH > 7, and a neutral solution has pH $= 7$.

In aqueous solution, the following are classified as strong acids because they ionize completely: $HClO_4$, HI, HBr, HCl, H_2SO_4 (first stage of ionization), and HNO_3. Most acids are weak acids and, therefore, weak electrolytes. The following are classified as strong bases because they ionize completely in aqueous solution: hydroxides of alkali metals and of alkaline earth metals (except beryllium).

The acid ionization constant K_a increases with acid strength. K_b similarly expresses the strengths of bases. Percent ionization is another measure of the strength of acids. The more dilute a solution of a weak acid, the greater the percent ionization of the acid. Diprotic and polyprotic acids yield more than one proton on ionization per molecule of the acid. The product of the ionization constant of an acid and the ionization constant of its conjugate base is equal to the ion-product constant of water; that is, $K_aK_b = K_w$.

Most salts are strong electrolytes and dissociate completely into ions in solution. The reaction of these ions with water, called salt hydrolysis, can lead to acidic or basic solutions. In salt hydrolysis, the conjugate bases of weak acids yield basic solutions, and the conjugate acids of weak bases yield acidic solutions. Small, highly charged metal ions such as Al^{3+} and Fe^{3+} hydrolyze to yield acidic solutions. Most oxides can be classified as acidic, basic, or amphoteric.

Lewis acids accept pairs of electrons and Lewis bases donate pairs of electrons. The term "Lewis acid" is generally reserved for substances that can accept electron pairs but do not contain ionizable hydrogen atoms.

KEY WORDS

Acid ionization constant (K_a), p. 455
Base ionization constant (K_b), p. 464
Conjugate acid-base pair, p. 448
Ion-product constant, p. 450
Lewis acid, p. 473
Lewis base, p. 473
Percent ionization, p. 461
pH, p. 451
Salt hydrolysis, p. 467
Strong acid, p. 453
Strong base, p. 453
Weak acid, p. 453
Weak base, p. 453

QUESTIONS AND PROBLEMS

BRØNSTED ACIDS AND BASES

Review Questions

16.1 Define Brønsted acids and bases. How do the Brønsted definitions differ from Arrhenius's definitions of acids and bases?

16.2 In order for a species to act as a Brønsted base, an atom in the species must possess a lone pair of electrons. Explain why this is so.

Problems

16.3 Classify each of the following species as a Brønsted acid or base, or both: (a) H_2O, (b) OH^-, (c) H_3O^+, (d) NH_3, (e) NH_4^+, (f) NH_2^-, (g) NO_3^-, (h) CO_3^{2-}, (i) HBr, (j) HCN.

16.4 What are the names and formulas of the conjugate bases of the following acids? (a) HNO_2, (b) H_2SO_4, (c) H_2S, (d) HCN, (e) HCOOH (formic acid).

16.5 Identify the acid-base conjugate pairs in each of the following reactions:

(a) $CH_3COO^- + HCN \rightleftharpoons CH_3COOH + CN^-$

(b) $HCO_3^- + HCO_3^- \rightleftharpoons H_2CO_3 + CO_3^{2-}$

(c) $H_2PO_4^- + NH_3 \rightleftharpoons HPO_4^{2-} + NH_4^+$

(d) $HClO + CH_3NH_2 \rightleftharpoons CH_3NH_3^+ + ClO^-$

(e) $CO_3^{2-} + H_2O \rightleftharpoons HCO_3^- + OH^-$

(f) $CH_3COO^- + H_2O \rightleftharpoons CH_3COOH + OH^-$

16.6 Give the conjugate acid of each of the following bases: (a) HS^-, (b) HCO_3^-, (c) CO_3^{2-}, (d) $H_2PO_4^-$, (e) HPO_4^{2-}, (f) PO_4^{3-}, (g) HSO_4^-, (h) SO_4^{2-}, (i) HSO_3^-, (j) SO_3^{2-}.

16.7 Give the conjugate base of each of the following acids: (a) $CH_2ClCOOH$, (b) HIO_4, (c) H_3PO_4, (d) $H_2PO_4^-$, (e) HPO_4^{2-}, (f) H_2SO_4, (g) HSO_4^-, (h) H_2SO_3, (i) HSO_3^-, (j) NH_4^+, (k) H_2S, (l) HS^-, (m) HClO.

16.8 Oxalic acid ($C_2H_2O_4$) has the following structure:

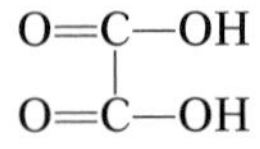

An oxalic acid solution contains the following species in varying concentrations: $C_2H_2O_4$, $C_2HO_4^-$, $C_2O_4^{2-}$, and H^+. (a) Draw Lewis structures of $C_2HO_4^-$ and $C_2O_4^{2-}$. (b) Which of the above four species can act only as acids, which can act only as bases, and which can act as both acids and bases?

PH AND POH CALCULATIONS

Review Questions

16.9 What is the ion-product constant for water?

16.10 Write an equation relating $[H^+]$ and $[OH^-]$ in solution at 25°C.

16.11 The ion-product constant for water is 1.0×10^{-14} at 25°C and 3.8×10^{-14} at 40°C. Is the process

$$H_2O(l) \rightleftharpoons H^+(aq) + OH^-(aq)$$

endothermic or exothermic?

16.12 Define pH. Why do chemists normally choose to discuss the acidity of a solution in terms of pH rather than hydrogen ion concentration, $[H^+]$?

16.13 The pH of a solution is 6.7. From this statement alone, can you conclude that the solution is acidic? If not, what additional information would you need? Can the pH of a solution be zero or negative? If so, give examples to illustrate these values.

16.14 Define pOH. Write an equation relating pH and pOH.

Problems

16.15 Calculate the hydrogen ion concentration for solutions with the following pH values: (a) 2.42, (b) 11.21, (c) 6.96, (d) 15.00.

16.16 Calculate the hydrogen ion concentration in moles per liter for each of the following solutions: (a) a solution whose pH is 5.20, (b) a solution whose pH is 16.00, (c) a solution whose hydroxide concentration is 3.7×10^{-9} *M*.

16.17 Calculate the pH of each of the following solutions: (a) 0.0010 *M* HCl, (b) 0.76 *M* KOH, (c) 2.8×10^{-4} *M* $Ba(OH)_2$, (d) 5.2×10^{-4} *M* HNO_3.

16.18 Calculate the pH of water at 40°C, given that K_w is 3.8×10^{-14} at this temperature.

16.19 Complete the following table for a solution:

pH	$[H^+]$	Solution is
<7		
	$<1.0 \times 10^{-7}$ *M*	
		Neutral

16.20 Fill in the word "acidic," "basic," or "neutral" for the following solutions:
(a) pOH > 7; solution is ______.
(b) pOH = 7; solution is ______.
(c) pOH < 7; solution is ______.

16.21 The pOH of a solution is 9.40. Calculate the hydrogen ion concentration of the solution.

16.22 Calculate the number of moles of KOH in 5.50 mL of a 0.360 *M* KOH solution. What is the pOH of the solution?

16.23 A solution is made by dissolving 18.4 g of HCl in 662 mL of water. Calculate the pH of the solution. (Assume that the volume of the solution is also 662 mL.)

16.24 How much NaOH (in grams) is needed to prepare 546 mL of solution with a pH of 10.00?

STRENGTHS OF ACIDS AND BASES

Review Questions

16.25 Explain what is meant by the strength of an acid.

16.26 Without referring to the text, write the formulas of four strong acids and four weak acids.

16.27 What are the strongest acid and strongest base that can exist in water?

16.28 H_2SO_4 is a strong acid, but HSO_4^- is a weak acid. Account for the difference in strength of these two related species.

Problems

16.29 Classify each of the following species as a weak or strong acid: (a) HNO_3, (b) HF, (c) H_2SO_4, (d) HSO_4^-, (e) H_2CO_3, (f) HCO_3^-, (g) HCl, (h) HCN, (i) HNO_2.

16.30 Classify each of the following species as a weak or strong base: (a) LiOH, (b) CN^-, (c) H_2O, (d) ClO_4^-, (e) NH_2^-.

16.31 Which of the following statements is/are true regarding a 0.10 *M* solution of a weak acid HA?
(a) The pH is 1.00.
(b) $[H^+] \gg [A^-]$
(c) $[H^+] = [A^-]$

(d) The pH is less than 1.

16.32 Which of the following statements is/are true regarding a 1.0 *M* solution of a strong acid HA?
(a) $[A^-] > [H^+]$
(b) The pH is 0.00.
(c) $[H^+] = 1.0\ M$
(d) $[HA] = 1.0\ M$

16.33 Predict the direction that predominates in this reaction:

$$F^-(aq) + H_2O(l) \rightleftharpoons HF(aq) + OH^-(aq)$$

16.34 Predict whether the following reaction will proceed from left to right to any measurable extent:

$$CH_3COOH(aq) + Cl^-(aq) \longrightarrow$$

16.35 Which of the following acids is the stronger: CH_3COOH or $CH_2ClCOOH$? Explain your choice. (*Hint:* Chlorine is more electronegative than hydrogen. How would its presence affect the ionization of the carboxyl group?)

16.36 Consider the following compounds:

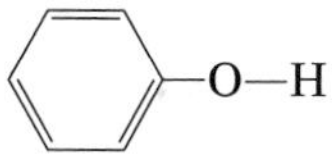

phenol

$CH_3—O—H$

methanol

Experimentally, phenol is found to be a stronger acid than methanol. Explain this difference in terms of the structures of the conjugate bases. (*Hint:* A more stable conjugate base favors ionization. Only one of the conjugate bases can be stabilized by resonance.)

WEAK ACID IONIZATION CONSTANTS

Review Questions

16.37 What does the ionization constant tell us about the strength of an acid?

16.38 List the factors on which the K_a of a weak acid depends.

16.39 Why do we normally not quote K_a values for strong acids such as HCl and HNO_3? Why is it necessary to specify temperature when giving K_a values?

16.40 Which of the following solutions has the highest pH? (a) 0.40 *M* HCOOH, (b) 0.40 *M* $HClO_4$, (c) 0.40 *M* CH_3COOH.

Problems

16.41 The K_a for benzoic acid is 6.5×10^{-5}. Calculate the concentrations of all the species (C_6H_5COOH, $C_6H_5COO^-$, H^+, and OH^-) in a 0.10 *M* benzoic acid solution.

16.42 A 0.0560-g quantity of acetic acid is dissolved in enough water to make 50.0 mL of solution. Calculate the concentrations of H^+, CH_3COO^-, and CH_3COOH at equilibrium. (K_a for acetic acid = 1.8×10^{-5}.)

16.43 The pH of a HF solution is 6.20. Calculate the ratio [conjugate base]/[acid] for HF at this pH.

16.44 What is the original molarity of a solution of formic acid (HCOOH) whose pH is 3.26 at equilibrium?

16.45 Calculate the percent ionization of benzoic acid at the following concentrations: (a) 0.20 *M*, (b) 0.00020 *M*.

16.46 Calculate the percent ionization of hydrofluoric acid at the following concentrations: (a) 0.60 *M*, (b) 0.080 *M*, (c) 0.0046 *M*, (d) 0.00028 *M*. Comment on the trends.

16.47 A 0.040 *M* solution of a monoprotic acid is 14 percent ionized. Calculate the ionization constant of the acid.

16.48 (a) Calculate the percent ionization of a 0.20 *M* solution of the monoprotic acetylsalicylic acid (aspirin). ($K_a = 3.0 \times 10^{-4}$.) (b) The pH of gastric juice in the stomach of a certain individual is 1.00. After a few aspirin tablets have been swallowed, the concentration of acetylsalicylic acid in the stomach is 0.20 *M*. Calculate the percent ionization of the acid under these conditions.

DIPROTIC AND POLYPROTIC ACIDS

Review Questions

16.49 Malonic acid is a diprotic acid. Explain what that means.

16.50 Write all the species (except water) that are present in a phosphoric acid solution. Indicate which species can act as a Brønsted acid, which as a Brønsted base, and which as both a Brønsted acid and a Brønsted base.

Problems

16.51 What are the concentrations of HSO_4^-, SO_4^{2-}, and H^+ in a 0.20 *M* $KHSO_4$ solution? (*Hint:* H_2SO_4 is a strong acid: K_a for $HSO_4^- = 1.3 \times 10^{-2}$.)

16.52 Calculate the concentrations of H^+, HCO_3^-, and CO_3^{2-} in a 0.025 *M* H_2CO_3 solution.

WEAK BASE IONIZATION CONSTANTS; K_a–K_b RELATIONSHIP

Review Questions

16.53 Use NH_3 to illustrate what we mean by the strength of a base.

16.54 Write the equation relating K_a for a weak acid and K_b for its conjugate base. Use NH_3 and its conjugate acid NH_4^+ to derive the relationship between K_a and K_b.

Problems

16.55 Calculate the pH for each of the following solutions: (a) 0.10 *M* NH_3, (b) 0.050 *M* pyridine.

16.56 The pH of a 0.30 *M* solution of a weak base is 10.66. What is the K_b of the base?

16.57 What is the original molarity of a solution of ammonia whose pH is 11.22?

16.58 In a 0.080 *M* NH_3 solution, what percent of the NH_3 is present as NH_4^+?

ACID-BASE PROPERTIES OF SALT SOLUTIONS

Review Questions

16.59 Define salt hydrolysis. Categorize salts according to how they affect the pH of a solution.

16.60 Explain why small, highly charged metal ions are able to undergo hydrolysis.

16.61 Al^{3+} is not a Brønsted acid, but $Al(H_2O)_6^{3+}$ is. Explain.

16.62 Specify which of the following salts will undergo hydrolysis: KF, $NaNO_3$, NH_4NO_2, $MgSO_4$, KCN, C_6H_5COONa, RbI, Na_2CO_3, $CaCl_2$, HCOOK.

16.63 Predict the pH (>7, <7, or ≈7) of the aqueous solutions containing the following salts: (a) KBr, (b) $Al(NO_3)_3$, (c) $BaCl_2$, (d) $Bi(NO_3)_3$.

16.64 Which ion of the alkaline earth metals is most likely to undergo hydrolysis?

Problems

16.65 A certain salt, MX (containing the M^+ and X^- ions), is dissolved in water, and the pH of the resulting solution is 7.0. Can you say anything about the strengths of the acid and the base from which the salt is derived?

16.66 In a certain experiment a student finds that the pHs of 0.10 *M* solutions of three potassium salts KX, KY, and KZ are 7.0, 9.0, and 11.0, respectively. Arrange the acids HX, HY, and HZ in the order of increasing acid strength.

16.67 Calculate the pH of a 0.36 *M* CH_3COONa solution.

16.68 Calculate the pH of a 0.42 *M* NH_4Cl solution.

16.69 Predict whether a solution containing the salt K_2HPO_4 will be acidic, neutral, or basic. (*Hint:* You need to consider both the ionization and hydrolysis of HPO_4^{2-}.)

16.70 Compare the pH of a 0.040 *M* HCl solution with that of a 0.040 *M* H_2SO_4 solution.

16.71 The first and second ionization constants of a diprotic acid H_2A are K_{a_1} and K_{a_2} at a certain temperature. Under what conditions will $[A^{2-}] = K_{a_2}$?

16.72 The K_a of formic acid is 1.7×10^{-4} at 25°C. Will the acid become stronger or weaker at 40°C? Explain.

LEWIS ACIDS AND BASES

Review Questions

16.73 What are the Lewis definitions of an acid and a base? In what way are they more general than the Brønsted definitions?

16.74 In terms of orbitals and electron arrangements, what must be present for a molecule or an ion to act as a Lewis acid (use H^+ and BF_3 as examples)? What must be present for a molecule or ion to act as a Lewis base (use OH^- and NH_3 as examples)?

Problems

16.75 Classify each of the following species as a Lewis acid or a Lewis base: (a) CO_2, (b) H_2O, (c) I^-, (d) SO_2, (e) NH_3, (f) OH^-, (g) H^+, (h) BCl_3.

16.76 Describe the following reaction according to the Lewis theory of acids and bases:

$$AlCl_3(s) + Cl^-(aq) \longrightarrow AlCl_4^-(aq)$$

16.77 Which would be considered a stronger Lewis acid: (a) BF_3 or BCl_3, (b) Fe^{2+} or Fe^{3+}? Explain.

16.78 All Brønsted acids are Lewis acids, but the reverse is not true. Give two examples of Lewis acids that are not Brønsted acids.

MISCELLANEOUS PROBLEMS

16.79 Classify the following oxides as acidic, basic, amphoteric, or neutral: (a) CO_2, (b) K_2O, (c) CaO, (d) N_2O_5, (e) CO, (f) NO, (g) SnO_2, (h) SO_3, (i) Al_2O_3, (j) BaO.

16.80 A typical reaction between an antacid and the hydrochloric acid in gastric juice is

$$NaHCO_3(aq) + HCl(aq) \longrightarrow NaCl(aq) + H_2O(l) + CO_2(g)$$

Calculate the volume (in liters) of CO_2 generated from 0.350 g of $NaHCO_3$ and excess gastric juice at 1.00 atm and 37.0°C.

16.81 To which of the following would addition of an equal volume of 0.60 *M* NaOH lead to a solution having a lower pH? (a) Water, (b) 0.30 *M* HCl, (c) 0.70 *M* KOH, (d) 0.40 *M* $NaNO_3$.

16.82 The pH of a 0.0642 *M* solution of a monoprotic acid is 3.86. Is this a strong acid?

16.83 Like water, ammonia undergoes autoionization in liquid ammonia:

$$NH_3 + NH_3 \rightleftharpoons NH_4^+ + NH_2^-$$

(a) Identify the Brønsted acids and Brønsted bases in this reaction. (b) What species correspond to H^+ and OH^-, and what is the condition for a neutral solution?

16.84 HA and HB are both weak acids although HB is the stronger of the two. Will it take more volume of a 0.10 *M* NaOH solution to neutralize 50.0 mL of 0.10 *M* HB than 50.0 mL of 0.10 *M* HA?

16.85 A 1.87-g sample of Mg reacts with 80.0 mL of a HCl solution whose pH is −0.544. What is the pH of the solution after all the Mg has reacted? Assume volume of solution is constant.

16.86 The three common chromium oxides are CrO, Cr_2O_3, and CrO_3. If Cr_2O_3 is amphoteric, what can you say about the acid-base properties of CrO and CrO_3?

16.87 Most of the hydrides of Group 1A and Group 2A metals are ionic (the exceptions are BeH_2 and MgH_2, which are covalent compounds). (a) Describe the reaction between the hydride ion (H^-) and water in terms of a Brønsted acid-base reaction. (b) The same reaction can also be classified as a redox reaction. Identify the oxidizing and reducing agents.

16.88 Use the data in Table 16.3 to calculate the equilibrium constant for the following reaction:

$$CH_3COOH(aq) + NO_2^-(aq) \rightleftharpoons CH_3COO^-(aq) + HNO_2(aq)$$

16.89 Calculate the pH of a 0.20 *M* ammonium acetate (CH_3COONH_4) solution.

16.90 Novocaine, used as a local anesthetic by dentists, is a weak base ($K_b = 8.91 \times 10^{-6}$). What is the ratio of the concentration of the base to that of its acid in the blood plasma (pH = 7.40) of a patient?

Answers to Practice Exercises: 16.1 (1) H_2O (acid) and OH^- (base); (2) HCN (acid) and CN^- (base); **16.2** 7.7×10^{-15} *M*; **16.3** 0.12; **16.4** 4.7×10^{-4} *M*; **16.5** 12.48; **16.6** $[H^+] = 7.3 \times 10^{-3}$ *M*, $[A^-] = 7.3 \times 10^{-3}$ *M*, [HA] = 0.19 *M*; **16.7** 2.09; **16.8** 2.2×10^{-6}; **16.9** $[C_2H_2O_4] = 0.11$ *M*, $[C_2HO_4^-] = 0.086$ *M*, $[C_2O_4^{2-}] = 6.1 \times 10^{-5}$ *M*, $[H^+] = 0.086$ *M*; **16.10** 12.03; **16.11** (a) ≃7, (b) >7, (c) <7, (d) >7.

CHAPTER 17

ACID-BASE EQUILIBRIA AND SOLUBILITY EQUILIBRIA

Chicken eggs.

◆ How is the shell of a hen's egg formed? The eggshell is largely composed of calcite, which is a crystalline form of calcium carbonate ($CaCO_3$).

An average eggshell weighs about 5 g and is 40 percent calcium. Most of the calcium in an eggshell is laid down within a 16-hour period. This means that it is deposited at a rate of about 125 milligrams per hour. No hen can consume calcium fast enough to meet this demand. Instead, it is supplied by special bony masses in the hen's long bones, which accumulate large reserves of calcium for eggshell formation. [The inorganic calcium component of the bone is calcium phosphate, $Ca_3(PO_4)_2$, another insoluble compound.] If a hen is fed a low-calcium diet, her eggshell becomes progressively thinner; she may have to mobilize 10 percent of the total amount of calcium in her bones just to lay one egg! When the food supply is constantly low in calcium, egg production eventually stops.

Normally, the raw materials, the Ca^{2+} and CO_3^{2-} ions, are supplied by the blood to the shell gland. The calcification process is a precipitation reaction:

$$Ca^{2+}(aq) + CO_3^{2-}(aq) \rightleftharpoons CaCO_3(s)$$

In the blood, free Ca^{2+} ions are in equilibrium with calcium ions bound to proteins. As the free ions are taken up by the shell gland, more are provided by the dissociation of the protein-bound calcium.

The carbonate ions necessary for eggshell formation are a metabolic by-product. Carbon dioxide produced during metabolism is converted to carbonic acid (H_2CO_3) by the enzyme carbonic anhydrase (CA):

$$CO_2(g) + H_2O(l) \overset{CA}{\rightleftharpoons} H_2CO_3(aq)$$

Carbonic acid ionizes stepwise to produce carbonate ions:

$$H_2CO_3(aq) \rightleftharpoons H^+(aq) + HCO_3^-(aq)$$

$$HCO_3^-(aq) \rightleftharpoons H^+(aq) + CO_3^{2-}(aq)$$

Chickens do not perspire and so must pant to cool themselves. Panting expels more CO_2 from the chicken's body than normal respiration does. According to Le Chatelier's principle, panting will shift the CO_2–H_2CO_3 equilibrium shown above from right to left, thereby lowering the concentration of the CO_3^{2-} ions in solution and resulting in thin eggshells. One remedy to this problem is to give chickens carbonated water to drink in hot weather. The CO_2 dissolved in the water adds CO_2 to the chicken's body fluids and shift the CO_2–H_2CO_3 equilibrium to the right. ◆

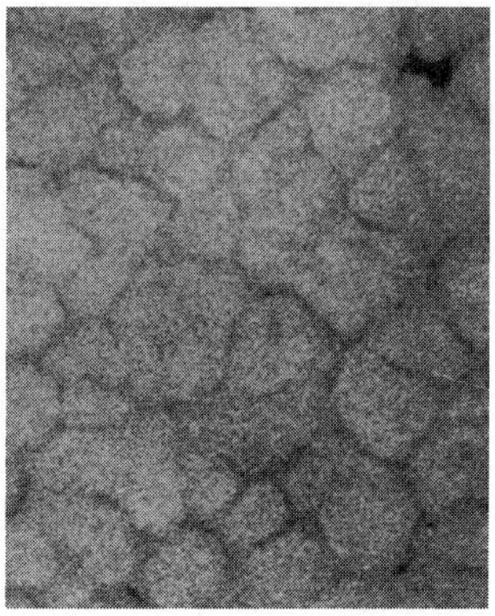

X-ray micrograph of an eggshell, showing columns of calcite.

17.1 HOMOGENEOUS VERSUS HETEROGENEOUS SOLUTION EQUILIBRIA

In Chapter 16 we saw that weak acids and weak bases never ionize completely in water. Thus at equilibrium a weak acid solution, for example, contains nonionized acid as well as H^+ ions and the conjugate base. Nevertheless, all of these species are dissolved, so that the system is an example of homogeneous equilibrium (see Chapter 15).

Another important type of equilibrium, which we will consider in the second half of the chapter, involves the dissolution and precipitation of slightly soluble substances. These processes are examples of heterogeneous equilibria; that is, they pertain to reactions in which the components are in more than one phase.

17.2 BUFFER SOLUTIONS

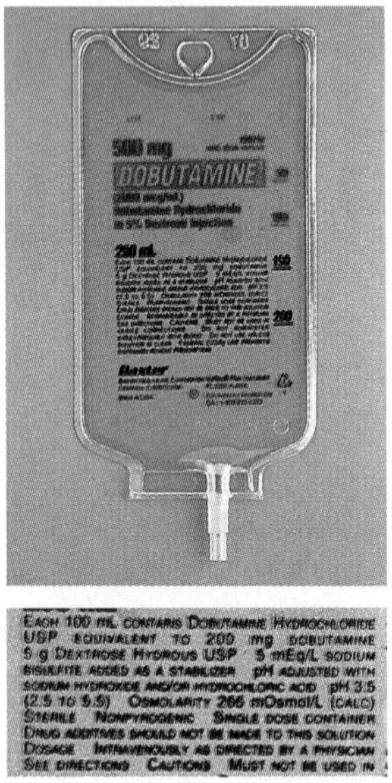

Fluids for intravenous injection must include buffer systems to maintain the proper blood pH.

A ***buffer solution*** is *a solution of (1) a weak acid or a weak base and (2) its salt; both components must be present. The solution has the ability to resist changes in pH upon the addition of small amounts of either acid or base.* Buffers are very important to chemical and biological systems. The pH in the human body varies greatly from one fluid to another; for example, the pH of blood is about 7.4, whereas the gastric juice in our stomachs has a pH of about 1.5. These pH values, which are crucial for the proper functioning of enzymes and the balance of osmotic pressure, are maintained by buffers in most cases.

A buffer solution must contain a relatively large concentration of acid to react with any OH^- ions that may be added to it and must contain a similar concentration of base to react with any added H^+ ions. Furthermore, the acid and the base components of the buffer must not consume each other in a neutralization reaction. These requirements are satisfied by an acid-base conjugate pair (a weak acid and its conjugate base or a weak base and its conjugate acid).

A simple buffer solution can be prepared by adding comparable amounts of acetic acid (CH_3COOH) and sodium acetate (CH_3COONa) to water. The equilibrium concentrations of both the acid and the conjugate base (from CH_3COONa) are assumed to be the same as the starting concentrations. This is so because (1) CH_3COOH is a weak acid and the extent of hydrolysis of the CH_3COO^- ion is very small and (2) the presence of CH_3COO^- ions suppresses the ionization of CH_3COOH, and the presence of CH_3COOH suppresses the hydrolysis of the CH_3COO^- ions.

A solution containing these two substances has the ability to neutralize either added acid or added base. Sodium acetate, a strong electrolyte, dissociates completely in water:

$$CH_3COONa(s) \xrightarrow{H_2O} CH_3COO^-(aq) + Na^+(aq)$$

If an acid is added, the H^+ ions will be consumed by the conjugate base in the buffer, CH_3COO^-, according to the equation

$$CH_3COO^-(aq) + H^+(aq) \longrightarrow CH_3COOH(aq)$$

If a base is added to the buffer system, the OH^- ions will be neutralized by the acid in the buffer:

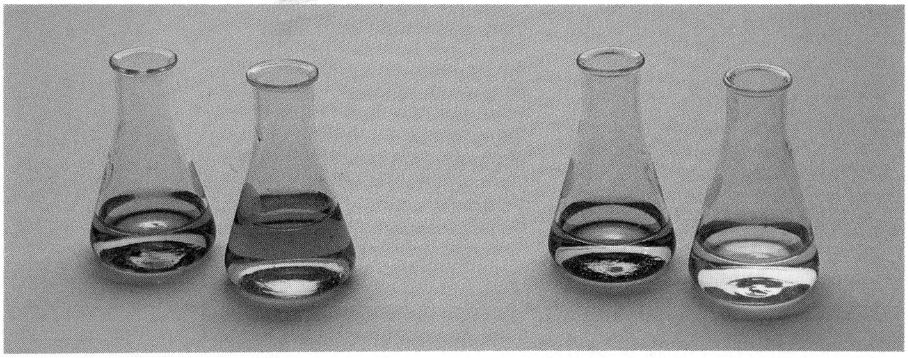

FIGURE 17.1

The acid-base indicator bromophenol blue (added to all solutions) is used to illustrate buffer action. The indicator's color is blue-purple above pH 4.6 and changes to yellow below pH 3.0. (a) A buffer solution made up of 50 mL of 0.1 M CH_3COOH and 50 mL of 0.1 M CH_3COONa. The solution has a pH of 4.7 and turns the indicator blue-purple. (b) After the addition of 40 mL of 0.1 M HCl solution to the solution in (a), the color remains blue-purple. (c) A 100-mL CH_3COOH solution whose pH is 4.7. (d) After the addition of 6 drops (about 0.3 mL) of 0.1 M HCl solution, the color turns yellow. Without buffer action, the pH of the solution decreases rapidly to below 3.0 upon the addition of 0.1 M HCl.

$$CH_3COOH(aq) + OH^-(aq) \longrightarrow CH_3COO^-(aq) + H_2O(l)$$

The *buffering capacity,* that is, the effectiveness of the buffer solution, depends on the amounts of acid and conjugate base from which the buffer is made. The larger the amount, the greater the buffering capacity.

In general, a buffer system can be represented as salt/acid or conjugate base/acid. Thus the sodium acetate–acetic acid buffer system can be written as CH_3COONa/CH_3COOH or CH_3COO^-/CH_3COOH. Figure 17.1 shows this buffer system in action.

EXAMPLE 17.1

Identifying Buffer Systems

Which of the following solutions are buffer systems? (a) KH_2PO_4/H_3PO_4, (b) $NaClO_4/HClO_4$, (c) C_5H_5N/C_5H_5NHCl (C_5H_5N is pyridine; its K_b is given in Table 16.4). Explain your answer.

Answer: (a) H_3PO_4 is a weak acid, and its conjugate base, $H_2PO_4^-$, is a weak base (see Table 16.4). Therefore, this is a buffer system.

(b) Because $HClO_4$ is a strong acid, its conjugate base, ClO_4^-, is an extremely weak base. This means that the ClO_4^- ion will not combine with a H^+ ion in solution to form $HClO_4$. Thus the system cannot act as a buffer system.

(c) As Table 16.4 shows, C_5H_5N is a weak base and its conjugate acid, $C_5H_5\overset{+}{N}H$ (the cation of the salt C_5H_5NHCl), is a weak acid. Therefore, this is a buffer system.

PRACTICE EXERCISE

Which of the following solutions are buffer systems? (a) KF/HF, (b) KBr/HBr, (c) $Na_2CO_3/NaHCO_3$.

EXAMPLE 17.2

Calculating the pH of a Buffer System

(a) Calculate the pH of a buffer system containing 1.0 *M* CH_3COOH and 1.0 *M* CH_3COONa. (b) What is the pH of the buffer system after the addition of 0.10 mole of gaseous HCl to 1 L of the solution? Assume that the volume of the solution does not change when the HCl is added.

Answer: (a) Assuming negligible ionization of the acetic acid and hydrolysis of the acetate ions, we have, at equilibrium,

$$[CH_3COOH] = 1.0\ M \qquad \text{and} \qquad [CH_3COO^-] = 1.0\ M$$

$$K_a = \frac{[H^+][CH_3COO^-]}{[CH_3COOH]} = 1.8 \times 10^{-5}$$

$$[H^+] = \frac{K_a[CH_3COOH]}{[CH_3COO^-]}$$

$$= \frac{(1.8 \times 10^{-5})(1.0)}{(1.0)}$$

$$= 1.8 \times 10^{-5}\ M$$

$$pH = -\log (1.8 \times 10^{-5})$$

$$= 4.74$$

(b) After the addition of HCl, complete ionization of HCl acid occurs:

$$\underset{0.10\ \text{mol}}{HCl(aq)} \longrightarrow \underset{0.10\ \text{mol}}{H^+(aq)} + \underset{0.10\ \text{mol}}{Cl^-(aq)}$$

Originally, there were 1.0 mol CH_3COOH and 1.0 mol CH_3COO^- present in 1 L of the solution. After neutralization of the HCl acid by CH_3COO^-, which we write as

$$\underset{0.10\ \text{mol}}{CH_3COO^-(aq)} + \underset{0.10\ \text{mol}}{H^+(aq)} \longrightarrow \underset{0.10\ \text{mol}}{CH_3COOH(aq)}$$

the number of moles of acetic acid and the number of moles of acetate ions present are

$$CH_3COOH: \qquad (1.0 + 0.1)\ \text{mol} = 1.1\ \text{mol}$$

$$CH_3COO^-: \qquad (1.0 - 0.1)\ \text{mol} = 0.90\ \text{mol}$$

Next we calculate the hydrogen ion concentration:

$$[H^+] = \frac{K_a[CH_3COOH]}{[CH_3COO^-]}$$

$$[H^+] = \frac{(1.8 \times 10^{-5})(1.1)}{0.90}$$

$$= 2.2 \times 10^{-5}\ M$$

The pH of the solution becomes

$$pH = -\log (2.2 \times 10^{-5})$$

$$= 4.66$$

PRACTICE EXERCISE

Calculate the pH of the 0.30 M NH_3/0.36 M NH_4Cl buffer system. What is the pH after the addition of 20.0 mL of 0.050 M NaOH to 80.0 mL of the buffer solution?

We see that in the buffer solution examined in Example 17.2 there is a decrease in pH of 0.08 unit (becoming more acidic) as a result of the addition of HCl. We can also compare the change in H^+ ion concentrations as follows:

Before addition of HCl: $[H^+] = 1.8 \times 10^{-5}\ M$

After addition of HCl: $[H^+] = 2.2 \times 10^{-5}\ M$

Thus the H^+ ion concentration increases by a factor of

$$\frac{2.2 \times 10^{-5}\ M}{1.8 \times 10^{-5}\ M} = 1.2$$

To appreciate the effectiveness of the CH_3COONa/CH_3COOH buffer, let us find out what would happen if 0.10 mol HCl were added to 1 L of water, and compare the increase in H^+ ion concentration.

Before addition of HCl: $[H^+] = 1.0 \times 10^{-7}\ M$

After addition of HCl: $[H^+] = 0.10\ M$

Thus, as a result of the addition of HCl, the H^+ ion concentration increases by a factor of

$$\frac{0.10\ M}{1.0 \times 10^{-7}\ M} = 1.0 \times 10^6$$

amounting to a millionfold increase! This comparison shows that a properly chosen buffer solution can maintain a fairly constant H^+ ion concentration, or pH.

Preparing a Buffer Solution with a Specific pH

Now suppose we want to prepare a buffer solution with a specific pH. How do we go about it? Referring to the acetic acid–sodium acetate buffer system, we can write the equilibrium constant as

$$K_a = \frac{[CH_3COO^-][H^+]}{[CH_3COOH]}$$

Note that this expression holds whether we have only acetic acid or a mixture of acetic acid and sodium acetate in solution. Rearranging the above equation gives

$$[H^+] = \frac{K_a[CH_3COOH]}{[CH_3COO^-]}$$

Taking the negative logarithm of both sides, we obtain

$$-\log [H^+] = -\log K_a - \log \frac{[CH_3COOH]}{[CH_3COO^-]}$$

or

$$-\log [H^+] = -\log K_a + \log \frac{[CH_3COO^-]}{[CH_3COOH]}$$

So

$$pH = pK_a + \log \frac{[CH_3COO^-]}{[CH_3COOH]} \tag{17.1}$$

where

$$pK_a = -\log K_a \tag{17.2}$$

pK_a is related to K_a as pH is related to $[H^+]$. Remember that the stronger the acid (that is, the larger the K_a), the smaller the pK_a.

Equation (17.1) is called the *Henderson-Hasselbalch equation.* In a more general form it can be expressed as

$$pH = pK_a + \log \frac{[\text{conjugate base}]}{[\text{acid}]} \tag{17.3}$$

If the molar concentrations of the acid and its conjugate base are approximately equal, that is, [acid] ≈ [conjugate base], then

$$\log \frac{[\text{conjugate base}]}{[\text{acid}]} \approx 0$$

or

$$pH \approx pK_a$$

Thus, to prepare a buffer solution, we choose a weak acid whose pK_a is close to the desired pH. This choice not only gives the correct pH value of the buffer system, but also ensures that we have *comparable* amounts of the acid and its conjugate base present; both are prerequisites for the buffer system to function effectively.

EXAMPLE 17.3
Preparing a Buffer Solution with a Specific pH

Describe how you would prepare a "phosphate buffer" at a pH of about 7.40.

Answer: We write three stages of ionization of phosphoric acid as follows. The K_a values are obtained from Table 16.4, and the pK_a values are found by applying Equation (17.2):

$$H_3PO_4(aq) \rightleftharpoons H^+(aq) + H_2PO_4^-(aq) \qquad K_{a_1} = 7.5 \times 10^{-3};\ pK_{a_1} = 2.12$$

$$H_2PO_4^-(aq) \rightleftharpoons H^+(aq) + HPO_4^{2-}(aq) \qquad K_{a_2} = 6.2 \times 10^{-8};\ pK_{a_2} = 7.21$$

$$HPO_4^{2-}(aq) \rightleftharpoons H^+(aq) + PO_4^{3-}(aq) \qquad K_{a_3} = 4.8 \times 10^{-13};\ pK_{a_3} = 12.32$$

The most suitable of the three buffer systems is $HPO_4^{2-}/H_2PO_4^-$, because the pK_a of the acid $H_2PO_4^-$ is closest to the desired pH. From the Henderson-Hasselbalch equation we write

$$pH = pK_a + \log \frac{[\text{conjugate base}]}{[\text{acid}]}$$

$$7.40 = 7.21 + \log \frac{[HPO_4^{2-}]}{[H_2PO_4^-]}$$

$$\log \frac{[HPO_4^{2-}]}{[H_2PO_4^-]} = 0.19$$

Taking the antilog, we obtain

$$\frac{[HPO_4^{2-}]}{[H_2PO_4^-]} = 1.5$$

Thus one way to prepare a phosphate buffer with a pH of 7.40 is to dissolve disodium hydrogen phosphate (Na_2HPO_4) and sodium dihydrogen phosphate (NaH_2PO_4) in a mole ratio of 1.5:1.0 in water. For example, we could dissolve 1.5 moles of Na_2HPO_4 and 1.0 mole of NaH_2PO_4 in enough water to make up a 1-L solution.

PRACTICE EXERCISE

How would you prepare a liter of "carbonate buffer" at a pH of 10.10? You are provided with carbonic acid (H_2CO_3), sodium hydrogen carbonate ($NaHCO_3$), and sodium carbonate (Na_2CO_3).

17.3 A CLOSER LOOK AT ACID-BASE TITRATIONS

Having discussed buffer solutions, we can now look in more detail at the quantitative aspects of acid-base titrations (see Section 4.6). We will consider three types of reactions: (1) titrations involving a strong acid and a strong base, (2) titrations involving a weak acid and a strong base, and (3) titrations involving a strong acid and a weak base. Titrations involving a weak acid and a weak base are complicated by the hydrolysis of both the cation and the anion of the salt formed. These titrations will not be dealt with here. Figure 17.2 shows the arrangement for monitoring the pH during the course of a titration.

STRONG ACID–STRONG BASE TITRATIONS

The reaction between a strong acid (say, HCl) and a strong base (say, NaOH) can be represented by

$$NaOH(aq) + HCl(aq) \longrightarrow NaCl(aq) + H_2O(l)$$

or in terms of the net ionic equation

$$H^+(aq) + OH^-(aq) \longrightarrow H_2O(l)$$

Consider the addition of a 0.100 M NaOH solution (from a buret) to an Erlenmeyer

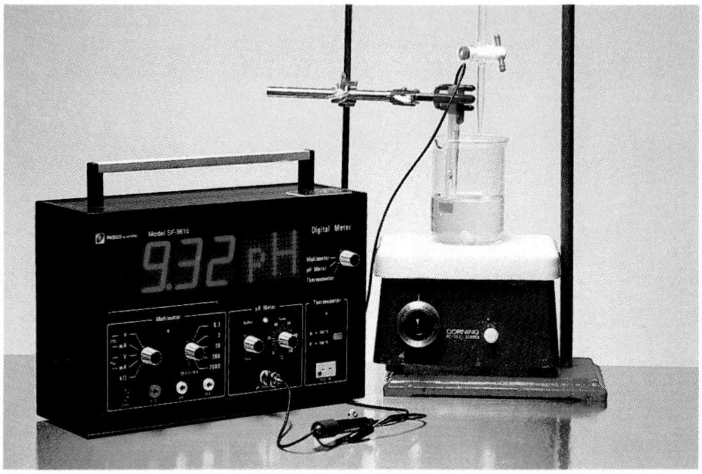

FIGURE 17.2

A pH meter is used to monitor an acid-base titration.

flask containing 25.0 mL of 0.100 *M* HCl. For convenience, we will use only three significant figures for volume and concentration and two significant figures for pH. Figure 17.3 shows the pH profile of the titration (also known as the titration curve). Before the addition of NaOH, the pH of the acid is given by −log (0.100), or 1.00. When NaOH is added, the pH of the solution increases slowly at first. Near the equivalence point the pH begins to rise steeply, and at the equivalence point (that is, the point at which equimolar amounts of acid and base have reacted) the curve rises almost vertically. In a strong acid–strong base titration both the hydrogen ion and hydroxide ion concentrations are very small at the equivalence point (approximately 1×10^{-7} *M*); consequently, the addition of a single drop of the base can cause a large increase in $[OH^-]$ and in the pH of the solution. Beyond the equivalence point, the pH again increases slowly with the addition of NaOH.

It is possible to calculate the pH of the solution at every stage of titration. Here are three sample calculations:

1. *After the addition of 10.0 mL of 0.100 M NaOH to 25.0 mL of 0.100 M HCl.* The total volume of the solution is 35.0 mL. The number of moles of NaOH in 10.0 mL is

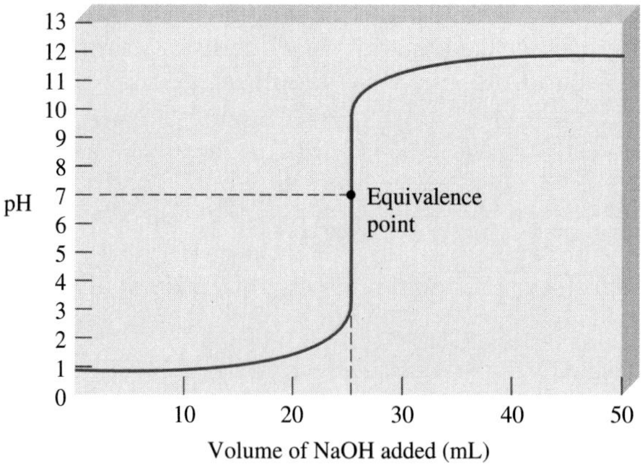

FIGURE 17.3

pH profile of a strong acid–strong base titration. A 0.100 M NaOH solution is added from a buret to 25.0 mL of a 0.100 M HCl solution in an Erlenmeyer flask (see Figure 4.16). This curve is sometimes referred to as a titration curve.

$$10.0 \text{ mL} \times \frac{0.100 \text{ mol NaOH}}{1 \text{ L NaOH}} \times \frac{1 \text{ L}}{1000 \text{ mL}} = 1.00 \times 10^{-3} \text{ mol}$$

The number of moles of HCl originally present in 25.0 mL of solution is

$$25.0 \text{ mL} \times \frac{0.100 \text{ mol HCl}}{1 \text{ L HCl}} \times \frac{1 \text{ L}}{1000 \text{ mL}} = 2.50 \times 10^{-3} \text{ mol}$$

Thus the amount of HCl left after partial neutralization is $(2.50 \times 10^{-3}) - (1.00 \times 10^{-3})$, or 1.50×10^{-3} mol. Next, the concentration of H^+ ions in 35.0 mL of solution is found as follows:

$$\frac{1.50 \times 10^{-3} \text{ mol HCl}}{35.0 \text{ mL}} \times \frac{1000 \text{ mL}}{1 \text{ L}} = 0.0429 \text{ mol HCl/L}$$

$$= 0.0429\ M \text{ HCl}$$

Thus $[H^+] = 0.0429\ M$, and the pH of the solution is

$$\text{pH} = -\log 0.0429 = 1.37$$

2. *After the addition of 25.0 mL of 0.100 M NaOH to 25.0 mL of 0.100 M HCl.* This is a simple calculation, because it involves a complete neutralization reaction and the salt (NaCl) does not undergo hydrolysis. At the equivalence point, $[H^+] = [OH^-]$ and the pH of the solution is 7.00.

3. *After the addition of 35.0 mL of 0.100 M NaOH to 25.0 mL of 0.100 M HCl.* The total volume of the solution is now 60.0 mL. The number of moles of NaOH added is

$$35.0 \text{ mL} \times \frac{0.100 \text{ mol NaOH}}{1 \text{ L NaOH}} \times \frac{1 \text{ L}}{1000 \text{ mL}} = 3.50 \times 10^{-3} \text{ mol}$$

The number of moles of HCl in 25.0 mL solution is 2.50×10^{-3}. After complete neutralization of HCl, the number of moles of NaOH left is $(3.50 \times 10^{-3}) - (2.50 \times 10^{-3})$, or 1.00×10^{-3} mol. The concentration of NaOH in 60.0 mL of solution is

$$\frac{1.00 \times 10^{-3} \text{ mol NaOH}}{60.0 \text{ mL}} \times \frac{1000 \text{ mL}}{1 \text{ L}} = 0.0167 \text{ mol NaOH/L}$$

$$= 0.0167\ M \text{ NaOH}$$

Thus $[OH^-] = 0.0167\ M$ and $\text{pOH} = -\log 0.0167 = 1.78$. Hence, the pH of the solution is

$$\begin{aligned} \text{pH} &= 14.00 - \text{pOH} \\ &= 14.00 - 1.78 \\ &= 12.22 \end{aligned}$$

Table 17.1 shows a more complete set of data for the titration.

WEAK ACID–STRONG BASE TITRATIONS

Consider the neutralization reaction between acetic acid (a weak acid) and sodium hydroxide (a strong base):

$$CH_3COOH(aq) + NaOH(aq) \longrightarrow CH_3COONa(aq) + H_2O(l)$$

TABLE 17.1
Titration Data for HCl versus NaOH*

Volume of NaOH added (mL)	Total volume (mL)	Mole excess of acid or base	pH of solution
0.0	25.0	2.50×10^{-3} H^+	1.00
10.0	35.0	1.50×10^{-3}	1.37
20.0	45.0	5.00×10^{-4}	1.95
24.0	49.0	1.00×10^{-4}	2.69
24.5	49.5	5.00×10^{-5}	3.00
25.0	50.0	0.00	7.00
25.5	50.5	5.00×10^{-5} OH^-	11.00
26.0	51.0	1.00×10^{-4}	11.29
30.0	55.0	5.00×10^{-4}	11.96
40.0	65.0	1.50×10^{-3}	12.36
50.0	75.0	2.50×10^{-3}	12.52

*The titration is carried out by slowly adding a 0.100 *M* NaOH solution from a buret to 25.0 mL of a 0.100 *M* HCl solution in an Erlenmeyer flask.

This equation can be simplified to

$$CH_3COOH(aq) + OH^-(aq) \longrightarrow CH_3COO^-(aq) + H_2O(l)$$

The acetate ion undergoes hydrolysis as follows:

$$CH_3COO^-(aq) + H_2O(l) \rightleftharpoons CH_3COOH(aq) + OH^-(aq)$$

Therefore, at the equivalence point the pH will be *greater than* 7 as a result of the excess OH^- ions formed [Figure 17.4(a)].

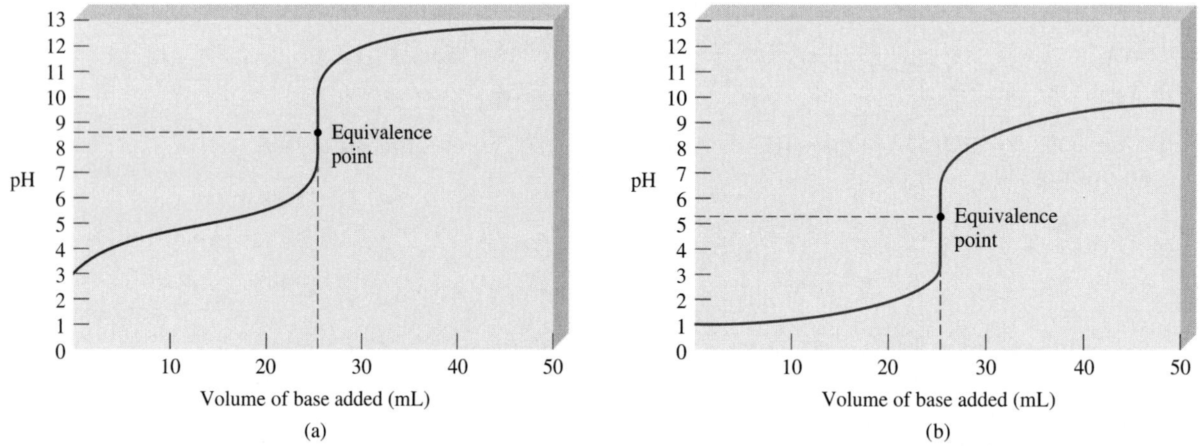

FIGURE 17.4

Titration curves. (a) Weak acid versus strong base. (b) Strong acid versus weak base. In each case, a 0.100 M base solution is added from a buret to 25.0 mL of a 0.100 M acid solution in an Erlenmeyer flask. As a result of salt hydrolysis, the pH at the equivalence point is greater than 7 for (a) and less than 7 for (b).

Strong Acid–Weak Base Titrations

Consider the titration of hydrochloric acid (a strong acid) and ammonia (a weak base):

$$HCl(aq) + NH_3(aq) \longrightarrow NH_4Cl(aq)$$

or simply

$$H^+(aq) + NH_3(aq) \longrightarrow NH_4^+(aq)$$

The pH at the equivalence point is *less than* 7 as a result of salt hydrolysis [Figure 17.4(b)]:

$$NH_4^+(aq) + H_2O(l) \rightleftharpoons NH_3(aq) + H_3O^+(aq)$$

or simply

$$NH_4^+(aq) \rightleftharpoons NH_3(aq) + H^+(aq)$$

17.4 Acid-Base Indicators

The end point or equivalence point in an acid-base titration is often signaled by a change in the color of an acid-base indicator, usually a weak organic acid or base that has distinctly different colors in its nonionized and ionized forms. These two forms are related to the pH of the solution in which the indicator is dissolved, as we will soon see. A change in the color of an indicator can be used to gauge the progress of a titration.

Not all indicators change color at the same pH, so the choice of indicator for a particular titration depends on the nature of the acid and base used in the titration (that is, whether they are strong or weak). Let us consider a weak monoprotic acid, HIn, which acts as an indicator. In solution

$$HIn(aq) \rightleftharpoons H^+(aq) + In^-(aq)$$

If the indicator is in a sufficiently acidic medium, the equilibrium, according to Le Chatelier's principle, shifts to the left and the predominant color of the indicator is that of the nonionized form (HIn). On the other hand, in a basic medium the equilibrium shifts to the right and the color of the conjugate base (In^-) predominates. Roughly speaking, we can use the following concentration ratios to predict the perceived color of the indicator:

$$\frac{[HIn]}{[In^-]} \geq 10 \qquad \text{color of acid (HIn) predominates}$$

$$\frac{[In^-]}{[HIn]} \geq 10 \qquad \text{color of conjugate base (In}^-\text{) predominates}$$

If $[HIn] \simeq [In^-]$, then the indicator color is a combination of the colors of HIn and In^-.

In Section 4.6, we mentioned that phenolphthalein is a suitable indicator for the titration of NaOH and HCl. Phenolphthalein is colorless in acidic and neutral solutions, but reddish pink in basic solutions. Measurements show that at $pH < 8.3$ the indicator is colorless but that it begins to turn reddish pink when the pH exceeds 8.3. Because of the steepness of the pH curve, near the equivalence point in Figure 17.3, the addition of a very small quantity of NaOH (say, 0.05 mL, which

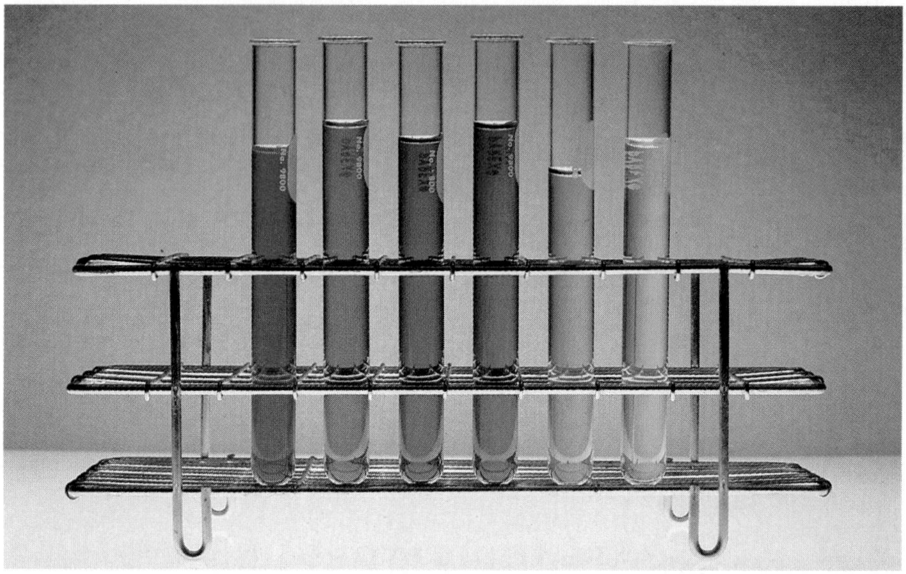

FIGURE 17.5

Solutions containing extracts of red cabbage (obtained by boiling the cabbage in water) develop different colors when treated with an acid and a base. The pH of the solutions increases from left to right.

is about the volume of a drop from the buret) brings about a large increase in the pH of the solution. What is important, however, is the fact that the steep portion of the pH profile includes the range over which phenolphthalein changes from colorless to reddish pink. Whenever such a matching occurs, the indicator can be used to locate the equivalence point of the titration.

Many acid-base indicators are plant pigments. For example, by boiling chopped red cabbage in water we can extract pigments that exhibit many different colors at various pHs (Figure 17.5). Table 17.2 lists a number of indicators commonly used in acid-base titrations. The criterion for choosing an appropriate indicator for a given titration is whether the pH range over which the indicator changes color corresponds with the steep portion of the titration curves shown in Figures 17.3 and 17.4. If not, the indicator will not accurately identify the equivalence point.

TABLE 17.2
Some Common Acid-Base Indicators

	Color		
Indicator	**In acid**	**In base**	**pH range***
Thymol blue	Red	Yellow	1.2–2.8
Bromophenol blue	Yellow	Bluish purple	3.0–4.6
Methyl orange	Orange	Yellow	3.1–4.4
Methyl red	Red	Yellow	4.2–6.3
Chlorophenol blue	Yellow	Red	4.8–6.4
Bromothymol blue	Yellow	Blue	6.0–7.6
Cresol red	Yellow	Red	7.2–8.8
Phenolphthalein	Colorless	Reddish pink	8.3–10.0

*The pH range is defined as the range over which the indicator changes from the acid color to the base color.

EXAMPLE 17.4
Choosing Suitable Acid-Base Indicators

Which indicator or indicators listed in Table 17.2 would you use for the acid-base titrations shown in (a) Figure 17.3, (b) Figure 17.4(a), and (c) Figure 17.4(b)?

Answer: (a) Near the equivalence point, the pH of the solution changes abruptly from 4 to 10. Therefore all the indicators except thymol blue, bromophenol blue, and methyl orange are suitable for use in the titration.

(b) Here the steep portion covers the pH range between 7 and 10; therefore, the suitable indicators are cresol red and phenolphthalein.

(c) Here the steep portion of the pH curve covers the pH range between 3 and 7; therefore, the suitable indicators are bromophenol blue, methyl orange, methyl red, and chlorophenol blue.

PRACTICE EXERCISE

Referring to Table 17.2, specify which indicator or indicators you would use for the following titrations: (a) HBr versus NH_3, (b) HNO_3 versus NaOH, (c) HNO_2 versus KOH.

17.5 SOLUBILITY EQUILIBRIA

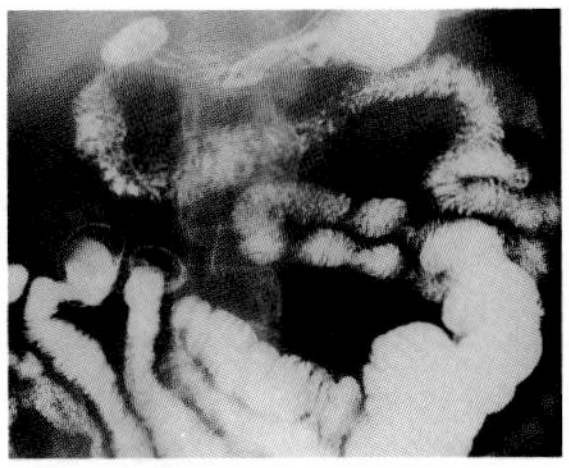

An aqueous suspension of $BaSO_4$ is used to examine the digestive tract.

Precipitation reactions are important in industry, medicine, and everyday life. For example, the preparation of many essential industrial chemicals such as sodium carbonate (Na_2CO_3) makes use of precipitation reactions. The dissolving of tooth enamel, which is mainly made of hydroxyapatite [$Ca_5(PO_4)_3OH$], in an acidic medium leads to tooth decay. Barium sulfate ($BaSO_4$), an insoluble compound that is opaque to X rays, is used to diagnose ailments of the digestive tract. Stalactites and stalagmites, which consist of calcium carbonate ($CaCO_3$), are produced by a precipitation reaction, and so are many foods, such as fudge.

The general rules for predicting the solubility of ionic compounds in water were introduced in Section 4.2. Although useful, these solubility rules do not allow us to make quantitative predictions about how much of a given ionic compound will dissolve in water. To develop a quantitative approach, we start with what we already know about chemical equilibrium.

Downward-growing stalactites and upward-growing stalagmites.

SOLUBILITY PRODUCT

Consider a saturated solution of silver chloride that is in contact with solid silver chloride. The solubility equilibrium can be represented as

$$AgCl(s) \rightleftharpoons Ag^+(aq) + Cl^-(aq)$$

Because salts such as AgCl are treated as strong electrolytes, all the AgCl that dissolves in water is assumed to dissociate completely into Ag^+ and Cl^- ions. We know from Chapter 15 that for heterogeneous reactions the concentration of the solid is a constant. Thus we can write the equilibrium constant for the dissolution of AgCl as

$$K_{sp} = [Ag^+][Cl^-]$$

where $\mathbf{K_{sp}}$ is called the solubility product constant or simply the solubility product. In general, the ***solubility product*** of a compound is *the product of the molar concentrations of the constituent ions, each raised to the power of its stoichiometric coefficient in the equilibrium equation.*

Because each AgCl unit contains only one Ag^+ ion and one Cl^- ion, its solubility product expression is particularly simple to write. The following cases are more complex.

- MgF_2

$$MgF_2(s) \rightleftharpoons Mg^{2+}(aq) + 2F^-(aq) \qquad K_{sp} = [Mg^{2+}][F^-]^2$$

- Ag_2CO_3

$$Ag_2CO_3(s) \rightleftharpoons 2Ag^+(aq) + CO_3^{2-}(aq) \qquad K_{sp} = [Ag^+]^2[CO_3^{2-}]$$

- $Ca_3(PO_4)_2$

$$Ca_3(PO_4)_2(s) \rightleftharpoons 3Ca^{2+}(aq) + 2PO_4^{3-}(aq) \qquad K_{sp} = [Ca^{2+}]^3[PO_4^{3-}]^2$$

Table 17.3 lists the solubility products for a number of salts of low solubility. Soluble salts such as NaCl and KNO_3, which have very large K_{sp} values, are not listed in the table.

For equilibrium reactions involving an ionic solid in aqueous solution, any one of the following conditions may exist: (1) The solution is unsaturated, (2) the solution is saturated, or (3) the solution is supersaturated. Following the procedure in Section 15.3, we use Q, called the *ion product,* to represent the product of the molar concentrations of the ions raised to the power of their stoichiometric coefficients. Thus for an aqueous solution containing Ag^+ and Cl^- ions at 25°C

TABLE 17.3

Solubility Products of Some Slightly Soluble Ionic Compounds at 25°C

Compound	K_{sp}	Compound	K_{sp}
Aluminum hydroxide [$Al(OH)_3$]	1.8×10^{-33}	Lead(II) chromate ($PbCrO_4$)	2.0×10^{-14}
Barium carbonate ($BaCO_3$)	8.1×10^{-9}	Lead(II) fluoride (PbF_2)	4.1×10^{-8}
Barium fluoride (BaF_2)	1.7×10^{-6}	Lead(II) iodide (PbI_2)	1.4×10^{-8}
Barium sulfate ($BaSO_4$)	1.1×10^{-10}	Lead(II) sulfide (PbS)	3.4×10^{-28}
Bismuth sulfide (Bi_2S_3)	1.6×10^{-72}	Magnesium carbonate ($MgCO_3$)	4.0×10^{-5}
Cadmium sulfide (CdS)	8.0×10^{-28}	Magnesium hydroxide [$Mg(OH)_2$]	1.2×10^{-11}
Calcium carbonate ($CaCO_3$)	8.7×10^{-9}	Manganese(II) sulfide (MnS)	3.0×10^{-14}
Calcium fluoride (CaF_2)	4.0×10^{-11}	Mercury(I) chloride (Hg_2Cl_2)	3.5×10^{-18}
Calcium hydroxide [$Ca(OH)_2$]	8.0×10^{-6}	Mercury(II) sulfide (HgS)	4.0×10^{-54}
Calcium phosphate [$Ca_3(PO_4)_2$]	1.2×10^{-26}	Nickel(II) sulfide (NiS)	1.4×10^{-24}
Chromium(III) hydroxide [$Cr(OH)_3$]	3.0×10^{-29}	Silver bromide (AgBr)	7.7×10^{-13}
Cobalt(II) sulfide (CoS)	4.0×10^{-21}	Silver carbonate (Ag_2CO_3)	8.1×10^{-12}
Copper(I) bromide (CuBr)	4.2×10^{-8}	Silver chloride (AgCl)	1.6×10^{-10}
Copper(I) iodide (CuI)	5.1×10^{-12}	Silver iodide (AgI)	8.3×10^{-17}
Copper(II) hydroxide [$Cu(OH)_2$]	2.2×10^{-20}	Silver sulfate (Ag_2SO_4)	1.4×10^{-5}
Copper(II) sulfide (CuS)	6.0×10^{-37}	Silver sulfide (Ag_2S)	6.0×10^{-51}
Iron(II) hydroxide [$Fe(OH)_2$]	1.6×10^{-14}	Strontium carbonate ($SrCO_3$)	1.6×10^{-9}
Iron(III) hydroxide [$Fe(OH)_3$]	1.1×10^{-36}	Strontium sulfate ($SrSO_4$)	3.8×10^{-7}
Iron(II) sulfide (FeS)	6.0×10^{-19}	Tin(II) sulfide (SnS)	1.0×10^{-26}
Lead(II) carbonate ($PbCO_3$)	3.3×10^{-14}	Zinc hydroxide [$Zn(OH)_2$]	1.8×10^{-14}
Lead(II) chloride ($PbCl_2$)	2.4×10^{-4}	Zinc sulfide (ZnS)	3.0×10^{-23}

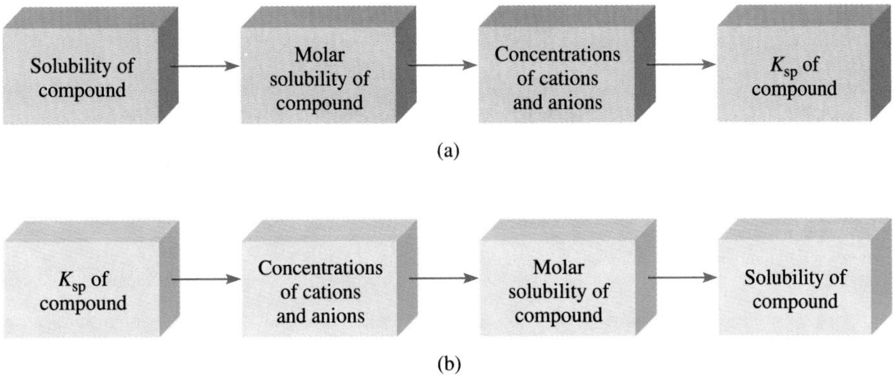

FIGURE 17.6

Sequence of steps (a) for calculating K_{sp} from solubility data and (b) for calculating solubility from K_{sp} data.

$$Q = [Ag^+]_0[Cl^-]_0$$

The subscript 0 reminds us that these are initial concentrations and do not necessarily correspond to those at equilibrium. The possible relationships between Q and K_{sp} are

$Q < K_{sp}$ $[Ag^+]_0[Cl^-]_0 < 1.6 \times 10^{-10}$	Unsaturated solution
$Q = K_{sp}$ $[Ag^+][Cl^-] = 1.6 \times 10^{-10}$	Saturated solution
$Q > K_{sp}$ $[Ag^+]_0[Cl^-]_0 > 1.6 \times 10^{-10}$	Supersaturated solution; AgCl will precipitate out until the product of the ionic concentrations is equal to 1.6×10^{-10}

MOLAR SOLUBILITY AND SOLUBILITY

The value of K_{sp} indicates the solubility of an ionic compound—the smaller the value, the less soluble the compound in water. However, in using K_{sp} values to compare solubilities, you should choose compounds that have similar formulas, such as AgCl and ZnS, or CaF_2 and $Fe(OH)_2$. There are two other quantities that express a substance's solubility: ***molar solubility,*** which is *the number of moles of solute in 1 L of a saturated solution (moles per liter), and* ***solubility,*** which is *the number of grams of solute in 1 L of a saturated solution (grams per liter).* Note that all these expressions refer to the concentration of saturated solutions at some given temperature (usually 25°C). Figure 17.6 shows the relationships among solubility, molar solubility, and K_{sp}.

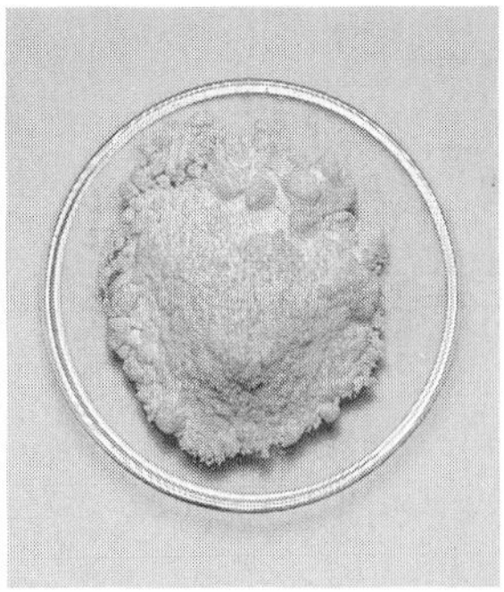

Silver sulfate.

EXAMPLE 17.5

Calculating K_{sp} from Molar Solubility

The molar solubility of silver sulfate is 1.5×10^{-2} mol/L. Calculate the solubility product of the salt.

Answer: First we write the solubility equilibrium equation:

$$Ag_2SO_4(s) \rightleftharpoons 2Ag^+(aq) + SO_4^{2-}(aq)$$

From the stoichiometry we see that 1 mole of Ag_2SO_4 produces 2 moles of Ag^+

and 1 mole of SO_4^{2-} in solution. Therefore, when 1.5×10^{-2} mol Ag_2SO_4 is dissolved in 1 L of solution, the concentrations are

$$[Ag^+] = 2(1.5 \times 10^{-2}\,M) = 3.0 \times 10^{-2}\,M$$
$$[SO_4^{2-}] = 1.5 \times 10^{-2}\,M$$

Now we can calculate the solubility product:

$$\begin{aligned} K_{sp} &= [Ag^+]^2[SO_4^{2-}] \\ &= (3.0 \times 10^{-2})^2(1.5 \times 10^{-2}) \\ &= 1.4 \times 10^{-5} \end{aligned}$$

PRACTICE EXERCISE

The molar solubility of barium fluoride (BaF_2) is 7.5×10^{-3} mol/L. What is the solubility product of the compound?

EXAMPLE 17.6

Calculating Molar Solubility from K_{sp}

Copper(I) iodide.

Using the data in Table 17.3, calculate the molar solubility of copper(I) iodide; that is, calculate the number of moles of CuI dissolved in 1 L of solution.

Answer: To convert K_{sp} to the concentrations of anions and cations, we use the same procedure outlined in Section 15.3 for calculating equilibrium concentrations from the equilibrium constant.

Step 1: Let s be the molar solubility (in moles per liter) of CuI. Since one unit of CuI yields one Cu^+ ion and one I^- ion, at equilibrium both $[Cu^+]$ and $[I^-]$ are equal to s. We summarize the changes in concentrations as follows:

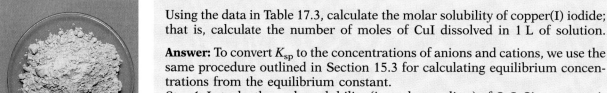

	$CuI(s) \rightleftharpoons$	$Cu^+(aq)$	+ $I^-(aq)$
Initial (M):		0.00	0.00
Change (M):		$+s$	$+s$
Equilibrium (M):		s	s

Step 2:

$$K_{sp} = [Cu^+][I^-]$$
$$5.1 \times 10^{-12} = (s)(s)$$
$$s = (5.1 \times 10^{-12})^{1/2} = 2.3 \times 10^{-6}\,M$$

Step 3: Therefore, at equilibrium

$$[Cu^+] = 2.3 \times 10^{-6}\,M$$
$$[I^-] = 2.3 \times 10^{-6}\,M$$

Because 1 mole of CuI yields 1 mole of Cu^+ and 1 mole of I^-, the concentration of Cu^+ or I^- ions is equal to that of CuI dissociated. Thus the molar solubility of CuI also is $2.3 \times 10^{-6}\,M$.

PRACTICE EXERCISE

Calculate the molar solubility of lead carbonate ($PbCO_3$).

EXAMPLE 17.7

Calculating Solubility from K_{sp}

Using the data in Table 17.3, calculate the solubility of copper hydroxide, $Cu(OH)_2$, in grams per liter.

Copper(II) hydroxide.

Answer:

Step 1: Let s be the molar solubility of $Cu(OH)_2$. Since one unit of $Cu(OH)_2$ yields one Cu^{2+} ion and two OH^- ions, at equilibrium $[Cu^{2+}]$ is s and $[OH^-]$ is $2s$. We summarize the changes in concentrations as follows:

	$Cu(OH)_2(s) \rightleftharpoons$	$Cu^{2+}(aq)$	$+\ 2OH^-(aq)$
Initial (M):		0.00	0.00
Change (M)		$+s$	$+2s$
Equilibrium (M):		s	$2s$

Step 2:

$$K_{sp} = [Cu^{2+}][OH^-]^2$$

$$2.2 \times 10^{-20} = (s)(2s)^2$$

$$s^3 = \frac{2.2 \times 10^{-20}}{4} = 5.5 \times 10^{-21}$$

Solving for s, we get

$$s = 1.8 \times 10^{-7}\ M$$

Knowing that the molar mass of $Cu(OH)_2$ is 97.57 g/mol and knowing its molar solubility, we can calculate the solubility in grams per liter as follows:

$$\text{solubility of } Cu(OH)_2 = \frac{1.8 \times 10^{-7} \text{ mol } Cu(OH)_2}{1 \text{ L soln}} \times \frac{97.57 \text{ g } Cu(OH)_2}{1 \text{ mol } Cu(OH)_2}$$

$$= 1.8 \times 10^{-5} \text{ g/L}$$

PRACTICE EXERCISE

Calculate the solubility of silver chloride (AgCl) in grams per liter.

As Examples 17.5–17.7 show, solubility and solubility product are related. If we know one, we can calculate the other, but each quantity provides different information. Table 17.4 shows the relationship between molar solubility and solubility product for a number of ionic compounds.

When carrying out solubility and/or solubility product calculations, keep in mind the following important points:

- The solubility is the quantity of a substance that dissolves in a certain quantity of water. In solubility equilibria calculations, it is usually expressed as *grams*

TABLE 17.4
Relationship between K_{sp} and Molar Solubility (s)

Compound	K_{sp} expression	Equilibrium concentration (M) Cation	Anion	Relation between K_{sp} and s
AgCl	$[Ag^+][Cl^-]$	s	s	$K_{sp} = s^2;\ s = (K_{sp})^{1/2}$
$BaSO_4$	$[Ba^{2+}][SO_4^{2-}]$	s	s	$K_{sp} = s^2;\ s = (K_{sp})^{1/2}$
Ag_2CO_3	$[Ag^+]^2[CO_3^{2-}]$	$2s$	s	$K_{sp} = 4s^3;\ s = \left(\frac{K_{sp}}{4}\right)^{1/3}$
PbF_2	$[Pb^{2+}][F^-]^2$	s	$2s$	$K_{sp} = 4s^3;\ s = \left(\frac{K_{sp}}{4}\right)^{1/3}$
$Al(OH)_3$	$[Al^{3+}][OH^-]^3$	s	$3s$	$K_{sp} = 27s^4;\ s = \left(\frac{K_{sp}}{27}\right)^{1/4}$
$Ca_3(PO_4)_2$	$[Ca^{2+}]^3[PO_4^{3-}]^2$	$3s$	$2s$	$K_{sp} = 108s^5;\ s = \left(\frac{K_{sp}}{108}\right)^{1/5}$

of solute per liter of solution. Molar solubility is the number of *moles* of solute per liter of solution.

- The solubility product is an equilibrium constant.
- Molar solubility, solubility, and solubility product all refer to a *saturated solution.*

PREDICTING PRECIPITATION REACTIONS

From a knowledge of the solubility rules (see Section 4.2) and the solubility products listed in Table 17.3, we can predict whether a precipitate will form when we mix two solutions.

EXAMPLE 17.8
Predicting a Precipitation Reaction

Exactly 200 mL of 0.0040 M $BaCl_2$ are added to exactly 600 mL of 0.0080 M K_2SO_4. Will a precipitate form?

Answer: According to solubility rule 3 on page 91, the only precipitate that might form is $BaSO_4$:

$$Ba^{2+}(aq) + SO_4^{2-}(aq) \longrightarrow BaSO_4(s)$$

The number of moles of Ba^{2+} present in the original 200 mL of solution is

$$200 \text{ mL} \times \frac{0.0040 \text{ mol } Ba^{2+}}{1 \text{ L soln}} \times \frac{1 \text{ L}}{1000 \text{ mL}} = 8.0 \times 10^{-4} \text{ mol } Ba^{2+}$$

The total volume after combining the two solutions is 800 mL. The concentration of Ba^{2+} in the 800 mL volume is

$$[Ba^{2+}] = \frac{8.0 \times 10^{-4}\ \text{mol}}{800\ \text{mL}} \times \frac{1000\ \text{mL}}{1\ \text{L soln}}$$

$$= 1.0 \times 10^{-3}\ M$$

The number of moles of SO_4^{2-} in the original 600 mL solution is

$$600\ \text{mL} \times \frac{0.0080\ \text{mol}\ SO_4^{2-}}{1\ \text{L soln}} \times \frac{1\ \text{L}}{1000\ \text{mL}} = 4.8 \times 10^{-3}\ \text{mol}\ SO_4^{2-}$$

The concentration of SO_4^{2-} in the 800 mL of the combined solution is

$$[SO_4^{2-}] = \frac{4.8 \times 10^{-3}\ \text{mol}}{800\ \text{mL}} \times \frac{1000\ \text{mL}}{1\ \text{L soln}}$$

$$= 6.0 \times 10^{-3}\ M$$

Now we must compare Q and K_{sp}. From Table 17.3, the K_{sp} for $BaSO_4$ is 1.1×10^{-10}. As for Q,

$$Q = [Ba^{2+}]_0[SO_4^{2-}]_0$$

$$= (1.0 \times 10^{-3})(6.0 \times 10^{-3})$$

$$= 6.0 \times 10^{-6}$$

Therefore,

$$Q > K_{sp}$$

The solution is supersaturated because the value of Q indicates that the concentrations of the ions are too large. Thus some of the $BaSO_4$ will precipitate out of solution until

$$[Ba^{2+}][SO_4^{2-}] = 1.1 \times 10^{-10}$$

PRACTICE EXERCISE

If 2.00 mL of 0.200 M NaOH are added to 1.00 L of 0.100 M $CaCl_2$, will precipitation occur?

17.6 THE COMMON ION EFFECT AND SOLUBILITY

As we have noted, the solubility product is an equilibrium constant; precipitation of an ionic compound from solution occurs whenever the ion product exceeds K_{sp} for that substance. In a saturated solution of AgCl, for example, the ion product $[Ag^+][Cl^-]$ is, of course, equal to K_{sp}. Furthermore, simple stoichiometry tells us that $[Ag^+] = [Cl^-]$. But this equality does not hold in all situations.

Suppose we study a solution containing two dissolved substances that share a common ion, say, AgCl and $AgNO_3$. In addition to the dissociation of AgCl, the following process also contributes to the total concentration of the common silver ions in solution:

$$AgNO_3(s) \xrightarrow{H_2O} Ag^+(aq) + NO_3^-(aq)$$

If $AgNO_3$ is added to a saturated AgCl solution, the increase in $[Ag^+]$ will make the ion product greater than the solubility product:

$$Q = [Ag^+]_0[Cl^-]_0 > K_{sp}$$

To reestablish equilibrium, some AgCl will precipitate out of the solution, as Le Chatelier's principle would predict, until the ion product is once again equal to K_{sp}. The effect of adding a common ion, then, is a *decrease* in the solubility of the salt (AgCl) in solution. Note that in this case $[Ag^+]$ is no longer equal to $[Cl^-]$ at equilibrium; rather, $[Ag^+] > [Cl^-]$.

EXAMPLE 17.9

Effect of a Common Ion on Solubility

Calculate the solubility of silver chloride (in grams per liter) in a $6.5 \times 10^{-3}\ M$ silver nitrate solution.

Answer:

Step 1: Since $AgNO_3$ is a soluble strong electrolyte, it dissociates completely:

$$\begin{array}{lccc} AgNO_3(s) \xrightarrow{H_2O} & Ag^+(aq) & + & NO_3^-(aq) \\ & 6.5 \times 10^{-3}\ M & & 6.5 \times 10^{-3}\ M \end{array}$$

Let s be the molar solubility of AgCl in $AgNO_3$ solution. We summarize the changes in concentrations as follows:

	$AgCl(s) \rightleftharpoons$	$Ag^+(aq)$	$+\ Cl^-(aq)$
Initial (M):		6.5×10^{-3}	0.0
Change (M)		$+s$	$+s$
Equilibrium (M):		$(6.5 \times 10^{-3} + s)$	s

Step 2:

$$K_{sp} = [Ag^+][Cl^-]$$

$$1.6 \times 10^{-10} = (6.5 \times 10^{-3} + s)(s)$$

Since AgCl is quite insoluble and the presence of Ag^+ ions from $AgNO_3$ further lowers the solubility of AgCl, s must be very small compared with 6.5×10^{-3}. Therefore, applying the approximation $6.5 \times 10^{-3} + s \simeq 6.5 \times 10^{-3}$, we obtain

$$1.6 \times 10^{-10} = 6.5 \times 10^{-3}s$$

$$s = 2.5 \times 10^{-8}\ M$$

Step 3: At equilibrium

$$[Ag^+] = (6.5 \times 10^{-3} + 2.5 \times 10^{-8})\ M \simeq 6.5 \times 10^{-3}\ M$$

$$[Cl^-] = 2.5 \times 10^{-8}\ M$$

and so our approximation $6.5 \times 10^{-3} + 2.5 \times 10^{-8} \simeq 6.5 \times 10^{-3}$ was justified in step 2. Since all the Cl^- ions must come from AgCl, the amount of AgCl dissolved in $AgNO_3$ solution also is $2.5 \times 10^{-8}\ M$. Then, knowing the molar mass of AgCl (143.4 g), we can calculate the solubility of the AgCl as follows:

$$\text{solubility of AgCl in AgNO}_3\text{ solution} = \frac{2.5 \times 10^{-8}\text{ mol AgCl}}{1\text{ L soln}} \times \frac{143.3\text{ g AgCl}}{1\text{ mol AgCl}}$$

$$= 3.6 \times 10^{-6}\text{ g/L}$$

The solubility of AgCl in pure water is 1.9×10^{-3} g/L (see the Practice Exercise in Example 17.7). Therefore, the answer is reasonable.

PRACTICE EXERCISE

Calculate the solubility in grams per liter of AgBr in (a) pure water and (b) 0.0010 *M* NaBr.

17.7 COMPLEX ION EQUILIBRIA AND SOLUBILITY

Lewis acid-base reactions in which a metal cation (electron-pair acceptor) combines with a Lewis base (electron-pair donor) result in the formation of complex ions:

$$\underset{\text{acid}}{Ag^+(aq)} + \underset{\text{base}}{2NH_3(aq)} \rightleftharpoons Ag(NH_3)_2^+(aq)$$

Thus, we can define a ***complex ion*** as *an ion containing a central metal cation bonded to one or more molecules or ions.* Complex ions are crucial to many chemical and biological processes. Here we will consider the effect of complex ion formation on solubility. In Chapter 18 we will discuss the chemistry of complex ions in more detail.

Transition metals have a particular tendency to form complex ions. For example, a solution of cobalt(II) chloride is pink because of the presence of the $Co(H_2O)_6^{2+}$ ions (Figure 17.7). When HCl is added, the solution turns blue as a result of the formation of the complex ion $CoCl_4^{2-}$:

$$Co^{2+}(aq) + 4Cl^-(aq) \rightleftharpoons CoCl_4^{2-}(aq)$$

FIGURE 17.7

(Left) An aqueous cobalt(II) chloride solution. The pink color is due to the $Co(H_2O)_6^{2+}$ ions. (Right) After the addition of HCl solution, the solution turned blue because of the formation of the complex $CoCl_4^{2-}$ ions.

FIGURE 17.8

(Left) A beaker containing an aqueous solution of copper(II) sulfate. (Center) After the addition of a few drops of a concentrated aqueous ammonia solution, a light-blue precipitate of $Cu(OH)_2$ is formed. (Right) When an excess of concentrated aqueous ammonia solution is added, the $Cu(OH)_2$ precipitate dissolves to form the dark-blue complex ion $Cu(NH_3)_4^{2+}$.

Copper(II) sulfate ($CuSO_4$) dissolves in water to produce a blue solution. The hydrated copper(II) ions are responsible for this color; many other sulfates (Na_2SO_4, for example) are colorless. Adding a *few drops* of concentrated ammonia solution to a $CuSO_4$ solution causes a light-blue precipitate, copper(II) hydroxide, to form:

$$Cu^{2+}(aq) + 2OH^-(aq) \longrightarrow Cu(OH)_2(s)$$

where the OH^- ions are supplied by the ammonia solution. If an *excess* of NH_3 is then added, the blue precipitate redissolves to produce a beautiful dark-blue solution, this time due to the formation of the complex ion $Cu(NH_3)_4^{2+}$ (Figure 17.8):

$$Cu(OH)_2(s) + 4NH_3(aq) \rightleftharpoons Cu(NH_3)_4^{2+}(aq) + 2OH^-(aq)$$

Thus the formation of the complex ion $Cu(NH_3)_4^{2+}$ increases the solubility of $Cu(OH)_2$.

A measure of the tendency of a metal ion to form a particular complex ion is given by the ***formation constant K_f*** (also called the *stability constant*), which is *the equilibrium constant for complex ion formation.* The larger K_f is, the more stable the complex ion is. Table 17.5 lists the formation constants of a number of complex ions.

The formation of the $Cu(NH_3)_4^{2+}$ ion can be expressed as

$$Cu^{2+}(aq) + 4NH_3(aq) \rightleftharpoons Cu(NH_3)_4^{2+}(aq)$$

TABLE 17.5

Formation Constants of Selected Complex Ions in Water at 25°C

Complex ion	Equilibrium expression	Formation constant (K_f)
$Ag(NH_3)_2^+$	$Ag^+ + 2NH_3 \rightleftharpoons Ag(NH_3)_2^+$	1.5×10^7
$Ag(CN)_2^-$	$Ag^+ + 2CN^- \rightleftharpoons Ag(CN)_2^-$	1.0×10^{21}
$Cu(CN)_4^{2-}$	$Cu^{2+} + 4CN^- \rightleftharpoons Cu(CN)_4^{2-}$	1.0×10^{25}
$Cu(NH_3)_4^{2+}$	$Cu^{2+} + 4NH_3 \rightleftharpoons Cu(NH_3)_4^{2+}$	5.0×10^{13}
$Cd(CN)_4^{2-}$	$Cd^{2+} + 4CN^- \rightleftharpoons Cd(CN)_4^{2-}$	7.1×10^{16}
CdI_4^{2-}	$Cd^{2+} + 4I^- \rightleftharpoons CdI_4^{2-}$	2.0×10^6
$HgCl_4^{2-}$	$Hg^{2+} + 4Cl^- \rightleftharpoons HgCl_4^{2-}$	1.7×10^{16}
HgI_4^{2-}	$Hg^{2+} + 4I^- \rightleftharpoons HgI_4^{2-}$	2.0×10^{30}
$Hg(CN)_4^{2-}$	$Hg^{2+} + 4CN^- \rightleftharpoons Hg(CN)_4^{2-}$	2.5×10^{41}
$Co(NH_3)_6^{3+}$	$Co^{3+} + 6NH_3 \rightleftharpoons Co(NH_3)_6^{3+}$	5.0×10^{31}
$Zn(NH_3)_4^{2+}$	$Zn^{2+} + 4NH_3 \rightleftharpoons Zn(NH_3)_4^{2+}$	2.9×10^9

for which the formation constant is

$$K_f = \frac{[Cu(NH_3)_4^{2+}]}{[Cu^{2+}][NH_3]^4}$$
$$= 5.0 \times 10^{13}$$

The very large value of K_f in this case indicates the great stability of the complex ion in solution and accounts for the very low concentration of copper(II) ions at equilibrium.

EXAMPLE 17.10

Complex Ion Formation

A 0.20-mole quantity of $CuSO_4$ is added to a liter of 1.20 *M* NH_3 solution. What is the concentration of Cu^{2+} ions at equilibrium?

Answer: The addition of $CuSO_4$ to the NH_3 solution results in the reaction

$$Cu^{2+}(aq) + 4NH_3(aq) \rightleftharpoons Cu(NH_3)_4^{2+}(aq)$$

Since K_f is very large (5.0×10^{13}), the reaction lies mostly to the right. As a good approximation, we can assume that essentially all the dissolved Cu^{2+} ions end up as $Cu(NH_3)_4^{2+}$ ions. Thus the amount of NH_3 consumed in forming the complex ions is 4×0.20 mol, or 0.80 mol. (Note that 0.20 mol Cu^{2+} is initially present in solution and four NH_3 molecules are needed to "complex" one Cu^{2+} ion.) The concentration of NH_3 at equilibrium therefore is (1.20 − 0.80) *M*, or 0.40 *M*, and that of $Cu(NH_3)_4^{2+}$ is 0.20 *M*, the same as the initial concentration of Cu^{2+}. Since $Cu(NH_3)_4^{2+}$ does dissociate to a slight extent, we call the concentration of Cu^{2+} ions at equilibrium x and write

$$K_f = \frac{[Cu(NH_3)_4^{2+}]}{[Cu^{2+}][NH_3]^4} = 5.0 \times 10^{13}$$

$$\frac{0.20}{x(0.40)^4} = 5.0 \times 10^{13}$$

Solving for x, we obtain

$$x = 1.6 \times 10^{-13}\ M = [Cu^{2+}]$$

The small value of $[Cu^{2+}]$ at equilibrium, compared with 0.20 *M*, certainly justifies our approximation.

PRACTICE EXERCISE

If 2.50 g of $CuSO_4$ are dissolved in 9.0×10^2 mL of 0.30 *M* NH_3, what are the concentrations of Cu^{2+}, $Cu(NH_3)_4^{2+}$, and NH_3 at equilibrium?

Finally, we note that there is a class of hydroxides, called *amphoteric hydroxides,* which can react with both acids and bases. Examples are $Al(OH)_3$, $Pb(OH)_2$, $Cr(OH)_3$, $Zn(OH)_2$, and $Cd(OH)_2$. For example, aluminum hydroxide reacts with acids and bases as follows:

$$Al(OH)_3(aq) + 3H^+(aq) \longrightarrow Al^{3+}(aq) + 3H_2O(l)$$
$$Al(OH)_3(aq) + OH^-(aq) \rightleftharpoons Al(OH)_4^-(aq)$$

FIGURE 17.9

(Left) The formation of $Al(OH)_3$ precipitate when a NaOH solution is added to a solution of $Al(NO_3)_3$. (Right) The addition of still more NaOH solution causes the $Al(OH)_3$ precipitate to dissolve due to complex ion formation $[Al(OH)_4^-]$.

The increase in solubility of $Al(OH)_3$ in a basic medium is the result of the formation of the complex ion $[Al(OH)_4^-]$ in which $Al(OH)_3$ acts as the Lewis acid and OH^- acts as the Lewis base (Figure 17.9). Other amphoteric hydroxides behave in a similar manner.

17.8 APPLICATION OF THE SOLUBILITY PRODUCT PRINCIPLE TO QUALITATIVE ANALYSIS

In Section 4.6, we discussed the principle of gravimetric analysis, by which we measure the amount of an ion in an unknown sample. Here we will briefly discuss ***qualitative analysis,*** *the determination of the types of ions present in a solution.* We will focus on the cations.

Twenty common cations can be analyzed readily in aqueous solution. These cations can be divided into five groups according to the solubility products of their insoluble salts (Table 17.6). Since an unknown solution may contain any one or up to all 20 ions, analysis must be carried out systematically from group 1 through group 5. Let us consider the general procedure for separating these ions by adding precipitating reagents to an unknown solution.

- *Group 1 cations.* When dilute HCl is added to the unknown solution, only the Ag^+, Hg_2^{2+}, and Pb^{2+} ions precipitate as insoluble chlorides. The other ions, whose chlorides are soluble, remain in solution.

- *Group 2 cations.* After the chloride precipitates have been removed by filtration, hydrogen sulfide is reacted with the unknown acidic solution. Under this condition, the concentration of the S^{2-} ion in solution is negligible. Therefore the precipitation of metal sulfides is best represented as

$$M^{2+}(aq) + H_2S(aq) \rightleftharpoons MS(s) + 2H^+(aq)$$

  Adding acid to the solution shifts this equilibrium to the left so that only the least soluble metal sulfides, that is, those with the smallest K_{sp} values, will precipitate out of solution. These are Bi_2S_3, CdS, CuS, and SnS.

- *Group 3 cations.* At this stage, sodium hydroxide is added to the solution to make it basic. In a basic solution, the above equilibrium shifts to the right. Therefore, the more soluble sulfides (CoS, FeS, MnS, NiS, ZnS) now precipitate out of solution. Note that the Al^{3+} and Cr^{3+} ions actually precipitate as the

TABLE 17.6

Separation of Cations into Groups According to Their Precipitation Reactions with Various Reagents

Group	Cation	Precipitating reagents	Insoluble compound	K_{sp}
1	Ag^+	HCl	AgCl	1.6×10^{-10}
	Hg_2^{2+}	↓	Hg_2Cl_2	3.5×10^{-18}
	Pb^{2+}	↓	$PbCl_2$	2.4×10^{-4}
2	Bi^{3+}	H_2S	Bi_2S_3	1.6×10^{-72}
	Cd^{2+}	in acidic	CdS	8.0×10^{-28}
	Cu^{2+}	solutions	CuS	6.0×10^{-37}
	Sn^{2+}	↓	SnS	1.0×10^{-26}
3	Al^{3+}	H_2S	$Al(OH)_3$	1.8×10^{-33}
	Co^{2+}	in basic	CoS	4.0×10^{-21}
	Cr^{3+}	solutions	$Cr(OH)_3$	3.0×10^{-29}
	Fe^{2+}	↓	FeS	6.0×10^{-19}
	Mn^{2+}	↓	MnS	3.0×10^{-14}
	Ni^{2+}	↓	NiS	1.4×10^{-24}
	Zn^{2+}	↓	ZnS	3.0×10^{-23}
4	Ba^{2+}	Na_2CO_3	$BaCO_3$	8.1×10^{-9}
	Ca^{2+}	↓	$CaCO_3$	8.7×10^{-9}
	Sr^{2+}	↓	$SrCO_3$	1.6×10^{-9}
5	K^+	No precipitating	None	
	Na^+	reagent	None	
	NH_4^+		None	

hydroxides $Al(OH)_3$ and $Cr(OH)_3$, rather than as the sulfides, because the hydroxides are less soluble. The solution is then filtered to remove the insoluble sulfides and hydroxides.

- *Group 4 cations.* After all the group 1, 2, and 3 cations have been removed from solution, sodium carbonate is added to the basic solution to precipitate Ba^{2+}, Ca^{2+}, and Sr^{2+} ions as $BaCO_3$, $CaCO_3$, and $SrCO_3$. These precipitates too are removed from solution by filtration.

- *Group 5 cations.* At this stage, the only cations possibly remaining in solution are Na^+, K^+, and NH_4^+. The presence of NH_4^+ can be determined by adding sodium hydroxide:

$$NaOH(aq) + NH_4^+(aq) \longrightarrow Na^+(aq) + H_2O(l) + NH_3(g)$$

The ammonia gas is detected either by noting its characteristic odor or by observing a piece of wet red litmus paper turning blue when placed above (not in contact with) the solution. To confirm the presence of Na^+ and K^+ ions, we usually use a flame test, as follows: A piece of platinum wire (chosen because platinum is inert) is moistened with the solution and is then held over a Bunsen burner flame. Each type of metal ion gives a characteristic color when heated in this manner. For example, the color emitted by Na^+ ions is yellow, that of K^+ ions is violet, and that of Cu^{2+} ions is green (Figure 17.10).

Figure 17.11 summarizes this scheme for separating metal ions.

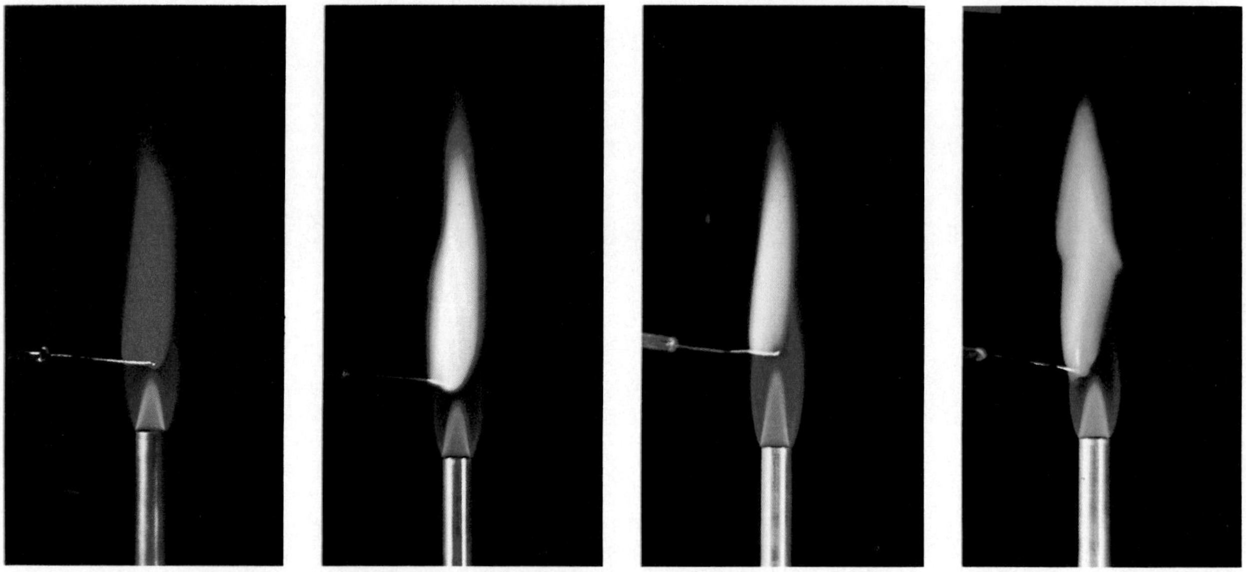

FIGURE 17.10

Left to right: Flame colors of lithium, sodium, potassium, and copper.

FIGURE 17.11

A flow chart for the separation of cations in qualitative analysis.

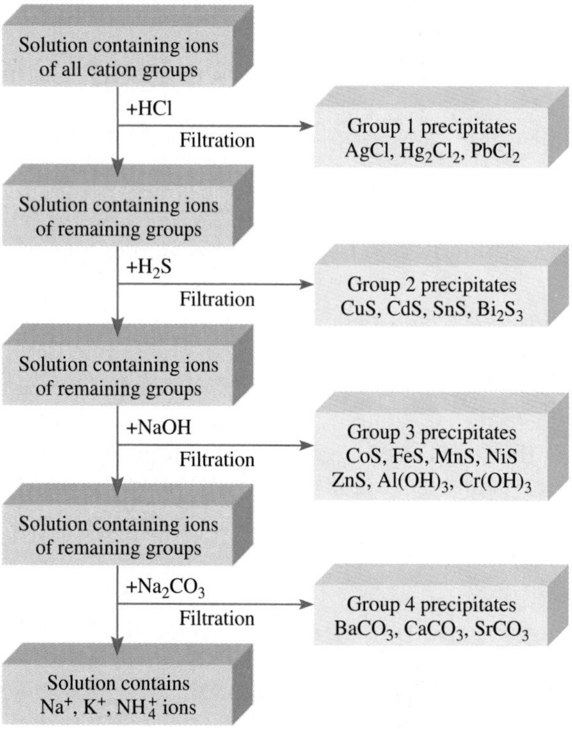

Two points regarding qualitative analysis must be mentioned. First, the separation of the cations into groups is made as selective as possible; that is, the anions that are added as reagents must be such that they will precipitate the fewest types of cations. For example, all the cations in group 1 form insoluble sulfides. Thus, if H_2S were reacted with the solution at the start, as many as seven different sulfides might precipitate out of solution (group 1 *and* group 2 sulfides), an undesirable outcome. Second, the separation of cations at each step must be carried out as completely as possible. For example, if we do not add enough HCl to the unknown solution to remove all the group 1 cations, they will precipitate with the group 2 cations as insoluble sulfides; this too would interfere with further chemical analysis and lead us to draw erroneous conclusions.

SUMMARY

Equilibria involving weak acids or weak bases in aqueous solution are homogeneous. Solubility equilibria are examples of heterogeneous equilibria.

A buffer solution is a combination of a weak acid and its weak conjugate base; the solution reacts with small amounts of added acid or base in such a way that the pH of the solution remains nearly constant. Buffer systems play a vital role in maintaining the pH of body fluids.

The pH at the equivalence point of an acid-base titration depends on the hydrolysis of the salt formed in the neutralization reaction. For strong acid–strong base titrations, the pH at the equivalence point is 7; for weak acid–strong base titrations, the pH at the equivalence point is greater than 7; for strong acid–weak base titrations, the pH at the equivalence point is less than 7. Acid-base indicators are weak organic acids or bases that change color at the equivalence point in an acid-base neutralization reaction.

The solubility product K_{sp} expresses the equilibrium between a solid and its ions in solution. Solubility can be found from K_{sp} and vice versa. The presence of a common ion decreases the solubility of a salt.

Complex ions are formed in solution by the combination of a metal cation with a Lewis base. The formation constant K_f measures the tendency toward the formation of a specific complex ion. Complex ion formation can increase the solubility of an insoluble substance.

Qualitative analysis is the identification of cations and anions in solution. It is based largely on the principles of solubility equilibria.

KEY WORDS

Buffer solution, p. 482
Complex ion, p. 501
Formation constant (K_f), p. 502
Molar solubility, p. 495
Qualitative analysis, p. 504
Solubility, p. 495
Solubility product (K_{sp}), p. 494

QUESTIONS AND PROBLEMS

BUFFER SOLUTIONS

Review Questions

17.1 Define buffer solution.

17.2 Define pK_a for a weak acid and explain the relationship between the value of the pK_a and the strength of the acid. Do the same for pK_b and a weak base.

17.3 The pK_as of two monoprotic acids HA and HB are 5.9 and 8.1, respectively. Which of the two is the stronger acid?

17.4 The pK_bs for the bases X^-, Y^-, and Z^- are 2.72, 8.66, and 4.57, respectively. Arrange the following acids in order of increasing strength: HX, HY, HZ.

Problems

17.5 Specify which of the following systems can be

classified as a buffer system: (a) KCl/HCl, (b) NH_3/NH_4NO_3, (c) Na_2HPO_4/NaH_2PO_4.

17.6 Specify which of the following systems can be classified as a buffer system: (a) KNO_2/HNO_2, (b) $KHSO_4/H_2SO_4$, (c) HCOOK/HCOOH.

17.7 The pH of a bicarbonate–carbonic acid buffer is 8.00. Calculate the ratio of the concentration of carbonic acid to that of the bicarbonate ion.

17.8 Calculate the pH of the following two buffer solutions: (a) 2.0 *M* CH_3COONa/2.0 *M* CH_3COOH, (b) 0.20 *M* CH_3COONa/0.20 *M* CH_3COOH. Which is the more effective buffer? Why?

17.9 Calculate the pH of the buffer system 0.15 *M* NH_3/0.35 *M* NH_4Cl.

17.10 What is the pH of the buffer 0.10 *M* Na_2HPO_4/0.15 *M* KH_2PO_4?

17.11 The pH of a sodium acetate–acetic acid buffer is 4.50. Calculate the ratio $[CH_3COO^-]/[CH_3COOH]$.

17.12 The pH of blood plasma is 7.40. Assuming the principal buffer system is HCO_3^-/H_2CO_3, calculate the ratio $[HCO_3^-]/[H_2CO_3]$. Is this buffer more effective against an added acid or an added base?

17.13 Calculate the pH of a buffer solution prepared by adding 20.5 g of CH_3COOH and 17.8 g of CH_3COONa to enough water to make 5.00×10^2 mL of solution.

17.14 Calculate the pH of 1.00 L of the buffer 1.00 *M* CH_3COONa/1.00 *M* CH_3COOH before and after the addition of (a) 0.080 mol NaOH and (b) 0.12 mol HCl. (Assume that there is no change in volume.)

17.15 A diprotic acid, H_2A, has the following ionization constants: $K_{a_1} = 1.1 \times 10^{-3}$ and $K_{a_2} = 2.5 \times 10^{-6}$. In order to make up a buffer solution of pH 5.80, which combination would you choose: NaHA/H_2A or Na_2A/NaHA?

17.16 A student wishes to prepare a buffer solution at pH = 8.60. Which of the following weak acids should she choose and why? HA ($K_a = 2.7 \times 10^{-3}$), HB ($K_a = 4.4 \times 10^{-6}$), or HC ($K_a = 2.6 \times 10^{-9}$).

ACID-BASE TITRATIONS

Problems

17.17 A 0.2688-g sample of a monoprotic acid neutralizes 16.4 mL of 0.08133 *M* KOH solution. Calculate the molar mass of the acid.

17.18 A 5.00-g quantity of a diprotic acid is dissolved in water and made up to exactly 250 mL. Calculate the molar mass of the acid if 25.0 mL of this solution required 11.1 mL of 1.00 *M* KOH for neutralization. Assume that both protons of the acid are titrated.

17.19 Calculate the pH at the equivalence point for the following titrations: (a) 0.10 *M* HCl versus 0.10 *M* NH_3, (b) 0.10 *M* CH_3COOH versus 0.10 *M* NaOH.

17.20 A sample of 0.1276 g of an unknown monoprotic acid was dissolved in 25.0 mL of water and titrated with 0.0633 *M* NaOH solution. The volume of base required to reach the equivalence point was 18.4 mL. (a) Calculate the molar mass of the acid. (b) After 10.0 mL of base had been added to the titration, the pH was determined to be 5.87. What is the K_a of the unknown acid?

ACID-BASE INDICATORS

Review Questions

17.21 Explain how an acid-base indicator works in a titration.

17.22 What are the criteria for choosing an indicator for a particular acid-base titration?

17.23 The amount of indicator used in an acid-base titration must be small. Why?

17.24 A student carried out an acid-base titration by adding NaOH solution from a buret to an Erlenmeyer flask containing HCl solution and using phenolphthalein as indicator. At the equivalence point, he observed a faint reddish-pink color. However, after a few minutes, the solution gradually turned colorless. What do you suppose happened?

Problems

17.25 Referring to Table 17.2, specify which indicator or indicators you would use for the following titrations: (a) HCOOH versus NaOH, (b) HCl versus KOH, (c) HNO_3 versus NH_3.

17.26 The ionization constant K_a of an indicator HIn is 1.0×10^{-6}. The color of the nonionized form is red and that of the ionized form is yellow. What is the color of this indicator in a solution whose pH is 4.00? (*Hint:* The color of an indicator can be estimated by considering the ratio $[HIn]/[In^-]$. If the ratio is equal to or greater than 10, the color will be that of the nonionized form. If the ratio is equal to or smaller than 0.1, the color will be that of the ionized form.)

SOLUBILITY AND SOLUBILITY PRODUCT

Review Questions

17.27 Define solubility, molar solubility, and solubility product. Explain the difference between solubility and the solubility product of a slightly soluble substance such as $BaSO_4$.

17.28 Why do we usually not quote the K_{sp} values for soluble ionic compounds?

17.29 Write balanced equations and solubility product expressions for the solubility equilibria of the following compounds: (a) CuBr, (b) ZnC_2O_4, (c) Ag_2CrO_4, (d) Hg_2Cl_2, (e) $AuCl_3$, (f) $Mn_3(PO_4)_2$.

17.30 Write the solubility product expression for the ionic compound A_xB_y.

17.31 How can we predict whether a precipitate will form when two solutions are mixed?

17.32 Silver chloride has a larger K_{sp} than silver carbonate (see Table 17.3). Does this mean that the former also has a larger molar solubility than the latter?

Problems

17.33 Calculate the concentration of ions in the following saturated solutions:
(a) $[I^-]$ in AgI solution with $[Ag^+] = 9.1 \times 10^{-9}$ *M*
(b) $[Al^{3+}]$ in $Al(OH)_3$ with $[OH^-] = 2.9 \times 10^{-9}$ *M*

17.34 From the solubility data given, calculate the solubility products for the following compounds:
(a) SrF_2, 7.3×10^{-2} g/L
(b) Ag_3PO_4, 6.7×10^{-3} g/L

17.35 The molar solubility of $MnCO_3$ is 4.2×10^{-6} *M*. What is K_{sp} for this compound?

17.36 The solubility of an ionic compound MX (molar mass = 346 g) is 4.63×10^{-3} g/L. What is K_{sp} for the compound?

17.37 The solubility of an ionic compound M_2X_3 (molar mass = 288 g) is 3.6×10^{-17} g/L. What is K_{sp} for the compound?

17.38 Using data from Table 17.3, calculate the molar solubility of CaF_2.

17.39 What is the pH of a saturated zinc hydroxide solution?

17.40 The pH of a saturated solution of a metal hydroxide MOH is 9.68. Calculate the K_{sp} for the compound.

17.41 A sample of 20.0 mL of 0.10 *M* $Ba(NO_3)_2$ is added to 50.0 mL of 0.10 *M* Na_2CO_3. Will $BaCO_3$ precipitate?

17.42 A volume of 75 mL of 0.060 *M* NaF is mixed with 25 mL of 0.15 *M* $Sr(NO_3)_2$. Calculate the concentrations in the final solution of NO_3^-, Na^+, Sr^{2+}, and F^-. (K_{sp} for $SrF_2 = 2.0 \times 10^{-10}$.)

THE COMMON ION EFFECT

Review Questions

17.43 How does a common ion affect solubility? Use Le Chatelier's principle to explain the decrease in solubility of $CaCO_3$ in a Na_2CO_3 solution.

17.44 The molar solubility of AgCl in 6.5×10^{-3} *M* $AgNO_3$ is 2.5×10^{-8} *M*. In deriving K_{sp} from these data, which of the following assumptions are reasonable?
(a) K_{sp} is the same as solubility.
(b) K_{sp} of AgCl is the same in 6.5×10^{-3} *M* $AgNO_3$ as in pure water.
(c) Solubility of AgCl is independent of the concentration of $AgNO_3$.
(d) $[Ag^+]$ in solution does not change significantly upon the addition of AgCl to 6.5×10^{-3} *M* $AgNO_3$.
(e) $[Ag^+]$ in solution after the addition of AgCl to 6.5×10^{-3} *M* $AgNO_3$ is the same as it would be in pure water.

Problems

17.45 How many grams of $CaCO_3$ will dissolve in 3.0×10^2 mL of 0.050 *M* $Ca(NO_3)_2$?

17.46 The solubility product of $PbBr_2$ is 8.9×10^{-6}. Determine the molar solubility (a) in pure water, (b) in 0.20 *M* KBr solution, (c) in 0.20 *M* $Pb(NO_3)_2$ solution.

17.47 Calculate the molar solubility of AgCl in a solution made by dissolving 10.0 g of $CaCl_2$ in 1.00 L of solution.

17.48 Calculate the molar solubility of $BaSO_4$ (a) in water and (b) in a solution containing 1.0 *M* SO_4^{2-} ions.

COMPLEX IONS

Review Questions

17.49 Explain the formation of complexes in Table 17.5 in terms of Lewis acid-base theory.

17.50 Give an example to illustrate the general effect of complex ion formation on solubility.

Problems

17.51 Write the formation constant expressions for

these complex ions: (a) $Zn(OH)_4^{2-}$, (b) $Co(NH_3)_6^{3+}$, (c) HgI_4^{2-}.

17.52 Explain, with balanced ionic equations, why (a) CuI_2 dissolves in ammonia solution, (b) AgBr dissolves in NaCN solution, (c) Hg_2Cl_2 dissolves in KCl solution.

17.53 If 2.50 g of $CuSO_4$ are dissolved in 9.0×10^2 mL of 0.30 *M* NH_3, what are the concentrations of Cu^{2+}, $Cu(NH_3)_4^{2+}$, and NH_3 at equilibrium?

17.54 Calculate the concentrations of Cd^{2+}, $Cd(CN)_4^{2-}$, and CN^- at equilibrium when 0.50 g of $Cd(NO_3)_2$ dissolves in 5.0×10^2 mL of 0.50 *M* NaCN.

17.55 If NaOH is added to 0.010 *M* Al^{3+}, which will be the predominant species at equilibrium: $Al(OH)_3$ or $Al(OH)_4^-$? The pH of the solution is 14.00. [K_f for $Al(OH)_4^- = 2.0 \times 10^{33}$.]

17.56 Calculate the molar solubility of AgI in a 1.0 *M* NH_3 solution.

QUALITATIVE ANALYSIS

Review Questions

17.57 Outline the general principle of qualitative analysis.

17.58 Give two examples of metal ions in each group (1 through 5) in the qualitative analysis scheme.

Problems

17.59 In a group 1 analysis, a student obtained a precipitate containing both AgCl and $PbCl_2$. Suggest one reagent that would allow her to separate AgCl(*s*) from $PbCl_2(s)$.

17.60 In a group 1 analysis, a student adds hydrochloric acid to the unknown solution to make $[Cl^-] = 0.15$ *M*. Some $PbCl_2$ precipitates. Calculate the concentration of Pb^{2+} remaining in solution.

17.61 Both KCl and NH_4Cl are white solids. Suggest one reagent that would allow you to distinguish between these two compounds.

17.62 Describe a simple test that would allow you to distinguish between $AgNO_3(s)$ and $Cu(NO_3)_2(s)$.

MISCELLANEOUS PROBLEMS

17.63 A quantity of 0.560 g of KOH is added to 25.0 mL of 1.00 *M* HCl. Excess Na_2CO_3 is then added to the solution. What mass (in grams) of CO_2 is formed?

17.64 A volume of 25.0 mL of 0.100 *M* HCl is titrated against a 0.100 *M* NH_3 solution added to it from a buret. Calculate the pH values of the solution (a) after 10.0 mL of NH_3 solution have been added, (b) after 25.0 mL of NH_3 solution have been added, (c) after 35.0 mL of NH_3 solution have been added.

17.65 The buffer range is defined by the equation pH = $pK_a \pm 1$. Calculate the range of the ratio [conjugate base]/[acid] that corresponds to this equation.

17.66 The pK_a of the indicator methyl orange is 3.46. Over what pH range does this indicator change from 90% HIn to 90% In^-?

17.67 Sketch the titration curve of a weak acid versus a strong base such that shown in Figure 17.4(a). On your graph indicate the volume of base used at the equivalence point and also at the half-equivalence point, that is, the point at which half of the base has been added. Show how you can measure the pH of the solution at the half-equivalence point. Using Equation (17.3), explain how you can determine the pK_a of the acid by this procedure.

17.68 A 200-mL volume of NaOH solution was added to 400 mL of a 2.00 *M* HNO_2 solution. The pH of the mixed solution was 1.50 units greater than that of the original acid solution. Calculate the molarity of the NaOH solution.

17.69 The pK_a of butyric acid (HBut) is 4.7. Calculate K_b for the butyrate ion (But^-).

17.70 A solution is made by mixing exactly 500 mL of 0.167 *M* NaOH with exactly 500 mL 0.100 *M* CH_3COOH. Calculate the equilibrium concentrations of H^+, CH_3COOH, CH_3COO^-, OH^-, and Na^+.

17.71 $Cd(OH)_2$ is an insoluble compound. It dissolves in excess NaOH in solution. Write a balanced ionic equation for this reaction. What type of reaction is this?

17.72 Calculate the pH of the 0.20 *M* NH_3/0.20 *M* NH_4Cl buffer. What is the pH of the buffer after the addition of 10.0 mL of 0.10 *M* HCl to 65.0 mL of the buffer?

17.73 For which of the following reactions is the equilibrium constant called a solubility product?
(a) $Zn(OH)_2(s) + 2OH^-(aq) \rightleftharpoons Zn(OH)_4^{2-}(aq)$
(b) $3Ca^{2+}(aq) + 2PO_4^{3-}(aq) \rightleftharpoons Ca_3(PO_4)_2(s)$
(c) $CaCO_3(s) + 2H^+(aq) \rightleftharpoons Ca^{2+}(aq) + H_2O(l) + CO_2(g)$
(d) $PbI_2(s) \rightleftharpoons Pb^{2+}(aq) + 2I^-(aq)$

17.74 A student mixes 50.0 mL of 1.00 *M* $Ba(OH)_2$ with

86.4 mL of 0.494 *M* H_2SO_4. Calculate the mass of $BaSO_4$ formed and the pH of the mixed solution.

17.75 A 2.0-L kettle contains 116 g of calcium carbonate as boiler scale. How many times would the kettle have to be completely filled with distilled water to remove all of the deposit at 25°C?

17.76 Equal volumes of 0.12 *M* $AgNO_3$ and 0.14 *M* $ZnCl_2$ solution are mixed. Calculate the equilibrium concentrations of Ag^+, Cl^-, Zn^{2+}, and NO_3^-.

17.77 Calculate the solubility (in grams per liter) of Ag_2CO_3.

17.78 Find the approximate pH range suitable for the separation of Fe^{3+} and Zn^{2+} by precipitation of $Fe(OH)_3$ from a solution that is initially 0.010 *M* in Fe^{3+} and Zn^{2+}.

17.79 Which of the following ionic compounds will be more soluble in acid solution than in water? (a) $BaSO_4$, (b) $PbCl_2$, (c) $Fe(OH)_3$, (d) $CaCO_3$.

17.80 Which of the following substances will be more soluble in acid solution than in pure water? (a) CuI, (b) Ag_2SO_4, (c) $Zn(OH)_2$, (d) BaC_2O_4, (e) $Ca_3(PO_4)_2$.

17.81 What is the pH of a saturated solution of aluminum hydroxide?

17.82 The molar solubility of $Pb(IO_3)_2$ in a 0.10 *M* $NaIO_3$ solution is 2.4×10^{-11} mol/L. What is K_{sp} for $Pb(IO_3)_2$?

17.83 The solubility product of $Mg(OH)_2$ is 1.2×10^{-11}. What minimum OH^- concentration must be attained (for example, by adding NaOH) to make the Mg^{2+} concentration in a solution of $Mg(NO_3)_2$ less than 1.0×10^{-10} *M*?

17.84 Calculate whether or not a precipitate will form if 2.00 mL of 0.60 *M* NH_3 are added to 1.0 L of 1.0×10^{-3} *M* of $FeSO_4$.

17.85 Both Ag^+ and Zn^{2+} form complex ions with NH_3. Write balanced equations for the reactions. However, $Zn(OH)_2$ is soluble in 6 *M* NaOH, and AgOH is not. Explain.

17.86 When a KI solution was added to a solution of mercury(II) chloride, a precipitate [mercury(II) iodide] was formed. A student plotted the mass of the precipitate formed versus the volume of the KI solution added and obtained the following graph. Explain the appearance of the graph.

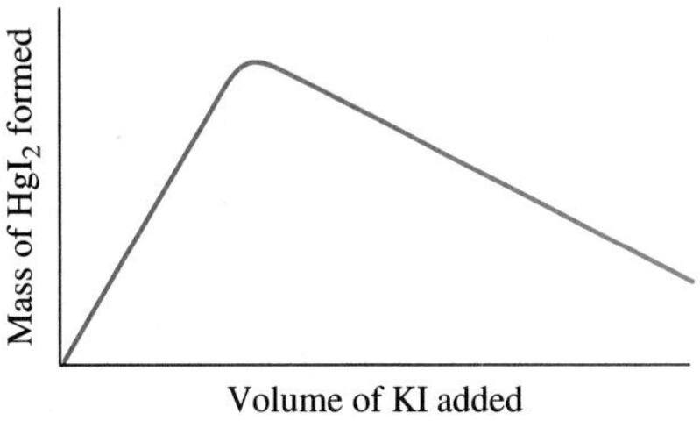

17.87 Barium is a toxic substance that can cause serious deterioration of the heart's function. In a barium enema procedure, a patient drinks an aqueous suspension of 20 g $BaSO_4$. If this substance were to equilibrate with the 5.0 L of the blood in the patient's body, how many grams of $BaSO_4$ will dissolve in the blood? For a good estimate, we may assume that the temperature is at 25°C. Why is $Ba(NO_3)_2$ not chosen for this procedure?

17.88 The pK_a of phenolphthalein is 9.10. Over what pH range does this indicator change from 95% HIn to 95% In^-?

17.89 Look up the K_{sp} values for $BaSO_4$ and $SrSO_4$ in Table 17.3. Calculate the values of $[Ba^{2+}]$, $[Sr^{2+}]$, and $[SO_4^{2-}]$ in a solution that is saturated with both compounds.

Answers to Practice Exercises: 17.1 (a) and (c); **17.2** 9.17, 9.20; **17.3** Weigh out Na_2CO_3 and $NaHCO_3$ in a mole ratio of 0.60 to 1.0. Dissolve in enough water to make up a 1-L solution; **17.4** (a) Bromophenol blue, methyl orange, methyl red, and chlorophenol blue; (b) all except thymol blue, bromophenol blue, and methyl orange; (c) cresol red and phenolphthalein; **17.5** 1.7×10^{-6}; **17.6** 1.8×10^{-7} *M*; **17.7** 1.9×10^{-3} g/L; **17.8** No; **17.9** (a) 1.7×10^{-4} g/L, (b) 1.4×10^{-7} g/L; **17.10** $[Cu^{2+}] = 1.2 \times 10^{-13}$ *M*, $[Cu(NH_3)_4^{2+}] = 0.017$ *M*, $[NH_3] = 0.23$ *M*.

CHAPTER 18
THE CHEMISTRY OF COORDINATION COMPOUNDS

◆ Luck often plays a role in major scientific breakthroughs, but it takes an alert and well-trained person to recognize the significance of an accidental discovery and to take full advantage of it. Such was the case when in 1964 the biophysicist Barnett Rosenberg and his research group at Michigan State University were studying the effect of an electric field on the growth of bacteria. They suspended a bacterial culture between two platinum electrodes and passed an electric current through it. To their surprise, they found that after an hour or so the bacteria cells ceased dividing. It did not take long for the group to determine that a platinum-containing substance that was extracted from the bacterial culture inhibited cell division.

Rosenberg reasoned that the platinum compound might be useful as an anticancer agent, because cancer involves uncontrolled division of cancerous cells, so he set out to identify the substance. Given the presence of ammonia and chloride ions in solution during electrolysis, Rosenberg synthesized a number of platinum compounds containing ammonia and chlorine. The one that proved most effective at inhibiting cell division was *cis*-diamminedichloroplatinum(II) $[Pt(NH_3)_2Cl_2]$, also called *cisplatin.* Cisplatin is one of a group of transition metal compounds called *coordination compounds,* which have some unique properties.

Interestingly, cisplatin has been known since 1845, but its ability to block cell division was not established until 1967. Today it is one of the most widely used anti-cancer drugs. Presumably, cisplatin binds to DNA in a specific manner which leads to a mistake (or mutation) in the latter's replication and the eventual destruction of the cancerous cell. ◆

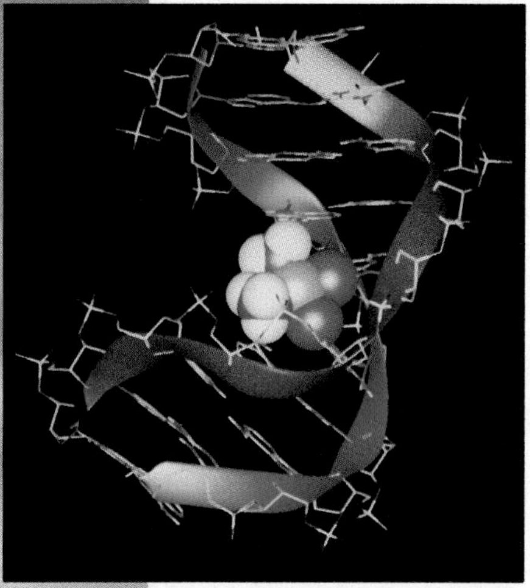

Cisplatin disrupts the replication of DNA by binding to the double helix.

18.1 PROPERTIES OF THE TRANSITION METALS

Transition metals characteristically have incompletely filled *d* subshells or readily give rise to ions with incompletely filled *d* subshells (Figure 18.1). This attribute is responsible for several notable properties, including distinctive coloring, formation of paramagnetic compounds, catalytic activity, and especially a great tendency to form complex ions. In this chapter we focus on the first-row elements from scandium to copper, the most common transition metals (Figure 18.2).

As we read across any period from left to right, atomic numbers increase, electrons are added to the outer shell, and the nuclear charge increases by the addition of protons. In the third-period elements—sodium to argon—the outer electrons weakly shield one another from the extra nuclear charge. Consequently, atomic radii decrease rapidly from sodium to argon and the electronegativities and ionization energies increase steadily (see Figures 8.5, 8.10, and 9.4).

For the transition metals, the trends are different. In Table 18.1, we see, reading across from scandium to copper, that the nuclear charge, of course, increases, but electrons are being added to the inner 3*d* subshell. These 3*d* electrons shield the 4*s* electrons from the increasing nuclear charge somewhat more effectively than outer-shell electrons can shield one another, so the atomic radii decrease less rapidly. For the same reason, electronegativities and ionization energies increase only slightly from scandium across to copper compared with the increases from sodium to argon.

Most transition metals are inert toward acids or react slowly because of a protective layer of oxide. A case in point is chromium: it is quite inert chemically because of the formation on its surface of chromium(III) oxide, Cr_2O_3. Consequently, chromium is commonly used as a protective and noncorrosive plating on other metals. On automobile bumpers and trim, chromium plating serves a decorative as well as a functional purpose.

1A	2A	3B	4B	5B	6B	7B	8B	8B	8B	1B	2B	3A	4A	5A	6A	7A	8A
1 H																	2 He
3 Li	4 Be											5 B	6 C	7 N	8 O	9 F	10 Ne
11 Na	12 Mg											13 Al	14 Si	15 P	16 S	17 Cl	18 Ar
19 K	20 Ca	21 Sc	22 Ti	23 V	24 Cr	25 Mn	26 Fe	27 Co	28 Ni	29 Cu	30 Zn	31 Ga	32 Ge	33 As	34 Se	35 Br	36 Kr
37 Rb	38 Sr	39 Y	40 Zr	41 Nb	42 Mo	43 Tc	44 Ru	45 Rh	46 Pd	47 Ag	48 Cd	49 In	50 Sn	51 Sb	52 Te	53 I	54 Xe
55 Cs	56 Ba	57 La	72 Hf	73 Ta	74 W	75 Re	76 Os	77 Ir	78 Pt	79 Au	80 Hg	81 Tl	82 Pb	83 Bi	84 Po	85 At	86 Rn
87 Fr	88 Ra	89 Ac	104 Unq	105 Unp	106 Unh	107 Uns	108 Uno	109 Une									

FIGURE 18.1

The transition metals. Note that although the Group 2B elements (Zn, Cd, Hg) are described as transition metals by some chemists, neither the metals nor their ions have incompletely filled d subshells.

FIGURE 18.2

The first-row transition metals.

TABLE 18.1

Electron Configuration and Other Properties of the First-Row Transition Metals

	Sc	Ti	V	Cr	Mn	Fe	Co	Ni	Cu
Electron configuration									
M	$4s^23d^1$	$4s^23d^2$	$4s^23d^3$	$4s^13d^5$	$4s^23d^5$	$4s^23d^6$	$4s^23d^7$	$4s^23d^8$	$4s^13d^{10}$
M^{2+}	—	$3d^2$	$3d^3$	$3d^4$	$3d^5$	$3d^6$	$3d^7$	$3d^8$	$3d^9$
M^{3+}	[Ar]	$3d^1$	$3d^2$	$3d^3$	$3d^4$	$3d^5$	$3d^6$	$3d^7$	$3d^8$
Electronegativity	1.3	1.5	1.6	1.6	1.5	1.8	1.9	1.9	1.9
Ionization energy (kJ/mol)									
First	631	658	650	652	717	759	760	736	745
Second	1235	1309	1413	1591	1509	1561	1645	1751	1958
Third	2389	2650	2828	2986	3250	2956	3231	3393	3578
Radius (pm)									
M	162	147	134	130	135	126	125	124	128
M^{2+}	—	90	88	85	80	77	75	69	72
M^{3+}	81	77	74	64	66	60	64	—	—

ELECTRON CONFIGURATIONS

The electron configurations of the first-row transition metals were discussed in Section 7.10. Calcium has the electron configuration $[Ar]4s^2$. From scandium across to copper, electrons are added to the 3*d* orbitals. Thus, the outer electron configuration of scandium is $4s^23d^1$, that of titanium is $4s^23d^2$, and so on. The two exceptions are chromium and copper, whose outer electron configurations are $4s^13d^5$ and $4s^13d^{10}$, respectively. These irregularities are the result of the extra stability associated with half-filled and completely filled 3*d* subshells.

When the first-row transition metals form cations, electrons are removed first from the 4*s* orbitals and then from the 3*d* orbitals. (This is the opposite of the order in which orbitals are filled in neutral atoms.) For example, the outer electron configuration of Fe^{2+} is $3d^6$, not $4s^23d^4$.

OXIDATION STATES

As noted in Chapter 4, the transition metals exhibit variable oxidation states in their compounds. Figure 18.3 shows the oxidation states from scandium to copper. Note that the common oxidation states for each element include +2, +3, or both. The +3 oxidation states are more stable at the beginning of the series, whereas toward the end the +2 oxidation states are more stable. The reason is that the ionization energies increase gradually from left to right. However, the third ionization energy (when an electron is removed from the 3*d* orbital) increases more sharply than the first and second ionization energies. Because it takes more energy to remove the third electron from the metals near the end of the row than from those near the beginning, the metals near the end tend to form M^{2+} ions rather than M^{3+} ions.

The highest oxidation state is +7, for manganese ($4s^23d^5$). Transition metals usually exhibit their highest oxidation states in compounds with very electronegative elements such as oxygen and fluorine—for example, VF_5, CrO_3, and Mn_2O_7.

Sc	Ti	V	Cr	Mn	Fe	Co	Ni	Cu
				+7				
			+6	+6	+6			
		+5	+5	+5	+5			
	+4	+4	+4	+4	+4	+4		
+3	+3	+3	+3	+3	+3	+3	+3	+3
	+2	+2	+2	+2	+2	+2	+2	+2
								+1

FIGURE 18.3

Oxidation states of the first-row transition metals. The most stable oxidation numbers are shown in color. The zero oxidation state is encountered in some compounds, such as $Ni(CO)_4$ *and* $Fe(CO)_5$.

18.2 COORDINATION COMPOUNDS

A ***coordination compound*** is *a neutral species in which a small number of molecules or ions surround a central metal atom or ion.* An example of a coordination compound is $[Co(NH_3)_6]Cl_3$, which is formed when the complex ion $[Co(NH_3)_6]^{3+}$ (see Table 17.5) combines with chloride ions (Cl^-). Placing the chloride ions outside the brackets shows that they are not part of the complex ion but are held to it by ionic forces. Most, but not all, of the metals in coordination compounds are transition metals. *The molecules or ions that surround the metal in a complex ion* are called ***ligands*** (Table 18.2). The interactions between a metal atom and its ligands can be thought of as Lewis acid-base reactions. As we saw in Section 16.10, a Lewis base is a substance capable of donating an electron pair. Every ligand has at least one lone pair of electrons on the ***donor atom,*** that is, *the atom in a ligand that is bonded directly to the metal atom,* as these examples show:

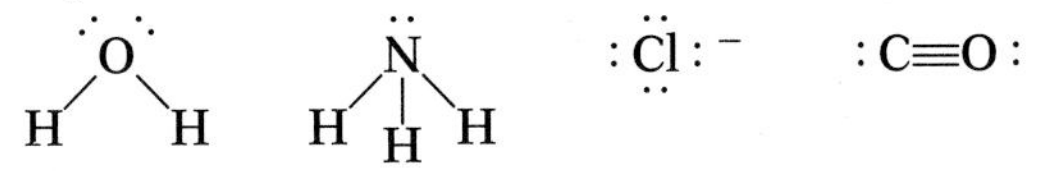

TABLE 18.2
Some Common Ligands

Name	Structure
	Monodentate ligands
Ammonia	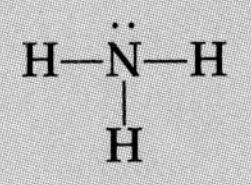
Carbon monoxide	:C≡O:
Chloride ion	$:\ddot{Cl}:^-$
Cyanide ion	$[:C\equiv N:]^-$
Thiocyanate ion	$[:\ddot{S}-C\equiv N:]^-$
	Bidentate ligands
Ethylenediamine	$H_2\ddot{N}-CH_2-CH_2-\ddot{N}H_2$
Oxalate ion	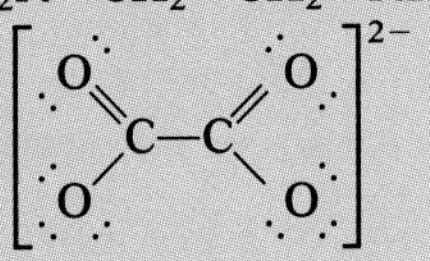
	Polydendate ligand
Ethylenediaminetetraacetate ion (EDTA)	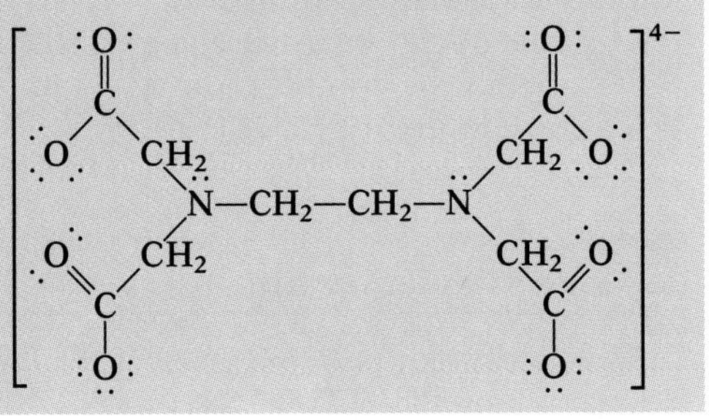

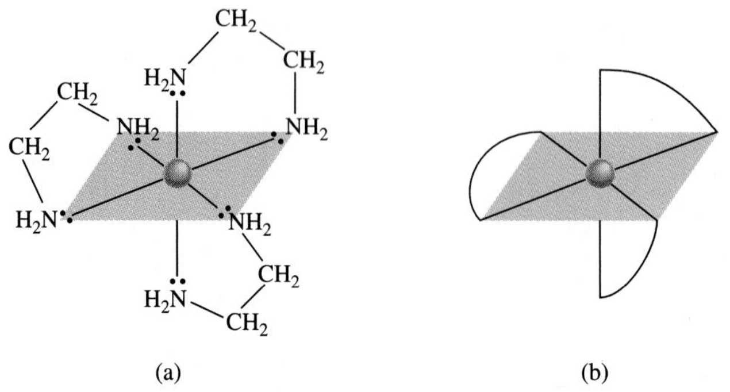

FIGURE 18.4

(a) Structure of a metal-ethylenediamine complex. Each ethylenediamine molecule provides two N donor atoms and is therefore a bidentate ligand. (b) Simplified structure of the same complex.

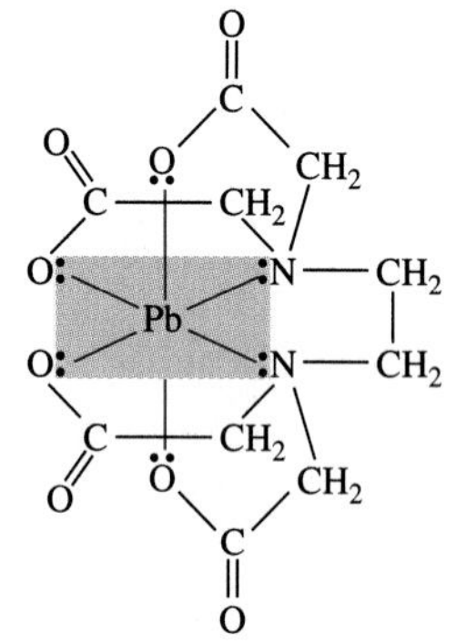

FIGURE 18.5

EDTA complex of lead. The complex bears a net charge of −2, since each O donor atom has one negative charge and the lead ion carries two positive charges. Note the octahedral geometry around the Pb^{2+} ion.

For example, the N atom (in NH_3) is the donor atom in $[Co(NH_3)_6]^{3+}$ since it is bonded to Co. Therefore, ligands play the role of Lewis bases. On the other hand, a transition metal atom (in either its neutral or positively charged state) acts as a Lewis acid, accepting (and sharing) pairs of electrons from the Lewis bases. Thus the metal-ligand bonds are usually coordinate covalent bonds (see Section 9.7).

The ***coordination number*** in coordination compounds is defined as *the number of donor atoms surrounding the central metal atom.* For example, the coordination number of Ag^+ in $[Ag(NH_3)_2]^+$ is 2, that of Cu^{2+} in $[Cu(NH_3)_4]^{2+}$ is 4, and that of Fe^{3+} in $[Fe(CN)_6]^{3-}$ is 6. The most common coordination numbers are 4 and 6, but coordination numbers such as 2 and 5 are also known.

Depending on the number of donor atoms present, ligands are classified as *monodentate, bidentate,* or *polydentate* (see Table 18.2). H_2O and NH_3 are monodentate ligands with only one donor atom each. One bidentate ligand is ethylenediamine (commonly abbreviated "en"):

$$H_2\ddot{N}—CH_2—CH_2—\ddot{N}H_2$$

The two nitrogen atoms can coordinate with a metal atom as shown in Figure 18.4.

Bidentate and polydentate ligands are also called ***chelating agents*** because of *their ability to hold the metal atom like a claw* (from the Greek *chele,* meaning "claw"). One example is ethylenediaminetetraacetate ion (EDTA), a polydentate ligand used to treat metal poisoning (Figure 18.5). Six donor atoms enable EDTA to form a very stable complex ion with lead, which is removed from the blood and excreted from the body. EDTA is also used to clean up spills of radioactive metals.

Oxidation Number of Metals in Coordination Compounds

The net charge of a complex ion is the sum of the charges on the central metal atom and its surrounding ligands. In the $[PtCl_6]^{2-}$ ion, for example, each chloride ion has an oxidation number of −1, so the oxidation number of Pt must be +4. If the ligands do not bear net charges, the oxidation number of the metal is equal to the charge of the complex ion. Thus in $[Cu(NH_3)_4]^{2+}$ each NH_3 is neutral, so the oxidation number of Cu is +2.

EXAMPLE 18.1

Assigning Oxidation Numbers

Specify the oxidation number of the central metal atom in each of the following compounds: (a) $[Ru(NH_3)_5(H_2O)]Cl_2$, (b) $[Cr(NH_3)_6](NO_3)_3$, (c) $[Fe(CO)_5]$, (d) $K_4[Fe(CN)_6]$.

Answer: (a) Both NH_3 and H_2O are neutral species. Since each chloride ion carries a -1 charge, and there are two Cl^- ions, the oxidation number of Ru must be $+2$.

(b) Each nitrate ion has a charge of -1; therefore, the cation must be $[Cr(NH_3)_6]^{3+}$. Because NH_3 is neutral, the oxidation number of Cr is $+3$.

(c) Since the CO species are neutral, the oxidation number of Fe is zero.

(d) Each potassium ion has a charge of $+1$; therefore, the anion is $[Fe(CN)_6]^{4-}$. Next, we know that each cyanide group bears a charge of -1, so Fe must have an oxidation number of $+2$.

PRACTICE EXERCISE

Write the oxidation numbers of the metals in $K[Au(OH)_4]$.

NAMING COORDINATION COMPOUNDS

Now that we have discussed the various types of ligands and the oxidation number of metals, our next step is to learn what to call these coordination compounds. The rules for naming coordination compounds are as follows:

- The cation is named before the anion, as is the case for other ionic compounds. The rule holds regardless of whether the complex ion bears a net positive or a negative charge. For example, in $K_3[Fe(CN)_6]$ and $[Co(NH_3)_4Cl_2]Cl$, we name the K^+ and $[Co(NH_3)_4Cl_2]^+$ cations first, respectively.
- Within a complex ion the ligands are named first, in alphabetical order, and the metal ion is named last.
- The names of anionic ligands end with the letter *o*, whereas a neutral ligand is usually called by the name of the molecule. The exceptions are H_2O (aquo), CO (carbonyl), and NH_3 (ammine). Table 18.3 lists some common ligands.

TABLE 18.3
Names of Common Ligands in Coordination Compounds

Ligand	Name of ligand in coordination compound
Bromide, Br^-	Bromo
Chloride, Cl^-	Chloro
Cyanide, CN^-	Cyano
Hydroxide, OH^-	Hydroxo
Oxide, O^{2-}	Oxo
Carbonate, CO_3^{2-}	Carbonato
Nitrite, NO_2^-	Nitro
Oxalate, $C_2O_4^{2-}$	Oxalato
Ammonia, NH_3	Ammine
Carbon monoxide, CO	Carbonyl
Water, H_2O	Aquo
Ethylenediamine	Ethylenediamine
Ethylenediaminetetraacetate	Ethylenediaminetetraacetato

TABLE 18.4
Names of Anions Containing Metal Atoms

Metal	Name of metal in anionic complex
Aluminum	Aluminate
Chromium	Chromate
Cobalt	Cobaltate
Copper	Cuprate
Gold	Aurate
Iron	Ferrate
Lead	Plumbate
Manganese	Manganate
Molybdenum	Molybdate
Nickel	Nickelate
Silver	Argentate
Tin	Stannate
Tungsten	Tungstate
Zinc	Zincate

- When several ligands of a particular kind are present, we use the Greek prefixes "di-," "tri-," "tetra-," "penta-," and "hexa-" to name them. Thus the ligands in $[Co(NH_3)_4Cl_2]^+$ are "tetraamminedichloro." (Note that prefixes play no role in the alphabetical order of ligands.) If the ligand itself contains a Greek prefix, we use the prefixes "bis" (2), "tris" (3), and "tetrakis" (4) to indicate the number of ligands present. For example, the ligand ethylenediamine already contains "di"; therefore, if two such ligands are present the name is bis(ethylenediamine).
- The oxidation number of the metal is written in Roman numerals following the name of the metal. For example, the Roman numeral III is used to indicate the +3 oxidation state of chromium in $[Cr(NH_3)_4Cl_2]^+$, which is called tetraamminedichlorochromium(III) ion.
- If the complex is an anion, its name ends in "-ate." For example, in $K_4[Fe(CN)_6]$ the anion $[Fe(CN)_6]^{4-}$ is called hexacyanoferrate(II) ion. Note that the Roman numeral II indicates the oxidation state of iron. Table 18.4 gives the names of anions containing metal atoms.

EXAMPLE 18.2
Naming Coordination Compounds

Write the systematic names of the following compounds: (a) $Ni(CO)_4$, (b) $[Co(NH_3)_4Cl_2]Cl$, (c) $K_3[Fe(CN)_6]$, (d) $[Cr(en)_3]Cl_3$.

Answer: (a) The CO ligands are neutral species and the nickel atom bears no net charge, so the compound is called tetracarbonylnickel(0), or, more commonly, nickel tetracarbonyl.

(b) Starting with the cation, each of the two chloride ligands bears a negative charge and the ammonia molecules are neutral. Thus the cobalt atom must have an oxidation number of +3 (to balance the chloride anion). The compound is called tetraamminedichlorocobalt(III) chloride.

(c) The complex ion is the anion and it bears three negative charges. Thus the iron atom must have an oxidation number of +3. The compound is potassium hexacyanoferrate(III). This compound is commonly called potassium ferricyanide.

(d) As we noted earlier, "en" is the abbreviation for the ligand ethylenediamine. Since there are three en groups present and the name of the ligand already contains "di," the name of the compound is tris(ethylenediamine)-chromium(III) chloride.

PRACTICE EXERCISE

What is the systematic name of $[Cr(H_2O)_4Cl_2]Cl$?

EXAMPLE 18.3

Writing Formulas for Coordination Compounds

Write the formulas for the following compounds: (a) pentaamminechlorocobalt(III) chloride, (b) dichlorobis(ethylenediamine)platinum(IV) nitrate, (c) sodium hexanitrocobaltate(III).

Answer: (a) The complex cation contains five NH_3 groups, a chloride ion, and a cobalt ion with a +3 oxidation number. The net charge on the cation must be 2+. Therefore, the formula for the compound is $[Co(NH_3)_5Cl]Cl_2$.

(b) There are two chloride ions, two ethylenediamine groups, and a platinum ion with an oxidation number of +4 in the complex cation. Therefore, the formula for the compound is $[Pt(en)_2Cl_2](NO_3)_2$.

(c) The complex anion contains six nitro groups and a cobalt ion with an oxidation number of +3. Therefore, the formula for the compound is $Na_3[Co(NO_2)_6]$.

PRACTICE EXERCISE

Write the formula for the following compound: tris(ethylenediamine)-cobalt(III) sulfate.

18.3 GEOMETRY OF COORDINATION COMPOUNDS

Figure 18.6 shows four different geometric arrangements for metal atoms with monodentate ligands. In these diagrams we see that the structure and the coordination number of the metal atom relate to each other as follows:

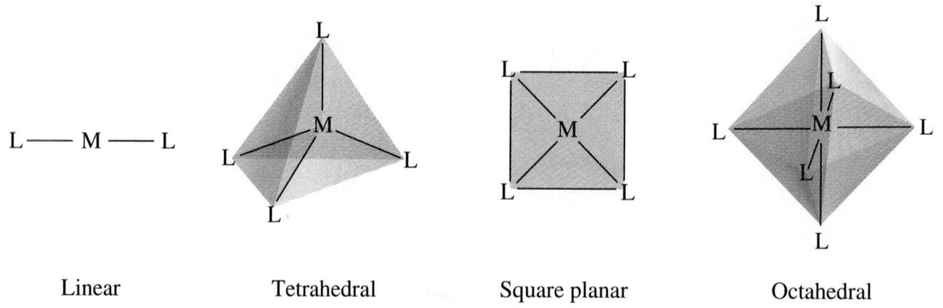

FIGURE 18.6

Common geometries of complex ions. In each case M is a metal and L is a monodentate ligand.

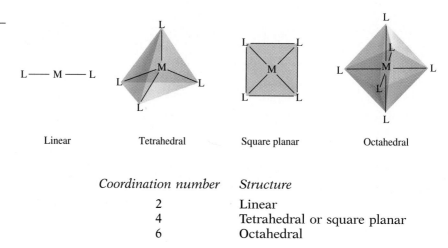

Coordination number	*Structure*
2	Linear
4	Tetrahedral or square planar
6	Octahedral

Coordination Number = 2

The $Ag(NH_3)_2^+$ complex ion, formed by the reaction between Ag^+ ions and ammonia (see Table 17.5), has a coordination number of 2 and a linear geometry. Other examples are $[CuCl_2]^-$ and $[Au(CN)_2]^-$.

Coordination Number = 4

There are two types of geometry with a coordination number of 4. The $[Zn(NH_3)_4]^{2+}$ and $[CoCl_4]^{2-}$ ions have tetrahedral geometry, while the $[Pt(NH_3)_4]^{2+}$ ion has the square planar geometry. In Chapter 13 we discussed geometric isomers of alkenes (see p. 374). Square planar complex ions with two different monodentate ligands can also exhibit geometric isomerism. Figure 18.7 shows the cis and trans isomers of diamminedichloroplatinum(II). Note that although the types of bonds are the same in both isomers (two Pt—N and two Pt—Cl bonds), the spatial arrangements are different. These two isomers have different properties (melting point, boiling point, color, solubility in water, and dipole moment).

Coordination Number = 6

Complex ions with a coordination number of 6 all have octahedral geometry (see Section 10.1). Geometric isomerism is possible in octahedral complexes when two or more different ligands are present. An example is the tetraamminedichlorocobalt(III) ion shown in Figure 18.8. The two geometric isomers have different colors and other properties even though they have the same ligands and the same number and types of bonds (Figure 18.9).

In addition to geometric isomerism, certain octahedral complex ions can also exhibit optical isomerism (that is, give rise to optical isomers, as discussed in

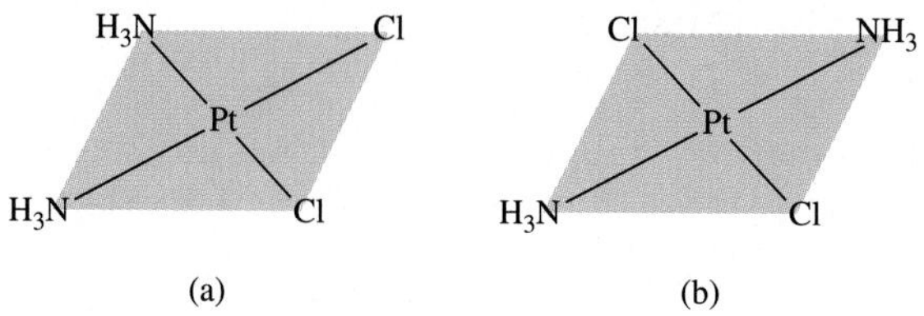

FIGURE 18.7

The (a) cis and (b) trans isomers of diamminedichloroplatinum(II). Note that the two Cl atoms are adjacent to each other in the cis isomer and diagonally across from each other in the trans isomer.

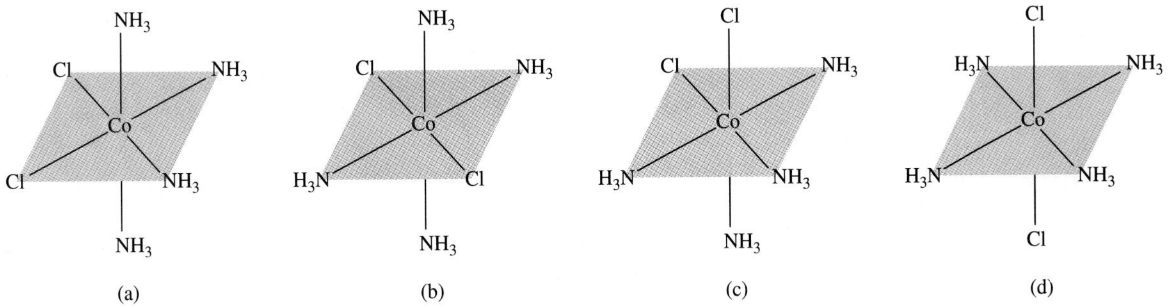

FIGURE 18.8

The (a) cis and (b) trans isomers of tetraamminedichlorocobalt(III) ion. The structure shown in (c) can be generated by rotating that in (a), and the structure shown in (d) can be generated by rotating that in (b). The ion has only two geometric isomers, (a) [or (c)] and (b) [or (d)].

Chapter 13). Figure 18.10 shows the cis and trans isomers of dichlorobis-(ethylenediamine)cobalt(III) ion and their mirror images. Careful examination reveals that the trans isomer and its mirror image are superimposable, but the cis isomer and its mirror image are not. Therefore, the cis isomer and its mirror images are optical isomers, or enantiomers. It is interesting to note that unlike the case in most organic compounds, there are no asymmetric carbon atoms in these compounds.

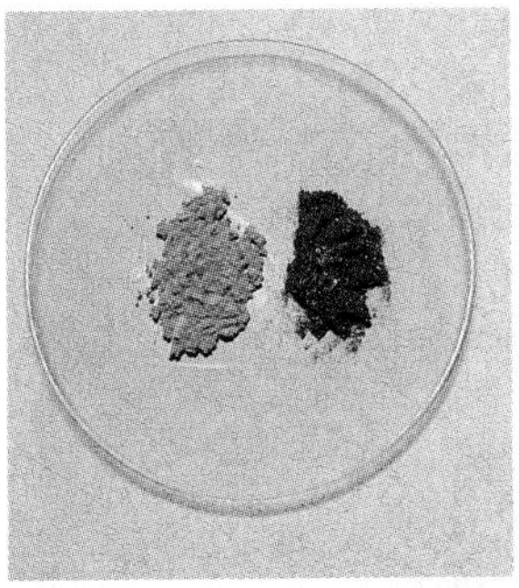

FIGURE 18.9

Cis-tetraammine-dichlorocobalt(III) chloride (left) and trans-tetraamminedichloro-cobalt(III) chloride

18.4 BONDING IN COORDINATION COMPOUNDS: CRYSTAL FIELD THEORY

A satisfactory theory of bonding in coordination compounds must account for properties such as color and magnetism, as well as isomerism and bond strength. No single theory as yet does all this for us. Rather, several different approaches have been applied to transition metal complexes. We will consider only one of

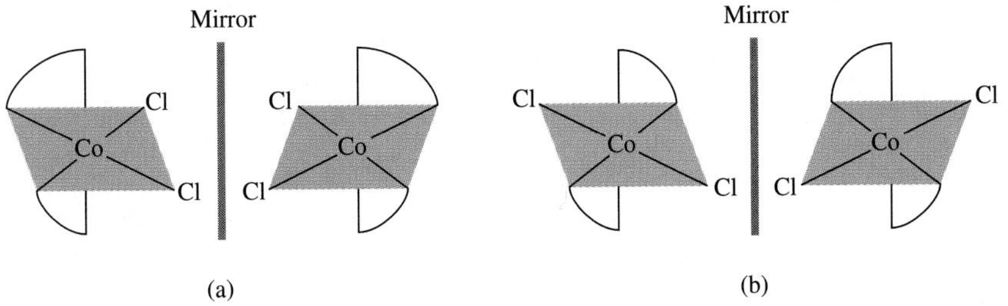

FIGURE 18.10

The (a) cis and (b) trans isomers of dichlorobis(ethylenediamine)cobalt(III) ion and their mirror images. If you could rotate the mirror image of the trans isomer 90° clockwise about the vertical position and place the ion over the trans isomer, you would find that the two are superimposable. No matter how you rotated the cis isomer and its mirror image, however, you could not superimpose one on the other.

them here—crystal field theory—because it accounts for both the color and magnetic properties of many coordination compounds. (The name crystal field theory is derived from its use in explaining the properties of solid, crystalline substances.)

We will begin our discussion of crystal field theory with the most straightforward case, namely, complex ions with octahedral geometry. Then we will see how it is applied to tetrahedral and square planar complexes.

Crystal Field Splitting in Octahedral Complexes

Crystal field theory explains the bonding in complex ions purely in terms of electrostatic forces. In a complex ion, two types of electrostatic interaction come into play. One is the attraction between the positive metal ion and the negatively charged ligand or the negatively charged end of a polar ligand. This is the force that binds the ligands to the metal. The second type of interaction is electrostatic repulsion between the lone pairs on the ligands and the electrons in the *d* orbitals of the metals.

As we saw in Chapter 7, *d* orbitals have different orientations, but in the absence of external disturbance they all have the same energy. In an octahedral complex, a central metal atom is surrounded by six lone pairs of electrons (on the six ligands), so all five *d* orbitals experience electrostatic repulsion. The magnitude of this repulsion depends on the orientation of the *d* orbital that is involved. Take the $d_{x^2-y^2}$ orbital as an example. In Figure 18.11, we see that the lobes of this orbital point toward corners of the octahedron along the *x*- and *y*-axes, where the lone-pair electrons are positioned. Thus an electron residing in this orbital would experience a greater repulsion from the ligands than an electron would in, say, the d_{xy} orbital. For this reason, the energy of the $d_{x^2-y^2}$ orbital is increased relative to the d_{xy} and d_{xz} orbitals. The d_{z^2} orbital's energy is also greater, because its lobes are pointed at the ligands along the *z*-axis.

As a result of these metal-ligand interactions, the five *d* orbitals in an octahedral complex are *split* between two energy levels: a higher level with two orbitals

FIGURE 18.11

The five d orbitals in an octahedral environment. The metal atom (or ion) is at the center of the octahedron, and the six lone pairs on the donor atoms of the ligands are at the corners.

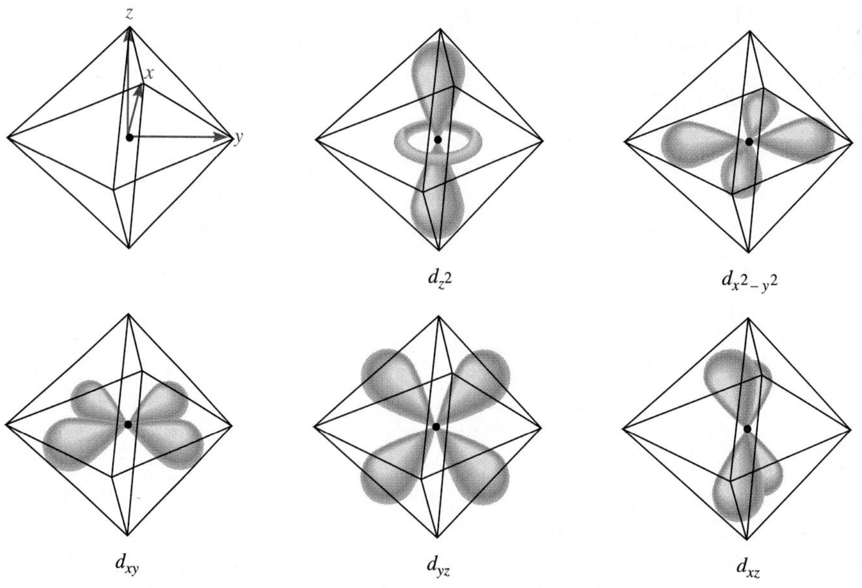

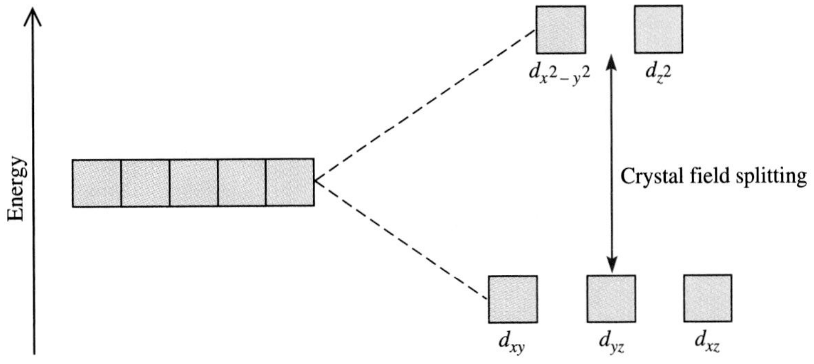

FIGURE 18.12

Crystal field splitting between d orbitals in an octahedral complex.

$d_{x^2-y^2}$ and d_{z^2} having the same energy and a lower level with three equal-energy orbitals (d_{xy}, d_{yz}, and d_{xz}), as shown in Figure 18.12. *The energy difference between these two sets of d orbitals* is called the ***crystal field splitting*** (Δ). The magnitude of Δ depends on the metal and the nature of the ligands; it has a direct effect on the color and magnetic properties of complex ions.

COLOR

In Chapter 7 we learned that white light, such as sunlight, is a combination of all colors. A substance appears black if it absorbs all the visible light that strikes it. If it absorbs no visible light, it is white or colorless. An object appears green if it absorbs all the light but reflects the green component. Alternatively, the object also appears green if it reflects all colors except red, the *complementary* color of green. Figure 18.13 shows the "color wheel," which is a quick reference for determining complementary colors.

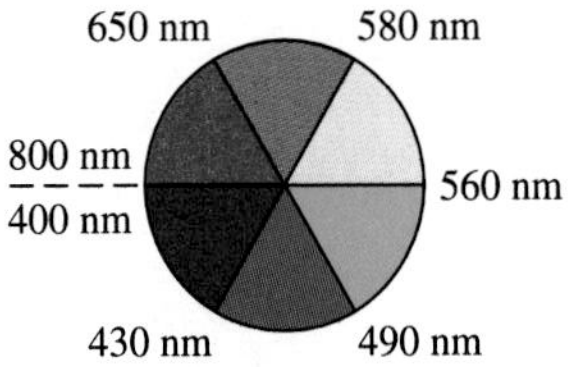

FIGURE 18.13

A color wheel showing the colors that are complementary to one another.

What has been said of reflected light also applies to transmitted light. For example,the hydrated cupric ion, $[Cu(H_2O)_6]^{2+}$, absorbs light in the orange region of the spectrum so that it appears blue to us. When the energy of a photon striking an ion (or compound) is equal to the difference between the lower and higher *d* orbital energy levels, absorption occurs; that is, an electron is promoted from a lower to a higher level. With this knowledge we can calculate the energy change involved in the electron transition that occurs in the cupric ion. Recall from Section 7.1 that the energy of a photon is given by Equation (7.2):

$$E = h\nu$$

where h represents Planck's constant (6.63×10^{-34} J s) and ν is the frequency of the radiation. Here $E = \Delta$, so we have

$$\begin{aligned}\Delta &= h\nu \\ &= (6.63 \times 10^{-34}\ \text{J s})(5 \times 10^{14}/\text{s}) \\ &= 3 \times 10^{-19}\ \text{J}\end{aligned}$$

(Note that this is the energy absorbed by *one* ion.) If the wavelength of the photon absorbed by an ion lies outside the visible region, then the transmitted light looks the same (to us) as the incident light—white—and the ion appears colorless.

Spectroscopic analysis offers the best means of measuring crystal field splitting. The $[Ti(H_2O)_6]^{3+}$ ion provides a particularly simple example, because Ti^{3+} has only one 3*d* electron (Figure 18.14). The $[Ti(H_2O)_6]^{3+}$ ion absorbs light in the visible region (Figure 18.15). The wavelength corresponding to maximum absorp-

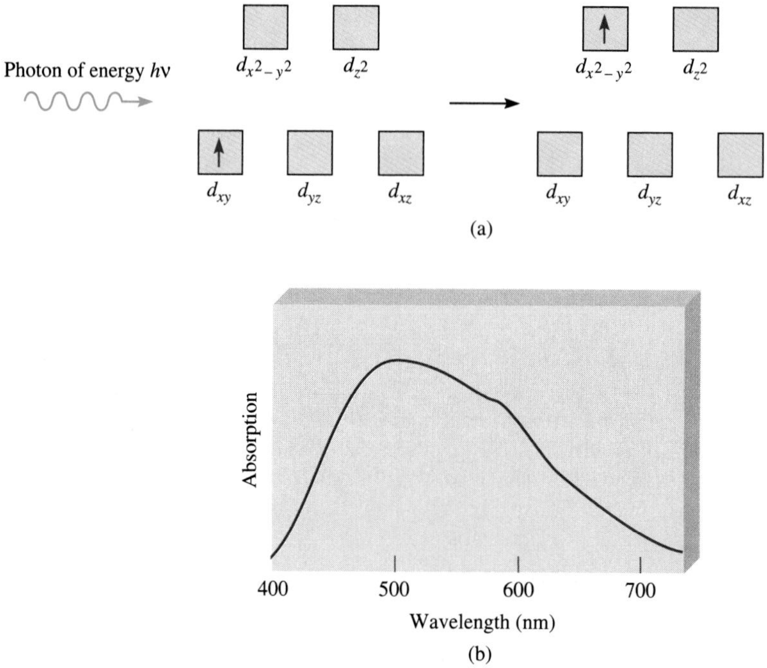

FIGURE 18.14

(a) The absorption process of a photon and (b) a graph of the absorption spectrum of $[Ti(H_2O)_6]^{3+}$. The energy of the incoming photon is equal to the crystal field splitting. The maximum absorption peak in the visible region occurs at 498 nm.

FIGURE 18.15

Colors of some of the first-row transition metal ions in solution. From left to right: Ti^{3+}, Cr^{3+}, Mn^{2+}, Fe^{3+}, Co^{2+}, Ni^{2+}, Cu^{2+}. The Sc^{3+} and V^{5+} ions are colorless.

tion is 498 nm [Figure 18.14(b)]. This information enables us to calculate the crystal field splitting as follows. We start by writing

$$\Delta = h\nu$$

Also

$$\nu = \frac{c}{\lambda}$$

where c is the velocity of light and λ is the wavelength. Therefore

$$\Delta = \frac{hc}{\lambda} = \frac{(6.63 \times 10^{-34}\ \text{J s})(3.00 \times 10^{8}\ \text{m/s})}{(498\ \text{nm})(1 \times 10^{-9}\ \text{m/1 nm})}$$

$$= 3.99 \times 10^{-19}\ \text{J}$$

This is the energy required to excite one $[Ti(H_2O)_6]^{3+}$ ion. To express this energy difference in the more convenient units of kilojoules per mole, we write

$$\begin{aligned}\Delta &= (3.99 \times 10^{-19}\ \text{J/ion})(6.02 \times 10^{23}\ \text{ions/mol})\\ &= 240{,}000\ \text{J/mol}\\ &= 240\ \text{kJ/mol}\end{aligned}$$

Aided by spectroscopic data for a number of complexes, all having the same metal ion but different ligands, chemists calculated the crystal splitting for each ligand and established a ***spectrochemical series,*** which is *a list of ligands arranged in order of their abilities to split the d orbital energies:*

$$I^- < Br^- < Cl^- < OH^- < F^- < H_2O < NH_3 < \text{en} < CN^- < CO$$

These ligands are arranged in the order of increasing value of Δ. CO and CN^- are called *strong-field ligands,* because they cause a large splitting of the *d* orbital energy levels. The halide ions and hydroxide ion are *weak-field ligands,* because they split the *d* orbitals to a lesser extent.

Magnetic Properties

The magnitude of the crystal field splitting also determines the magnetic properties of a complex ion. The $[Ti(H_2O)_6]^{3+}$ ion, having only one *d* electron, is always paramagnetic. However, in an ion where several *d* electrons are present, such as Fe^{3+} complexes, the situation becomes more involved. Consider the octahedral complexes $[FeF_6]^{3-}$ and $[Fe(CN)_6]^{3-}$ (Figure 18.16). The electron configuration of Fe^{3+} is $[Ar]3d^5$, and there are two possible ways to distribute the five *d* electrons among the *d* orbitals. According to Hund's rule (see Section 7.9), maximum stability is reached when the electrons are placed in five separate orbitals with parallel spins. But this arrangement can be achieved only at a cost; two of the five electrons must be energetically promoted to the high-lying $d_{x^2-y^2}$ and d_{z^2} orbitals. No such energy investment is needed if all five electrons enter the d_{xy}, d_{yz}, and d_{xz} orbitals. According to Pauli's exclusion principle (p. 200), there will be only one unpaired electron present in this case.

Figure 18.17 shows the distribution of electrons among *d* orbitals for low- and high-spin complexes. The actual arrangement of the electrons is determined by the amount of stability gained by having maximum parallel spins versus the investment in energy required to promote electrons to higher *d* orbitals. Because F^- is a weak-field ligand, the five *d* electrons enter five separate *d* orbitals with parallel

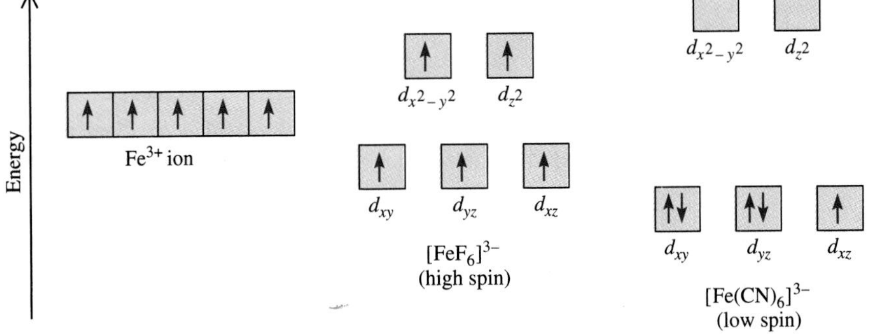

FIGURE 18.16

Energy-level diagrams for the Fe^{3+} ion and $[FeF_6]^{3-}$ and $[Fe(CN)_6]^{3-}$ complex ions.

FIGURE 18.17

Orbital diagrams for the high-spin and low-spin octahedral complexes corresponding to the electron configurations d^4, d^5, d^6, and d^7. No such distinctions can be made for d^1, d^2, d^3, d^8, d^9, and d^{10}.

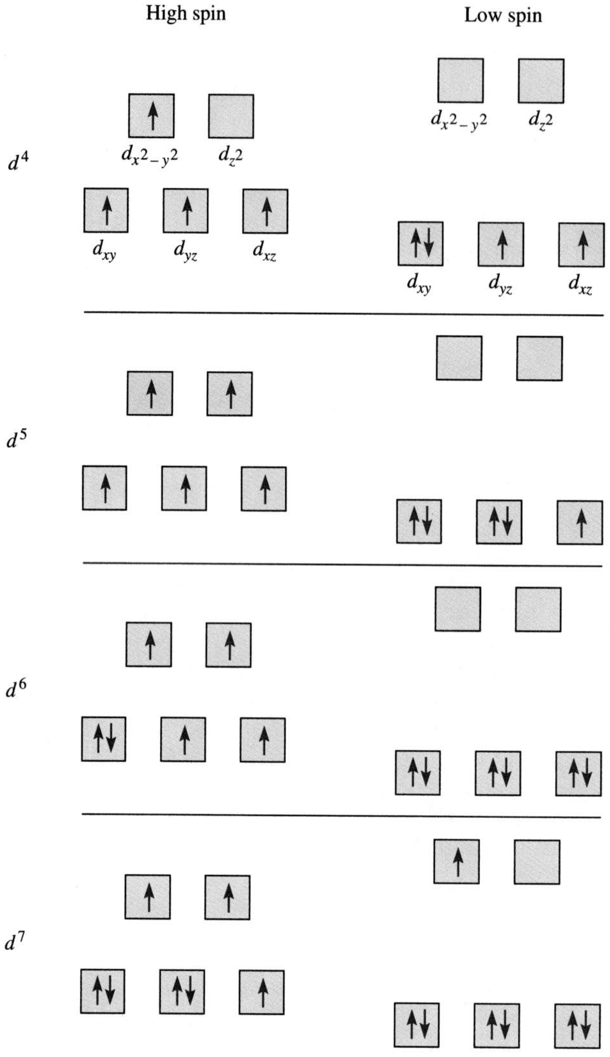

spins to create a *high-spin* complex (see Figure 18.16). On the other hand, the cyanide ion is a strong-field ligand, so it is energetically preferable for all five electrons to be in the lower orbitals and therefore a *low-spin* complex is formed. High-spin complexes are more paramagnetic than low-spin complexes.

The actual number of unpaired electrons (or spins) in a complex ion can be found by magnetic measurements, and in general, experimental findings support predictions based on crystal field splitting. However, a distinction between low- and high-spin complexes can be made only if the metal ion contains more than three and fewer than eight *d* electrons, as shown in Figure 18.17.

EXAMPLE 18.4

Predicting the Number of Unpaired Spins

Predict the number of unpaired spins in the $[Cr(en)_3]^{2+}$ ion.

Answer: The electron configuration of Cr^{2+} is [Ar]$3d^4$. Since en is a strong-field ligand, we expect $[Cr(en)_3]^{2+}$ to be a low-spin complex. According to Figure 18.17, all four electrons will be placed in the lower *d* orbitals (d_{xy}, d_{yz}, and d_{xz}) and there will be a total of two unpaired spins.

PRACTICE EXERCISE

How many unpaired spins are in $[Mn(H_2O)_6]^{2+}$? (H_2O is a weak-field ligand.)

TETRAHEDRAL AND SQUARE PLANAR COMPLEXES

The splitting of the *d* orbital energy levels in tetrahedral and square planar complexes can also be accounted for satisfactorily by the crystal field theory. In fact, the splitting pattern for a tetrahedral ion is just the reverse of that for octahedral complexes. In this case, the d_{xy}, d_{yz}, and d_{xz} orbitals are more closely directed at the ligands and therefore have higher energy than the $d_{x^2-y^2}$ and d_{z^2} orbitals (Figure 18.18). Most tetrahedral complexes are high-spin complexes. Presumably, the tetrahedral arrangement reduces the magnitude of metal-ligand interactions, resulting in a smaller Δ value. This is a reasonable assumption since the number of ligands is smaller in a tetrahedral complex.

As Figure 18.19 shows, the splitting pattern for square planar complexes is the

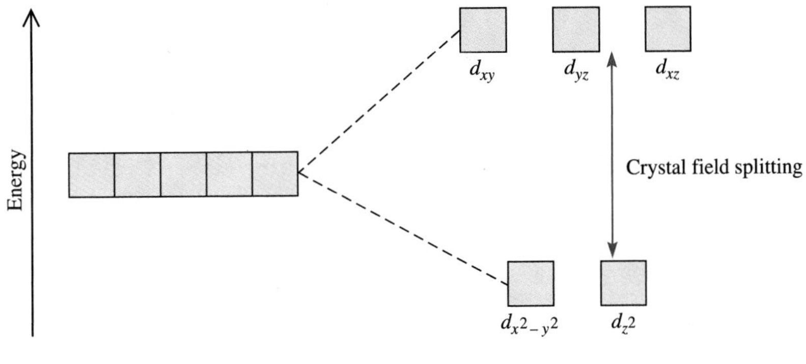

FIGURE 18.18

Crystal field splitting between d orbitals in a tetrahedral complex.

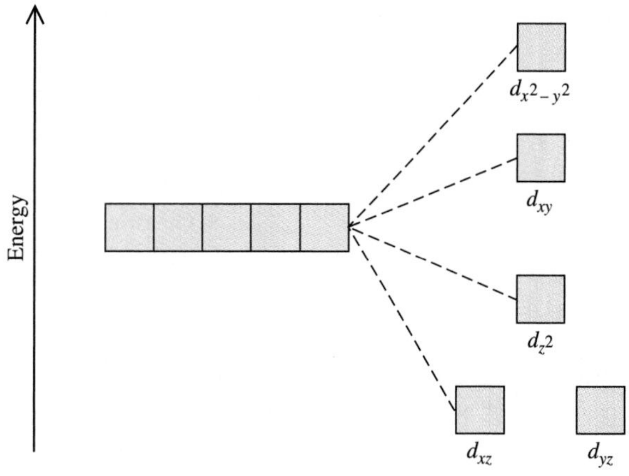

FIGURE 18.19

Energy-level diagram for a square planar complex. Because there are more than two energy levels, we cannot define crystal field splitting as we can for octahedral and tetrahedral complexes.

most complicated. Clearly, the $d_{x^2-y^2}$ orbital possesses the highest energy (as in the octahedral case), and the d_{xy} orbital the next highest. However, the relative placement of the d_{z^2} and the d_{xz} and d_{yz} orbitals cannot be determined simply by inspection and must be calculated.

SUMMARY

Transition metals usually have incompletely filled *d* orbitals and have a pronounced tendency to form complexes. The first-row transition metals (scandium to copper) are the most common transition metals; their chemistry is characteristic, in many ways, of the entire group.

A coordination compound is a neutral species in which a small number of molecules or ions surround a central metal atom or ion. The donor atoms in the ligands each contribute an electron pair to the central metal ion in a complex. Coordination compounds may display geometric and/or optical isomerism.

Crystal field theory explains bonding in complexes in terms of electrostatic interactions. According to crystal field theory, the *d* orbitals are split into two higher-energy and three lower-energy orbitals in an octahedral complex. The energy difference between these two sets of *d* orbitals is the crystal field splitting. Strong-field ligands cause a large splitting, and weak-field ligands cause a small splitting. Electron spins tend to be parallel with weak-field ligands and paired with strong-field ligands, whereas a greater investment of energy is required to promote electrons into the high-lying *d* orbitals. The splitting of *d* orbitals in a tetrahedral complex is the opposite of that in an octahedral complex, and that in a square planar complex is the most complicated.

KEY WORDS

Chelating agent, p. 518
Coordination compound, p. 517
Coordination number, p. 518
Crystal field splitting (Δ), p. 525
Donor atom, p. 517
Ligand, p. 517
Spectrochemical series, p. 527

QUESTIONS AND PROBLEMS

PROPERTIES OF TRANSITION METALS

Review Questions

18.1 What distinguishes a transition metal from a representative metal?

18.2 Why is zinc not considered a transition metal?

18.3 Explain why atomic radii decrease very gradually from scandium to copper.

18.4 Without referring to the text, write the ground-state electron configurations of the first-row transition metals. Explain any irregularities.

18.5 Write the electron configurations of the following ions: V^{5+}, Cr^{3+}, Mn^{2+}, Fe^{3+}, Cu^{2+}, Sc^{3+}, Ti^{4+}.

18.6 Why do transition metals have more oxidation states than other elements? Give the highest oxidation states for scandium to copper.

18.7 As we read across the first-row transition metals from left to right, the +2 oxidation state becomes more stable in comparison with the +3 state. Why is this so?

18.8 Chromium exhibits several oxidation states in its compounds, whereas aluminum exhibits only the +3 oxidation state. Explain.

COORDINATION COMPOUNDS: NOMENCLATURE; OXIDATION NUMBER

Review Questions

18.9 Define the following terms: coordination compound, ligand, donor atom, coordination number, chelating agent.

18.10 Describe the interaction between a donor atom and a metal atom in terms of a Lewis acid-base reaction.

Problems

18.11 Complete the following statements for the complex ion $[Co(en)_2(H_2O)CN]^{2+}$. (a) The term "en" is the abbreviation for ______. (b) The oxidation number of Co is ______. (c) The coordination number of Co is ______. (d) ______ is a bidentate ligand.

18.12 Complete the following statements for the complex ion $[Cr(C_2O_4)_2(H_2O)_2]^-$. (a) The oxidation number of Cr is ______. (b) The coordination number of Cr is ______. (c) ______ is a bidentate ligand.

18.13 Give the oxidation numbers of the metals in the following species: (a) $K_3[Fe(CN)_6]$, (b) $K_3[Cr(C_2O_4)_3]$, (c) $[Ni(CN)_4]^{2-}$.

18.14 Give the oxidation numbers of the metals in the following species: (a) Na_2MoO_4, (b) $MgWO_4$, (c) $K_4[Fe(CN)_6]$.

18.15 What are the systematic names for the following ions and compounds? (a) $[Co(NH_3)_4Cl_2]^+$, (b) $Cr(NH_3)_3Cl_3$, (c) $[Co(en)_2Br_2]^+$, (d) $Fe(CO)_5$.

18.16 What are the systematic names for the following ions and compounds? (a) [*cis*-$Co(en)_2Cl_2]^+$, (b) $[Pt(NH_3)_5Cl]Cl_3$, (c) $[Co(NH_3)_6]Cl_3$, (d) $[Co(NH_3)_5Cl]Cl_2$, (e) *trans*-$Pt(NH_3)_2Cl_2$.

18.17 Write the formulas for each of the following ions and compounds: (a) tetrahydroxozincate(II), (b) pentaaquochlorochromium(III) chloride, (c) tetrabromocuprate(II), (d) ethylenediaminetetraacetatoferrate(II).

18.18 Write the formulas for each of the following ions and compounds: (a) bis(ethylenediamine) dichlorochromium(III), (b) pentacarbonyliron(0), (c) potassium tetracyanocuprate(II), (d) tetraammineaquochlorocobalt(III) chloride.

STRUCTURE OF COORDINATION COMPOUNDS

Problems

18.19 Draw structures of all the geometric and optical isomers of each of the following cobalt complexes:
(a) $[Co(NH_3)_6]^{3+}$
(b) $[Co(NH_3)_5Cl]^{2+}$
(c) $[Co(NH_3)_4Cl_2]^+$
(d) $[Co(en)_3]^{3+}$
(e) $[Co(C_2O_4)_3]^{3-}$

18.20 How many geometric isomers are in the following species? (a) $[Co(NH_3)_2Cl_4]^-$, (b) $[Co(NH_3)_3Cl_3]$.

18.21 A student prepared a cobalt complex that has one of the following structures: $[Co(NH_3)_6]Cl_3$, $[Co(NH_3)_5Cl]Cl_2$, or $[Co(NH_3)_4Cl_2]Cl$. Explain how the student would distinguish among these possibilities by an electrical conductance experiment. At the student's disposal are three strong electrolytes: NaCl, $MgCl_2$, and $FeCl_3$, which may be used for comparison purposes.

18.22 The complex ion $[Ni(CN)_2Br_2]^{2-}$ has a square planar geometry. Draw the structures of the geometric isomers of this complex.

BONDING, COLOR, MAGNETISM

Review Questions

18.23 Briefly describe the crystal field theory. Define the following terms: crystal field splitting, high-spin complex, low-spin complex, spectrochemical series.

18.24 What is the origin of color in a compound?

18.25 Compounds containing the Sc^{3+} ion are colorless, whereas those containing the Ti^{3+} ion are colored. Explain.

18.26 What factors determine whether a given complex will be diamagnetic or paramagnetic?

Problems

18.27 For the same type of ligands, explain why the crystal field splitting for an octahedral complex is always greater than that for a tetrahedral complex.

18.28 Transition metal complexes containing CN^- ligands are often yellow in color, whereas those containing H_2O ligands are often green or blue. Explain.

18.29 The $[Ni(CN)_4]^{2-}$ ion, which has a square planar geometry, is diamagnetic, whereas the $[NiCl_4]^{2-}$ ion, which has a tetrahedral geometry, is paramagnetic. Show the crystal field splitting diagrams for those two complexes.

18.30 Predict the number of unpaired electrons in the following complex ions: (a) $[Cr(CN]_6]^{4-}$, (b) $[Cr(H_2O)_6]^{2+}$.

18.31 The absorption maximum for the complex ion $[Co(NH_3)_6]^{3+}$ occurs at 470 nm. (a) Predict the color of the complex and (b) calculate the crystal field splitting in kilojoules per mole.

18.32 A solution made by dissolving 0.875 g of $Co(NH_3)_4Cl_3$ in 25.0 g of water freezes 0.56°C below the freezing point of pure water. Calculate the number of moles of ions produced when 1 mole of $Co(NH_3)_4Cl_3$ is dissolved in water, and suggest a structure for the complex ion present in this compound.

MISCELLANEOUS PROBLEMS

18.33 Explain the following facts: (a) Copper and iron have several oxidation states, whereas zinc exists in only one. (b) Copper and iron form colored ions, whereas zinc does not.

18.34 The formation constant for the reaction $Ag^+ + 2NH_3 \rightleftharpoons [Ag(NH_3)_2]^+$ is 1.5×10^7 and that for the reaction $Ag^+ + 2CN^- \rightleftharpoons [Ag(CN)_2]^-$ is 1.0×10^{21} at 25°C (see Table 17.5). Calculate the equilibrium constant at 25°C for the reaction

$$[Ag(NH_3)_2]^+ + 2CN^- \rightleftharpoons [Ag(CN)_2]^- + 2NH_3$$

18.35 Hemoglobin is the oxygen-carrying protein. In each hemoglobin molecule, there are four heme groups. In each heme group an Fe(II) ion is octahedrally bound to five N atoms and to either a water molecule (called deoxyhemoglobin) or an oxygen molecule (called oxyhemoglobin). Oxyhemoglobin is bright red, whereas deoxyhemoglobin is purple. Show that the difference in color can be accounted for qualitatively on the basis of high-spin and low-spin complexes. (*Hint:* O_2 is a strong-field ligand.)

18.36 Hydrated Mn^{2+} ions are practically colorless (see Figure 18.15) even though they possess five 3*d* electrons. Explain. (*Hint:* Electronic transitions in which there is a change in the number of unpaired electrons do not occur readily.)

18.37 Which of the following hydrated cations are colorless? $Fe^{2+}(aq)$, $Zn^{2+}(aq)$, $Cu^+(aq)$, $Cu^{2+}(aq)$, $V^{5+}(aq)$, $Ca^{2+}(aq)$, $Co^{2+}(aq)$, $Sc^{3+}(aq)$, $Pb^{2+}(aq)$. Explain your choice.

18.38 In each of the following pairs of complexes, choose the one that absorbs light at a longer wavelength: (a) $[Co(NH_3)_6]^{2+}$, $[Co(H_2O)_6]^{2+}$; (b) $[FeF_6]^{3-}$, $[Fe(CN)_6]^{3-}$; (c) $[Cu(NH_3)_4]^{2+}$, $[CuCl_4]^{2-}$.

18.39 A student in 1895 prepared three chromium coordination compounds having the same formulas of $CrCl_3(H_2O)_6$ with the following properties:

Color	Cl^- ions in solution per formula unit
Violet	3
Light green	2
Dark green	1

Write modern formulas for these compounds and suggest a method for confirming the number of Cl^- ions present in solution in each case. (*Hint:* Some of the compounds may exist as hydrates, which are compounds that have a specific number of water molecules attached to them.)

18.40 Complex ion formation has been used to extract gold which exists in nature in the uncombined state. To separate it from other solid impurities, the ore is treated with a sodium cyanide (NaCN) solution in the presence of air to dissolve the gold by forming the soluble complex ion $[Au(CN)_2]^-$. (a) Balance the following equation:

$$Au + CN^- + O_2 + H_2O \longrightarrow [Au(CN)_2]^- + OH^-$$

(b) The gold is obtained by reducing the complex ion with zinc metal. Write a balanced ionic equation for this process. (c) What is the geometry and coordination number of the $[Au(CN)_2]^-$ ion?

18.41 The following molecules (AX_4Y_2) have octahedral geometry. Group the molecules that are equivalent to each other.

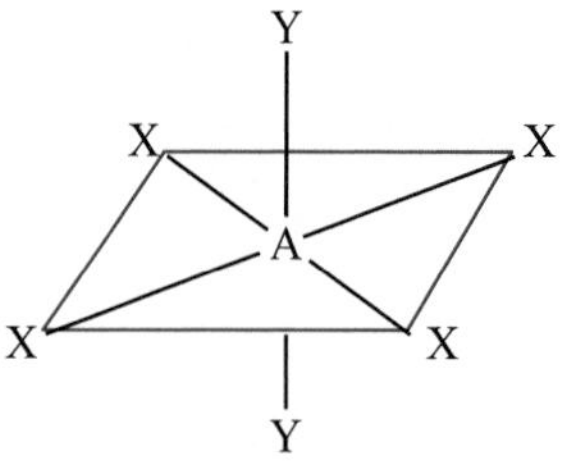

(a)

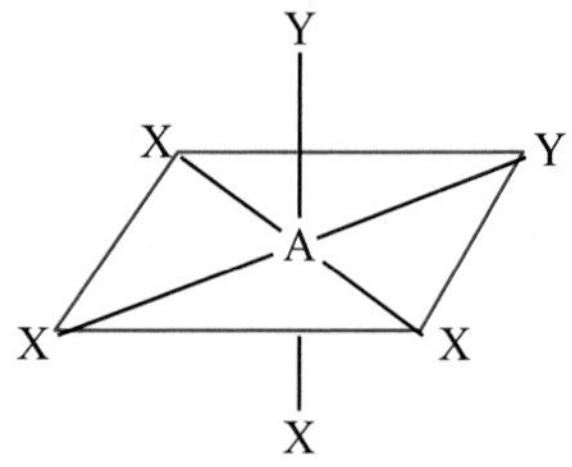

(b)

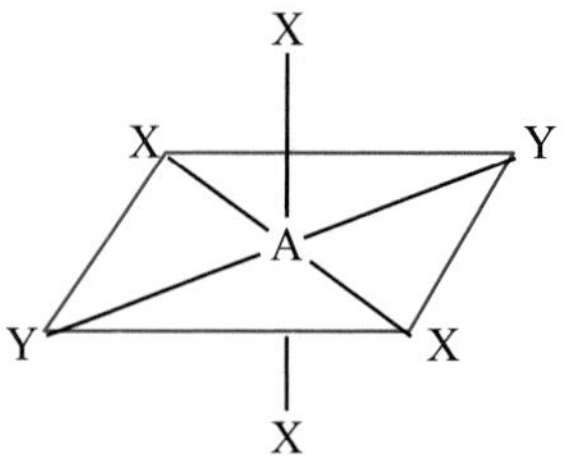

(c)

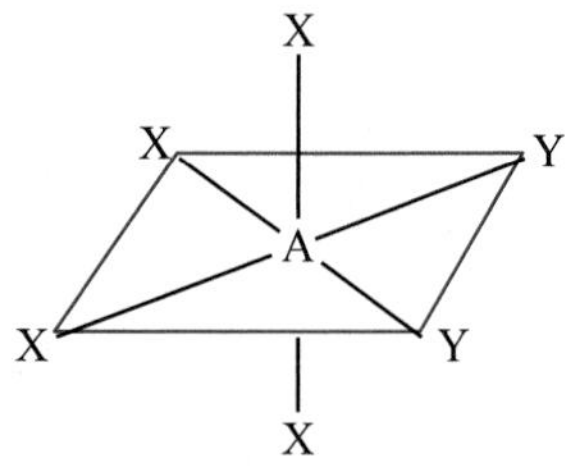

(d)

Answers to Practice Exercises: **18.1** K: +1, Au: +3; **18.2** Tetraaquodichlorochromium(III) chloride; **18.3** $[Co(en)_3]_2(SO_4)_3$; **18.4** 5.

CHAPTER 19

THERMODYNAMICS

◆ The second law of thermodynamics was developed by the French engineer Sadi Carnot around 1820. Carnot's formulation has to do with the efficiency of heat engines, machines that convert heat to mechanical work. It helps us answer the question, for a given amount of heat input, how much useful work can we get out of the engine?

Consider the operation of a steam turbine, a typical heat engine, at an electric power station. Superheated steam at about 560°C (or 833 K) is used to drive the turbine to run the generator. Eventually, the steam is discharged to the surroundings at about 38°C (or 311 K). Carnot showed that the maximum efficiency of this and all other heat engines is given by $(T_2 - T_1)/T_2$, where T_2 and T_1 are the upper and lower temperatures (in kelvins). Since T_1 cannot be zero and T_2 cannot be infinite, the efficiency can *never* approach 100 percent. (For the steam turbine described above, the maximum efficiency would be 0.63, or 63 percent.) Thus it is impossible to build a heat engine that would convert heat totally into mechanical work; some heat must inevitably be discharged to the surroundings. This is the essence of the second law.

Carnot's work is significant but not directly applicable to chemistry. The foundation of "chemical thermodynamics" was laid by an American scientist, Josiah Willard Gibbs (1839–1903). In 1876 Gibbs published a 323-page paper entitled "On the Equilibrium of Heterogeneous Substances," in which he presented the basic principles of chemical equilibrium and phase equilibrium, introduced a new concept called *free energy,* and explained the relations governing energy changes in electrochemical cells.

The son of a Yale professor and the first person to be awarded a Ph.D. in science from an American university, Gibbs spent all of his professional life as a professor of mathematical physics at Yale. Although he is arguably the most brilliant native-born American scientist, Gibbs was a modest and private individual, and he never gained the eminence that his contemporary and admirer James Maxwell did in Europe. Even today, very few people outside of chemistry and physics have ever heard of Gibbs. ◆

The efficiency of converting heat to work, as in the case of a steam locomotive, is governed by the second law of thermodynamics.

19.1 LAWS OF THERMODYNAMICS

Thermodynamics is an extensive and far-reaching scientific discipline that enables us to use information gained from experiments on a system to draw conclusions about other aspects of the same system without further experimentation. In Chapter 6 we discussed the first law of thermodynamics, which says that energy can be converted from one form to another, but it cannot be created or destroyed. One measure of these changes is the amount of heat given off or absorbed by a system during a constant-pressure process, which we define as a change in enthalpy (ΔH).

This chapter introduces the second law of thermodynamics and the Gibbs free-energy function. The second law, which is based on entropy change of the universe, explains why chemical processes tend to favor one direction. More conveniently, we can study the spontaneity of a reaction by using the Gibbs free energy, which focuses only on the system.

19.2 SPONTANEOUS PROCESSES

One of the main objectives of thermodynamics is to predict whether or not a reaction will occur when reactants are brought together under specific conditions (for example, at a certain temperature, pressure, and concentration). A reaction that *does* occur under the given set of conditions is called a *spontaneous reaction.*

We observe spontaneous physical and chemical processes every day, including many of the following examples:

- A waterfall runs downhill, but never up, spontaneously.
- A lump of sugar spontaneously dissolves in a cup of coffee, but dissolved sugar does not spontaneously reappear in its original form.
- Water freezes spontaneously below 0°C, and ice melts spontaneously above 0°C (at 1 atm).
- Heat flows from a hotter object to a colder one, but the reverse never happens spontaneously.
- The expansion of a gas in an evacuated bulb is a spontaneous process [Figure 19.1(a)]. The reverse process, that is, the gathering of all the molecules into one bulb, is not spontaneous [Figure 19.1(b)].
- A piece of sodium metal reacts violently with water to form sodium hydroxide and hydrogen gas. However, hydrogen gas does not react with sodium hydroxide to form water and sodium.
- Iron exposed to water and oxygen forms rust, but rust does not spontaneously change back to iron.

These examples show that processes that occur spontaneously in one direction cannot, *under the same conditions,* also take place spontaneously in the opposite direction. Note that "spontaneous" in this context does *not* mean "fast." It simply means that a reaction is certain to occur. For instance, the formation of rust can take months, but under the right conditions, iron will inevitably rust.

How can we predict whether or not a process will occur spontaneously? If we assume that spontaneous processes take place because they decrease the energy of

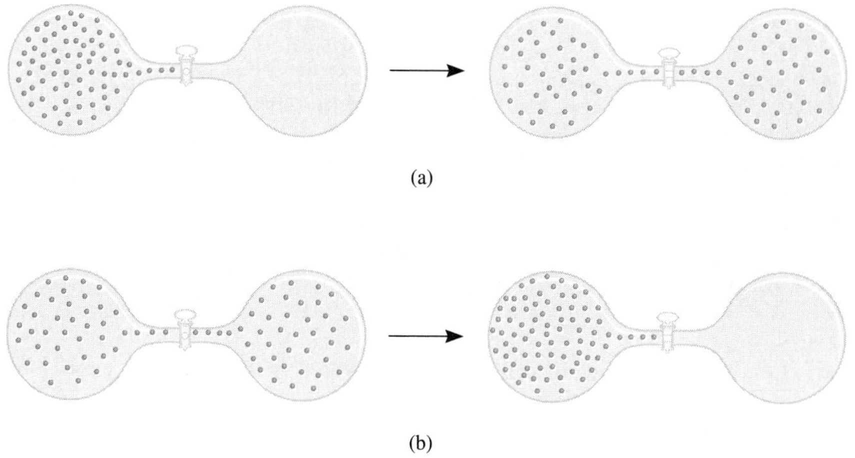

FIGURE 19.1

(a) A spontaneous process. (b) A non-spontaneous process.

a system, we can explain why a ball rolls downhill and why springs in a clock unwind. In chemistry we find that a large number of exothermic reactions are spontaneous. An example is the combustion of methane:

$$CH_4(g) + 2O_2(g) \longrightarrow CO_2(g) + 2H_2O(l) \qquad \Delta H° = -890.4 \text{ kJ}$$

Another example is the acid-base neutralization reaction:

$$H^+(aq) + OH^-(aq) \longrightarrow H_2O(l) \qquad \Delta H° = -56.2 \text{ kJ}$$

But the assumption that spontaneous processes always decrease a system's energy fails in a number of cases. Consider a solid-to-liquid phase transition such as

$$H_2O(s) \longrightarrow H_2O(l) \qquad \Delta H° = 6.01 \text{ kJ}$$

Experience tells us that ice melts spontaneously above 0°C even though the process is endothermic. Now we consider the cooling that results when ammonium nitrate dissolves in water:

$$NH_4NO_3(s) \xrightarrow{H_2O} NH_4^+(aq) + NO_3^-(aq) \qquad \Delta H° = 25 \text{ kJ}$$

The dissolution process is spontaneous; yet it is also endothermic. The decomposition of mercury(II) oxide is an endothermic reaction that is nonspontaneous at room temperature but becomes spontaneous when the temperature is raised:

$$2HgO(s) \longrightarrow 2Hg(l) + O_2(g) \qquad \Delta H° = 90.7 \text{ kJ}$$

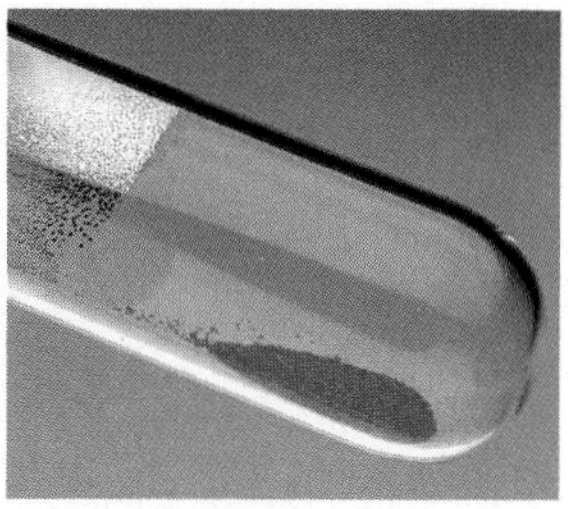

On heating, HgO decomposes to give Hg and O_2.

From a study of the examples mentioned and many more cases, chemists have come to the following conclusion: Exothermicity favors the spontaneity of a reaction but does not guarantee it. Enthalpy changes alone do not enable us to predict spontaneity or nonspontaneity. Thus, we will set aside the discussion of enthalpy changes and spontaneous reactions for the moment and look at another thermodynamic function that can be used to predict the direction of a reaction.

19.3 ENTROPY

In order to predict the spontaneity of a process, we need to know the changes in both the enthalpy and the disorder of the system. ***Entropy (S)*** is *a direct measure of* *the randomness or disorder of a system.* In other words, entropy describes the extent to which atoms, molecules, or ions are distributed in a disorderly fashion in a given region in space. The greater the disorder of a system, the greater its entropy. Conversely, the more ordered a system, the lower its entropy.

At the macroscopic level, we can think of order and disorder in terms of a deck of playing cards. A new deck of cards is arranged in order (the cards are in sequence from ace to king, and the order of the suits is spades, hearts, diamonds, clubs), but once the deck has been shuffled, the cards are disordered with respect to both number and suit. It is possible, but extremely unlikely, that reshuffling the cards will restore the original order. Moreover, there are many ways for the cards to be out of sequence but only one way for them to be ordered according to our definition.

The notion of order and disorder is related to probability. A probable event is one that can happen in many ways, and an improbable event is one that can happen in only one or a few ways. The idea of probability can be applied to the situation shown in Figure 19.1. Referring to Figure 19.1(a), suppose that initially we have only one molecule in the left flask. Since the two flasks have equal volume, the probability of finding the molecule in either flask after the stopcock is opened is $\frac{1}{2}$. If the number of molecules is increased to two, the probability of finding both molecules in the *same* flask becomes $\frac{1}{2} \times \frac{1}{2}$, or $\frac{1}{4}$. Now $\frac{1}{4}$ is an appreciable quantity, so it would not be surprising to find both molecules in the same flask after opening the stopcock. However, it is not difficult to see that as the number of molecules increases, the probability (P) of finding all the molecules in the *same* flask becomes progressively smaller:

$$P = \left(\frac{1}{2}\right) \times \left(\frac{1}{2}\right) \times \left(\frac{1}{2}\right) \cdots$$
$$= \left(\frac{1}{2}\right)^N$$

where N is the total number of molecules present. If $N = 100$, we have

$$P = \left(\frac{1}{2}\right)^{100} = 8 \times 10^{-31}$$

If N is of the order of 6×10^{23}, the number of molecules in 1 mole of a gas, the probability becomes $(\frac{1}{2})^{6\times10^{23}}$, which is such a small number that, for all practical purposes, it can be regarded as zero. On the basis of probability considerations, we would expect the gas to fill both flasks uniformly. By the same token, we now understand why the situation depicted in Figure 19.1(b) is not spontaneous: It represents a highly improbable event. In other words, an ordered state has a low probability of occurring and a low entropy, while a disordered state has a high probability of occurring and a high entropy.

Unlike energy and enthalpy, it is possible to determine the absolute entropy of a substance. Absolute entropies measured at 1 atm and 25°C, called *standard* entropies ($S°$), are generally used. (Recall that the standard state refers only to 1 atm. We choose 25°C because many processes are carried out at room temperature.) Table 19.1 lists standard entropies of a few elements and compounds; Appendix 2 provides a more extensive listing. The units of entropy are J/K or J/K · mol for 1 mole of the substance. (We use joules rather than kilojoules because entropy values are typically quite small.) Entropies of elements and compounds are positive (that is, $S° > 0$). By contrast, the standard enthalpy of formation (ΔH_f°) for elements in their stable form is zero, and for compounds it may be positive or negative.

TABLE 19.1
Standard Entropy Values ($S°$) for Selected Substances

Substance	$S°$ (J/K · mol)
$H_2O(l)$	69.9
$H_2O(g)$	188.7
$Br_2(l)$	152.3
$Br_2(g)$	245.3
$I_2(s)$	116.7
$I_2(g)$	260.6
C(diamond)	2.44
C(graphite)	5.69
He(g)	126.1
Ne(g)	146.2

For any substance, the particles in the solid are more ordered than those in the liquid state, which in turn are more ordered than those in the gaseous state. Water vapor has a greater entropy value than liquid water, because a mole of water has a much smaller volume in the liquid state than it does in the gaseous state. Consequently, the water molecules are more ordered in the liquid state because there are fewer positions for them to occupy. Similarly, bromine vapor has a greater entropy value than liquid bromine, and iodine vapor has a greater entropy value than solid iodine. Thus for any substance, the entropy for the same molar amount always increases in the following order:

$$S_{\text{solid}} < S_{\text{liquid}} < S_{\text{gas}}$$

Both diamond and graphite are solid forms of carbon. Diamond has a smaller entropy value than graphite because diamond has a more ordered structure. Neon has a higher entropy value than helium, because atoms that have more electrons and greater mass have greater entropy than atoms with fewer electrons and smaller mass.

Like energy and enthalpy, entropy is a state function. Consider a certain process in which a system changes from some initial state to some final state. The entropy change for the process, ΔS, is

$$\Delta S = S_{\text{f}} - S_{\text{i}}$$

where S_{f} and S_{i} are the entropies of the system in the final and initial states, respectively. If the change results in an increase in randomness, or disorder, then $S_{\text{f}} > S_{\text{i}}$, and $\Delta S > 0$. Figure 19.2 shows several processes that lead to an increase in entropy. In each case the system changes from a more ordered state to a less

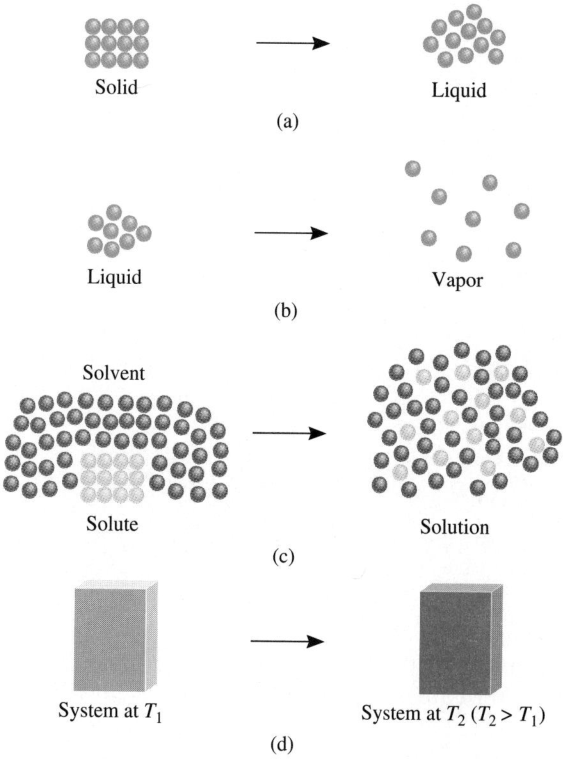

FIGURE 19.2

Processes that lead to an increase in entropy of the system: (a) melting: $S_{liquid} > S_{solid}$; (b) vaporization: $S_{vapor} > S_{liquid}$; (c) dissolving: $S_{soln} > (S_{solvent} + S_{solute})$; (d) heating: $S_{T_2} > S_{T_1}$.

FIGURE 19.3

(a) A diatomic molecule can rotate about the y- and z-axes (the x-axis is along the bond). (b) Vibrational motion of a diatomic molecule. Chemical bonds can be stretched and compressed like a spring.

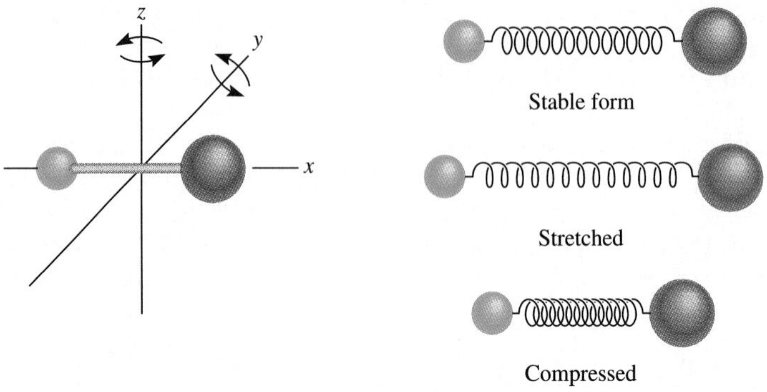

ordered one. It is clear that for melting and vaporization processes alike, $\Delta S > 0$.

When a sugar crystal dissolves in water, the highly ordered structure of the solid and part of the water are destroyed. Consequently, the solution possesses greater disorder than the pure solute and pure solvent. (The hydration of sugar molecules actually results in a decrease in entropy because the water molecules become more ordered around the solute molecules. But this decrease in not enough to offset the increase due to the mixing of two different compounds.)

Heating also increases the entropy of a system because it increases the random motion of atoms and molecules. In addition to translational motion (that is, movement from one point to another), molecules can also experience rotational and vibrational motion (Figure 19.3). The higher the temperature, the greater the kinetic energy of the system. This means more randomness at the molecular level, which corresponds to an increase in entropy.

Bromine is a fuming liquid at room temperature.

EXAMPLE 19.1

Predicting Entropy Changes

Predict whether the entropy change is greater than or less than zero for each of the following processes: (a) freezing ethanol, (b) evaporating a beaker of bromine at room temperature, (c) dissolving urea in water, (d) cooling nitrogen gas from 80°C to 20°C.

Answer: (a) This is a liquid-to-solid phase transition. The system becomes more ordered, so that $\Delta S < 0$.

(b) This is a liquid-to-vapor phase transition. The system becomes more disordered, and $\Delta S > 0$.

(c) A solution is invariably more disordered than its components (the solute and solvent). Therefore, $\Delta S > 0$.

(d) Cooling decreases molecular motion; therefore, $\Delta S < 0$.

PRACTICE EXERCISE

How does the entropy of a system change for each of the following processes? (a) Condensing water vapor, (b) crystallizing sucrose, (c) heating hydrogen gas from 60°C to 80°C, (d) subliming dry ice.

19.4 THE SECOND LAW OF THERMODYNAMICS

The connection between entropy and the spontaneity of a reaction is expressed by the ***second law of thermodynamics:*** *The entropy of the universe increases in a spontaneous process and remains unchanged in an equilibrium process.* Since the universe is made up of the system and the surroundings, the entropy change in the universe (ΔS_{univ}) for any process is the *sum* of the entropy changes in the system (ΔS_{sys}) and in the surroundings (ΔS_{surr}). Mathematically, we can express the second law of thermodynamics as follows:

A spontaneous process: $\Delta S_{univ} = \Delta S_{sys} + \Delta S_{surr} > 0$ (19.1)

An equilibrium process: $\Delta S_{univ} = \Delta S_{sys} + \Delta S_{surr} = 0$ (19.2)

For a spontaneous process this law says that ΔS_{univ} must be greater than zero, but it does not place a restriction on either ΔS_{sys} or ΔS_{surr}. Thus it is possible for either ΔS_{sys} or ΔS_{surr} to be negative, as long as the sum of these two quantities is greater than zero. For an equilibrium process, ΔS_{univ} is zero. In this case ΔS_{sys} and ΔS_{surr} must be equal in magnitude, but opposite in sign. What if for some process we find that S_{univ} is negative? What this means is that the process is not spontaneous in the direction described. Rather, the process is spontaneous in the *opposite* direction.

To calculate ΔS_{univ}, we need to know both ΔS_{sys} and ΔS_{surr}. Here we will focus on ΔS_{sys}. Let us suppose that the system is represented by the following reaction:

$$a\text{A} + b\text{B} \longrightarrow c\text{C} + d\text{D}$$

As is the case for the enthalpy of a reaction [see Equation (6.8)], the standard entropy change $\Delta S°$ is given by

$$\Delta S°_{rxn} = [cS°(\text{C}) + dS°(\text{D})] - [aS°(\text{A}) + bS°(\text{B})] \qquad (19.3)$$

or, in general, using Σ to represent summation and m and n for the stoichiometric coefficients in the reaction,

$$\Delta S°_{rxn} = \Sigma nS°(\text{products}) - \Sigma mS°(\text{reactants}) \qquad (19.4)$$

Using the standard entropy values ($S°$) listed in Appendix 2, we can calculate $\Delta S°_{rxn}$, which corresponds to ΔS_{sys}.

EXAMPLE 19.2
Calculating Entropy Changes of a System

From the absolute entropy values in Appendix 2, calculate the standard entropy changes for the following reactions at 25°C.

(a) $CaCO_3(s) \longrightarrow CaO(s) + CO_2(g)$
(b) $N_2(g) + 3H_2(g) \longrightarrow 2NH_3(g)$
(c) $H_2(g) + Cl_2(g) \longrightarrow 2HCl(g)$

Answer: We can calculate $\Delta S°$ by using Equation (19.4).

(a)
$$\begin{aligned}\Delta S°_{rxn} &= [S°(CaO) + S°(CO_2)] - [S°(CaCO_3)] \\ &= [(1\text{ mol})(39.8\text{ J/K}\cdot\text{mol}) + (1\text{ mol})(213.6\text{ J/K}\cdot\text{mol})] \\ &\qquad - (1\text{ mol})(92.9\text{ J/K}\cdot\text{mol}) \\ &= 160.5\text{ J/K}\end{aligned}$$

Thus, when 1 mole of $CaCO_3$ decomposes to form 1 mole of CaO and 1 mole of gaseous CO_2, there is an increase in entropy equal to 160.5 J/K.

(b) $$\begin{aligned}\Delta S^\circ_{rxn} &= [2S^\circ(NH_3)] - [S^\circ(N_2) + 3S^\circ(H_2)]\\ &= (2\text{ mol})(193\text{ J/K}\cdot\text{mol}) - [(1\text{ mol})(192\text{ J/K}\cdot\text{mol}) + (3\text{ mol})(131\text{ J/K}\cdot\text{mol})]\\ &= -199\text{ J/K}\end{aligned}$$

This result shows that when 1 mole of gaseous nitrogen reacts with 3 moles of gaseous hydrogen to form 2 moles of gaseous ammonia, there is a decrease in entropy equal to -199 J/K.

(c) $$\begin{aligned}\Delta S^\circ_{rxn} &= [2S^\circ(HCl)] - [S^\circ(H_2) + S^\circ(Cl_2)]\\ &= (2\text{ mol})(187\text{ J/K}\cdot\text{mol}) - [(1\text{ mol})(131\text{ J/K}\cdot\text{mol}) + (1\text{ mol})(223\text{ J/K}\cdot\text{mol})]\\ &= 20\text{ J/K}\end{aligned}$$

Thus the formation of 2 moles of gaseous HCl from 1 mole of gaseous H_2 and 1 mole of gaseous Cl_2 results in a small increase in entropy equal to 20 J/K.

PRACTICE EXERCISE

Calculate the standard entropy change for the following reactions at 25°C:

(a) $2CO(g) + O_2(g) \longrightarrow 2CO_2(g)$
(b) $3O_2(g) \longrightarrow 2O_3(g)$
(c) $2NaHCO_3(s) \longrightarrow Na_2CO_3(s) + H_2O(l) + CO_2(g)$

The results of Example 19.2 are consistent with the results observed for many other reactions. Taken together, they support the following general rules:

- If a reaction produces more gas molecules than it consumes [Example 19.2(a)], ΔS° is positive. If the total number of gas molecules diminishes [Example 19.2(b)], ΔS° is negative.
- If there is no net change in the total number of gas molecules [Example 19.2(c)], then ΔS° may be positive or negative, but it will be relatively small numerically.

These conclusions make sense, given that gases invariably have greater entropy than liquids and solids. For reactions involving only liquids and solids prediction is more difficult, but the general rule still holds in many cases that an increase in the total number of molecules and/or ions is accompanied by an increase in entropy.

EXAMPLE 19.3

Predicting Entropy Changes of a System

Predict whether the entropy change of the system in each of the following reactions is positive or negative.

(a) $Ag^+(aq) + Cl^-(aq) \longrightarrow AgCl(s)$
(b) $NH_4Cl(s) \longrightarrow NH_3(g) + HCl(g)$
(c) $H_2(g) + Br_2(g) \longrightarrow 2HBr(g)$

Answer: (a) The Ag^+ and Cl^- ions are free to move in solution, whereas AgCl is a solid. Furthermore, the number of particles decreases from left to right. Therefore, ΔS is negative.

(b) Since the solid is converted to two gaseous products, ΔS is positive.

(c) We see that the same number of moles of gas is involved in the reactants as in the product. Therefore we cannot predict the sign of ΔS, but we know the change must be quite small.

PRACTICE EXERCISE

Discuss qualitatively the sign of the entropy change expected for each of the following processes:

(a) $I_2(g) \longrightarrow 2I(g)$
(b) $2Zn(s) + O_2(g) \longrightarrow 2ZnO(s)$
(c) $N_2(g) + O_2(g) \longrightarrow 2NO(g)$

19.5 GIBBS FREE ENERGY

The second law of thermodynamics tells us that a spontaneous reaction increases the entropy of the universe; that is, $\Delta S_{univ} > 0$. In order to determine the sign of ΔS_{univ} for a reaction, however, we would need to calculate both ΔS_{sys} and ΔS_{surr}. But the calculation of ΔS_{surr} can be quite difficult, and besides, we are usually concerned only with what happens in a particular system rather than with what happens in the entire universe. Therefore, we rely on another thermodynamic function to help us determine whether a reaction will occur spontaneously if we consider only the system itself. This function, called ***Gibbs free energy (G),*** or simply *free energy,* brings the entropy and enthalpy together in an equation that shows the spontaneity of a reaction more directly:

Josiah Willard Gibbs (1839–1903)

$$G = H - TS \tag{19.5}$$

Remember that *all* the quantities in Equation (19.5) pertain to the system, and T is the temperature (in kelvins) of the system. You can see that G has units of energy (both H and TS have energy units), and, like H and S, it is a state function.

The change in free energy (ΔG) of a system for a process *at constant temperature* is

$$\Delta G = \Delta H - T\,\Delta S \tag{19.6}$$

We can summarize the conditions for spontaneity and equilibrium at constant temperature and pressure in terms of ΔG as follows:

$\Delta G < 0$ A spontaneous process in the forward direction.

$\Delta G > 0$ A nonspontaneous reaction. The reaction is spontaneous in the reverse direction.

$\Delta G = 0$ The system is at equilibrium. There is no net change.

"Free" energy in this context is simply the energy available to do work. Thus, if a particular reaction is accompanied by a release of usable energy (that is, if ΔG is negative), this fact alone guarantees that it must be spontaneous, and there is no need to worry about what happens to the rest of the universe.

STANDARD FREE-ENERGY CHANGES

For a reaction carried out under standard-state conditions, that is, reactants in their standard states are converted to products in their standard states, the free-energy change is called the ***standard free-energy change, $\Delta G°$.*** Table 19.2 summarizes the conventions used by chemists to define the standard states of pure substances as well as solutions. To calculate $\Delta G°$ we start with the equation

$$a\text{A} + b\text{B} \longrightarrow c\text{C} + d\text{D}$$

The standard free-energy change for this reaction is given by

$$\Delta G°_{\text{rxn}} = [c\Delta G°_f(\text{C}) + d\Delta G°_f(\text{D})] - [a\Delta G°_f(\text{A}) + b\Delta G°_f(\text{B})]$$

or, in general,

$$\Delta G°_{\text{rxn}} = \Sigma n\Delta G°_f(\text{products}) - \Sigma m\Delta G°_f(\text{reactants}) \qquad (19.7)$$

where m and n are stoichiometric coefficients. The term $\boldsymbol{\Delta G°_f}$ is the ***standard free energy of formation*** of a compound, *the free-energy change that occurs when 1 mole of the compound is synthesized from its elements in their standard states.* For the combustion of graphite:

$$\text{C(graphite)} + \text{O}_2(g) \longrightarrow \text{CO}_2(g)$$

the standard free-energy change [from Equation (19.7)] is

$$\Delta G°_{\text{rxn}} = \Delta G°_f(\text{CO}_2) - [\Delta G°_f(\text{C, graphite}) + \Delta G°_f(\text{O}_2)]$$

As in the case of the standard enthalpy of formation (p. 160), we define the standard free energy of formation of any element in its stable form as zero. Thus

$$\Delta G°_f(\text{C, graphite}) = 0 \qquad \text{and} \qquad \Delta G°_f(\text{O}_2) = 0$$

Therefore the standard free-energy change for the reaction in this case is numerically equal to the standard free energy of formation of CO_2:

$$\Delta G°_{\text{rxn}} = \Delta G°_f(\text{CO}_2)$$

Note that $\Delta G°_{\text{rxn}}$ is in kilojoules, but $\Delta G°_f$ is in kilojoules per mole. The equation holds because the coefficient in front of $\Delta G°_f$ (1 in this case) has the unit "mol." Appendix 2 lists the values of $\Delta G°_f$ for a number of compounds.

TABLE 19.2
Conventions for Standard States

State of matter	Standard state
Gas	1 atm pressure
Liquid	Pure liquid
Solid	Pure solid
Elements*	$\Delta G°_f(\text{element}) = 0$
Solution	1 molar concentration

*The most stable allotropic form at 25°C and 1 atm.

EXAMPLE 19.4
Calculating Standard Free-Energy Changes

Calculate the standard free-energy changes for the following reactions at 25°C.

(a) $CH_4(g) + 2O_2(g) \longrightarrow CO_2(g) + 2H_2O(l)$
(b) $2MgO(s) \longrightarrow 2Mg(s) + O_2(g)$

Answer: (a) According to Equation (19.7), we write

$$\Delta G^\circ_{rxn} = [\Delta G^\circ_f(CO_2) + 2\Delta G^\circ_f(H_2O)] - [\Delta G^\circ_f(CH_4) + 2\Delta G^\circ_f(O_2)]$$

We insert the values from Appendix 2:

$$\Delta G^\circ_{rxn} = [(1\text{ mol})(-394.4\text{ kJ/mol}) + (2\text{ mol})(-237.2\text{ kJ/mol})] - [(1\text{ mol})(-50.8\text{ kJ/mol}) + (2\text{ mol})(0\text{ kJ/mol})]$$
$$= -818.0\text{ kJ}$$

(b) The equation is

$$\Delta G^\circ_{rxn} = [2\Delta G^\circ_f(Mg) + \Delta G^\circ_f(O_2)] - [2\Delta G^\circ_f(MgO)]$$

From data in Appendix 2 we write

$$\Delta G^\circ_{rxn} = [(2\text{ mol})(0\text{ kJ/mol}) + (1\text{ mol})(0\text{ kJ/mol})] - [(2\text{ mol})(-569.6\text{ kJ/mol})]$$
$$= 1139\text{ kJ}$$

PRACTICE EXERCISE

Calculate the standard free-energy changes for the following reactions at 25°C:

(a) $H_2(g) + Br_2(g) \longrightarrow 2HBr(g)$
(b) $2C_2H_6(g) + 7O_2(g) \longrightarrow 4CO_2(g) + 6H_2O(l)$

In the preceding example the large negative value of ΔG° for the combustion of methane in (a) means that the reaction is a spontaneous process under standard-state conditions, whereas the decomposition of MgO in (b) is nonspontaneous because ΔG° is a large, positive quantity. Remember, however, that a large, negative ΔG° does not tell us anything about the actual *rate* of the spontaneous process; a mixture of CH_4 and O_2 at 25°C could sit unchanged for quite some time in the absence of a spark or flame.

Applications of Equation (19.6)

In order to predict the sign of ΔG, according to Equation (19.6), we need to know both ΔH and ΔS. A negative ΔH (an exothermic reaction) and a positive ΔS (a reaction that results in an increase in disorder of the system) tend to make ΔG negative, although temperature may influence the *direction* of a spontaneous reaction. The four possible outcomes of this relationship are:

- If both ΔH and ΔS are positive, then ΔG will be negative only when the $T\Delta S$ term is greater in magnitude than ΔH. This condition is met when T is large.
- If ΔH is positive and ΔS is negative, ΔG will always be positive, regardless of temperature.

TABLE 19.3
Factors Affecting the Sign of ΔG in the Relationship $\Delta G = \Delta H - T\Delta S$

ΔH	ΔS	ΔG	Example
+	+	Reaction proceeds spontaneously at high temperatures. At low temperatures, reaction is spontaneous in the reverse direction.	$H_2(g) + I_2(g) \longrightarrow 2HI(g)$
+	−	ΔG is always positive. Reaction is spontaneous in the reverse direction at all temperatures.	$3O_2(g) \longrightarrow 2O_3(g)$
−	+	ΔG is always negative. Reaction proceeds spontaneously at all temperatures.	$2H_2O_2(l) \longrightarrow 2H_2O(l) + O_2(g)$
−	−	Reaction proceeds spontaneously at low temperatures. At high temperatures, the reverse reaction becomes spontaneous.	$NH_3(g) + HCl(g) \longrightarrow NH_4Cl(s)$

- If ΔH is negative and ΔS is positive, then ΔG will always be negative regardless of temperature.
- If ΔH is negative and ΔS is negative, then ΔG will be negative only when $T\Delta S$ is smaller in magnitude than ΔH. This condition is met when T is small.

What temperatures will cause ΔG to be negative for the first and last cases depends on the actual values of ΔH and ΔS of the system. Table 19.3 summarizes the effects of the possibilities just described.

We will now consider two specific applications of Equation (19.6).

Temperature and Chemical Reactions

Calcium oxide (CaO), also called quicklime, is an extremely valuable inorganic substance used in steelmaking, production of calcium metal, the paper industry, water treatment, and pollution control. It is prepared by decomposing limestone ($CaCO_3$) in a kiln at high temperatures (Figure 19.4):

$$CaCO_3(s) \rightleftharpoons CaO(s) + CO_2(g)$$

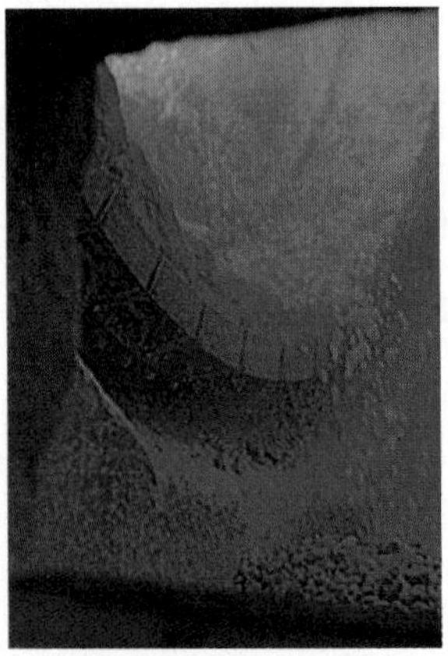

FIGURE 19.4

The production of CaO in a rotary kiln.

The reaction is reversible, and CaO readily combines with CO_2 to form $CaCO_3$. The pressure of CO_2 in equilibrium with $CaCO_3$ and CaO increases with temperature. In actual production the system is never maintained at equilibrium; rather, CO_2 is constantly removed from the kiln to shift the equilibrium from left to right, promoting the formation of calcium oxide.

The important information for the practical chemist is the temperature at which the decomposition of $CaCO_3$ becomes appreciable (that is, at what temperature the reaction becomes spontaneous). We can make a reliable estimate of that temperature as follows. First we calculate $\Delta H°$ and $\Delta S°$ for the reaction at 25°C, using the data in Appendix 2. To determine $\Delta H°$, we apply Equation (6.8):

$$\begin{aligned}\Delta H° &= [\Delta H_f°(CaO) + \Delta H_f°(CO_2)] - [\Delta H_f°(CaCO_3)] \\ &= [(1\ \text{mol})(-635.6\ \text{kJ/mol}) + (1\ \text{mol})(-393.5\ \text{kJ/mol})] \\ &\quad - [(1\ \text{mol})(-1206.9\ \text{kJ/mol})] \\ &= 177.8\ \text{kJ}\end{aligned}$$

Next we apply Equation (19.4) to find $\Delta S°$:

$$\Delta S^\circ = [S^\circ(\text{CaO}) + S^\circ(\text{CO}_2)] - [S^\circ(\text{CaCO}_3)]$$
$$= [(1 \text{ mol})(39.8 \text{ J/K} \cdot \text{mol}) + (1 \text{ mol})(213.6 \text{ J/K} \cdot \text{mol})] - [(1 \text{ mol})(92.9 \text{ J/K} \cdot \text{mol})]$$
$$= 160.5 \text{ J/K}$$

For reactions carried out under standard-state conditions, Equation (19.6) takes the form

$$\Delta G^\circ = \Delta H^\circ - T\Delta S^\circ$$

so we obtain

$$\Delta G^\circ = 177.8 \text{ kJ} - (298 \text{ K})(160.5 \text{ J/K})\left(\frac{1 \text{ kJ}}{1000 \text{ J}}\right)$$
$$= 130.0 \text{ kJ}$$

Since ΔG° is a large positive quantity, we conclude that the reaction is not favored at 25°C (or 298 K). In order to make ΔG° negative, we first have to find the temperature at which ΔG° is zero; that is,

$$0 = \Delta H^\circ - T\Delta S^\circ$$

or

$$T = \frac{\Delta H^\circ}{\Delta S^\circ}$$
$$= \frac{(177.8 \text{ kJ})(1000 \text{ J/1 kJ})}{160.5 \text{ J/K}}$$
$$= 1108 \text{ K}$$
$$= 835°\text{C}$$

At a temperature higher than 835°C, ΔG° becomes negative, indicating that the decomposition is spontaneous. For example, at 840°C, or 1113 K,

$$\Delta G^\circ = \Delta H^\circ - T\Delta S^\circ$$
$$= 177.8 \text{ kJ} - (1113 \text{ K})(160.5 \text{ J/K})\left(\frac{1 \text{ kJ}}{1000 \text{ J}}\right)$$
$$= -0.8 \text{ kJ}$$

Two points are worth making about such a calculation. First, we used the ΔH° and ΔS° values at 25°C to calculate changes that occur at a much higher temperature. Since both ΔH° and ΔS° change with temperature, this approach will not give us an accurate value of ΔG°, but it is good enough for "ball park" estimates. Second, we should not be misled into thinking that nothing happens below 835°C and that at 835°C $CaCO_3$ suddenly begins to decompose. Far from it. The fact that ΔG° is a positive value at some temperature below 835°C does not mean that no CO_2 is produced. It only says that the pressure of the CO_2 gas formed at that temperature will be below 1 atm (its standard-state value; see Table 19.2). As Figure 19.5 shows, the pressure of CO_2 at first increases very slowly with temperature; it becomes easily measurable above 700°C. The significance of 835°C is that this is the temperature at which the equilibrium pressure of CO_2 reaches 1 atm. Above 835°C, the equilibrium pressure of CO_2 exceeds 1 atm. (In Section 19.5 we will see how ΔG° is related to the equilibrium constant of a given reaction.)

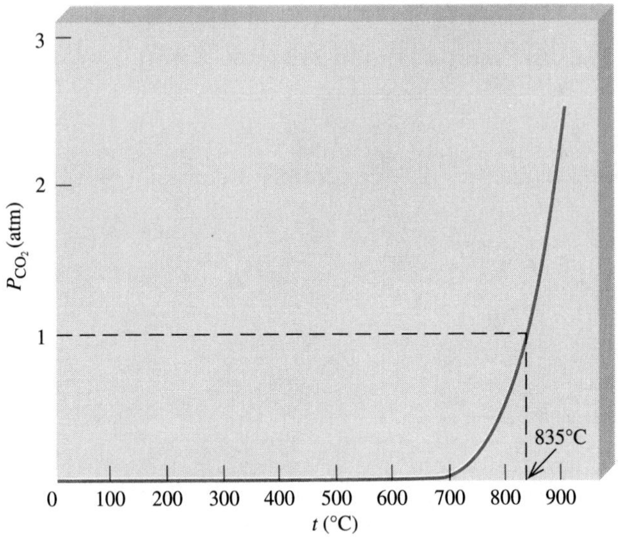

FIGURE 19.5

Equilibrium pressure of CO_2 from the decomposition of $CaCO_3$, as a function of temperature. This curve is calculated by assuming that ΔH and ΔS of the reaction do not change with temperature.

Phase Transitions

At the transition temperature (that is, at the melting point or the boiling point) a system is at equilibrium ($\Delta G = 0$), so Equation (19.6) becomes

$$0 = \Delta H - T\Delta S$$

$$\Delta S = \frac{\Delta H}{T}$$

Let us first consider the ice-water equilibrium. For the ice → water transition, ΔH is the molar heat of fusion (see Table 11.7) and T is the melting point. The entropy change is therefore

$$\Delta S_{\text{ice}\rightarrow\text{water}} = \frac{6010 \text{ J/mol}}{273 \text{ K}}$$
$$= 22.0 \text{ J/K} \cdot \text{mol}$$

Thus when 1 mole of ice melts at 0°C, there is an increase in entropy of 22.0 J/K. The increase in entropy is consistent with the increase in disorder from solid to liquid. Conversely, for the water → ice transition, the decrease in entropy is given by

$$\Delta S_{\text{water}\rightarrow\text{ice}} = \frac{-6010 \text{ J/mol}}{273 \text{ K}}$$
$$= -22.0 \text{ J/K} \cdot \text{mol}$$

In the laboratory we normally carry out unidirectional changes, that is, either ice to water or water to ice. We can calculate the entropy change in each case using the equation $\Delta S = \Delta H/T$ as long as the temperature remains at 0°C. The same procedure can be applied to the water-steam transition. In this case ΔH is the heat of vaporization and T is the boiling point of water.

EXAMPLE 19.5
Entropy Changes Due to Phase Transitions

The molar heats of fusion and vaporization of benzene are 10.9 kJ/mol and 31.02 kJ/mol, respectively. Calculate the entropy changes for the solid → liquid and liquid → vapor transitions for benzene. At 1 atm pressure, benzene melts at 5.5°C and boils at 80.1°C.

Answer: At the melting point, the system is at equilibrium. Therefore $\Delta G = 0$ and the entropy of fusion is given by

$$0 = \Delta H_{\text{fus}} - T_{\text{f}}\,\Delta S_{\text{fus}}$$

$$\Delta S_{\text{fus}} = \frac{\Delta H_{\text{fus}}}{T_{\text{f}}} = \frac{(10.9\ \text{kJ/mol})(1000\ \text{J/1 kJ})}{(5.5 + 273)\ \text{K}} = 39.1\ \text{J/K} \cdot \text{mol}$$

Similarly, at the boiling point, $\Delta G = 0$ and we have

$$\Delta S_{\text{vap}} = \frac{\Delta H_{\text{vap}}}{T_{\text{bp}}} = \frac{(31.0\ \text{kJ/mol})(1000\ \text{J/1 kJ})}{(80.1 + 273)\ \text{K}} = 87.8\ \text{J/K} \cdot \text{mol}$$

PRACTICE EXERCISE

The molar heats of fusion and vaporization of argon are 1.3 kJ/mol and 6.3 kJ/mol, and argon's melting point and boiling point are −190°C and −186°C, respectively. Calculate the entropy changes for fusion and vaporization.

Liquid and solid benzene in equilibrium.

19.6 FREE ENERGY AND CHEMICAL EQUILIBRIUM

In this section we take a closer look at the difference between ΔG and $\Delta G°$ and at the relationship between free-energy changes and equilibrium constants.

The equations for free-energy change and standard free-energy change are, respectively,

$$\Delta G = \Delta H - T\Delta S$$

$$\Delta G° = \Delta H° - T\Delta S°$$

It is important to understand the conditions under which these equations apply and what kind of information we can obtain from ΔG and $\Delta G°$. Let us illustrate with the following reaction:

$$\text{reactants} \longrightarrow \text{products}$$

The standard free-energy change for the reaction is given by

$$\Delta G^\circ = G^\circ(\text{products}) - G^\circ(\text{reactants})$$

The quantity ΔG° represents the change when the reactants in their standard states are converted to products in their standard states (see Table 19.2 for definitions of standard states). We have already seen in Example 19.4 how ΔG° for a reaction is calculated from the known standard free energies of formation of reactants and products. Suppose we start the reaction in solution with all the reactants in their standard states (that is, all at 1 *M* concentration). As soon as the reaction starts, the standard-state condition no longer exists for the reactants or the products since neither remains at 1 *M* concentration. Under conditions that are *not* standard state, we must use ΔG rather than ΔG° to predict the direction of the reaction. The relationship between ΔG and ΔG° is

$$\Delta G = \Delta G^\circ + RT \ln Q \tag{19.8}$$

where R is the gas constant (8.314 J/K · mol), T the absolute temperature of the reaction, ln the natural logarithm (see Appendix 3), and Q the reaction quotient defined on page 427. We see that ΔG depends on two quantities: ΔG° and $RT \ln Q$. For a given reaction at temperature T the value of ΔG° is fixed but that of $RT \ln Q$ is not, because Q varies according to the composition of the reacting mixture. Let us consider two special cases:

Case 1: If ΔG° is a large negative value, the $RT \ln Q$ term will not be positive enough to match the ΔG° term until significant product formation has occurred.

Case 2: If ΔG° is a large positive value, the $RT \ln Q$ term will be more negative than ΔG° is positive only as long as very little product has formed and the concentration of the reactant is high relative to that of the product.

At equilibrium, by definition, $\Delta G = 0$ and $Q = K$, where K is the equilibrium constant. Thus

$$0 = \Delta G^\circ + RT \ln K$$

or

$$\Delta G^\circ = -RT \ln K \tag{19.9}$$

In this equation K_P is used for gases and K_c for reactions in solution. Note that the larger K is, the more negative is ΔG°.

For chemists, Equation (19.9) is one of the most important equations in thermodynamics because it relates the equilibrium constant of a reaction to the change in standard free energy. Thus from a knowledge of K we can calculate ΔG°, and vice versa. Figure 19.6 shows plots of free energy of a reacting system versus the extent of the reaction for two types of reactions. As you can see, if $\Delta G^\circ < 0$, the products are favored over reactants at equilibrium. Conversely, if $\Delta G^\circ > 0$, there will be more reactants than products at equilibrium. Table 19.4 summarizes the three possible relations between ΔG° and K, as predicted by Equation (19.9).

For reactions having very large or very small equilibrium constants, it is generally very difficult, if not impossible, to measure K by monitoring the concentrations of all the reacting species. Consider, for example, the formation of nitric oxide from molecular nitrogen and molecular oxygen:

$$N_2(g) + O_2(g) \rightleftharpoons 2NO(g)$$

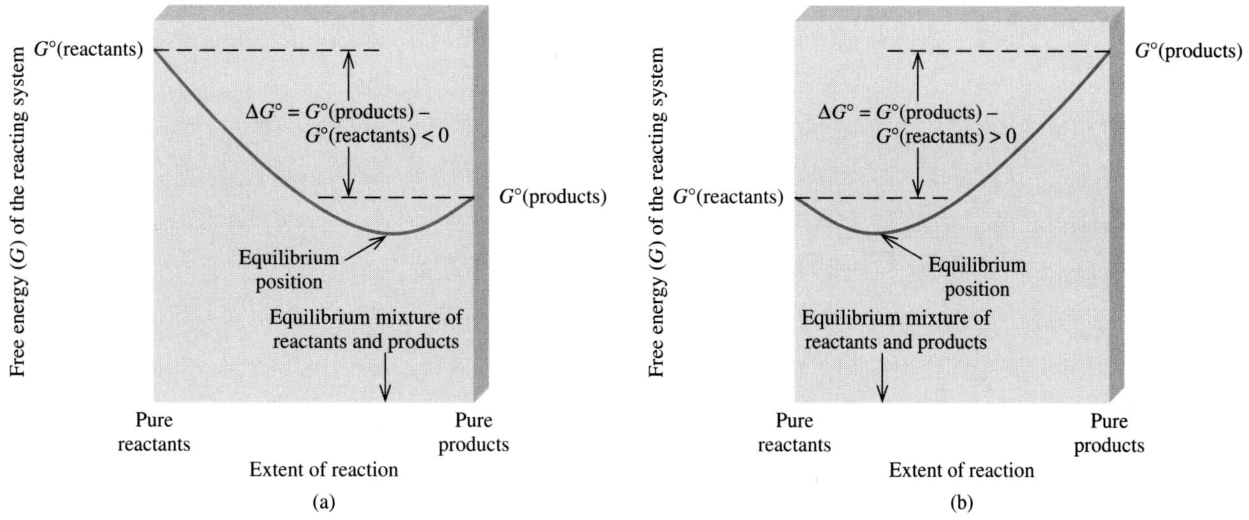

FIGURE 19.6

(a) $G^\circ_{products} < G^\circ_{reactants}$, so $\Delta G^\circ < 0$. At equilibrium, there is a significant conversion of reactants to products. (b) $G^\circ_{products} > G^\circ_{reactants}$, so $\Delta G^\circ > 0$. At equilibrium, reactants are favored over products.

TABLE 19.4

Relation between ΔG° and K as Predicted by the Equation $\Delta G^\circ = -RT \ln K$

K	$\ln K$	ΔG°	Comments
>1	Positive	Negative	Products are favored over reactants at equilibrium.
=1	0	0	Products and reactants are equally favored at equilibrium.
<1	Negative	Positive	Reactants are favored over products at equilibrium.

At 25°C, the equilibrium constant K_c is

$$K_c = \frac{[NO]^2}{[N_2][O_2]} = 4.0 \times 10^{-31}$$

The very small value of K_c means that the concentration of NO at equilibrium will be exceedingly low. In such a case the equilibrium constant is more conveniently obtained from the ΔG° of the reaction. (As we have seen, ΔG° can be calculated from the ΔH° and ΔS° values.) On the other hand, the equilibrium constant for the formation of hydrogen iodide from molecular hydrogen and molecular iodine is near unity at room temperature:

$$H_2(g) + I_2(g) \rightleftharpoons 2HI(g)$$

For this reason it is easier to measure K and then calculate ΔG° using Equation (19.9) than to measure ΔH° and ΔS°.

EXAMPLE 19.6
Using $\Delta G°$ to Calculate the Equilibrium Constant

Using data listed in Appendix 2, calculate the equilibrium constant (K_P) for the following reaction at 25°C:

$$2H_2O(l) \rightleftharpoons 2H_2(g) + O_2(g)$$

Answer: According to Equation (19.7)

$$\Delta G°_{rxn} = [2\Delta G°_f(H_2) + \Delta G°_f(O_2)] - [2\Delta G°_f(H_2O)]$$

$$= [2\text{ mol})(0\text{ kJ/mol}) + (1\text{ mol})(0\text{ kJ/mol})] - [(2\text{ mol})(-237.2\text{ kJ/mol})]$$

$$= 474.4\text{ kJ}$$

Using Equation (19.9)

$$\Delta G°_{rxn} = -RT\ln K_P$$

$$474.4\text{ kJ} \times \frac{1000\text{ J}}{1\text{ kJ}} = -(8.314\text{ J/K}\cdot\text{mol})(298\text{ K})\ln K_P$$

$$\ln K_P = -191.5$$

$$K_P = e^{-191.5}$$

$$= 7 \times 10^{-84}$$

PRACTICE EXERCISE

Calculate the equilibrium constant (K_P) for the reaction

$$2O_3(g) \longrightarrow 3O_2(g)$$

at 25°C.

To apply Equation (19.8), recall that the reaction quotient Q has the same form as the equilibrium constant K except that the concentration terms in Q do not refer to those at equilibrium (see p. 427).

EXAMPLE 19.7
Using ΔG to Predict the Direction of a Reaction

The standard free-energy change for the reaction

$$N_2(g) + 3H_2(g) \rightleftharpoons 2NH_3(g)$$

is -33.2 kJ at 25°C. In a certain experiment, the initial pressures are P_{H_2} = 0.250 atm, P_{N_2} = 0.870 atm, and P_{NH_3} = 12.9 atm. Calculate ΔG for the reaction at these pressures, and predict the direction of reaction.

Answer: Equation (19.8) can be written as

$$\Delta G = \Delta G° + RT\ln Q_P$$

$$= \Delta G° + RT\ln\frac{P^2_{NH_3}}{P^3_{H_2}P_{N_2}}$$

$$\Delta G = -33.2 \times 1000 \text{ J} + (8.314 \text{ J/K} \cdot \text{mol})(298 \text{ K}) \times \ln \frac{(12.9)^2}{(0.250)^3(0.870)}$$

$$= -33.2 \times 10^3 \text{ J} + 23.3 \times 10^3 \text{ J}$$

$$= -9.9 \times 10^3 \text{ J} = -9.9 \text{ kJ}$$

Since ΔG is negative, the net reaction proceeds from left to right.

PRACTICE EXERCISE

The $\Delta G°$ for the reaction

$$H_2(g) + I_2(g) \rightleftharpoons 2HI(g)$$

is 2.60 kJ at 25°C. In one experiment, the initial pressures are $P_{H_2} = 4.26$ atm, $P_{I_2} = 0.024$ atm, and $P_{HI} = 0.23$ atm. Calculate ΔG for the reaction and predict the direction of reaction.

19.7 THERMODYNAMICS IN LIVING SYSTEMS

Many biochemical reactions are nonspontaneous (that is, they have a positive $\Delta G°$ value), yet they are essential to the maintenance of life. These reactions are made to occur by coupling them to an energetically favorable process, one that has a negative $\Delta G°$ value. The principle of *coupled reactions* is based on a simple concept, although the necessary machinery for carrying out these reactions in the human body took billions of years to develop.

During metabolism, food molecules, represented by glucose ($C_6H_{12}O_6$), are converted to carbon dioxide and water with a substantial release of free energy:

$$C_6H_{12}O_6(s) + 6O_2(g) \longrightarrow 6CO_2(g) + 6H_2O(l) \qquad \Delta G° = -2880 \text{ kJ}$$

In a living cell, this reaction does not take place in a single step (as burning glucose in a flame would); rather, the glucose molecule, in the presence of enzymes, is broken down in many steps to yield the final products. Much of the energy released along the way is captured (stored) by synthesizing adenosine triphosphate (ATP) from adenosine diphosphate (ADP) and phosphoric acid (Figure 19.7):

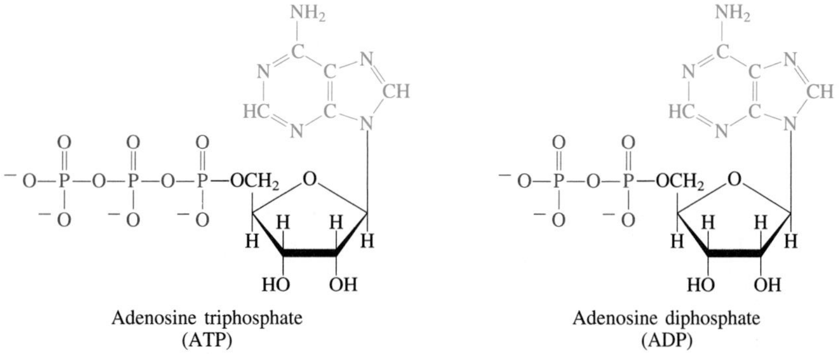

FIGURE 19.7

Structure of ATP and ADP in ionized forms. The adenine group is in blue, the ribose group in black, and the phosphate group in red. Note that ADP has one fewer phosphate group than ATP.

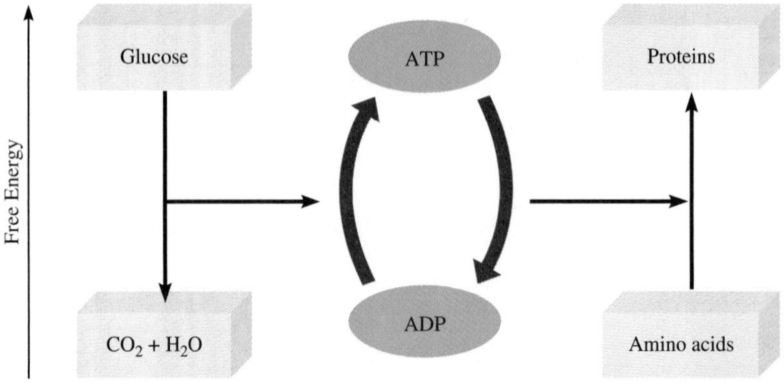

FIGURE 19.8

Schematic representation of ATP synthesis and coupled reactions in living systems. The conversion of glucose to carbon dioxide and water during metabolism releases free energy. This release of free energy is used to convert ADP into ATP. The ATP molecules are then used as an energy source to drive nonspontaneous reactions, such as protein synthesis from amino acids.

$$\text{ADP} + \text{H}_3\text{PO}_4 \longrightarrow \text{ATP} + \text{H}_2\text{O} \qquad \Delta G^\circ = +31 \text{ kJ}$$

Under appropriate conditions, ATP undergoes hydrolysis to give ADP and phosphoric acid, with a release of 31 kJ of free energy which can be used to drive energetically unfavorable reactions.

Consider the synthesis of the dipeptide analyglycine shown in Figure 14.13:

$$\text{alanine} + \text{glycine} \longrightarrow \text{alanylglycine} \qquad \Delta G^\circ = +29 \text{ kJ}$$

This nonspontaneous process is the first step in the formation of a polypeptide chain (a protein molecule). With the aid of an enzyme, the reaction is coupled to the hydrolysis of ATP as follows:

$$\text{ATP} + \text{H}_2\text{O} + \text{alanine} + \text{glycine} \longrightarrow \text{ADP} + \text{H}_3\text{PO}_4 + \text{alanylglycine}$$

The overall free-energy change is given by $\Delta G^\circ = -31 \text{ kJ} + 29 \text{ kJ} = -2 \text{ kJ}$, which means that the coupled reaction is spontaneous. Figure 19.8 shows the ATP-ADP interconversions that result in energy storage (from metabolism) and free-energy release (from ATP hydrolysis) to drive life-sustaining reactions.

SUMMARY

Entropy is a measure of the disorder of a system. Any spontaneous process must lead to a net increase in entropy in the universe (second law of thermodynamics) or, stated mathematically, $\Delta S_{\text{univ}} = \Delta S_{\text{sys}} + \Delta S_{\text{surr}} > 0$. The standard entropy change of a chemical reaction ($\Delta S^\circ_{\text{rxn}}$) can be calculated from the absolute entropies of reactants and products.

The equation $\Delta G = \Delta H - T\Delta S$ can be used to predict the spontaneity of a process. Under conditions of constant temperature and pressure, the free-energy change ΔG is less than zero for a spontaneous process and greater than zero for a nonspontaneous process. For an equilibrium process, $\Delta G = 0$. The standard free-energy change for a reaction, ΔG°, can be found from the standard free energies of formation of reactants and products. The equilibrium constant of a reaction and the standard free-energy change of the reaction are related by the equation $\Delta G^\circ = -RT \ln K$.

Most biological reactions are nonspontaneous. They are driven by the hydrolysis of ATP, which has a negative ΔG° value.

KEY WORDS

Entropy (S), p. 537
Gibbs free energy, p. 543
Second law of thermodynamics, p. 541
Standard free-energy change ($\Delta G°$), p. 544
Standard free energy of formation ($\Delta G_f°$), p. 544

QUESTIONS AND PROBLEMS

SPONTANEOUS PROCESSES

Review Questions

19.1 Explain what a spontaneous process is. Give two examples each of spontaneous and nonspontaneous processes.

19.2 Which of the following processes are spontaneous and which are nonspontaneous? (a) Dissolving table salt (NaCl) in hot soup, (b) climbing Mt. Everest, (c) spreading fragrance in a room by removing the cap from a perfume bottle.

19.3 Which of the following processes are spontaneous and which are nonspontaneous at a given temperature?

(a) $NaNO_3(s) \xrightarrow{H_2O} NaNO_3(aq)$ saturated soln

(b) $NaNO_3(s) \xrightarrow{H_2O} NaNO_3(aq)$ unsaturated soln

(c) $NaNO_3(s) \xrightarrow{H_2O} NaNO_3(aq)$ supersaturated soln

Problem

19.4 Referring to Figure 19.1(a), calculate the probability of all the molecules in the same flask if the number is (a) 6, (b) 60, (c) 600.

ENTROPY; SECOND LAW OF THERMODYNAMICS

Review Questions

19.5 Define entropy. What are the units of entropy? State the second law of thermodynamics in words and express it mathematically.

19.6 How does the entropy of a system change for each of the following processes?
(a) A solid melts.
(b) A liquid freezes.
(c) A liquid boils.
(d) A vapor is converted to a solid.
(e) A vapor condenses to a liquid.
(f) A solid sublimes.

Problems

19.7 For each pair of substances listed here, choose the one having the larger standard entropy at 25°C. The same molar amount is used in the comparison. Explain the basis for your choice. (a) $Li(s)$ or $Li(l)$; (b) $C_2H_5OH(l)$ or $CH_3OCH_3(l)$ (*Hint:* Which molecule can form hydrogen bonds?); (c) $Ar(g)$ or $Xe(g)$; (d) $CO(g)$ or $CO_2(g)$; (e) graphite or diamond; (f) $NO_2(g)$ or $N_2O_4(g)$.

19.8 Arrange the following substances (1 mole each) in order of increasing entropy at 25°C: (a) $Ne(g)$, (b) $SO_2(g)$, (c) $Na(s)$, (d) $NaCl(s)$, (e) $NH_3(g)$. Give the reasons for your arrangement.

19.9 Using the data in Appendix 2, calculate the standard entropy change for the following reactions at 25°C:
(a) $S(s) + O_2(g) \rightarrow SO_2(g)$
(b) $MgCO_3(s) \rightarrow MgO(s) + CO_2(g)$

19.10 Using the data in Appendix 2, calculate the standard entropy change for the following reactions at 25°C:
(a) $H_2(g) + CuO(s) \rightarrow Cu(s) + H_2O(g)$
(b) $2Al(s) + 3ZnO(s) \rightarrow Al_2O_3(s) + 3Zn(s)$
(c) $CH_4(g) + 2O_2(g) \rightarrow CO_2(g) + 2H_2O(l)$

19.11 Without consulting Appendix 2, predict whether the entropy change is positive or negative for the following reactions. Explain your predictions.
(a) $2KClO_4(s) \rightarrow 2KClO_3(s) + O_2(g)$
(b) $H_2O(g) \rightarrow H_2O(l)$
(c) $S(s) + O_2(g) \rightarrow SO_2(g)$
(d) $2Na(s) + 2H_2O(l) \rightarrow 2NaOH(aq) + H_2(g)$
(e) $CH_4(g) + 2O_2(g) \rightarrow CO_2(g) + 2H_2O(l)$
(f) $N_2(g) \rightarrow 2N(g)$
(g) $2LiOH(aq) + CO_2(g) \rightarrow Li_2CO_3(aq) + H_2O(l)$

19.12 State whether the sign of the entropy change expected for each of the following processes will be positive or negative and explain your predictions.
(a) $PCl_3(l) + Cl_2(g) \rightarrow PCl_5(s)$
(b) $2HgO(s) \rightarrow 2Hg(l) + O_2(g)$
(c) $H_2(g) \rightarrow 2H(g)$
(d) $U(s) + 3F_2(g) \rightarrow UF_6(s)$

FREE ENERGY

Review Questions

19.13 Define free energy. What are its units?

19.14 Why is it more convenient to predict the direction of a reaction in terms of ΔG_{sys} rather than ΔS_{univ}? Under what conditions can ΔG_{sys} be used to predict the spontaneity of a reaction?

Problems

19.15 Calculate $\Delta G°$ for the following reactions at 25°C:
(a) $N_2(g) + O_2(g) \rightarrow 2NO(g)$
(b) $H_2O(l) \rightarrow H_2O(g)$
(c) $2C_2H_2(g) + 5O_2(g) \rightarrow 4CO_2(g) + 2H_2O(l)$
(*Hint:* Look up the standard free energies of formation of the reactants and products in Appendix 2.)

19.16 Calculate $\Delta G°$ for the following reactions at 25°C:
(a) $2Mg(s) + O_2(g) \rightarrow 2MgO(s)$
(b) $2SO_2(g) + O_2(g) \rightarrow 2SO_3(g)$
(c) $2C_2H_6(g) + 7O_2(g) \rightarrow 4CO_2(g) + 6H_2O(l)$
See Appendix 2 for thermodynamic data.

19.17 From the following ΔH and ΔS values, predict whether each of the reactions would be spontaneous at 25°C. If not, at what temperature might the reaction become spontaneous? Reaction A: ΔH = 10.5 kJ, ΔS = 30 J/K; reaction B: ΔH = 1.8 kJ, ΔS = −113 J/K.

19.18 At what temperatures would reactions with the following ΔH and ΔS values become spontaneous? (a) ΔH = −126 kJ, ΔS = 84 J/K; (b) ΔH = −11.7 kJ, ΔS = −105 J/K.

FREE ENERGY AND CHEMICAL EQUILIBRIUM

Review Questions

19.19 Explain why Equation (19.9) is of great importance in chemistry.

19.20 Explain clearly the difference between ΔG and $\Delta G°$.

Problems

19.21 For the reaction

$$H_2(g) + I_2(g) \rightleftharpoons 2HI(g)$$

$\Delta G°$ = 2.60 kJ. Calculate K_P for the reaction at 25°C.

19.22 For the autoionization of water

$$H_2O(l) \rightleftharpoons H^+(aq) + OH^-(aq)$$

K_w is 1.0×10^{-14}. What is $\Delta G°$ for the process?

19.23 Consider the following reaction at 25°C:

$$Fe(OH)_2(s) \rightleftharpoons Fe^{2+}(aq) + 2OH^-(aq)$$

Calculate $\Delta G°$ for the reaction. K_{sp} for $Fe(OH)_2$ is 1.6×10^{-14}.

19.24 Calculate $\Delta G°$ and K_P for the equilibrium reaction

$$2H_2O(g) \rightleftharpoons 2H_2(g) + O_2(g)$$

at 25°C.

19.25 (a) Calculate $\Delta G°$ and K_P for the following equilibrium reaction at 25°C. The $\Delta G_f°$ values are 0 for $Cl_2(g)$, −286 kJ/mol for $PCl_3(g)$, and −325 kJ/mol for $PCl_5(g)$.

$$PCl_5(g) \rightleftharpoons PCl_3(g) + Cl_2(g)$$

(b) Calculate ΔG for the reaction if the partial pressures of the initial mixture are P_{PCl_5} = 0.0029 atm, P_{PCl_3} = 0.27 atm, and P_{Cl_2} = 0.40 atm.

19.26 The equilibrium constant (K_P) for the reaction

$$H_2(g) + CO_2(g) \rightleftharpoons H_2O(g) + CO(g)$$

is 4.40 at 2000 K. (a) Calculate $\Delta G°$ for the reaction. (b) Calculate ΔG for the reaction when the partial pressures are P_{H_2} = 0.25 atm, P_{CO_2} = 0.78 atm, P_{H_2O} = 0.66 atm, and P_{CO} = 1.20 atm.

19.27 Consider the decomposition of calcium carbonate:

$$CaCO_3(s) \rightleftharpoons CaO(s) + CO_2(g)$$

Calculate the pressure in atmospheres of CO_2 in an equilibrium process (a) at 25°C and (b) at 800°C. Assume that $\Delta H°$ = 177.8 kJ and $\Delta S°$ = 160.5 J/K.

19.28 The equilibrium constant K_P for the reaction

$$CO(g) + Cl_2(g) \rightleftharpoons COCl_2(g)$$

is 5.62×10^{35} at 25°C. Calculate $\Delta G_f°$ for $COCl_2$ at 25°C.

19.29 At 25°C, $\Delta G°$ for the process

$$H_2O(l) \rightleftharpoons H_2O(g)$$

is 8.6 kJ. Calculate the "equilibrium constant" for the process.

19.30 Calculate $\Delta G°$ for the process

$$C(\text{diamond}) \longrightarrow C(\text{graphite})$$

Is the reaction spontaneous at 25°C? If so, why don't diamonds become graphite on standing?

MISCELLANEOUS PROBLEMS

19.31 Explain the following nursery rhyme in terms of

the second law of thermodynamics.

Humpty Dumpty sat on a wall;
Humpty Dumpty had a great fall.
All the King's horses and all the King's men
Couldn't put Humpty together again.

19.32 Calculate ΔG for the reaction

$$H_2O(l) \rightleftharpoons H^+(aq) + OH^-(aq)$$

at 25°C for the following conditions:
(a) $[H^+] = 1.0 \times 10^{-7}\ M$, $[OH^-] = 1.0 \times 10^{-7}\ M$
(b) $[H^+] = 1.0 \times 10^{-3}\ M$, $[OH^-] = 1.0 \times 10^{-4}\ M$
(c) $[H^+] = 1.0 \times 10^{-12}\ M$, $[OH^-] = 2.0 \times 10^{-8}\ M$
(d) $[H^+] = 3.5\ M$, $[OH^-] = 4.8 \times 10^{-4}\ M$

19.33 Which of the following thermodynamic functions are associated only with the first law of thermodynamics: S, E, G, or H?

19.34 A student placed 1 g of each of three compounds A, B, and C in a container and found that after one week no change had occurred. Offer some possible explanations for the fact that no reactions took place. Assume that A, B, and C are totally miscible liquids.

19.35 Give a detailed example of each of the following, with an explanation: (a) a thermodynamically spontaneous process; (b) a process that would violate the first law of thermodynamics; (c) a process that would violate the second law of thermodynamics; (d) an irreversible process; (e) an equilibrium process.

19.36 Predict the signs of ΔH, ΔS, and ΔG of the system for the following processes at 1 atm: (a) Ammonia melts at −60°C, (b) ammonia melts at −77.7°C, (c) ammonia melts at −100°C. (The normal melting point of ammonia is −77.7°C.)

19.37 Consider the following facts: Water freezes spontaneously at −5°C and 1 atm, and ice has a more ordered structure than liquid water. Explain how a spontaneous process can lead to a decrease in entropy.

19.38 Ammonium nitrate (NH_4NO_3) dissolves spontaneously and endothermically in water. What can you deduce about the sign of ΔS for the solution process?

19.39 Calculate the equilibrium pressure due to the decomposition of ammonium chloride (NH_4Cl) at 25°C.

19.40 (a) Trouton's rule states that the ratio of the molar heat of vaporization of a liquid (ΔH_{vap}) to its boiling point in kelvins is approximately 90 J/K · mol. Show that this is the case with the following data:

	t_{bp} (°C)	ΔH_{vap} (kJ/mol)
Benzene	80.1	31.0
Hexane	68.7	30.8
Mercury	357	59.0
Toluene	110.6	35.2

and explain why this rule holds true. (b) Use the values in Table 11.5 to calculate the same ratio for ethanol and water. Explain why Trouton's rule does not apply to these two substances as well as it does to other liquids.

19.41 Carbon monoxide (CO) and nitric oxide (NO) are undesirable gases emitted in an automobile's exhaust. Under suitable conditions, these gases can be made to react to form nitrogen (N_2) and the less harmful carbon dioxide (CO_2). (a) Write an equation for this reaction. (b) Identify the oxidizing and reducing agents. (c) Calculate the K_P for the reaction at 25°C. (d) Under normal atmospheric conditions, the partial pressures are $P_{N_2} = 0.80$ atm, $P_{CO_2} = 3.0 \times 10^{-4}$ atm, $P_{CO} = 5.0 \times 10^{-5}$ atm, and $P_{NO} = 5.0 \times 10^{-7}$ atm. Calculate Q_P and predict the direction toward which the reaction will proceed. (e) Will raising the temperature favor the formation of N_2 and CO_2?

19.42 For reactions carried out under standard-state conditions, Equation (19.6) takes the form $\Delta G° = \Delta H° - T\Delta S°$. (a) Assuming $\Delta H°$ and $\Delta S°$ are independent of temperature, derive the equation

$$\ln \frac{K_2}{K_1} = \frac{\Delta H°}{R}\left(\frac{T_2 - T_1}{T_1 T_2}\right)$$

where K_1 and K_2 are the equilibrium constants at T_1 and T_2, respectively. (b) Given that K_c for the reaction

$$N_2O_4(g) \rightleftharpoons 2NO_2(g) \qquad \Delta H° = 58.0 \text{ kJ}$$

is 4.63×10^{-3} at 25°C, calculate the equilibrium constant at 65°C.

19.43 Some copper(II) sulfate pentahydrate is placed in a closed container, and the following equilibrium is established at 25°C:

$$CuSO_4 \cdot 5H_2O(s) \rightleftharpoons CuSO_4(s) + 5H_2O(g)$$

Is the pressure of the water vapor measurable? ($\Delta G_f°$ for $CuSO_4 \cdot 5H_2O$ is −1879.7 kJ/mol.)

19.44 The K_{sp} of AgCl is given in Table 17.3. What is its value at 60°C? (*Hint:* You need the result in Problem 19.42 and the data in Appendix 2 to calculate $\Delta H°$.)

19.45 Under what conditions does a substance have a standard entropy of zero? Can a substance ever have a negative standard entropy?

19.46 Water gas is a fuel made by reacting steam with red-hot coke (a by-product of coal distillation):

$$H_2O(g) + C(s) \rightleftharpoons CO(g) + H_2(g)$$

From the data in Appendix 2, estimate the temperature at which the reaction becomes favorable.

19.47 Consider the following Bronsted acid-base reaction at 25°C:

$$HF(aq) + Cl^-(aq) \rightleftharpoons HCl(aq) + F^-(aq)$$

(a) Predict whether K will be greater or smaller than unity. (b) Does $\Delta S°$ or $\Delta H°$ make a greater contribution to $\Delta G°$? (c) Is $\Delta H°$ likely to be positive or negative?

19.48 Crystallization of sodium acetate from a supersaturated solution occurs spontaneously (see p. 341). What can you deduce about the signs of ΔS and ΔH?

19.49 Consider the thermal decomposition of $CaCO_3$:

$$CaCO_3(s) \rightleftharpoons CaO(s) + CO_2(g)$$

The equilibrium vapor pressures of CO_2 are 22.6 mmHg at 700°C and 1829 mmHg at 950°C. Calculate the standard enthalpy of the reaction. (*Hint:* See Problem 19.42.)

19.50 A certain reaction is spontaneous at 72°C. If the enthalpy change for the reaction is 19 kJ, what is the *minimum* value of ΔS (in joules per kelvin) for the reaction?

19.51 Predict whether the entropy change is positive or negative for each of the following reactions:
(a) $Zn(s) + 2HCl(aq) \rightleftharpoons ZnCl_2(aq) + H_2(g)$
(b) $O(g) + O(g) \rightleftharpoons O_2(g)$
(c) $NH_4NO_3(s) \rightleftharpoons N_2O(g) + 2H_2O(g)$
(d) $Ba^{2+}(aq) + SO_4^{2-}(aq) \rightleftharpoons BaSO_4(s)$
(e) $2H_2O_2(l) \rightleftharpoons 2H_2O(l) + O_2(g)$

19.52 The reaction $NH_3(g) + HCl(g) \rightarrow NH_4Cl(s)$ proceeds spontaneously at 25°C even though there is a decrease in disorder in the system (gases are converted to a solid). Explain.

19.53 Use the following data to determine the normal boiling point, in kelvins, of mercury. What assumptions must you make in order to do the calculation?

Hg(l): $\Delta H_f^\circ = 0$ (by definition)
$S° = 77.4$ J/K · mol

Hg(g): $\Delta H_f^\circ = 60.78$ kJ/mol
$S° = 174.7$ J/K · mol

19.54 The molar heat of vaporization of ethanol is 39.3 kJ/mol, and the boiling point of ethanol is 78.3°C. Calculate ΔS for the vaporization of 0.50 mole ethanol.

19.55 A certain reaction is known to have a $\Delta G°$ value of −122 kJ. Will the reaction necessarily occur if the reagents are mixed together?

19.56 Calculate K_P for the following reaction at 25°C:

$$N_2(g) + O_2(g) \rightleftharpoons 2NO(g)$$

From your result, what can you conclude about the stability of a mixture of O_2 and N_2 gases in the atmosphere?

19.57 Calculate $\Delta G°$ and K_P for the following processes at 25°C:
(a) $H_2(g) + Br_2(l) \rightleftharpoons 2HBr(g)$
(b) $\frac{1}{2}H_2(g) + \frac{1}{2}Br_2(l) \rightleftharpoons HBr(g)$
Account for the differences in $\Delta G°$ and K_P obtained for (a) and (b).

19.58 Calculate the pressure of O_2 (in atmospheres) over a sample of NiO at 25°C if $\Delta G° = 212$ kJ for the reaction

$$NiO(s) \rightleftharpoons Ni(s) + \tfrac{1}{2}O_2(g)$$

19.59 Comment on the statement: Just talking about entropy increases its value in the universe.

19.60 Which of the following statements applies to a reaction with a negative $\Delta G°$ value: (a) The equilibrium constant K is greater than 1; (b) the reaction is spontaneous when all the reactants and products are in their standard states; (c) the reaction is always exothermic.

19.61 Consider the reaction

$$N_2(g) + O_2(g) \rightleftharpoons 2NO(g)$$

Given that the $\Delta G°$ for the reaction at 25°C is 173.4 kJ: (a) Calculate the standard free energy of formation of NO. (b) Calculate K_P of the reaction. (c) One of the starting substances in smog formation is NO. Assuming that the temperature in a running car's engine is 1100°C, estimate the K_P for the reaction. (d) Some farmers know that lightning helps to produce a better crop. Why?

19.62 In the Mond process for the purification of nickel, carbon monoxide is reacted with heated nickel to produce $Ni(CO)_4$, which is a gas and can therefore be separated from solid impurities:

$$Ni(s) + 4CO(g) \rightleftharpoons Ni(CO)_4(g)$$

Given that the standard free energies of formation of $CO(g)$ and $Ni(CO)_4(g)$ are -137.3 kJ/mol and -587.4 kJ/mol, respectively, calculate the equilibrium constant of the reaction at 80°C. Assume the ΔG_f°'s to be temperature independent.

19.63 Carry out the following experiments: Quickly stretch a rubber band (at least 0.5 cm wide) and press it against your lips. You will feel a warming effect. Next, carry out the reverse procedure; that is, first stretch a rubber band and hold it in position for a few seconds. Now quickly release the tension and press the rubber band against your lips. You will feel a cooling effect. (a) Apply Equation (19.6) to these processes to determine the signs of ΔG, ΔH, and hence ΔS in each case. (b) From the signs of ΔS, what can you conclude about the structure of rubber molecules? (*Hint:* You may need to consult Chapter 14 for the structure of rubber.)

19.64 The standard free energies of formation (ΔG_f°) of the three isomers of pentane (see p. 370) are *n*-pentane: -8.37 kJ/mol; 2-methylbutane: -14.8 kJ/mol; 2,2-dimethylpropane: -15.2 kJ/mol. (a) Determine the mole percent of these molecules in an equilibrium mixture at 25°C. (b) How does the stability of these molecules depend on the extent of branching?

19.65 In the metabolism of glucose, the first step is the conversion of glucose to glucose 6-phosphate:

$$\text{glucose} + H_3PO_4 \longrightarrow \text{glucose 6-phosphate} + H_2O \qquad \Delta G^\circ = 13.4 \text{ kJ}$$

Since ΔG° is positive, this reaction is not spontaneous. Show how this reaction can be made to proceed by coupling it with the hydrolysis of ATP. Write an equation for the coupled reaction and estimate the equilibrium constant for the coupled process.

19.66 As the following ΔG° shows, heating copper(I) oxide at 400°C does not produce any appreciable amount of Cu:

$$Cu_2O(s) \rightleftharpoons 2Cu(s) + \tfrac{1}{2}O_2(g) \qquad \Delta G^\circ = 125 \text{ kJ}$$

However, if this reaction is coupled to the conversion of graphite to carbon monoxide, it becomes spontaneous. Write an equation for the coupled process and calculate the equilibrium constant for the coupled reaction.

Answers to Practice Exercises: **19.1** (a) Entropy decreases, (b) entropy decreases, (c) entropy increases, (d) entropy increases; **19.2** (a) -173.6 J/K, (b) -139.8 J/K, (c) 215.3 J/K; **19.3** (a) $\Delta S > 0$, (b) $\Delta S < 0$, (c) $\Delta S \simeq 0$; **19.4** (a) -106.4 kJ, (b) -2935.0 kJ; **19.5** $\Delta S_{fus} =$ 16 J/K · mol, $\Delta S_{vap} = 72$ J/K · mol; **19.6** 1.93×10^{57}; **19.7** $\Delta G = 0.97$ kJ, direction is from right to left.

CHAPTER 20

REDOX REACTIONS AND ELECTROCHEMISTRY

◆ "Prometheus brought fire to mankind; for electricity we owe it to Faraday."*

Michael Faraday is regarded as the greatest experimentalist of the nineteenth century and one of the greatest scientists of all time. This is all the more remarkable considering that he had no formal schooling beyond the primary grades. Born in England in 1791, Faraday was one of ten children of a poor blacksmith. At the age of 14, he was apprenticed to a bookbinder. He became interested in science after reading a book on chemistry. Faraday attended a series of lectures by the renowned chemist Sir Humphry Davy and took neat and copious notes on them. Later, when he applied to Davy for a job at the Royal Institution, he submitted his lecture notes as proof of his earnestness. Davy promptly hired Faraday as his laboratory assistant. On a tour through Europe, Faraday and Davy used a lens to focus the sun's rays on a diamond to convert it to carbon dioxide, thus proving that diamond, like the more common substance graphite, is made of carbon atoms.

During his career, Faraday made an amazing number of important contributions. He discovered benzene and determined its composition. He was the first to liquefy many gases and to produce the first stainless steel. He discovered the rotation of polarized light in a magnetic field (now known as the Faraday effect) and demonstrated electromagnetic induction, using a moving magnet to generate electric current in a wire. He also developed the laws of electrolysis and coined the terms "anion," "cation," "electrode," and "electrolyte."

In 1822 Faraday replaced Davy as director of the Royal Institution. When Faraday was nominated for membership in the Royal Society, Davy cast the only dissenting vote, possibly out of spite. Nevertheless, Faraday was elected in 1824.

Faraday could have made a fortune from some of his discoveries, but he deliberately dropped every project when it acquired commercial value. Born in poverty, he also died in poverty, in 1867. His consuming work, he felt, was sufficient reward. ◆

*Quotation attributed to Sir Lawrence Bragg, a British physicist.

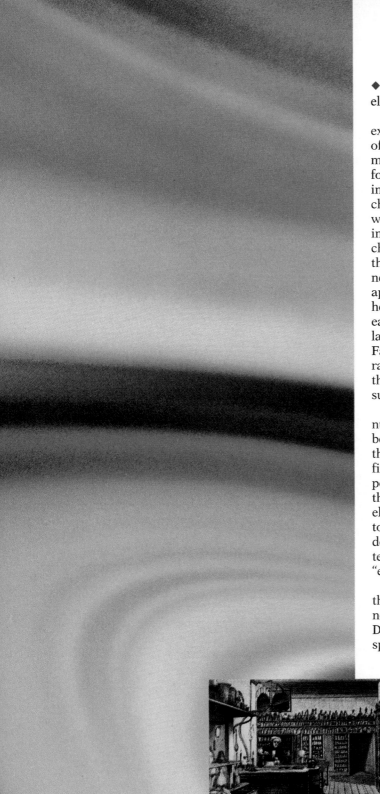

Michael Faraday at work in his laboratory.

20.1 REDOX REACTIONS

***Electrochemistry** is the branch of chemistry that deals with the interconversion of electrical energy and chemical energy.* Electrochemical processes are redox (oxidation-reduction) reactions in which the energy released by a spontaneous reaction is converted to electricity or in which electrical energy is used to cause a nonspontaneous reaction to occur. Although redox reactions were discussed in Chapter 4, it is helpful to review some of the basic concepts that will come up in this chapter.

Redox reactions involve the transfer of electrons from one substance to another. The reaction between magnesium metal and hydrochloric acid is an example of a redox reaction:

$$\overset{0}{Mg}(s) + 2\overset{+1}{H}Cl(aq) \longrightarrow \overset{+2}{Mg}Cl_2(aq) + \overset{0}{H_2}(g)$$

Recall that the numbers written above the elements are the oxidation numbers of the elements. The loss of electrons that occurs during oxidation is marked by an increase in the oxidation number of an element in a reaction. In reduction, there is a gain of electrons, indicated by a decrease in oxidation number of an element in a reaction. In the above reaction Mg metal is oxidized and H^+ ions are reduced; the Cl^- ions are spectator ions.

BALANCING REDOX EQUATIONS

Equations we have discussed so far for the redox reactions here and in Chapter 4 are relatively easy to balance. However, in the laboratory we often encounter more complex redox reactions, reactions that invariably involve oxoanions such as chromate (CrO_4^{2-}), dichromate ($Cr_2O_7^{2-}$), permanganate (MnO_4^-), nitrate (NO_3^-), and sulfate (SO_4^{2-}). In principle we can balance any redox equation using the procedure outlined in Section 3.7, but there are some special techniques for handling redox reactions, techniques that also give us insight into electron transfer processes. Here we will discuss one such procedure, called the *ion-electron method*. In this approach, the overall reaction is divided into two half-reactions, one for oxidation and one for reduction. The equations for the two half-reactions are balanced separately and then added together to give the overall balanced equation.

Suppose we are asked to balance the equation showing the oxidation of Fe^{2+} ions to Fe^{3+} ions by dichromate ions ($Cr_2O_7^{2-}$) in an acidic medium. The $Cr_2O_7^{2-}$ ions are reduced to Cr^{3+} ions. The following steps will help us accomplish this task.

Step 1: Write the unbalanced equation for the reaction in ionic form.

$$Fe^{2+} + Cr_2O_7^{2-} \longrightarrow Fe^{3+} + Cr^{3+}$$

Step 2: Separate the equation into two half-reactions.

$$\text{Oxidation:}\quad \overset{+2}{Fe^{2+}} \longrightarrow \overset{+3}{Fe^{3+}}$$

$$\text{Reduction:}\quad \overset{+6}{Cr_2}O_7^{2-} \longrightarrow \overset{+3}{Cr^{3+}}$$

Step 3: Balance the atoms other than O and H in each half-reaction separately.

The oxidation half-reaction is already balanced for Fe atoms. For the reduction half-reaction we multiply the Cr^{3+} by 2 to balance the Cr atoms.

$$Cr_2O_7^{2-} \longrightarrow 2Cr^{3+}$$

Step 4: For reactions in an acidic medium, add H_2O to balance the O atoms and H^+ to balance the H atoms.

Since the reaction takes place in an acidic medium, we add seven H_2O molecules to the right-hand side of the reduction half-reaction to balance the O atoms:

$$Cr_2O_7^{2-} \longrightarrow 2Cr^{3+} + 7H_2O$$

To balance the H atoms, we add fourteen H^+ ions on the left-hand side:

$$14H^+ + Cr_2O_7^{2-} \longrightarrow 2Cr^{3+} + 7H_2O$$

Step 5: Add electrons to one side of each half-reaction to balance the charges. If necessary, equalize the number of electrons in the two half-reactions by multiplying one or both half-reactions by appropriate coefficients.

For the oxidation half-reaction we write

$$Fe^{2+} \longrightarrow Fe^{3+} + e^-$$

We added an electron to the right-hand side so that there is a 2+ charge on each side of the half-reaction.

In the reduction half-reaction there are twelve net positive charges on the left-hand side and only six positive charges on the right-hand side. Therefore, we add six electrons on the left.

$$14H^+ + Cr_2O_7^{2-} + 6e^- \longrightarrow 2Cr^{3+} + 7H_2O$$

To equalize the number of electrons in both half-reactions, we multiply the oxidation half-reaction by 6:

$$6Fe^{2+} \longrightarrow 6Fe^{3+} + 6e^-$$

Step 6: Add the two half-reactions and balance the final equation by inspection. The electrons on both sides must cancel.

The two half-reactions are added to give

$$14H^+ + Cr_2O_7^{2-} + 6Fe^{2+} + \cancel{6e^-} \longrightarrow 2Cr^{3+} + 6Fe^{3+} + 7H_2O + \cancel{6e^-}$$

The electrons on both sides cancel, and we are left with the balanced net ionic equation:

$$14H^+ + Cr_2O_7^{2-} + 6Fe^{2+} \longrightarrow 2Cr^{3+} + 6Fe^{3+} + 7H_2O$$

Step 7: Verify that the equation contains the same types and numbers of atoms and the same charges on both sides of the equation.

A final check shows that the resulting equation is "atomically" and "electrically" balanced. For a reaction in a basic medium, first balance the atoms as you would in an acidic medium (step 4). Then, for every H^+ ion, add an equal number of OH^- ions to *both* sides of the equation. Where H^+ and OH^- appear on the same side of the equation, combine the ions to give H_2O. The following example illustrates the use of this procedure.

EXAMPLE 20.1

Balancing a Redox Equation

Write a balanced ionic equation to represent the oxidation of iodide ion (I^-) by permanganate ion (MnO_4^-) in basic solution to yield molecular iodine (I_2) and manganese(IV) oxide (MnO_2).

Answer

Step 1: The unbalanced equation is

$$MnO_4^- + I^- \longrightarrow MnO_2 + I_2$$

Step 2: The two half-reactions are

$$\text{Oxidation:} \quad \overset{-1}{I^-} \longrightarrow \overset{0}{I_2}$$

$$\text{Reduction:} \quad \overset{+7}{MnO_4^-} \longrightarrow \overset{+4}{MnO_2}$$

Step 3: To balance the I atoms in the oxidation half-reaction, we write

$$2I^- \longrightarrow I_2$$

Step 4: In the reduction half-reaction, to balance the O atoms we add two H_2O molecules on the right:

$$MnO_4^- \longrightarrow MnO_2 + 2H_2O$$

To balance the H atoms, we add four H^+ ions on the left:

$$MnO_4^- + 4H^+ \longrightarrow MnO_2 + 2H_2O$$

Since the reaction occurs in a basic medium and there are four H^+ ions, we add four OH^- ions to both sides of the equation:

$$MnO_4^- + 4H^+ + 4OH^- \longrightarrow MnO_2 + 2H_2O + 4OH^-$$

Combining the H^+ and OH^- ions to form H_2O and subtracting $2H_2O$ from both sides, we write

$$MnO_4^- + 2H_2O \longrightarrow MnO_2 + 4OH^-$$

Step 5: Next, we balance the charges of the two half-reactions as follows:

$$2I^- \longrightarrow I_2 + 2e^-$$

$$MnO_4^- + 2H_2O + 3e^- \longrightarrow MnO_2 + 4OH^-$$

To equalize the number of electrons, we multiply the oxidation half-reaction by 3 and the reduction half-reaction by 2:

$$6I^- \longrightarrow 3I_2 + 6e^-$$

$$2MnO_4^- + 4H_2O + 6e^- \longrightarrow 2MnO_2 + 8OH^-$$

Step 6: The two half-reactions are added together to give

$$6I^- + 2MnO_4^- + 4H_2O + \cancel{6e^-} \longrightarrow 3I_2 + 2MnO_2 + 8OH^- + \cancel{6e^-}$$

After canceling the electrons on both sides, we obtain

$$6I^- + 2MnO_4^- + 4H_2O \longrightarrow 3I_2 + 2MnO_2 + 8OH^-$$

Step 7: A final check shows that the equation is balanced in terms of both atoms and charges.

PRACTICE EXERCISE

Balance the following equation for the reaction in an acidic medium by the ion-electron method:

$$Fe^{2+} + MnO_4^- \longrightarrow Fe^{3+} + Mn^{2+}$$

20.2 GALVANIC CELLS

In Section 4.4 we saw that when a piece of zinc metal is placed in a $CuSO_4$ solution, Zn is oxidized to Zn^{2+} ions while Cu^{2+} ions are reduced to metallic copper (see Figure 4.10):

$$Zn(s) + Cu^{2+}(aq) \longrightarrow Zn^{2+}(aq) + Cu(s)$$

The electrons are transferred directly from the reducing agent (Zn) to the oxidation agent (Cu^{2+}) in solution. However, if we physically separate the oxidizing agent from the reducing agent, the transfer of electrons can take place via an external conducting medium. As the reaction progresses, it sets up a constant flow of electrons and hence generates electricity (that is, it produces electrical work).

The experimental apparatus for generating electricity through the use of a redox reaction is called an *electrochemical cell.* Figure 20.1 shows the essential components of one type of electrochemical cell, called a ***galvanic cell*** or *voltaic cell, in which electricity is produced by means of a spontaneous redox reaction.* A zinc bar is dipped into a $ZnSO_4$ solution, and a bar of copper is dipped into a $CuSO_4$ solution. The galvanic cell operates on the principle that the oxidation of Zn to Zn^{2+} and the reduction of Cu^{2+} to Cu can be made to take place simultaneously in separate locations with the transfer of electrons through the external wire. The zinc and copper bars are called *electrodes.* By definition, *the electrode at which oxidation occurs* is called the ***anode;*** *the electrode at which reduction occurs* is the ***cathode.***

For the system shown in Figure 20.1, *the oxidation and reduction reactions at the electrodes,* which are called ***half-cell reactions,*** are

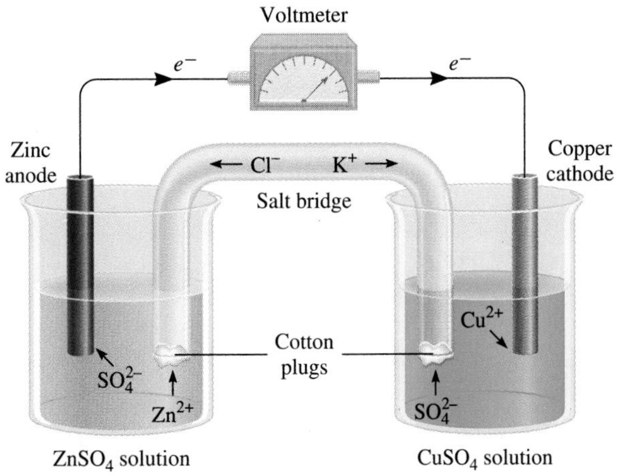

FIGURE 20.1

A galvanic cell. The salt bridge (an inverted U tube) containing a KCl solution provides an electrically conducting medium between two solutions. The openings of the U tube are loosely plugged with cotton balls to prevent the KCl solution from flowing into the containers while allowing the anions and cations to move across. Electrons flow externally from the Zn electrode (anode) to the Cu electrode (cathode).

$$\text{Zn electrode (anode):} \qquad Zn(s) \longrightarrow Zn^{2+}(aq) + 2e^-$$

$$\text{Cu electrode (cathode):} \qquad Cu^{2+}(aq) + 2e^- \longrightarrow Cu(s)$$

Note that unless the two solutions are separated from each other, the Cu^{2+} ions will react directly with the zinc bar:

$$Cu^{2+}(aq) + Zn(s) \longrightarrow Cu(s) + Zn^{2+}(aq)$$

and no useful electrical work will be obtained.

To complete the electric circuit, the solutions must be connected by a conducting medium through which the cations and anions can move. This requirement is satisfied by a *salt bridge,* which, in its simplest form, is an inverted U tube containing an inert electrolyte such as KCl or NH_4NO_3, whose ions will not react with other ions in solution or with the electrodes (see Figure 20.1). During the course of the overall redox reaction, electrons flow externally from the anode (Zn electrode) through the wire and voltmeter to the cathode (Cu electrode). In the solution, the cations (Zn^{2+}, Cu^{2+}, and K^+) move toward the cathode, while the anions (SO_4^{2-} and Cl^-) move in the opposite direction, toward the anode.

The fact that electrons flow from one electrode to the other indicates that there is a voltage difference between the two electrodes. This *voltage difference between the electrodes,* called the ***electromotive force,*** or ***emf*** ($\mathscr{E}$), can be measured by connecting a voltmeter to both electrodes (Figure 20.2). The emf of a galvanic cell is usually measured in volts; it is also referred to as *cell voltage* or *cell potential.* We will see that the emf of a cell depends not only on the nature of the electrodes and the ions, but also on the concentrations of the ions and the temperature at which the cell is operated.

The conventional notation for representing galvanic cells is the *cell diagram.* For the galvanic cell just described, using KCl as the electrolyte in the salt bridge and assuming 1 *M* concentrations, the cell diagram is

$$Zn(s)|Zn^{2+}(aq,\ 1\ M)|KCl(\text{sat'd})|Cu^{2+}(aq,\ 1\ M)|Cu(s)$$

where the vertical lines represent phase boundaries. For example, the zinc electrode is a solid and the Zn^{2+} ions (from $ZnSO_4$) are in solution. Thus we draw a line between Zn and Zn^{2+} to show the phase boundaries. Note that there is also a line between the $ZnSO_4$ solution and the KCl solution in the salt bridge because these two solutions are not mixed and therefore constitute two separate phases. By

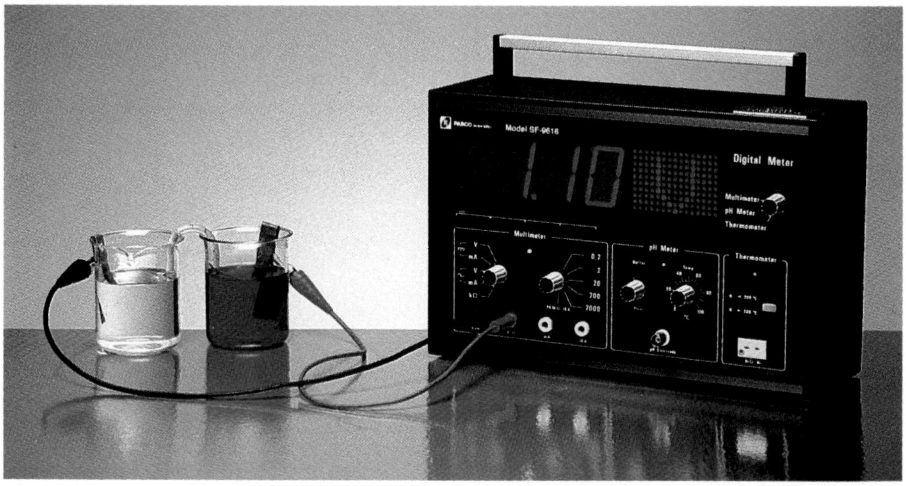

FIGURE 20.2

Practical setup of the galvanic cell described in Figure 20.1. Note the U tube (salt bridge) connecting the two beakers. When the concentrations of $ZnSO_4$ and $CuSO_4$ are 1 molar (1 M) at 25°C, the cell voltage is 1.10 V.

convention, the anode is written first at the left and the other components appear in the order in which we would encounter them in moving from the anode to the cathode.

20.3 STANDARD ELECTRODE POTENTIALS

When the concentrations of the Cu^{2+} and Zn^{2+} ions are both 1.0 *M*, we find that the emf of the cell shown in Figure 20.1 is 1.10 V at 25°C (see Figure 20.2). This voltage must be related directly to the redox reactions, but how? Just as the overall cell reaction can be thought of as the sum of two half-cell reactions, the measured emf of the cell can be treated as the sum of the electrical potentials at the Zn and Cu electrodes. Knowing one of these electrode potentials, we could obtain the other by subtraction (from 1.10 V). It is impossible to measure the potential of just a single electrode, but if we arbitrarily set the potential value of a particular electrode at zero, we can use it to determine the relative potentials of other electrodes. The hydrogen electrode, shown in Figure 20.3, serves as the standard for this purpose. Hydrogen gas is bubbled into a hydrochloric acid solution at 25°C. The platinum electrode has two functions. First, it provides a surface on which the dissociation of hydrogen molecules can take place:

$$H_2 \longrightarrow 2H^+ + 2e^-$$

Second, it serves as an electrical conductor to the external circuit.

In Chapter 19 we saw that the standard-state conditions for gases refer to 1 atm pressure and for solutes refer to 1 *M* concentration (see Table 19.2). When the hydrogen electrode is operated under standard-state conditions, that is, when the pressure of H_2 is 1 atm and the concentration of the HCl solution or $[H^+]$ is 1 *M*, the potential for the reduction of H^+ at 25°C is taken to be *exactly* zero:

$$2H^+(aq,\ 1\ M) + 2e^- \longrightarrow H_2(g,\ 1\ \text{atm}) \qquad \mathscr{E}^\circ = 0\ \text{V}$$

The superscript o denotes standard-state conditions. *For a reduction reaction at an electrode when all solutes are 1 M and all gases are at 1 atm, the voltage* is called the ***standard reduction potential.*** Thus, the standard reduction potential of the hydrogen electrode is defined as zero. The electrode is called the *standard hydrogen electrode (SHE)*.

We can use the SHE to measure the potentials of other kinds of electrodes. Figure 20.4(a) shows a galvanic cell with a zinc electrode and a SHE. Experiments

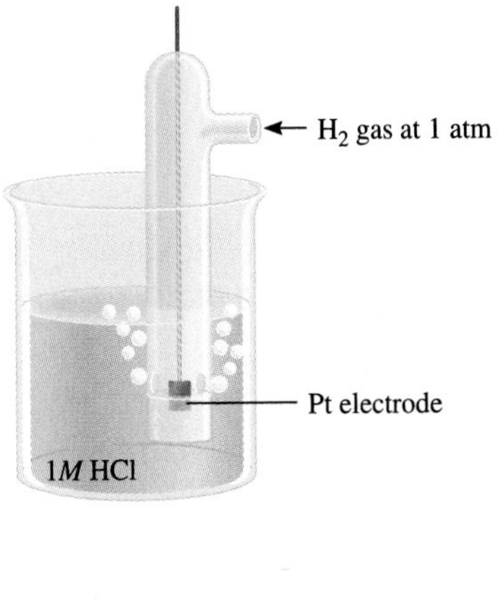

FIGURE 20.3

A hydrogen electrode operating under standard-state conditions. Hydrogen gas at 1 atm is bubbled through a 1 M HCl solution. The platinum electrode is part of the hydrogen electrode.

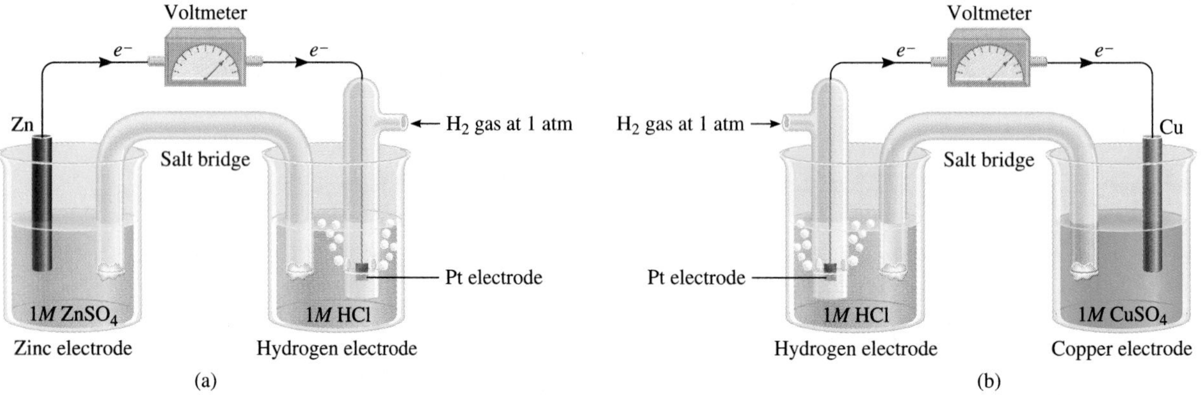

FIGURE 20.4

(a) A cell consisting of a zinc electrode and a hydrogen electrode. (b) A cell consisting of a copper electrode and a hydrogen electrode. Both cells are operating under standard-state conditions. Note that in (a) the SHE acts as the cathode, but in (b) the SHE is the anode.

show that the zinc electrode is the anode and the SHE is the cathode. We deduce this because the mass of the zinc electrode decreases during the operation of the cell, a fact consistent with the loss of zinc to the solution by the oxidation reaction:

$$\text{Zn}(s) \longrightarrow \text{Zn}^{2+}(aq) + 2e^-$$

The cell diagram is

$$\text{Zn}(s)|\text{Zn}^{2+}(aq,\ 1\ M)|\text{KCl(sat'd)}|\text{H}^+(aq,\ 1\ M)|\text{H}_2(g,\ 1\ \text{atm})|\text{Pt}(s)$$

When all the reactants are in their standard states (that is, H_2 at 1 atm, H^+ and Zn^{2+} ions at 1 M), the emf of the cell is 0.76 V. We can write the half-cell reactions as follows:

$$\begin{array}{llll} \text{Anode:} & \text{Zn}(s) \longrightarrow & \text{Zn}^{2+}(aq,\ 1\ M) + 2e^- & \mathscr{E}^\circ_{\text{Zn/Zn}^{2+}} \\ \text{Cathode:} & 2\text{H}^+(aq,\ 1\ M) + 2e^- \longrightarrow & \text{H}_2(g,\ 1\ \text{atm}) & \mathscr{E}^\circ_{\text{H}^+/\text{H}_2} \\ \hline \text{Overall:} & \text{Zn}(s) + 2\text{H}^+(aq,\ 1\ M) \longrightarrow & \text{Zn}^{2+}(aq,\ 1\ M) + \text{H}_2(g,\ 1\ \text{atm}) & \mathscr{E}^\circ_{\text{cell}} \end{array}$$

where $\mathscr{E}^\circ_{\text{Zn/Zn}^{2+}}$ is the ***standard oxidation potential,*** or *voltage under standard-state conditions, for the oxidation reaction:*

$$\text{Zn}(s) \longrightarrow \text{Zn}^{2+}(aq,\ 1\ M) + 2e^-$$

and $\mathscr{E}^\circ_{\text{H}^+/\text{H}_2}$ is the standard reduction potential previously defined for the SHE. (Zn/Zn^{2+} means $Zn \rightarrow Zn^{2+} + 2e^-$, and H^+/H_2 means $2H^+ + 2e^- \rightarrow H_2$.) The ***standard emf*** of the cell, $\mathscr{E}^\circ_{\text{cell}}$, is *the sum of the standard oxidation potential and the standard reduction potential:*

$$\mathscr{E}^\circ_{\text{cell}} = \mathscr{E}^\circ_{\text{ox}} + \mathscr{E}^\circ_{\text{red}} \tag{20.1}$$

where the subscripts "ox" and "red" indicate oxidation and reduction, respectively. For our galvanic cell,

$$\mathscr{E}^\circ_{\text{cell}} = \mathscr{E}^\circ_{\text{Zn/Zn}^{2+}} + \mathscr{E}^\circ_{\text{H}^+/\text{H}_2}$$

$$0.76\ \text{V} = \mathscr{E}^\circ_{\text{Zn/Zn}^{2+}} + 0$$

Thus the standard oxidation potential of zinc is 0.76 V.

We can evaluate the standard reduction potential of zinc, $\mathscr{E}^\circ_{\text{Zn}^{2+}/\text{Zn}}$, by reversing the oxidation half-reaction:

$$\text{Zn}^{2+}(aq,\ 1\ M) + 2e^- \longrightarrow \text{Zn}(s) \qquad \mathscr{E}^\circ_{\text{Zn}^{2+}/\text{Zn}} = -0.76\ \text{V}$$

Remember this: Whenever we reverse a half-cell reaction, $\mathscr{E}^\circ$ changes sign.

The standard electrode potentials of copper can similarly be obtained using a galvanic cell with a copper electrode and a SHE [Figure 20.4(b)]. In this case, the copper electrode is the cathode because its mass increases during the operation of the cell, as is consistent with the reduction reaction:

$$Cu^{2+}(aq) + 2e^- \longrightarrow Cu(s)$$

The cell diagram is

$$Pt(s)|H_2(g,\ 1\ atm)|H^+(aq,\ 1\ M)|KCl(sat'd)|Cu^{2+}(aq,\ 1\ M)|Cu(s)$$

and the half-cell reactions are

Anode:	$H_2(g,\ 1\ atm) \longrightarrow 2H^+(aq,\ 1\ M) + 2e^-$	$\mathscr{E}^\circ_{H_2/H^+}$
Cathode:	$Cu^{2+}(aq,\ 1\ M) + 2e^- \longrightarrow Cu(s)$	$\mathscr{E}^\circ_{Cu^{2+}/Cu}$
Overall:	$H_2(g\ 1\ atm) + Cu^{2+}(aq,\ 1\ M) \longrightarrow 2H^+(aq,\ 1\ M) + Cu(s)$	$\mathscr{E}^\circ_{cell}$

Under standard-state conditions and at 25°C, the emf of the cell is 0.34 V, so we write

$$\mathscr{E}^\circ_{cell} = \mathscr{E}^\circ_{H_2/H^+} + \mathscr{E}^\circ_{Cu^{2+}/Cu}$$

$$0.34\ V = 0 + \mathscr{E}^\circ_{Cu^{2+}/Cu}$$

Thus the standard reduction potential of copper is 0.34 V and its standard oxidation potential, $\mathscr{E}^\circ_{Cu/Cu^{2+}}$, is −0.34 V.

For the cell shown in Figure 20.1, we can now write

Anode:	$Zn(s) \longrightarrow Zn^{2+}(aq,\ 1\ M) + 2e^-$	$\mathscr{E}^\circ_{Zn/Zn^{2+}}$
Cathode:	$Cu^{2+}(aq,\ 1\ M) + 2e^- \longrightarrow Cu(s)$	$\mathscr{E}^\circ_{Cu^{2+}/Cu}$
Overall:	$Zn(s) + Cu^{2+}(aq,\ 1\ M) \longrightarrow Zn^{2+}(aq,\ 1\ M) + Cu(s)$	$\mathscr{E}^\circ_{cell}$

The emf of the cell is

$$\begin{aligned}\mathscr{E}^\circ_{cell} &= \mathscr{E}^\circ_{Zn/Zn^{2+}} + \mathscr{E}^\circ_{Cu^{2+}/Cu} \\ &= 0.76\ V + 0.34\ V \\ &= 1.10\ V\end{aligned}$$

This example illustrates how we can use the *sign* of the emf of the cell to predict the spontaneity of a redox reaction. Under standard-state conditions for reactants and products, the redox reaction is spontaneous in the forward direction if the standard emf of the cell is positive. If it is negative, the reaction is spontaneous in the opposite direction. It is important to keep in mind that a negative $\mathscr{E}^\circ_{cell}$ does *not* mean that a reaction will not occur if the reactants are mixed at 1 *M* concentrations. It merely means that the equilibrium of a redox reaction, when reached, will lie to the left. We will examine the relationships among $\mathscr{E}^\circ_{cell}$, ΔG°, and K later in this chapter.

Table 20.1 lists the standard reduction potentials of half-reactions. By definition the SHE has an $\mathscr{E}^\circ$ value of 0.00 V. The negative standard reduction potentials increase above it, and the positive standard reduction potentials increase below it. The more positive the value of $\mathscr{E}^\circ$, the greater the tendency for the substance to be reduced and therefore the stronger its tendency to act as an oxidizing agent. In Table 20.1 we see that F_2 is the strongest oxidizing agent and Li^+ the weakest.

EXAMPLE 20.2

Comparing Strengths of Oxidizing Agents

Arrange the following species in order of increasing strength as oxidizing agents under standard-state conditions: Sn^{2+}, $Cr_2O_7^{2-}$ (in acid solution), and $Br_2(l)$.

TABLE 20.1
Standard Reduction Potentials at 25°C*

Half-reaction	$\mathscr{E}°$(V)
$Li^+(aq) + e^- \longrightarrow Li(s)$	−3.05
$K^+(aq) + e^- \longrightarrow K(s)$	−2.93
$Ba^{2+}(aq) + 2e^- \longrightarrow Ba(s)$	−2.90
$Sr^{2+}(aq) + 2e^- \longrightarrow Sr(s)$	−2.89
$Ca^{2+}(aq) + 2e^- \longrightarrow Ca(s)$	−2.87
$Na^+(aq) + e^- \longrightarrow Na(s)$	−2.71
$Mg^{2+}(aq) + 2e^- \longrightarrow Mg(s)$	−2.37
$Be^{2+}(aq) + 2e^- \longrightarrow Be(s)$	−1.85
$Al^{3+}(aq) + 3e^- \longrightarrow Al(s)$	−1.66
$Mn^{2+}(aq) + 2e^- \longrightarrow Mn(s)$	−1.18
$2H_2O + 2e^- \longrightarrow H_2(g) + 2OH^-(aq)$	−0.83
$Zn^{2+}(aq) + 2e^- \longrightarrow Zn(s)$	−0.76
$Cr^{3+}(aq) + 3e^- \longrightarrow Cr(s)$	−0.74
$Fe^{2+}(aq) + 2e^- \longrightarrow Fe(s)$	−0.44
$Cd^{2+}(aq) + 2e^- \longrightarrow Cd(s)$	−0.40
$PbSO_4(s) + 2e^- \longrightarrow Pb(s) + SO_4^{2-}(aq)$	−0.31
$Co^{2+}(aq) + 2e^- \longrightarrow Co(s)$	−0.28
$Ni^{2+}(aq) + 2e^- \longrightarrow Ni(s)$	−0.25
$Sn^{2+}(aq) + 2e^- \longrightarrow Sn(s)$	−0.14
$Pb^{2+}(aq) + 2e^- \longrightarrow Pb(s)$	−0.13
$2H^+(aq) + 2e^- \longrightarrow H_2(g)$	0.00
$Sn^{4+}(aq) + 2e^- \longrightarrow Sn^{2+}(aq)$	+0.13
$Cu^{2+}(aq) + e^- \longrightarrow Cu^+(aq)$	+0.15
$SO_4^{2-}(aq) + 4H^+(aq) + 2e^- \longrightarrow SO_2(g) + 2H_2O$	+0.20
$AgCl(s) + e^- \longrightarrow Ag(s) + Cl^-(aq)$	+0.22
$Cu^{2+}(aq) + 2e^- \longrightarrow Cu(s)$	+0.34
$O_2(g) + 2H_2O + 4e^- \longrightarrow 4OH^-(aq)$	+0.40
$I_2(s) + 2e^- \longrightarrow 2I^-(aq)$	+0.53
$MnO_4^-(aq) + 2H_2O + 3e^- \longrightarrow MnO_2(s) + 4OH^-(aq)$	+0.59
$O_2(g) + 2H^+(aq) + 2e^- \longrightarrow H_2O_2(aq)$	+0.68
$Fe^{3+}(aq) + e^- \longrightarrow Fe^{2+}(aq)$	+0.77
$Ag^+(aq) + e^- \longrightarrow Ag(s)$	+0.80
$Hg_2^{2+}(aq) + 2e^- \longrightarrow 2Hg(l)$	+0.85
$2Hg^{2+}(aq) + 2e^- \longrightarrow Hg_2^{2+}(aq)$	+0.92
$NO_3^-(aq) + 4H^+(aq) + 3e^- \longrightarrow NO(g) + 2H_2O$	+0.96
$Br_2(l) + 2e^- \longrightarrow 2Br^-(aq)$	+1.07
$O_2(g) + 4H^+(aq) + 4e^- \longrightarrow 2H_2O$	+1.23
$MnO_2(s) + 4H^+(aq) + 2e^- \longrightarrow Mn^{2+}(aq) + 2H_2O$	+1.23
$Cr_2O_7^{2-}(aq) + 14H^+(aq) + 6e^- \longrightarrow 2Cr^{3+}(aq) + 7H_2O$	+1.33
$Cl_2(g) + 2e^- \longrightarrow 2Cl^-(aq)$	+1.36
$Au^{3+}(aq) + 3e^- \longrightarrow Au(s)$	+1.50
$MnO_4^-(aq) + 8H^+(aq) + 5e^- \longrightarrow Mn^{2+}(aq) + 4H_2O$	+1.51
$Ce^{4+}(aq) + e^- \longrightarrow Ce^{3+}(aq)$	+1.61
$PbO_2(s) + 4H^+(aq) + SO_4^{2-}(aq) + 2e^- \longrightarrow PbSO_4(s) + 2H_2O$	+1.70
$H_2O_2(aq) + 2H^+(aq) + 2e^- \longrightarrow 2H_2O$	+1.77
$Co^{3+}(aq) + e^- \longrightarrow Co^{2+}(aq)$	+1.82
$O_3(g) + 2H^+(aq) + 2e^- \longrightarrow O_2(g) + H_2O(l)$	+2.07
$F_2(g) + 2e^- \longrightarrow 2F^-(aq)$	+2.87

Increasing strength as oxidizing agent ↓ (down the table); Increasing strength as reducing agent ↑ (up the table)

*For all half-reactions the concentration is 1 *M* for dissolved species and the pressure is 1 atm for gases. These are the standard-state values.

Answer: Consulting Table 20.1, we write the half-reactions in the order of increasing $\mathscr{E}°$ value:

$$Sn^{2+}(aq,\ 1\ M) + 2e^- \longrightarrow Sn(s) \qquad \mathscr{E}° = -0.14\ V$$

$$Br_2(l) + 2e^- \longrightarrow 2Br^-(aq,\ 1\ M) \qquad \mathscr{E}° = +1.07\ V$$

$$Cr_2O_7^{2-}(aq,\ 1\ M) + 14H^+(aq,\ 1\ M) + 6e^- \longrightarrow 2Cr^{3+}(aq,\ 1\ M) + 7H_2O \qquad \mathscr{E}° = +1.33\ V$$

Since the reduction of $Cr_2O_7^{2-}$ has the highest positive potential and Br_2 the next highest, the order of increasing strength of oxidizing agents is

$$Sn^{2+} < Br_2 < Cr_2O_7^{2-}$$

PRACTICE EXERCISE

Which of the following is the strongest reducing agent under standard-state conditions: Cl^-, I^-, Pb?

EXAMPLE 20.3

Calculating the $\mathscr{E}°$ of a Galvanic Cell

A galvanic cell consists of a Mg electrode in a 1.0 *M* $Mg(NO_3)_2$ solution and a Ag electrode in a 1.0 *M* $AgNO_3$ solution. Calculate the standard emf of this electrochemical cell at 25°C.

Answer: Table 20.1 gives the standard reduction potentials of the two electrodes:

$$Mg^{2+}(aq,\ 1\ M) + 2e^- \longrightarrow Mg(s) \qquad \mathscr{E}° = -2.37\ V$$

$$Ag^+(aq,\ 1\ M) + e^- \longrightarrow Ag(s) \qquad \mathscr{E}° = 0.80\ V$$

It follows that Mg is a stronger reducing agent than Ag. We can write the half-reactions as follows:

$$\begin{array}{lrcl} \text{Anode:} & Mg(s) & \longrightarrow & Mg^{2+}(aq,\ 1\ M) + 2e^- \\ \text{Cathode:} & 2Ag^+(aq,\ 1\ M) + 2e^- & \longrightarrow & 2Ag(s) \\ \hline \text{Overall:} & Mg(s) + 2Ag^+(aq,\ 1\ M) & \longrightarrow & Mg^{2+}(aq,\ 1\ M) + 2Ag(s) \end{array}$$

Note that in order to balance the overall equation we multiplied the reduction of Ag^+ by 2. We can do so because $\mathscr{E}°$ is an intensive property and is not affected by this procedure. We find the emf of the cell by using Equation (20.1) and Table 20.1:

$$\begin{aligned} \mathscr{E}°_{cell} &= \mathscr{E}°_{Mg/Mg^{2+}} + \mathscr{E}°_{Ag^+/Ag} \\ &= 2.37\ V + 0.80\ V \\ &= 3.17\ V \end{aligned}$$

PRACTICE EXERCISE

What is the standard emf of a galvanic cell made of a Cd electrode in a 1.0 *M* $Cd(NO_3)_2$ solution and a Cr electrode in a 1.0 *M* $Cr(NO_3)_3$ solution?

20.4 SPONTANEITY OF REDOX REACTIONS

Table 20.1 enables us to predict the outcome of redox reactions under standard-state conditions, whether they take place in an electrochemical cell, where the reducing agent and oxidizing agent are physically separated from each other, or in a beaker, where the reactants are all mixed together. Our next step is to see how $\mathscr{E}^\circ_{cell}$ is related to other thermodynamic quantities such as ΔG° and K.

In Chapter 19 we saw that the free-energy change (decrease) in a spontaneous process is the energy available to do work. In fact, for a process carried out at constant temperature and pressure

$$\Delta G = w_{max}$$

where w_{max} is the maximum amount of work that can be done.

In a galvanic cell, chemical energy is converted into electrical energy. Electrical energy in this case is the product of the emf of the cell and the total electrical charge (in coulombs) that passes through the cell:

$$\begin{aligned} \text{electrical energy} &= \text{volts} \times \text{coulombs} \\ &= \text{joules} \end{aligned}$$

The total charge is determined by the number of moles of electrons (n) transferred from the reducing agent to the oxidizing agent in the overall redox equation. By definition

$$\text{total charge} = nF$$

where F, the Faraday constant, is the electrical charge contained in 1 mole of electrons. Experimentally, 1 ***faraday*** has been found to be *equivalent to 96,487 coulombs,* or 96,500 coulombs, rounded off to three significant figures. Thus

$$1\ F = 96{,}500\ \text{C/mol}$$

Since

$$1\ \text{J} = 1\ \text{C} \times 1\ \text{V}$$

we can also express the units of faraday as

$$1\ F = 96{,}500\ \text{J/V} \cdot \text{mol}$$

A common device for measuring the cell's emf is the *potentiometer,* which can precisely match the voltage of the cell without actually draining any current from the cell. Thus the measured emf is the *maximum* voltage that the cell can achieve. This value is used to calculate the maximum amount of electrical energy that can be obtained from the chemical reaction. This energy is used to do electrical work (w_{ele}), so

$$\begin{aligned} w_{max} &= w_{ele} \\ &= -nF\mathscr{E}_{cell} \end{aligned}$$

The negative sign on the right-hand side indicates that the electrical work is done by the system on the surroundings. Now, since

$$\Delta G = w_{max}$$

we obtain

$$\Delta G = -nF\mathscr{E}_{cell} \tag{20.2}$$

TABLE 20.2
Relationships among $\Delta G°$, K, and $\mathscr{E}^\circ_{cell}$

$\Delta G°$	K	$\mathscr{E}^\circ_{cell}$	Reaction under standard-state conditions
Negative	> 1	Positive	Spontaneous
0	$= 1$	0	At equilibrium
Positive	< 1	Negative	Nonspontaneous. Reaction is spontaneous in the reverse direction.

Both n and F are positive quantities and ΔG is negative for a spontaneous process, so $\mathscr{E}_{cell}$ is positive. For reactions in which reactants and products are in their standard states, Equation (20.2) becomes

$$\Delta G° = -nF\mathscr{E}^\circ_{cell} \qquad (20.3)$$

Here again, $\mathscr{E}^\circ_{cell}$ is positive for a spontaneous process.

In Section 19.6 we saw that the standard free-energy change $\Delta G°$ for a reaction is related to its equilibrium constant as follows [see Equation (19.9)]:

$$\Delta G° = -RT \ln K$$

Therefore, from Equations (20.3) and (19.9) we obtain

$$-nF\mathscr{E}^\circ_{cell} = -RT \ln K$$

Solving for $\mathscr{E}^\circ_{cell}$ we get

$$\mathscr{E}^\circ_{cell} = \frac{RT}{nF} \ln K \qquad (20.4)$$

When $T = 298$ K, Equation (20.4) can be simplified by substituting for R and F and converting the natural logarithm to the common (base-10) logarithm:

$$\mathscr{E}^\circ_{cell} = \frac{2.303(8.314 \text{ J/K} \cdot \text{mol})(298 \text{ K})}{n(96{,}500 \text{ J/V} \cdot \text{mol})} \log K$$

$$= \frac{0.0591 \text{ V}}{n} \log K \qquad (20.5)$$

Thus, if any one of the three quantities $\Delta G°$, K, or $\mathscr{E}^\circ_{cell}$ is known, the other two can be calculated using Equation (19.9), Equation (20.3), or Equation (20.4). We can summarize the relationships among $\Delta G°$, K, and $\mathscr{E}^\circ_{cell}$ and characterize the spontaneity of a redox reaction as shown in Table 20.2.

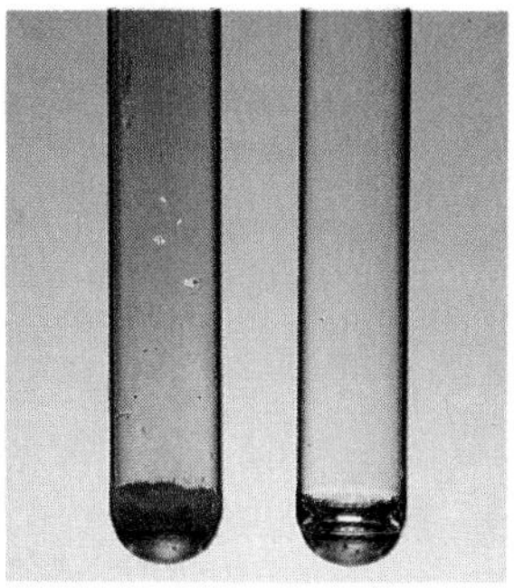

HNO_3 (left) can oxidize Hg, but HCl (right) cannot.

EXAMPLE 20.4
Comparing the Relative Oxidizing Strengths of HCl and HNO_3

Using data in Table 20.1, calculate $\mathscr{E}°$ for the reactions of mercury with (a) 1 M HCl and (b) 1 M HNO_3. Which acid will oxidize Hg to Hg_2^{2+} under standard-state conditions?

Answer: (a) HCl: First we write the half-reactions:

Oxidation:	$2Hg(l) \longrightarrow Hg_2^{2+}(aq, 1\ M) + 2e^-$
Reduction:	$2H^+(aq, 1\ M) + 2e^- \longrightarrow H_2(g, 1 \text{ atm})$
Overall:	$2Hg(l) + 2H^+(aq, 1\ M) \longrightarrow Hg_2^{2+}(aq, 1\ M) + H_2(g, 1 \text{ atm})$

The standard emf, $\mathscr{E}°$, is given by

$$\begin{aligned}\mathscr{E}° &= \mathscr{E}°_{Hg/Hg_2^{2+}} + \mathscr{E}°_{H^+/H_2}\\ &= -0.85\text{ V} + 0\\ &= -0.85\text{ V}\end{aligned}$$

(We omit the subscript "cell" because this reaction is not carried out in an electrochemical cell.) Since $\mathscr{E}°$ is negative, we conclude that mercury is not oxidized by hydrochloric acid under standard-state conditions.

(b) HNO_3: The reactions are

Oxidation: $3[2Hg(l) \longrightarrow Hg_2^{2+}(aq, 1\,M) + 2e^-]$

Reduction: $2[NO_3^-(aq, 1\,M) + 4H^+(aq, 1\,M) + 3e^- \longrightarrow NO(g, 1\text{ atm}) + 2H_2O(l)]$

Overall:
$6Hg(l) + 2NO_3^-(aq, 1\,M) + 8H^+(aq, 1\,M) \longrightarrow 3Hg_2^{2+}(aq, 1\,M) + 2NO(g, 1\text{ atm}) + 4H_2O(l)$

Thus

$$\begin{aligned}\mathscr{E}° &= \mathscr{E}°_{Hg/Hg_2^{2+}} + \mathscr{E}°_{NO_3^-/NO}\\ &= -0.85\text{ V} + 0.96\text{ V}\\ &= 0.11\text{ V}\end{aligned}$$

Since $\mathscr{E}°$ is positive, the reaction is spontaneous under standard-state conditions.

PRACTICE EXERCISE

Will H_2O_2 (in acid solution) oxidize Mn^{2+} to MnO_4^- under standard-state conditions?

EXAMPLE 20.5

Calculating $\Delta G°$ and K from $\mathscr{E}°$

Calculate the standard free-energy change and the equilibrium constant for the following reaction at 25°C:

$$Sn(s) + 2Cu^{2+}(aq) \rightleftharpoons Sn^{2+}(aq) + 2Cu^+(aq)$$

Answer: The two half-reactions for the overall process are

Oxidation: $Sn(s) \longrightarrow Sn^{2+}(aq) + 2e^-$

Reduction: $2Cu^{2+}(aq) + 2e^- \longrightarrow 2Cu^+(aq)$

In Table 20.1 we find that $\mathscr{E}°_{Sn^{2+}/Sn} = -0.14$ V and $\mathscr{E}°_{Cu^{2+}/Cu^+} = 0.15$ V. Thus

$$\begin{aligned}\mathscr{E}° &= \mathscr{E}°_{Sn/Sn^{2+}} + \mathscr{E}°_{Cu^{2+}/Cu^+}\\ &= 0.14\text{ V} + 0.15\text{ V}\\ &= 0.29\text{ V}\end{aligned}$$

Now we use Equation (20.3):

$$\Delta G^\circ = -nF\mathscr{E}^\circ$$

The overall reaction shows that $n = 2$, so

$$\begin{aligned}\Delta G^\circ &= -(2\text{ mol})(96{,}500\text{ J/V}\cdot\text{mol})(0.29\text{ V})\\ &= -5.6\times 10^4\text{ J}\\ &= -56\text{ kJ}\end{aligned}$$

To calculate the equilibrium constant we rearrange Equation (20.5) as follows:

$$\log K = \frac{n\mathscr{E}^\circ}{0.0591\text{ V}}$$

Since $n = 2$, we write

$$\log K = \frac{(2)(0.29\text{ V})}{0.0591\text{ V}} = 9.8$$

$$K = 6\times 10^9$$

PRACTICE EXERCISE

Calculate ΔG° and K for the following reaction at 25°C:

$$2Al^{3+}(aq) + 3Mg(s) \rightleftharpoons 2Al(s) + 3Mg^{2+}(aq)$$

20.5 EFFECT OF CONCENTRATION ON CELL EMF: THE NERNST EQUATION

So far we have focused on redox reactions in which the reactants and products are in their standard states, that is, gases at 1 atm and solutes at 1 M concentration. However, standard-state conditions are often difficult and sometimes impossible to maintain. The relationship between the emf of a cell and the concentrations of reactants and products under non-standard-state conditions in a redox reaction of the type

$$a\text{A} + b\text{B} \longrightarrow c\text{C} + d\text{D}$$

can be derived as follows. From Equation (19.8) we write

$$\Delta G = \Delta G^\circ + RT\ln Q$$

where Q is the reaction quotient (see Section 15.3). Since $\Delta G = -nF\mathscr{E}$ and $\Delta G^\circ = -nF\mathscr{E}^\circ$, the equation can be expressed as

$$-nF\mathscr{E} = -nF\mathscr{E}^\circ + RT\ln Q$$

Dividing the equation through by $-nF$ and converting to the common logarithm, we get

$$\mathscr{E} = \mathscr{E}^\circ - \frac{2.303RT}{nF}\log Q \qquad (20.6)$$

Equation (20.6) is known as the ***Nernst equation,*** after the German chemist Walter Hermann Nernst, who did much work in thermodynamics and electrochemistry. When $T = 298$ K, Equation (20.6) can be written as

$$\mathscr{E} = \mathscr{E}^\circ - \frac{0.0591\ \text{V}}{n} \log Q \qquad (20.7)$$

At equilibrium, there is no net transfer of electrons, so $\mathscr{E} = 0$ and $Q = K$, where K is the equilibrium constant of the redox reaction.

The Nernst equation enables us to calculate $\mathscr{E}$ as a function of reactant and product concentrations in a redox reaction. Referring to the galvanic cell in Figure 20.1, we get

$$\text{Zn}(s) + \text{Cu}^{2+}(aq) \longrightarrow \text{Zn}^{2+}(aq) + \text{Cu}(s)$$

The Nernst equation for this cell at $T = 298$ K is

$$\mathscr{E} = 1.10\ \text{V} - \frac{0.0591\ \text{V}}{2} \log \frac{[\text{Zn}^{2+}]}{[\text{Cu}^{2+}]}$$

If the ratio $[\text{Zn}^{2+}]/[\text{Cu}^{2+}]$ is less than 1, $\log [\text{Zn}^{2+}]/[\text{Cu}^{2+}]$ is a negative number, so that the second term on the right-hand side of the above equation is positive. Under this condition $\mathscr{E}$ is greater than the standard emf $\mathscr{E}^\circ$. If the ratio is greater than 1, $\mathscr{E}$ is smaller than $\mathscr{E}^\circ$.

EXAMPLE 20.6

Using the Nernst Equation to Predict the Spontaneity of a Redox Reaction

Predict whether the following reaction would proceed spontaneously as written as 298 K:

$$\text{Co}(s) + \text{Fe}^{2+}(aq) \longrightarrow \text{Co}^{2+}(aq) + \text{Fe}(s)$$

given that $[\text{Co}^{2+}] = 0.15\ M$ and $[\text{Fe}^{2+}] = 0.68\ M$.

Answer: The half-reactions are

$$\begin{aligned} &\text{Oxidation:} && \text{Co}(s) \longrightarrow \text{Co}^{2+}(aq) + 2e^- \\ &\text{Reduction:} && \text{Fe}^{2+}(aq) + 2e^- \longrightarrow \text{Fe}(s) \end{aligned}$$

In Table 20.1, we find that $\mathscr{E}^\circ_{\text{Co}^{2+}/\text{Co}} = -0.28$ V and $\mathscr{E}^\circ_{\text{Fe}^{2+}/\text{Fe}} = -0.44$ V. Therefore, the standard emf is

$$\begin{aligned} \mathscr{E}^\circ &= \mathscr{E}^\circ_{\text{Co}/\text{Co}^{2+}} + \mathscr{E}^\circ_{\text{Fe}^{2+}/\text{Fe}} \\ &= 0.28\ \text{V} + (-0.44\ \text{V}) \\ &= -0.16\ \text{V} \end{aligned}$$

From Equation (20.7) we write

$$\begin{aligned} \mathscr{E} &= \mathscr{E}^\circ - \frac{0.0591\ \text{V}}{n} \log \frac{[\text{Co}^{2+}]}{[\text{Fe}^{2+}]} \\ &= -0.16\ \text{V} - \frac{0.0591\ \text{V}}{2} \log \frac{0.15}{0.68} \\ &= -0.16\ \text{V} + 0.019\ \text{V} \\ &= -0.14\ \text{V} \end{aligned}$$

Since $\mathscr{E}$ is negative (or ΔG is positive), the reaction is *not* spontaneous in the direction written.

PRACTICE EXERCISE

Will the following reaction occur spontaneously at 25°C, given that $[Fe^{2+}] = 0.60\ M$ and $[Cd^{2+}] = 0.010\ M$?

$$Cd(s) + Fe^{2+}(aq) \longrightarrow Cd^{2+}(aq) + Fe(s)$$

It is interesting to determine at what ratio of $[Co^{2+}]$ to $[Fe^{2+}]$ the reaction in Example 20.6 will become spontaneous. We start with the Nernst equation as written for 298 K [Equation (20.7)]:

$$\mathscr{E} = \mathscr{E}° - \frac{0.0591\ V}{n} \log Q$$

We set $\mathscr{E}$ equal to zero since this corresponds to the equilibrium situation.

$$0 = -0.16\ V - \frac{0.0591\ V}{2} \log \frac{[Co^{2+}]}{[Fe^{2+}]}$$

$$\log \frac{[Co^{2+}]}{[Fe^{2+}]} = -5.4$$

Taking the antilog of both sides, we obtain

$$\frac{[Co^{2+}]}{[Fe^{2+}]} = 4 \times 10^{-6} = K$$

Thus for the reaction to be spontaneous, the ratio $[Co^{2+}]/[Fe^{2+}]$ must be smaller than 4×10^{-6}.

EXAMPLE 20.7

Using the Nernst Equation to Calculate Concentration

Consider the galvanic cell shown in Figure 20.4(a). In a certain experiment, the emf($\mathscr{E}$) of the cell is found to be 0.54 V at 25°C. Suppose that $[Zn^{2+}] = 1.0\ M$ and $P_{H_2} = 1.0$ atm. Calculate the molar concentration of H^+.

Answer: The overall cell reaction is

$$Zn(s) + 2H^+(aq,\ ?\ M) \longrightarrow Zn^{2+}(aq,\ 1\ M) + H_2(g,\ 1\ atm)$$

As we saw earlier (p. 568), the standard emf for the cell is 0.76 V. From Equation (20.7), we write

$$\mathscr{E} = \mathscr{E}° - \frac{0.0591\ V}{n} \log \frac{[Zn^{2+}]P_{H_2}}{[H^+]^2}$$

$$0.54\ V = 0.76\ V - \frac{0.0591\ V}{2} \log \frac{(1.0)(1.0)}{[H^+]^2}$$

$$-0.22\ V = -\frac{0.0591\ V}{2} \log \frac{1}{[H^+]^2}$$

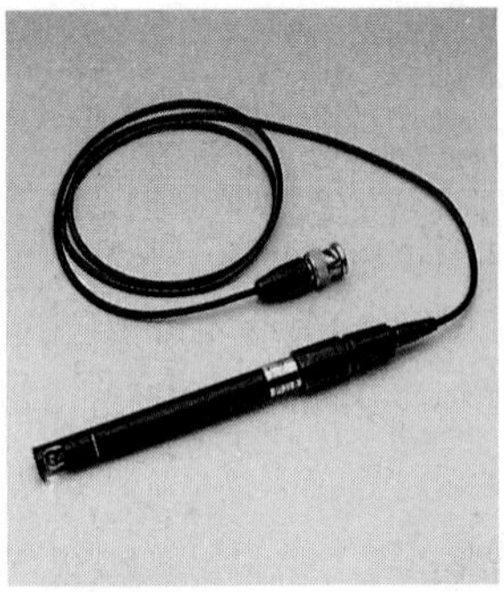

FIGURE 20.5

A glass electrode that is used in conjunction with a reference electrode in a pH meter.

$$7.4 = \log \frac{1}{[H^+]^2}$$
$$7.4 = -2 \log [H^+]$$
$$\log [H^+] = -3.7$$
$$[H^+] = 2 \times 10^{-4}\ M$$

PRACTICE EXERCISE

What is the emf of a cell consisting of a Cd/Cd^{2+} half-cell and a $Pt/H_2/H^+$ half-cell if $[Cd^{2+}] = 0.20\ M$, $[H^+] = 0.16\ M$, and $P_{H_2} = 0.80$ atm?

Example 20.7 shows that a galvanic cell whose cell reaction involves H^+ ions can be used to measure $[H^+]$ or pH. The pH meter described in Section 16.3 is based on this principle, but for practical reasons the electrodes used in a pH meter are quite different from the SHE and zinc electrode in the galvanic cell (Figure 20.5).

20.6 BATTERIES

A **battery** is *an electrochemical cell, or,* often, *several electrochemical cells connected in series, that can be used as a source of direct electric current at a constant voltage.* Although the operation of a battery is similar in principle to that of the galvanic cells described in Section 20.2, a battery has the advantage of being completely self-contained and requiring no auxiliary components such as salt bridges. We will describe several types of batteries that are in widespread use.

THE DRY CELL BATTERY

The most common dry cell, that is, a cell without fluid components, is the *Leclanché cell* used in flashlights and portable CD players. The anode of the cell consists of a zinc can or container that is in contact with manganese dioxide (MnO_2) and an electrolyte. The electrolyte consists of ammonium chloride and zinc chloride in water, with starch added to form a paste that is less likely to leak than a solution (Figure 20.6). The cathode is a carbon rod, which is immersed in the electrolyte in the center of the cell. The cell reactions are

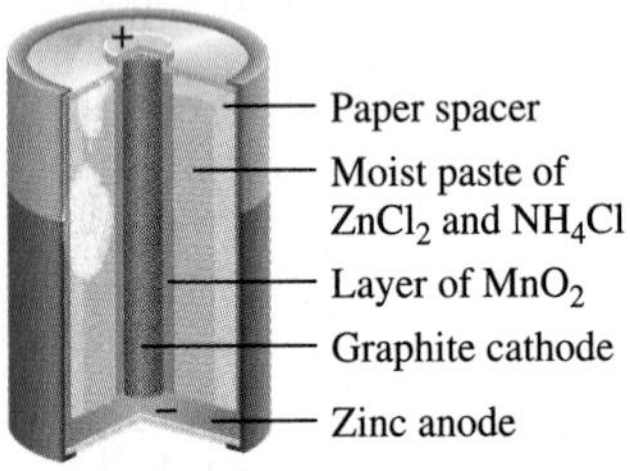

FIGURE 20.6

Interior section of a dry cell of the kind used in flashlights and portable CD players. Actually, the cell is not completely dry. It contains a moist electrolyte paste.

$$\begin{array}{ll} \text{Anode:} & Zn(s) \longrightarrow Zn^{2+}(aq) + 2e^- \\ \text{Cathode:} & 2NH_4^+(aq) + 2MnO_2(s) + 2e^- \longrightarrow Mn_2O_3(s) + 2NH_3(aq) + H_2O(l) \\ \hline \text{Overall:} & Zn(s) + 2NH_4^+(aq) + 2MnO_2(s) \longrightarrow \\ & \quad Zn^{2+}(aq) + 2NH_3(aq) + H_2O(l) + Mn_2O_3(s) \end{array}$$

Actually the reactions that occur in the cell are much more complex than the equations suggest. The voltage produced by a dry cell is about 1.5 V.

THE MERCURY BATTERY

The mercury battery is used extensively in medicine and electronics and is more expensive than the common dry cell. Contained in a stainless-steel cylinder, the mercury battery consists of a zinc anode (amalgamated with mercury) in contact with a strongly alkaline electrolyte containing zinc oxide and mercury(II) oxide (Figure 20.7). The cell reactions are

$$\begin{array}{ll} \text{Anode:} & Zn(Hg) + 2OH^-(aq) \longrightarrow ZnO(s) + H_2O(l) + 2e^- \\ \text{Cathode:} & HgO(s) + H_2O(l) + 2e^- \longrightarrow Hg(l) + 2OH^-(aq) \\ \hline \text{Overall:} & Zn(Hg) + HgO(s) \longrightarrow ZnO(s) + Hg(l) \end{array}$$

Because there is no change in electrolyte composition during operation—the overall cell reaction involves only solid substances—the mercury battery provides a more constant voltage (1.35 V) than the Leclanché cell. It also has a considerably higher capacity and longer life. These qualities make the mercury battery ideal for use in pacemakers, hearing aids, electric watches, and light meters.

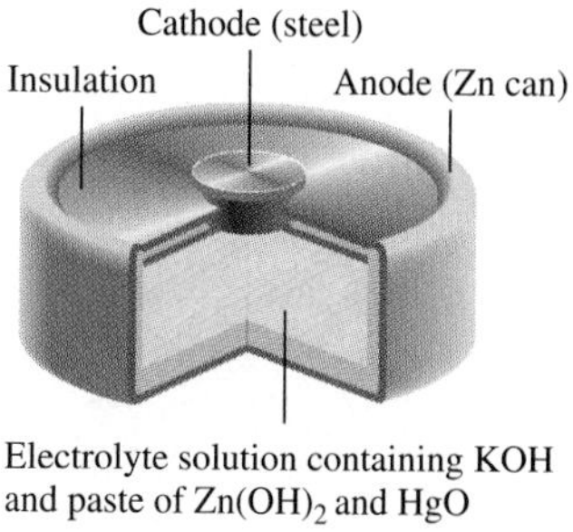

FIGURE 20.7

Interior section of a mercury battery.

THE LEAD STORAGE BATTERY

The lead storage battery commonly used in automobiles consists of six identical cells joined together in series. Each cell has a lead anode and a cathode made of lead dioxide (PbO_2) packed on a metal plate (Figure 20.8). Both the cathode and the anode are immersed in an aqueous solution of sulfuric acid, which acts as the electrolyte. The cell reactions are

$$\begin{array}{ll} \text{Anode:} & Pb(s) + SO_4^{2-}(aq) \longrightarrow PbSO_4(s) + 2e^- \\ \text{Cathode:} & PbO_2(s) + 4H^+(aq) + SO_4^{2-}(aq) + 2e^- \longrightarrow PbSO_4(s) + 2H_2O(l) \\ \hline \text{Overall:} & Pb(s) + PbO_2(s) + 4H^+(aq) + 2SO_4^{2-}(aq) \longrightarrow 2PbSO_4(s) + 2H_2O(l) \end{array}$$

Under normal operating conditions, each cell produces 2 V; a total of 12 V from the six cells is used to power the ignition circuit of the automobile and its other electrical systems. The lead storage battery can deliver large amounts of current for a short time, such as the time it takes to start up the engine.

Unlike the Leclanché cell and the mercury battery, the lead storage battery is rechargeable. Recharging the battery means reversing the normal electrochemical reaction by applying an external voltage at the cathode and the anode. (This process is called *electrolysis;* see p. 586.) The reactions that replenish the original materials are

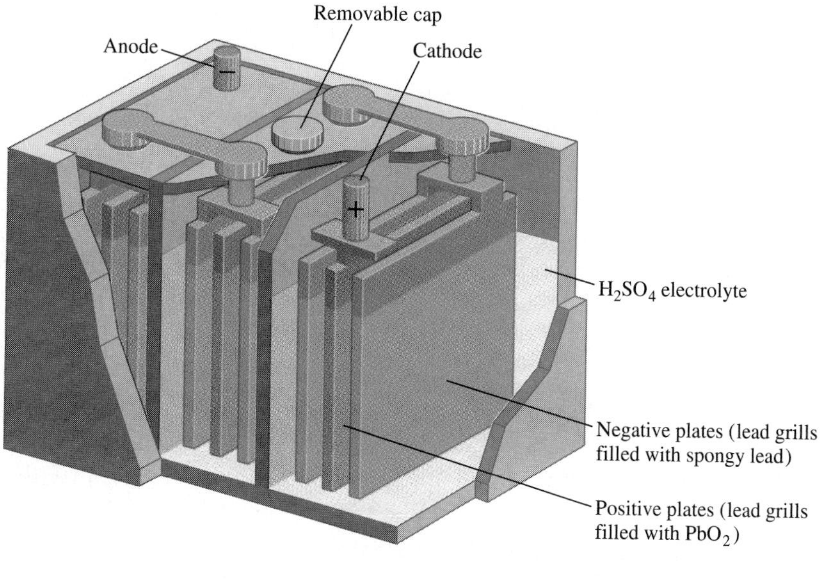

FIGURE 20.8

Interior section of a lead storage battery. Under normal operating conditions, the concentration of the sulfuric acid solution is about 38 percent by mass.

$$
\begin{array}{lrcl}
\text{Anode:} & PbSO_4(s) + 2e^- & \longrightarrow & Pb(s) + SO_4^{2-}(aq) \\
\text{Cathode:} & PbSO_4(s) + 2H_2O(l) & \longrightarrow & PbO_2(s) + 4H^+(aq) + SO_4^{2-}(aq) + 2e^- \\
\hline
\text{Overall:} & 2PbSO_4(s) + 2H_2O(l) & \longrightarrow & Pb(s) + PbO_2(s) + 4H^+(aq) + 2SO_4^{2-}(aq)
\end{array}
$$

The overall reaction is exactly the opposite of the normal cell reaction.

Two aspects of the operation of a lead storage battery are worth noting. First, because the electrochemical reaction consumes sulfuric acid, the degree to which the battery has been discharged can be checked by measuring the density of the electrolyte with a hydrometer, as is usually done at gas stations. The density of the fluid in a "healthy," fully charged battery should be equal to or greater than 1.2 g/mL. Second, people living in cold climates sometimes have trouble starting their cars because the battery has "gone dead." Thermodynamic calculations show that the emf of many electrochemical cells decreases with decreasing temperature. However, for a lead storage battery, the temperature coefficient is about 1.5×10^{-4} V/°C; that is, there is a decrease in voltage of 1.5×10^{-4} V for every degree drop in temperature. Thus, even allowing for a 40°C change in temperature, the decrease in voltage amounts to only 6×10^{-3} V, which is about

$$\frac{6 \times 10^{-3}\ \text{V}}{12\ \text{V}} \times 100\% = 0.05\%$$

of the operating voltage, an insignificant change. The real cause of a battery's apparent breakdown is an increase in the viscosity of the electrolyte as the temperature decreases. For the battery to function properly, the electrolyte must be fully conducting. However, the ions move much more slowly in a viscous medium, so the resistance of the fluid increases, leading to a decrease in the power output of the battery. If an apparently "dead battery" is warmed to near room temperature, it recovers its ability to deliver normal power.

FUEL CELLS

Fossil fuels are a major source of energy, but conversion of fossil fuel into electrical energy is a highly inefficient process. Consider the combustion of methane:

$$CH_4(g) + 2O_2(g) \longrightarrow CO_2(g) + 2H_2O(l) + \text{energy}$$

To generate electricity, heat produced by the reaction is first used to convert water to steam, which then drives a turbine that drives a generator. An appreciable fraction of the energy released in the form of heat is lost to the surroundings at each step; the most efficient power plant now in existence converts only about 40 percent of the original chemical energy into electricity. Because combustion reactions are redox reactions, it is more desirable to carry them out directly by electrochemical means, thereby greatly increasing the efficiency of power production. This objective can be accomplished by devices known as fuel cells.

In its simplest form, a hydrogen-oxygen fuel cell consists of an electrolyte solution, such as potassium hydroxide solution, and two inert electrodes. Hydrogen and oxygen gases are bubbled through the anode and cathode compartments (Figure 20.9), where the following reactions take place:

$$
\begin{array}{lrcl}
\text{Anode:} & 2H_2(g) + 4OH^-(aq) & \longrightarrow & 4H_2O(l) + 4e^- \\
\text{Cathode:} & O_2(g) + 2H_2O(l) + 4e^- & \longrightarrow & 4OH^-(aq) \\
\hline
\text{Overall:} & 2H_2(g) + O_2(g) & \longrightarrow & 2H_2O(l)
\end{array}
$$

The standard emf of the cell can be calculated as follows, with data from Table 20.1:

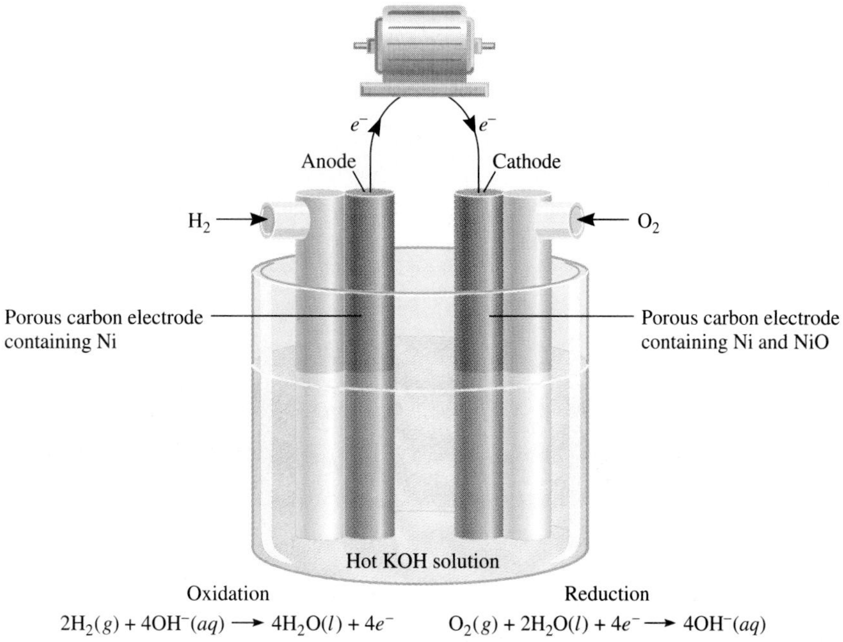

FIGURE 20.9

A hydrogen-oxygen fuel cell. The Ni and NiO embedded in the porous carbon electrodes are electrocatalysts.

$$
\begin{aligned}
\mathscr{E}^\circ &= \mathscr{E}^\circ_{\text{ox}} + \mathscr{E}^\circ_{\text{red}} \\
&= 0.83\ \text{V} + 0.40\ \text{V} \\
&= 1.23\ \text{V}
\end{aligned}
$$

Thus the cell reaction is spontaneous under standard-state conditions. Note that the reaction is the same as the hydrogen combustion reaction, but the oxidation and reduction are carried out separately at the anode and the cathode. Like platinum in the standard hydrogen electrode, the electrodes have a twofold function. They serve as electrical conductors, and they provide the necessary surfaces for the initial decomposition of the molecules into atomic species, prior to electron transfer. They are *electrocatalysts.* Metals such as platinum, nickel, and rhodium are good electrocatalysts.

In addition to the H_2–O_2 system, a number of other fuel cells have been developed. Among these is the propane-oxygen fuel cell. The half-cell reactions are

$$
\begin{array}{lrcl}
\text{Anode:} & C_3H_8(g) + 6H_2O(l) & \longrightarrow & 3CO_2(g) + 20H^+(aq) + 20e^- \\
\text{Cathode:} & 5O_2(g) + 20H^+(aq) + 20e^- & \longrightarrow & 10H_2O(l) \\
\hline
\text{Overall:} & C_3H_8(g) + 5O_2(g) & \longrightarrow & 3CO_2(g) + 4H_2O(l)
\end{array}
$$

The overall reaction is identical to the burning of propane in oxygen.

Unlike batteries, fuel cells do not store chemical energy. Reactants must be constantly resupplied, and products must be constantly removed from a fuel cell. In this respect, a fuel cell resembles an engine more than it does a battery. Properly designed fuel cells may be as much as 70 percent efficient, about twice as efficient as an internal combustion engine. In addition, fuel-cell generators are free of the noise, vibration, heat transfer, thermal pollution, and other problems normally associated with conventional power plants. Nevertheless, fuel cells are not yet in widespread use. The most successful application of fuel cells to date has been in space vehicles (Figure 20.10).

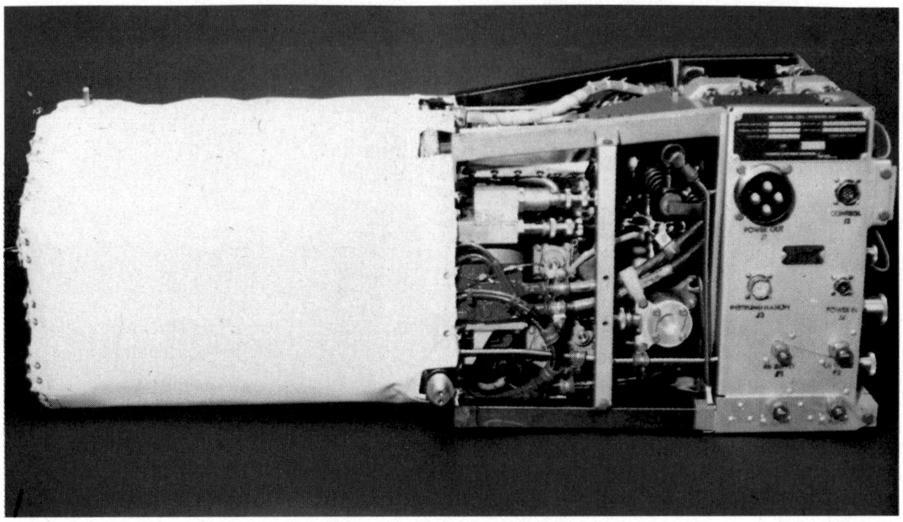

FIGURE 20.10

A hydrogen-oxygen fuel cell used in the space program. The pure water produced by the cell is drunk by the astronauts.

20.7 CORROSION

Corrosion is the term usually applied to *the deterioration of metals by an electrochemical process.* Some of the examples of corrosion we see around us are iron rust, silver tarnish, and the green patina formed on copper and brass (Figure 20.11). Corrosion causes enormous damage to buildings, bridges, ships, and cars. The cost of metallic corrosion to the U.S. economy is approximately 100 billion dollars a year! This section discusses some of the fundamental processes that occur in corrosion and methods used to protect metals against it.

By far the most familiar example of corrosion is the formation of rust on iron. Oxygen gas and water must be present in order for iron to rust. Although the reactions involved are quite complex and not completely understood, the main steps are believed to be as follows. A region of the metal's surface serves as the anode, where oxidation occurs:

$$Fe(s) \longrightarrow Fe^{2+}(aq) + 2e^-$$

The electrons given up by iron reduce atmospheric oxygen to water at the cathode, which is another region of the same metal's surface:

$$O_2(g) + 4H^+(aq) + 4e^- \longrightarrow 2H_2O(l)$$

The overall redox reaction is

$$2Fe(s) + O_2(g) + 4H^+(aq) \longrightarrow 2Fe^{2+}(aq) + 2H_2O(l)$$

With data from Table 20.1, we find the standard emf for this process:

$$\begin{aligned}\mathscr{E}^\circ &= \mathscr{E}^\circ_{ox} + \mathscr{E}^\circ_{red}\\ &= 0.44\ \text{V} + 1.23\ \text{V}\\ &= 1.67\ \text{V}\end{aligned}$$

Note that this reaction occurs in an acidic medium; the H^+ ions are supplied in part by the reaction of atmospheric carbon dioxide with water to form H_2CO_3.

FIGURE 20.11

Examples of corrosion: (a) a rusted ship, (b) a half-tarnished silver dish, and (c) the Statue of Liberty coated with patina before its restoration in 1986.

The Fe^{2+} ions formed at the anode are further oxidized by oxygen:

$$4Fe^{2+}(aq) + O_2(g) + (4 + 2x)H_2O(l) \longrightarrow 2Fe_2O_3 \cdot xH_2O(s) + 8H^+(aq)$$

This hydrated form of iron(III) oxide is known as rust. The amount of water associated with the iron oxide varies, so we represent the formula as $Fe_2O_3 \cdot xH_2O$.

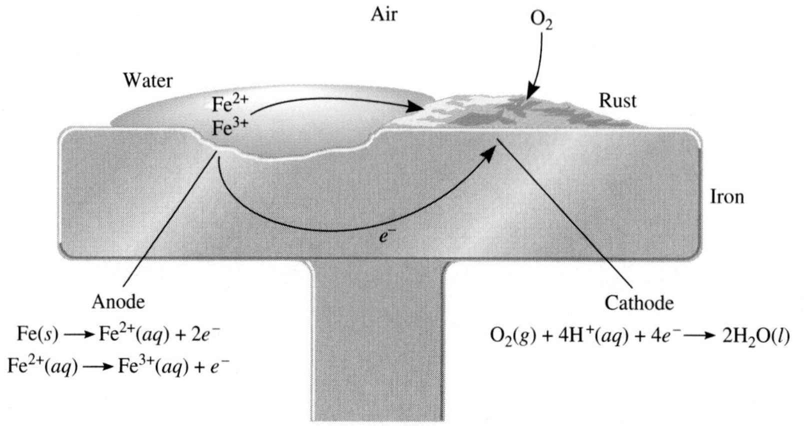

FIGURE 20.12

The electrochemical process involved in rust formation. The H^+ ions are supplied by H_2CO_3, which forms when CO_2 dissolves in water.

Figure 20.12 shows the mechanism of rust formation. The electric circuit is completed by the migration of electrons and ions; this is the reason that rusting occurs so rapidly in salt water. In cold climates, salts (NaCl or $CaCl_2$) spread on roadways to melt ice and snow are a major cause of rust formation on automobiles.

Metallic corrosion is not limited to iron. Another corrosion-prone metal is aluminum, a component of many things, including airplanes and beverage cans. Aluminum has a much greater tendency to oxidize than does iron; in Table 20.1 we see that Al has a more negative standard reduction potential than Fe. Based on this fact alone, we might expect to see airplanes slowly corrode away and soda cans transformed into piles of corroded aluminum. These processes do not occur because the layer of insoluble aluminum oxide (Al_2O_3) that forms on its surface when the metal is exposed to air serves to protect the aluminum underneath from further corrosion. The rust that forms on the surface of iron, however, is too porous to protect the underlying metal.

Coinage metals such as copper and silver also corrode, but much more slowly.

$$Cu(s) \longrightarrow Cu^{2+}(aq) + 2e^-$$

$$Ag(s) \longrightarrow Ag^+(aq) + e^-$$

In normal atmospheric exposure, copper forms a layer of copper carbonate ($CuCO_3$), a green substance also called patina, that protects the metal underneath from further corrosion. Likewise, silverware that comes into contact with foodstuffs develops a layer of silver sulfide (Ag_2S).

A number of methods have been devised to protect metals from corrosion. Most of them are aimed at preventing rust formation. The most obvious approach is to coat the metal surface with paint. However, if the paint is scratched, pitted, or dented to expose even the smallest area of bare metal, rust will form under the paint layer. An alternative is to render the surface of iron inactive by a process called *passivation,* in which the metal is treated with a strong oxidizing agent such as concentrated nitric acid, to produce a thin oxide layer. To prevent rust formation in cooling systems and radiators, a solution of sodium chromate is often added.

The tendency for iron to oxidize is greatly reduced by alloying it with certain other metals. For example, stainless steel, which is an alloy of iron, chromium, and nickel, is protected from corrosion by an outer layer of chromium(III) oxide.

An iron container can be coated with a layer of another metal such as tin or zinc. For instance, a "tin" can is made by applying a thin layer of tin over iron. As long as the tin layer remains intact, no rust will form. However, once the surface has been scratched, rusting occurs rapidly. If we look up the standard reduction potentials, we find that iron acts as the anode and tin as the cathode in the corrosion process:

$$Fe(s) \longrightarrow Fe^{2+}(aq) + 2e^- \qquad \mathscr{E}^\circ = +0.44\ V$$

$$Sn^{2+}(aq) + 2e^- \longrightarrow Sn(s) \qquad \mathscr{E}^\circ = -0.14\ V$$

The protective process is different for zinc-plated, or *galvanized,* iron. Zinc is more easily oxidized than iron (see Table 20.1):

$$Zn(s) \longrightarrow Zn^{2+}(aq) + 2e^- \qquad \mathscr{E}^\circ = +0.76\ V$$

So even if a scratch exposes the iron, the zinc is still attacked. In this case, the zinc metal serves as the anode and the iron is the cathode.

Cathodic protection is a process in which the metal that is to be protected from corrosion is made the cathode in what amounts to an electrochemical cell. Figure 20.13 shows how an iron nail can be protected from rusting by connecting the nail to a piece of zinc. Without such protection, an iron nail quickly rusts in water. Rusting of underground iron pipes and iron storage tanks can be prevented or greatly reduced by connecting them to metals such as zinc and magnesium, which oxidize more readily than iron (Figure 20.14).

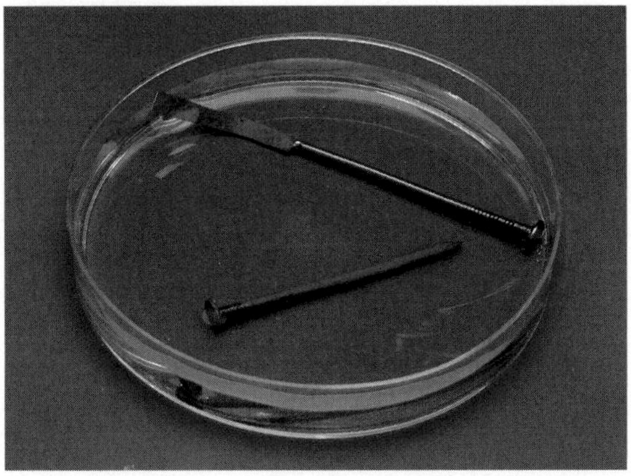

FIGURE 20.13

An iron nail that is cathodically protected by a strip of zinc does not rust in water, while an iron nail without such protection rusts readily.

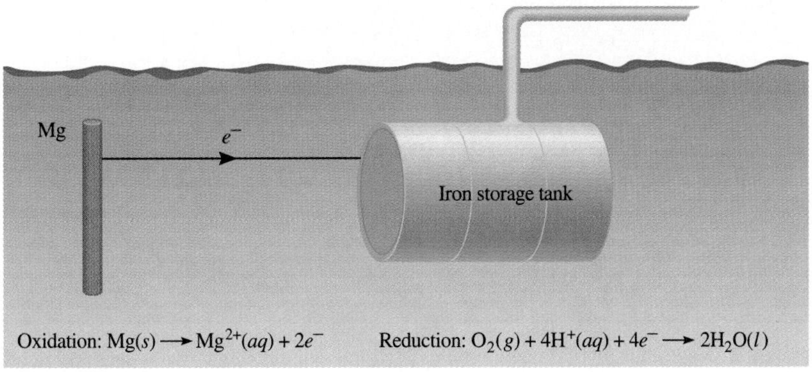

FIGURE 20.14

Cathodic protection of an iron storage tank (cathode) by magnesium, a more electropositive metal (anode). Since only the magnesium is depleted in the electrochemical process, it is sometimes called the sacrificial anode.

20.8 ELECTROLYSIS

In contrast to spontaneous redox reactions, which result in the conversion of chemical energy into electrical energy, ***electrolysis*** is the process in which *electrical energy is used to cause a nonspontaneous chemical reaction to occur.* The same principles underlie electrolysis and the processes that take place in galvanic cells. Here we will discuss three examples of electrolysis based on those principles. Then we will look at the quantitative aspects of electrolysis.

ELECTROLYSIS OF MOLTEN SODIUM CHLORIDE

In its molten state, sodium chloride, an ionic compound, can be electrolyzed to form sodium metal and chlorine. Figure 20.15(a) is a diagram of a *Downs cell,* which is used for large-scale electrolysis of NaCl. In molten NaCl, the cations and anions are the Na^+ and Cl^- ions, respectively. Figure 20.15(b) is a simplified diagram showing the reactions that occur at the electrodes. The *electrolytic cell* contains a pair of electrodes connected to the battery. The battery serves as an "electron pump," driving electrons to the cathode, where reduction occurs, and withdrawing electrons from the anode, where oxidation occurs. The reactions at the electrodes are

$$\begin{array}{lrcl} \text{Anode (oxidation):} & 2Cl^- & \longrightarrow & Cl_2(g) + 2e^- \\ \text{Cathode (reduction):} & 2Na^+ + 2e^- & \longrightarrow & 2Na(l) \\ \hline \text{Overall:} & 2Na^+ + 2Cl^- & \longrightarrow & 2Na(l) + Cl_2(g) \end{array}$$

This process is a major source of pure sodium metal and chlorine gas.

ELECTROLYSIS OF WATER

Water in a beaker under atmospheric conditions (1 atm and 25°C) will not spontaneously decompose to form hydrogen and oxygen gas because the standard free-energy change for the reaction is a large positive quantity:

$$2H_2O(l) \longrightarrow 2H_2(g) + O_2(g) \qquad \Delta G° = 474.4 \text{ kJ}$$

However, this reaction can be induced by electrolyzing water in a cell like the one

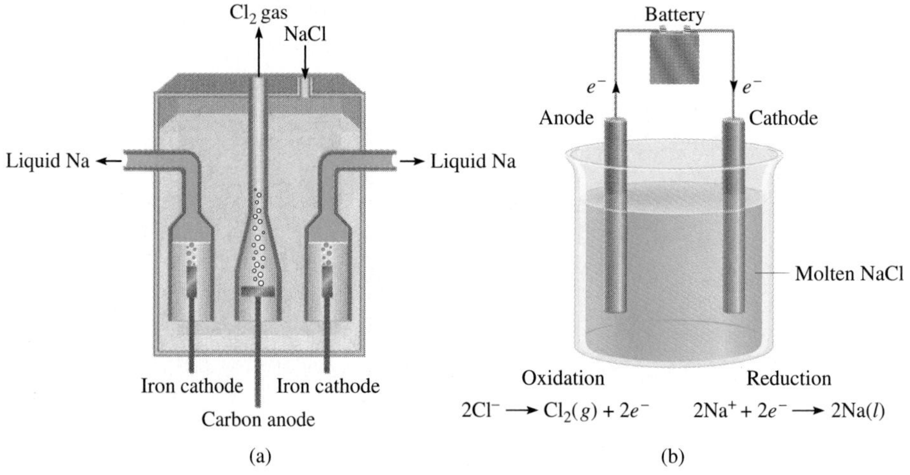

FIGURE 20.15

(a) A practical arrangement called a Downs cell for the electrolysis of molten NaCl (m.p. = 801°C). The sodium metal formed at the cathodes is in the liquid state. Since liquid sodium metal is lighter than molten NaCl, the sodium floats to the surface, as shown, and is collected. Chlorine gas forms at the anode and is collected at the top. (b) A simplified diagram showing the electrode reactions during the electrolysis of molten NaCl. The battery is needed to drive the nonspontaneous reactions.

shown in Figure 20.16. This electrolytic cell consists of a pair of electrodes made of a nonreactive metal, such as platinum, immersed in water. When the electrodes are connected to the battery, nothing happens because there are not enough ions in pure water to carry much of an electric current. (Remember that at 25°C, pure water has only 1×10^{-7} *M* H^+ ions and 1×10^{-7} *M* OH^- ions.)

On the other hand, the reaction occurs readily in a 0.1 *M* H_2SO_4 solution because there are sufficient ions to conduct electricity. Immediately, gas bubbles begin to appear at both electrodes. The process is diagrammed in Figure 20.17. The reaction at the anode is

$$2H_2O(l) \longrightarrow O_2(g) + 4H^+(aq) + 4e^-$$

At the cathode we have

$$H^+(aq) + e^- \longrightarrow \tfrac{1}{2}H_2(g)$$

The overall reaction is given by

$$\begin{array}{lrcl} \text{Anode (oxidation):} & 2H_2O(l) & \longrightarrow & O_2(g) + 4H^+(aq) + 4e^- \\ \text{Cathode (reduction):} & 4[H^+(aq) + e^- & \longrightarrow & \tfrac{1}{2}H_2(g)] \\ \hline \text{Overall:} & 2H_2O(l) & \longrightarrow & 2H_2(g) + O_2(g) \end{array}$$

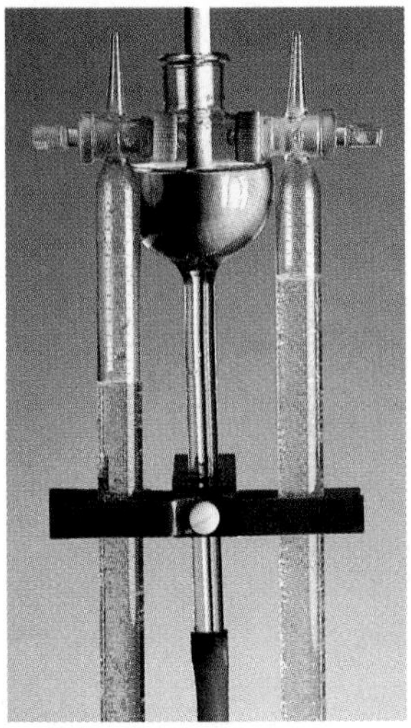

FIGURE 20.16

Apparatus for small-scale electrolysis of water. The volume of hydrogen gas generated (left column) is twice that of oxygen gas (right column). Also see page 8.

ELECTROLYSIS OF AN AQUEOUS SODIUM CHLORIDE SOLUTION

This is the most complicated of the three examples of electrolysis considered here because aqueous sodium chloride solution contains several species that could be oxidized and reduced. The oxidation reactions that might occur at the anode are

$$(1) \quad 2H_2O(l) \longrightarrow O_2(g) + 4H^+(aq) + 4e^-$$

$$(2) \quad 2Cl^-(aq) \longrightarrow Cl_2(g) + 2e^-$$

Referring to Table 20.1, we find

$$O_2(g) + 4H^+(aq) + 4e^- \longrightarrow 2H_2O(l) \qquad \mathscr{E}° = 1.23\text{ V}$$

$$Cl_2(g) + 2e^- \longrightarrow 2Cl^-(aq) \qquad \mathscr{E}° = 1.36\text{ V}$$

The standard reduction potentials of (1) and (2) are not very different, but the values do suggest that H_2O should be preferentially oxidized at the anode. However, by experiment we find that the gas liberated at the anode is Cl_2, not O_2! In

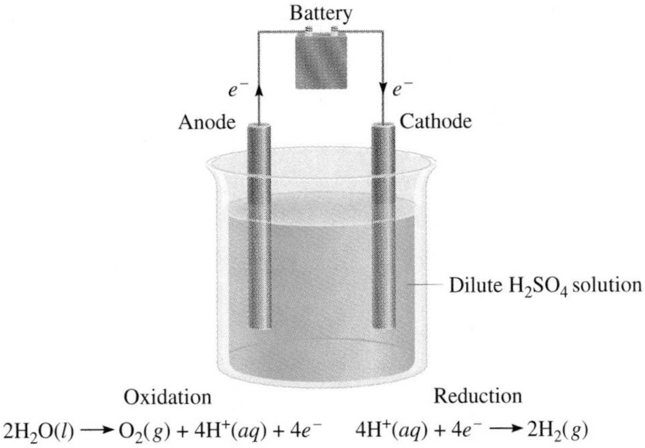

FIGURE 20.17

A diagram showing the electrode reactions during the electrolysis of water.

studying electrolytic processes we sometimes find that the voltage required for a reaction is considerably higher than the electrode potential indicates. *The additional voltage required to cause electrolysis* is called the ***overvoltage.*** The overvoltage for O_2 formation is quite high. Therefore, under normal operating conditions Cl_2 gas is actually formed at the anode instead of O_2.

The reductions that might occur at the cathode are

$$(3)\quad Na^+(aq) + e^- \longrightarrow Na(s) \qquad \mathscr{E}^\circ = -2.71\ V$$

$$(4)\quad 2H_2O(l) + 2e^- \longrightarrow H_2(g) + 2OH^-(aq) \qquad \mathscr{E}^\circ = -0.83\ V$$

$$(5)\quad 2H^+(aq) + 2e^- \longrightarrow H_2(g) \qquad \mathscr{E}^\circ = 0.00\ V$$

Reaction (3) is ruled out because it has a very negative standard reduction potential. Reaction (5) is preferred over (4) under standard-state conditions. At a pH of 7 (as is the case for a NaCl solution), however, they are equally probable. We generally use (4) to describe the cathode reaction because the concentration of H^+ ions is too low (about $1 \times 10^{-7}\ M$) to make (5) a reasonable choice.

Thus, the reactions in the electrolysis of aqueous sodium chloride are

Anode (oxidation):	$2Cl^-(aq) \longrightarrow Cl_2(g) + 2e^-$
Cathode (reduction):	$2H_2O(l) + 2e^- \longrightarrow H_2(g) + 2OH^-(aq)$
Overall:	$2H_2O(l) + 2Cl^-(aq) \longrightarrow H_2(g) + Cl_2(g) + 2OH^-(aq)$

As the overall reaction shows, the concentration of the Cl^- ions decreases during electrolysis and that of the OH^- ions increases. Therefore, in addition to H_2 and Cl_2, the useful by-product NaOH can be obtained by evaporating the aqueous solution at the end of the electrolysis.

EXAMPLE 20.8

Predicting the Products of Electrolysis

An aqueous Na_2SO_4 solution is electrolyzed, using the apparatus shown in Figure 20.16. If the products formed at the anode and cathode are oxygen gas and hydrogen gas, respectively, describe the electrolysis in terms of the reactions at the electrodes.

Answer: Before we look at the electrode reactions, we should consider the following facts: (1) Since Na_2SO_4 does not hydrolyze in water, the pH of the solution is close to 7. (2) The Na^+ ions are not reduced at the cathode, and the SO_4^{2-} ions are not oxidized at the anode. These conclusions are drawn from the electrolysis of water in the presence of sulfuric acid and in aqueous sodium chloride solution. Therefore, the electrode reactions are

$$\text{Anode:}\quad 2H_2O(l) \longrightarrow O_2(g) + 4H^+(aq) + 4e^-$$

$$\text{Cathode:}\quad 2H_2O(l) + 2e^- \longrightarrow H_2(g) + 2OH^-$$

The overall reaction, obtained by doubling the cathode reaction coefficients and adding the result to the anode reaction, is

$$6H_2O(l) \longrightarrow 2H_2(g) + O_2(g) + 4H^+(aq) + 4OH^-(aq)$$

If the H^+ and OH^- ions are allowed to mix, then

$$4H^+(aq) + 4OH^-(aq) \longrightarrow 4H_2O(l)$$

and the overall reaction becomes

$$2H_2O(l) \longrightarrow 2H_2(g) + O_2(g)$$

PRACTICE EXERCISE

An aqueous solution of $Mg(NO_3)_2$ is electrolyzed. What are the gaseous products at the anode and cathode?

Electrolysis has many important applications in industry, mainly in the extraction and purification of metals. We will discuss some of them in Section 20.9.

QUANTITATIVE ASPECTS OF ELECTROLYSIS

The quantitative treatment of electrolysis was developed primarily by Faraday. He observed that the mass of product formed (or reactant consumed) at an electrode is proportional to both the amount of electricity transferred at the electrode and the molar mass of the substance in question. For example, in the electrolysis of molten NaCl, the cathode reaction tells us that one Na atom is produced when one Na^+ ion accepts an electron from the electrode. To reduce 1 mole of Na^+ ions, we must supply Avogadro's number (6.02×10^{23}) of electrons to the cathode. On the other hand, the stoichiometry of the anode reaction shows that oxidation of two Cl^- ions yields one chlorine molecule. Therefore, the formation of 1 mole of Cl_2 results in the transfer of 2 moles of electrons from the Cl^- ions to the anode. Similarly, it takes 2 moles of electrons to reduce 1 mole of Mg^{2+} ions and 3 moles of electrons to reduce 1 mole of Al^{3+} ions:

$$Mg^{2+} + 2e^- \longrightarrow Mg$$

$$Al^{3+} + 3e^- \longrightarrow Al$$

Therefore

$$2\,F \simeq 1 \text{ mol } Mg^{2+}$$

$$3\,F \simeq 1 \text{ mol } Al^{3+}$$

where F is the faraday.

In an electrolysis experiment, we generally measure the current (in amperes, A) that passes through an electrolytic cell in a given period of time. The relationship between charge (in coulombs, C) and current is

$$1 \text{ C} = 1 \text{ A} \times 1 \text{ s}$$

that is, a coulomb is the quantity of electrical charge passing any point in the circuit in 1 second when the current is 1 ampere.

Figure 20.18 shows the steps involved in calculating the quantities of substances produced in electrolysis.

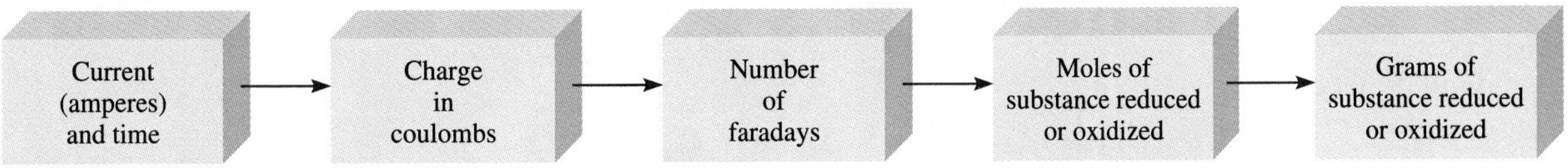

FIGURE 20.18

Steps involved in calculating amounts of substances reduced or oxidized in electrolysis.

EXAMPLE 20.9

Calculating the Quantity of Products in Electrolysis

A current of 0.452 A is passed through an electrolytic cell containing molten $CaCl_2$ for 1.50 hours. Write the electrode reactions and calculate the quantity of products (in grams) formed at the electrodes.

Answer: Since the only ions present in molten $CaCl_2$ are Ca^{2+} and Cl^-, the reactions are

$$\begin{array}{lrcl} \text{Anode:} & 2Cl^- & \longrightarrow & Cl_2(g) + 2e^- \\ \text{Cathode:} & Ca^{2+} + 2e^- & \longrightarrow & Ca(l) \\ \hline \text{Overall:} & Ca^{2+} + 2Cl^- & \longrightarrow & Ca(l) + Cl_2(g) \end{array}$$

The quantities of Ca metal and chlorine gas formed depend on the number of electrons that pass through the electrolytic cell, which in turn depends on the current and time, or charge:

$$? \text{ C} = 0.452 \text{ A} \times 1.50 \text{ h} \times \frac{3600 \text{ s}}{1 \text{ h}} \times \frac{1 \text{ C}}{1 \text{ A} \cdot \text{s}} = 2.44 \times 10^3 \text{ C}$$

Since 1 F = 96,500 C and 2 F are required to reduce 1 mole of Ca^{2+} ions, the mass of Ca metal formed at the cathode is calculated as follows:

$$? \text{ g Ca} = 2.44 \times 10^3 \text{ C} \times \frac{1 F}{96{,}500 \text{ C}} \times \frac{1 \text{ mol Ca}}{2 F} \times \frac{40.08 \text{ g Ca}}{1 \text{ mol Ca}} = 0.507 \text{ g Ca}$$

The anode reaction indicates that 1 mole of chlorine is produced per 2 F of electricity. Hence the mass of chlorine gas formed is

$$? \text{ g } Cl_2 = 2.44 \times 10^3 \text{ C} \times \frac{1 F}{96{,}500 \text{ C}} \times \frac{1 \text{ mol } Cl_2}{2 F} \times \frac{70.90 \text{ g } Cl_2}{1 \text{ mol } Cl_2} = 0.896 \text{ g } Cl_2$$

PRACTICE EXERCISE

A constant current is passed through an electrolytic cell containing molten $MgCl_2$ for 18 hours. If 4.8×10^5 g of Cl_2 are obtained, what is the current in amperes?

Charles Hall (1863–1914). Hall became interested in finding an inexpensive way to extract aluminum while an undergraduate at Oberlin College. Shortly after graduation, when he was only 22 years old, he succeeded in obtaining aluminum from aluminum oxide in a backyard woodshed.

20.9 ELECTROMETALLURGY

Electrolysis methods are useful for obtaining a pure metal from its ores or for refining (purifying) the metal. Collectively, these processes are called *electrometallurgy*. In the last section we saw how an active metal, sodium, can be obtained by electrolytically reducing its cation in the molten NaCl salt (p. 586). Here we will consider two other examples.

PRODUCTION OF ALUMINUM METAL

Aluminum is usually prepared from bauxite ore ($Al_2O_3 \cdot 2H_2O$). The ore is first treated to remove various impurities and then heated to obtain the anhydrous Al_2O_3. The oxide is dissolved in molten cryolite (Na_3AlF_6) in a Hall electrolytic cell

(Figure 20.19). The cell contains a series of carbon anodes; the cathode is also made of carbon and constitutes the lining inside the cell. The solution is electrolyzed to produce aluminum and oxygen gas:

$$\begin{array}{lrcl} \text{Anode:} & 3[2O^{2-} & \longrightarrow & O_2(g) + 4e^-] \\ \text{Cathode:} & 4[Al^{3+} + 3e^- & \longrightarrow & Al(l)] \\ \hline \text{Overall:} & 2Al_2O_3 & \longrightarrow & 4Al(l) + 3O_2(g) \end{array}$$

Oxygen gas reacts with the carbon anodes at 1000°C (the melting point of cryolite) to form carbon monoxide, which escapes as a gas. The liquid aluminum metal (m.p. 660°C) sinks to the bottom of the vessel, from which it can be drained.

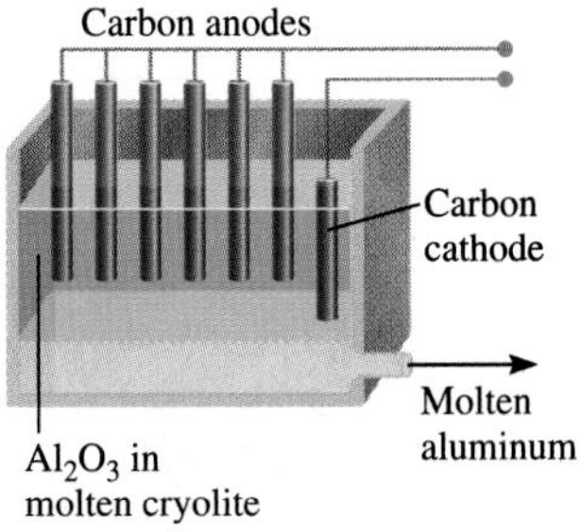

FIGURE 20.19

Electrolytic production of aluminum based on the Hall process.

PURIFICATION OF COPPER METAL

The copper metal obtained from its ores usually contains a number of impurities such as zinc, iron, silver, and gold. The more electropositive metals are removed by an electrolysis process in which the impure copper acts as the anode and *pure* copper acts as the cathode in a sulfuric acid solution containing Cu^{2+} ions (Figure 20.20). The reactions are

$$\begin{array}{lrcl} \text{Anode:} & Cu(s) & \longrightarrow & Cu^{2+}(aq) + 2e^- \\ \text{Cathode:} & Cu^{2+}(aq) + 2e^- & \longrightarrow & Cu(s) \end{array}$$

Reactive metals in the copper anode, such as iron and zinc, are also oxidized at the anode and enter the solution as Fe^{2+} and Zn^{2+} ions. They are not reduced at the cathode, however. The less electropositive metals, such as gold and silver, are not oxidized at the anode. Eventually, as the copper anode dissolves, these metals fall to the bottom of the cell. Thus the net result of this electrolysis process is the transfer of copper from the anode to the cathode. Copper prepared this way has a purity greater than 99.5 percent. It is interesting to note that the metal impurities (mostly silver and gold) from the copper anode are valuable by-products, the sale of which often pays for the electricity used to drive the electrolysis.

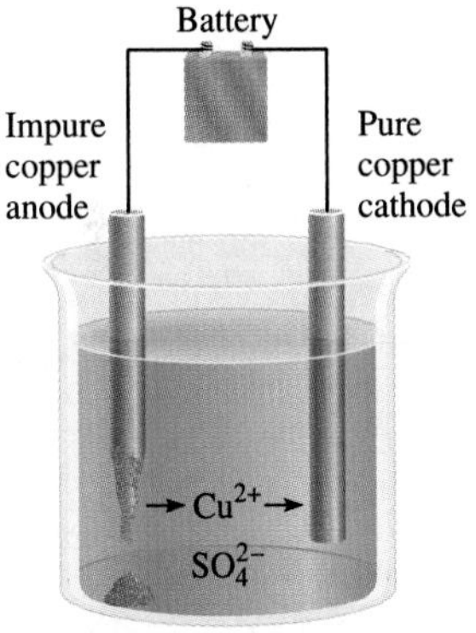

FIGURE 20.20

Electrolytic purification of copper.

SUMMARY

Redox reactions involve the transfer of electrons. Equations representing redox processes can be balanced using the ion-electron method.

In a galvanic cell, electricity is produced by a spontaneous chemical reaction. The oxidation at the anode and the reduction at the cathode take place separately, and the electrons flow through an external circuit. The two parts of a galvanic cell are the half-cells, and the reactions at the electrodes are the half-cell reactions. A salt bridge allows ions to flow between the half-cells.

The electromotive force (emf) of a cell is the voltage difference between the two electrodes. In the external circuit, electrons flow from the anode to the cathode in a galvanic cell. In solution, the anions move toward the anode and the cations move toward the cathode. The quantity of electricity carried by 1 mole of electrons is called a faraday, which is equal to 96,500 coulombs.

Standard reduction potentials show the relative likelihood of half-cell reduction reactions and can be used to predict the products, direction, and spontaneity of redox reactions between various substances. The decrease in free energy of the system in a spontaneous redox reaction is equal to the electrical work done by the system on the surroundings, or $\Delta G = -nF\mathscr{E}$. The equilibrium constant for a redox reaction can be found from the standard electromotive force of a cell.

The Nernst equation gives the relationship between the cell emf and the concentrations of the reactants and products under conditions other than the standard state.

Batteries, which consist of one or more electrochemical cells, are used widely as self-contained power sources. Some of the better-known batteries are the dry cell, such as the Leclanché cell; the mercury battery; the nickel-cadmium battery; and the lead storage battery used in automobiles. Fuel cells produce electrical energy from a continuous supply of reactants.

The corrosion of metals, the most familiar example of which is the rusting of iron, is an electrochemical phenomenon.

In electrolysis, electric current from an external source is used to drive a nonspontaneous chemical reaction. The amount of product formed or reactant consumed in an electrolytic cell depends on the quantity of electricity transferred at the electrode. Electrolysis plays an important role in obtaining pure metals from their ores and in purifying metals.

KEY WORDS

Anode, p. 565
Battery, p. 578
Cathode, p. 565
Corrosion, p. 582
Electrochemistry, p. 562
Electrolysis, p. 586
Electromotive force (emf), p. 566
Faraday (F), p. 572
Galvanic cell, p. 565
Half-cell reaction, p. 565
Nernst equation, p. 576
Overvoltage, p. 588
Standard emf, p. 568
Standard oxidation potential, p. 568
Standard reduction potential, p. 567

QUESTIONS AND PROBLEMS

BALANCING REDOX EQUATIONS

Problems

20.1 Balance the following redox equations by the ion-electron method:
(a) $H_2O_2 + Fe^{2+} \rightarrow Fe^{3+} + H_2O$ (in acidic solution)
(b) $Cu + HNO_3 \rightarrow Cu^{2+} + NO + H_2O$ (in acidic solution)
(c) $CN^- + MnO_4^- \rightarrow CNO^- + MnO_2$ (in basic solution)
(d) $Br_2 \rightarrow BrO_3^- + Br^-$ (in basic solution)
(e) $S_2O_3^{2-} + I_2 \rightarrow I^- + S_4O_6^{2-}$ (in acidic solution)

20.2 Balance the following redox equations by the ion-electron method:
(a) $Mn^{2+} + H_2O_2 \rightarrow MnO_2 + H_2O$ (in basic solution)
(b) $Bi(OH)_3 + SnO_2^{2-} \rightarrow SnO_3^{2-} + Bi$ (in basic solution)
(c) $Cr_2O_7^{2-} + C_2O_4^{2-} \rightarrow Cr^{3+} + CO_2$ (in acidic solution)
(d) $ClO_3^- + Cl^- \rightarrow Cl_2 + ClO_2$ (in acidic solution)

GALVANIC CELLS

Review Questions

20.3 Define the following terms: anode, cathode, electromotive force, standard oxidation potential, standard reduction potential.

20.4 Describe the basic features of a galvanic cell. Why are the two components in a galvanic cell separated from each other? What is the function of a salt bridge in a galvanic cell?

20.5 What is the difference between the half-reactions that characterize redox processes described in Section 4.4 and the half-cell reactions discussed in Section 20.2?

20.6 After operating a galvanic cell like the one shown in Figure 20.1 for a few minutes, a student notices that the cell emf begins to drop. Why?

Problems

20.7 Calculate the standard emf of a cell that uses the Mg/Mg^{2+} and Cu/Cu^{2+} half-cell reactions at 25°C. Write the equation for the cell reaction occurring under standard-state conditions.

20.8 Calculate the standard emf of a cell that uses

Ag/Ag^+ and Al/Al^{3+} half-cell reactions. Write the cell reaction occurring under standard-state conditions.

SPONTANEITY OF REDOX REACTIONS

Problems

20.9 Predict whether Fe^{3+} can oxidize I^- to I_2 under standard-state conditions.

20.10 Which of the following reagents can oxidize H_2O to $O_2(g)$ under standard-state conditions? $H^+(aq)$, $Cl^-(aq)$, $Cl_2(g)$, $Cu^{2+}(aq)$, $Pb^{2+}(aq)$, $MnO_4^-(aq)$ (in acid).

20.11 Consider the following half-reactions:

$$MnO_4^-(aq) + 8H^+(aq) + 5e^- \longrightarrow Mn^{2+}(aq) + 4H_2O(l)$$

$$NO_3^-(aq) + 4H^+(aq) + 3e^- \longrightarrow NO(g) + 2H_2O(l)$$

Predict whether NO_3^- ions will oxidize Mn^{2+} to MnO_4^- under standard-state conditions.

20.12 Predict whether the following reactions would occur spontaneously in aqueous solution at 25°C. Assume that the initial concentrations of dissolved species are all 1.0 *M*.
(a) $Ca(s) + Cd^{2+}(aq) \rightarrow Ca^{2+}(aq) + Cd(s)$
(b) $2Br^-(aq) + Sn^{2+}(aq) \rightarrow Br_2(l) + Sn(s)$
(c) $2Ag(s) + Ni^{2+} \rightarrow 2Ag^+(aq) + Ni(s)$
(d) $Cu^+(aq) + Fe^{3+}(aq) \rightarrow Cu^{2+}(aq) + Fe^{2+}(aq)$

20.13 Which species in each pair is a better oxidizing agent under standard-state conditions? (a) Br_2 or Au^{3+}, (b) H_2 or Ag^+, (c) Cd^{2+} or Cr^{3+}, (d) O_2 in acidic media or O_2 in basic media.

20.14 Which species in each pair is a better reducing agent under standard-state conditions? (a) Na or Li, (b) H_2 or I_2, (c) Fe^{2+} or Ag, (d) Br^- or Co^{2+}.

RELATIONSHIP AMONG $\mathscr{E}°$, $\Delta G°$, AND K

Review Questions

20.15 Write the equations relating $\Delta G°$ and K to the standard emf of a cell. Define all the terms.

20.16 Compare the ease of measuring the equilibrium constant electrochemically with that of finding K by chemical means [see Equation (19.9)].

Problems

20.17 What is the equilibrium constant for the following reaction at 25°C?

$$Mg(s) + Zn^{2+}(aq) \rightleftharpoons Mg^{2+}(aq) + Zn(s)$$

20.18 The equilibrium constant for the reaction

$$Sr(s) + Mg^{2+}(aq) \rightleftharpoons Sr^{2+}(aq) + Mg(s)$$

is 2.69×10^{12} at 25°C. Calculate $\mathscr{E}°$ for a cell made up of the Sr/Sr^{2+} and Mg/Mg^{2+} half-cells.

20.19 Use the standard reduction potentials to find the equilibrium constant for each of the following reactions at 25°C:
(a) $Br_2(l) + 2I^-(aq) \rightleftharpoons 2Br^-(aq) + I_2(s)$
(b) $2Ce^{4+}(aq) + 2Cl^-(aq) \rightleftharpoons Cl_2(g) + 2Ce^{3+}(aq)$
(c) $5Fe^{2+}(aq) + MnO_4^-(aq) + 8H^+(aq) \rightleftharpoons Mn^{2+}(aq) + 4H_2O + 5Fe^{3+}(aq)$

20.20 Calculate $\Delta G°$ and K_c for the following reactions at 25°C:
(a) $Mg(s) + Pb^{2+}(aq) \rightleftharpoons Mg^{2+}(aq) + Pb(s)$
(b) $Br_2(l) + 2I^-(aq) \rightleftharpoons 2Br^-(aq) + I_2(s)$
(c) $O_2(g) + 4H^+(aq) + 4Fe^{2+}(aq) \rightleftharpoons 2H_2O(l) + 4Fe^{3+}(aq)$
(d) $2Al(s) + 3I_2(s) \rightleftharpoons 2Al^{3+}(aq) + 6I^-(aq)$

20.21 What spontaneous reaction will occur in aqueous solution, under standard-state conditions, among the ions Ce^{4+}, Ce^{3+}, Fe^{3+}, and Fe^{2+}? Calculate $\Delta G°$ and K_c for the reaction.

20.22 Given that $\mathscr{E}° = 0.52$ V for the reduction $Cu^+(aq) + e^- \rightarrow Cu(s)$, calculate $\mathscr{E}°$, $\Delta G°$, and K for the following reaction at 25°C:

$$2Cu^+(aq) \longrightarrow Cu^{2+}(aq) + Cu(s).$$

THE NERNST EQUATION

Review Questions

20.23 Write the Nernst equation and explain all the terms.

20.24 Write the Nernst equation for the following processes at some temperature T:
(a) $Mg(s) + Sn^{2+}(aq) \rightarrow Mg^{2+}(aq) + Sn(s)$
(b) $2Cr(s) + 3Pb^{2+}(aq) \rightarrow 2Cr^{3+}(aq) + 3Pb(s)$

Problems

20.25 What is the potential of a cell made up of Zn/Zn^{2+} and Cu/Cu^{2+} half-cells at 25°C if $[Zn^{2+}] = 0.25$ *M* and $[Cu^{2+}] = 0.15$ *M*?

20.26 Calculate $\mathscr{E}°$, $\mathscr{E}$, and ΔG for the following cell reactions:
(a) $Mg(s) + Sn^{2+}(aq) \rightarrow Mg^{2+}(aq) + Sn(s)$
$[Mg^{2+}] = 0.045$ *M*, $[Sn^{2+}] = 0.035$ *M*
(b) $3Zn(s) + 2Cr^{3+}(aq) \rightarrow 3Zn^{2+}(aq) + 2Cr(s)$
$[Cr^{3+}] = 0.010$ *M*, $[Zn^{2+}] = 0.0085$ *M*

20.27 Calculate the standard potential of the cell consisting of the Zn/Zn^{2+} half-cell and the SHE. What will the emf of the cell be if $[Zn^{2+}] = 0.45$ *M*, $P_{H_2} = 2.0$ atm, and $[H^+] = 1.8$ *M*?

20.28 What is the emf of a cell consisting of a Pb/Pb^{2+}

half-cell and a $Pt/H_2/H^+$ half-cell if $[Pb^{2+}] = 0.10\ M$, $[H^+] = 0.050\ M$, and $P_{H_2} = 1.0$ atm?

20.29 Referring to Figure 20.1, calculate the $[Cu^{2+}]/[Zn^{2+}]$ ratio at which the following reaction will become spontaneous at 25°C:

$$Cu(s) + Zn^{2+}(aq) \longrightarrow Cu^{2+}(aq) + Zn(s)$$

20.30 Calculate the emf of the following concentration cell:

$$Mg(s)|Mg^{2+}(0.24\ M)|KCl(\text{sat'd})|Mg^{2+}(0.53\ M)|Mg(s)$$

BATTERIES AND FUEL CELLS

Review Questions

20.31 Explain clearly the differences between a primary galvanic cell—one that is not rechargeable—and a storage cell (for example, the lead storage battery), which is rechargeable.

20.32 Discuss the advantages and disadvantages of fuel cells over conventional power plants in producing electricity.

Problems

20.33 The hydrogen-oxygen fuel cell is described in Section 20.6. (a) What volume of $H_2(g)$, stored at 25°C at a pressure of 155 atm, would be needed to run an electric motor drawing a current of 8.5 A for 3.0 h? (b) What volume (in liters) of air at 25°C and 1.00 atm will have to pass into the cell per minute to run the motor? Assume that air is 20% O_2 by volume and that all the O_2 is consumed in the cell. The other components of air do not affect the fuel-cell reactions. Assume ideal gas behavior.

20.34 Calculate the standard emf of the propane fuel cell discussed on p. 581 at 25°C, given that ΔG_f° for propane is -23.5 kJ/mol.

CORROSION

Review Questions

20.35 Steel hardware, including nuts and bolts, is often coated with a thin cadmium plating. Explain the function of the cadmium layer.

20.36 "Galvanized iron" is steel sheet coated with zinc; "tin" cans are made of steel sheet coated with tin. Discuss the functions of these coatings and the electrochemistry of the corrosion reactions that occur if an electrolyte contacts the scratched surface of a galvanized iron sheet or a tin can.

20.37 Tarnished silver contains Ag_2S. The tarnish can be removed by placing silverware in an aluminum pan containing an inert electrolyte solution, such as NaCl. Explain the electrochemical principle for this procedure. [The standard reduction potential for the half-cell reaction $Ag_2S(s) + 2e^- \rightarrow 2Ag(s) + S^{2-}(aq)$ is -0.71 V.]

20.38 How does the tendency of iron to rust depend on the pH of solution?

ELECTROLYSIS

Review Questions

20.39 What is the difference between an electrochemical cell (such as a galvanic cell) and an electrolytic cell?

20.40 What is Faraday's contribution to quantitative electrolysis?

Problems

20.41 The half-reaction at an electrode is

$$Mg^{2+}(\text{molten}) + 2e^- \longrightarrow Mg(s)$$

Calculate the number of grams of magnesium that can be produced by passing 1.00 F through the electrode.

20.42 Consider the electrolysis of molten calcium chloride, $CaCl_2$. (a) Write the electrode reactions. (b) How many grams of calcium metal can be produced by passing 0.50 A for 30 min?

20.43 Considering only the cost of electricity, would it be cheaper to produce a ton of sodium or a ton of aluminum by electrolysis?

20.44 If the cost of electricity to produce magnesium by the electrolysis of molten magnesium chloride is \$155 per ton of metal, what is the cost (in dollars) of electricity to produce (a) 10.0 tons of aluminum, (b) 30.0 tons of sodium, (c) 50.0 tons of calcium?

20.45 In the electrolysis of water one of the half-reactions is

$$2H_2O(l) \longrightarrow O_2(g) + 4H^+(aq) + 4e^-$$

Suppose 0.076 L of O_2 is collected at 25°C and 755 mmHg. How many faradays of electricity must have passed through the solution?

20.46 How many faradays of electricity are required to produce (a) 0.84 L of O_2 at exactly 1 atm and 25°C from aqueous H_2SO_4 solution, (b) 1.50 L of Cl_2 at 750 mmHg and 20°C from molten NaCl, (c) 6.0 g of Sn from molten $SnCl_2$?

20.47 Calculate the amounts of Cu and Br_2 produced at inert electrodes by passing a current of 4.50 A through a solution of $CuBr_2$ for 1.0 h.

20.48 In the electrolysis of an aqueous $AgNO_3$ solution, 0.67 g of Ag is deposited after a certain period of time. (a) Write the half-reaction for the reduction of Ag^+. (b) What is the probable oxidation half-reaction? (c) Calculate the quantity of electricity used, in coulombs.

20.49 A steady current was passed through molten $CoSO_4$ until 2.35 g of metallic cobalt were produced. Calculate the number of coulombs of electricity used.

20.50 A constant electric current flows for 3.75 h through two electrolytic cells connected in series. One contains a solution of $AgNO_3$ and the second a solution of $CuCl_2$. During this time 2.00 g of silver are deposited in the first cell. (a) How many grams of copper are deposited in the second cell? (b) What is the current flowing, in amperes?

20.51 What is the hourly production rate of chlorine gas (in kilograms) from an electrolytic cell using aqueous NaCl electrolyte and carrying a current of 1.500×10^3 A? The anode efficiency for the oxidation of Cl^- is 93.0 percent.

20.52 Chromium plating is applied by electrolysis to objects suspended in a dichromate solution, according to the following (unbalanced) half-reaction:

$$Cr_2O_7^{2-}(aq) + e^- + H^+ \longrightarrow Cr(s) + H_2O(l)$$

How long (in hours) would it take to apply a chromium plating of thickness 1.0×10^{-2} mm to a car bumper of surface area 0.25 m^2 in an electrolysis cell carrying a current of 25.0 A? (The density of chromium is 7.19 g/cm^3.)

20.53 The passage of a current of 0.750 A for 25.0 min deposited 0.369 g of copper from a $CuSO_4$ solution. From this information, calculate the molar mass of copper.

20.54 A quantity of 0.300 g of copper was deposited from a $CuSO_4$ solution by passing a current of 3.00 A for 304 s. Calculate the value of the faraday constant.

20.55 In a certain electrolysis experiment, 1.44 g of Ag were deposited in one cell (containing an aqueous $AgNO_3$ solution), while 0.120 g of an unknown metal X was deposited in another cell (containing an aqueous XCl_3 solution) in series. Calculate the molar mass of X.

20.56 In the electrolysis of water one of the half-reactions is

$$2H^+(aq) + 2e^- \longrightarrow H_2(g)$$

Suppose 0.854 L of H_2 is collected at 25°C and 782 mmHg. How many faradays of electricity must have passed through the solution?

MISCELLANEOUS PROBLEMS

20.57 For each of the following redox reactions (i) write the half-reactions; (ii) write a balanced equation for the whole reaction; (iii) determine in which direction the reaction will proceed spontaneously under standard-state conditions.

(a) $H_2(g) + Ni^{2+}(aq) \rightarrow H^+(aq) + Ni(s)$

(b) $MnO_4^-(aq) + Cl^-(aq) \rightarrow Mn^{2+}(aq) + Cl_2(g)$ (in acid solution)

(c) $Cr(s) + Zn^{2+}(aq) \rightarrow Cr^{3+}(aq) + Zn(s)$

20.58 Oxidation of 25.0 mL of a solution containing Fe^{2+} requires 26.0 mL of 0.0250 *M* $K_2Cr_2O_7$ in acidic solution. Balance the following equation and calculate the molar concentration of Fe^{2+}:

$$Cr_2O_7^{2-} + Fe^{2+} + H^+ \longrightarrow Cr^{3+} + Fe^{3+}$$

20.59 The SO_2 present in air is mainly responsible for the acid rain phenomenon. Its concentration can be determined by titrating against a standard permanganate solution as follows:

$$5SO_2 + 2MnO_4^- + 2H_2O \longrightarrow 5SO_4^{2-} + 2Mn^{2+} + 4H^+$$

Calculate the number of grams of SO_2 in a sample of air if 7.37 mL of 0.00800 *M* $KMnO_4$ solution are required for the titration.

20.60 A sample of iron ore weighing 0.2792 g was dissolved in dilute acid solution, and all the Fe(II) was converted to Fe(III) ions. The solution required 23.30 mL of 0.0194 *M* $KMnO_4$ for titration. Calculate the percent by mass of iron in the ore.

20.61 The concentration of a hydrogen peroxide solution can be conveniently determined by titration against a standardized potassium permanganate solution in acidic medium according to the following unbalanced equation:

$$MnO_4^- + H_2O_2 \longrightarrow O_2 + Mn^{2+}$$

(a) Balance the above equation. (b) If 36.44 mL of a 0.01652 *M* $KMnO_4$ solution are required to completely oxidize 25.00 mL of a H_2O_2 solution, calculate the molarity of the H_2O_2 solution.

20.62 Oxalic acid ($H_2C_2O_4$) is present in many plants and vegetables. (a) Balance the following equation in acid solution:

$$MnO_4^- + C_2O_4^{2-} \longrightarrow Mn^{2+} + CO_2$$

(b) If a 1.00-g sample of $H_2C_2O_4$ requires 24.0 mL of 0.0100 M $KMnO_4$ solution to reach an equivalence point, what is the percent by mass of $H_2C_2O_4$ in the sample?

20.63 Complete the following table.

$\mathscr{E}$	ΔG	Cell reaction
>0		
	>0	
=0		

State whether the cell reaction is spontaneous, nonspontaneous, or at equilibrium.

20.64 Calcium oxalate (CaC_2O_4) is insoluble in water. This property has been used to determine the amount of Ca^{2+} ions in fluids such as blood. The calcium oxalate isolated from blood is dissolved in acid and titrated against a standardized $KMnO_4$ solution as described in Problem 20.62. In one test it is found that the calcium oxalate isolated from a 10.0-mL sample of blood requires 24.2 mL of 9.56×10^{-4} M $KMnO_4$ for titration. Calculate the number of milligrams of calcium per milliliter of blood.

20.65 From the following information, calculate the solubility product of AgBr.

$$AgBr(s) + e^- \longrightarrow Ag(s) + Br^-(aq) \qquad \mathscr{E}° = 0.07\ V$$

$$Ag^+(aq) + e^- \longrightarrow Ag(s) \qquad \mathscr{E}° = 0.80\ V$$

20.66 Consider the cell composed of the SHE and a half-cell using the reaction $Ag^+(aq) + e^- \rightarrow Ag(s)$. (a) Calculate the standard cell potential. (b) What is the spontaneous cell reaction under standard-state conditions? (c) Calculate the cell potential when $[H^+]$ in the hydrogen electrode is changed to (i) 1.0×10^{-2} M and (ii) 1.0×10^{-5} M, all other reagents being held at standard-state conditions. (d) Based on this cell arrangement, suggest a design for a pH meter.

20.67 A galvanic cell consists of a silver electrode in contact with 346 mL of 0.100 M $AgNO_3$ solution and a magnesium electrode in contact with 288 mL of 0.100 M $Mg(NO_3)_2$ solution. (a) Calculate $\mathscr{E}$ for the cell at 25°C. (b) A current is drawn from the cell until 1.20 g of silver have been deposited at the silver electrode. Calculate $\mathscr{E}$ for the cell at this stage of operation.

20.68 Explain why chlorine gas can be prepared by electrolyzing an aqueous solution of NaCl but fluorine gas cannot be prepared by electrolyzing an aqueous solution of NaF.

20.69 Calculate the emf of the following concentration cell at 25°C:

$$Cu(s)|Cu^{2+}(0.080\ M)|KCl(sat'd)|Cu^{2+}(1.2\ M)|Cu(s)$$

20.70 The cathode reaction in the Leclanché cell is given by

$$2MnO_2(s) + Zn^{2+} + 2e^- \longrightarrow ZnMn_2O_4(s)$$

If a Leclanché cell produces a current of 0.0050 A, calculate how many hours this current supply will last if initially 4.0 g of MnO_2 are present in the cell. Assume that there is an excess of Zn^{2+} ions.

20.71 Suppose you are asked to verify experimentally the electrode reactions shown in Example 20.8. In addition to the apparatus and the solution, you are also given two pieces of litmus paper, one blue and the other red. Describe what steps you would take in this experiment.

20.72 For a number of years it was not clear whether mercury(I) ions existed in solution as Hg^+ or as Hg_2^{2+}. To distinguish between these two possibilities, we could set up the following system:

$$Hg(l)|soln\ A|KCl(sat'd)|soln\ B|Hg(l)$$

where soln A contained 0.263 g mercury(I) nitrate per liter and soln B contained 2.63 g mercury(I) nitrate per liter. If the measured emf of such a cell is 0.0289 V at 18°C, what can you deduce about the nature of the mercury ions?

20.73 An aqueous KI solution to which a few drops of phenophthalein have been added is electrolyzed using an apparatus similar to that shown in Figure 20.17. Describe what you would observe at the anode and the cathode. (*Hint:* Molecular iodine is only slightly soluble in water, but in the presence of I^- ions, it forms the brown color of I_3^- ions according to the equation $I^- + I_2 \rightarrow I_3^-$.)

20.74 A piece of magnesium metal weighing 1.56 g is placed in 100.0 mL of 0.100 M $AgNO_3$ at 25°C. Calculate $[Mg^{2+}]$ and $[Ag^+]$ in solution at equilibrium. What is the mass of the magnesium left? The volume remains constant.

20.75 Describe an experiment that would enable you to determine which is the cathode and which is the anode in a galvanic cell using copper and zinc electrodes.

20.76 An acidified solution was electrolyzed using copper electrodes. After passing a constant current of 1.18 A for 1.52×10^3 s, the anode was found to have lost 0.584 g. (a) What gas is produced at the cathode and what is its volume at STP? (b) Given that the charge of an electron is 1.6022×10^{-19} C, calculate Avogadro's number. Assume that copper is oxidized to Cu^{2+} ions.

20.77 In a certain electrolysis experiment involving Al^{3+} ions, 60.2 g of Al are recovered when a current of 0.352 A is used. How long in minutes did the electrolysis take?

20.78 Consider the oxidation of ammonia:

$$4NH_3(g) + 3O_2(g) \longrightarrow 2N_2(g) + 6H_2O(l)$$

(a) Calculate the $\Delta G°$ for the reaction. (b) If this reaction were used in a fuel cell, what would be the standard cell potential?

20.79 A galvanic cell is constructed by immersing a piece of copper wire in 25.0 mL of a 0.20 *M* $CuSO_4$ solution and a zinc strip in 25.0 mL of a 0.20 *M* $ZnSO_4$ solution. (a) Calculate the emf of the cell at 25°C and predict what would happen if a small amount of concentrated NH_3 solution is added to (i) the $CuSO_4$ solution and (ii) the $ZnSO_4$ solution. Assume the volume in each compartment remains 25.00 mL. (b) In a separate experiment, 25.0 mL of 3.00 *M* NH_3 are added to the $CuSO_4$ solution. If the emf of the cell is 0.68 V, calculate the formation constant (K_f) of $Cu(NH_3)_4^{2+}$.

20.80 In an electrolysis experiment a student passes the same quantity of electricity through two electrolytic cells containing silver and gold salts, respectively. Over a certain period of time, she finds that 2.64 g of Ag and 1.61 g of Au are deposited at the cathodes. What is the oxidation state of gold in the gold salt?

20.81 People living in cold-climate countries where there is plenty of snow are advised not to heat their garages in the winter. What is the electrochemical basis for this action?

20.82 Given that

$$2Hg^{2+}(aq) + 2e^- \longrightarrow Hg_2^{2+}(aq) \qquad \mathscr{E}° = 0.92\text{ V}$$

$$Hg_2^{2+}(aq) + 2e^- \longrightarrow 2Hg(l) \qquad \mathscr{E}° = 0.85\text{ V}$$

calculate $\Delta G°$ and K for the process at 25°C:

$$Hg_2^{2+}(aq) \longrightarrow Hg^{2+}(aq) + Hg(l)$$

(The above reaction is an example of *disproportionation reaction* in which an element in one oxidation state is both oxidized and reduced.)

20.83 Fluorine (F_2) is obtained by the electrolysis of liquid hydrogen fluoride (HF) containing potassium fluoride (KF). (a) Write the half-cell reactions and the overall reaction for the process. (b) What is the purpose of KF? (c) Calculate the volume of F_2 (in liters) collected at 24.0°C and 1.2 atm after electrolyzing the solution for 15 h at a current of 502 A.

20.84 A 300-mL solution of NaCl was electrolyzed for 6.00 min. If the pH of the final solution was 12.24, calculate the average current used.

20.85 Industrially, copper is purified by electrolysis. The impure copper acts as the anode, and the cathode is made of pure copper. The electrodes are immersed in a $CuSO_4$ solution. During electrolysis, copper at the anode enters the solution as Cu^{2+} while Cu^{2+} ions are reduced at the cathode. (a) Write half-cell reactions and the overall reaction for the electrolytic process. (b) Suppose the anode was contaminated with Zn and Ag. Explain what happens to these impurities during electrolysis. (c) How many hours will it take to obtain 1.00 kg of Cu at a current of 18.9 A?

20.86 An aqueous solution of a platinum salt is electrolyzed by passing a current of 2.50 A for 2.00 h. As a result, 9.09 g of metallic Pt are formed at the cathode. Calculate the charge on the Pt ions in this solution.

20.87 Consider a galvanic cell consisting of a magnesium electrode in contact with 1.0 *M* $Mg(NO_3)_2$ and a cadmium electrode in contact with 1.0 *M* $Cd(NO_3)_2$. Calculate $\mathscr{E}°$ for the cell and draw a diagram showing the cathode, the anode, and the direction of electron flow.

20.88 A current of 6.00 A passes through an electrolytic cell containing dilute sulfuric acid for 3.40 h. If the volume of O_2 gas generated at the anode is 4.26 L (at STP), calculate the charge (in coulombs) on an electron.

20.89 (a) Gold will not dissolve in either concentrated nitric acid or concentrated hydrochloric acid. However, the metal does dissolve in a mixture of the acids (one part HNO_3 and three parts HCl by volume), called *aqua regia*. Write a balanced equation for this reaction. (*Hint:* Among the products are $HAuCl_4$ and NO_2.) (b) Use Le Chatelier's principle to explain why the oxidizing power of the NO_3^- ion is enhanced in a strongly acidic solution.

Answers to Practice Exercises: 20.1 $5Fe^{2+} + MnO_4^- + 8H^+ \rightarrow 5Fe^{3+} + Mn^{2+} + 4H_2O$; **20.2** Pb; **20.3** 0.34 V; **20.4** Yes; **20.5** $\Delta G° = -4.1 \times 10^2$ kJ, $K = 7 \times 10^{71}$; **20.6** Yes; **20.7** 0.38 V; **20.8** Anode: O_2, cathode: H_2; **20.9** 2.0×10^4 A.

CHAPTER 21

CHEMICAL KINETICS

◆ In recent years, advances in electronics and lasers have enabled chemists to probe reactions that take place extremely fast, on the order of picoseconds (10^{-12} s) or even femtoseconds (10^{-15} s). A number of chemical and biological processes occur on such a short time scale. Among them is the initial step in vision.

Lining the back of the eye are several layers of light-sensitive rod- and cone-shaped cells. These cells contain a substance called *retinal,* which is associated with proteins called *opsins.* The retinal-opsin complexes are called rhodopsin, which allows us to see black and white, and iodopsin, which is responsible for color vision.

In 1938 the German scientist Selig Hecht discovered that a human rod cell can be excited by a *single* photon. Twenty years later, the American biochemist George Wald showed that the primary event in visual excitation is an isomerization process.

Retinal is a complex compound with a number of geometric isomers, but only the all-*trans* and 11-*cis* isomers are involved in vision. In rhodopsin, the absorption of a photon converts the 11-*cis* retinal configuration to the all-*trans* form within a few picoseconds. This change in geometry triggers the visual process, by which an electrical impulse is transmitted to the optic nerve and ultimately to the brain. It is this final step that allows us to "see." ◆

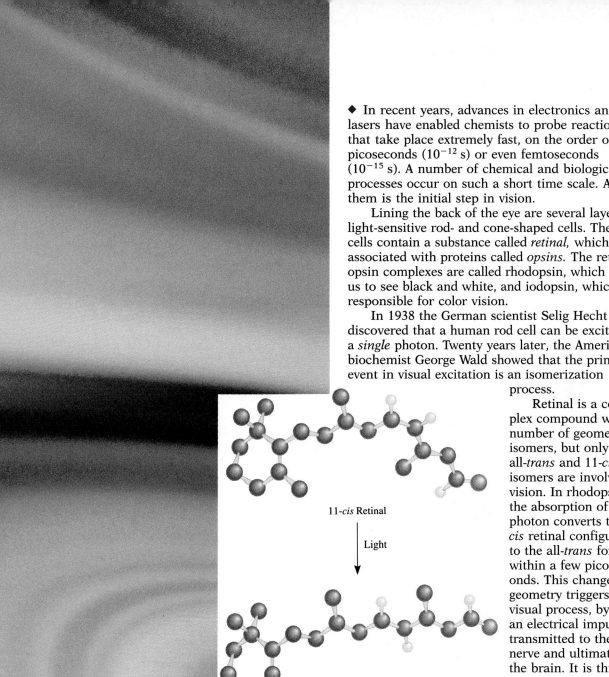

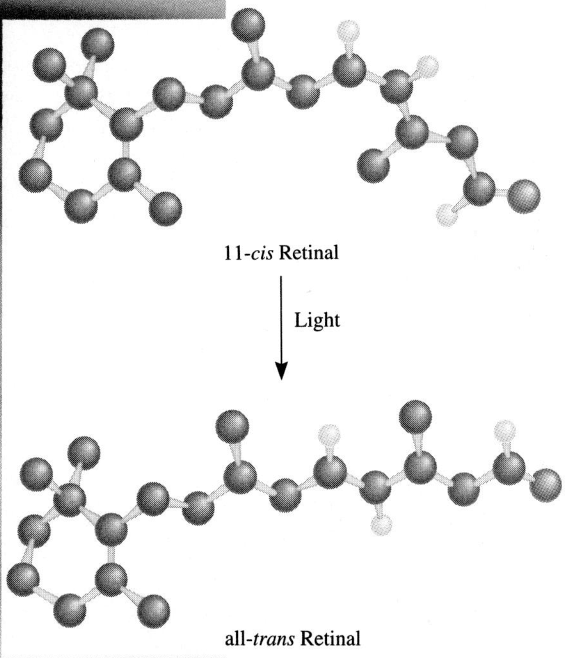

Under the influence of visible light, 11-cis retinal is converted to the all-trans isomer. The black spheres are C atoms, the yellow spheres H atoms, and the red spheres O atoms. Other H atoms are omitted for clarity. The retinal is bound to opsin at the O atom.

21.1 THE RATE OF A REACTION

The area of chemistry concerned with the speed, or rate, at which a chemical reaction occurs is called ***chemical kinetics.*** The word "kinetic" suggests movement or change; in Chapter 5 we defined kinetic energy as the energy available because of the motion of an object. Here kinetics refers to the rate of a reaction, or the ***reaction rate,*** which is *the change in concentration of a reactant or a product with time* (*M*/s).

We know that any reaction can be represented by the general equation

$$\text{reactants} \longrightarrow \text{products}$$

This equation tells us that, during the course of a reaction, reactant molecules are consumed while product molecules are formed. As a result, we can follow the progress of a reaction by monitoring either the decrease in concentration of the reactants or the increase in concentration of the products.

Figure 21.1 shows the progress of a simple reaction in which A molecules are converted to B molecules (for example, the conversion of *cis*-stilbene to *trans*-stilbene shown on p. 428):

$$\text{A} \longrightarrow \text{B}$$

The decrease in the number of A molecules and the increase in the number of B molecules with time are shown in Figure 21.2. In general, it is more convenient to express the rate in terms of change in concentration with time. Thus for the above reaction we can express the rate as

$$\text{rate} = -\frac{\Delta[\text{A}]}{\Delta t} \qquad \text{or} \qquad \text{rate} = \frac{\Delta[\text{B}]}{\Delta t}$$

where Δ[A] and Δ[B] are the changes in concentration (in molarity) over a time period Δt. Because the concentration of A *decreases* during the time interval, Δ[A] is a negative quantity. The rate of a reaction is a positive quantity, so a minus sign is needed in the rate expression to make the rate positive. On the other hand, the rate of product formation does not require a minus sign because Δ[B] is a positive quantity (the concentration of B *increases* with time).

For more complex reactions, we must be careful in writing the rate expression.

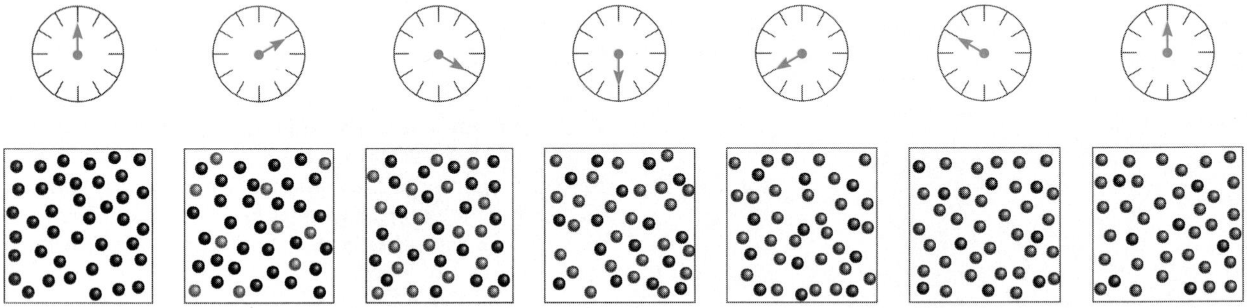

FIGURE 21.1

The progress of reaction A → B at 10-s intervals over a period of 60 s. Initially, only A molecules (black dots) are present. As time progresses, B molecules (colored dots) are formed.

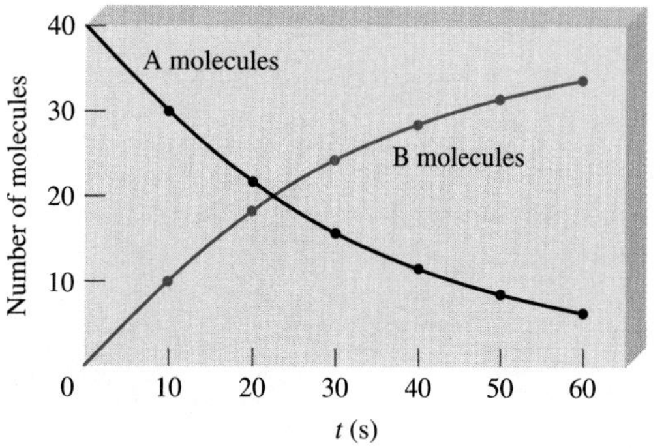

FIGURE 21.2

The rate of reaction A → B, represented as the decrease of A molecules with time and as the increase of B molecules with time.

Consider, for example, the reaction

$$2A \longrightarrow B$$

Two moles of A disappear for each mole of B that forms—that is, the rate of disappearance of A is twice as fast as the rate of appearance of B. We write the rate as either

$$\text{rate} = -\frac{1}{2}\frac{\Delta[A]}{\Delta t} \qquad \text{or} \qquad \text{rate} = \frac{\Delta[B]}{\Delta t}$$

for the reaction

$$aA + bB \longrightarrow cC + dD$$

the rate is given by

$$\text{rate} = -\frac{1}{a}\frac{\Delta[A]}{\Delta t} = -\frac{1}{b}\frac{\Delta[B]}{\Delta t} = \frac{1}{c}\frac{\Delta[C]}{\Delta t} = \frac{1}{d}\frac{\Delta[D]}{\Delta t}$$

EXAMPLE 21.1

Writing Rate Expressions

Write the rate expressions for the following reactions in terms of the disappearance of the reactants and the appearance of the products:

(a) $I^-(aq) + OCl^-(aq) \longrightarrow Cl^-(aq) + OI^-(aq)$

(b) $4NH_3(g) + 5O_2(g) \longrightarrow 4NO(g) + 6H_2O(g)$

Answer: (a) Since each of the stoichiometric coefficients equals 1,

$$\text{rate} = -\frac{\Delta[I^-]}{\Delta t} = -\frac{\Delta[OCl^-]}{\Delta t} = \frac{\Delta[Cl^-]}{\Delta t} = \frac{\Delta[OI^-]}{\Delta t}$$

(b) In this reaction

$$\text{rate} = -\frac{1}{4}\frac{\Delta[NH_3]}{\Delta t} = -\frac{1}{5}\frac{\Delta[O_2]}{\Delta t} = \frac{1}{4}\frac{\Delta[NO]}{\Delta t} = \frac{1}{6}\frac{\Delta[H_2O]}{\Delta t}$$

PRACTICE EXERCISE

Write the rate expression for the following reaction:

$$CH_4(g) + 2O_2(g) \longrightarrow CO_2(g) + 2H_2O(g)$$

FIGURE 21.3

The decrease in bromine concentration as time elapses shows up as a loss in color (from left to right).

Depending on the nature of the reaction, there are a number of ways in which to measure reaction rate. For example, in aqueous solutions, molecular bromine reacts with formic acid (HCOOH) as follows:

$$Br_2(aq) + HCOOH(aq) \longrightarrow 2H^+(aq) + 2Br^-(aq) + CO_2(g)$$

Molecular bromine is reddish brown in color. All other species in the reaction are colorless. As the reaction progresses, the concentration of Br_2 steadily decreases and its color fades (Figure 21.3). Thus the change in concentration (which is evident by the intensity of the color) with time can be followed with a spectrometer. We can determine the reaction rate graphically by plotting the concentration of bromine versus time, as Figure 21.4 shows. The rate of the reaction at a particular instant is given by the slope of the tangent (which is $\Delta[Br_2]/\Delta t$) at that instant. In a certain experiment we find that the rate is 2.96×10^{-5} *M*/s at 100 s after the start of the reaction, 2.09×10^{-5} *M*/s at 200 s, and so on. Because generally the rate is proportional to the concentration of the reactant, it is not surprising that its value falls as the concentration of bromine decreases.

If one of the products or reactants of a reaction is a gas, we can use a manome-

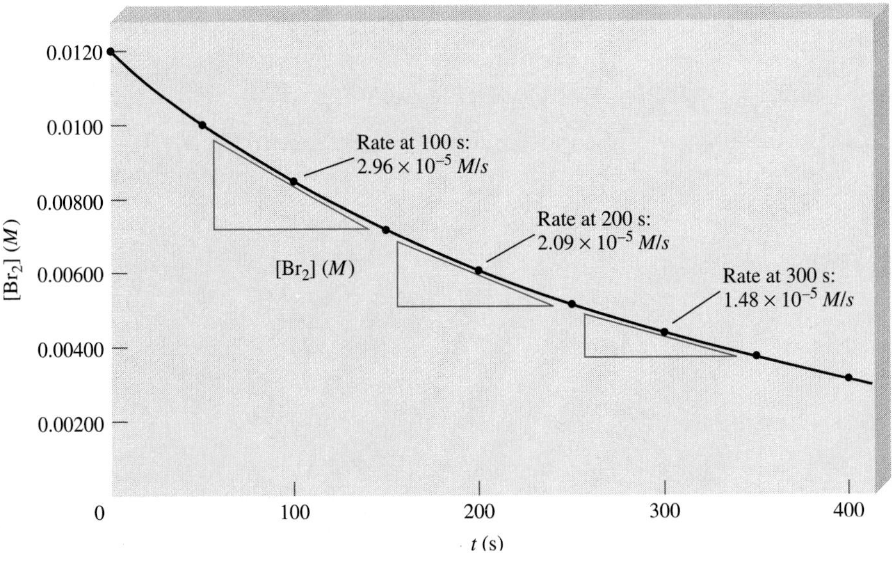

FIGURE 21.4

The instantaneous rates of the reaction between molecular bromine and formic acid at t = 100 s, 200 s, and 300 s are given by the slopes of the tangents (the straight lines touching the curve) at these times.

ter to find the reaction rate. To illustrate this method, let us consider the decomposition of hydrogen peroxide:

$$2H_2O_2(l) \longrightarrow 2H_2O(l) + O_2(g)$$

In this case, the rate of decomposition can be conveniently determined by measuring the rate of oxygen evolution with a manometer (Figure 21.5). The oxygen pressure can be readily converted to concentration by using the ideal gas equation [Equation (5.7)]:

$$PV = nRT$$

or

$$P = \frac{n}{V}RT = MRT$$

where n/V gives the molarity of oxygen gas. Rearranging the equation, we get

$$M = \frac{1}{RT}P$$

The reaction rate, which is given by the rate of oxygen production, can now be written as

$$\text{rate} = \frac{\Delta[O_2]}{\Delta t} = \frac{1}{RT}\frac{\Delta P}{\Delta t}$$

If a reaction either consumes or generates ions, its rate can be measured by monitoring electrical conductance. If H^+ ion is the reactant or product, we can determine the reaction rate by measuring the solution's pH as a function of time.

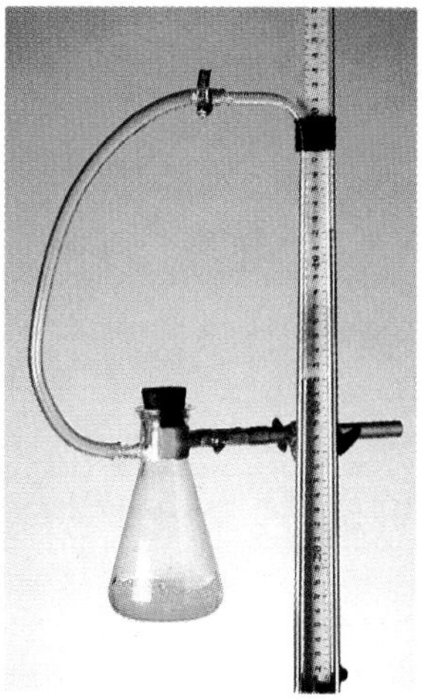

FIGURE 21.5

The rate of hydrogen peroxide decomposition can be measured with a manometer, which shows the increase in the oxygen gas pressure with time.

21.2 THE RATE LAWS

One way to study the effect of reactant concentration on reaction rate is to determine how the initial rate depends on the starting concentrations. In general it is preferable to measure the initial rate because as the reaction proceeds, the concentrations of the reactants decrease and it may become difficult to measure the changes accurately. Also, there may be a reverse reaction such that

$$\text{products} \longrightarrow \text{reactants}$$

which would introduce error in the rate measurement. Both of these complications are virtually absent during the early stages of the reaction.

Table 21.1 shows three rate measurements for the reaction

$$F_2(g) + 2ClO_2(g) \longrightarrow 2FClO_2(g)$$

TABLE 21.1

Rate Data for the Reaction between F_2 and ClO_2

$[F_2]$ (M)	$[ClO_2]$ (M)	Initial rate (M/s)
1. 0.10	0.010	1.2×10^{-3}
2. 0.10	0.040	4.8×10^{-3}
3. 0.20	0.010	2.4×10^{-3}

Looking at table entries 1 and 3, we see that if we double $[F_2]$ while holding $[ClO_2]$ constant, the rate doubles. Thus the rate is directly proportional to $[F_2]$. Similarly, the data in entries 1 and 2 show that when we quadruple $[ClO_2]$ at constant $[F_2]$, the rate increases by four times, so that the rate is also directly proportional to $[ClO_2]$. We can summarize these observations by writing

$$\begin{aligned} \text{rate} &\propto [F_2][ClO_2] \\ &= k[F_2][ClO_2] \end{aligned}$$

The term k is the ***rate constant,*** *a constant of proportionality between the reaction rate and the concentrations of the reactants.* The above equation is known as the ***rate law,*** *an expression relating the rate of a reaction to the rate constant and the concentrations of the reactants.* For a general reaction of the type

$$a\text{A} + b\text{B} \longrightarrow c\text{C} + d\text{D}$$

the rate law takes the form

$$\text{rate} = k[\text{A}]^x[\text{B}]^y \tag{21.1}$$

If we know the values of k, x, and y, as well as the concentrations of A and B, we can use the rate law to calculate the rate of the reaction. Like k, x and y must be determined experimentally. *The sum of the powers to which all reactant concentrations appearing in the rate law are raised* is called the overall ***reaction order.*** In the rate law expression shown above, the overall reaction order is given by $x + y$. For the reaction involving F_2 and ClO_2, the overall order is 1 + 1, or 2. We say that the reaction is first order in F_2 and first order in ClO_2, or second order overall. Note that reaction order is *always* determined by reactant concentrations and never by product concentrations.

Reaction order enables us to appreciate better the dependence of rate on reactant concentrations. Suppose, for example, that, for a certain reaction, $x = 1$ and $y = 2$. The rate law for the reaction is

$$\text{rate} = k[\text{A}][\text{B}]^2$$

This reaction is first order in A, second order in B, and third order overall (1 + 2 = 3). Let us assume that initially $[\text{A}] = 1.0\ M$ and $[\text{B}] = 1.0\ M$. The rate law tells us that if we double the concentration of A from 1.0 M to 2.0 M at constant [B], we also double the reaction rate:

$$\begin{aligned} \text{for } [\text{A}] = 1.0\ M \qquad \text{rate}_1 &= k(1.0\ M)(1.0\ M)^2 \\ &= k(1.0\ M^3) \\ \text{for } [\text{A}] = 2.0\ M \qquad \text{rate}_2 &= k(2.0\ M)(1.0\ M)^2 \\ &= k(2.0\ M^3) \end{aligned}$$

Hence

$$\text{rate}_2 = 2(\text{rate}_1)$$

On the other hand, if we double the concentration of B from 1.0 M to 2.0 M at constant [A], the reaction rate will increase by a factor of 4 because of the power 2 in the exponent:

$$\begin{aligned} \text{for } [\text{B}] = 1.0\ M \qquad \text{rate}_1 &= k(1.0\ M)(1.0\ M)^2 \\ &= k(1.0\ M^3) \\ \text{for } [\text{B}] = 2.0\ M \qquad \text{rate}_2 &= k(1.0\ M)(2.0\ M)^2 \end{aligned}$$

$$\text{rate}_2 = k(4.0\ M^3)$$

Hence

$$\text{rate}_2 = 4(\text{rate}_1)$$

If, for a certain reaction, $x = 0$ and $y = 1$, then the rate law is

$$\begin{aligned}\text{rate} &= k[\text{A}]^0[\text{B}] \\ &= k[\text{B}]\end{aligned}$$

This reaction is zero order in A, first order in B, and first order overall. Thus the rate of this reaction is *independent* of the concentration of A.

Experimental Determination of Rate Laws

If a reaction involves only one reactant, the rate law can be readily determined by measuring the initial rate of the reaction as a function of the reactant's concentration. For example, if the rate doubles when the concentration of the reactant doubles, then the reaction is first order in the reactant. If the rate quadruples when the concentration doubles, the reaction is second order in the reactant.

For a reaction involving more than one reactant, we can find the rate law by measuring the dependence of the reaction rate on the concentration of each reactant, one at a time. We fix the concentrations of all but one reactant and record the rate of the reaction as a function of the concentration of that reactant. Any changes in the rate must be due only to changes in that substance. The dependence thus observed gives us the order in that particular reactant. The same procedure is then applied to the next reactant, and so on. This approach is known as the *isolation method.*

EXAMPLE 21.2
Determining the Rate Law of a Reaction

The rate of the reaction $\text{A} + 2\text{B} \rightarrow \text{C}$ has been observed at 25°C. From the following data, determine the rate law for the reaction and calculate the rate constant.

Experiment	*Initial [A] (M)*	*Initial [B] (M)*	*Initial rate (M/s)*
1	0.100	0.100	5.50×10^{-6}
2	0.200	0.100	2.20×10^{-5}
3	0.400	0.100	8.80×10^{-5}
4	0.100	0.300	1.65×10^{-5}
5	0.100	0.600	3.30×10^{-5}

Answer: We assume that the rate law takes the form

$$\text{rate} = k[\text{A}]^x[\text{B}]^y$$

Experiments 1 and 2 show that when we double the concentration of A at constant [B], the rate increases fourfold. Thus the reaction is second order in A. Experiments 4 and 5 indicate that doubling [B] at constant [A] doubles the rate; the reaction is first order in B.

We can arrive at the same conclusion by taking the ratios of rates as follows. Using the data from experiments 1 and 2, we can write the ratio of the rates as

$$\frac{\text{rate}_1}{\text{rate}_2} = \frac{5.50 \times 10^{-6}\ M/\text{s}}{2.20 \times 10^{-5}\ M/\text{s}} = \frac{1}{4}$$

However, the ratio of the rates can also be expressed in terms of the rate law, as follows:

$$\frac{\text{rate}_1}{\text{rate}_2} = \frac{1}{4} = \frac{\cancel{k}(0.100\ M)^x\cancel{(0.100\ M)^y}}{\cancel{k}(0.200\ M)^x\cancel{(0.100\ M)^y}}$$

$$= \left(\frac{0.100\ M}{0.200\ M}\right)^x = \left(\frac{1}{2}\right)^x$$

Thus $x = 2$, and the reaction is second order in A.

Similarly, using the data for experiments 4 and 5, we have

$$\frac{\text{rate}_4}{\text{rate}_5} = \frac{1.65 \times 10^{-5}\ M/\text{s}}{3.30 \times 10^{-5}\ M/\text{s}} = \frac{1}{2}$$

From the rate law we write

$$\frac{\text{rate}_4}{\text{rate}_5} = \frac{1}{2} = \frac{k\cancel{(0.100\ M)^x}(0.300\ M)^y}{k\cancel{(0.100\ M)^x}(0.600\ M)^y}$$

$$= \left(\frac{0.300\ M}{0.600\ M}\right)^y = \left(\frac{1}{2}\right)^y$$

Therefore, $y = 1$, so the reaction is first order in B. The overall order is (2 + 1), or 3 (third order), and the rate law is

$$\text{rate} = k[\text{A}]^2[\text{B}]$$

The rate constant k can be calculated using the data from any one of the experiments.

Since

$$k = \frac{\text{rate}}{[\text{A}]^2[\text{B}]}$$

data from experiment 1 give us

$$k = \frac{5.50 \times 10^{-6}\ M/\text{s}}{(0.100\ M)^2(0.100\ M)}$$

$$= 5.50 \times 10^{-3}/M^2\text{s}$$

PRACTICE EXERCISE

The reaction of peroxydisulfate ion ($S_2O_8^{2-}$) with iodide ion (I^-) is

$$S_2O_8^{2-}(aq) + 3I^-(aq) \longrightarrow 2SO_4^{2-}(aq) + I_3^-(aq)$$

From the following data collected at a certain temperature, determine the rate law and calculate the rate constant.

Experiment	*[$S_2O_8^{2-}$]*	*[I^-]*	*Initial rate (M/s)*
1	0.080	0.034	2.2×10^{-4}
2	0.080	0.017	1.1×10^{-4}
3	0.16	0.017	2.2×10^{-4}

Example 21.2 brings out an important point about rate laws: There is no relationship between the exponents x and y and the stoichiometric coefficients in the balanced equation. As another example, consider the thermal decomposition of N_2O_5:

$$2N_2O_5(g) \longrightarrow 4NO_2(g) + O_2(g)$$

The rate law is

$$\text{rate} = k[N_2O_5]$$

and *not* rate $= k[N_2O_5]^2$, as we might have guessed from the balanced equation. *In general, the order of a reaction must be determined by experiment; it cannot be deduced from the overall balanced equation.*

21.3 RELATION BETWEEN REACTANT CONCENTRATIONS AND TIME: FIRST-ORDER REACTIONS

Rate laws enable us to calculate the rate of a reaction from the rate constant and reactant concentrations. They can also be converted into equations that allow us to determine the concentrations of reactants at any time during the course of a reaction. We will illustrate this application by considering one of the simplest kind of rate laws—that applying to reactions that are first order overall.

A ***first-order reaction*** is *a reaction whose rate depends on the reactant concentration raised to the first power.* In a first-order reaction of the type

$$A \longrightarrow \text{product}$$

the rate is

$$\text{rate} = -\frac{\Delta[A]}{\Delta t}$$

From the rate law, we also know that

$$\text{rate} = k[A]$$

Thus

$$-\frac{\Delta[A]}{\Delta t} = k[A] \qquad (21.2)$$

We can determine the units of the first-order rate constant k by transposing:

$$k = -\frac{\Delta[A]}{[A]}\frac{1}{\Delta t}$$

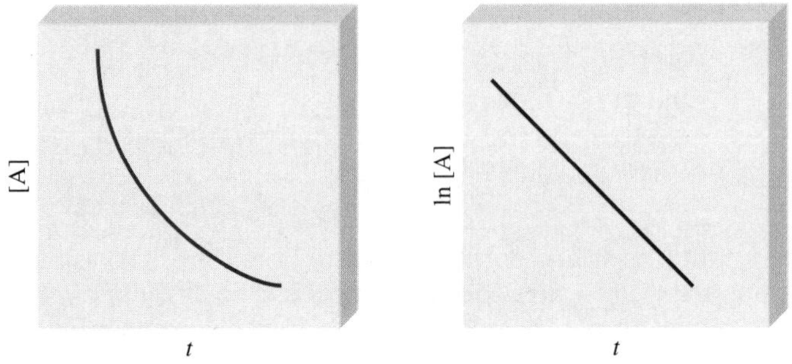

FIGURE 21.6

First-order reaction characteristics: (a) decrease of reactant concentration with time; (b) plot of the straight-line relationship to obtain the rate constant. The slope of the line is equal to $-k$.

Since the unit for Δ[A] and [A] is M and that for Δt is s, the unit for k is

$$\frac{M}{M\text{ s}} = \frac{1}{\text{s}} = \text{s}^{-1}$$

(The minus sign does not enter into the evaluation of units.) Using calculus, we can show from Equation (21.2) that

$$\ln \frac{[\text{A}]_0}{[\text{A}]} = kt \tag{21.3}$$

where ln is the natural logarithm, and $[\text{A}]_0$ and [A] are the concentrations of A at times $t = 0$ and $t = t$, respectively. It should be understood that $t = 0$ need not correspond to the beginning of the experiment; it can be any time when we choose to monitor the change in the concentration of A.

Equation (21.3) can be rearranged as follows:

$$\ln [\text{A}]_0 - \ln [\text{A}] = kt$$

or

$$\ln [\text{A}] = -kt + \ln [\text{A}]_0 \tag{21.4}$$

Equation (21.4) has the form of the linear equation $y = mx + b$, in which m is the slope of the line that is the graph of the equation:

$$\ln [\text{A}] = (-k)(t) + \ln [\text{A}]_0$$

$$\updownarrow \quad \updownarrow \quad \updownarrow \quad \updownarrow$$

$$y \quad = \quad m \quad x \quad + \quad b$$

Thus a plot of ln [A] versus t (or y versus x) gives a straight line with a slope of $-k$ (or m). This allows us to calculate the rate constant k. Figure 21.6 shows the characteristics of a first-order reaction.

There are many known first-order reactions. All nuclear decay processes are first order in nature (see Chapter 22). Another example is the decomposition of ethane (C_2H_6) into highly reactive methyl radicals (CH_3):

$$C_2H_6 \longrightarrow 2CH_3$$

The decomposition of N_2O_5, which we discussed earlier, is first order in N_2O_5.

EXAMPLE 21.3

Analyzing a First-Order Reaction

The decomposition of dinitrogen pentoxide is a first-order reaction with a rate constant of $5.1 \times 10^{-4}\ \text{s}^{-1}$ at 45°C:

$$2N_2O_5(g) \longrightarrow 4NO_2(g) + O_2(g)$$

(a) If the initial concentration of N_2O_5 was 0.25 *M*, what is the concentration after 3.2 min? (b) How long will it take for the concentration of N_2O_5 to decrease from 0.25 *M* to 0.15 *M*? (c) How long will it take to convert 62 percent of the starting material?

N_2O_5 decomposes to give NO_2.

Answer: (a) Applying Equation (21.3),

$$\ln \frac{[A]_0}{[A]} = kt$$

$$\ln \frac{0.25\ M}{[A]} = (5.1 \times 10^{-4}\ s^{-1})\left(3.2\ \text{min} \times \frac{60\ s}{1\ \text{min}}\right)$$

Solving the equation, we obtain

$$\ln \frac{0.25\ M}{[A]} = 0.098$$

$$\frac{0.25\ M}{[A]} = e^{0.098} = 1.1$$

$$[A] = 0.23\ M$$

(b) Again using Equation (21.3), we have

$$\ln \frac{0.25\ M}{0.15\ M} = (5.1 \times 10^{-4}\ s^{-1})t$$

$$t = 1.0 \times 10^3\ s$$

$$= 17\ \text{min}$$

(c) In a calculation of this type, we do not need to know the actual concentration of the starting material. If 62 percent of the starting material has reacted, then the amount left after time *t* is (100% − 62%), or 38%. Thus $[A]/[A]_0 =$ 38%/100%, or 0.38. From Equation (21.3), we write

$$t = \frac{1}{k} \ln \frac{[A]_0}{[A]}$$

$$= \frac{1}{51 \times 10^{-4}\ s^{-1}} \ln \frac{1.0}{0.38}$$

$$= 1.9 \times 10^3\ s$$

$$= 32\ \text{min}$$

PRACTICE EXERCISE

The reaction $2A \rightarrow B$ is first order in A with a rate constant of $2.8 \times 10^{-2}\ s^{-1}$ at 80°C. How long (in seconds) will it take for A to decrease from 0.88 *M* to 0.14 *M*?

The ***half-life*** of a reaction, $t_{\frac{1}{2}}$, is *the time required for the concentration of a reactant to decrease to half of its initial concentration.* We can obtain an expression for $t_{\frac{1}{2}}$ for a first-order reaction as follows. From Equation (21.3) we write

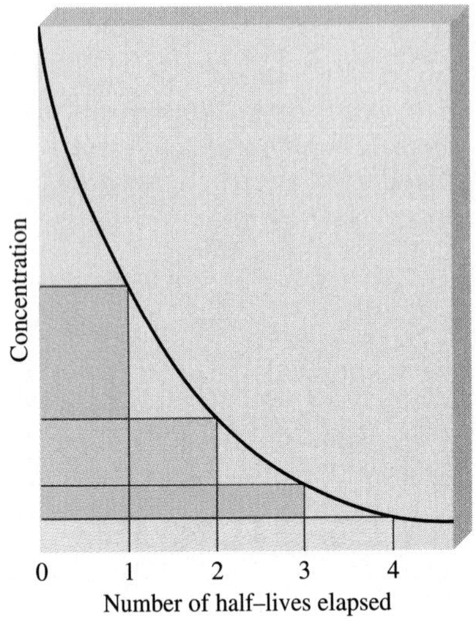

FIGURE 21.7

Change in concentration of a reactant with number of half-lives for a first-order reaction.

$$t = \frac{1}{k} \ln \frac{[A]_0}{[A]}$$

By the definition of half-life, when $t = t_{\frac{1}{2}}$, $[A] = [A]_0/2$, so

$$t_{\frac{1}{2}} = \frac{1}{k} \ln \frac{[A]_0}{[A]_0/2}$$

or

$$t_{\frac{1}{2}} = \frac{1}{k} \ln 2 = \frac{0.693}{k} \qquad (21.5)$$

Equation (21.5) tells us that the half-life of a first-order reaction is *independent* of the initial concentration of the reactant. Thus, it takes the same time for the concentration of the reactant to decrease from 1.0 *M* to 0.50 *M*, say, as it does for a decrease in concentration from 0.10 *M* to 0.050 *M* (Figure 21.7). Measuring the half-life of a reaction is one way to determine the rate constant of a first-order reaction.

EXAMPLE 21.4

Determining the Half-Life of a First-Order Reaction

The conversion of cyclopropane to propene in the gas phase

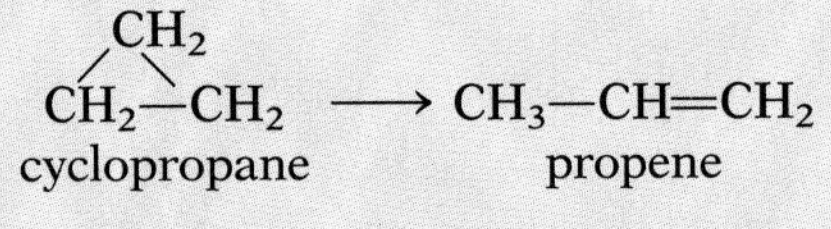

is a first-order reaction with a rate constant of $6.7 \times 10^{-4}\ s^{-1}$ at 500°C. Calculate the half-life of the reaction.

Answer: Applying Equation (21.5), we get

$$t_{\frac{1}{2}} = \frac{0.693}{k}$$

$$t_{\frac{1}{2}} = \frac{0.693}{6.7 \times 10^{-4}\ s^{-1}}$$
$$= 1.0 \times 10^3\ s$$

Thus it will take 1.0×10^3 s, or about 17 min, for *any* initial concentration of cyclopropane to decrease to half its value.

PRACTICE EXERCISE

The half-life of a first-order reaction is 84.1 min. Calculate the rate constant of the reaction.

Equations (21.3) and (21.4) can help us evaluate a first-order rate constant. From experimental data for concentration versus time, we can determine k graphically (as shown in Figure 21.6), or we can use the half-life method, as Example 21.5 will show.

For gas-phase reactions we can replace the concentration terms in Equation (21.3) with the pressures of the gaseous reactant. Consider the first-order reaction

$$A(g) \longrightarrow \text{product}$$

Using the ideal gas equation we write

$$PV = n_A RT$$

or

$$\frac{n_A}{V} = [A] = \frac{P}{RT}$$

Substituting $[A] = P/RT$ in Equation (21.3), we get

$$\ln\frac{[A]_0}{[A]} = \ln\frac{P_0/RT}{P/RT} = \ln\frac{P_0}{P} = kt$$

EXAMPLE 21.5
Analyzing First-Order Kinetics

The rate of decomposition of azomethane is studied by monitoring the partial pressure of the reactant as a function of time:

$$CH_3{-}N{=}N{-}CH_3(g) \longrightarrow N_2(g) + C_2H_6(g)$$

The data obtained at 300°C are shown in the following table:

Time (s)	*Partial pressure of azomethane (mmHg)*
0	284
100	220
150	193
200	170
250	150
300	132

(a) Are these values consistent with first-order kinetics? If so, determine the rate constant, (b) by plotting the data as shown in Figure 21.6 and (c) by using the half-life method.

Answer: (a) The partial pressure of azomethane at any time is directly proportional to its concentration (in moles per liter). Therefore Equation (21.4) can be written in terms of partial pressures as

$$\ln P = -kt + \ln P_0$$

where P_0 and P are the partial pressures of azomethane at time $t = 0$ and $t = t$. This equation has the form of the linear equation $y = mx + b$. Figure 21.8(a), which is based on the data given in the table below, shows that a plot of ln P versus t yields a straight line, so the reaction is indeed first order.

t (s)	*ln P*
0	5.649
100	5.394
150	5.263
200	5.136
250	5.011
300	4.883

(b) From Equation (21.4) we know that the slope of the line for a first-order reaction is equal to $-k$. In Figure 21.8(a) the slope is $-2.55 \times 10^{-3}\ s^{-1}$. Therefore

$$-k = \text{slope}$$
$$-k = -2.55 \times 10^{-3}\ s^{-1}$$
$$k = 2.55 \times 10^{-3}\ s^{-1}$$

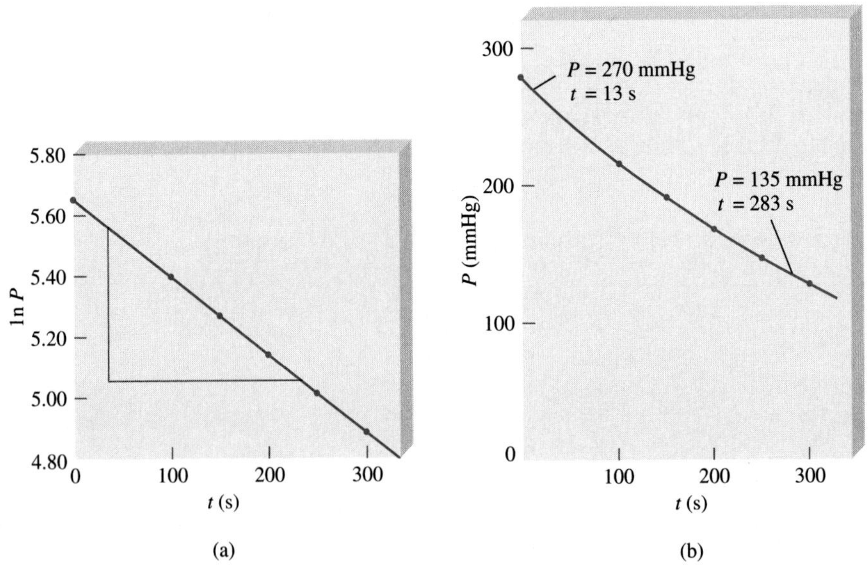

FIGURE 21.8

(a) Plot of ln P versus time. The slope of the line is calculated from two pairs of coordinates:

$$\text{slope} = \frac{5.05 - 5.56}{(233 - 33)s} = -2.55 \times 10^{-3}\ s^{-1}$$

According to Equation (21.4), the slope is equal to −k. (b) The half-life method for determining the rate constant of the first-order reaction. The time it takes for the pressure to drop from 270 mmHg (an arbitrary choice) to 135 mmHg is the half-life of the reaction. Here $t_{\frac{1}{2}}$ *= 283 s − 13 s, or 270 s.*

(c) To calculate the rate constant by the half-life method we need to plot the partial pressure versus time, as shown in Figure 21.8(b), because the given data do not show the time it takes for the partial pressure to decrease to half of its value. From the curve we find that $t_{\frac{1}{2}} = 270$ s, so the rate constant can be calculated by using Equation (21.5):

$$t_{\frac{1}{2}} = \frac{0.693}{k}$$

$$k = \frac{0.693}{270 \text{ s}}$$

$$= 2.57 \times 10^{-3} \text{ s}^{-1}$$

The small difference between the two rate constants calculated in (b) and (c) for the same reaction should not be surprising because they are arrived at by different ways of analyzing the data.

PRACTICE EXERCISE

Azomethane ($C_2H_6N_2$) decomposes at a certain temperature in the gas phase as follows

$$C_2H_6N_2(g) \longrightarrow C_2H_6(g) + N_2(g)$$

From the following data determine the order of the reaction and the rate constant.

Time (min)	*[$C_2H_6N_2$] (M)*
0	0.36
15	0.30
30	0.25
48	0.19
75	0.13

21.4 ACTIVATION ENERGY AND TEMPERATURE DEPENDENCE OF RATE CONSTANTS

With very few exceptions, reaction rates increase with increasing temperature. For example, much less time is required to hard-boil an egg at 100°C (about 10 min) than at 80°C (about 30 min). Conversely, an effective way to preserve foods is to store them at subzero temperatures, thereby slowing the rate of bacterial decay. Figure 21.9 shows a typical example of the relationship between the rate constant of a reaction and temperature. In order to explain this behavior, we must ask how reactions get started in the first place.

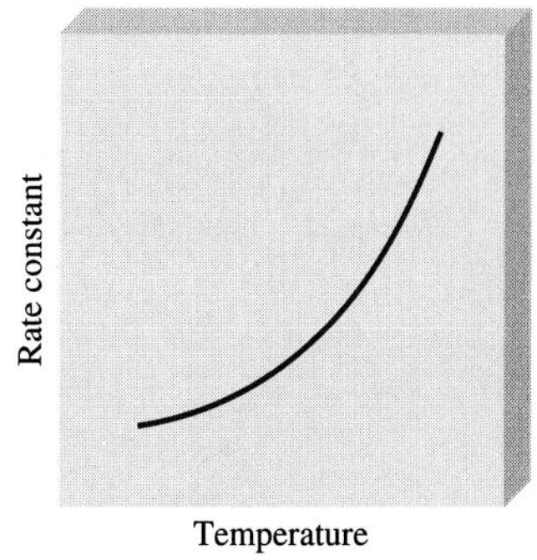

FIGURE 21.9

Dependence of rate constant on temperature. The rate constants of most reactions increase with increasing temperature.

THE COLLISION THEORY OF CHEMICAL KINETICS

The kinetic molecular theory of gases (p. 135) states that gas molecules frequently collide with one another. Therefore it seems logical to assume—and it is generally

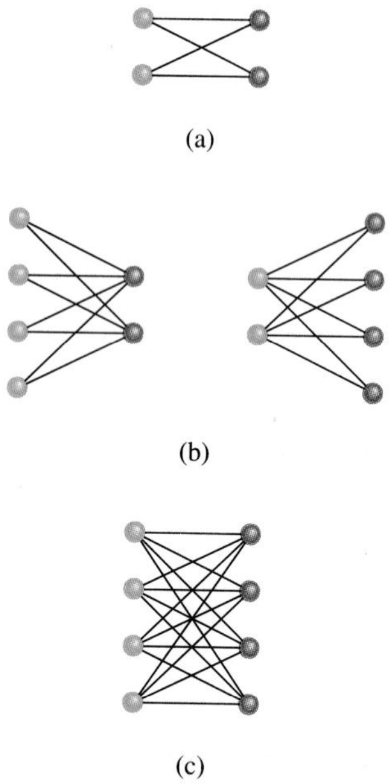

FIGURE 21.10

Dependence of number of collisions on concentration. We consider here only A-B collisions, which can lead to the formation of products. (a) There are four possible collisions among two A and two B molecules. (b) Doubling the number of either type of molecule (but not both) increases the number of collisions to eight. (c) Doubling both the A and B molecules increases the number of collisions to sixteen.

true—that chemical reactions occur as a result of collisions between reacting molecules. In terms of the *collision theory* of chemical kinetics, then, we expect the rate of a reaction to be directly proportional to the number of molecular collisions per second, or to the frequency of molecular collisions:

$$\text{rate} \propto \frac{\text{number of collisions}}{\text{s}}$$

This simple relationship explains the dependence of reaction rate on concentration.

Consider the reaction of A molecules with B molecules to form some product. Suppose that each product molecule is formed by the direct combination of an A molecule and a B molecule. If we doubled the concentration of A, say, then the number of A-B collisions would also double, because, in any given volume, there would be twice as many A molecules that could collide with B molecules (Figure 21.10). Consequently, the rate would increase by a factor of 2. Similarly, doubling the concentration of B molecules would increase the rate twofold. Thus we can express the rate law as

$$\text{rate} = k[\text{A}][\text{B}]$$

The reaction is first order in both A and B and obeys second-order kinetics.

The collision theory is intuitively appealing, but the relationship between rate and molecular collision is more complicated than you might expect. The implication of the collision theory is that a reaction always occurs when an A and a B molecule collide. However, not all collisions lead to reactions. Calculations based on the kinetic molecular theory show that, at ordinary pressures (say, 1 atm) and temperatures (say, 298 K), there are about 1×10^{27} binary collisions (collisions between two molecules) in 1 mL of volume every second, in the gas phase. Even more collisions per second occur in liquids. If every binary collision led to a product, then most reactions would be complete almost instantaneously. In practice, we find that the rates of reactions differ greatly. This means that, in many cases, collisions alone do not guarantee that a reaction will take place.

Any molecule in motion possesses kinetic energy; the faster it moves, the greater the kinetic energy. When molecules collide, part of their kinetic energy is converted to vibrational energy. If the initial kinetic energies are large, then the colliding molecules will vibrate so strongly as to break some of the chemical bonds. This bond fracture is the first step toward product formation. If the initial kinetic energies are small, the molecules will merely bounce off each other intact. Energetically speaking, there is some minimum collision energy below which no reaction occurs.

We postulate that, in order to react, the colliding molecules must have a total kinetic energy equal to or greater than the ***activation energy*** **(E_a),** which is *the minimum amount of energy required to initiate a chemical reaction.* Lacking this energy, the molecules remain intact, and no change results from the collision. *The species temporarily formed by the reactant molecules as a result of the collision before they form the product* is called the ***activated complex.***

Figure 21.11 shows two different potential energy profiles for the reaction

$$\text{A} + \text{B} \longrightarrow \text{C} + \text{D}$$

If the products are more stable than the reactants, then the reaction will be accompanied by a release of heat; that is, the reaction is exothermic [Figure 21.11(a)]. On the other hand, if the products are less stable than the reactants, then heat will be absorbed by the reacting mixture from the surroundings and we have an endother-

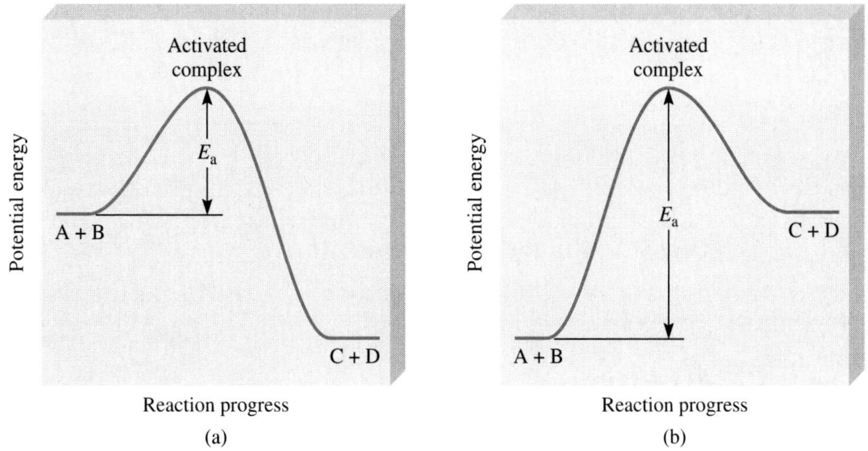

FIGURE 21.11

Potential energy profiles for (a) exothermic and (b) endothermic reactions. These plots show the change in potential energy as reactants A and B are converted to products C and D. The activated complex is a highly unstable species with a high potential energy. The activation energy is defined for the forward reaction in both (a) and (b). Note that the products C and D are more stable than the reactants in (a) and less stable than those in (b).

mic reaction [Figure 21.11(b)]. In both cases we plot the potential energy of the reacting system versus the progress of the reaction. Qualitatively, these plots show the potential energy changes as reactants are converted to products.

We can think of activation energy as a barrier that prevents less energetic molecules from reacting. Because the number of reactant molecules in an ordinary reaction is very large, the speeds, and hence also the kinetic energies of the molecules, vary greatly. Normally, only a small fraction of the colliding molecules—the fastest-moving ones—have enough kinetic energy to exceed the activation energy. These molecules can therefore take part in the reaction. The increase in the rate (or the rate constant) with temperature can now be explained: The speeds of the molecules obey the Maxwell distributions shown in Figure 5.14. Compare the speed distributions at two different temperatures. Since more high-energy molecules are present at the higher temperature, the rate of product formation is also greater at the higher temperature.

THE ARRHENIUS EQUATION

The dependence of the rate constant of a reaction on temperature can be expressed by the following equation, now known as the *Arrhenius equation:*

$$k = Ae^{-E_a/RT} \tag{21.6}$$

where E_a is the activation energy of the reaction (in kilojoules per mole), R the gas constant (8.314 J/K · mol), T the absolute temperature, and e the base of the natural logarithm scale (see Appendix 3). The quantity A represents the collision frequency and is called the *frequency factor.* It can be treated as a constant for a given reacting system over a fairly wide temperature range. Equation (21.6) shows that the rate constant is directly proportional to A and, therefore, to the collision frequency. Further, because of the minus sign associated with the exponent E_a/RT, the rate constant decreases with increasing activation energy and increases with increasing temperature. This equation can be expressed in a more useful form by taking the natural logarithm of both sides:

$$\ln k = \ln Ae^{-E_a/RT}$$

$$= \ln A - \frac{E_a}{RT} \tag{21.7}$$

Equation (21.7) can take the form of a linear equation:

$$\begin{array}{ccccccc} \ln k & = & \left(-\frac{E_a}{R}\right) & \left(\frac{1}{T}\right) & + & \ln A \\ \updownarrow & & \updownarrow & \updownarrow & & \updownarrow \\ y & = & m & x & + & b \end{array}$$

Thus, a plot of $\ln k$ versus $1/T$ gives a straight line whose slope m is equal to $-E_a/R$ and whose intercept b with the ordinate (the y-axis) is $\ln A$.

EXAMPLE 21.6

Applying the Arrhenius Equation

The rate constants for the decomposition of acetaldehyde

$$CH_3CHO(g) \longrightarrow CH_4(g) + CO(g)$$

were measured at five different temperatures. The data are shown below. Plot $\ln k$ versus $1/T$, and determine the activation energy (in kilojoules per mole) for the reaction. Note that the reaction is "$\frac{3}{2}$" order in CH_3CHO, so k has the units of $1/M^{\frac{1}{2}}$ s.

k ($1/M^{\frac{1}{2}}$ s)	T (K)
0.011	700
0.035	730
0.105	760
0.343	790
0.789	810

Answer: We need to plot $\ln k$ on the y-axis versus $1/T$ on the x-axis. From the given data we obtain

$\ln k$	$1/T$ (K^{-1})
−4.51	1.43×10^{-3}
−3.35	1.37×10^{-3}
−2.254	1.32×10^{-3}
−1.070	1.27×10^{-3}
−0.237	1.23×10^{-3}

These data, when plotted, yield the graph shown in Figure 21.12. The slope of the straight line is -2.09×10^4 K. From the linear form of Equation (21.7)

$$\text{slope} = -\frac{E_a}{R} = -2.09 \times 10^4 \text{ K}$$

$$E_a = (8.314 \text{ J/K} \cdot \text{mol})(2.09 \times 10^4 \text{ K})$$

$$= 1.74 \times 10^5 \text{ J/mol}$$

$$= 1.74 \times 10^2 \text{ kJ/mol}$$

It is important to note that although the rate constant itself has the units

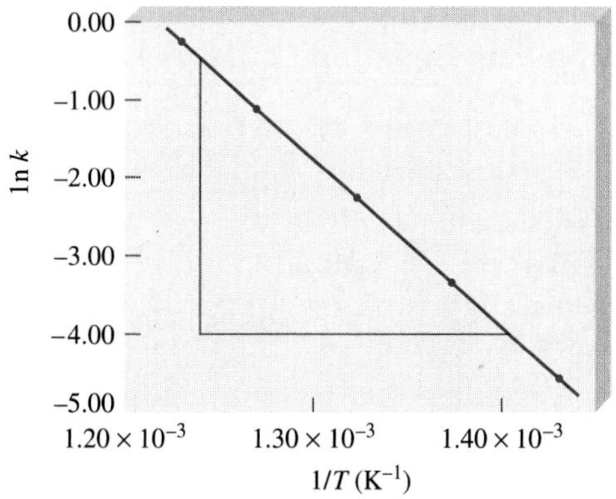

FIGURE 21.12

Plot of ln k versus 1/T. The slope of the line is calculated from two pairs of coordinates:

$$\text{slope} = \frac{-4.00 - (-0.45)}{(1.41 - 1.24) \times 10^{-3}\ K^{-1}} = -2.09 \times 10^4\ K$$

The slope is equal to $-E_a/R$.

$1/M^{\frac{1}{2}}$ s, the quantity ln k has no units (we cannot take the logarithm of any unit).

PRACTICE EXERCISE

The second-order rate constant for the decomposition of nitrous oxide (N_2O) into nitrogen molecules and oxygen atoms has been measured at a series of temperatures:

k ($1/M \cdot s$)	t (°C)
1.87×10^{-3}	600
0.0113	650
0.0569	700
0.244	750

Determine graphically the activation energy for the reaction.

An equation relating the rate constants k_1 and k_2 at temperatures T_1 and T_2 can be used to calculate the activation energy or to find the rate constant at another temperature if the activation energy is known. To derive such an equation we start with Equation (21.7):

$$\ln k_1 = \ln A - \frac{E_a}{RT_1}$$

$$\ln k_2 = \ln A - \frac{E_a}{RT_2}$$

Subtracting ln k_2 from ln k_1 gives

$$\ln k_1 - \ln k_2 = \frac{E_a}{R}\left(\frac{1}{T_2} - \frac{1}{T_1}\right)$$

$$\ln \frac{k_1}{k_2} = \frac{E_a}{R}\left(\frac{1}{T_2} - \frac{1}{T_1}\right)$$

$$= \frac{E_a}{R}\left(\frac{T_1 - T_2}{T_1 T_2}\right) \qquad (21.8)$$

EXAMPLE 21.7
Applying Equation (21.8)

The rate constant of a first-order reaction is $3.46 \times 10^{-2}\ s^{-1}$ at 298 K. What is the rate constant at 350 K if the activation energy for the reaction is 50.2 kJ/mol?

Answer: The data are

$$k_1 = 3.46 \times 10^{-2}\ s^{-1} \qquad k_2 = ?$$
$$T_1 = 298\ K \qquad T_2 = 350\ K$$

Substituting in Equation (21.8),

$$\ln \frac{3.46 \times 10^{-2}}{k_2} = \frac{50.2 \times 10^3\ J/mol}{8.314\ J/K \cdot mol}\left[\frac{298\ K - 350\ K}{(298\ K)(350\ K)}\right]$$

Solving the equation gives

$$\ln \frac{3.46 \times 10^{-2}}{k_2} = -3.01$$
$$\frac{3.46 \times 10^{-2}}{k_2} = e^{-3.01}$$
$$= 0.0493$$
$$k_2 = 0.702\ s^{-1}$$

The rate constant is expected to be greater at a higher temperature. Therefore, the answer is reasonable.

PRACTICE EXERCISE

The first-order rate constant for the reaction of methyl chloride (CH_3Cl) with water to produce methanol (CH_3OH) and hydrochloric acid (HCl) is $3.32 \times 10^{-10}\ s^{-1}$ at 25°C. Calculate the rate constant at 40°C if the activation energy is 116 kJ/mol.

For simple reactions (for example, those between atoms), we can equate the frequency factor (A) in the Arrhenius equation with the frequency of collisions between the reacting species. For more complex reactions, we must also consider the "orientation factor," that is, how reacting molecules are oriented relative to each other. The carefully studied reaction between potassium atoms (K) and methyl iodide (CH_3I) to form potassium iodide (KI) and a methyl radical (CH_3) illustrates this point:

$$K + CH_3I \longrightarrow KI + CH_3$$

This reaction is most favorable only when the K atom collides head-on with the I atom in CH_3I (Figure 21.13). Otherwise, a few or no products are formed. The

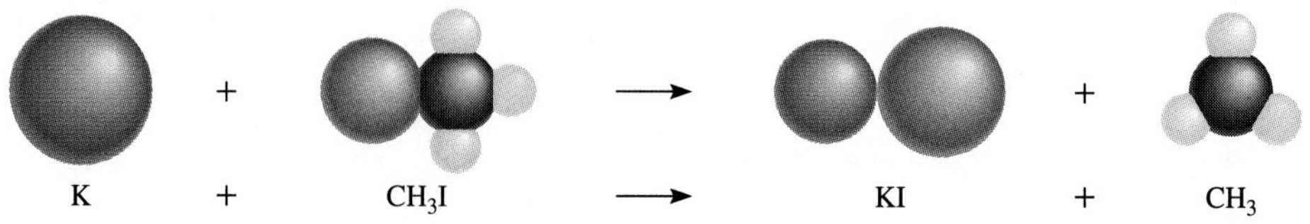

FIGURE 21.13

Relative orientation of reacting molecules. Only when the K atom collides directly with the I atom will the reaction most likely occur.

nature of the orientation factor is satisfactorily dealt with in a more advanced treatment of chemical kinetics.

21.5 REACTION MECHANISMS AND RATE LAWS

As we mentioned earlier, an overall balanced chemical equation does not tell us much about how a reaction actually takes place. In many cases, it merely represents the sum of *a series of simple reactions* that are often called the ***elementary steps*** (or *elementary reactions*) because they *represent the progress of the overall reaction at the molecular level. The sequence of elementary steps that leads to product formation* is called the ***reaction mechanism.*** As an example of a reaction mechanism, let us consider the reaction between nitric oxide and oxygen:

$$2NO(g) + O_2(g) \longrightarrow 2NO_2(g)$$

We know that the products are not formed directly from the collision of a NO molecule with an O_2 molecule because N_2O_2 is detected during the course of the reaction. Let us assume that the reaction actually takes place via two elementary steps as follows:

$$2NO(g) \longrightarrow N_2O_2(g)$$

$$N_2O_2(g) + O_2(g) \longrightarrow 2NO_2(g)$$

In the first elementary step, two NO molecules collide to form a N_2O_2 molecule. This event is followed by the reaction between N_2O_2 and O_2 to give two molecules of NO_2. The net chemical equation, which represents the overall change, is given by the sum of the elementary steps:

$$\begin{array}{ll} \text{Elementary step:} & NO + NO \longrightarrow N_2O_2 \\ \text{Elementary step:} & N_2O_2 + O_2 \longrightarrow 2NO_2 \\ \hline \text{Overall reaction:} & 2NO + \cancel{N_2O_2} + O_2 \longrightarrow \cancel{N_2O_2} + 2NO_2 \end{array}$$

Species such as N_2O_2 are called ***intermediates*** because they *appear in the mechanism of the reaction (that is, the elementary steps) but not in the overall balanced equation.* Keep in mind that an intermediate is always formed in an early elementary step and consumed in a later elementary step.

The number of molecules reacting in an elementary step determines the ***molecularity of a reaction.*** Each of the elementary steps discussed above is called a ***bimolecular reaction,*** *an elementary step that involves two molecules.* An example of a ***unimolecular reaction,*** *an elementary step in which only one reacting molecule participates,* is the conversion of cyclopropane to propene discussed in Example 21.4. Very few ***termolecular reactions,*** *reactions that involve the participation of three molecules in one elementary step,* are known. The reason is that in a termolecular reaction the product forms as a result of the simultaneous encounter of three molecules, which is a far less likely event than a bimolecular collision.

Knowing the elementary steps of a reaction enables us to deduce the rate law. Suppose we have the following unimolecular elementary step:

$$A \longrightarrow \text{products}$$

Because this *is* the process that occurs at the molecular level, the larger the number of A molecules present, the faster the rate of product formation. Thus we can write the rate law directly according to the elementary step:

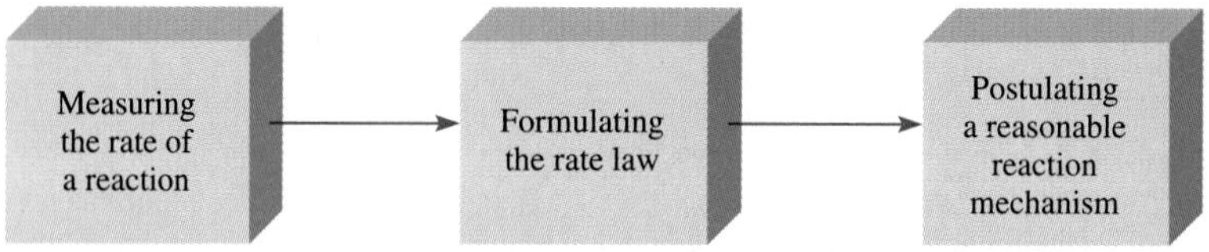

FIGURE 21.14

Sequence of steps in the study of a reaction mechanism.

$$\text{rate} = k[\text{A}]$$

For a bimolecular elementary step involving A and B molecules:

$$\text{A} + \text{B} \longrightarrow \text{product}$$

the rate of product formation depends on how frequently A and B collide, which in turn depends on the concentrations of A and B. In this case we can write the rate law as

$$\text{rate} = k[\text{A}][\text{B}]$$

Similarly, for a bimolecular elementary step of the type

$$\text{A} + \text{A} \longrightarrow \text{products}$$

or

$$2\text{A} \longrightarrow \text{products}$$

the rate law becomes

$$\text{rate} = k[\text{A}]^2$$

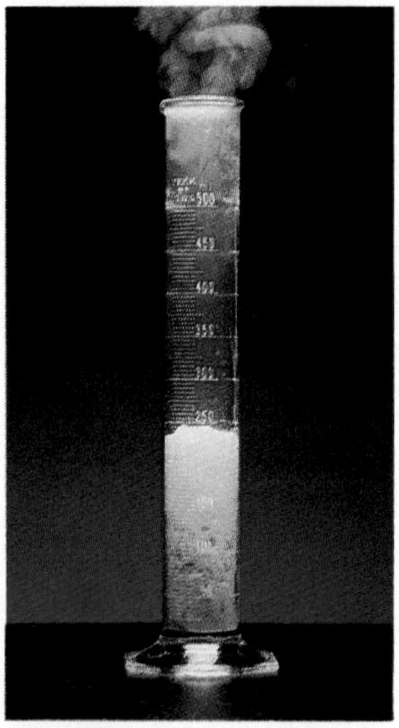

FIGURE 21.15

The decomposition of hydrogen peroxide is catalyzed by the iodide ion. A few drops of liquid soap have been added to the solution to dramatize the evolution of oxygen gas. Some of the iodide ions are oxidized to molecular iodine, which then reacts with iodide ions to form the brown triiodide ion, I_3^-.

The above examples show that the reaction order for each reactant in an elementary step is equal to its stoichiometric coefficient in the chemical equation for that step. In contrast, we cannot tell by merely looking at the overall balanced equation whether the reaction occurs as shown or in a series of elementary steps. This determination is made in the laboratory.

Experimental studies of reaction mechanisms begin with the collection of data (rate measurements). Next, we analyze these data to determine the rate constant and order of the reaction, and we write the rate law. Finally, we suggest a plausible mechanism for the reaction in terms of elementary steps (Figure 21.14). The elementary steps must satisfy two requirements:

- The sum of the elementary steps must give the overall balanced equation for the reaction.
- The ***rate-determining step,*** which is *the slowest step in the sequence of steps leading to product formation,* should predict the same rate law as is determined experimentally.

An analogy for the rate-determining step is the flow of traffic along a narrow road. Assuming the cars cannot pass one another on the road, the rate at which the cars travel is governed by the slowest-moving car. Keep in mind that for any proposed reaction scheme, we must be able to detect the presence of any intermediate(s) formed in one or more elementary steps.

The decomposition of hydrogen peroxide illustrates the elucidation of reaction mechanisms by experimental studies. This reaction is facilitated by iodide ions (Figure 21.15). The overall reaction is

$$2H_2O_2(aq) \longrightarrow 2H_2O(l) + O_2(g)$$

By experiment, the rate law is found to be

$$\text{rate} = k[H_2O_2][I^-]$$

Thus the reaction is first order with respect to both H_2O_2 and I^-. You can see that decomposition does not occur in a single elementary step corresponding to the overall balanced equation. If it did, the reaction would be second order in H_2O_2 (note the coefficient 2 in the equation). What's more, the I^- ion, which is not even in the overall equation, appears in the rate law expression. How can we reconcile these facts?

We can account for the observed rate law by assuming that the reaction takes place in two separate elementary steps, each of which is bimolecular:

Step 1: $$H_2O_2 + I^- \xrightarrow{k_1} H_2O + IO^-$$

Step 2: $$H_2O_2 + IO^- \xrightarrow{k_2} H_2O + O_2 + I^-$$

If we further assume that step 1 is the rate-determining step, then the rate of the reaction can be determined from the first step alone:

$$\text{rate} = k_1[H_2O_2][I^-]$$

Note that the IO^- ion is an intermediate because it does not appear in the overall balanced equation. Although the I^- ion also does not appear in the overall equation, I^- differs from IO^- in that the former is present at the start of the reaction and at its completion. The function of I^- is to speed up the reaction; that is, it is a catalyst. We will discuss catalysis in the next section. Finally, note that the sum of steps 1 and 2 gives the overall balanced equation shown above.

EXAMPLE 21.8

Studying Reaction Mechanisms

The gas-phase decomposition of nitrous oxide (N_2O) is believed to occur via two elementary steps:

Step 1: $$N_2O \xrightarrow{k_1} N_2 + O$$

Step 2: $$N_2O + O \xrightarrow{k_2} N_2 + O_2$$

Experimentally the rate law is found to be rate $= k[N_2O]$. (a) Write the equation for the overall reaction. (b) Which of the species are intermediates? (c) What can you say about the relative rates of steps 1 and 2?

Answer: (a) Adding the equations for steps 1 and 2 gives the overall reaction

$$2N_2O \longrightarrow 2N_2 + O_2$$

(b) Since the O atom is produced in the first elementary step and it does not appear in the overall balanced equation, it is an intermediate.

(c) If we assume that step 1 is the rate-determining step (that is, if $k_2 \gg k_1$), then the rate of the overall reaction is given by

$$\text{rate} = k_1[N_2O]$$

and $k = k_1$.

PRACTICE EXERCISE

The reaction between NO_2 and CO to produce NO and CO_2 is believed to occur via two steps:

Step 1: $NO_2 + NO_2 \longrightarrow NO + NO_3$
Step 2: $NO_3 + CO \longrightarrow NO_2 + CO_2$

The experimental rate law is rate = $k[NO_2]^2$. (a) Write the equation for the overall reaction. (b) Identify the intermediate. (c) What can you say about the relative rates of steps 1 and 2?

21.6 CATALYSIS

We saw in studying the decomposition of hydrogen peroxide that the reaction rate depends on the concentration of iodide ions even though I^- does not appear in the overall equation. We noted there that I^- acts as a catalyst for that reaction. A ***catalyst*** is *a substance that increases the rate of a chemical reaction without itself being consumed.* The catalyst may react to form an intermediate, but it is regenerated in a subsequent step of the reaction.

In the laboratory preparation of molecular oxygen, a sample of potassium chlorate is heated; the reaction is

$$2KClO_3(s) \longrightarrow 2KCl(s) + 3O_2(g)$$

However, this thermal decomposition is very slow in the absence of a catalyst. The rate of decomposition can be increased dramatically by adding a small amount of the catalyst manganese dioxide (MnO_2), a black powdery substance. All the MnO_2 can be recovered at the end of the reaction, just as all the I^- ions remain following H_2O_2 decomposition.

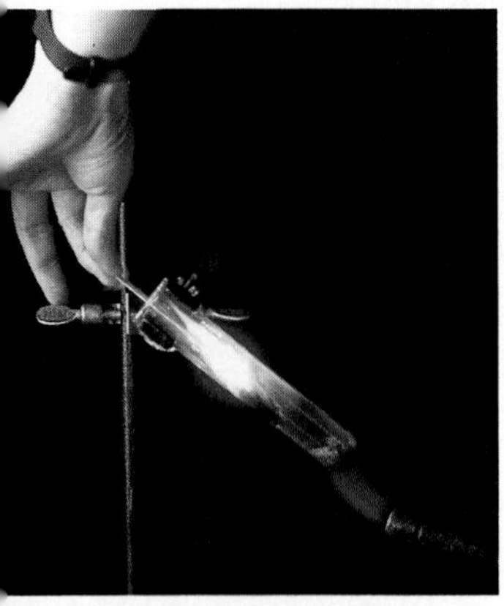

MnO_2 catalyzes the thermal decomposition of $KClO_3$. The oxygen gas produced supports the combustion of the wood splint.

A catalyst speeds up a reaction by providing a set of elementary steps with more favorable kinetics than those that exist in its absence. From Equation (21.7) we know that the rate constant k (and hence the rate) of a reaction depends on the frequency factor A and the activation energy E_a—the larger the A or the smaller the E_a, the greater the rate. In many cases, a catalyst increases the rate by lowering the activation energy for the reaction.

Let us assume that the following reaction has a certain rate constant k and an activation energy E_a.

$$A + B \xrightarrow{k} C + D$$

In the presence of a catalyst, however, the rate constant is k_c (called the *catalytic rate constant*):

$$A + B \xrightarrow{k_c} C + D$$

By the definition of a catalyst,

$$\text{rate}_{\text{catalyzed}} > \text{rate}_{\text{uncatalyzed}}$$

Figure 21.16 shows the potential energy profiles for both reactions. Note that the total energies of the reactants (A and B) and those of the products (C and D) are

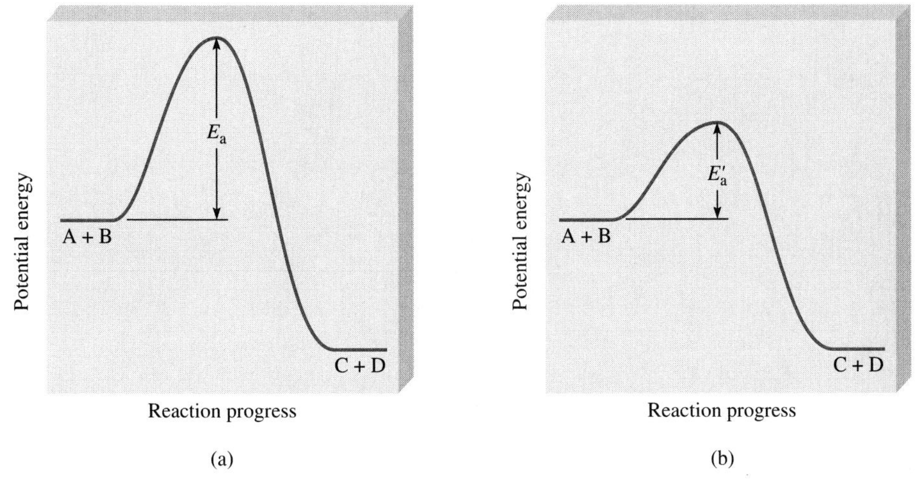

FIGURE 21.16

Comparison of the activation energy barriers of an uncatalyzed reaction and the same reaction with a catalyst. The catalyst lowers the energy barrier but does not affect the actual energies of the reactants or products. Although the reactants and products are the same in both cases, the reaction mechanisms and rate laws are different in (a) and (b). Therefore, two separate diagrams are used to show how the reaction progresses.

unaffected by the catalyst; the only difference between the two is a lowering of the activation energy from E_a to E'_a. Because the activation energy for the reverse reaction is also lowered, a catalyst enhances the rate of the reverse reaction to the same extent as it does the forward reaction rate.

There are three general types of catalysis, depending on the nature of the rate-increasing substance: heterogeneous catalysis, homogeneous catalysis, and enzyme catalysis.

Heterogeneous Catalysis

In *heterogeneous catalysis,* the reactants and the catalyst are in different phases. Usually the catalyst is a solid and the reactants are either gases or liquids. Heterogeneous catalysis is by far the most important type of catalysis in industrial chemistry, especially in the synthesis of many key chemicals. Here we describe two specific examples of heterogeneous catalysis.

The Manufacture of Nitric Acid

Nitric acid is one of the most important inorganic acids. It is used in the production of fertilizers, dyes, drugs, and explosives. The major industrial method of producing nitric acid is the *Ostwald process.* The starting materials, ammonia and molecular oxygen, are heated in the presence of a platinum-rhodium catalyst (Figure 21.17) to about 800°C:

$$4NH_3(g) + 5O_2(g) \longrightarrow 4NO(g) + 6H_2O(g)$$

The nitric oxide formed readily oxidizes (without catalysis) to nitrogen dioxide:

$$2NO(g) + O_2(g) \longrightarrow 2NO_2(g)$$

When dissolved in water, NO_2 forms both nitrous acid and nitric acid:

$$2NO_2(g) + H_2O(l) \longrightarrow HNO_2(aq) + HNO_3(aq)$$

On heating, nitrous acid is converted to nitric acid as follows:

$$3HNO_2(aq) \longrightarrow HNO_3(aq) + H_2O(l) + 2NO(g)$$

The NO generated can be recycled to produce NO_2 in the second step.

FIGURE 21.17

Platinum-rhodium catalyst used in the Ostwald process.

Catalytic Converters

At high temperatures inside a running car's engine, nitrogen and oxygen gases react to form nitric oxide:

$$N_2(g) + O_2(g) \rightleftharpoons 2NO(g)$$

When released into the atmosphere, NO rapidly combines with O_2 to form NO_2. Nitrogen dioxide and other gases emitted by an automobile, such as carbon monoxide (CO) and various unburned hydrocarbons, make automobile exhaust a major source of air pollution.

Most new cars are equipped with catalytic converters (Figure 21.18). An efficient catalytic converter serves two purposes: It oxidizes CO and unburned hydrocarbons to CO_2 and H_2O, and it reduces NO and NO_2 to N_2 and O_2. Hot exhaust gases into which air has been injected are passed through the first chamber of one converter to accelerate the complete burning of hydrocarbons and to decrease CO emission. (A cross section of the catalytic converter, containing Pt or Pd or a transition metal oxide such as CuO or Cr_2O_3, is shown in Figure 21.19). However, since high temperatures increase NO production, a second chamber containing a different catalyst (a transition metal or a transition metal oxide) and operating at a lower temperature is required to dissociate NO into N_2 and O_2 before the exhaust is discharged through the tailpipe.

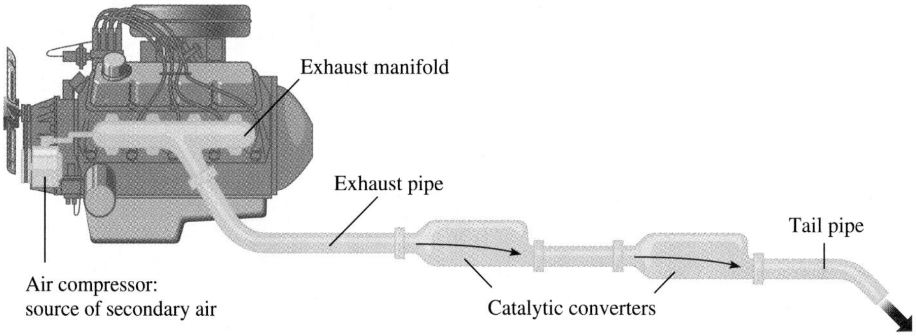

FIGURE 21.18

A two-stage catalytic converter for an automobile.

FIGURE 21.19

A cross-sectional view of a catalytic converter. The beads contain platinum, palladium, and rhodium, which catalyze the combustion of CO and hydrocarbons.

Homogeneous Catalysis

In *homogeneous catalysis* the reactants and catalyst are dispersed in a single phase, usually liquid. Acid and base catalyses are the most important type of homogeneous catalysis in liquid solution. For example, the reaction of ethyl acetate with water to form acetic acid and ethanol normally occurs too slowly to be measured.

$$\underset{\text{ethyl acetate}}{CH_3-\overset{\overset{\displaystyle O}{\|}}{C}-O-C_2H_5} + H_2O \longrightarrow \underset{\text{acetic acid}}{CH_3-\overset{\overset{\displaystyle O}{\|}}{C}-OH} + \underset{\text{ethanol}}{C_2H_5OH}$$

In the absence of the catalyst, the rate law is given by

$$\text{rate} = k[CH_3COOC_2H_5]$$

However, the reaction can be catalyzed by an acid. In the presence of hydrochloric acid, the rate is given by

$$\text{rate} = k_c[CH_3COOC_2H_5][H^+]$$

Enzyme Catalysis

Of all the intricate processes that have evolved in living systems, none is more striking or more essential than enzyme catalysis. ***Enzymes*** are *biological catalysts.* The amazing fact about enzymes is that not only can they increase the rate of biochemical reactions by factors ranging from 10^6 to 10^{12}, but they are also highly specific. An enzyme acts only on certain molecules, called *substrates* (that is, reactants), while leaving the rest of the system unaffected. It has been estimated that an average living cell may contain some 3000 different enzymes, each of them catalyzing a specific reaction in which a substrate is converted into the appropriate products. Enzyme catalysis is homogeneous because the substrate, enzyme, and products are all present in aqueous solution.

An enzyme is typically a large protein molecule that contains one or more *active sites* where interactions with substrates take place. These sites are structurally compatible with specific molecules, in much the same way as a key fits a particular lock (Figure 21.20). However, an enzyme molecule (or at least its active site) has a fair amount of structural flexibility and can modify its shape to accommodate more than one substrate.

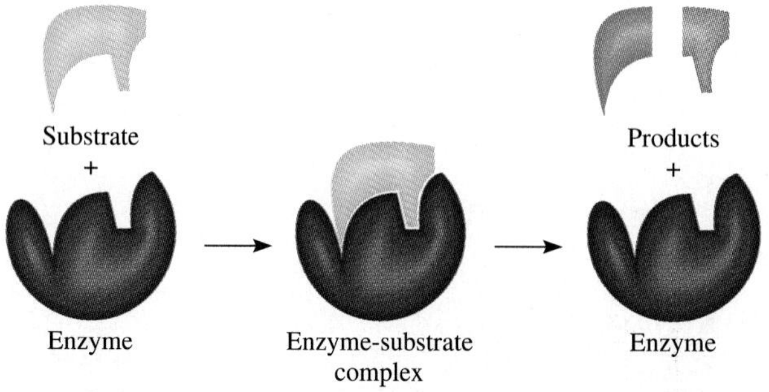

FIGURE 21.20

The lock-and-key model of an enzyme's specificity for substrate molecules.

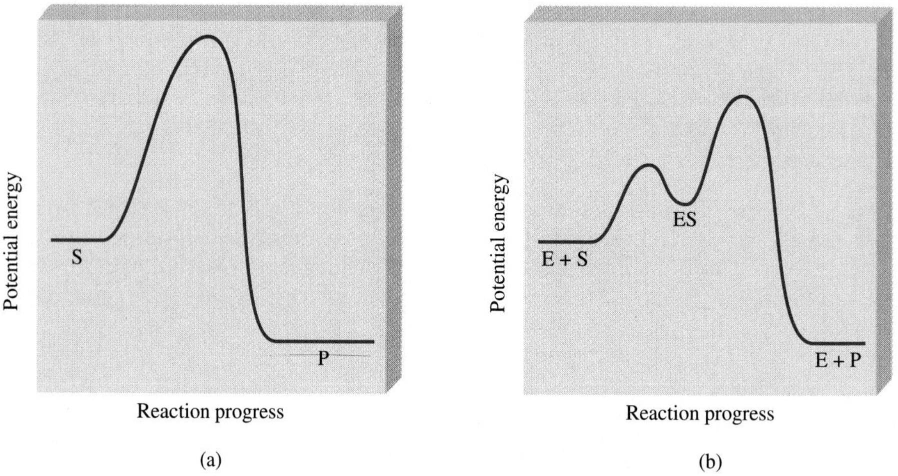

FIGURE 21.21

Comparison of (a) an uncatalyzed reaction and (b) the same reaction catalyzed by an enzyme. The plot in (b) assumes that the catalyzed reaction has a two-step mechanism, in which the second step (ES → E + P) is rate-determining.

The mathematical treatment of enzyme kinetics is quite complex, even when we know the basic steps involved in the reaction. A simplified scheme is as follows:

$$E + S \rightleftharpoons ES$$

$$ES \xrightarrow{k} P + E$$

where E, S, and P represent enzyme, substrate, and product, and ES is the enzyme-substrate intermediate. Figure 21.21 shows the potential energy profile for the reaction. It is often assumed that the formation of ES and its decomposition back to enzyme and substrate molecules occur rapidly and that the rate-determining step is the formation of product. In general, the rate of such a reaction is given by the equation

$$\text{rate} = \frac{\Delta[P]}{\Delta t} = k[ES]$$

The concentration of the ES intermediate is itself proportional to the amount of the substrate present, and a plot of the rate versus the concentration of substrate typically yields a curve such as that shown in Figure 21.22. Initially the rate rises rapidly with increasing substrate concentration. However, above a certain concentration all the active sites are occupied, and the reaction becomes zero order in the substrate. That is, the rate remains the same even though the substrate concentration increases. At and beyond this point, the rate of formation of product depends only on how fast the ES intermediate breaks down, not on the number of substrate molecules present.

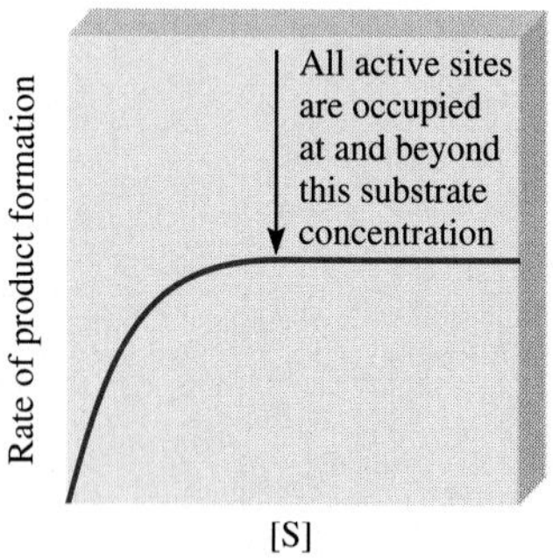

FIGURE 21.22

Plot of the rate of product formation versus substrate concentration in an enzyme-catalyzed reaction.

SUMMARY

The rate of a chemical reaction is shown by the change in the concentration of reactant or product molecules over time. The rate is not constant but varies continuously as concentration changes.

The rate law is an expression relating the rate of a reaction to the rate constant and the concentrations of the reactants raised to appropriate powers. The rate constant k for a given reaction changes only with temperature. Overall reaction order is the sum of the powers to which reactant concentrations are raised in the rate law. The rate law and the reaction order cannot be determined from the stoichiometry of the overall equation for a reaction; they must be determined by experiment. The half-life of a reaction (the time it takes for the concentration of a reactant to decrease by one-half) can be used to determine the rate constant of a first-order reaction.

According to collision theory, a reaction occurs when molecules collide with sufficient energy, called the activation energy, to break the bond and initiate the reaction. The rate constant and the activation energy are related by the Arrhenius equation, $k = Ae^{-E_a/RT}$.

The overall balanced equation for a reaction may be the sum of a series of simple reactions, called elementary steps. The complete series of elementary steps for a reaction is the reaction mechanism. The slowest step in a reaction mechanism is the rate-determining step.

A catalyst speeds up a reaction usually by lowering the value of E_a. A catalyst can be recovered unchanged at the end of a reaction. In heterogeneous catalysis, which is of great industrial importance, the catalyst is a solid and the reactants are gases or liquids. In homogeneous catalysis, the catalyst and the reactants are in the same phase. Enzymes are catalysts in living systems.

KEY WORDS

Activated complex, p. 614
Activation energy (E_a), p. 614
Bimolecular reaction, p. 619
Catalyst, p. 622
Chemical kinetics, p. 600
Elementary step, p. 619
Enzymes, p. 625
First-order reaction, p. 607
Half-life, p. 609
Intermediate, p. 619
Molecularity of a reaction, p. 619
Rate constant, p. 604
Rate-determining step, p. 620
Rate law, p. 604
Reaction mechanism, p. 619
Reaction order, p. 604
Reaction rate, p. 600
Termolecular reaction, p. 619
Unimolecular reaction, p. 619

QUESTIONS AND PROBLEMS

REACTION RATE

Review Questions

21.1 What is meant by the rate of a chemical reaction?

21.2 What are the units of the rate of a reaction?

21.3 What are the advantages of measuring the initial rate of a reaction?

21.4 Can you suggest two reactions that are very slow (take days or longer to complete) and two reactions that are very fast (are over in minutes or seconds)?

Problems

21.5 Write the reaction rate expressions for the following reactions in terms of the disappearance of the reactants and the appearance of products:
(a) $H_2(g) + I_2(g) \rightarrow 2HI(g)$
(b) $2H_2(g) + O_2(g) \rightarrow 2H_2O(g)$
(c) $5Br^-(aq) + BrO_3^-(aq) + 6H^+(aq) \rightarrow 3Br_2(aq) + 3H_2O(l)$

21.6 Consider the reaction

$$N_2(g) + 3H_2(g) \longrightarrow 2NH_3(g)$$

Suppose that at a particular moment during the reaction molecular hydrogen is reacting at the rate of 0.074 *M*/s. (a) At what rate is ammonia being formed? (b) At what rate is molecular nitrogen reacting?

RATE LAWS

Review Questions

21.7 Explain what is meant by the rate law of a reaction.

21.8 What is meant by the order of a reaction?

21.9 What are the units for the rate constants of first-order and second-order reactions?

21.10 Write an equation relating the concentration of a reactant A at $t = 0$ to that at $t = t$ for a first-order reaction. Define all the terms and give their units.

21.11 Consider the zero-order reaction $A \rightarrow$ product. (a) Write the rate law for the reaction. (b) What are the units for the rate constant? (c) Plot the rate of the reaction versus [A].

21.12 The rate constant of a first-order reaction is $66\ s^{-1}$. What is the rate constant in units of minutes?

21.13 On which of the following quantities does the rate constant of a reaction depend? (a) Concentrations of reactants, (b) nature of reactants, (c) temperature.

21.14 For each of the following pairs of reaction conditions, indicate which has the faster rate of formation of hydrogen gas: (a) sodium or potassium with water, (b) magnesium or iron with 1.0 *M* HCl, (c) magnesium rod or magnesium powder with 1.0 *M* HCl, (d) magnesium with 0.10 *M* HCl or magnesium with 1.0 *M* HCl.

Problems

21.15 The rate law for the reaction

$$NH_4^+(aq) + NO_2^-(aq) \longrightarrow N_2(g) + 2H_2O(l)$$

is given by rate $= k[NH_4^+][NO_2^-]$. At 25°C, the rate constant is $3.0 \times 10^{-4}/M \cdot s$. Calculate the rate of the reaction at this temperature if $[NH_4^+] = 0.26\ M$ and $[NO_2^-] = 0.080\ M$.

21.16 Starting with the data in Table 21.1, (a) deduce the rate law for the reaction, (b) calculate the rate constant, and (c) calculate the rate of the reaction at the time when $[F_2] = 0.010\ M$ and $[ClO_2] = 0.020\ M$.

21.17 Consider the reaction

$$A + B \longrightarrow \text{products}$$

From the following data obtained at a certain temperature, determine the order of the reaction and calculate the rate constant:

[A] (*M*)	[B] (*M*)	Rate (*M*/s)
1.50	1.50	3.20×10^{-1}
1.50	2.50	3.20×10^{-1}
3.00	1.50	6.40×10^{-1}

21.18 Consider the reaction

$$X + Y \longrightarrow Z$$

The following data are obtained at 360 K:

Initial rate of disappearance of X (*M*/s)	[X]	[Y]
0.147	0.10	0.50
0.127	0.20	0.30
4.064	0.40	0.60
1.016	0.20	0.60
0.508	0.40	0.30

(a) Determine the order of the reaction. (b) Determine the initial rate of disappearance of X when the concentration of X is 0.30 *M* and that of Y is 0.40 *M*.

21.19 Determine the overall orders of the reactions to which the following rate laws apply: (a) rate $= k[NO_2]^2$; (b) rate $= k$; (c) rate $= k[H_2][Br_2]^{\frac{1}{2}}$; (d) rate $= k[NO]^2[O_2]$.

21.20 Consider the reaction

$$A \longrightarrow B$$

The rate of the reaction is $1.6 \times 10^{-2}\ M/s$ when the concentration of A is 0.35 *M*. Calculate the rate constant if the reaction is (a) first order in A, (b) second order in A.

HALF-LIFE

Review Questions

21.21 Define the half-life of a reaction. Write the equation relating the half-life of a first-order reaction to the rate constant.

21.22 For a first-order reaction, how long will it take for the concentration of reactant to fall to one-eighth its original value? Express your answer in terms of the half-life ($t_{\frac{1}{2}}$) and in terms of the rate constant *k*.

Problems

21.23 What is the half-life of a compound if 75 percent of a given sample of the compound decomposes in 60 min? Assume first-order kinetics.

21.24 The thermal decomposition of phosphine (PH_3) into phosphorus and molecular hydrogen is a first-order reaction:

$$4PH_3(g) \longrightarrow P_4(g) + 6H_2(g)$$

The half-life of the reaction is 35.0 s at 680°C. Calculate (a) the first-order rate constant for the reaction and (b) the time required for 95 percent of the phosphine to decompose.

ACTIVATION ENERGY

Review Questions

21.25 Define activation energy. What role does activation energy play in chemical kinetics?

21.26 Write the Arrhenius equation and define all terms.

21.27 Use the Arrhenius equation to show why the rate constant of a reaction (a) decreases with increasing activation energy and (b) increases with increasing temperature.

21.28 As we know, methane burns readily in oxygen in a highly exothermic reaction. Yet a mixture of methane and oxygen gas can be kept indefinitely without any apparent change. Explain.

21.29 Sketch a potential-energy-versus-reaction-progress plot for the following reactions:
(a) $S(s) + O_2(g) \rightarrow SO_2(g)$ $\Delta H° = -296.06$ kJ
(b) $Cl_2(g) \rightarrow Cl(g) + Cl(g)$ $\Delta H° = 242.7$ kJ

21.30 The reaction $H + H_2 \rightarrow H_2 + H$ has been studied for many years. Sketch a potential-energy-versus-reaction-progress diagram for this reaction.

Problems

21.31 Variation of the rate constant with temperature for the first-order reaction

$$2N_2O_5(g) \longrightarrow 2N_2O_4(g) + O_2(g)$$

is given in the following table. Determine graphically the activation energy for the reaction.

T (K)	k (s^{-1})
273	7.87×10^3
298	3.46×10^5
318	4.98×10^6
338	4.87×10^7

21.32 Given the same concentrations, the reaction

$$CO(g) + Cl_2(g) \longrightarrow COCl_2(g)$$

at 250°C is 1.50×10^3 times as fast as the same reaction at 150°C. Calculate the energy of activation for this reaction. Assume that the frequency factor is constant.

21.33 For the reaction

$$NO(g) + O_3(g) \longrightarrow NO_2(g) + O_2(g)$$

the frequency factor A is 8.7×10^{12} s^{-1} and the activation energy is 63 kJ/mol. What is the rate constant for the reaction at 75°C?

21.34 The rate constant of a first-order reaction is 4.60×10^{-4} s^{-1} at 350°C. If the activation energy is 104 kJ/mol, calculate the temperature at which its rate constant is 8.80×10^{-4} s^{-1}.

21.35 The rate constants of some reactions double with every 10-degree rise in temperature. Assume a reaction takes place at 295 K and 305 K. What must the activation energy be for the rate constant to double as described?

21.36 The rate at which tree crickets chirp is 2.0×10^2 per minute at 27°C but only 39.6 per minute at 5°C. From these data, calculate the "energy of activation" for the chirping process. (*Hint:* The ratio of rates is equal to the ratio of rate constants.)

REACTION MECHANISMS

Review Questions

21.37 What do we mean by the mechanism of a reaction?

21.38 What is an elementary step?

21.39 What is the molecularity of a reaction?

21.40 Reactions can be classified as unimolecular, bimolecular, and so on. Why are there no zero-molecular reactions?

21.41 Explain why termolecular reactions are rare.

21.42 What is the rate-determining step of a reaction? Give an everyday analogy to illustrate the meaning of the term "rate determining."

21.43 The equation for the combustion of ethane (C_2H_6) is

$$2C_2H_6 + 7O_2 \longrightarrow 4CO_2 + 6H_2O$$

Explain why it is unlikely that this equation also represents the elementary step for the reaction.

21.44 Which of the following species cannot be isolated in a reaction? Activated complex, product, intermediate.

Problems

21.45 The rate law for the reaction

$$2NO(g) + Cl_2(g) \longrightarrow 2NOCl(g)$$

is given by rate = $k[NO][Cl_2]$. (a) What is the order of the reaction? (b) A mechanism involving the following steps has been proposed for the reaction

$$NO(g) + Cl_2(g) \longrightarrow NOCl_2(g)$$

$$NOCl_2(g) + NO(g) \longrightarrow 2NOCl(g)$$

If this mechanism is correct, what does it imply about the relative rates of these two steps?

21.46 For the reaction $X_2 + Y + Z \rightarrow XY + XZ$ it is found that doubling the concentration of X_2 doubles the reaction rate, tripling the concentration of Y triples the rate, and doubling the concentration of Z has no effect. (a) What is the rate law for this reaction? (b) Why is it that the change in the concentration of Z has no effect on the rate? (c) Suggest a mechanism for the reaction that is consistent with the rate law.

CATALYSIS

Review Questions

21.47 How does a catalyst increase the rate of a reaction?

21.48 What are the characteristics of a catalyst?

21.49 A certain reaction is known to proceed slowly at room temperature. Is it possible to make the reaction proceed at a faster rate without changing the temperature?

21.50 Distinguish between homogeneous catalysis and heterogeneous catalysis. Describe some important industrial processes that utilize heterogeneous catalysis.

21.51 Are enzyme-catalyzed reactions examples of homogeneous or heterogeneous catalysis?

21.52 The concentrations of enzymes in cells are usually quite small. What is the biological significance of this fact?

Problems

21.53 Most reactions, including enzyme-catalyzed reactions, proceed faster at higher temperatures. However, for a given enzyme, the rate drops off abruptly at a certain temperature. Account for this behavior.

21.54 Consider the following mechanism for the enzyme-catalyzed reaction

$$E + S \underset{k_{-1}}{\overset{k_1}{\rightleftharpoons}} ES \qquad \text{(fast equilibrium)}$$

$$ES \xrightarrow{k_2} E + P \qquad \text{(slow)}$$

Derive an expression for the rate law of the reaction in terms of the concentrations of E and S. (*Hint:* To solve for [ES], make use of the fact that, at equilibrium, the rate of the forward reaction is equal to the rate of the reverse reaction.)

MISCELLANEOUS PROBLEMS

21.55 Suggest experimental means by which the rates of the following reactions could be followed:
(a) $CaCO_3(s) \rightarrow CaO(s) + CO_2(g)$
(b) $Cl_2(g) + 2Br^-(aq) \rightarrow Br_2(aq) + 2Cl^-(aq)$
(c) $C_2H_6(g) \rightarrow C_2H_4(g) + H_2(g)$

21.56 List four factors that influence the rate of a reaction.

21.57 "The rate constant for the reaction

$$NO_2(g) + CO(g) \longrightarrow NO(g) + CO_2(g)$$

is $1.64 \times 10^{-6}/M \cdot s$." What is incomplete about this statement?

21.58 In a certain industrial process using a heterogeneous catalyst, the volume of the catalyst (in the shape of a sphere) is 10.0 cm^3. Calculate the surface area of the catalyst. If the sphere is broken down into eight spheres, each of which has a volume of 1.25 cm^3, what is the total surface area of the spheres? Which of the two geometric configurations of the catalyst is more effective? Explain. (The surface area of a sphere is $4\pi r^2$, where r is the radius of the sphere.)

21.59 When methyl phosphate is heated in acid solution, it reacts with water:

$$CH_3OPO_3H_2 + H_2O \longrightarrow CH_3OH + H_3PO_4$$

If the reaction is carried out in water enriched with ^{18}O, the oxygen-18 isotope is found in the phosphoric acid product but not in the methanol. What does this tell us about the mechanism of the reaction?

21.60 The rate of the reaction

$$CH_3COOC_2H_5(aq) + H_2O(l) \longrightarrow CH_3COOH(aq) + C_2H_5OH(aq)$$

shows first-order characteristics—that is, rate = $k[CH_3COOC_2H_5]$—even though this is a second-order reaction (first order in $CH_3COOC_2H_5$ and first order in H_2O). Explain.

21.61 Explain why most metals used in catalysis are transition metals.

21.62 The bromination of acetone is acid-catalyzed:

$$CH_3COCH_3 + Br_2 \xrightarrow[\text{catalyst}]{H^+} CH_3COCH_2Br + H^+ + Br^-$$

The rate of disappearance of bromine was measured for several different concentrations of acetone, bromine, and H^+ ions at a certain temperature:

	$[CH_3COCH_3]$	$[Br_2]$	$[H^+]$	Rate of disappearance of Br_2 (M/s)
(a)	0.30	0.050	0.050	5.7×10^{-5}
(b)	0.30	0.10	0.050	5.7×10^{-5}
(c)	0.30	0.050	0.10	1.2×10^{-4}
(d)	0.40	0.050	0.20	3.1×10^{-4}
(e)	0.40	0.050	0.050	7.6×10^{-5}

(a) What is the rate law for the reaction? (b) Determine the rate constant.

21.63 The reaction $2A + 3B \rightarrow C$ is first order with respect to A and B. When the initial concentrations are $[A] = 1.6 \times 10^{-2}\ M$ and $[B] = 2.4 \times 10^{-3}\ M$, the rate is $4.1 \times 10^{-4}\ M$/s. Calculate the rate constant of the reaction.

21.64 The decomposition of N_2O to N_2 and O_2 is a first-order reaction. At 730°C the half-life of the reaction is 3.58×10^3 min. If the initial pressure of N_2O is 2.10 atm at 730°C, calculate the total gas pressure after one half-life. Assume that the volume remains constant.

21.65 The reaction $S_2O_8^{2-} + 2I^- \rightarrow 2SO_4^{2-} + I_2$ proceeds slowly in aqueous solution, but it can be catalyzed by the Fe^{3+} ion. Given that Fe^{3+} can iodize I^- and Fe^{2+} can reduce $S_2O_8^{2-}$, write a plausible two-step mechanism for this reaction. Explain why the uncatalyzed reaction is slow.

21.66 What are the units of the rate constant for a third-order reaction?

21.67 Consider the zero-order reaction $A \rightarrow B$. Sketch the following plots: (a) rate versus [A] and (b) [A] versus t.

21.68 A flask contains a mixture of compounds A and B. Both compounds decompose by first-order kinetics. The half-lives are 50.0 min for A and 18.0 min for B. If the concentrations of A and B are equal initially, how long will it take for the concentration of A to be four times that of B?

21.69 Referring to Example 21.5, explain how you would measure experimentally the partial pressure of azomethane as a function of time.

21.70 The rate law for the reaction $2NO_2(g) \rightarrow N_2O_4(g)$ is rate $= k[NO_2]^2$. Which of the following changes will change the value of k? (a) The pressure of NO_2 is doubled. (b) The reaction is run in an organic solvent. (c) The volume of the container is doubled. (d) The temperature is decreased. (e) A catalyst is added to the container.

21.71 The reaction of G_2 with E_2 to form 2EG is exothermic, and the reaction of G_2 with X_2 to form 2XG is endothermic. The activation energy of the exothermic reaction is greater than that of the endothermic reaction. Sketch the potential energy profile diagrams for these two reactions on the same graph.

21.72 In the nuclear industry, workers use a rule of thumb that the radioactivity from any sample will be relatively harmless after ten half-lives. Calculate the fraction of a radioactive sample that remains after this time period. (*Hint:* Radioactive decays obey first-order kinetics.)

21.73 Briefly comment on the effect of a catalyst on each of the following: (a) activation energy, (b) reaction mechanism, (c) enthalpy of reaction, (d) rate of forward step, (e) rate of reverse step.

21.74 A quantity of 6 g of granulated Zn is added to a solution of 2 M HCl in a beaker at room temperature. Hydrogen gas is generated. For each of the following changes (at constant volume of the acid) state whether the rate of hydrogen gas evolution will be increased, decreased, or unchanged: (a) 6 g of powdered Zn is used; (b) 4 g of granulated Zn is used; (c) 2 M acetic acid is used instead of 2 M HCl; (d) temperature is raised to 40°C.

21.75 The following data were collected for the reaction between hydrogen and nitric oxide at 700°C:

$$2H_2(g) + 2NO(g) \longrightarrow 2H_2O(g) + N_2(g)$$

Experiment	$[H_2]$	[NO]	Initial rate (M/s)
1	0.010	0.025	2.4×10^{-6}
2	0.0050	0.025	1.2×10^{-6}
3	0.010	0.0125	0.60×10^{-6}

(a) Determine the order of the reaction. (b) Calculate the rate constant. (c) Suggest a plausible mechanism that is consistent with the rate law. (*Hint:* Assume the oxygen atom is the intermediate.)

21.76 A certain first-order reaction is 35.5 percent complete in 4.90 min at 25°C. What is its rate constant?

21.77 The decomposition of dinitrogen pentoxide has been studied in carbon tetrachloride solvent

(CCl_4) at a certain temperature:

$$2N_2O_5 \longrightarrow 4NO_2 + O_2$$

$[N_2O_5]$ (M)	Initial rate (M/s)
0.92	0.95×10^{-5}
1.23	1.20×10^{-5}
1.79	1.93×10^{-5}
2.00	2.10×10^{-5}
2.21	2.26×10^{-5}

Determine graphically the rate law for the reaction and calculate the rate constant.

21.78 The thermal decomposition of N_2O_5 obeys first-order kinetics. At 45°C, a plot of ln $[N_2O_5]$ versus t gives a slope of -6.18×10^{-4} min^{-1}. What is the half-life of the reaction?

21.79 When a mixture of methane and bromine is exposed to light, the following reaction occurs slowly:

$$CH_4(g) + Br_2(g) \longrightarrow CH_3Br(g) + HBr(g)$$

Suggest a reasonable mechanism for this reaction. (*Hint:* Bromine vapor is deep red; methane is colorless.)

21.80 Consider the following elementary step:

$$X + 2Y \longrightarrow XY_2$$

(a) Write a rate law for this reaction. (b) If the initial rate of formation of XY_2 is 3.8×10^{-3} M/s and the initial concentrations of X and Y are 0.26 M and 0.88 M, what is the rate constant of the reaction?

21.81 Consider the reaction

$$C_2H_5I(aq) + H_2O(l) \longrightarrow C_2H_5OH(aq) + H^+(aq) + I^-(aq)$$

How could you follow the progress of the reaction by measuring the electrical conductance of the solution?

21.82 A compound X undergoes two *simultaneous* first-order reactions as follows: X → Y with rate constant k_1 and X → Z with rate constant k_2. The ratio of k_1/k_2 at 40°C is 8.0. What is the ratio at 300°C? Assume that the frequency factor of the two reactions is the same.

21.83 In recent years ozone in the stratosphere has been depleted at an alarmingly fast rate by chlorofluorocarbons (CFCs). A CFC molecule such as $CFCl_3$ is first decomposed by UV radiation:

$$CFCl_3 \longrightarrow CFCl_2 + Cl$$

The chlorine radical then reacts with ozone as follows:

$$Cl + O_3 \longrightarrow ClO + O_2$$
$$ClO + O \longrightarrow Cl + O_2$$

(a) Write the overall reaction for the last two steps. (b) What are the roles of Cl and ClO? (c) Why is the fluorine radical not important in this mechanism? (d) One suggestion to reduce the concentration of chlorine radicals is to add hydrocarbons such as ethane (C_2H_6) to the stratosphere. How will this work?

21.84 Consider a car fitted with a catalytic converter. The first ten minutes or so after it is started is the most polluting. Why?

Answers to Practice Exercises: 21.1 rate $= -\frac{\Delta[CH_4]}{\Delta t} = -\frac{1}{2}\frac{\Delta[O_2]}{\Delta t} = \frac{\Delta[CO_2]}{\Delta t} = \frac{1}{2}\frac{\Delta[H_2O]}{\Delta t}$; **21.2** rate $= k[S_2O_8^{2-}][I^-]$, $k = 8.1 \times 10^{-2}/M$s; **21.3** 66 s; **21.4** 8.24×10^{-3} min^{-1}; **21.5** rate $= k[C_2H_6N_2]$, $k = 0.014$ min^{-1}; **21.6** 240 kJ/mol; **21.7** 3.13×10^{-9} s^{-1}; **21.8** (a) $NO_2 + CO \rightarrow NO + CO_2$, (b) NO_3, (c) The first step is rate-determining.

CHAPTER 22
NUCLEAR CHEMISTRY

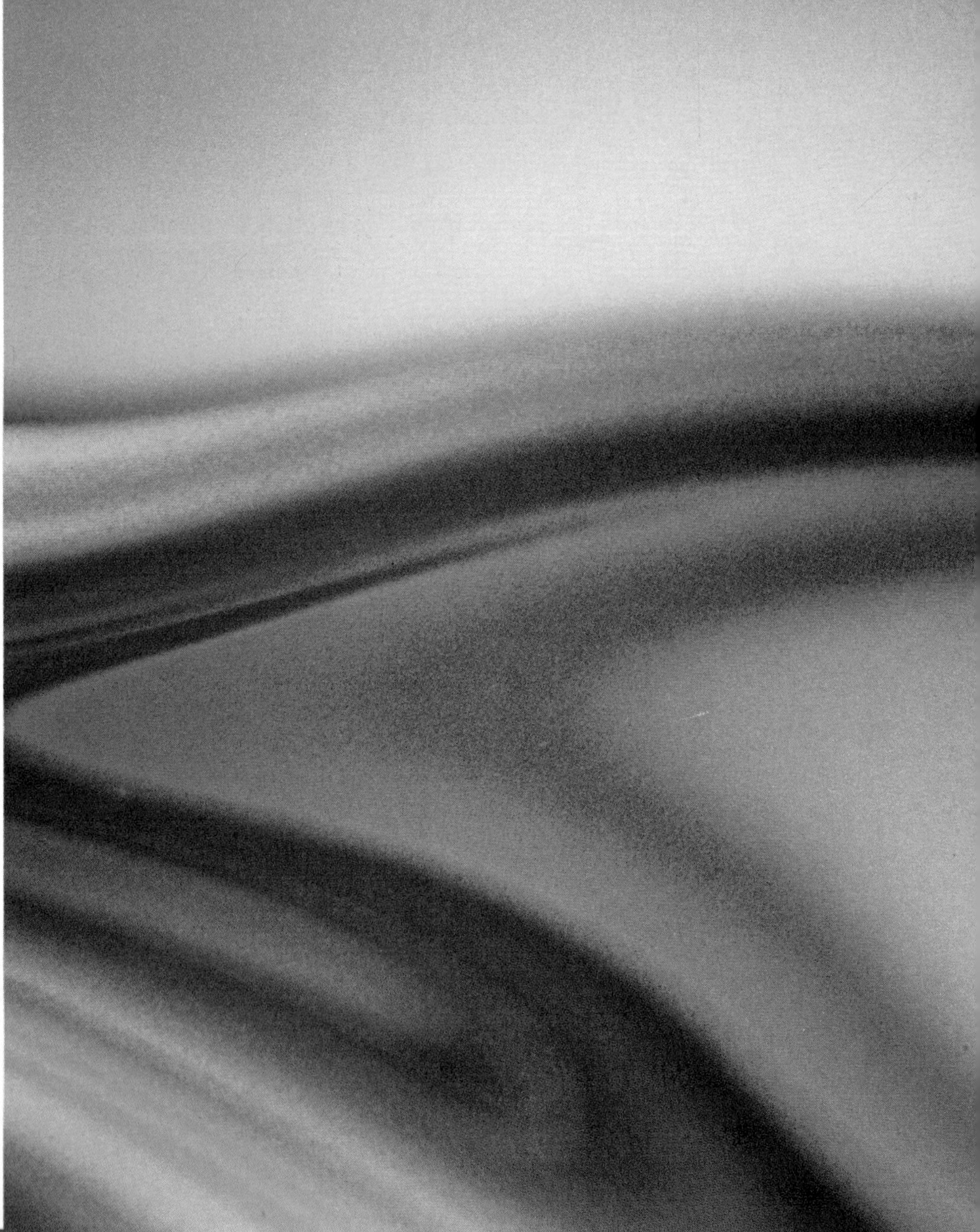

Lise Meitner and Otto Hahn in their laboratory.

◆ In December 1938, Lise Meitner, an Austrian physicist living in Sweden, received a puzzling letter from her long-time collaborator at the University of Berlin, Otto Hahn. The letter described an experiment in which Hahn and another colleague, Fritz Strassman, had bombarded uranium with neutrons in the hope of creating heavier elements. When they analyzed the products, the two chemists were astonished to find barium, which has roughly half the mass of uranium. On a walk through the Swedish countryside, Meitner and her nephew Otto Frisch, who was also a physicist, concluded that Hahn and Strassman had split apart the uranium nucleus. Borrowing the term "fission" from biology, where it applies to cell division, the scientists called the newly discovered phenomenon *nuclear fission.* They calculated that the process would release enormous energy.

Lise Meitner was a well-known and respected figure in nuclear physics; Einstein nicknamed her the "German Madam Curie." She was born in 1878 in Vienna, Austria, and showed a marked bent for mathematics and natural science from childhood. In 1906 she received a Ph.D. in physics from the University of Vienna, the second woman to do so in the university's 500-year history. Soon afterward she began a lifelong collaboration with Hahn. Being a woman, Meitner was given only a carpenter's workshop as her laboratory, and she worked for five years without pay! Gradually her situation improved, and in 1922 she was named a lecturer at the University of Berlin.

Meitner's research was interrupted in 1938 when Hitler seized Austria, and she lost her citizenship. Being Jewish, she fled to Sweden, where she continued to work with Hahn by mail.

In 1944 Hahn was awarded the Nobel Prize in chemistry for his work on nuclear fission. Meitner received no credit for her contribution. Today, Lise Meitner's statue is displayed among those of other prominent physicists at the Deutsche Museum in Munich, Germany, and element 109 is named meitnerium (Mt) in her honor. The inscription on her gravestone in Cambridge, England, reads: "Lise Meitner, 1878–1968, a great scientist who never forgot her humanity." ◆

22.1 THE NATURE OF NUCLEAR REACTIONS

With the exception of hydrogen (^1_1H), all nuclei contain two kinds of fundamental particles, called *protons* and *neutrons.* Some nuclei are unstable; they emit particles and/or electromagnetic radiation spontaneously (see Section 2.2). The name for this phenomenon is *radioactivity.* All elements having an atomic number greater than 83 are radioactive.

Nuclei can also undergo change as a result of bombardment by neutrons, protrons, or other nuclei. This process is known as ***nuclear transmutation.*** An example of nuclear transmutation is the conversion of atmospheric $^{14}_7N$ to $^{14}_6C$ and ^1_1H, which results when the nitrogen isotope captures a neutron (from the sun). In some cases, heavier elements are synthesized from lighter elements. This type of transmutation occurs naturally in outer space, but it can also be achieved artificially, as we will see in Section 22.4

Radioactive decay and nuclear transmutation are ***nuclear reactions,*** which *involve changes in atomic nuclei.* Nuclear chemistry is the study of nuclear reactions. Nuclear reactions differ significantly from ordinary chemical reactions, which involve only the rearrangement of electrons. Table 22.1 summarizes the differences.

BALANCING NUCLEAR EQUATIONS

In order to discuss nuclear reactions in any depth, we need to understand how to write and balance the equations. Writing a nuclear equation differs somewhat from writing equations for chemical reactions. In addition to writing the symbols for various chemical elements, we must also explicitly indicate protons, neutrons, and electrons. In fact, we must show the numbers of protons and neutrons present in *every* species in such an equation.

The symbols for elementary particles are as follows:

^1_1p or ^1_1H	^1_0n	$^{\ 0}_{-1}e$ or $^{\ 0}_{-1}\beta$	$^{\ 0}_{+1}e$ or $^{\ 0}_{+1}\beta$	^4_2He or $^4_2\alpha$
proton	neutron	electron	positron	α particle

TABLE 22.1
Comparison of Chemical Reactions and Nuclear Reactions

Chemical reactions	Nuclear reactions
1. Atoms are rearranged by the breaking and forming of chemical bonds.	1. Elements (or isotopes of the same elements) are converted from one to another.
2. Only electrons in atomic orbitals are involved in the breaking and forming of bonds.	2. Protons, neutrons, electrons, and other elementary particles may be involved.
3. Reactions are accompanied by absorption or release of relatively small amounts of energy.	3. Reactions are accompanied by absorption or release of tremendous amounts of energy.
4. Rates of reaction are influenced by temperature, pressure, concentration, and catalysts.	4. Rates of reaction normally are not affected by temperature, pressure, and catalysts.

In accordance with the notation used in Section 2.3, the superscript in each case denotes the mass number (the total number of neutrons and protons present) and the subscript is the atomic number (the number of protons). Thus, the "atomic number" of a proton is 1, because there is one proton present, and the "mass number" is also 1, because there is one proton but are no neutrons present. On the other hand, the "mass number" of a neutron is 1, but its "atomic number" is zero, because there are no protons present. For the electron, the "mass number" is zero (there are neither protons nor neutrons present), but the "atomic number" is −1, because the electron possesses a unit negative charge.

The symbol ${}_{-1}^{0}e$ represents an electron in or from an atomic orbital. The symbol ${}_{-1}^{0}\beta$ represents an electron that, although physically identical to any other electron, comes from a nucleus and not from an atomic orbital.

The ***positron*** *has the same mass as the electron but bears a +1 charge.* The α particle has two protons and two neutrons, so its atomic number is 2 and its mass number is 4.

In balancing any nuclear equation, we observe the following rules:

- The total number of protons plus neutrons in the products and in the reactants must be the same (conservation of mass number).
- The total number of nuclear charges in the products and in the reactants must be the same (conservation of atomic number).

If the atomic numbers and mass numbers of all the species but one in a nuclear equation are known, the unknown species can be identified by applying these rules, as shown in the following example, which illustrates balancing nuclear decay equations.

EXAMPLE 22.1

Balancing Nuclear Equations

Balance the following nuclear equations (that is, identify the product X):

(a) ${}_{84}^{212}\text{Po} \longrightarrow {}_{82}^{208}\text{Pb} + \text{X}$

(b) ${}_{55}^{137}\text{Cs} \longrightarrow {}_{-1}^{0}\beta + \text{X}$

Answer: (a) The mass number and atomic number are 212 and 84, respectively, on the left-hand side and 208 and 82, respectively, on the right-hand side. Thus X must have a mass number of 4 and an atomic number of 2, which means that it is an α particle. The balanced equation is

$$ {}_{84}^{212}\text{Po} \longrightarrow {}_{82}^{208}\text{Pb} + {}_{2}^{4}\text{He} $$

(b) In this case the mass number of X is the same as that for Cs-137 while the atomic number of X must be 56 so that the sum of protons is 55 on the right-hand side. Thus X is ${}_{56}^{137}\text{Ba}$, and the balanced equation is

$$ {}_{55}^{137}\text{Cs} \longrightarrow {}_{-1}^{0}\beta + {}_{56}^{137}\text{Ba} $$

PRACTICE EXERCISE

Identify X in the following nuclear equations: (a) ${}_{33}^{78}\text{As} \rightarrow {}_{-1}^{0}\beta + \text{X}$. (b) ${}_{11}^{20}\text{Na} \rightarrow {}_{10}^{20}\text{Ne} + \text{X}$.

22.2 NUCLEAR STABILITY

The nucleus occupies a very small portion of the total volume of an atom, but it contains most of the atom's mass because both the protons and neutrons reside there. We would expect the protons to strongly repel one another when they are packed so closely together. This indeed is so. However, in addition to the repulsion, there are short-range attractions between proton and proton, proton and neutron, and neutron and neutron. The stability of any nucleus is determined by the difference between the forces of repulsion and the short-range attraction. If repulsion outweighs attraction, the nucleus disintegrates, emitting particles and/or radiation. This is the phenomenon of radioactivity we discussed in Chapter 2. If attraction prevails, the nucleus will be stable.

The principal factor for determining whether a nucleus is stable is the *neutron-to-proton ratio* (n:p). For stable atoms of elements of low atomic number, the n:p value is close to 1. As the atomic number increases, the neutron-to-proton ratios of the stable nuclei become greater than 1. This deviation at higher atomic numbers arises because a larger number of neutrons is needed to stabilize the nucleus by counteracting the strong repulsion among the protons. The following rules are useful in predicting nuclear stability:

- Nuclei that contain 2, 8, 20, 50, 82, or 126 protons or neutrons are generally more stable than nuclei that do not possess these numbers. For example, there are ten stable isotopes of tin (Sn) with the atomic number 50 and only two stable isotopes of antimony (Sb) with the atomic number 51. The numbers 2, 8, 20, 50, 82, and 126 are called *magic numbers*. These numbers have a significance in nuclear stability similar to the numbers of electrons associated with the very stable noble gases (that is, 2, 10, 18, 36, 54, and 86 electrons).
- Nuclei with even numbers of both protons and neutrons are generally more stable than those with odd numbers of these particles (Table 22.2).
- All isotopes of the elements starting with polonium (Po, $Z = 84$) are radioactive. All isotopes of technetium (Tc, $Z = 43$) and promethium (Pm, $Z = 61$) are also radioactive.

Figure 22.1 shows a plot of the number of neutrons versus the number of protons in various isotopes. The stable nuclei are located in an area of the graph known as the *belt of stability*. Most of the radioactive nuclei lie outside this belt. Above the stability belt, the nuclei have higher neutron-to-proton ratios than those within the belt (for the same number of protons). Below the stability belt the nuclei

TABLE 22.2
Number of Stable Isotopes with Even and Odd Numbers of Protons and Neutrons

Protons	Neutrons	Number of stable isotopes
Odd	Odd	4
Odd	Even	50
Even	Odd	53
Even	Even	157

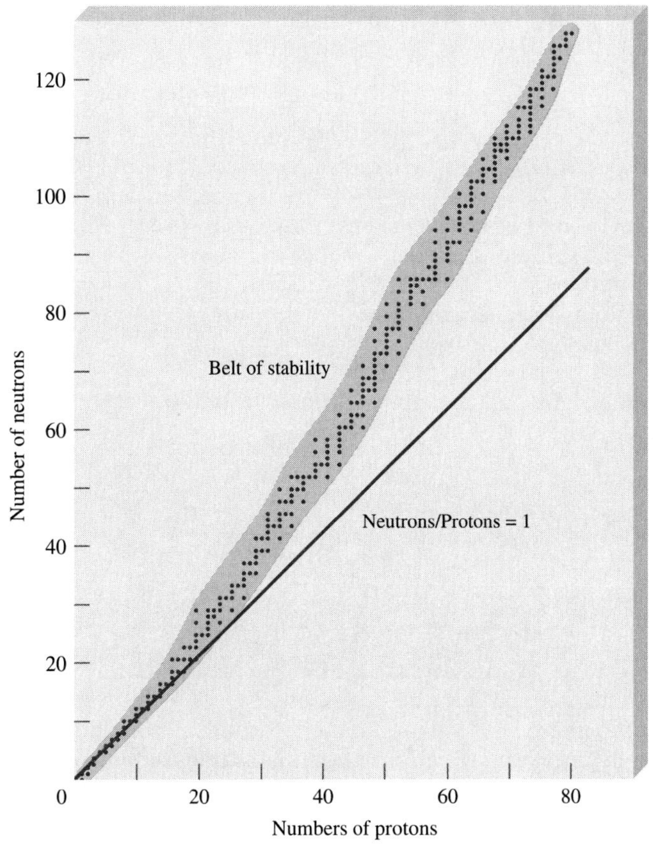

FIGURE 22.1

Plot of neutrons versus protons for various stable isotopes, represented by dots. The straight line represents the points at which the neutron-to-proton ratio equals 1. The shaded area represents the belt of stability.

have lower neutron-to-proton ratios than those in the belt (for the same number of protons).

Nuclear Binding Energy

A quantitative measure of nuclear stability is the ***nuclear binding energy,*** which is *the energy required to break up a nucleus into its component protons and neutrons.* This quantity represents the conversion of mass to energy that occurs during an exothermic nuclear reaction; that is, the synthesis of the nucleus from the protons and neutrons. As is true for chemical reactions, the more exothermic the reaction, the greater the stability of the product nucleus.

The concept of nuclear binding energy evolved from studies showing that the masses of nuclei are always less than the sum of the masses of the ***nucleons,*** that is, *both the protons and the neutrons in a nucleus.* For example, the $^{19}_{9}F$ isotope has an atomic mass of 18.9984 amu. The nucleus has 9 protons and 10 neutrons and therefore a total of 19 nucleons. Using the known mass of the $^{1}_{1}H$ atom (1.007825 amu) and the neutron (1.008665 amu), we can carry out the following analysis. The mass of 9 $^{1}_{1}H$ atoms (that is, the mass of 9 protons and 9 electrons) is

$$9 \times 1.007825 \text{ amu} = 9.070425 \text{ amu}$$

and the mass of 10 neutrons is

$$10 \times 1.008665 \text{ amu} = 10.08665 \text{ amu}$$

Therefore, the atomic mass of a $^{19}_{9}F$ atom calculated from the known numbers of electrons, protons, and neutrons is

$$9.070425 \text{ amu} + 10.08665 \text{ amu} = 19.15708 \text{ amu}$$

which is larger than 18.9984 amu (the measured mass of $^{19}_{9}F$) by 0.1587 amu.

The difference between the mass of an atom and the sum of the masses of its protons, neutrons, and electrons is called the ***mass defect.*** According to *Einstein's mass-energy equivalence relationship* ($E = mc^2$, where E is energy, m is mass, and c is the velocity of light), the loss in mass shows up as energy (heat) given off to the surroundings. Thus the formation of $^{19}_{9}F$ is exothermic. We can calculate the amount of energy released by writing

$$\Delta E = (\Delta m)c^2$$

where ΔE and Δm are defined as follows:

$$\Delta E = \text{energy of product} - \text{energy of reactants}$$

$$\Delta m = \text{mass of product} - \text{mass of reactants}$$

Thus we have for the change in mass

$$\begin{aligned}\Delta m &= 18.9984 \text{ amu} - 19.15708 \text{ amu}\\ &= -0.1587 \text{ amu}\end{aligned}$$

Because $^{19}_{9}F$ has a mass that is less than the mass calculated from the number of electrons and nucleons present, Δm is a negative quantity. (Note that there is no change in the electron's mass because it is not a nucleon.) Consequently, ΔE is also a negative quantity; that is, energy is released to the surroundings as a result of the formation of the fluorine-19 nucleus. So we calculate ΔE as follows:

$$\begin{aligned}\Delta E &= (-0.1587 \text{ amu})(3.00 \times 10^8 \text{ m/s})^2\\ &= -1.43 \times 10^{16} \text{ amu m}^2\text{/s}^2\end{aligned}$$

With the conversion factors

$$1 \text{ kg} = 6.022 \times 10^{26} \text{ amu}$$

$$1 \text{ J} = 1 \text{ kg m}^2\text{/s}^2$$

we obtain

$$\begin{aligned}\Delta E &= \left(-1.43 \times 10^{16} \frac{\text{amu m}^2}{\text{s}^2}\right) \times \left(\frac{1.00 \text{ kg}}{6.022 \times 10^{26} \text{ amu}}\right) \times \left(\frac{1 \text{ J}}{1 \text{ kg m}^2\text{/s}^2}\right)\\ &= -2.37 \times 10^{-11} \text{ J}\end{aligned}$$

This is the amount of energy released when one fluorine-19 nucleus is formed from 9 protons and 10 neutrons. The nuclear binding energy of the nucleus is 2.37×10^{-11} J, which is the amount of energy needed to decompose the nucleus into separate protons and neutrons. In general, the larger the mass defect, the greater the nuclear binding energy and the more stable the nucleus. In the formation of 1 mole of fluorine nuclei, for instance, the energy released is

$$\begin{aligned}\Delta E &= (-2.37 \times 10^{-11} \text{ J})(6.022 \times 10^{23}\text{/mol})\\ &= -1.43 \times 10^{13} \text{ J/mol}\\ &= -1.43 \times 10^{10} \text{ kJ/mol}\end{aligned}$$

The nuclear binding energy, therefore, is 1.43×10^{10} kJ for 1 mole of fluorine-19 nuclei, which is a tremendously large quantity when we consider that the enthal-

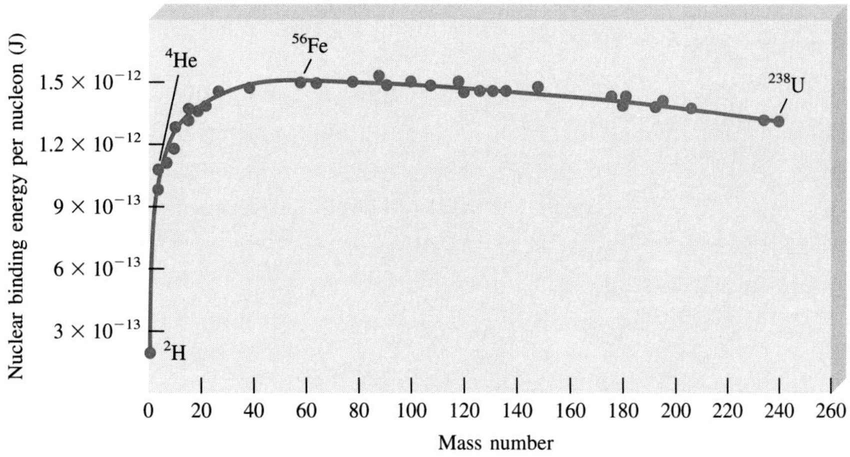

FIGURE 22.2
Plot of nuclear binding energy per nucleon versus mass number.

pies of ordinary chemical reactions are of the order of only 200 kJ. The procedure we have followed can be used to calculate the nuclear binding energy of any nucleus.

As we have noted, nuclear binding energy is an indication of the stability of a nucleus. However, in comparing the stability of any two nuclei we must account for the fact that they have different numbers of nucleons. For this reason it is more meaningful to use the *nuclear binding energy per nucleon,* defined as

$$\text{nuclear binding energy per nucleon} = \frac{\text{nuclear binding energy}}{\text{number of nucleons}}$$

For the fluorine-19 nucleus,

$$\text{nuclear binding energy per nucleon} = \frac{2.37 \times 10^{-11}\text{ J}}{19\text{ nucleons}}$$
$$= 1.25 \times 10^{-12}\text{ J/nucleon}$$

The nuclear binding energy per nucleon allows us to compare the stability of all nuclei on a common basis. Figure 22.2 shows the variation of nuclear binding energy per nucleon plotted against mass number. As you can see, the curve rises rather steeply; the highest binding energies per nucleon belong to elements with intermediate mass numbers—between 40 and 100—and are greatest for elements in the iron, cobalt, and nickel region (the Group 8B elements) of the periodic table. This result shows that the *net* attractive forces among the particles (protons and neutrons) are greatest for the nuclei of these elements.

EXAMPLE 22.2
Calculating Nuclear Binding Energy

The atomic mass of $^{127}_{53}I$ is 126.9004 amu. Calculate the nuclear binding energy of this nucleus and the corresponding nuclear binding energy per nucleon.

Answer: There are 53 protons and 74 neutrons in the nucleus. The mass of 53 $^{1}_{1}H$ atoms is

$$53 \times 1.007825\text{ amu} = 53.41473\text{ amu}$$

and the mass of 74 neutrons is

$$74 \times 1.008665 \text{ amu} = 74.64121 \text{ amu}$$

The predicted mass for $^{127}_{53}I$ is 53.41473 + 74.64121 = 128.05594 amu, and the mass defect is

$$\begin{aligned}\Delta m &= 126.9004 \text{ amu} - 128.05594 \text{ amu} \\ &= -1.1555 \text{ amu}\end{aligned}$$

The energy released is

$$\begin{aligned}\Delta E &= (\Delta m)c^2 \\ &= (-1.1555 \text{ amu})(3.00 \times 10^8 \text{ m/s})^2 \\ &= -1.04 \times 10^{17} \text{ amu m}^2/\text{s}^2 \\ &= \left(-1.04 \times 10^{17} \frac{\text{amu m}^2}{\text{s}^2}\right) \times \left(\frac{1.00 \text{ kg}}{6.022 \times 10^{26} \text{ amu}}\right) \times \left(\frac{1 \text{ J}}{1 \text{ kg m}^2/\text{s}^2}\right) \\ &= -1.73 \times 10^{-10} \text{ J}\end{aligned}$$

Thus the nuclear binding energy is 1.73×10^{-10} J. The nuclear binding energy per nucleon is obtained as follows:

$$\frac{1.73 \times 10^{-10} \text{ J}}{127 \text{ nucleons}} = 1.36 \times 10^{-12} \text{ J/nucleon}$$

PRACTICE EXERCISE

Calculate the nuclear binding energy (in joules) and the binding energy per nucleon of $^{209}_{83}Bi$ (208.9804 amu).

22.3 NATURAL RADIOACTIVITY

Nuclei outside the belt of stability, as well as nuclei with more than 83 protons, tend to be radioactive. The main types of radiation emitted by nuclei of radioactive elements are α particles (or doubly charged helium nuclei, He^{2+}); β particles (or electrons); γ rays, which are very-short-wavelength (0.1 nm to 10^{-4} nm) electromagnetic waves; positron emission; and electron capture.

When a radioactive nucleus disintegrates, the products formed may also be unstable and, if so, will undergo further disintegration. This process is repeated until a stable product finally is formed. Starting with the original radioactive nucleus, *the sequence of disintegration steps leading to a stable species* is called a ***decay series.*** Table 22.3 shows the decay series of naturally occurring uranium-238, which involves 14 steps. This decay scheme, known as the *uranium decay series,* also shows the half-lives of all the products.

It is important to be able to balance the nuclear reaction for each of the steps in a decay series. For example, the first step in the uranium decay series is the decay of uranium-238 to thorium-234, with the emission of an α particle. Hence, the reaction is

$$^{238}_{92}U \longrightarrow {}^{234}_{90}Th + {}^{4}_{2}He$$

TABLE 22.3
The Uranium Decay Series

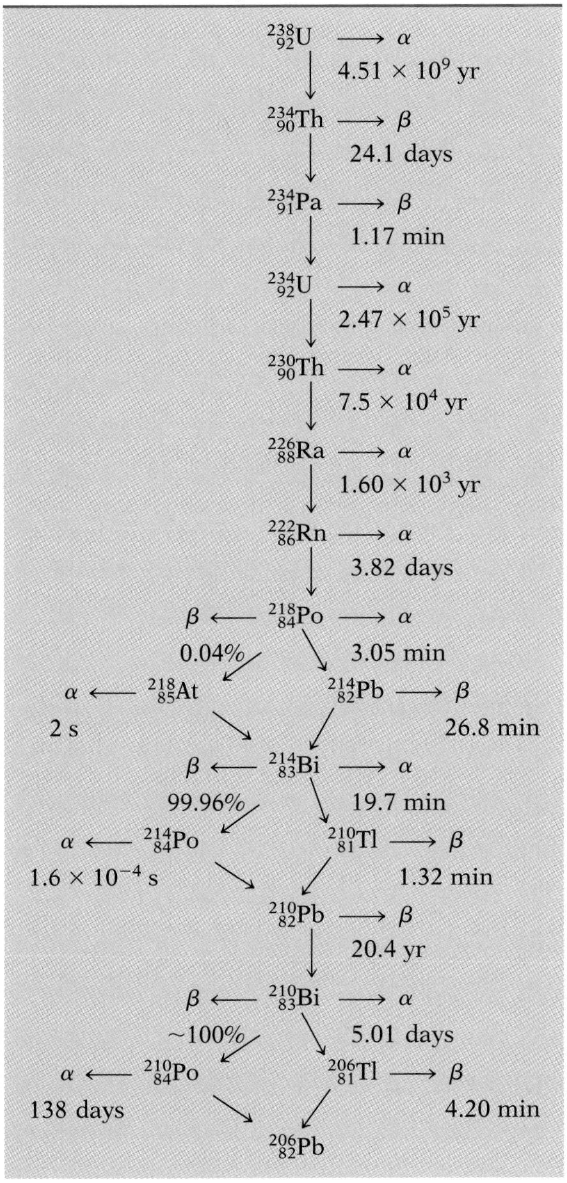

The next step is represented by

$$^{234}_{90}\text{Th} \longrightarrow\ ^{234}_{91}\text{Pa} +\ ^{0}_{-1}\beta$$

and so on. In a discussion of radioactive decay steps, the beginning radioactive isotope is called the *parent* and the product, the *daughter*.

Kinetics of Radioactive Decay

All radioactive decays obey first-order kinetics. This means that the rate of radioactive decay at any time t is given by

$$\text{rate of decay at time } t = kN$$

where k is the first-order rate constant and N is the number of radioactive nuclei present at time t. The units of the decay rate are number of disintegrations per unit time. According to Equation (21.3), the number of radioactive nuclei at time zero (N_0) and time t (N_t) is

$$\ln \frac{N_0}{N_t} = kt$$

The corresponding half-life of the reaction is given by Equation (21.5):

$$t_{\frac{1}{2}} = \frac{0.693}{k}$$

The half-lives (hence the rate constants) of radioactive isotopes vary greatly from nucleus to nucleus. For example, looking at Table 22.3, we find two extreme cases:

$$^{238}_{92}\text{U} \longrightarrow {}^{234}_{90}\text{Th} + {}^{4}_{2}\text{He} \qquad t_{\frac{1}{2}} = 4.51 \times 10^{9}\text{ yr}$$

$$^{214}_{84}\text{Po} \longrightarrow {}^{210}_{82}\text{Pb} + {}^{4}_{2}\text{He} \qquad t_{\frac{1}{2}} = 1.6 \times 10^{-4}\text{ s}$$

The ratio of these two rate constants after conversion to the same time unit is about 1×10^{21}, an enormously large number. Furthermore, the rate constants are unaffected by changes in environmental conditions such as temperature and pressure. These highly unusual features are not seen in ordinary chemical reactions (see Table 22.1).

Dating Based on Radioactive Decay

The half-lives of radioactive isotopes have been used as "atomic clocks" to determine the ages of certain objects. Here we will consider two methods for dating objects.

The age of the Shroud of Turin was shown by carbon-14 dating to be between A.D. 1260 and A.D. 1390, and therefore the shroud cannot be the burial cloth of Jesus Christ.

Radiocarbon Dating

The carbon-14 isotope is produced when atmospheric nitrogen is bombarded by cosmic rays:

$$^{14}_{7}\text{N} + {}^{1}_{0}\text{n} \longrightarrow {}^{14}_{6}\text{C} + {}^{1}_{1}\text{H}$$

The radioactive carbon-14 isotope decays according to the equation

$$^{14}_{6}\text{C} \longrightarrow {}^{14}_{7}\text{N} + {}^{0}_{-1}\beta \qquad t_{\frac{1}{2}} = 5730\text{ yr}$$

The carbon-14 isotopes enter the biosphere as CO_2, which is taken up in plant photosynthesis. Plant-eating animals in turn exhale carbon-14 in CO_2. Eventually, carbon-14 participates in many aspects of the carbon cycle. The ^{14}C lost by radioactive decay is constantly replenished by the production of new isotopes in the atmosphere until a dynamic equilibrium is established whereby the ratio of ^{14}C to ^{12}C remains constant in living matter. But when an individual plant or an animal dies, the carbon-14 isotope in it is no longer replenished, so the ratio decreases as ^{14}C decays. This same change occurs when carbon atoms are trapped in coal, petroleum, or wood preserved underground, and in mummified bodies. After a number of years, there are proportionately fewer ^{14}C nuclei in a mummy than in a living person.

The decreasing ratio of ^{14}C to ^{12}C can be used to estimate the age of a specimen. Using Equation (21.3), we can write

$$\ln \frac{N_0}{N_t} = kt$$

where N_0 and N_t are the number of ^{14}C nuclei present at $t = 0$ and $t = t$, and k is the first-order rate constant ($1.21 \times 10^{-4}\ yr^{-1}$). Since the decay rate is proportional to the amount of the radioactive isotope present, we have

$$t = \frac{1}{k} \ln \frac{N_0}{N}$$

$$= \frac{1}{1.21 \times 10^{-4}\ yr^{-1}} \ln \frac{\text{decay rate of fresh sample}}{\text{decay rate of old sample}}$$

Thus, by measuring the decay rates of the fresh sample and the old sample, we can calculate t, which is the age of the old sample. Radiocarbon dating is a valuable tool for estimating the age of objects (containing C atoms) dating back 1000 to 50,000 years.

Dating Using Uranium-238 Isotopes

Because some members of the uranium series have very long half-lives (see Table 22.3), this series is particularly suitable for estimating the age of rocks in the earth and of extraterrestrial objects. The half-life for the first step ($^{238}_{92}U$ to $^{234}_{90}Th$) is 4.51×10^9 yr. This is about 20,000 times the second largest value (that is, 2.47×10^5 yr), which is the half-life for $^{234}_{92}U$ to $^{230}_{90}Th$. Therefore, as a good approximation, we can assume that the half-life for the overall process (that is, from $^{238}_{92}U$ to $^{206}_{82}Pb$) is governed solely by the first step:

$$^{238}_{92}U \longrightarrow {}^{206}_{82}Pb + 8\,{}^{4}_{2}He + 6\,{}^{0}_{-1}\beta \qquad t_{\frac{1}{2}} = 4.51 \times 10^9\ yr$$

In naturally occurring uranium minerals we should and do find some lead-206 isotopes formed by radioactive decay. Assuming that no lead was present when the mineral was formed and that the mineral has not undergone chemical changes that would allow the lead-206 isotope to be separated from the parent uranium-238, it is possible to estimate the age of the rocks from the mass ratio of $^{206}_{82}Pb$ to $^{238}_{92}U$. The equation for the decay of $^{238}_{92}U$ to $^{206}_{82}Pb$ tells us that for every mole, or 238 g, of uranium that undergoes complete decay, 1 mole, or 206 g, of lead is formed. If only half a mole of uranium-238 has undergone decay, the mass ratio Pb-206/U-238 becomes

$$\frac{206\ g/2}{238\ g/2} = 0.866$$

and the process would have taken a half-life of 4.51×10^9 yr to complete (Figure 22.3). Ratios lower than 0.866 mean that the rocks are less than 4.51×10^9 yr old, and higher ratios suggest a greater age. Interestingly, studies based on the uranium series as well as other decay series put the age of the oldest rocks and, therefore, probably the age of Earth itself at 4.5×10^9, or 4.5 billion, years.

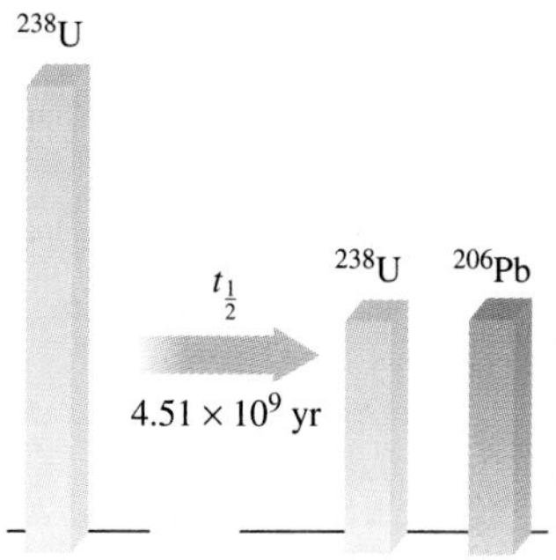

FIGURE 22.3

After one half-life, nearly half of the original uranium-238 is converted to lead-206.

22.4 NUCLEAR TRANSMUTATION

The scope of nuclear chemistry would be rather narrow if study had been limited to natural radioactive elements. An experiment performed by Rutherford in 1919, however, suggested the possibility of producing radioactivity artificially. When Rutherford bombarded a sample of nitrogen with α particles, the following reaction took place:

$$^{14}_{7}N + ^{4}_{2}He \longrightarrow ^{17}_{8}O + ^{1}_{1}p$$

An oxygen-17 isotope was produced with the emission of a proton. This reaction demonstrated for the first time the feasibility of nuclear transmutation.

The above reaction can be abbreviated as $^{14}_{7}N(\alpha,p)^{17}_{8}O$. Note that in the parentheses the bombarding particle is written first, followed by the ejected particle. Unlike radioactive decay, nuclear transmutation is *not* a spontaneous process; consequently, nuclear transmutation equations have more than one reactant on the left side of the equation.

EXAMPLE 22.3
Balancing Nuclear Transmutation Equations

Write the balanced equation for the nuclear reaction $^{56}_{26}Fe(d,\alpha)^{54}_{25}Mn$, where d represents the deuterium nucleus (that is, $^{2}_{1}H$).

Answer: The abbreviation tells us that when iron-56 is bombarded with a deuterium nucleus, it produces the manganese-54 nucleus plus an α particle. Thus, the equation for this reaction is

$$^{56}_{26}Fe + ^{2}_{1}H \longrightarrow ^{54}_{25}Mn + ^{4}_{2}He$$

PRACTICE EXERCISE

Write a balanced equation for $^{106}_{46}Pd(\alpha,p)^{109}_{47}Ag$.

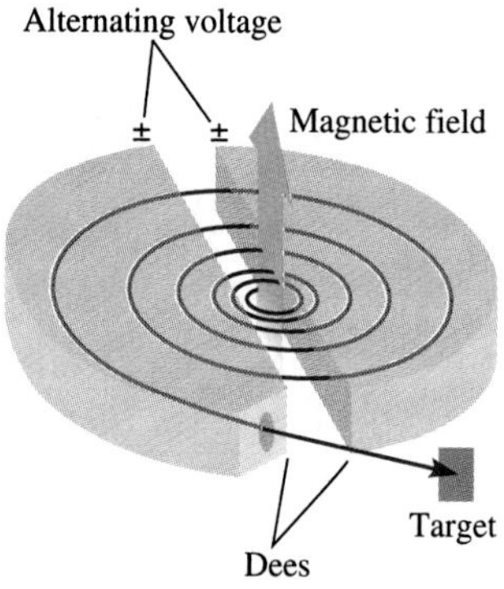

FIGURE 22.4
Schematic diagram of a cyclotron accelerator. The particle (an ion) to be accelerated starts at the center and is forced to move in a spiral path through the influence of electric and magnetic fields until it emerges at a high velocity. The magnetic fields are perpendicular to the plane of the dees, which are hollow and serve as electrodes.

Although light elements are generally not radioactive, they can be made so by bombarding their nuclei with appropriate particles. As we saw earlier, the radioactive carbon-14 isotope can be prepared by bombarding nitrogen-14 with neutrons. Another example is tritium, $^{3}_{1}H$, which is prepared according to the following bombardment:

$$^{6}_{3}Li + ^{1}_{0}n \longrightarrow ^{3}_{1}H + ^{4}_{2}He$$

Tritium decays with the emission of β particles:

$$^{3}_{1}H \longrightarrow ^{3}_{2}He + ^{0}_{-1}\beta \qquad t_{\frac{1}{2}} = 12.5 \text{ yr}$$

Many synthetic isotopes are prepared by using neutrons as projectiles. This approach is particularly convenient because neutrons carry no charges and therefore are not repelled by the targets—the nuclei. In contrast, when the projectiles are positively charged particles (for example, protons or α particles), they must have considerable kinetic energy in order to overcome the electrostatic repulsion between themselves and the target atoms. The synthesis of phosphorus from aluminum is one example:

$$^{27}_{13}Al + ^{4}_{2}He \longrightarrow ^{30}_{15}P + ^{1}_{0}n$$

In a *particle accelerator,* electric and magnetic fields are used to increase the kinetic energy of charged species so that a reaction will occur (Figure 22.4). Alternating the polarity (that is, + and −) on specially constructed plates causes the particles to accelerate along a spiral path. When they have the energy necessary to

FIGURE 22.5

A section of a linear particle accelerator.

carry out the desired nuclear reaction, they are guided out of the accelerator into a collision with a target substance.

Scientists have designed a variety of particle accelerators.The cyclotron shown in Figure 22.4 is one example. Another is a linear accelerator, which accelerates particles along a linear path of about 3 km (Figure 22.5). It is now possible to accelerate particles to a speed well above 90 percent of the speed of light. (According to Einstein's relativity theory, it is impossible for a particle to move *at* the speed of light. The only exception is the photon, which has a zero rest mass.) The extremely energetic particles produced in accelerators are employed by physicists to smash atomic nuclei to fragments. Studying the debris from such disintegrations provides valuable information about nuclear structure and binding forces.

THE TRANSURANIUM ELEMENTS

Particle accelerators made it possible to synthesize *elements with atomic numbers greater than 92,* called ***transuranium elements.*** Neptunium ($Z = 93$) was first prepared in 1940. Since then, 16 other transuranium elements have been synthesized. All isotopes of these elements are radioactive. Table 22.4 lists the transuranium elements and the reactions through which they are formed.

22.5 NUCLEAR FISSION

In ***nuclear fission*** *a heavy nucleus (mass number >200) divides to form smaller nuclei of intermediate mass and one or more neutrons.* Because the heavy nucleus is less stable than its products (see Figure 22.2), this process releases a large amount of energy.

TABLE 22.4
The Transuranium Elements

Atomic number	Name	Symbol	Preparation
93	Neptunium	Np	${}^{238}_{92}\text{U} + {}^{1}_{0}\text{n} \longrightarrow {}^{239}_{93}\text{Np} + {}^{0}_{-1}\beta$
94	Plutonium	Pu	${}^{239}_{93}\text{Np} \longrightarrow {}^{239}_{94}\text{Pu} + {}^{0}_{-1}\beta$
95	Americium	Am	${}^{239}_{94}\text{Pu} + {}^{1}_{0}\text{n} \longrightarrow {}^{240}_{95}\text{Am} + {}^{0}_{-1}\beta$
96	Curium	Cm	${}^{239}_{94}\text{Pu} + {}^{4}_{2}\text{He} \longrightarrow {}^{242}_{96}\text{Cm} + {}^{1}_{0}\text{n}$
97	Berkelium	Bk	${}^{241}_{95}\text{Am} + {}^{4}_{2}\text{He} \longrightarrow {}^{243}_{97}\text{Bk} + 2{}^{1}_{0}\text{n}$
98	Californium	Cf	${}^{242}_{96}\text{Cm} + {}^{4}_{2}\text{He} \longrightarrow {}^{245}_{98}\text{Cf} + {}^{1}_{0}\text{n}$
99	Einsteinium	Es	${}^{238}_{92}\text{U} + 15\,{}^{1}_{0}\text{n} \longrightarrow {}^{253}_{99}\text{Es} + 7\,{}^{0}_{-1}\beta$
100	Fermium	Fm	${}^{238}_{92}\text{U} + 17\,{}^{1}_{0}\text{n} \longrightarrow {}^{255}_{100}\text{Fm} + 8\,{}^{0}_{-1}\beta$
101	Mendelevium	Md	${}^{253}_{99}\text{Es} + {}^{4}_{2}\text{He} \longrightarrow {}^{256}_{101}\text{Md} + {}^{1}_{0}\text{n}$
102	Nobelium	No	${}^{246}_{96}\text{Cm} + {}^{12}_{6}\text{C} \longrightarrow {}^{254}_{102}\text{No} + 4\,{}^{1}_{0}\text{n}$
103	Lawrencium	Lr	${}^{252}_{98}\text{Cf} + {}^{10}_{5}\text{B} \longrightarrow {}^{257}_{103}\text{Lr} + 5\,{}^{1}_{0}\text{n}$
104	Unnilquadium	Unq	${}^{249}_{98}\text{Cf} + {}^{12}_{6}\text{C} \longrightarrow {}^{257}_{104}\text{Unq} + 4\,{}^{1}_{0}\text{n}$
105	Unnilpentium	Unp	${}^{249}_{98}\text{Cf} + {}^{15}_{7}\text{N} \longrightarrow {}^{260}_{105}\text{Unp} + 4\,{}^{1}_{0}\text{n}$
106	Unnilhexium	Unh	${}^{249}_{98}\text{Cf} + {}^{18}_{8}\text{O} \longrightarrow {}^{263}_{106}\text{Unh} + 4\,{}^{1}_{0}\text{n}$
107	Unnilseptium	Uns	${}^{209}_{83}\text{Bi} + {}^{54}_{24}\text{Cr} \longrightarrow {}^{262}_{107}\text{Uns} + {}^{1}_{0}\text{n}$
108	Unniloctium	Uno	${}^{208}_{82}\text{Pb} + {}^{58}_{26}\text{Fe} \longrightarrow {}^{265}_{108}\text{Uno} + {}^{1}_{0}\text{n}$
109	Unnilennium	Une	${}^{209}_{83}\text{Bi} + {}^{58}_{26}\text{Fe} \longrightarrow {}^{266}_{109}\text{Une} + {}^{1}_{0}\text{n}$

The first nuclear fission reaction studied was that of uranium-235 bombarded with slow neutrons, whose speed is comparable to that of air molecules at room temperature. Under these conditions, uranium-235 undergoes fission, as shown in Figure 22.6. Actually, this reaction is very complex: More than 30 different elements have been found among the fission products. A representative reaction is

$${}^{235}_{92}\text{U} + {}^{1}_{0}\text{n} \longrightarrow {}^{90}_{38}\text{Sr} + {}^{143}_{54}\text{Xe} + 3\,{}^{1}_{0}\text{n}$$

Although many heavy nuclei can be made to undergo fission, only the fission of naturally occurring uranium-235 and of the artificial plutonium-239 has any practical importance. As Figure 22.2 shows, the binding energy per nucleon for

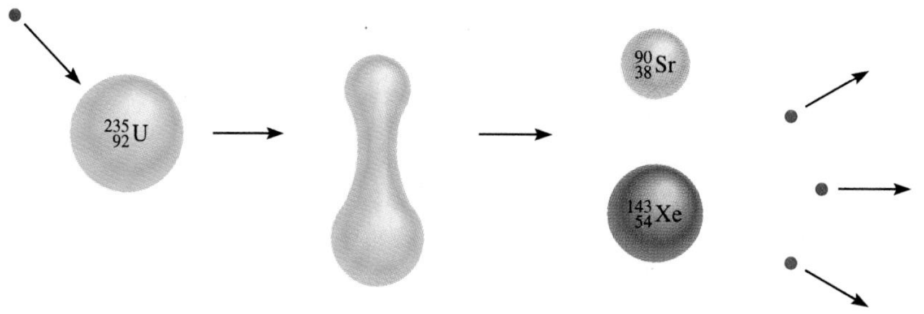

FIGURE 22.6

Nuclear fission of U-235. When a U-235 nucleus captures a neutron (red sphere), it undergoes fission to yield two smaller nuclei. On the average, three neutrons are emitted for every U-235 nucleus that divides.

uranium-235 is less than the sum of the binding energies for strontium-90 and xenon-143. Therefore, when a uranium-235 nucleus is split into two smaller nuclei, a certain amount of energy is released. Let us estimate the magnitude of this energy, using the preceding reaction as an example. Table 22.5 shows the nuclear binding energies of uranium-235 and its fission products. The difference between the binding energies of the reactants and products is $(1.23 \times 10^{-10} + 1.92 \times 10^{-10})$ J − (2.82×10^{-10}) J, or 3.3×10^{-11} J per uranium-235 nucleus. For 1 mole of uranium-235, the energy released would be $(3.3 \times 10^{-11})(6.02 \times 10^{23})$, or 2.0×10^{13} J. This is an extremely exothermic reaction, considering that the heat of combustion of 1 ton of coal is only about 8×10^{7} J.

TABLE 22.5
Nuclear Binding Energies of ^{235}U and Its Fission Products

	Nuclear binding energy
^{235}U	2.82×10^{-10} J
^{90}Sr	1.23×10^{-10} J
^{143}Xe	1.92×10^{-10} J

The significant feature of uranium-235 fission is not just the enormous amount of energy released, but the fact that more neutrons are produced than are originally captured in the process. This property makes possible a ***nuclear chain reaction,*** which is *a self-sustaining sequence of nuclear fission reactions.* The neutrons generated during the initial stages of fission can induce fission in other uranium-235 nuclei, which in turn produce more neutrons, and so on. In less than a second, the reaction can become uncontrollable, liberating a tremendous amount of heat to the surroundings.

Figure 22.7 shows two types of fission reactions. For a chain reaction to occur, enough uranium-235 must be present in the sample to capture the neutrons. Otherwise, many of the neutrons will escape from the sample and the chain reaction will not occur [Figure 22.7(a)]. In this situation the mass of the sample is said to be *subcritical.* Figure 22.7(b) shows what happens when the amount of the fissionable material is equal to or greater than the ***critical mass,*** *the minimum mass of fissionable material required to generate a self-sustaining nuclear chain reaction.* In this case most of the neutrons will be captured by uranium-235 and a chain reaction will occur.

THE ATOMIC BOMB

The first application of nuclear fission was in the development of the atomic bomb. How is such a bomb made and detonated? The crucial factor in the bomb's design is the determination of the critical mass for the bomb. A small atomic bomb is equivalent to 20,000 tons of TNT. Since 1 ton of TNT releases about 4×10^{9} J of

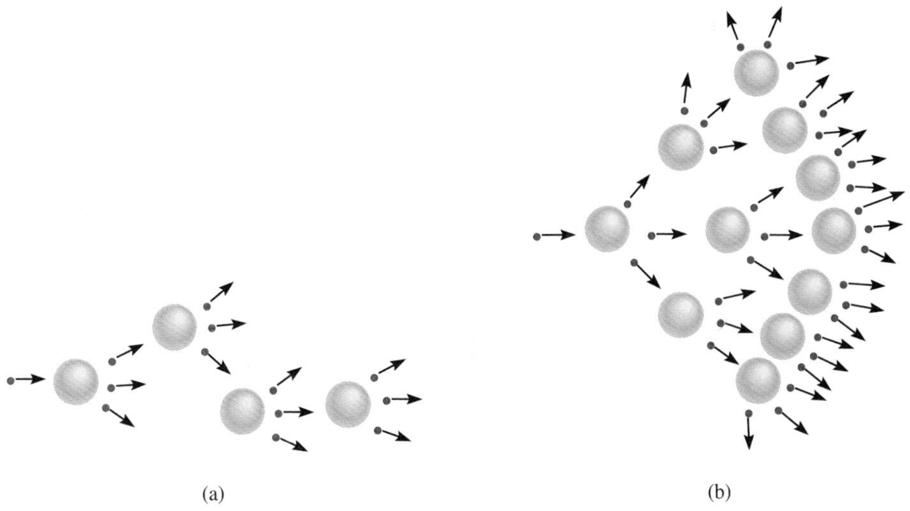

FIGURE 22.7
Two types of nuclear fission. (a) If the mass of U-235 is subcritical, no chain reaction will result. Many of the neutrons produced will escape to the surroundings. (b) If the mass is critical, most of the neutrons will be captured by U-235 nuclei and an uncontrollable chain reaction will occur.

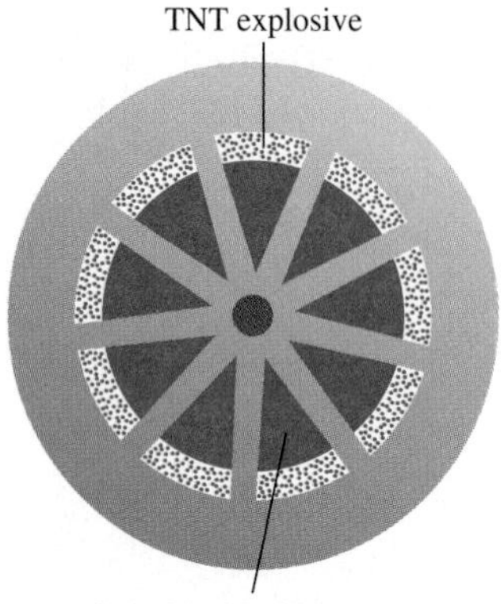

FIGURE 22.8

Schematic cross section of an atomic bomb. The TNT explosives are set off first. The explosion forces the sections of fissionable material together to form an amount considerably larger than the critical mass.

energy, 20,000 tons would produce 8×10^{13} J. Earlier we saw that 1 mole, or 235 g, of uranium-235 liberates 2.0×10^{13} J of energy when it undergoes fission. Thus the mass of the isotope present in a small bomb must be at least

$$235 \text{ g} \times \frac{8 \times 10^{13} \text{ J}}{2.0 \times 10^{13} \text{ J}} \simeq 1 \times 10^3 \text{ g} = 1 \text{ kg}$$

For obvious reasons, an atomic bomb is never assembled with the critical mass already present. Instead, the critical mass is formed by using a conventional explosive, such as TNT, to force the fissionable sections together, as shown in Figure 22.8. Neutrons from a source at the center of the device trigger the nuclear chain reaction. Uranium-235 was the fissionable material in the bomb dropped on Hiroshima, Japan, on August 6, 1945. Plutonium-239 was used in the bomb exploded over Nagasaki three days later. The fission reactions generated were similar in these two cases, as was the extent of the destruction.

Nuclear Reactors

A peaceful but controversial application of nuclear fission is the generation of electricity using heat from a controlled chain reaction in a nuclear reactor (Figure 22.9). Currently, nuclear reactors provide about 8 percent of the electrical energy in the United States. This is a small but by no means negligible contribution to the nation's energy production. Several different types of nuclear reactors are in operation; we will briefly discuss the main features of three of them, along with their advantages and disadvantages.

FIGURE 22.9

Schematic diagram of a nuclear fission reactor. The fission process is controlled by cadmium or boron rods. The heat generated by the process is used to produce steam for the generation of electricity via a heat exchange system.

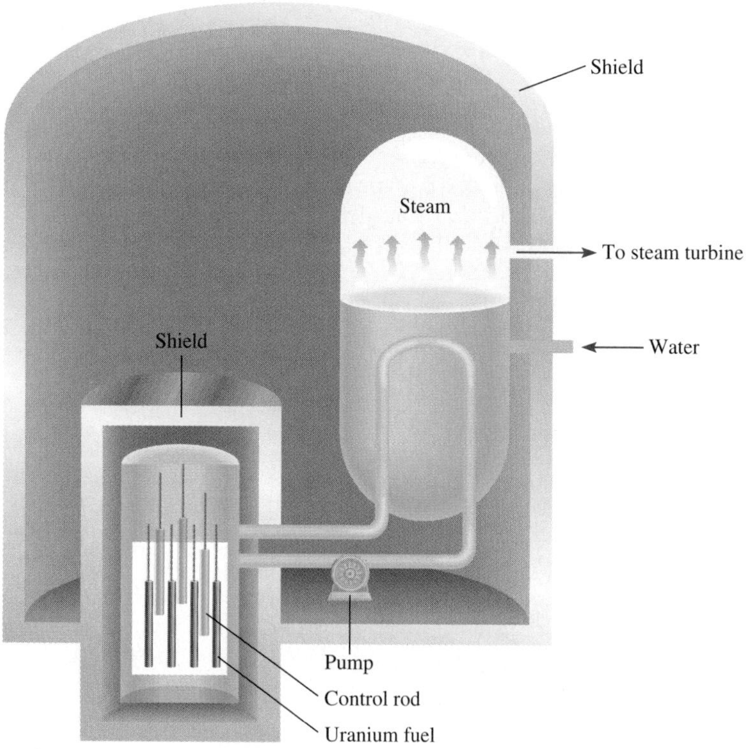

Light Water Reactors

Most of the nuclear reactors in the United States are *light water reactors.* Figure 22.9 is a schematic diagram of such a reactor, and Figure 22.10 shows the refueling process in the core of a nuclear reactor.

An important aspect of the fission process is the speed of the neutrons. Slow neutrons split uranium-235 nuclei more efficiently than do fast ones. Because fission reactions are so exothermic, the neutrons produced usually move at high velocities. For greater efficiency they must be slowed down before they can be used to induce nuclear disintegration. To accomplish this goal, scientists use ***moderators,*** which are *substances that can reduce the kinetic energy of neutrons.* A good moderator must satisfy several requirements: It must be a fluid so it can be used also as a coolant; it should possess a high specific heat; it should be nontoxic and inexpensive (as very large quantities of it are necessary); and it should resist conversion into a radioactive substance by neutron bombardment. No substance fits all these requirements, although water comes closer than many others that have been considered. Nuclear reactors using light water (H_2O) as a moderator are called light water reactors because ^1_1H is the lightest isotope of the element hydrogen.

FIGURE 22.10

Refueling the core of a nuclear reactor.

The nuclear fuel consists of uranium, usually in the form of its oxide, U_3O_8 (Figure 22.11). Naturally occurring uranium contains about 0.7 percent of the uranium-235 isotope, which is too low a concentration to sustain a small-scale chain reaction. For effective operation of a light water reactor, uranium-235 must be enriched to a concentration of 3 or 4 percent. In principle, the main difference between an atomic bomb and a nuclear reactor is that the chain reaction that takes place in a nuclear reactor is kept under control at all times. The factor limiting the rate of the reaction is the number of neutrons present. This can be controlled by lowering cadmium or boron rods between the fuel elements. These rods capture neutrons according to the equations

$$^{113}_{48}\text{Cd} + ^1_0\text{n} \longrightarrow ^{114}_{48}\text{Cd} + \gamma$$

$$^{10}_{5}\text{B} + ^1_0\text{n} \longrightarrow ^{7}_{3}\text{Li} + ^4_2\text{He}$$

where γ denotes gamma rays. Without the control rods the heat generated would melt down the reactor core, releasing radioactive materials into the environment.

FIGURE 22.11

Uranium oxide, U_3O_8.

Nuclear reactors have rather elaborate cooling systems that absorb the heat given off by the nuclear reaction and transfer it outside the reactor core, where it is used to produce enough steam to drive an electric generator. In this respect, a nuclear power plant is similar to a conventional power plant that burns fossil fuel. In both cases, large quantities of cooling water are needed to condense steam for reuse. Thus most nuclear power plants are built near a river or a lake. Unfortunately, this method of cooling causes thermal pollution (see Section 12.4).

Heavy Water Reactors

Another type of nuclear reactor uses D_2O, or heavy water, as the moderator, rather than H_2O. Heavy water slows down the neutrons emerging from the fission reaction less efficiently than does light water. Consequently, there is no need to use enriched uranium for fission in a *heavy water reactor.* The faster-moving neutrons travel greater distances, and so the probability that they will strike the proper targets—uranium-235 isotopes—is correspondingly high. Eventually most of the uranium-235 isotopes will take part in the fission process.

The main advantage of a heavy water reactor is that it eliminates the need for building expensive uranium enrichment facilities. However, D_2O must be prepared by either fractional distillation or electrolysis of ordinary water, which can be very expensive, considering the quantity of water used in a nuclear reactor. In countries where hydroelectric power is abundant, the cost of producing D_2O by electrolysis can be reasonably low. At present, Canada is the only nation successfully using heavy water nuclear reactors. The fact that no enriched uranium is required in a heavy water reactor allows a country to enjoy the benefits of nuclear power without undertaking work that is closely associated with weapons technology.

Breeder Reactors

A ***breeder reactor*** *produces more fissionable materials than it uses.* We know that when uranium-238 is bombarded with fast neutrons, the following reactions take place:

$$^{238}_{92}U + ^{1}_{0}n \longrightarrow ^{239}_{92}U$$
$$^{239}_{92}U \longrightarrow ^{239}_{93}Np + ^{0}_{-1}\beta \qquad t_{\frac{1}{2}} = 23.4 \text{ min}$$
$$^{239}_{93}Np \longrightarrow ^{239}_{94}Pu + ^{0}_{-1}\beta \qquad t_{\frac{1}{2}} = 2.35 \text{ days}$$

In this manner the nonfissionable uranium-238 is transmuted into the fissionable isotope plutonium-239 (Figure 22.12).

It turns out that more neutrons are generated when fission is induced by fast neutrons, with the most neutrons obtained from fast-neutron fission in Pu-239. In a typical breeder reactor, nuclear fuel containing plutonium-239 is mixed with uranium-238 so that breeding takes place within the core, and it has no moderator to slow the neutrons. Since fission is less likely with fast neutrons, a breeder reactor needs more highly enriched fuel than ordinary, slow-neutron reactors. During the operation of a breeder reactor, for every uranium-235 (or plutonium-239) nucleus undergoing fission, more than one neutron is captured by uranium-238 to generate plutonium-239. Thus the stockpile of fissionable material can be steadily increased as the starting nuclear fuels are consumed. It takes about 7 to 10 years to regenerate the sizable amount of material needed to refuel the original reactor and to fuel another reactor of comparable size. This interval is called the *doubling time.*

Another fertile isotope is $^{232}_{90}Th$. Upon capturing slow neutrons, thorium is transmuted to uranium-233, which, like uranium-235, is a fissionable isotope:

$$^{232}_{90}Th + ^{1}_{0}n \longrightarrow ^{233}_{90}Th$$
$$^{233}_{90}Th \longrightarrow ^{233}_{91}Pa + ^{0}_{-1}\beta \qquad t_{\frac{1}{2}} = 22 \text{ min}$$
$$^{233}_{91}Pa \longrightarrow ^{233}_{92}U + ^{0}_{-1}\beta \qquad t_{\frac{1}{2}} = 27.4 \text{ days}$$

Uranium-233 is stable enough for long-term storage. The amounts of uranium-238 and thorium-232 in Earth's crust are relatively plentiful (4 ppm and 12 ppm by mass, respectively).

Despite the promising prospects, the development of breeder reactors has been very slow. To date, the United States does not have a single operating breeder reactor, and only a few have been built in other countries, such as France and Russia. One problem is economics; breeder reactors are more expensive to build than conventional reactors. There are also more technical difficulties associated with the construction of such reactors. As a result, the future of breeder reactors, in the United States at least, is rather uncertain.

FIGURE 22.12

The red glow of the radioactive plutonium-239 isotope. The orange color is due to the presence of its oxide.

Hazards of Nuclear Energy

Many people, including environmentalists, regard nuclear fission as a highly undesirable method of energy production. Many fission products such as strontium-90 are dangerous radioactive isotopes with long half-lives. Plutonium-239, used as a nuclear fuel and produced in breeder reactors, is one of the most toxic substances known. It is an α emitter with a half-life of 24,400 yr.

Accidents, too, present many dangers. The accident at the Three Mile Island reactor in Pennsylvania in 1979 first brought the potential hazards of nuclear plants to public attention. In this instance very little radiation escaped the reactor, but the plant remained closed for more than a decade while repairs were made and safety issues addressed. Only a few years later, the disaster at the Chernobyl nuclear plant in Belarus, on April 26, 1986, was a tragic reminder of just how catastrophic a runaway nuclear reaction can be. On that day, a reactor at the plant surged out of control. The fire and explosion that followed released much radioactive material into the environment. People working near the plant died within weeks as a result of the exposure to the intense radiation. The long-term effect of the radioactive fallout from this incident has not yet been clearly assessed, although agriculture and dairy farming were affected by the fallout. The number of potential cancer deaths attributable to the radiation contamination is estimated to be between a few thousand and more than 100,000.

Furthermore, the problem of radioactive waste disposal has not been satisfactorily resolved even for safely operated nuclear plants. Many suggestions have been made for the storage or disposal of nuclear waste, including burial underground, burial beneath the ocean floor, and storage in deep geologic formations. But none of these sites has proved absolutely safe in the long run. Leakage of radioactive wastes into underground water, for example, can endanger nearby communities. The ideal disposal site would seem to be the sun, where a bit more radiation would make little difference, but this kind of operation requires 100 percent reliability in space technology.

Because of the hazards, what was once hailed as the solution to our future energy needs is now being debated by scientists and the general public alike. The controversy surrounding nuclear reactors seems likely to continue for some time.

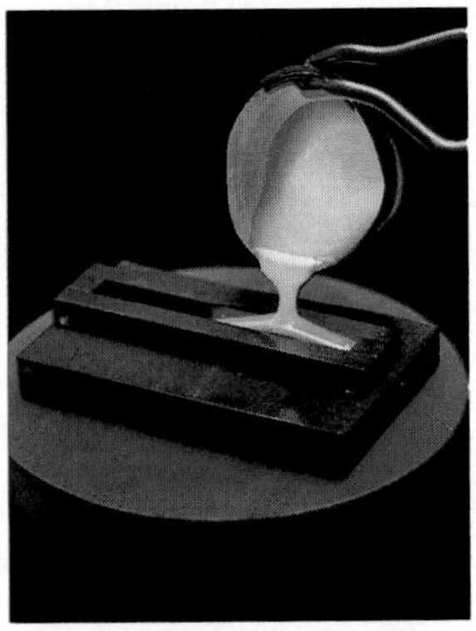

Molten glass is poured over nuclear waste before burial.

22.6 NUCLEAR FUSION

In contrast to the nuclear fission process, ***nuclear fusion,*** *the combining of small nuclei into larger ones,* is largely exempt from the waste disposal problem.

Figure 22.2 showed that for the lightest elements, nuclear stability increases with increasing mass number. This behavior suggests that if two light nuclei combine or fuse together to form a larger, more stable nucleus, an appreciable amount of energy will be released in the process. This is the basis for ongoing research into the harnessing of nuclear fusion for the production of energy.

Nuclear fusion occurs constantly in the sun (Figure 22.13). The sun is made up mostly of hydrogen and helium. In its interior, where temperatures reach about 30 million degrees Celsius, the following fusion reactions are believed to take place:

$$^{1}_{1}\text{H} + {}^{2}_{1}\text{H} \longrightarrow {}^{3}_{2}\text{He}$$

$$^{3}_{2}\text{He} + {}^{3}_{2}\text{He} \longrightarrow {}^{4}_{2}\text{He} + 2\,{}^{1}_{1}\text{H}$$

$$^{1}_{1}\text{H} + {}^{1}_{1}\text{H} \longrightarrow {}^{2}_{1}\text{H} + {}^{0}_{+1}\beta$$

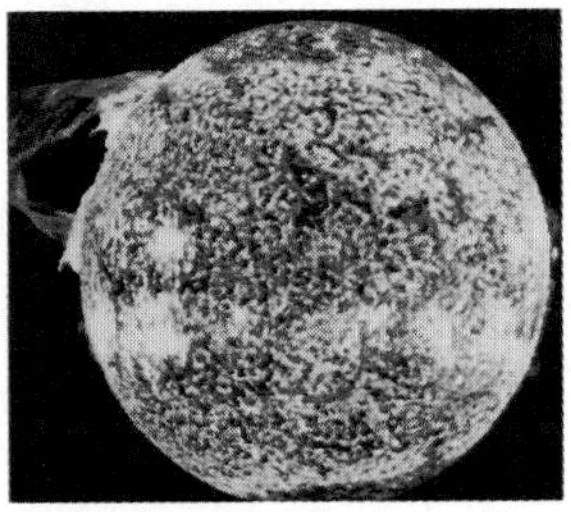

FIGURE 22.13

Nuclear fusion keeps the temperature in the interior of the sun at about 30 million °C.

Because *fusion reactions take place only at very high temperatures,* they are often called ***thermonuclear reactions.***

Fusion Reactors

A major concern in choosing the proper nuclear fusion process for energy production is the temperature necessary to carry out the process. Some promising reactions are

Reaction	*Energy released*
${}^{2}_{1}H + {}^{2}_{1}H \longrightarrow {}^{3}_{1}H + {}^{1}_{1}H$	6.3×10^{-13} J
${}^{2}_{1}H + {}^{3}_{1}H \longrightarrow {}^{4}_{2}He + {}^{1}_{0}n$	2.8×10^{-12} J
${}^{6}_{3}Li + {}^{2}_{1}H \longrightarrow 2\,{}^{4}_{2}He$	3.6×10^{-12} J

These reactions take place at extremely high temperatures, of the order of 100 million degrees Celsius, to overcome the repulsive forces between the nuclei. The first reaction is particularly attractive because the world's supply of deuterium is virtually inexhaustible: The total volume of water on Earth is about 1.5×10^{21} L. Since the natural abundance of deuterium is 1.5×10^{-2} percent, the total amount of deuterium present is roughly 4.5×10^{21} g, or 5.0×10^{15} tons. The cost of preparing deuterium is minimal compared with the value of the energy released by the reaction.

In contrast to the fission process, nuclear fusion looks like a very promising energy source, at least "on paper." Although thermal pollution would be a problem, fusion has the following advantages: (1) The fuels are cheap and almost inexhaustible and (2) the process produces little radioactive waste. If a fusion machine were turned off, it would shut down completely and instantly, without any danger of a meltdown.

If nuclear fusion is so great, why isn't there even one fusion reactor producing energy? Although we command the scientific knowledge to design such a reactor, the technical difficulties have not yet been solved. The basic problem is finding a way to hold the nuclei together long enough, and at the appropriate temperature, for fusion to occur. At temperatures of about 100 million degrees Celsius, molecules cannot exist, and most or all of the atoms are stripped of their electrons. This *state of matter, a gaseous mixture of positive ions and electrons,* is called ***plasma.*** The problem of containing this plasma is a formidable one. What solid container can exist at such temperatures? None, unless the amount of plasma is small; but then the solid surface would immediately cool the sample and quench the fusion reaction. One approach to solving this problem is to use *magnetic confinement.* Since a plasma consists of charged particles moving at high speeds, a magnetic field will exert force on it. As Figure 22.14 shows, the plasma moves through a doughnut-shaped tunnel, confined by a complex magnetic field. Thus the plasma never comes in contact with the walls of the container.

Another promising development employs high-power lasers to initiate the fusion reaction. In test runs a number of laser beams transfer energy to a small fuel pellet, heating it and causing it to *implode,* that is, to collapse inward from all sides and compress into a small volume (Figure 22.15). Consequently, fusion occurs. Like the magnetic confinement approach, laser fusion presents a number of technical difficulties that still need to be overcome before it can be put to practical use on a large scale.

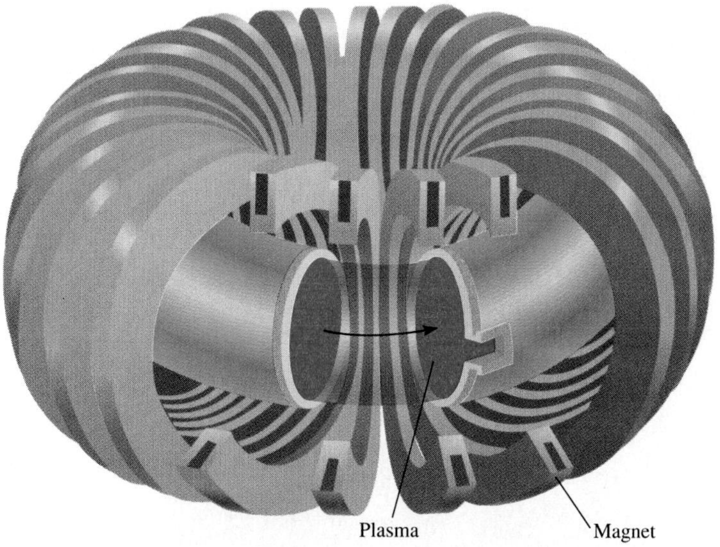

FIGURE 22.14

A magnetic confinement design called tokamak.

FIGURE 22.15

This small-scale fusion reaction was created at the Lawrence Livermore National Laboratory using the world's most powerful laser, Nova.

THE HYDROGEN BOMB

The technical problems inherent in the design of a nuclear fusion reactor do not affect the production of a hydrogen bomb, also called a thermonuclear bomb. In this case the objective is all power and no control. Hydrogen bombs do not contain gaseous hydrogen or gaseous deuterium; they contain solid lithium deuteride (LiD), which can be packed very tightly. The detonation of a hydrogen bomb occurs in two stages—first a fission reaction and then a fusion reaction. The required temperature for fusion is derived from an atomic bomb. Immediately after the atomic bomb explodes, the following fusions occur, releasing vast amounts of energy (Figure 22.16):

$$^{6}_{3}\text{Li} + ^{2}_{1}\text{H} \longrightarrow 2\,^{4}_{2}\text{He}$$

$$^{2}_{1}\text{H} + ^{2}_{1}\text{H} \longrightarrow ^{3}_{1}\text{H} + ^{1}_{1}\text{H}$$

FIGURE 22.16

Explosion of a thermonuclear bomb.

There is no critical mass in a fusion bomb, and the force of explosion is limited only by the quantity of reactants present. Thermonuclear bombs are described as being "cleaner" than atomic bombs because they do not produce radioactive isotopes except for tritium, which is a weak β-particle emitter ($t_{\frac{1}{2}} = 12.5$ yr), and the products from the fission starter. Their damaging effects on the environment can be aggravated, however, by incorporating in the construction some nonfissionable material such as cobalt. Upon bombardment by neutrons, cobalt-59 is converted to cobalt-60, which is a very strong γ-ray emitter with a half-life of 5.2 yr. The presence of these radioactive cobalt isotopes in the debris or fallout from a thermonuclear explosion would be fatal to those who survived the initial blast.

22.7 USES OF ISOTOPES

Radioactive and stable isotopes alike have many applications in science and medicine. We have previously described the use of radioactive isotopes in dating (see Section 22.3). In this section we discuss a few more examples.

DETERMINATION OF MOLECULAR STRUCTURE

The formula of the thiosulfate ion is $S_2O_3^{2-}$. For some years chemists were uncertain about whether the two sulfur atoms occupied equivalent positions in the ion. The thiosulfate ion is prepared by treatment of the sulfite ion with elemental sulfur:

$$SO_3^{2-}(aq) + S(s) \longrightarrow S_2O_3^{2-}(aq) \qquad (22.1)$$

When thiosulfate is treated with dilute acid, the reaction is reversed. The sulfite ion is re-formed and elemental sulfur precipitates:

$$S_2O_3^{2-}(aq) \xrightarrow{H^+} SO_3^{2-}(aq) + S(s) \qquad (22.2)$$

If this sequence is started with elemental sulfur enriched in the radioactive sulfur-

35 isotope, the isotope acts as a "label" for S atoms. All the labels are found in the sulfur precipitate in Equation (22.2); none of them appears in the final sulfite ions. Clearly, then, the two atoms in sulfur in $S_2O_3^{2-}$ are not structurally equivalent, as would be the case if the structure were

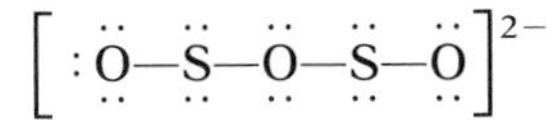

Otherwise, the radioactive isotope would be present in both the elemental sulfur precipitate and the sulfite ion. Based on spectroscopic studies, we know that the structure of the thiosulfate ion is

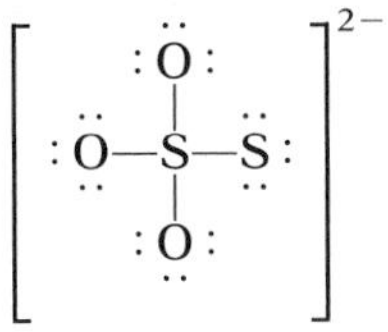

Study of Photosynthesis

The study of photosynthesis is also rich with isotope applications. The overall photosynthesis reaction can be represented as

$$6CO_2 + 6H_2O \longrightarrow C_6H_{12}O_6 + 6O_2$$

The radioactive carbon-14 isotope helped to determine the path of carbon in photosynthesis. Starting with $^{14}CO_2$, it was possible to isolate the intermediate products during photosynthesis and measure the amount of radioactivity of each carbon-containing compound. In this manner the path from CO_2 through various intermediate compounds to carbohydrate could be clearly charted. *Isotopes, especially radioactive isotopes that are used to trace the path of the atoms of an element in a chemical or biological process,* are called ***tracers.***

Photosynthesis takes place in the green leaves of plants.

Isotopes in Medicine

Tracers are used also for diagnosis in medicine. Sodium-24 (a β emitter with a half-life of 14.8 h) injected into the bloodstream as a salt solution can be monitored to trace the flow of blood and detect possible constrictions or obstructions in the circulatory system. Iodine-131 (a β emitter with a half-life of 8 days) has been used to test the activity of the thyroid gland. A malfunctioning thyroid can be detected by giving the patient a drink of a solution containing a known amount of $Na^{131}I$ and measuring the radioactivity just above the thyroid to see if the iodine is absorbed at the normal rate. Of course, the amounts of radioisotope used in the human body must always be kept small; otherwise, the patient might suffer permanent damage from the high-energy radiation. Another radioactive isotope of iodine, iodine-123 (a γ-ray emitter), is used to image the brain (Figure 22.17).

Technetium is one of the most useful elements in nuclear medicine. Although technetium is a transition metal, all its isotopes are radioactive. Therefore, technetium does not occur naturally on Earth. In the laboratory it is prepared by the nuclear reactions

$$^{98}_{42}Mo + ^{1}_{0}n \longrightarrow ^{99}_{42}Mo$$

$$^{99}_{42}Mo \longrightarrow ^{99m}_{43}Tc + ^{0}_{-1}\beta$$

where the superscript m denotes that the technetium-99 isotope is produced in its

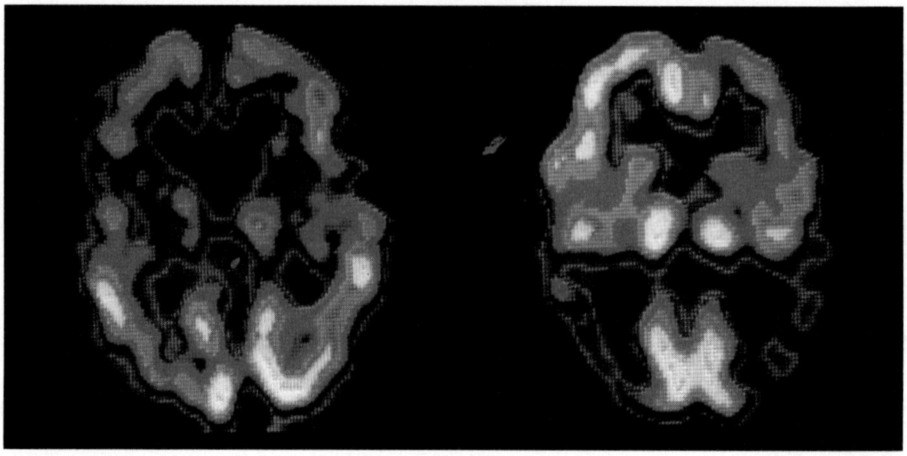

FIGURE 22.17

A compound labeled with iodine-123 is used to image the brain. (Left) A normal brain. (Right) The brain of a patient with Alzheimer's disease.

excited nuclear state. This isotope has a half-life of about 6 h, decaying by γ radiation to technetium-99 in its nuclear ground state. Thus it is a valuable diagnostic tool. The patient either drinks or is injected with a solution containing ^{99m}Tc. By detecting the γ rays emitted by ^{99m}Tc, doctors can obtain images of organs such as the heart, liver, and lungs.

A major advantage of using radioactive isotopes as tracers is that they are easy to detect. Their presence even in very small amounts can be detected by photographic techniques or by devices known as counters. Figure 22.18 is a diagram of a Geiger counter, an instrument widely used in scientific work and medical laboratories to detect radiation.

22.8 BIOLOGICAL EFFECTS OF RADIATION

In this section we will examine briefly the effects of radiation on biological systems. But first let us define quantitative measures of radiation. The fundamental unit of radioactivity is the *curie* (Ci); 1 Ci corresponds to exactly 3.70×10^{10} nu-

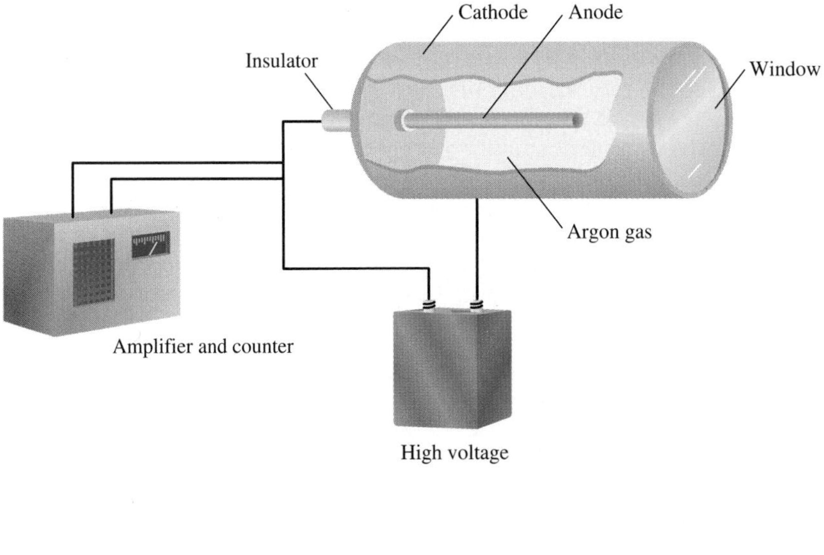

FIGURE 22.18

Schematic diagram of a Geiger counter. Radiation (α, β, or γ rays) entering through the window causes ionization of the argon gas, which allows a small current to flow between the electrodes. This current is amplified and is used to flash a light or operate a counter with a clicking sound.

clear disintegrations per second. This decay rate is equivalent to that of 1 g of radium. A *millicurie* (mCi) is one-thousandth of a curie. Thus, 10 mCi of a carbon-14 sample is the quantity that undergoes

$$(10 \times 10^{-3})(3.70 \times 10^{10}) = 3.70 \times 10^{8}$$

disintegrations per second. The intensity of radiation depends on the number of disintegrations as well as on the energy and type of radiation emitted. One common unit for the absorbed dose of radiation is the *rad* (*r*adiation *a*bsorbed *d*ose), which is the amount of radiation that results in the absorption of 1×10^{-5} J per gram of irradiated material. The biological effect of radiation depends on the part of the body irradiated and the type of radiation. For this reason the rad is often multiplied by a factor called *RBE* (*r*elative *b*iological *e*ffectiveness). The product is called a *rem* (*r*oentgen *e*quivalent for *m*an):

$$1 \text{ rem} = 1 \text{ rad} \times 1 \text{ RBE}$$

Of the three types of nuclear radiation, α particles usually have the least penetrating power. Beta particles are more penetrating than α particles but less so than γ rays. Gamma rays have very short wavelengths and high energies. Furthermore, since they carry no charge, they cannot be stopped by shielding materials as easily as α and β particles. However, if α or β emitters are ingested, their damaging effects are greatly aggravated because the organs will be constantly subject to damaging radiation at close range. For example, strontium-90, a β emitter, can replace calcium in bones, where it does the greatest damage.

Table 22.6 lists the average amounts of radiation an American receives every year. It should be pointed out that for short-term exposures to radiation, a dosage of 50–200 rem will cause a decrease in white blood cell counts and other complications while a dosage of 500 rem or greater will result in death within weeks.

The chemical basis of radiation damage is that of ionizing radiation. Radiation of either particles or γ rays can remove electrons from atoms and molecules in its path, leading to the formation of ions and radicals. For example, when water is irradiated with γ rays, the following reactions take place:

$$H_2O \xrightarrow{\text{radiation}} H_2O^+ + e^-$$

$$H_2O^+ + H_2O \longrightarrow H_3O^+ + \underset{\text{hydroxyl radical}}{\cdot OH}$$

TABLE 22.6
Average Yearly Radiation Doses for Americans

Source	Dose (mrem/yr)*
Cosmic rays	20–50
Ground and surroundings	25
Human body†	26
Medical and dental X rays	50–75
Air travel	5
Fallout from weapons tests	5
Nuclear waste	2
Total	133–188

*1 mrem = 1 millirem = 1×10^{-3} rem.
†The radioactivity in the body comes from food and air.

The electron (in the hydrated form) can subsequently react with water or with a hydrogen ion to form atomic hydrogen, and with oxygen to produce the superoxide ion, O_2^- (a radical):

$$e^- + O_2 \longrightarrow \cdot O_2^-$$

In the tissues the superoxide ions and other free radicals attack cell membranes and a host of organic compounds, such as enzymes and DNA molecules. Organic compounds can themselves be directly ionized and destroyed by high-energy radiation.

It has long been known that exposure to high-energy radiation can induce cancer in humans and other animals. Cancer is characterized by uncontrolled cellular growth. On the other hand, it is also well established that cancer cells can be destroyed by proper radiation treatment. In radiation therapy, a compromise is sought. The radiation to which the patient is exposed must be sufficient to destroy cancer cells without killing too many normal cells and, it is hoped, without inducing another form of cancer.

Radiation damage to living systems is generally classified as *somatic* or *genetic*. Somatic injuries are those that affect the organism during its own lifetime. Sunburn, skin rash, cancer, and cataracts are examples of somatic damage. Genetic damage means inheritable changes or gene mutations. For example, a person whose chromosomes have been damaged or altered by radiation may have deformed offspring.

SUMMARY

Nuclear chemistry is the study of changes in the atomic nuclei. Such changes are termed nuclear reactions. Radioactive decay and nuclear transmutation are nuclear reactions.

For stable nuclei of low atomic number, the neutron-to-proton ratio is close to 1. For heavier stable nuclei, the ratio becomes greater than 1. All nuclei with 84 or more protons are unstable and radioactive. Nuclei with even atomic numbers are more stable than those with odd atomic numbers. A quantitative measure of nuclear stability is the nuclear binding energy, which can be calculated from a knowledge of the mass defect of the nucleus.

Radioactive nuclei emit α particles, β particles, positrons, or γ rays. The equation for a nuclear reaction includes the particles emitted, and both the mass numbers and the atomic numbers must balance. Uranium-238 is the parent of a natural radioactive decay series. A number of radioactive isotopes, such as U-238 and C-14, can be used to date objects. Artificially radioactive elements are created by the bombardment of other elements by accelerated neutrons, protons, or α particles.

Nuclear fission is the splitting of a large nucleus into two smaller nuclei plus neutrons. When these neutrons are captured efficiently by other nuclei, an uncontrollable chain reaction can occur. Nuclear reactions use the heat from a controlled nuclear fission reaction to produce power. The three important types of reactors are light water reactors, heavy water reactors, and breeder reactors.

Nuclear fusion, the type of reaction that occurs in the sun, is the combination of two light nuclei to form one heavy nucleus. Fusion takes place only at very high temperatures—so high that controlled large-scale nuclear fusion has so far not been achieved.

Radioactive isotopes are easy to detect and thus make excellent tracers in chemical reactions and in medical practice. High-energy radiation damages living systems by causing ionization and the formation of free radicals.

KEY WORDS

Breeder reactor, p. 652
Critical mass, p. 649
Decay series, p. 642
Mass defect, p. 640
Moderators, p. 651
Nuclear binding energy, p. 639
Nuclear chain reaction, p. 649
Nuclear fission, p. 647
Nuclear fusion, p. 653
Nuclear reaction, p. 636
Nuclear transmutation, p. 636
Nucleon, p. 639
Plasma, p. 654
Positron, p. 637
Thermonuclear reaction, p. 654
Tracer, p. 657
Transuranium elements, p. 647

QUESTIONS AND PROBLEMS

NUCLEAR REACTIONS

Review Questions

22.1 How do nuclear reactions differ from ordinary chemical reactions?

22.2 What are the steps in balancing nuclear equations?

22.3 What is the difference between ${}^{0}_{-1}e$ and ${}^{0}_{-1}\beta$?

22.4 What is the difference between an electron and a positron?

Problems

22.5 Complete the following nuclear equations and identify X in each case:
(a) ${}^{26}_{12}\text{Mg} + {}^{1}_{1}\text{p} \rightarrow {}^{4}_{2}\text{He} + \text{X}$
(b) ${}^{59}_{27}\text{Co} + {}^{2}_{1}\text{H} \rightarrow {}^{60}_{27}\text{Co} + \text{X}$
(c) ${}^{235}_{92}\text{U} + {}^{1}_{0}\text{n} \rightarrow {}^{94}_{36}\text{Kr} + {}^{139}_{56}\text{Ba} + 3\text{X}$
(d) ${}^{53}_{24}\text{Cr} + {}^{4}_{2}\text{He} + {}^{1}_{0}\text{n} + \text{X}$
(e) ${}^{20}_{8}\text{O} \rightarrow {}^{20}_{9}\text{F} + \text{X}$

22.6 Complete the following nuclear equations and identify X in each case:
(a) ${}^{135}_{53}\text{I} \rightarrow {}^{135}_{54}\text{Xe} + \text{X}$
(b) ${}^{40}_{19}\text{K} \rightarrow {}^{0}_{-1}\beta + \text{X}$
(c) ${}^{59}_{27}\text{Co} + {}^{1}_{0}\text{n} \rightarrow {}^{56}_{25}\text{Mn} + \text{X}$
(d) ${}^{235}_{92}\text{U} + {}^{1}_{0}\text{n} \rightarrow {}^{99}_{40}\text{Sr} + {}^{135}_{52}\text{Te} + 2\text{X}$

22.7 Write balanced nuclear equations for the following reactions and identify X: (a) $\text{X}(\text{p},\alpha){}^{12}_{6}\text{C}$, (b) ${}^{27}_{13}\text{Al}(\text{d},\alpha)\text{X}$, (c) ${}^{55}_{25}\text{Mn}(\text{n},\gamma)\text{X}$

22.8 Write balanced nuclear equations for the following reactions and identify X: (a) ${}^{80}_{34}\text{Se}(\text{d},\text{p})\text{X}$, (b) $\text{X}(\text{d},2\text{p}){}^{9}_{3}\text{Li}$, (c) ${}^{10}_{5}\text{B}(\text{n},\alpha)\text{X}$

22.9 Describe how you would prepare astatine-211, starting with bismuth-209.

22.10 A long-cherished dream of alchemists was to produce gold from cheaper and more abundant elements. This dream was finally realized when ${}^{198}_{80}\text{Hg}$ was converted into gold by neutron bombardment. Write a balanced equation for this reaction.

NUCLEAR STABILITY

Review Questions

22.11 State the general rules for predicting nuclear stability.

22.12 What is the belt of stability?

22.13 Why is it impossible for the isotope ${}^{2}_{2}\text{He}$ to exist?

22.14 Define nuclear binding energy, mass defect, and nucleon.

22.15 How does Einstein's equation, $E = mc^2$, allow us to calculate nuclear binding energy?

22.16 Why is it preferable to compare the stability of nuclei in terms of nuclear binding energy per nucleon?

Problems

22.17 The radius of a uranium-235 nucleus is about 7.0×10^{-3} pm. Calculate the density of the nucleus in grams per cubic centimeter. (Assume the atomic mass is 235 amu.)

22.18 For each pair of isotopes listed, predict which one is less stable: (a) ${}^{6}_{3}\text{Li}$ or ${}^{9}_{3}\text{Li}$, (b) ${}^{23}_{11}\text{Na}$ or ${}^{25}_{11}\text{Na}$, (c) ${}^{48}_{20}\text{Ca}$ or ${}^{48}_{21}\text{Sc}$.

22.19 For each pair of elements listed, predict which one has more stable isotopes: (a) Co or Ni, (b) F or Se, (c) Ag or Cd.

22.20 In each pair of isotopes shown, indicate which one you would expect to be radioactive: (a) ${}^{20}_{10}\text{Ne}$

and $^{17}_{10}Ne$, (b) $^{40}_{20}Ca$ and $^{45}_{20}Ca$, (c) $^{95}_{42}Mo$ and $^{92}_{43}Tc$, (d) $^{195}_{80}Hg$ and $^{196}_{80}Hg$, (e) $^{209}_{83}Bi$ and $^{242}_{96}Cm$.

22.21 Given that

$$H(g) + H(g) \longrightarrow H_2(g) \qquad \Delta H° = -436.4 \text{ kJ}$$

calculate the change in mass (in kilograms) per mole of H_2 formed.

22.22 Estimates show that the total energy output of the sun is 5×10^{26} J/s. What is the corresponding mass loss in kilograms per second of the sun?

22.23 Calculate the nuclear binding energy (in joules) and the binding energy per nucleon of the following isotopes: (a) $^{7}_{3}Li$ (7.01600 amu) and (b) $^{35}_{17}Cl$ (34.95952 amu).

22.24 Calculate the nuclear binding energy (in joules) and the binding energy per nucleon of the following isotopes: (a) $^{4}_{2}He$ (4.0026 amu) and (b) $^{184}_{74}W$ (183.9510 amu).

RADIOACTIVE DECAY; DATING

Problems

22.25 Complete the following radioactive decay series:

(a) $^{232}Th \xrightarrow{\alpha}$ ____ $\xrightarrow{\beta}$ ____ $\xrightarrow{\beta}$ ^{228}Th

(b) $^{235}U \xrightarrow{\alpha}$ ____ $\xrightarrow{\beta}$ ____ $\xrightarrow{\alpha}$ ^{227}Ac

(c) ____ $\xrightarrow{\alpha}$ $^{233}Pa \xrightarrow{\beta}$ ____ $\xrightarrow{\alpha}$ ____

22.26 A radioactive substance decays as follows:

Time (days)	*Mass (g)*
0	500
1	389
2	303
3	236
4	184
5	143
6	112

Calculate the first-order decay constant and the half-life of the reaction.

22.27 The radioactive decay of Tl-206 to Pb-206 has a half-life of 4.20 min. Starting with 5.00×10^{22} atoms of Tl-206, calculate the number of such atoms left after 42.0 min.

22.28 A freshly isolated sample of ^{90}Y was found to have an activity of 9.8×10^5 disintegrations per minute at 1:00 P.M. on December 3, 1995. At 2:15 P.M. on December 17, 1995, its activity was redetermined and found to be 2.6×10^4 disintegrations per minute. Calculate the half-life of ^{90}Y.

22.29 Why do radioactive decay series obey first-order kinetics?

22.30 In the thorium decay series, thorium-232 loses a total of six α particles and four β particles in a 10-stage process. What is the final isotope produced?

22.31 Strontium-90 is one of the products of the fission of uranium-235. This isotope of strontium is radioactive, with a half-life of 28.1 yr. Calculate how long (in years) it will take for 1.00 g of the isotope to be reduced to 0.200 g by decay.

22.32 Consider the decay series

$$A \longrightarrow B \longrightarrow C \longrightarrow D$$

where A, B, and C are radioactive with half-lives of 4.50 s, 15.0 days, and 1.00 s, respectively, and D is nonradioactive. Starting with 1.00 mole of A, and none of B, C, or D, calculate the number of moles of A, B, C, and D left after 30 days.

NUCLEAR FISSION; NUCLEAR REACTORS

Review Questions

22.33 Define nuclear fission, nuclear chain reaction, and critical mass.

22.34 Which isotopes can undergo nuclear fission?

22.35 Explain how an atomic bomb works.

22.36 Explain the functions of a moderator and a control rod in a nuclear reactor.

22.37 Discuss the differences between light water and heavy water nuclear fission reactors. What are the advantages of a breeder reactor over a conventional nuclear fission reactor?

22.38 No form of energy production is without risk. Make a list of the risks to society involved in fueling and operating a conventional coal-fired electric power plant, and compare them with the risks of fueling and operating a nuclear fission-powered electric plant.

NUCLEAR FUSION

Review Questions

22.39 Define nuclear fusion, thermonuclear reaction, and plasma.

22.40 Why do heavy elements such as uranium undergo fission while light elements such as hydrogen and lithium undergo fusion?

22.41 How does a hydrogen bomb work?

22.42 What are the advantages of a fusion reactor over a fission reactor? What are the practical difficulties in operating a large-scale fusion reactor?

APPLICATIONS OF ISOTOPES

Problems

22.43 Describe how you would use a radioactive iodine isotope to demonstrate that the following process is in dynamic equilibrium:

$$PbI_2(s) \rightleftharpoons Pb^{2+}(aq) + 2I^-(aq)$$

22.44 Consider the following redox reaction:

$$IO_4^-(aq) + 2I^-(aq) + H_2O(l) \longrightarrow I_2(s) + IO_3^-(aq) + 2OH^-(aq)$$

When KIO_4 is added to a solution containing iodide ions labeled with radioactive iodine-128, all the radioactivity appears in I_2 and none in the IO_3^- ion. What can you deduce about the mechanism for the redox process?

22.45 Explain how you might use a radioactive tracer to show that ions are not completely motionless in crystals.

22.46 Each molecule of hemoglobin, the oxygen carrier ion blood, contains four Fe atoms. Explain how you would use the radioactive $^{59}_{26}Fe$ ($t_{\frac{1}{2}}$ = 46 days) to show that the iron in a certain diet is converted into hemoglobin.

MISCELLANEOUS PROBLEMS

22.47 How does a Geiger counter work?

22.48 Nuclei with an even number of protons and an even number of neutrons are more stable than those with an odd number of protons and/or odd number of neutrons. What is the significance of the even number of protons and of neutrons in this case?

22.49 Tritium, ^3H, is radioactive and decays by electron emission. Its half-life is 12.5 yr. In ordinary water the ratio of ^1H to ^3H atoms is 1.0×10^{17} to 1. (a) Write a balanced nuclear equation for tritium decay. (b) How many disintegrations will be observed per minute in a 1.00-kg sample of water?

22.50 (a) What is the activity, in millicuries, of a 0.500-g sample of $^{237}_{93}Np$? (This isotope decays by α-particle emission and has a half-life of 2.20×10^6 yr.) (b) Write a balanced nuclear equation for the decay of $^{237}_{93}Np$.

22.51 The following equations are for nuclear reactions that are known to occur in the explosion of an atomic bomb. Identify X.

(a) $^{235}_{92}U + ^1_0n \rightarrow ^{140}_{56}Ba + 3^1_0n + X$
(b) $^{235}_{92}U + ^1_0n \rightarrow ^{144}_{55}Cs + ^{90}_{37}Rb + 2X$
(c) $^{235}_{92}U + ^1_0n \rightarrow ^{87}_{35}Br + 3^1_0n + X$
(d) $^{235}_{92}U + ^1_0n \rightarrow ^{160}_{62}Sm + ^{72}_{30}Zn + 4X$

22.52 Calculate the nuclear binding energies, in joules per nucleon, for the following species: (a) ^{10}B (10.0129 amu), (b) ^{11}B (11.00931 amu), (c) ^{14}N (14.00307 amu), (d) ^{56}Fe (55.9349 amu).

22.53 Write complete nuclear equations for the following processes: (a) Tritium, ^3H, undergoes β decay; (b) ^{242}Pu undergoes α-particle emission; (c) ^{131}I undergoes β decay; (d) ^{251}Cf emits an α particle.

22.54 The nucleus of nitrogen-18 lies above the stability belt. Write an equation for a nuclear reaction by which nitrogen-18 can achieve stability.

22.55 Why is strontium-90 a particularly dangerous isotope for humans?

22.56 How are scientists able to tell the age of a fossil?

22.57 After the Chernobyl accident, people living close to the nuclear reactor site were urged to take large amounts of potassium iodide as a safety precaution. What is the chemical basis for this action?

22.58 Astatine, the last member of Group 7A, can be prepared by bombarding bismuth-209 with α particles. (a) Write an equation for the reaction. (b) Represent the equation in the abbreviated form as discussed in Section 22.4.

22.59 To detect bombs that may be smuggled onto airplanes, the Federal Aviation Administration requires all major airports in the United States to install thermal neutron analyzers. A thermal neutron analyzer bombards baggages with low-energy neutrons, converting some of the nitrogen-14 nuclei to nitrogen-15, with simultaneous emission of γ rays. Because nitrogen content is usually high in explosives, detection of a high dosage of γ rays will suggest that a bomb may be present. (a) Write an equation for the nuclear process. (b) Compare this technique with the conventional X-ray detection method.

22.60 Explain why, to achieve nuclear fusion in the laboratory, the temperature must reach about 100 million degrees Celsius, which is much higher than that in the interior of the sun (30 million degrees Celsius).

22.61 Tritium contains one proton and two neutrons. There is no proton-proton repulsion present in the nucleus. Why, then, is tritium radioactive?

22.62 The carbon-14 decay rate of a sample obtained

from a young tree is 0.260 disintegration per second per gram of the sample. Another wood sample prepared from an object recovered at an archaeological excavation gives a decay rate of 0.186 disintegration per second per gram of the sample. What is the age of the object?

22.63 The usefulness of radiocarbon dating is limited to objects no older than 50,000 years. What percent of the carbon-14 originally present in the sample remains after this period of time?

22.64 The radioactive potassium-40 isotope decays to argon-40 with a half-life of 1.2×10^9 yr. (a) Write a balanced equation for the reaction. (b) A sample of moon rock is found to contain 18 percent potassium-40 and 82 percent argon by mass. Calculate the age of the rock in years.

22.65 Both barium (Ba) and radium (Ra) are members of Group 4A and are expected to exhibit similar chemical properties. However, Ra is not found in barium ores. Instead, it is found in uranium ores. Explain.

22.66 Nuclear waste disposal is one of the major concerns of the nuclear industry. In choosing a safe and stable environment to store nuclear wastes, consideration must be given to the heat released during nuclear decay. As an example, consider the β decay of ^{90}Sr (89.907738 amu):

$$^{90}_{38}\text{Sr} \longrightarrow {}^{90}_{39}\text{Y} + {}^{0}_{-1}\beta \qquad t_{\frac{1}{2}} = 28.1 \text{ yr}$$

The ^{90}Y (89.907152 amu) formed further decays as follows:

$$^{90}_{39}\text{Y} \longrightarrow {}^{90}_{40}\text{Zr} + {}^{0}_{-1}\beta \qquad t_{\frac{1}{2}} = 64 \text{ h}$$

Zirconium-90 (89.904703 amu) is a stable isotope. (a) Use the mass defect to calculate the energy released (in joules) in each of the above two decays. (The mass of the electron is 5.4857×10^{-4} amu.) (b) Starting with one mole of ^{90}Sr, calculate the number of moles of ^{90}Sr that will decay in a year. (c) Calculate the amount of heat released (in kilojoules) corresponding to the number of moles of ^{90}Sr decayed to ^{90}Zr in (b).

22.67 Which of the following poses a greater health hazard: A radioactive isotope with a short half-life or a radioactive isotope with a long half-life? Explain.

22.68 For generations there was controversy about whether the Shroud of Turin, a piece of linen bearing the image of a man, was the burial cloth of Christ. In 1988 scientists carried out a carbon-14 dating experiment on the shroud, using less than 50 mg of the material. In one measurement the decay rate of the shroud was found to be 13.0 cpm/g C (13.0 counts per minute per gram of carbon). Given that the decay rate of wood from a living tree is 15.3 cpm/g C, what can you conclude about the origin of the shroud?

22.69 As a result of being exposed to the radiation released during the Chernobyl nuclear accident, the dose of iodine-131 in a person's body is 7.4 mC ($1 \text{ mC} = 1 \times 10^{-3}$ Ci). Using the relationship rate = kN, calculate the number of atoms of iodine-131 to which this dose corresponds. (The half-life of I-131 is 8.1 days.)

Answers to Practice Exercises 22.1 (a) $^{78}_{34}\text{Se}$, (b) $^{0}_{+1}\beta$; **22.2** 2.63×10^{-10} J, 1.26×10^{-12} J/nucleon; **22.3** $^{106}_{46}\text{Pd} + {}^{4}_{2}\text{He} \rightarrow {}^{109}_{47}\text{Ag} + {}^{1}_{1}\text{p}$.

APPENDIX 1
UNITS FOR THE GAS CONSTANT

In this appendix we will see how the gas constant R can be expressed in units $\mathrm{J/K \cdot mol}$. Our first step is to derive a relationship between atm and pascal. We start with

$$\begin{aligned}
\text{pressure} &= \frac{\text{force}}{\text{area}} \\
&= \frac{\text{mass} \times \text{acceleration}}{\text{area}} \\
&= \frac{\text{volume} \times \text{density} \times \text{acceleration}}{\text{area}} \\
&= \text{length} \times \text{density} \times \text{acceleration}
\end{aligned}$$

By definition, the standard atmosphere is the pressure exerted by a column of mercury exactly 76 cm high of density 13.5951 $\mathrm{g/cm^3}$, in a place where acceleration due to gravity is 980.665 $\mathrm{cm/s^2}$. However, to express pressure in $\mathrm{N/m^2}$ it is necessary to write

$$\begin{aligned}
\text{density of mercury} &= 1.35951 \times 10^4\ \mathrm{kg/m^3} \\
\text{acceleration due to gravity} &= 9.80665\ \mathrm{m/s^2}
\end{aligned}$$

The standard atmosphere is given by

$$\begin{aligned}
1\ \mathrm{atm} &= (0.76\ \mathrm{m\ Hg})(1.35951 \times 10^4\ \mathrm{kg/m^3})(9.80665\ \mathrm{m/s^2}) \\
&= 101325\ \mathrm{kg\ m/m^2 \cdot s^2} \\
&= 101325\ \mathrm{N/m^2} \\
&= 101325\ \mathrm{Pa}
\end{aligned}$$

From Section 5.5 we see that the gas constant R is given by 0.082057 $\mathrm{L \cdot atm/K \cdot mol}$. Using the conversion factors

$$\begin{aligned}
1\ \mathrm{L} &= 1 \times 10^{-3}\ \mathrm{m^3} \\
1\ \mathrm{atm} &= 101325\ \mathrm{N/m^2}
\end{aligned}$$

we write

$$R = \left(0.082057 \frac{\text{L atm}}{\text{K mol}}\right)\left(\frac{1 \times 10^{-3}\ \text{m}^3}{1\ \text{L}}\right)\left(\frac{101325\ \text{N/m}^2}{1\ \text{atm}}\right)$$

$$= 8.314 \frac{\text{N m}}{\text{K mol}}$$

$$= 8.314 \frac{\text{J}}{\text{K mol}}$$

$$1\ \text{L} \cdot \text{atm} = (1 \times 10^{-3}\ \text{m}^3)(101325\ \text{N/m}^2)$$

$$= 101.3\ \text{N m}$$

and

$$= 101.3\ \text{J}$$

APPENDIX 2
SELECTED THERMODYNAMIC DATA AT 1 ATM AND 25°C

Inorganic Substances

Substance	**ΔH_f° (kJ/mol)**	**ΔG_f° (kJ/mol)**	**S° (J/K · mol)**
$Ag(s)$	0	0	42.7
$Ag^+(aq)$	105.9	77.1	73.9
$AgCl(s)$	−127.0	−109.7	96.1
$AgBr(s)$	−99.5	−95.9	107.1
$AgI(s)$	−62.4	−66.3	114.2
$AgNO_3(s)$	−123.1	−32.2	140.9
$Al(s)$	0	0	28.3
$Al^{3+}(aq)$	−524.7	−481.2	−313.38
$Al_2O_3(s)$	−1669.8	−1576.4	50.99
$As(s)$	0	0	35.15
$AsO_4^{3-}(aq)$	−870.3	−635.97	−144.77
$AsH_3(g)$	171.5		
$H_3AsO_4(s)$	−900.4		
$Au(s)$	0	0	47.7
$Au_2O_3(s)$	80.8	163.2	125.5
$AuCl(s)$	−35.2		
$AuCl_3(s)$	−118.4		
$B(s)$	0	0	6.5
$B_2O_3(s)$	−1263.6	−1184.1	54.0
$H_3BO_3(s)$	−1087.9	−963.16	89.58
$H_3BO_3(aq)$	−1067.8	−963.3	159.8
$Ba(s)$	0	0	66.9
$Ba^{2+}(aq)$	−538.4	−560.66	12.55
$BaO(s)$	−558.2	−528.4	70.3
$BaCl_2(s)$	−860.1	−810.86	125.5
$BaSO_4(s)$	−1464.4	−1353.1	132.2
$BaCO_3(s)$	−1218.8	−1138.9	112.1

(Continued)

Substance	ΔH_f° (kJ/mol)	ΔG_f° (kJ/mol)	S° (J/K · mol)
$Be(s)$	0	0	9.5
$BeO(s)$	−610.9	−581.58	14.1
$Br_2(l)$	0	0	152.3
$Br^-(aq)$	−120.9	−102.8	80.7
$HBr(g)$	−36.2	−53.2	198.48
C(graphite)	0	0	5.69
C(diamond)	1.90	2.87	2.4
$CO(g)$	−110.5	−137.3	197.9
$CO_2(g)$	−393.5	−394.4	213.6
$CO_2(aq)$	−412.9	−386.2	121.3
$CO_3^{2-}(aq)$	−676.3	−528.1	−53.1
$HCO_3^-(aq)$	−691.1	−587.1	94.98
$H_2CO_3(aq)$	−699.7	−623.2	187.4
$CS_2(g)$	115.3	65.1	237.8
$CS_2(l)$	87.9	63.6	151.0
$HCN(aq)$	105.4	112.1	128.9
$CN^-(aq)$	151.0	165.69	117.99
$(NH_2)_2CO(s)$	−333.19	−197.15	104.6
$(NH_2)_2CO(aq)$	−319.2	−203.84	173.85
$Ca(s)$	0	0	41.6
$Ca^{2+}(aq)$	−542.96	−553.0	−55.2
$CaO(s)$	−635.6	−604.2	39.8
$Ca(OH)_2(s)$	−986.6	−896.8	76.2
$CaF_2(s)$	−1214.6	−1161.9	68.87
$CaCl_2(s)$	−794.96	−750.19	113.8
$CaSO_4(s)$	−1432.69	−1320.3	106.69
$CaCO_3(s)$	−1206.9	−1128.8	92.9
$Cd(s)$	0	0	51.46
$Cd^{2+}(aq)$	−72.38	−77.7	−61.09
$CdO(s)$	−254.6	−225.06	54.8
$CdCl_2(s)$	−389.1	−342.59	118.4
$CdSO_4(s)$	−926.17	−820.2	137.2
$Cl_2(g)$	0	0	223.0
$Cl^-(aq)$	−167.2	−131.2	56.5
$HCl(g)$	−92.3	−95.27	187.0
$Co(s)$	0	0	28.45
$Co^{2+}(aq)$	−67.36	−51.46	155.2
$CoO(s)$	−239.3	−213.38	43.9
$Cr(s)$	0	0	23.77
$Cr^{2+}(aq)$	−138.9		
$Cr_2O_3(s)$	−1128.4	−1046.8	81.17
$CrO_4^{2-}(aq)$	−863.16	−706.26	38.49
$CrO_7^{2-}(aq)$	−1460.6	−1257.29	213.8

(Continued)

Substance	ΔH_f° (kJ/mol)	ΔG_f° (kJ/mol)	S° (J/K · mol)
$Cs(s)$	0	0	82.8
$Cs^+(aq)$	−247.69	−282.0	133.05
$Cu(s)$	0	0	33.3
$Cu^+(aq)$	51.88	50.2	−26.36
$Cu^{2+}(aq)$	64.39	64.98	98.7
$CuO(s)$	−155.2	−127.2	43.5
$Cu_2O(s)$	−166.69	−146.36	100.8
$CuCl(s)$	−134.7	−118.8	91.6
$CuCl_2(s)$	−205.85		
$CuS(s)$	−48.5	−49.0	66.5
$CuSO_4(s)$	−769.86	−661.9	113.39
$F_2(g)$	0	0	203.34
$F^-(aq)$	−329.1	−276.48	−9.6
$HF(g)$	−271.6	−270.7	173.5
$Fe(s)$	0	0	27.2
$Fe^{2+}(aq)$	−87.86	−84.9	−113.39
$Fe^{3+}(aq)$	−47.7	−10.5	−293.3
$Fe_2O_3(s)$	−822.2	−741.0	90.0
$Fe(OH)_2(s)$	−568.19	−483.55	79.5
$Fe(OH)_3(s)$	−824.25		
$H(g)$	218.2	203.2	114.6
$H_2(g)$	0	0	131.0
$H^+(aq)$	0	0	0
$OH^-(aq)$	−229.94	−157.30	−10.5
$H_2O(g)$	−241.8	−228.6	188.7
$H_2O(l)$	−285.8	−237.2	69.9
$H_2O_2(l)$	−187.6	−118.1	?
$Hg(l)$	0	0	77.4
$Hg^{2+}(aq)$		−164.38	
$HgO(s)$	−90.7	−58.5	72.0
$HgCl_2(s)$	−230.1		
$Hg_2Cl_2(s)$	−264.9	−210.66	196.2
$HgS(s)$	−58.16	−48.8	77.8
$HgSO_4(s)$	−704.17		
$Hg_2SO_4(s)$	−741.99	−623.92	200.75
$I_2(s)$	0	0	116.7
$I^-(aq)$	55.9	51.67	109.37
$HI(g)$	25.9	1.30	206.3
$K(s)$	0	0	63.6
$K^+(aq)$	−251.2	−282.28	102.5
$KOH(s)$	−425.85		
$KCl(s)$	−435.87	−408.3	82.68

(Continued)

Substance	ΔH_f° (kJ/mol)	ΔG_f° (kJ/mol)	S° (J/K · mol)
$KClO_3(s)$	−391.20	−289.9	142.97
$KClO_4(s)$	−433.46	−304.18	151.0
$KBr(s)$	−392.17	−379.2	96.4
$KI(s)$	−327.65	−322.29	104.35
$KNO_3(s)$	−492.7	−393.1	132.9
$Li(s)$	0	0	28.0
$Li^+(aq)$	−278.46	−293.8	14.2
$Li_2O(s)$	−595.8	?	?
$LiOH(s)$	−487.2	−443.9	50.2
$Mg(s)$	0	0	32.5
$Mg^{2+}(aq)$	−461.96	−456.0	−117.99
$MgO(s)$	−601.8	−569.6	26.78
$Mg(OH)_2(s)$	−924.66	−833.75	63.1
$MgCl_2(s)$	−641.8	−592.3	89.5
$MgSO_4(s)$	−1278.2	−1173.6	91.6
$MgCO_3(s)$	−1112.9	−1029.3	65.69
$Mn(s)$	0	0	31.76
$Mn^{2+}(aq)$	−218.8	−223.4	−83.68
$MnO_2(s)$	−520.9	−466.1	53.1
$N_2(g)$	0	0	191.5
$N_3^-(aq)$	245.18	?	?
$NH_3(g)$	−46.3	−16.6	193.0
$NH_4^+(aq)$	−132.80	−79.5	112.8
$NH_4Cl(s)$	−315.39	−203.89	94.56
$NH_3(aq)$	−366.1	−263.76	181.17
$N_2H_4(l)$	50.4		
$NO(g)$	90.4	86.7	210.6
$NO_2(g)$	33.85	51.8	240.46
$N_2O_4(g)$	9.66	98.29	304.3
$N_2O(g)$	81.56	103.6	219.99
$HNO_2(aq)$	−118.8	−53.6	
$HNO_3(l)$	−173.2	−79.9	155.6
$NO_3^-(aq)$	−206.57	−110.5	146.4
$Na(s)$	0	0	51.05
$Na^+(aq)$	−239.66	−261.87	60.25
$Na_2O(s)$	−415.89	−376.56	72.8
$NaCl(s)$	−411.0	−384.0	72.38
$NaI(s)$	−288.0		
$Na_2SO_4(s)$	−1384.49	−1266.8	149.49
$NaNO_3(s)$	−466.68	−365.89	116.3
$Na_2CO_3(s)$	−1130.9	−1047.67	135.98
$NaHCO_3(s)$	−947.68	−851.86	102.09
$Ni(s)$	0	0	30.1

(Continued)

Substance	ΔH_f° (kJ/mol)	ΔG_f° (kJ/mol)	S° (J/K · mol)
$Ni^{2+}(aq)$	−64.0	−46.4	159.4
$NiO(s)$	−244.35	−216.3	38.58
$Ni(OH)_2(s)$	−538.06	−453.1	79.5
$O(g)$	249.4	230.1	160.95
$O_2(g)$	0	0	205.0
$O_3(aq)$	−12.09	16.3	110.88
$O_3(g)$	142.2	163.4	237.6
P(white)	0	0	44.0
P(red)	−18.4	13.8	29.3
$PO_4^{3-}(aq)$	−1284.07	−1025.59	−217.57
$P_4O_{10}(s)$	−3012.48		
$PH_3(g)$	9.25	18.2	210.0
$HPO_4^{2-}(aq)$	−1298.7	−1094.1	−35.98
$H_2PO_4^-(aq)$	−1302.48	−1135.1	89.1
$Pb(s)$	0	0	64.89
$Pb^{2+}(aq)$	1.6	24.3	21.3
$PbO(s)$	−217.86	−188.49	69.45
$PbO_2(s)$	−276.65	−218.99	76.57
$PbCl_2(s)$	−359.2	−313.97	136.4
$PbS(s)$	−94.3	−92.68	91.2
$PbSO_4(s)$	−918.4	−811.2	147.28
$Pt(s)$	0	0	41.84
$PtCl_4^{2-}(aq)$	−516.3	−384.5	175.7
$Rb(s)$	0	0	69.45
$Rb^+(aq)$	−246.4	−282.2	124.27
S(rhombic)	0	0	31.88
S(monoclinic)	0.30	0.10	32.55
$SO_2(g)$	−296.1	−300.4	248.5
$SO_3(g)$	−395.2	−370.4	256.2
$SO_3^{2-}(aq)$	−624.25	−497.06	43.5
$SO_4^{2-}(aq)$	−907.5	−741.99	17.15
$H_2S(g)$	−20.15	−33.0	205.64
$HSO_3^-(aq)$	−627.98	−527.3	132.38
$HSO_4^-(aq)$	−885.75	−752.87	126.86
$H_2SO_4(l)$	−811.3	?	?
$SF_6(g)$	−1096.2	?	?
$Se(s)$	0	0	42.44
$SeO_2(s)$	−225.35		
$H_2Se(g)$	29.7	15.90	218.9
$Si(s)$	0	0	18.70
$SiO_2(s)$	−859.3	−805.0	41.84

(Continued)

Substance	ΔH_f° (kJ/mol)	ΔG_f° (kJ/mol)	S° (J/K · mol)
Sr(*s*)	0	0	54.39
Sr^{2+}(*aq*)	−545.5	−557.3	39.33
$SrCl_2$(*s*)	−828.4	−781.15	117.15
$SrSO_4$(*s*)	−1444.74	−1334.28	121.75
$SrCO_3$(*s*)	−1218.38	−1137.6	97.07
W(*s*)	0	0	33.47
WO_3(*s*)	−840.3	−763.45	83.26
WO_4^-(*aq*)	−1115.45		
Zn(*s*)	0	0	41.6
Zn^{2+}(*aq*)	−152.4	−147.2	106.48
ZnO(*s*)	−348.0	−318.2	43.9
$ZnCl_2$(*s*)	−415.89	−369.26	108.37
ZnS(*s*)	−202.9	−198.3	57.7
$ZnSO_4$(*s*)	−978.6	−871.6	124.7

Organic Substances

Substance	Formula	ΔH_f° (kJ/mol)	ΔG_f° (kJ/mol)	S° (J/K · mol)
Acetic acid(*l*)	CH_3COOH	−484.2	−389.45	159.83
Acetaldehyde(*g*)	CH_3CHO	−166.35	−139.08	264.2
Acetone(*l*)	CH_3COCH_3	−246.8	−153.55	198.74
Acetylene(*g*)	C_2H_2	226.6	209.2	200.8
Benzene(*l*)	C_6H_6	49.04	124.5	124.5
Ethanol(*l*)	C_2H_5OH	−276.98	−174.18	161.04
Ethane(*g*)	C_2H_6	−84.7	−32.89	229.49
Ethylene(*g*)	C_2H_4	52.3	68.1	219.45
Formic acid(*l*)	HCOOH	−409.2	−346.0	128.95
Glucose(*s*)	$C_6H_{12}O_6$	−1274.5	−910.56	212.1
Methane(*g*)	CH_4	−74.85	−50.8	186.19
Methanol(*l*)	CH_3OH	−238.7	−166.3	126.78
Sucrose(*s*)	$C_{12}H_{22}O_{11}$	−2221.7	−1544.3	360.24

APPENDIX 3
MATHEMATICAL OPERATIONS

Logarithms

Common Logarithms. The concept of the logarithm is an extension of the concept of exponents, which is discussed in Chapter 1. The *common,* or base-10, logarithm of any number is the power to which 10 must be raised to equal the number. The following examples illustrate this relationship:

Logarithm	Exponent
$\log 1 = 0$	$10^0 = 1$
$\log 10 = 1$	$10^1 = 10$
$\log 100 = 2$	$10^2 = 100$
$\log 10^{-1} = -1$	$10^{-1} = 0.1$
$\log 10^{-2} = -2$	$10^{-2} = 0.01$

In each case the logarithm of the number can be obtained by inspection.

Since the logarithms of numbers are exponents, they have the same properties as exponents. Thus, we have

Logarithm	Exponent
$\log AB = \log A + \log B$	$10^A \times 10^B = 10^{A+B}$
$\log \dfrac{A}{B} = \log A - \log B$	$\dfrac{10^A}{10^B} = 10^{A-B}$

Furthermore, $\log A^n = n \log A$.

Now suppose we want to find the common logarithm of 6.7×10^{-4}. On most electronic calculators, the number is entered first and then the log key is punched. This operation gives us

$$\log 6.7 \times 10^{-4} = -3.17$$

Note that there are as many digits *after* the decimal point as there are significant figures in the original number. The original number has two significant figures and the "17" in -3.17 tells us that the log has two significant figures. Other examples are

Number	Common Logarithm
62	1.79
0.872	−0.0595
1.0×10^{-7}	−7.00

Sometimes (as in the case of pH calculations) it is necessary to obtain the number whose logarithm is known. This procedure is known as taking the antilogarithm; it is simply the reverse of taking the logarithm of a number. Suppose in a certain calculation we have pH = 1.46 and are asked to calculate $[H^+]$. From the definition of pH ($pH = -\log [H^+]$) we can write

$$[H^+] = 10^{-1.46}$$

Many calculators have a key labeled $\log^{-1}$ or INV log to obtain antilogs. Other calculators have a 10^x or y^x key (where x corresponds to -1.46 in our example and y is 10 for base-10 logarithm). Therefore, we find that $[H^+] = 0.035\ M$.

Natural Logarithms. Logarithms taken to the base e instead of 10 are known as natural logarithms (denoted by ln or $\log_e$); e is equal to 2.7183. The relationship between common logarithms and natural logarithms is as follows:

$$\log 10 = 1 \qquad 10^1 = 10$$

$$\ln 10 = 2.303 \qquad e^{2.303} = 10$$

Thus

$$\ln x = 2.303 \log x$$

To find the natural logarithm of 2.27, say, we first enter the number on the electronic calculator and then punch the ln key to get

$$\ln 2.27 = 0.820$$

If no ln key is provided, we can proceed as follows:

$$2.303 \log 2.27 = 2.303 \times 0.356$$
$$= 0.820$$

Sometimes we may be given the natural logarithm and asked to find the number it represents. For example

$$\ln x = 59.7$$

On many calculators, we simply enter the number and punch the e key:

$$e^{59.7} = 8.46 \times 10^{25}$$

The Quadratic Equation

A quadratic equation takes the form

$$ax^2 + bx + c = 0$$

If coefficients a, b, and c are known, then x is given by

$$x = \frac{-b \pm \sqrt{b^2 - 4ac}}{2a}$$

Suppose we have the following quadratic equation:

$$2x^2 + 5x - 12 = 0$$

Solving for x, we write

$$x = \frac{-5 \pm \sqrt{(5)^2 - 4(2)(-12)}}{2(2)}$$
$$= \frac{-5 \pm \sqrt{25 + 96}}{4}$$

Therefore

$$x = \frac{-5 + 11}{4} = \frac{3}{2}$$

and

$$x = \frac{-5 - 11}{4} = -4$$

APPENDIX 4
THE ELEMENTS AND THE DERIVATION OF THEIR NAMES AND SYMBOLS*

Element	Symbol	Atomic No.	Atomic Mass†	Date of Discovery	Discoverer and Nationality‡	Derivation
Actinium	Ac	89	(227)	1899	A. Debierne (Fr.)	Gr. *aktis,* beam or ray
Aluminum	Al	13	26.98	1827	F. Wochler (Ge.)	Alum, the aluminum compound in which it was discovered; derived from L. *alumen,* astringent taste
Americium	Am	95	(243)	1944	A. Ghiorso (USA) R. A. James (USA) G. T. Seaborg (USA) S. G. Thompson (USA)	The Americas
Antimony	Sb	51	121.8	Ancient		L. *antimonium* (*anti,* opposite of; *monium,* isolated condition), so named because it is a tangible (metallic) substance which combines readily; symbol, L. *stibium,* mark
Argon	Ar	18	39.95	1894	Lord Raleigh (GB) Sir William Ramsay (GB)	Gr. *argos,* inactive

(Continued)

Source: From "The Elements and Derivation of Their Names and Symbols," G. P. Dinga, *Chemistry* **41** (2), 20–22 (1968). Copyright by the American Chemical Society.

*At the time this table was drawn up, only 103 elements were known to exist.

†The atomic masses given here correspond to the 1961 values of the Commission on Atomic Weights. Masses in parentheses are those of the most stable or most common isotopes.

‡The abbreviations are (Ar.) Arabic; (Au.) Austrian; (Du.) Dutch; (Fr.) French; (Ge.) German; (GB) British; (Gr.) Greek; (H.) Hungarian; (I.) Italian; (L.) Latin; (P.) Polish; (R.) Russian; (Sp.) Spanish; (Swe.) Swedish; (USA) American.

Element	Symbol	Atomic No.	Atomic Mass†	Date of Discovery	Discoverer and Nationality‡	Derivation
Arsenic	As	33	74.92	1250	Albertus Magnus(Ge.)	Gr. *aksenikon,* yellow pigment; L. *arsenicum,* orpiment; the Greeks once used arsenic trisulfide as a pigment
Astatine	At	85	(210)	1940	D. R. Corson (USA) K. R. MacKenzie (USA) E. Segre (USA)	Gr. *astatos,* unstable
Barium	Ba	56	137.3	1808	Sir Humphry Davy (GB)	barite, a heavy spar, derived from Gr. *barys,* heavy
Berkelium	Bk	97	(247)	1950	G. T. Seaborg (USA) S. G. Thompson (USA) A. Ghiorso (USA)	Berkeley, Calif.
Beryllium	Be	4	9.012	1828	F. Woehler (Ge.) A. A. B. Bussy (Fr.)	Fr. L. *beryl,* sweet
Bismuth	Bi	83	209.0	1753	Claude Geoffroy (Fr.)	Ge. *bismuth,* probably a distortion of *weisse masse* (white mass) in which it was found
Boron	B	5	10.81	1808	Sir Humphry Davy (GB) J. L. Gay-Lussac (Fr.) L. J. Thenard (Fr.)	The compound borax, derived from Ar. *buraq.* white
Bromine	Br	35	79.90	1826	A. J. Balard (Fr.)	Gr. *bromos,* stench
Cadmium	Cd	48	112.4	1817	Fr. Stromeyer (Ge.)	Gr. *kadmia,* earth; L. *cadmia,* calamine (because it is found along with calamine)
Calcium	Ca	20	40.08	1808	Sir Humphry Davy (GB)	L. *calx,* lime
Californium	Cf	98	(249)	1950	G. T. Seaborg (USA) S. G. Thompson (USA) A. Ghiorso (USA) K. Street, Jr. (USA)	California
Carbon	C	6	12.01	Ancient		L. *carbo,* charcoal
Cerium	Ce	58	140.1	1803	J. J. Berzelius (Swe.) William Hisinger (Swe.) M. H. Klaproth (Ge.)	Asteroid Ceres
Cesium	Cs	55	132.9	1860	R. Bunsen (Ge.) G. R. Kirchhoff (Ge.)	L. *caesium,* blue (cesium was discovered by its spectral lines, which are blue)
Chlorine	Cl	17	35.45	1774	K. W. Scheele (Swe.)	Gr. *chloros,* light green
Chromium	Cr	24	52.00	1797	L. N. Vauquelin (Fr.)	Gr. *chroma,* color (because it is used in pigments)
Cobalt	Co	27	58.93	1735	G. Brandt (Ge.)	Ge. *Kobold,* goblin (because the ore yielded cobalt instead of the expected metal, copper, it was attributed to goblins)

(Continued)

Element	Symbol	Atomic No.	Atomic Mass†	Date of Discovery	Discoverer and Nationality‡	Derivation
Copper	Cu	29	63.55	Ancient		L. *cuprum,* copper, derived from *cyprium,* Island of Cyprus, the main source of ancient copper
Curium	Cm	96	(247)	1944	G. T. Seaborg (USA) R. A. James (USA) A. Ghiorso (USA)	Pierre and Marie Curie
Dysprosium	Dy	66	162.5	1886	Lecoq de Boisbaudran (Fr.)	Gr. *dysprositos,* hard to get at
Einsteinium	Es	99	(254)	1952	A. Ghiorso (USA)	Albert Einstein
Erbium	Er	68	167.3	1843	C. G. Mosander (Swe.)	Ytterby, Sweden, where many rare earths were discovered
Europium	Eu	63	152.0	1896	E. Demarcay (Fr.)	Europe
Fermium	Fm	100	(253)	1953	A. Ghiorso (USA)	Enrico Fermi
Fluorine	F	9	19.00	1886	H. Moissan (Fr.)	Mineral fluorspar, from L. *fluere,* flow (because fluorspar was used as a flux)
Francium	Fr	87	(223)	1939	Marguerite Perey (Fr.)	France
Gadolinium	Gd	64	157.3	1880	J. C. Marignac (Fr.)	Johan Gadolin, Finnish rare earth chemist
Gallium	Ga	31	69.72	1875	Lecoq de Boisbaudran (Fr.)	L. *Gallia,* France
Germanium	Ge	32	72.59	1886	Clemens Winkler (Ge.)	L. *Germania,* Germany
Gold	Au	79	197.0	Ancient		L. *aurum,* shining dawn
Hafnium	Hf	72	178.5	1923	D. Coster (Du.) G. von Hevesey (H.)	L. *Hafnia,* Copenhagen
Helium	He	2	4.003	1868	P. Janssen (spectr) (Fr.) Sir William Ramsay (isolated) (GB)	Gr. *helios,* sun (because it was first discovered in the sun's spectrum)
Holmium	Ho	67	164.9	1879	P. T. Cleve (Swe.)	L. *Holmia,* Stockholm
Hydrogen	H	1	1.008	1766	Sir Henry Cavendish (GB)	Gr. *hydro,* water; *genes,* forming (because it produces water when burned with oxygen)
Indium	In	49	114.8	1863	F. Reich (Ge.) T. Richter (Ge.)	Indigo, because of its indigo blue lines in the spectrum
Iodine	I	53	126.9	1811	B. Courtois (Fr.)	Gr. *iodes,* violet
Iridium	Ir	77	192.2	1803	S. Tennant (GB)	L. *iris,* rainbow
Iron	Fe	26	55.85	Ancient		L. *ferrum,* iron
Krypton	Kr	36	83.80	1898	Sir William Ramsay (GB) M. W. Travers (GB)	Gr. *kryptos,* hidden
Lanthanum	La	57	138.9	1839	C. G. Mosander (Swe.)	Gr. *lanthanein,* concealed
Lawrencium	Lr	103	(257)	1961	A. Ghiorso (USA) T. Sikkeland (USA) A. E. Larsh (USA) R. M. Latimer (USA)	E. O. Lawrence (USA), inventor of the cyclotron

(Continued)

Element	Symbol	Atomic No.	Atomic Mass†	Date of Discovery	Discoverer and Nationality‡	Derivation
Lead	Pb	82	207.2	Ancient		Symbol, L. *plumbum,* lead, meaning heavy
Lithium	Li	3	6.941	1817	A. Arfvedson (Swe.)	Gr. *lithos,* rock (because it occurs in rocks)
Lutetium	Lu	71	175.0	1907	G. Urbain (Fr.) C. A. von Welsbach (Au.)	*Lutetia,* ancient name for Paris
Magnesium	Mg	12	24.31	1808	Sir Humphry Davy (GB)	*Magnesia,* a district in Thessaly; possibly derived from L. *magnesia*
Manganese	Mn	25	54.94	1774	J. G. Gahn (Swe.)	L. *magnes,* magnet
Mendelevium	Md	101	(256)	1955	A. Ghiorso (USA) G. R. Choppin (USA) G. T. Seaborg (USA) B. G. Harvey (USA) S. G. Thompson (USA)	Mendeleev, Russian chemist who prepared the periodic chart and predicted properties of undiscovered elements
Mercury	Hg	80	200.6	Ancient		Symbol, L. *hydrargyrum,* liquid silver
Molybdenum	Mo	42	95.94	1778	G. W. Scheele (Swe.)	Gr. *molybdos,* lead
Neodymium	Nd	60	144.2	1885	C. A. von Welsbach (Au.)	Gr. *neos,* new; *didymos,* twin
Neon	Ne	10	20.18	1898	Sir William Ramsay (GB) M. W. Travers (GB)	Gr. *neos,* new
Neptunium	Np	93	(237)	1940	E. M. McMillan (USA) P. H. Abelson (USA)	Planet Neptune
Nickel	Ni	28	58.69	1751	A. F. Cronstedt (Swe.)	Swe. *kopparnickel,* false copper; also Ge. *nickel,* referring to the devil that prevented copper from being extracted from nickel ores
Niobium	Nb	41	92.91	1801	Charles Hatchett (GB)	Gr. *Niobe,* daughter of Tantalus (niobium was considered identical to tantalum, named after *Tantalus,* until 1884; originally called columbium, with symbol Cb)
Nitrogen	N	7	14.01	1772	Daniel Rutherford (GB)	Fr. *nitrogene,* derived from L. *nitrum,* native soda, or Gr. *nitron,* native soda, and Gr. *genes,* forming
Nobelium	No	102	(253)	1958	A. Ghiorso (USA) T. Sikkeland (USA) J. R. Walton (USA) G. T. Seaborg (USA)	Alfred Nobel
Osmium	Os	76	190.2	1803	S. Tennant (GB)	Gr. *osme,* odor

(Continued)

Element	Symbol	Atomic No.	Atomic Mass†	Date of Discovery	Discoverer and Nationality‡	Derivation
Oxygen	O	8	16.00	1774	Joseph Priestley (GB) C. W. Scheele (Swe.)	Fr. *oxygene,* generator of acid, derived from Gr. *oxys,* acid, and L. *genes,* forming (because it was once thought to be a part of all acids)
Palladium	Pd	46	106.4	1803	W. H. Wollaston (GB)	Asteroid Pallas
Phosphorus	P	15	30.97	1669	H. Brandt (Ge.)	Gr. *phosphoros,* light bearing
Platinum	Pt	78	195.1	1735 1741	A. de Ulloa (Sp.) Charles Wood (GB)	Sp. *platina,* silver
Plutonium	Pu	94	(242)	1940	G. T. Seaborg (USA) E. M. McMillan (USA) J. W. Kennedy (USA) A. C. Wahl (USA)	Planet Pluto
Polonium	Po	84	(210)	1898	Marie Curie (P.)	Poland
Potassium	K	19	39.10	1807	Sir Humphry Davy (GB)	Symbol, L. *kalium,* potash
Praseodymium	Pr	59	140.9	1885	C. A. von Welsbach (Au.)	Gr. *prasios,* green; *didymos,* twin
Promethium	Pm	61	(147)	1945	J. A. Marinsky (USA) *L. E. Glendenin (USA)* C. D. Coryell (USA)	Gr. mythology, *Prometheus,* the Greek Titan who stole fire from heaven
Protactinium	Pa	91	(231)	1917	O. Hahn (Ge.) L. Meitner (Au.)	Gr. *protos,* first; *actinium* (because it disintegrates into actinium)
Radium	Ra	88	(226)	1898	Pierre and Marie Curie (Fr.; P.)	L. *radius,* ray
Radon	Rn	86	(222)	1900	F. E. Dorn (Ge.)	Derived from radium with suffix "on" common to inert gases (once called nitron, meaning shining, with symbol Nt)
Rhenium	Re	75	186.2	1925	W. Noddack (Ge.) I. Tacke (Ge.) Otto Berg (Ge.)	L. *Rhenus,* Rhine
Rhodium	Rh	45	102.9	1804	W. H. Wollaston, (GB)	Gr. *rhodon,* rose (because some of its salts are rose-colored)
Rubidium	Rb	37	85.47	1861	R. W. Bunsen (Ge.) G. Kirchoff (Ge.)	L. *rubidius,* dark red (discovered with the spectroscope, its spectrum shows red lines)
Ruthenium	Ru	44	101.1	1844	K. K. Klaus (R.)	L. *Ruthenia,* Russia
Samarium	Sm	62	150.4	1879	Lecoq de Boisbaudran (Fr.)	Samarskite, after Samarski, a Russian engineer
Scandium	Sc	21	44.96	1879	L. F. Nilson (Swe.)	Scandinavia
Selenium	Se	34	78.96	1817	J. J. Berzelius (Swe.)	Gr. *selene,* moon (because it resembles tellurium, named for the earth)

(Continued)

Element	Symbol	Atomic No.	Atomic Mass†	Date of Discovery	Discoverer and Nationality‡	Derivation
Silicon	Si	14	28.09	1824	J. J. Berzelius (Swe.)	L. *silex, silicis,* flint
Silver	Ag	47	107.9	Ancient		Symbol, L. *argentum,* silver
Sodium	Na	11	22.99	1807	Sir Humphry Davy (GB)	L. *sodanum,* headache remedy; symbol, L. *natrium,* soda
Strontium	Sr	38	87.62	1808	Sir Humphry Davy (GB)	Strontian, Scotland, derived from mineral strontionite
Sulfur	S	16	32.07	Ancient		L. *sulphurium* (Sanskrit, *sulvere*)
Tantalum	Ta	73	180.9	1802	A. G. Ekeberg (Swe.).	Gr. mythology, *Tantalus,* because of difficulty in isolating it (Tantalus, son of Zeus, was punished by being forced to stand up to his chin in water which receded whenever he tried to drink)
Technetium	Tc	43	(99)	1937	C. Perrier (I.)	Gr. *technetos,* artificial (because it was the first artificial element)
Tellurium	Te	52	127.6	1782	F. J. Müller (Au.)	L. *tellus,* earth
Terbium	Tb	65	158.9	1843	C. G. Mosander (Swe.)	Ytterby, Sweden
Thallium	Tl	81	204.4	1861	Sir William Crookes (GB)	Gr. *thallos,* a budding twig (because its spectrum shows a bright green line)
Thorium	Th	90	232.0	1828	J. J. Berzelius (Swe.)	Mineral thorite, derived from *Thor,* Norse god of war
Thulium	Tm	69	168.9	1879	P. T. Cleve (Swe.)	*Thule,* early name for Scandinavia
Tin	Sn	50	118.7	Ancient		Symbol, L. *stannum,* tin
Titanium	Ti	22	47.88	1791	W. Gregor (GB)	Gr. giants, the Titans, and L. *titans,* giant deities
Tungsten	W	74	183.9	1783	J. J. and F. de Elhuyar (Sp.)	Swe. *tung sten,* heavy stone; symbol, wolframite, a mineral
Uranium	U	92	238.0	1789 1841	M. H. Klaproth (Ge.) E. M. Peligot (Fr.)	Planet Uranus
Vanadium	V	23	50.94	1801 1830	A. M. del Rio (Sp.) N. G. Sefstrom (Swe.)	*Vanadis,* Norse goddess of love and beauty
Xenon	Xe	54	131.3	1898	Sir William Ramsay (GB) M. W. Travers (GB)	Gr. *xenos,* stranger
Ytterbium	Yb	70	173.0	1907	G. Urbain (Fr.)	Ytterby, Sweden
Yttrium	Y	39	88.91	1843	C. G. Mosander (Swe.)	Ytterby, Sweden
Zinc	Zn	30	65.39	1746	A. S. Marggraf (Ge.)	Ge. *zink,* of obscure origin
Zirconium	Zr	40	91.22	1789	M. H. Klaproth (Ge.)	Zircon, in which it was found, derived from *Ar. zargum,* gold color

GLOSSARY*

Absolute temperature scale. A temperature scale on which absolute zero (0 K) is the lowest temperature (also called the Kelvin temperature scale). (5.3)

Absolute zero. Theoretically the lowest attainable temperature. (5.3)

Accuracy. The closeness of a measurement to the true value of the quantity that is being measured. (1.5)

Acid. A substance that yields hydrogen ions (H^+) when dissolved in water. (2.7)

Acid ionization constant. The equilibrium constant for acid ionization. (16.5)

Actinide series. Elements that have incompletely filled $5f$ subshells or readily give rise to cations that have incompletely filled $5f$ subshells. (7.10)

Activated complex. The species temporarily formed by reactant molecules as a result of a collision before they form the product. (21.4)

Activation energy. The minimum amount of energy required to initiate a chemical reaction. (21.4)

Activity series. A summary of the results of many possible displacement reactions (4.4)

Actual yield. The amount of product actually obtained in a reaction. (3.9)

Addition reaction. A reaction in which one molecule is added to another. (13.2)

Adhesion. Attraction between unlike molecules. (11.3)

Alcohol. An organic compound containing the hydroxyl group (—OH). (13.3)

Aldehydes. Compounds with a carbonyl functional group and the general formula RCHO, where R is an H atom, an alkyl, or an aryl group. (13.2)

Aliphatic hydrocarbons. Hydrocarbons that do not contain the benzene group or the benzene ring. (13.2)

Alkali metals. The Group 1A elements (Li, Na, K, Rb, Cs, and Fr). (2.4)

Alkaline earth metals. The Group 2A elements (Be, Mg, Ca, Sr, Ba, and Ra). (2.4)

Alkanes. Hydrocarbons having the general formula C_nH_{2n+2}, where $n = 1, 2, \ldots$ (13.2)

Alkenes. Hydrocarbons that contain one or more carbon-carbon double bonds. They have the general formula C_nH_{2n}, where $n = 2, 3, \ldots$ (13.2)

Alkynes. Hydrocarbons that contain one or more carbon-carbon triple bonds. They have the general formula C_nH_{2n-2}, where $n = 2, 3, \ldots$ (13.2)

Allotropes. Two or more forms of the same element that differ significantly in chemical and physical properties. (2.6)

Alpha particles. See alpha rays.

Alpha (α) rays. Helium ions with a charge of +2. (2.2)

Amines. Organic bases that have the functional group $—NR_2$, where R may be H, an alkyl group, or an aromatic hydrocarbon group. (13.2)

Amino acid. A special kind of carboxylic acid that contains at least one carboxyl group (—COOH) and at least one amino group ($—NH_2$). (13.3)

Amphoteric oxide. An oxide that exhibits both acidic and basic properties. (8.6)

Amplitude. The vertical distance from the middle of a wave to the peak or trough. (7.1)

Anion. An ion with a net negative charge. (2.5)

Anode. The electrode at which oxidation occurs. (20.2)

Aqueous solution. A solution in which the solvent is water. (4.1)

Aromatic hydrocarbon. A hydrocarbon that contains one or more benzene rings. (13.2)

Atmospheric pressure. The pressure exerted by Earth's atmosphere. (5.2)

Atom. The basic unit of an element that can enter into chemical combination. (2.2)

*The number in parentheses is the number of the section in which the term first appears.

Atomic mass. The mass of an atom in atomic mass units. (3.1)
Atomic mass unit. A mass exactly equal to one-twelfth the mass of one carbon-12 atom. (3.1)
Atomic number The number of protons in the nucleus of an atom. (2.3)
Atomic orbital. The wave function of an electron in an atom. (7.6)
Atomic radius. One-half the distance between the nuclei in two adjacent atoms of the same element in a metal. For elements that exist as diatomic units, the atomic radius is one-half the distance between the nuclei of two atoms in a particular molecule. (8.3)
Aufbau principle. As protons are added one by one to the nucleus to build up the elements, electrons similarly are added to the atomic orbitals. (7.10)
Avogadro's law. At constant pressure and temperature, the volume of a gas is directly proportional to the number of moles of the gas present. (5.3)
Avogadro's number. 6.022×10^{23}; the number of particles in a mole. (3.2)
Barometer. An instrument that measures atmospheric pressure. (5.2)
Base. A substance that yields hydroxide ions (OH^-) when dissolved in water. (2.7)
Base ionization constant. The equilibrium constant for the ionization of a base. (16.2)
Battery. An electrochemical cell or a series of several connected electrochemical cells that can be used as a source of direct electric current at a constant voltage. (20.6)
Beta (β) rays. Streams of electrons emitted during the decay of certain radioactive substances. (2.2)
Bimolecular reaction. An elementary step involving two molecules that is part of a reaction mechanism. (21.5)
Binary acid. An acid that contains only two elements. (15.4)
Binary compounds. Compounds containing just two elements. (2.7)
Boiling point. The temperature at which the vapor pressure of a liquid is equal to the external atmospheric pressure. (11.6)
Bond dissociation energy. The enthalpy change required to break a bond in a mole of gaseous molecules. (9.8)
Bond energy. See bond dissociation energy.
Bond length. The distance between the centers of two bonded atoms in a molecule. (9.2)
Boundary surface diagram. Diagram of the region containing about 90 percent the electron density in an atomic orbital. (7.8)
Boyle's law. The volume of a fixed amount of gas is inversely proportional to the gas pressure at constant temperature. (5.3)
Breeder reactor. A nuclear reactor that produces more fissionable material than it uses. (22.5)
Brønsted acid. A substance capable of donating a proton in a reaction. (4.3)
Brønsted base. A substance capable of accepting a proton in a reaction. (4.3)
Buffer solution. A solution of (a) a weak acid or base and (b) its salt; both components must be present. A buffer solution has the ability to resist changes in pH when small amounts of either acid or base are added to it. (17.2)
Calorimetry. The measurement of heat changes. (6.4)
Carboxylic acids. Acids that contain the carboxyl group (—COOH). (13.3)
Catalyst. A substance that increases the rate of a chemical reaction without being consumed during the reaction. (21.6)
Cathode. The electrode at which reduction occurs. (20.2)
Cation. An ion with a net positive charge. (2.5)
Charles' and Gay-Lussac's law. See Charles' law.
Charles' law. The volume of a fixed amount of gas is directly proportional to the absolute temperature of the gas when the pressure is held constant. (5.3)
Chelating agent. A substance that forms complex ions with metal ions in solution. (18.2)
Chemical energy. Energy stored within the structural units of chemical substances. (6.1)
Chemical equation. An equation that uses chemical symbols to show what happens during a chemical reaction. (3.7)
Chemical equilibrium. A state in which the rates of the forward and reverse reactions are equal and no net changes can be observed. (4.1)
Chemical formula. An expression showing the chemical composition of a compound in terms of the symbols for the atoms of the elements involved. (2.6)
Chemical kinetics. The area of chemistry concerned with the speeds, or rates, at which chemical reactions occur. (21.1)
Chemical property. Any property of a substance that cannot be studied without converting the substance into some other substance. (1.3)
Chemical reaction. Chemical change. (3.7)
Chemistry. The science that studies the properties of substances and how substances react with one another. (1.1)
Chiral. Compounds or ions that are not superimposable with their mirror images. (13.4)

Closed system. A system that allows the exchange of energy (usually in the form of heat) but not mass with its surroundings. (6.2)

Closest packing. The most efficient arrangements for packing atoms, molecules, or ions in a crystal. (11.4)

Cohesion. The intermolecular attraction between like molecules. (11.3)

Colligative properties. Properties of solutions that depend on the number of solute particles in solution and not on the nature of the solute. (12.6)

Complex ion. An ion containing a central metal cation bonded to one or more molecules or ions. (17.7)

Compound. A substance composed of atoms of two or more elements chemically united in fixed proportions. (1.2)

Concentration of a solution. The amount of solute present in a given quantity of solution. (4.5)

Condensation. The phenomenon of going from the gaseous state to the liquid state. (11.8)

Condensation reaction. A reaction in which two smaller molecules combine to form a larger molecule. Water is invariably one of the products of such a reaction. (13.3)

Conjugate acid-base pair. An acid and its conjugate base or a base and its conjugate acid. (16.1)

Coordinate covalent bond. A bond in which the pair of electrons is supplied by one of the two bonded atoms. (9.7)

Coordination compound. A neutral species containing a complex ion. (18.2)

Coordination number. In a crystal lattice it is defined as the number of atoms (or ions) surrounding an atom (or ion) (11.4). In coordination compounds it is defined as the number of donor atoms surrounding the central metal atom in a complex. (18.2)

Copolymerization. The formation of a polymer that contains two or more different monomers. (14.2)

Corrosion. The deterioration of metals by an electrochemical process. (20.7)

Covalent bond. A bond in which two electrons are shared by two atoms. (9.2)

Critical mass. The minimum mass of fissionable material required to generate a self-sustaining nuclear chain reaction. (22.5)

Critical pressure. The minimum pressure necessary to bring about liquefaction at the critical temperature. (11.6)

Critical temperature. The temperature above which a gas will not liquefy. (11.6)

Crystal-field splitting. The energy difference between two sets of *d* orbitals of a metal atom in the presence of ligands. (18.4)

Crystalline solid. A solid that possesses rigid and long-range structural order; its atoms, molecules, or ions occupy specific positions. (11.4)

Crystallization. The process in which dissolved solute comes out of solution and forms crystals. (12.1)

Cycloalkanes. Hydrocarbons having the general formula C_nH_{2n} where $n = 3, 4, \ldots$. (13.2)

Dalton's law of partial pressures. The total pressure of a mixture of gases is just the sum of the pressures that each gas would exert if it were present alone. (5.5)

Decay series. The sequence of disintegration steps leading from a radioactive species to a stable product. (22.3)

Denatured protein. Protein that does not exhibit normal biological activities. (14.3)

Density. The mass of a substance divided by its volume. (1.4)

Deposition. The process in which vapor molecules are converted directly to the solid phase. (11.8)

Diagonal relationship. Similarities between pairs of elements in different groups and periods of the periodic table. (8.6)

Diamagnetic. Repelled by a magnet; a diamagnetic substance contains only paired electrons. (7.9)

Diatomic molecule. A molecule that consists of two atoms. (2.5)

Dilution. A procedure for preparing a less concentrated solution from a more concentrated solution. (4.5)

Dipole moment. The product of charge and the distance between the charges in a molecule. (10.2)

Dipole-dipole forces. Forces that act between polar molecules. (11.2)

Diprotic acid. Each unit of the acid yields two hydrogen ions. (4.3)

Dispersion forces. The attractive forces that arise as a result of temporary dipoles induced in the atoms or molecules. (11.2)

Displacement reaction. A reaction in which an atom or an ion in a compound is replaced by an atom of another element. (4.4)

Donor atom. The atom in a ligand that is bonded directly to the metal atom. (18.3)

Double bond. A covalent bond in which two atoms share two pairs of electrons. (9.2)

Dynamic equilibrium. The condition in which the rate of a forward process is exactly balanced by the rate of the reverse process. (11.6)

Electrochemistry. The branch of chemistry that deals with the interconversion of electrical energy and chemical energy. (20.1)

Electrolysis. A process in which electrical energy is used to bring about a nonsponstaneous chemical reaction. (20.8)

Electrolyte. A substance that, when dissolved in water, results in a solution that can conduct electricity. (4.1)

Electromagnetic wave. A wave that has an electric field component and a magnetic field component. (7.1)

Electromotive force (emf). The voltage difference between electrodes. (20.2)

Electromotive series. See activity series.

Electron. A subatomic particle that has a very low mass and carries a single negative electric charge. (2.2)

Electron affinity. The energy change that takes place when an electron is accepted by an atom (or an ion) in the gaseous state. (8.5)

Electron configuration. The distribution of electrons among the various orbitals in an atom or molecule. (7.9)

Electron density. The probability that an electron will be found at a particular region in an atomic orbital. (7.6)

Electronegativity. The ability of an atom to attract electrons toward itself in a chemical bond. (9.3)

Element. A substance that cannot be separated into simpler substances by chemical means. (1.2)

Elementary steps. A series of simple reactions that represent the overall progress of a reaction at the molecular level. (21.5)

Emission spectrum. The continuous or line spectrum of electromagnetic radiation emitted by a substance. (7.3)

Empirical formula. An expression using chemical symbols to show the types of elements in a substance and the ratios of the different kinds of atoms. (2.6)

Enantiomers. Optical isomers, that is, compounds and their nonsuperimposable mirror images. (13.4)

Endothermic processes. Processes that absorb heat from the surroundings. (6.2)

Energy. The capacity to do work or to produce change. (1.7)

Enthalpy. A thermodynamic quantity used to describe heat changes taking place at constant pressure. (6.3)

Enthalpy of reaction. The difference between the enthalpies of the products and the enthalpies of the reactants. (6.3)

Entropy. A direct measure of the randomness or disorder of a system. (19.3)

Enzyme. A biological catalyst. (21.6)

Equilibrium. A state in which there are no observable changes as time goes by. (14.1)

Equilibrium constant. A number equal to the ratio of the equilibrium concentrations of products to the equilibrium concentrations of reactants, each raised to the power of its stoichiometric coefficient. (14.2)

Equilibrium vapor pressure. The vapor pressure measured for a dynamic equilibrium of condensation and evaporation. (11.6)

Equivalence point. The point at which an acid is completely reacted with or neutralized by a base. (4.6)

Esters. Compounds that have the general formula RCOOR′, where R can be H or an alkyl group or an aromatic hydrocarbon group and R′ is an alkyl group or an aromatic hydrocarbon group. (13.3)

Ether. An organic compound containing the R—O—R′ linkage, where R and R′ are alkyl and/or aromatic hydrocarbon groups. (13.3)

Evaporation. The escape of molecules from the surface of a liquid; also called vaporization. (11.6)

Excess reagent. A reactant present in a quantity greater than necessary to react with the amount of the limiting reagent present. (3.9)

Excited state. A state that has higher energy than the ground state of the system. (7.3)

Exothermic processes. Processes that give off heat to the surroundings. (6.2)

Extensive property. A property that depends on how much matter is being considered. (1.3)

Family. The elements in a vertical column of the periodic table. (2.4)

Faraday. Change contained in 1 mole of electrons, equivalent to 96,487 coulombs. (20.4)

First law of thermodynamics. Energy can be converted from one form to another, but cannot be created or destroyed. (6.6)

First-order reaction. A reaction whose rate depends on reactant concentration raised to the first power. (21.2)

Formal charge. The difference between the valence electrons in an isolated atom and the number of electrons assigned to that atom in a Lewis structure. (9.5)

Formation constant. The equilibrium constant for the complex ion formation. (17.7)

Free energy. The energy available to do useful work. (19.3)

Free radical. A species containing an unpaired electron. (13.2)

Frequency. The number of waves that pass through a particular point per unit time. (7.1)

Functional group. That part of a molecule characterized by a special arrangement of atoms that is largely responsible for the chemical behavior of the parent molecule. (13.1)

Galvanic cell. An electrochemical cell that generates electricity by means of a spontaneous redox reaction. (20.2)

Gamma (γ) rays. High-energy radiation. (2.2)

Gas constant (*R*). The constant that appears in the ideal gas equation ($PV = nRT$). It is usually expressed as 0.08206 L · atm/K · mol, or 8.314 J/K · mol. (5.4)

Geometric isomers. Compounds with the same type and number of atoms and the same chemical bonds but different spatial arrangements; such isomers cannot be interconverted without breaking a chemical bond. (13.2)

Gibbs free energy. See free energy.

Gravimetric analysis. An experimental procedure that involves the measurement of mass to identify an unknown component of a substance. (4.6)

Ground state. The lowest energy state of a system. (7.3)

Group. The elements in a vertical column of the periodic table. (2.4)

Half-cell reactions. Oxidation and reduction reactions that occur at the electrodes. (20.2)

Half-life. The time required for the concentration of a reactant to decrease to half its initial concentration. (21.3)

Half-reaction. A reaction that explicitly shows electrons involved in either oxidation or reduction. (4.4)

Halogens. The nonmetallic elements in Group 7A (F, Cl, Br, I, and At). (2.4)

Heat. Transfer of energy between two bodies that are at different temperatures. (6.2)

Heat capacity. The amount of heat required to raise the temperature of a given quantity of a substance by one degree Celsius. (6.4)

Heat content. See enthalpy.

Heisenberg uncertainty principle. It is impossible to know simultaneously both the momentum and the position of a particle with certainty. (7.5)

Henry's law. The solubility of a gas in a liquid is proportional to the pressure of the gas over the solution. (12.5)

Hess's law. When reactants are converted to products, the change in enthalpy is the same whether the reaction takes place in one step or in a series of steps. (6.5)

Heterogeneous equilibrium. An equilibrium state in which the reacting species are not all in the same phase. (14.3)

Heterogeneous mixture. The individual components of such a mixture remain physically separate and can be seen as separate components. (1.2)

Homogeneous equilibrium. An equilibrium condition in which all reacting species are in the same phase. (14.3)

Homogeneous mixture. The composition of the mixture is the same throughout the solution. (1.2)

Homopolymer. A polymer consisting of only one type of monomer. (14.2)

Hund's rule. The most stable arrangement of electrons in atomic subshells is the one with the greatest number of parallel spins. (7.9)

Hybrid orbitals. Atomic orbitals obtained when two or more nonequivalent orbitals of the same atom combine prior to covalent bond formation. (10.3)

Hybridization. The process of mixing the atomic orbitals in an atom (usually the central atom) to generate a set of new atomic orbitals prior to covalent bond formation. (10.3)

Hydration. A process in which an ion or a molecule is surrounded by water molecules arranged in a specific manner. (4.1)

Hydrocarbons. Compounds made up of only carbon and hydrogen. (13.2)

Hydrogen bond. A special type of dipole-dipole interaction between the hydrogen atom bonded to an atom of a very electronegative element (F, N, O) and another atom of one of the three electronegative elements. (11.2)

Hydrogenation. The addition of hydrogen, especially to compounds with double and triple carbon-carbon bonds. (13.2)

Hydronium ion. H_3O^+. (4.3)

Ideal gas. A hypothetical gas whose pressure-volume-temperature behavior can be completely accounted for by the ideal gas equation. (5.4)

Ideal gas equation. An equation expressing the relationships among pressure, volume, temperature, and amount of gas ($PV = nRT$, where R is the gas constant). (5.4)

Ideal solution. Any solution that obeys Raoult's law. (12.8)

Indicators. Substances that have distinctly different colors in acidic and basic media. (4.6)

Induced dipole. The separation of positive and negative charges in a neutral atom (or a nonpolar molecule) caused by the proximity of an ion or a polar molecule. (11.2)

Intensive property. A property that does not depend on how much matter is being considered. (1.3)

Intermediate. A species that appears in the mechanism of the reaction (that is, in the elementary steps) but not in the overall balanced equation. (21.5)

Intermolecular forces. Attractive forces that exist among molecules. (11.2)

International System of Units. A revised metric system (abbreviated SI) that is widely used in scientific research. (1.4)

Intramolecular forces. Forces that hold atoms together in a molecule. (11.2)

Ion. A charged particle formed when a neutral atom or group of atoms gain or lose one or more electrons. (2.5)

Ionic bond. The electrostatic force that holds ions together in an ionic compound. (9.3)

Ionic compound. Any neutral compound containing cations and anions. (2.5)

Ionic equation. An equation that shows dissolved ionic compounds in terms of their free ions. (4.2)

Ionic radium. The radius of a cation or an anion as measured in an ionic compound. (8.3)

Ionization energy. The minimum energy required to remove an electron from an isolated atom (or an ion) in its ground state. (8.4)

Ion pair. A species made up of at least one cation and at least one anion held together by electrostatic forces. (12.6)

Ion-dipole forces. Forces that operate between an ion and a dipole. (11.2)

Ion-produce constant. Product of hydrogen ion concentration and hydroxide ion concentration (both in molarity) at a particular temperature. (16.2)

Isoelectronic. Ions, or atoms and ions, that possess the same number of electrons, and hence the same ground-state electron configuration, are said to be isoelectronic. (8.2)

Isolated system. A system that does not allow the transfer of either mass or energy to or from its surroundings. (6.2)

Isotopes. Atoms having the same atomic number but different mass numbers. (2.3)

Joule. Unit of energy given by newtons × meters. (5.6)

Kelvin temperature scale. See absolute temperature scale.

Ketones. Compounds with a carbonyl functional group and the general formula RR′CO, where R and R′ are alkyl and/or aromatic hydrocarbon groups. (13.3)

Kinetic energy. Energy available because of the motion of an object. (5.6)

Kinetic molecular theory of gases. A theory that describes the physical behavior of gases at the molecular level. (5.6)

Lanthanide series. Elements that have incompletely filled 4*f* subshells or readily give rise to cations that have incompletely filled 4*f* subshells. (7.10)

Lattice points. The positions occupied by atoms, molecules, or ions that define the geometry of a unit cell. (11.4)

Law of conservation of energy. The total quantity of energy in the universe is constant. (6.1)

Law of conservation of mass. Matter can be neither created nor destroyed. (2.1)

Law of definite proportions. Different samples of the same compound always contain its constituent elements in the same proportions by mass. (2.1)

Law of multiple proportions. If two elements can combine to form more than one type of compound, the masses of one element that combine with a fixed mass of the other element are in ratios of small whole numbers. (2.1)

Le Chatelier's principle. If an external stress is applied to a system at equilibrium, the system will adjust itself in such a way as to partially offset the stress. (14.6)

Lewis acid. A substance that can accept a pair of electrons. (16.10)

Lewis base. A substance that can donate a pair of electrons. (16.10)

Lewis dot symbol. The symbol of an element with one or more dots that represent the number of valence electrons in an atom of the element. (9.1)

Lewis structure. A representation of covalent bonding using Lewis symbols. Shared electron pairs are shown either as lines or as pairs of dots between two atoms, and lone pairs are shown as pairs of dots on individual atoms. (9.2)

Ligand. A molecule or an ion that is bonded to the metal ion in a complex ion. (18.2)

Limiting reagent. The reactant used up first in a reaction. (3.9)

Line spectrum. Spectrum produced when radiation is absorbed or emitted by a substance only at some wavelengths. (7.3)

Liter. The volume occupied by one cubic decimeter. (1.4)

Lone pairs. Valence electrons that are not involved in covalent bond formation. (9.2)

Macroscopic properties. Properties that can be measured directly. (1.6)

Many-electron atoms. Atoms that contain two or more electrons. (7.6)

Mass. A measure of the quantity of matter contained in an object. (1.4)

Mass defect. The difference between the mass of an atom and the sum of the masses of its protons, neutrons, and electrons. (22.2)

Mass number. The total number of neutrons and protons present in the nucleus of an atom. (2.3)

Matter. Anything that occupies space and possesses mass. (1.3)

Melting point. The temperature at which solid and liquid phases coexist in equilibrium. (11.6)

Metalloid. An element with properties intermediate between those of metals and nonmetals. (2.4)

Metals. Elements that are good conductors of heat and electricity and have the tendency to form positive ions in ionic compounds. (2.4)

Microscopic properties. Properties that must be measured indirectly with the aid of a microscope or other special instrument. (1.4)

Miscible. Two liquids that are completely soluble in each other in all proportions are said to be miscible. (12.2)

Mixture. A combination of two or more substances in which the substances retain their identities. (1.2)

Moderator. A substance that can reduce the kinetic energy of neutrons. (22.5)

Molality. The number of moles of solute dissolved in one kilogram of solvent. (12.3)

Molar concentration. See molarity.

Molar heat of fusion. The energy (in kilojoules) required to melt one mole of a solid. (11.6)

Molar heat of sublimation. The energy (in kilojoules) required to sublime one mole of a solid (11.6)

Molar heat of vaporization. The energy (in kilojoules) required to vaporize one mole of a liquid. (11.6)

Molar mass. The mass (in grams or kilograms) of one mole of atoms, molecules, or other particles. (3.2)

Molar solubility. The number of moles of solute in one liter of a saturated solution (mol/L). (17.5)

Molarity. The number of moles of solute in one liter of solution. (4.5)

Mole. The amount of substance that contains as many elementary entities (atoms, molecules, or other particles) as there are atoms in exactly 12 grams (or 0.012 kilograms) of the carbon-12 isotope. (3.2)

Mole fraction. Ratio of the number of moles of one component of a mixture to the total number of moles of all components in the mixture. (5.5)

Mole method. An approach for determining the amount of product formed in a reaction. (3.8)

Molecular equations. Equations in which the formulas of the compounds are written as though all species existed as molecules or whole units. (4.2)

Molecular formula. An expression showing the exact numbers of atoms of each element in a molecule. (2.6)

Molecular mass. The sum of the atomic masses (in amu) present in a given molecule. (3.3)

Molecularity of a reaction. The number of molecules reacting in an elementary step. (21.5)

Molecule. An aggregate of at least two atoms in a definite arrangement held together by special forces. (2.5)

Monatomic ion. An ion that contains only one atom. (2.5)

Monomer. The single repeating unit of a polymer. (14.2)

Monoprotic acid. Each unit of the acid yields one hydrogen ion. (4.3)

Multiple bonds. Bonds formed when two atoms share two or more pairs of electrons. (9.2)

Net ionic equation. An equation that includes only the ionic species that actually take part in the reaction. (4.2)

Neutralization reaction. A reaction between an acid and a base. (4.3)

Neutron. A subatomic particle that bears no net electric charge. Its mass is slightly greater than a proton's. (2.2)

Newton. The SI unit for force. (5.2)

Noble gas core. The noble gas that most nearly precedes the element being considered; used in writing electron configurations. (7.10)

Noble gases. Nonmetallic elements in Group 8A (He, Ne, Ar, Kr, Xe, and Rn). (2.4)

Nonbonding electrons. Valence electrons that are not involved in covalent bond formation. (9.2)

Nonelectrolyte. A substance that, when dissolved in water, gives a solution that is not electrically conducting. (4.1)

Nonmetals. Elements that are usually poor conductors of heat and electricity. (2.4)

Nonpolar molecule. A molecule that does not possess a dipole moment. (10.2)

Nonvolatile. Does not have a measurable vapor pressure. (12.6)

Nuclear binding energy. The energy required to break up a nucleus into protons and neutrons. (22.2)

Nuclear chain reaction. A self-sustaining sequence of nuclear fission reactions. (22.5)

Nuclear fission. The process in which a heavy nucleus (mass number > 200) divides to form small nuclei of intermediate mass and one or more neutrons. (22.5)

Nuclear fusion. The combining of small nuclei into larger ones. (22.6)

Nuclear reaction. A reaction involving change in an atomic nucleus. (22.1)

Nuclear transmutation. The change undergone by a nucleus as a result of bombardment by neutrons or other particles. (22.1)

Nucleons. Protons and neutrons in a nucleus. (22.2)

Nucleotide. The repeating unit in each strand of a DNA molecule which consists of a base-deoxyribose-phosphate linkage. (14.4)

Nucleus. The central core of an atom. (2.2)

Octet rule. An atom other than hydrogen tends to form bonds until it is surrounded by eight valence electrons. (9.2)

Open system. A system that can exchange mass and energy (usually in the form of heat) with its surroundings. (6.2)

Optical isomers. Compounds that are nonsuperimposable mirror images. (13.4)

Orbital. See atomic orbital.

Organic chemistry. The branch of chemistry that deals with carbon compounds. (13.1)

Osmosis. The net movement of solvent molecules through a semipermeable membrane from a pure solvent or from a dilute solution to a more concentrated solution. (12.6)

Osmotic pressure. The pressure required to stop osmosis. (12.6)

Overvoltage. The additional voltage required to cause electrolysis. (20.8)

Oxidation number. The number of charges an atom would have in a molecule if electrons were transferred completely in the direction indicated by the difference in electronegativity. (4.4)

Oxidation reaction. The half-reaction that represents the loss of electrons in a redox process. (4.4)

Oxidation state. See oxidation number.

Oxidizing agent. A substance that can accept electrons from another substance or increase the oxidation numbers of another substance. (4.4)

Oxoacid. An acid containing hydrogen, oxygen, and another element (the central element). (2.7)

Oxoanion. An anion derived from an oxoacid. (2.7)

Paramagnetic. Attracted to a magnet. A paramagnetic substance contains one or more unpaired electrons. (7.9)

Partial pressure. The pressure of one component in a mixture of gases. (5.5)

Pascal. A pressure of one newton per square meter ($1\ N/m^2$). (5.2)

Pauli exclusion principle. No two electrons in an atom can have the same four quantum numbers. (7.9)

Percent composition. The percent by mass of each element in a compound. (3.5)

Percent ionization. The ratio of ionized acid concentration at equilibrium to the initial concentration of acid. (16.5)

Percent by mass. The ratio of the mass of a solute to the mass of the solution, multiplied by 100%. (12.3)

Percent yield. The ratio of the actual yield of a reaction to the theoretical yield, multiplied by 100%. (3.9)

Period. A horizontal row of the periodic table. (2.4)

Periodic table. A tabular arrangement of the elements by similarities in properties and by increasing atomic number. (2.4)

pH. The negative logarithm of the hydrogen ion concentration in an aqueous solution. (16.3)

Phase. A homogeneous part of a system that is in contact with other parts of the system but separated from them by a well-defined boundary. (11.6)

Phase change. Transformation from one phase to another. (11.6)

Phase diagram. A diagram showing the conditions at which a substance exists as a solid, liquid, and vapor. (11.7)

Photon. A particle of light. (7.2)

Physical equilibrium. An equilibrium in which only physical properties change. (14.2)

Physical property. Any property of a substance that can be observed without transforming the substance into some other substance. (1.3)

Pi bond. A covalent bond formed by sideways overlapping orbitals; its electron density is concentrated above and below the plane of the nuclei of the bonding atoms. (10.4)

Plasma. A gaseous state of matter consisting of positive ions and electrons. (22.6)

Polar molecule. A molecules that possesses a dipole moment. (10.2)

Polarimeter. The instrument for studying interaction between plane-polarized light and chiral molecules. (13.4)

Polarizability. The ease with which the electron density in a neutral atom (or molecule) can be distorted. (11.2)

Polyatomic ion. An ion that contains more than one atom. (2.5)

Polyatomic molecule. A molecule that consists of more than two atoms. (2.5)

Polymer. A chemical species distinguished by a high molar mass, ranging into thousands and millions of grams. (14.1)

Positron. A particle that has the same mass as the electron but bears a +1 charge. (22.1)

Potential energy. Energy available by virtue of an object's position. (6.1)

Precipitate. An insoluble solid that separates from a solution. (4.2)

Precipitation reaction. A reaction characterized by the formation of a precipitate. (4.2)

Precision. The closeness of agreement of two or more measurements of the same quantity. (1.5)

Pressure. Force applied per unit area. (5.2)

Product. The substance formed as a result of a chemical reaction. (3.7)

Proton. A subatomic particle having a single positive electric charge. The mass of a proton is about 1840 times that of an electron. (2.2)

Qualitative analysis. The determination of the types of ions present in a solution. (17.8)

Quantum. The smallest quantity of energy that can be emitted or absorbed in the form of electromagnetic radiation. (7.1)

Quantum numbers. Numbers that describe the distribution of electrons in atoms. (7.7)

Racemic mixture. An equimolar mixture of two enantiomers. (13.4)

Radiant energy. Energy transmitted in the form of waves. (6.1)
Radiation. The emission and transmission of energy through space in the form of particles and/or waves. (2.2)
Radioactivity. The spontaneous breakdown of an atom by the emission of particles and/or radiation. (2.2)
Raoult's law. The partial pressure of the solvent over a solution is given by the product of the vapor pressure of the pure solvent and the mole fraction of the solvent in the solution. (12.6)
Rare earth series. See lanthanide series.
Rate constant. Proportionality constant relating reaction rate to the concentrations of reactants. (21.1)
Rate law. An expression relating the rate of a reaction to the rate constant and the concentrations of the reactants. (21.2)
Rate-determining step. The slowest step in the sequence of steps leading to the formation of products. (21.5)
Reactants. The starting substances in a chemical reaction. (3.7)
Reaction mechanism. The sequence of elementary steps that leads to product formation. (21.5)
Reaction order. The sum of the powers to which all reactant concentrations appearing in the rate law are raised. (21.2)
Reaction quotient. A number equal to the ratio of product concentrations to reactant concentrations, each raised to the power of its stoichiometric coefficient at some point other than equilibrium. (14.5)
Reaction rate. The change in concentration of reactant or product with time. (21.1)
Redox reaction. A reaction in which there is either a transfer of electrons or a change in the oxidation numbers of the substances taking part in the reaction. Also called oxidation-reduction reaction. (4.4)
Reducing agent. A substance that can donate electrons to another substance or decrease the oxidation numbers in another substance. (4.4)
Reduction reaction. The half-reaction that represents the gain of electrons in a redox process. (4.4)
Representative elements. Elements in Groups 1A through 7A, all of which have at least one incompletely filled *s* or *p* subshell of the highest principal quantum number. (8.2)
Resonance. The use of two or more Lewis structures to represent a particular molecule. (9.6)
Resonance structure. One of two or more alternative Lewis structures for a single molecule that cannot be described fully with a single Lewis structure. (9.6)
Reversible reaction. A reaction that can occur in both directions. (4.1)
Salt. An ionic compound made up of a cation other than H^+ and an anion other than OH^- or O^{2-}. (4.3)
Salt hydrolysis. The reaction of the anion or cation, or both, of a salt with water. (16.8)
Saponification. Soapmaking. (13.3)
Saturated hydrocarbons. Hydrocarbons that contain only single covalent bonds. (13.2)
Saturated solution. At a given temperature, the solution that results when the maximum amount of a substance has dissolved in a solvent. (12.1)
Second law of thermodynamics. The entropy of the universe increases in a spontaneous process and remains unchanged in an equilibrium process. (19.4)
Semipermeable membrane. A membrane that allows solvent molecules to pass through, but blocks the movement of solute molecules. (12.6)
Sigma bond. A covalent bond formed by orbitals overlapping end-to-end; its electron density is concentrated between the nuclei of the bonding atoms. (10.4)
Significant figures. The number of meaningful digits in a measured or calculated quantity. (1.5)
Solubility. The maximum amount of solute that can be dissolved in a given quantity of solvent at a specific temperature. (4.2)
Solubility product. The product of the molar concentrations of constituent ions, each raised to the power of its stoichiometric coefficient in the equilibrium equation. (17.5)
Solute. The substance present in the smaller amount in a solution. (4.1)
Solution. A homogeneous mixture of two or more substances. (4.1)
Solvation. The process in which an ion or molecule is surrounded by solvent molecules arranged in an ordered manner. (12.2)
Solvent. The substance present in the larger amount in a solution. (4.1)
Specific heat. The amount of heat energy required to raise the temperature of one gram of a substance by one degree Celsius. (6.4)
Spectator ions. Ions that are not involved in the overall reaction. (4.2)
Spectrochemical series. A list of ligands arranged in order of their abilities to split the *d*-orbital energies. (18.4)
Standard atmospheric pressure (1 atm). The pressure that supports a column of mercury exactly 76 cm high at 0°C at sea level. (5.2)
Standard emf. The sum of the standard reduction potential of the substance that undergoes reduction and the oxidation potential of the substance that undergoes oxidation in a redox process. (20.3)
Standard enthalpy of formation. The heat change that

results when one mole of a compound is formed from its elements in their standard states. (6.5)

Standard enthalpy of reaction. The enthalpy change that occurs when a reaction is carried out under standard-state conditions. (6.5)

Standard free-energy change. The free-energy change that occurs when reactants in their standard states are converted to products in their standard states. (19.5)

Standard free-energy of formation. The free-energy change when 1 mole of a compound is synthesized from its elements in their standard states. (19.5)

Standard oxidation potential. The voltage measured as oxidation occurs at the electrode when all solutes are 1 *M* and all gases are at 1 atm. (20.3)

Standard reduction potential. The voltage measured as a reduction reaction occurs at the electrode when all solutes are 1 *M* and all gases are at 1 atm. (20.3)

Standard solution. A solution of accurately known concentration. (4.6)

Standard state. The condition of 1 atm of pressure. (6.5)

Standard temperature and pressure (STP). 0°C and 1 atm. (5.4)

State function. A property that is determined by the state of the system. (6.6)

State of a system. The values of all pertinent macroscopic variables (for example, composition, volume, pressure, and temperature) of a system. (6.6)

Stoichiometric amounts. The exact molar amounts of reactants and products that appear in a balanced chemical equation. (3.9)

Stoichiometry. The mass relationships among reactants and products in chemical reactions. (3.8)

Strong acid. An acid that is a strong electrolyte. (16.4)

Strong base. A base that is a strong electrolyte. (16.4)

Structural isomers. Molecules that have the same molecular formula but different structures. (13.2)

Sublimation. The process in which molecules go directly from the solid phase into the vapor phase. (11.6)

Substance. A form of matter that has a definite or constant composition (the number and type of basic units present) and distinct properties. (1.2)

Supersaturated solution. A solution that contains more of the solute than is present in a saturated solution. (12.1)

Surface tension. The amount of energy required to stretch or increase the surface of a liquid by a unit area. (11.3)

Surroundings. The rest of the universe outside a system. (6.2)

System. Any specific part of the universe that is of interest to us. (6.2)

Termolecular reaction. An elementary step involving three molecules that is part of a reaction mechanism. (21.5)

Ternary compounds. Compounds consisting of three elements. (2.7)

Theoretical yield. The amount of product predicted by the balanced equation when all of the limiting reagent has reacted. (3.9)

Thermal energy. Energy associated with the random motion of atoms and molecules. (6.1)

Thermal pollution. The heating of the environment to temperatures that are harmful to its living inhabitants. (12.4)

Thermochemical equation. An equation that shows both the mass and enthalpy relations. (6.3)

Thermochemistry. The study of heat changes in chemical reactions. (6.2)

Thermodynamics. The scientific study of the interconversion of heat and other forms of energy. (6.6)

Thermonuclear reactions. Nuclear fusion reactions that occur at very high temperatures. (22.6)

Titration. The gradual addition of a solution of accurately known concentration to another solution of unknown concentration until the chemical reaction between the two solutions is complete. (4.6)

Tracers. Isotopes, especially radioactive isotopes, that are used to trace the path of the atoms of an element in a chemical or biological process. (22.7)

Transition metals. Elements that have incompletely filled *d* subshells or readily give rise to cations that have incompletely filled *d* subshells. (7.10)

Transuranium elements. Elements with atomic numbers greater than 92. (22.4)

Triple bond. A covalent bond in which two atoms share three pairs of electrons. (9.2)

Triple point. The point at which the vapor, liquid, and solid states of a substance are in equilibrium. (11.7)

Triprotic acid. Each unit of the acid yields three hydrogen ions. (4.3)

Unimolecular reaction. An elementary step in a reaction mechanism in which only one reacting molecule participates. (21.5)

Unit cell. The basic repeating unit of the arrangement of atoms, molecules, or ions in a crystalline solid. (11.4)

Unsaturated hydrocarbons. Hydrocarbons that contain carbon-carbon double bonds or carbon-carbon triple bonds. (13.2)

Unsaturated solution. A solution that contains less solute than it has the capacity to dissolve. (12.1)

Valence electrons. The outer electrons of an atom, which are the ones involved in chemical bonding. (8.2)

Valence shell. The outermost electron-occupied shell of an atom, which holds the electrons that are usually involved in bonding. (10.1)

Valence-shell electron-pair repulsion (VSEPR) model. A model that accounts for the geometrical arrangements of shared and unshared electron pairs around a central atom in terms of the repulsive forces between electron pairs. (10.1)

Valence-shell expansion. The use of *d* orbitals in addition to *s* and *p* orbitals to form covalent bonds. (10.3)

Van der Waals equation. An equation that describes the relationship among *P*, *V*, and *T* for a nonideal gas. (5.7)

Van der Waals forces. The collective name for certain attractive forces between atoms and molecules, namely, dipole-dipole, dipole-induced dipole, and dispersion forces. (11.2)

Vaporization. The escape of molecules from the surface of a liquid; also called evaporation. (11.6)

Viscosity. A measure of a fluid's resistance to flow. (11.3)

Volatile. Having a measurable vapor pressure. (12.6)

Volume. Length cubed. (1.4)

Wave. A vibrating disturbance by which energy is transmitted. (7.1)

Wavelength. The distance between identical points on successive waves. (7.1)

Weak acid. An acid that is a weak electrolyte. (16.4)

Weak base. An base that is a weak electrolyte. (16.4)

Weight. The force that gravity exerts on an object. (1.4)

Work. Directed energy change resulting from a process. (6.1)

ANSWERS TO EVEN-NUMBERED PROBLEMS

CHAPTER 1

1.8 **(a)** Physical change, **(b)** chemical change, **(c)** physical change, **(d)** chemical change, **(e)** physical change.
1.10 **(a)** Extensive, **(b)** intensive, **(c)** intensive.
1.12 **(a)** Compound, **(b)** element, **(c)** compound, **(d)** element.
1.18 1.30×10^3 g.
1.20 **(a)** 41°C, **(b)** 11.3°F, **(c)** 1.1×10^4°F.
1.22 **(a)** 7.49×10^{-1}, **(b)** 8.026×10^2, **(c)** 6.21×10^{-7}.
1.24 **(a)** 0.00003256, **(b)** 6030000.
1.26 **(a)** 1.8×10^{-2}, **(b)** 1.14×10^{10}, **(c)** -5×10^4, **(d)** 1.3×10^3.
1.28 **(a)** Three, **(b)** one, **(c)** one or two, **(d)** two.
1.30 **(a)** 1.28, **(b)** 3.18×10^{-3} mg, **(c)** 8.14×10^7 dm.
1.32 **(a)** 1.10×10^8 mg, **(b)** 6.83×10^{-5} m^3.
1.34 3.1557×10^7 s.
1.36 **(a)** 81 in/s, **(b)** 1.2×10^2 m/min, **(c)** 7.4 km/h.
1.38 **(a)** 8.35×10^{12} mi, **(b)** 2.96×10^3 cm, **(c)** 9.8×10^8 ft/s, **(d)** 8.6°C, **(e)** −459.67°F, **(f)** 7.12×10^{-5} m^3, **(g)** 7.2×10^3 L.
1.40 6.25×10^{-4} g/cm^3.
1.42 4.33×10^7 ton.
1.44 2.6 g/cm^3.
1.46 0.882 cm.
1.48 10.5 g/cm^3.
1.50 767 mph.
1.52 75.6°F ± 0.2°F.
1.54 500 mL.
1.56 5.4×10^{10} L.
1.58 −40°.
1.60 **(a)** 0.5%, **(b)** 3.1%.

CHAPTER 2

2.8 0.62 mi.
2.12 145.
2.14 N-15: 7 protons, 7 electrons, and 8 neutrons; S-33; 16 protons, 16 electrons and 17 neutrons; Cu-63: 29 protons, 29 electrons, and 34 neutrons; Sr-84: 38 protons, 38 electrons, and 46 neutrons; Ba-130: 56 protons, 56 electrons and 74 neutrons; W-186: 74 protons, 74 electrons, and 112 neutrons; Hg-202: 80 protons, 80 electrons, and 122 neutrons.
2.16 **(a)** ${}^{186}_{74}W$, **(b)** ${}^{201}_{80}Hg$.
2.34 **(a)** $AlBr_3$, **(b)** $NaSO_2$, **(c)** N_2O_5, **(d)** $K_2Cr_2O_7$.
2.38 K^+(19,18), Mg^{2+}(12,10), Fe^{3+}(26,23), Br^-(35,36), Mn^{2+}(25,23), C^{4-}(6,10), Cu^{2+}(29,27).
2.40 Ionic: NaBr, BaF_2, CsCl; Molecular: CH_4, CCl_4, ICl, NF_3.
2.42 **(a)** Potassium hypochlorite, **(b)** silver carbonate, **(c)** iron(II) chloride, **(d)** potassium permanganate, **(e)** cesium chlorate, **(f)** potassium ammonium sulfate, **(g)** iron(II) oxide, **(h)** iron(III) oxide, **(i)** titanium(IV) chloride, **(j)** sodium hydride, **(k)** lithium nitride, **(l)** sodium oxide, **(m)** sodium peroxide.
2.44 **(a)** CuCN, **(b)** $Sr(ClO_2)_2$, **(c)** $HBrO_4$, **(d)** HI, **(e)** $Na_2NH_4PO_4$, **(f)** $PbCO_3$, **(g)** SnF_2, **(h)** P_4S_{10}, **(i)** HgO, **(j)** Hg_2I_2.
2.46 (c).
2.48 **(a)** H or H_2?, **(b)** NaCl is an ionic compound.
2.50 **(a)** Molecule and compound, **(b)** element and molecule, **(c)** element, **(d)** molecule and compound, **(e)** element, **(f)** element and molecule, **(g)** element and molecule, **(h)** molecule and compound, **(i)** compound, **(j)** element, **(k)** element and molecule, **(l)** compound.
2.52 **(a)** CO_2(solid), **(b)** NaCl, **(c)** N_2O, **(d)** $CaCO_3$, **(e)** CaO, **(f)** $Ca(OH)_2$, **(g)** $NaHCO_3$, **(h)** $Mg(OH)_2$.
2.54 **(a)** Metals in groups 1A, 2A, and aluminum and nonmetals such as nitrogen, oxygen, and the halogens, **(b)** the transition metals.
2.56 ${}^{23}Na$.
2.58 Mercury (Hg) and bromine (Br_2).
2.60 N_2, O_2, O_3, F_2, Cl_2, He, Ne, Ar, Kr, Xe, Rn.
2.62 He, Ne, and Ar are chemically inert; Kr, Xe, and Rn only form a few compounds.

CHAPTER 3

3.10 Li-6: 7.493%; Li-7: 92.507%.
3.12 5.82×10^3 light years.
3.14 5.1×10^{24} amu.
3.16 9.96×10^{-15} mol.

3.18 3.01×10^3 g Au.
3.20 **(a)** 1.244×10^{-22} g/atom, **(b)** 9.746×10^{-23} g/atom.
3.22 2.98×10^{22} Cu atoms.
3.24 Lead.
3.26 **(a)** 256.6 g, **(b)** 76.15 g, **(c)** 119.37 g, **(d)** 176.12 g.
3.28 1.11×10^{-2} mol.
3.30 1.68×10^{26} C atoms; 6.72×10^{26} H atoms; 3.36×10^{26} N atoms; 1.68×10^{26} O atoms.
3.32 8.56×10^{22} molecules.
3.36 7 peaks.
3.42 **(a)** Na: 54.75%; F: 45.25%, **(b)** Na: 39.34%; Cl: 60.66%, **(c)** Na: 22.34%; Br: 77.66%, **(d)** Na: 15.34%; I: 84.66%.
3.44 Both are $C_6H_{10}S_2O$.
3.46 Ammonia.
3.48 39.3 g.
3.50 5.97 g F.
3.52 O: 66.1%; $C_2H_3NO_5$.
3.54 $C_5H_8O_4NNa$.
3.60 **(a)** $2KClO_3 \longrightarrow 2KCl + 3O_2$
(b) $2KNO_3 \longrightarrow 2KNO_2 + O_2$
(c) $NH_4NO_3 \longrightarrow N_2O + 2H_2O$
(d) $NH_4NO_2 \longrightarrow N_2 + 2H_2O$
(e) $2NaHCO_3 \longrightarrow Na_2CO_3 + H_2O + CO_2$
(f) $P_4O_{10} + 6H_2O \longrightarrow 4H_3PO_4$
(g) $2HCl + CaCO_3 \longrightarrow CaCl_2 + H_2O + CO_2$
(h) $2Al + 3H_2SO_4 \longrightarrow Al_2(SO_4)_3 + 3H_2$
(i) $CO_2 + 2KOH \longrightarrow K_2CO_3 + H_2O$
(j) $CH_4 + 2O_2 \longrightarrow CO_2 + 2H_2O$
3.66 1.01 mol.
3.68 **(a)** $2NaHCO_3 \longrightarrow Na_2CO_3 + H_2O + CO_2$. **(b)** 78.3 g
3.70 0.324 L.
3.72 0.294 mol.
3.74 **(a)** $NH_4NO_3 \longrightarrow N_2O + 2H_2O$. **(b)** 2.0×10 g.
3.76 18.0 g.
3.80 0.709 g NO_2; O_3; 6.9×10^{-3} mol NO.
3.82 HCl; 23.4 g.
3.86 **(a)** 7.05 g. **(b)** 92.9%
3.88 3.48×10^3 g.
3.90 **(a)** 0.212 mol. **(b)** 0.424 mol.
3.92 18.
3.94 2.4×10^{23} atoms.
3.96 65.4 g; Zn.
3.98 89.6%.
3.100 $C_6H_{12}O_6$.

CHAPTER 4

4.6 **(a)** Strong electrolyte. **(b)** Nonelectrolyte. **(c)** Weak electrolyte. **(d)** Strong electrolyte.
4.8 **(a)** Solid NaCl does not conduct electricity. **(b)** Molten NaCl conducts. **(c)** Aqueous solution of NaCl conducts.
4.10 In water HCl breaks up as H^+ and Cl^- ions; in benzene it remains as a molecule.
4.14 **(a)** Insoluble. **(b)** Soluble. **(c)** Soluble. **(d)** Insoluble. **(e)** Soluble.
4.16 **(a)** Ionic: $2Na^+(aq)+S^{2-}(aq) + Zn^{2+}(aq) + 2Cl^-(aq) \longrightarrow ZnS(s) + 2Na^+(aq) + 2Cl^-(aq)$; Net ionic: $Zn^{2+}(aq) + S^{2-}(aq) \longrightarrow ZnS(s)$. **(b)** Ionic: $6K^+(aq) + 2PO_4^{3-}(aq) + 3Sr^{2+}(aq) + 6NO_3^-(aq) \longrightarrow Sr_3(PO_4)_2(s) + 6K^+(aq) + 6NO_3^-(aq)$; Net ionic: $3Sr^{2+}(aq) + 2PO_4^{3-}(aq) \longrightarrow Sr_3(PO_4)_2(s)$. **(c)** Ionic: $Mg^{2+}(aq) + 2NO_3^-(aq) + 2Na^+(aq) + 2OH^-(aq) \longrightarrow Mg(OH)_2(s) + 2Na^+(aq) + 2NO_3^-(aq)$; Net ionic: $Mg^{2+}(aq) + 2OH^-(aq) \longrightarrow Mg(OH)_2(s)$.
4.18 **(a)** Add chloride ions. KCl is soluble; AgCl is not. **(b)** Add sulfate ions. Ag_2SO_4 is moderately soluble, but $PbSO_4$ is insoluble. **(c)** Add carbonate ions. $(NH_4)_2CO_3$ is soluble, but $CaCO_3$ is insoluble. **(d)** Add sulfate ions. $CuSO_4$ is soluble, but $BaSO_4$ is insoluble.
4.26 **(a)** Brønsted base. **(b)** Brønsted base. **(c)** Brønsted acid. **(d)** Brønsted acid and Brønsted base.
4.28 **(a)** Ionic: $CH_3COOH(aq) + K^+(aq) + OH^-(aq) \longrightarrow CH_3COO^-(aq) + K^+(aq) + H_2O(l)$; net ionic: $CH_3COOH(aq) + OH^-(aq) \longrightarrow CH_3COO^-(aq) + H_2O(l)$. **(b)** Ionic: $H_2CO_3(aq) + 2Na^+(aq) + 2OH^-(aq) \longrightarrow 2Na^+(aq) + CO_3^{2-}(aq) + 2H_2O(l)$; net ionic: $H_2CO_3(aq) + 2OH^-(aq) \longrightarrow CO_3^{2-}(aq) + 2H_2O(l)$. **(c)** Ionic: $2H^+(aq) + 2NO_3^-(aq) + Ba^{2+}(aq) + 2OH^-(aq) \longrightarrow Ba^{2+}(aq) + 2NO_3^-(aq) + 2H_2O(l)$; net ionic: $2H^+(aq) + 2OH^-(aq) \longrightarrow 2H_2O(l)$.
4.32 **(a)** Reducing agent: $Fe \longrightarrow Fe^{3+} + 3e^-$; oxidizing agent: $O_2 + 4e^- \longrightarrow 2O^{2-}$. **(b)** Reducing agent: $2Br^- \longrightarrow Br_2 + 2e^-$; oxidizing agent: $Cl_2 + 2e^- \longrightarrow 2Cl^-$. **(c)** Reducing agent: $Si \longrightarrow Si^{4+} + 4e^-$; oxidizing agent: $F_2 + 2e^- \longrightarrow 2F^-$. **(d)** Reducing agent: $H_2 \longrightarrow 2H^+ + 2e^-$; oxidizing agent: $Cl_2 + 2e^- \longrightarrow 2Cl^-$.
4.36 **(a)** +5. **(b)** +1. **(c)** +3. **(d)** +5. **(e)** +5. **(f)** +5.
4.38 **(a)** +1. **(b)** −1. **(c)** +3. **(d)** +3. **(e)** +4. **(f)** +6. **(g)** +2. **(h)** +4. **(i)** +2. **(j)** +3. **(k)** +5.
4.40 **(a)** −3. **(b)** $-\frac{1}{2}$. **(c)** −1. **(d)** +4. **(e)** +3. **(f)** −2. **(g)** +3. **(h)** +6.
4.42 SO_3 (S is in its maximum oxidation state of +6).
4.44 Li and Ca.
4.48 0.131 *M*.
4.50 10.8 g.
4.52 **(a)** 1.37 *M*. **(b)** 0.426 *M*. **(c)** 0.716 *M*.
4.54 **(a)** 6.50 g. **(b)** 2.45 g. **(c)** 2.65 g. **(d)** 7.36 g. **(e)** 3.95 g.
4.58 0.0433 *M*.
4.60 126 mL.
4.62 1.09 *M*.
4.66 35.72%
4.68 2.31×10^{-4} *M*.
4.72 **(a)** 6.0 mL. **(b)** 8.0 mL. **(c)** 23 mL.
4.74 The Ba^{2+} ions combine with the SO_4^{2-} ions to form $BaSO_4$ precipitate.
4.76 Physical test: Only the NaCl solution would conduct electricity; chemical test: Add $AgNO_3$ solution. Only the NaCl

solution would give AgCl precipitate.

4.78 Mg, Na, Ca, Ba, K, or Li.

4.80 Oxidation number of C is +2 in CO and +4 (maximum) in CO_2.

4.82 1.3 *M*.

4.84 0.171 *M*.

4.86 0.115 *M*.

4.88 0.80 L.

CHAPTER 5

5.14 0.797 atm.

5.18 53 atm.

5.20 **(a)** 0.69 L. **(b)** 61 atm.

5.22 −196°C; −268.8°C; 5.7×10^3 °C

5.24 1.3×10^2 K.

5.26 ClF_3.

5.32 6.2 atm.

5.34 472°C.

5.36 1.9 atm.

5.38 0.82 L.

5.40 33.6 mL.

5.42 6.1×10^{-3} atm.

5.44 35.1 g/mol.

5.46 N_2: 2.1×10^{22} molecules; O_2: 5.7×10^{21} molecules; Ar: 2.7×10^{20} atoms.

5.48 2.98 g/L.

5.50 SF_4.

5.54 **(a)** 0.89 atm. **(b)** 1.4 L.

5.56 349 mmHg.

5.58 19.8 g.

5.60 N_2: 217 mmHg; H_2: 650 mmHg.

5.68 N_2: 471.7 m/s; O_2: 441.4 m/s; O_3: 360.4 m/s.

5.74 Not ideal. Ideal gas pressure is 164 atm.

5.76 Ideal.

5.78 169 L. CO_2: 0.50 atm; N_2: 0.25 atm; H_2O: 0.41 atm; O_2: 0.041 atm.

5.80 **(a)** $NH_4NO_2(s) \longrightarrow N_2(g) + 2H_2O(l)$. **(b)** 0.273 g.

5.82 No. An ideal gas would not condense.

5.84 Higher in the winter.

5.86 **(a)** 0.86 L. **(b)** $NH_4HCO_3(s) \longrightarrow NH_3(g) + CO_2(g) + H_2O(l)$. Advantage: more gases (CO_2 and NH_3) generated; disadvantage: odor of ammonia!

5.88 3.88 L.

5.90 **(a)** $C_3H_8(g) + 5O_2(g) \longrightarrow 3CO_2(g) + 4H_2O(g)$. **(b)** 11.4 L.

5.92 O_2: 0.167 atm; NO_2: 0.333 atm.

CHAPTER 6

6.20 728 kJ.

6.22 50.8°C.

6.26 25.03°C.

6.30 O_2.

6.32 Greater: **(a)** $Cl_2(l)$; **(b)** $Br_2(g)$; **(c)** $I_2(g)$.

6.34 Ag_2O: Measure the standard enthalpy of reaction for $2Ag(s) + \frac{1}{2}O_2(g) \longrightarrow Ag_2O(s)$. $CaCl_2$: Measure the standard enthalpy of reaction for $Ca(s) + Cl_2(g) \longrightarrow CaCl_2(s)$.

6.38 −84.6 kJ.

6.40 −780 kJ.

6.44 −56.2 kJ.

6.46 **(a)** −1411 kJ. **(b)** −1124 kJ.

6.48 218.2 kJ/mol.

6.50 −4.51 kJ/g.

6.52 2.70×10^2 kJ.

6.58 48 J.

6.60 −3.1 kJ.

6.62 $\Delta H_2 - \Delta H_1$.

6.64 **(a)** −336.5 kJ. **(b)** $N_2H_4(l) + O_2(g) \longrightarrow N_2(g) + 2H_2O(l)$; $\Delta H^\circ_{rxn} = -622.0$ kJ. $4NH_3(g) + 3O_2(g) \longrightarrow 2N_2(g) + 6H_2O(l)$; $\Delta H^\circ_{rxn} = -1529.6$ kJ. Ammonia is a better fuel.

6.66 **(a)** Measure ΔH°_{rxn} for the reverse reaction, which is −2801.3 kJ. Therefore ΔH°_{rxn} for photosynthesis is +2801.3 kJ. **(b)** 1.1×10^{19} kJ.

6.68 −553.8 kJ.

6.70 −367 kJ.

6.72 0.492 J/g · °C.

6.74 The first reaction is exothermic (−110.5 kJ) which can be used to promote the second reaction which is endothermic (131.3 kJ).

6.76 1.09×10^4 L.

6.78 4.10 L.

6.80 5.60 kJ/mol.

6.82 −0.16 J.

6.84 1.9 kJ.

6.86 −66.5 kJ.

6.88 1.64×10^3 g.

6.90 −1.06 kJ.

CHAPTER 7

7.8 **(a)** 6.58×10^{14} Hz. **(b)** 1.36×10^8 nm.

7.10 1.5×10^2 s.

7.12 4.95×10^{14}/s.

7.16 **(a)** 4.0×10^2 nm. **(b)** 5.0×10^{-19} J.

7.18 1.2×10^2 nm; UV.

7.20 **(a)** 3.70×10^2 nm. **(b)** UV. **(c)** 5.38×10^{-19} J.

7.26 Use a prism.

7.28 Excited atoms emit the same characteristic frequencies or lines on Earth as they do from a distant star.

7.30 3.027×10^{-19} J.

7.32 6.17×10^{14}/s; 4.86×10^2 nm.

7.34 5.

7.40 1.37×10^{-6} nm.

7.42 1.7×10^{-23} nm.

7.54 $l = 2$: $m_l = -2, -1, 0, 1, 2$; $l = 1$: $m_l = -1, 0, 1$; $l = 0$: $m_l = 0$.

7.56 **(a)** $n = 4, l = 1, m_l = -1, 0, 1$. **(b)** $n = 3, l = 2, m_l = -2, -1, 0, 1, 2$. **(c)** $n = 3, l = 0, m_l = 0$. **(d)** $n = 5, l = 3, m_l = -3, -2, -1, 0, 1, 2, 3$.

7.58 Different orientations in space.

7.60 Values of l: 0 (6*s*, one subshell), 1 (6*p*, three subshells), 2 (6*d*, five subshells), 3 (6*f*, seven subshells), 4 (6*g*, nine subshells), 5 (6*h*, eleven subshells).

7.62 $2n^2$.

7.64 **(a)** 3, **(b)** 6, **(c)** 0.

7.66 Shielding in a many-electron atom.

7.68 **(a)** 2*s*. **(b)** 3*p*. **(c)** 3*s*. **(d)** 4*d*.

7.78 Al: $1s^22s^22p^63s^23p^1$; B: $1s^22s^22p^1$; F: $1s^22s^22p^5$.

7.80 B(1), Ne(0), P(3), Sc(1), Mn(5), Se(2), Zr(2), Ru(4), Cd(0), I(1), W(4), Pb(2), Ce(2), Ho(3).

7.82 Ge: $[Ar]4s^23d^{10}4p^2$; Fe: $[Ar]4s^23d^6$; Zn: $[Ar]4s^23d^{10}$; Ru: $[Kr]5s^14d^7$; W: $[Xe]6s^24f^{14}5d^4$; Tl: $[Xe]6s^24f^{14}5d^{10}6p^1$.

7.84 S^+.

7.86 **(a)** Incorrect, **(b)** correct, **(c)** incorrect.

7.88 **(a)** 4 (one in 2*s* and three in 2*p*). **(b)** 6 (2 in 4*p*, 2 in 4*d*, and 2 in 4*f*). **(c)** 10 (2 in each of 3*d*). **(d)** 1 (in 2*s*). **(e)** 2 (in 4*f*).

7.90 Wave properties.

7.92 **(a)** 8.76×10^{-35} m. **(b)** 7.39×10^{-9} m.

CHAPTER 8

8.18 (a) and (d); (b) and (e); (c) and (f).

8.20 **(a)** Group 1A, **(b)** Group 5A, **(c)** Group 8A, **(d)** Group 8B.
8.22 Fe^{3+}.
8.28 **(a)** [Ne], **(b)** [Ne], **(c)** [Ar], **(d)** [Ar], **(e)** [Ar], **(f)** $[Ar]3d^6$, **(g)** $[Ar]3d^9$, **(h)** $[Ar]3d^{10}$.
8.30 **(a)** Cr^{3+}, **(b)** Sc^{3+}, **(c)** Rh^{3+}, **(d)** Ir^{3+}.
8.32 Be^{2+} and He; F^- and N^{3-}; Fe^{2+} and Co^{3+}; S^{2-} and Ar.
8.38 $Na > Mg > Al > P > Cl$.
8.40 F.
8.42 The effective nuclear charge on the outermost electrons increases from left to right as a result of incomplete shielding by the inner electrons.
8.44 $Mg^{2+} < Na^+ < F^- < O^{2-} < N^{3-}$.
8.46 Te^{2-}.
8.48 −199.4°C.
8.52 The $3p^1$ electron of Al is effectively shielded by the inner electrons and $3s^2$ electrons.
8.54 2080 kJ/mol is for $1s^22s^22p^6$.
8.56 8.43×10^6 kJ/mol.
8.60 Cl.
8.62 Alkali metals have ns^1 and so can accept another electron.
8.66 Low ionization energy, reacts with water to form FrOH, and with oxygen to form oxide and superoxide.
8.68 The ns^1 electron of Group 1B metals is incompletely shielded by the inner *d* electrons and therefore they have much higher first ionization energies.
8.70 **(a)** $Li_2O(s) + H_2O(l) \longrightarrow 2LiOH(aq)$. **(b)** $CaO(s) + H_2O(l) \longrightarrow Ca(OH)_2(aq)$. **(c)** $CO_2(g) + H_2O(l) \longrightarrow H_2CO_3(aq)$.
8.72 BaO.
8.74 **(a)** Bromine, **(b)** nitrogen, **(c)** rubidium, **(d)** magnesium.
8.76 **(a)** $Mg^{2+} < Na^+ < F^- < O^{2-}$. **(b)** $O^{2-} < F^- < Na^+ < Mg^{2+}$.
8.78 M is potassium (K) and X is bromine (Br_2).
8.80 O^+ and N; Ar and S^{2-}; Ne and N^{3-}; Zn and As^{3+}; Cs^+ and Xe.
8.82 (a) and (d).
8.84 First: bromine; second: iodine; third: chlorine; fourth: fluorine.
8.86 Fluorine.
8.88 H^-.
8.90 Li_2O (basic), BeO (amphoteric), B_2O_3 (acidic), CO_2 (acidic), N_2O_5 (acidic).

CHAPTER 9

9.14 C—H < Br—H < F—H < Na—I < Li—Cl < K—F.
9.16 Cl—Cl < Br—Cl < Si—C < Cs—F.
9.18 **(a)** Covalent, **(b)** polar covalent, **(c)** ionic, **(d)** polar covalent.
9.22 **(a)** $[:\ddot{O}-\ddot{O}:]^{2-}$
(b) $[:C\equiv C:]^{2-}$
(c) $[:N\equiv O:]^+$
(d) 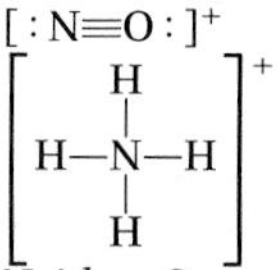
9.24 **(a)** Neither O atom has a complete octet; one H atom forms a double bond.
(b) 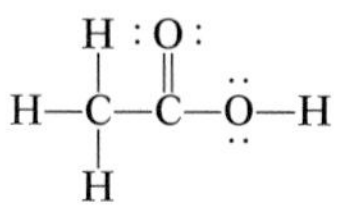
9.30 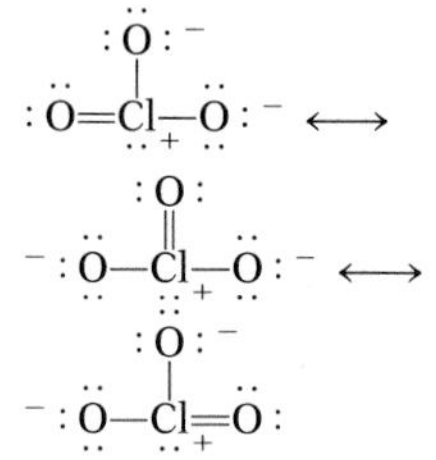
9.32 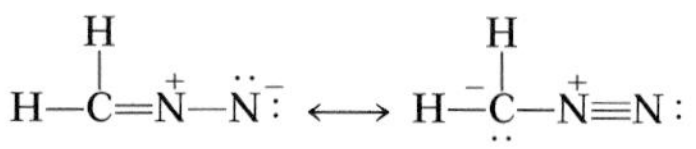
9.34 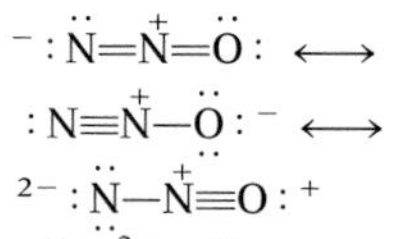
9.40 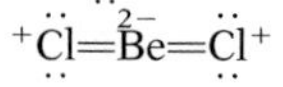
9.42 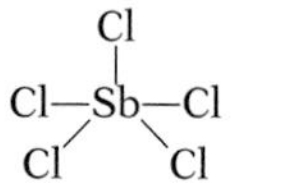
9.44 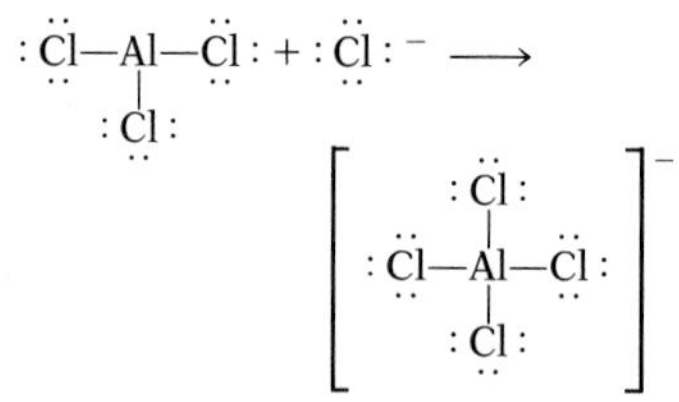
9.48 303.0 kJ/mol.
9.50 **(a)** −2759 kJ. **(b)** −3119 kJ.
9.52 Covalent: SiF_4, PF_5, SF_6, ClF_3; Ionic: NaF, MgF_2, AlF_3.
9.54 KF is a solid; has a high melting point; is an electrolyte. CO_2 is a gas; is a molecular compound.
9.56 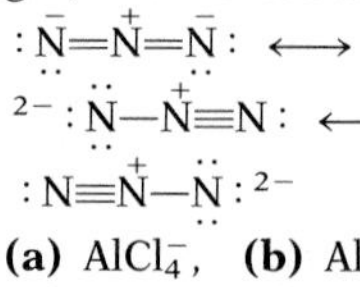
9.58 **(a)** $AlCl_4^-$, **(b)** AlF_6^{3-}, **(c)** $AlCl_3$.
9.60 C has incomplete octet in CF_2; C has an expanded octet in CH_5; F and H can only form a single bond; the I atoms are too large to surround the P atom.
9.62 **(a)** False, **(b)** true, **(c)** false, **(d)** false.
9.64 −67 kJ/mol.
9.66 N_2.
9.68 NH_4^+ and CH_4; CO and N_2; $B_3N_3H_6$ and C_6H_6.
9.70 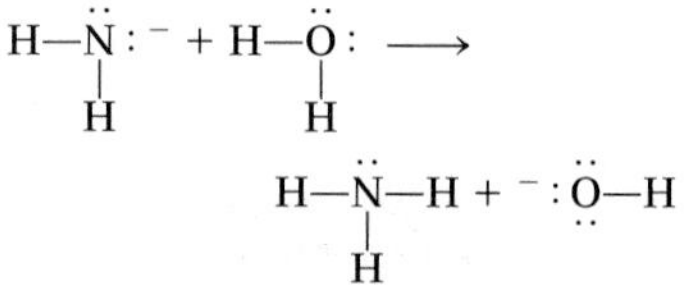
9.72 F cannot form an expanded octet.
9.74 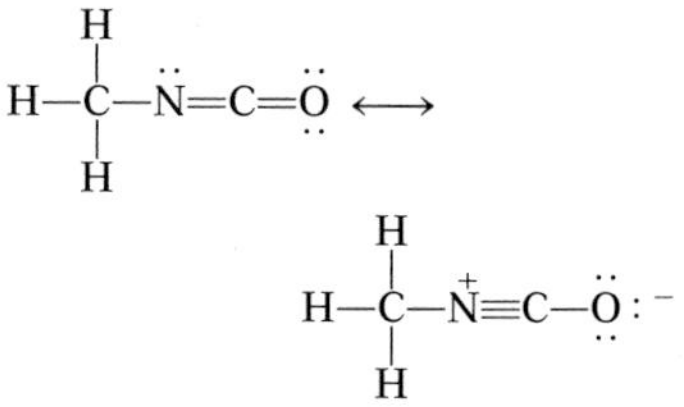
9.76 **(a)** Important, **(b)** important, **(c)** not important, **(d)** not important.
9.78 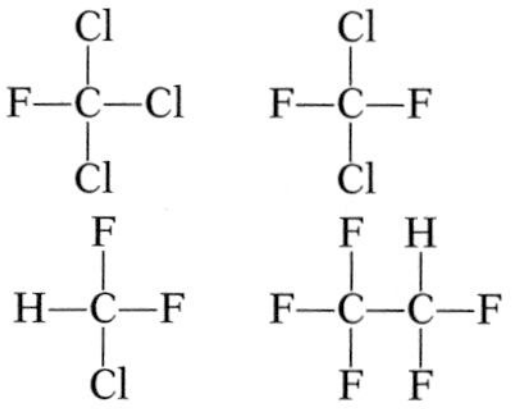
9.80 **(a)** −9.2 kJ. **(b)** −9.2 kJ.
9.82 **(a)** $^-{:}C\equiv O{:}^+$ **(b)** $:N\equiv O{:}^+$ **(c)** $^-{:}C\equiv N:$ **(d)** $:N\equiv N:$
9.84 True.
9.86 **(a)** 114 kJ. **(b)** Extra electron repulsion in F_2^- weakens the bond.

CHAPTER 10

10.8 **(a)** Trigonal planar, **(b)** linear, **(c)** tetrahedral.

10.10 **(a)** Tetrahedral, **(b)** bent, **(c)** trigonal planar, **(d)** linear, **(e)** square planar, **(f)** tetrahedral, **(g)** trigonal bipyramid, **(h)** trigonal pyramid, **(i)** tetrahedral.
10.12 $SiCl_4$, CI_4, $CdCl_4^{2-}$.
10.18 The electronegativity of the halogens decreases from F to I.
10.20 Higher.
10.22 (b) = (d) < (c) < (a).
10.32 sp^3.
10.34 Before the reaction: B is sp^2 and N is sp^3; after the reaction: both B and N are sp^3.
10.36 **(a)** Both C atoms are sp^3. **(b)** First C atom is sp^3; second and third C atoms are sp^2. **(c)** Both C atoms are sp^3. **(d)** First C atom is sp^3; second C atom is sp^2. **(e)** First C atom is sp^3; second C atom is sp^2.
10.38 The molecule is linear and the central N atom is *sp* hybridized.
10.40 sp^3d.
10.42 9 pi bonds and 9 sigma bonds.
10.44 Br-Hg-Br. Linear.
10.46 The large Si atoms prevent the effective sideways overlap of the 3*p* orbitals to form pi bonds.
10.48 XeF_3^+: T-shaped; XeF_5^+: square pyramid; SbF_6^-: octahedral.
10.50 **(a)** 180°, **(b)** 120°, **(c)** 109.5°, **(d)** roughly 109.5°, **(e)** 180°, **(f)** roughly 120°, **(g)** roughly 109.5°, **(h)** roughly 109.5°, **(i)** 109.5°.
10.52 sp^3d.
10.54 ICl_2^- and $CdBr_2$.
10.56 **(a)** sp^2. **(b)** Molecule on the right.
10.58 Rotation in *cis*-dichloroethylene would break a pi bond; there is only a sigma bond in 1,2-dichloroethane so rotation is free.
10.60 The repulsion between the bulky S=O double bonds spread out the angle close to 120°.
10.62 O_3, CO, CO_2, NO_2, $CFCl_3$.

CHAPTER 11

11.8 Methane; it has the lowest boiling point.
11.10 **(a)** Dispersion forces, **(b)** dispersion and dipole-dipole forces, **(c)** same as (b), **(d)** ionic and dispersion forces, **(e)** dispersion forces.
11.12 (e).
11.14 Only 1-butanol can form hydrogen bonds so it has a higher boiling point.
11.16 **(a)** Xe (greater dispersion forces), **(b)** CS_2 (greater dispersion forces), **(c)** Cl_2 (greater dispersion forces), **(d)** LiF (ionic compound), **(e)** NH_3 (hydrogen bond).
11.18 **(a)** Hydrogen bonding, dipole-dipole forces, dispersion forces, **(b)** dispersion forces, **(c)** dispersion forces, **(d)** the attractive forces of the covalent bond.
11.20 The compound on the left can form hydrogen bonds with itself (intramolecular hydrogen bonding).
11.32 Its viscosity is between that of ethanol and glycerol.
11.44 Simple cubic: 1 sphere; body-centered cubic: 2 spheres; face-centered cubic: 4 spheres.
11.46 6.21×10^{23} atom/mol.
11.48 458 pm.
11.50 XY_3.
11.52 Each C atom in diamond is covalently bonded to three other C atoms.
11.72 2670 kJ.
11.74 47.03 kJ/mol.
11.76 First step: freezing; second step: sublimation.
11.78 Additional heat is liberated when steam condenses at 100°C.
11.82 Initially, the ice melts because of the increase in pressure. As the wire sinks into the ice, the water above the wire refreezes. Eventually the wire moves completely through the ice block without cutting it in half.
11.84 **(a)** Ice melts and eventually water boils. **(b)** Ice forms. **(c)** Water boils.
11.86 (d).
11.88 Covalent crystal.
11.90 XYZ_3.
11.92 760 mmHg.
11.94 It has reached the critical temperature.

CHAPTER 12

12.8 Cyclohexane cannot form hydrogen bonds with ethanol.
12.10 The longer the chain gets, the more nonpolar the molecule becomes. The —OH group can form hydrogen bonds with water, but the rest of the molecule cannot.
12.14 **(a)** 25.9 g, **(b)** 1.72×10^3 g.
12.16 **(a)** 2.68 *m*. **(b)** 7.82 *m*.
12.20 5.0×10^2 *m*; 18.3 *M*.
12.22 **(a)** 2.41 *m*, **(b)** 2.13 *M*, **(c)** 0.0587 L.
12.26 45.9 g.
12.34 At the bottom of the mine the carbon dioxide pressure is greater and the dissolved gas is not released from the solution. Coming up, the pressure decreases and carbon dioxide is released from the solution.
12.36 0.28 L.
12.50 1.3×10^3 g.
12.52 Ethanol: 30.0 mmHg; propanol: 26.3 mmHg.
12.54 128 g.
12.56 0.59 *m*.
12.58 120 g/mol; $C_4H_8O_4$.
12.60 −8.6°C.
12.62 4.3×10^2 g/mol; $C_{24}H_{20}P_4$.
12.64 1.75×10^4 g/mol.
12.66 342 g/mol.
12.72 Boiling-point elevation, vapor-pressure lowering, osmotic pressure.
12.74 0.50 *m* glucose > 0.50 *m* acetic acid > 0.50 *m* HCl.
12.76 0.9420 *m*.
12.78 7.6 atm.
12.80 1.6 atm.
12.82 3.5 atm.
12.84 **(a)** 104 mmHg, **(b)** 116 mmHg.
12.86 2.95×10^3 g/mol.
12.88 No. Compound is assumed to be pure, monomeric, and a nonelectrolyte.
12.90 12.3 *M*.

CHAPTER 13

13.8 $CH_3CH_2CH_2CH_2CH_2Cl$
$CH_3CH_2CH_2CHClCH_3$
$CH_3CH_2CHClCH_2CH_3$

13.10 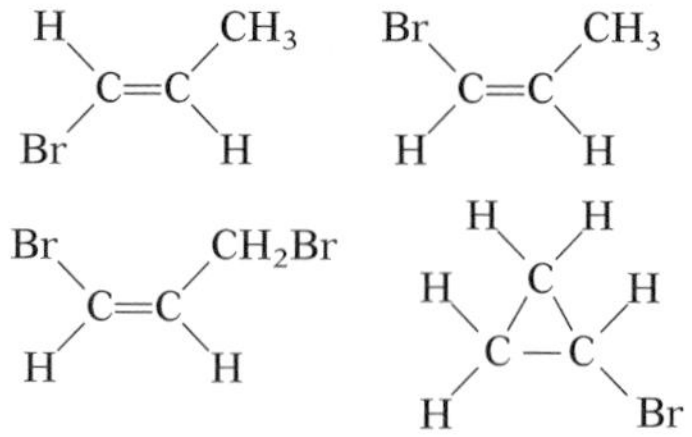

13.12 (a) Alkene or cycloalkane, **(b)** alkyne, **(c)** alkane, **(d)** alkene or cycloalkane, **(e)** alkyne.

13.14 The alkene (right) undergoes addition reactions with hydrogen, halogens, and hydrogen halides. The alkane (left) does not.

13.16 $\Delta H° = -630.8$ kJ.

13.18 (a) *cis*-1,2-dichlorocyclopropane, **(b)** *trans*-1,2-dichlorocyclopropane.

13.22 (a) Ether, **(b)** amine, **(c)** aldehyde, **(d)** ketone, **(e)** carboxylic acid, **(f)** alcohol, **(g)** amino acid.

13.24 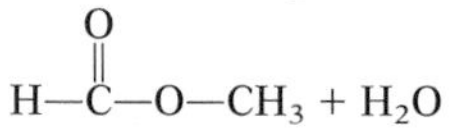

methyl formate

13.26 Two possible alcohols:

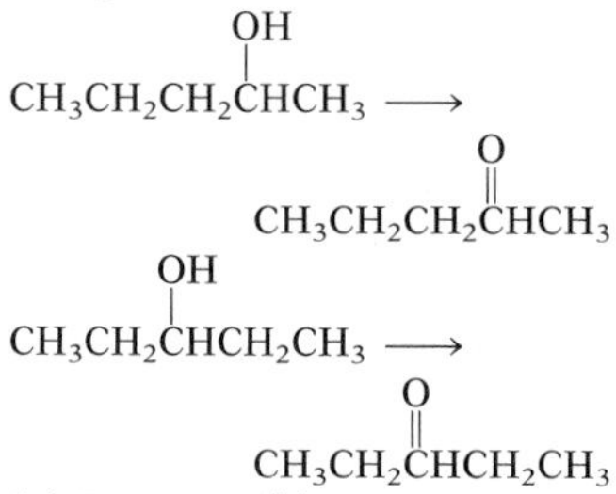

13.28 (a) Ketone, **(b)** ester, **(c)** ether.

13.32 (a) and (c).

13.34 (a)

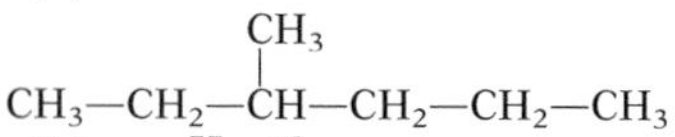

(b) 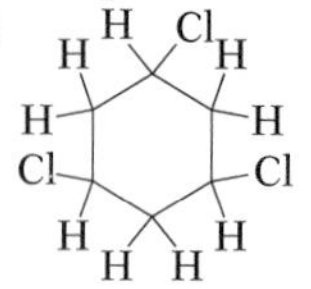

(c) 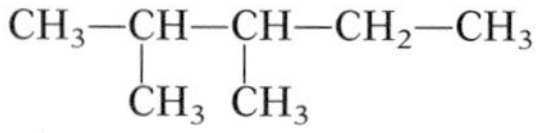

(d) 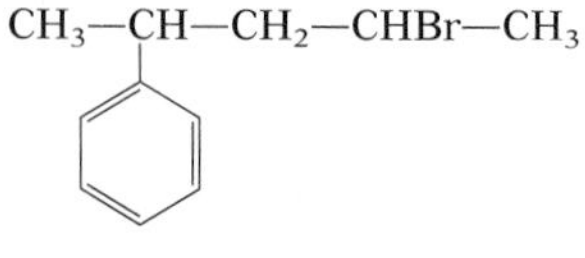

(e)

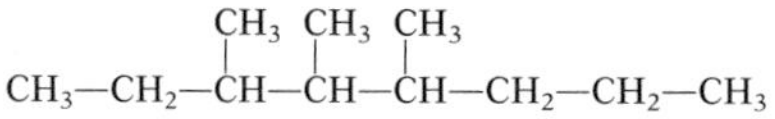

13.36 (a) 1,3-Dichloro-4-methylbenzene, **(b)** 1,4-dinitro-2-ethylbenzene, **(c)** 1,2,4,5-tetramethylbenzene.

13.38 −174 kJ.

13.40 (a), (c), (d), (f).

13.42 8.4×10^3 L.

13.44 (a) 6, **(b)** 4, **(c)** 8.

13.46 (a) 15.81 mg C; 1.33 mg H; 3.49 mg O. **(b)** C_6H_6O. **(c)** Phenol (see Figure 13.7).

13.48 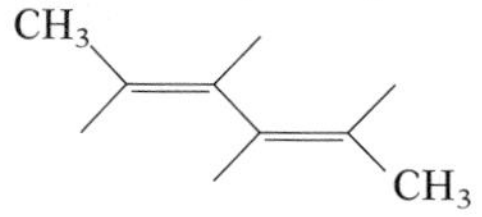

trans, trans

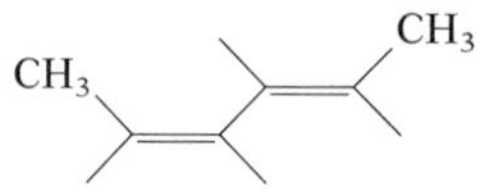

cis, trans

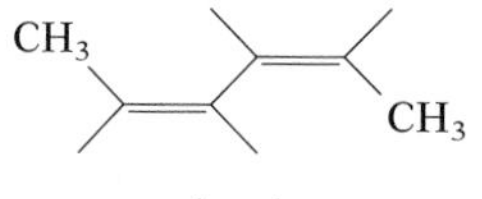

cis, cis

13.50 (a) React with HBr, **(b)** react with Br_2, **(c)** react with H_2.

13.52 (a) Alkyne, **(b)** alcohol, **(c)** ether, **(d)** aldehyde, **(e)** carboxylic acid, **(f)** amine.

13.54 The acids in lemon convert the amines to ammonium salts which have a low vapor pressure.

13.56 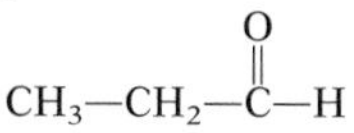

13.58 0.2525%.

13.60 (a) One, **(b)** two, **(c)** 5, **(d)** 1.

CHAPTER 14

14.8 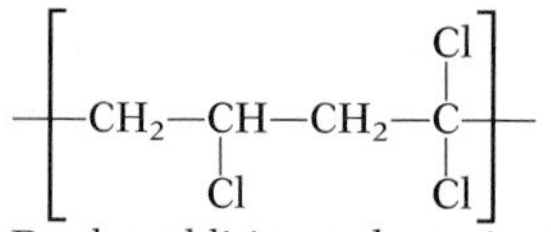

14.10 By the addition polymerization of styrene.

14.12 (a) $H_2C{=}CH{-}CH{=}CH_2$
(b) $HO_2C{-}(CH_2)_6{-}NH_2$

14.18 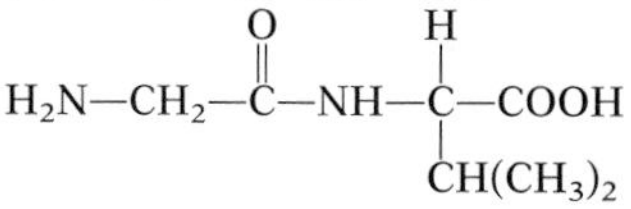

and

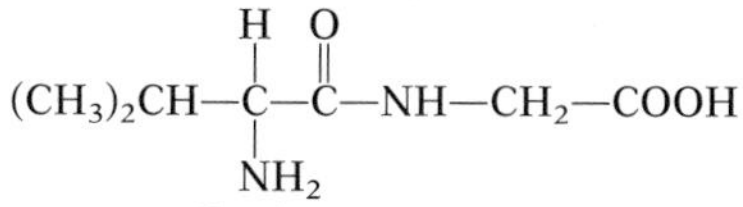

14.20 Initially the rate increases with temperature. At 35° C the enzyme begins to denature. At 45°C the enzyme has largely denatured.

14.26 DNAs and RNAs are made up of only four different building blocks (A, C, G, and T) whereas proteins are made up of twenty amino acids. Thus proteins have more varied structure.

14.28 Sample that has a higher C-G base pairs. A C-G pair has 3 hydrogen bonds and is harder to break up than an A-T pair which only has 2 hydrogen bonds.

14.30 Salt and vinegar denature the proteins, resulting in gel formation which seals the cracks.

14.32 Insects have blood that contains no hemoglobins. Without hemoglobin, oxygen needed for metabolism has to be transported by diffusion which would be too inefficient for a human-size insect.

14.34 4 Fe per hemoglobin molecule.

14.36 Dispersion forces.

CHAPTER 15

15.12 1.08×10^7.

15.14 1.1×10^{-5}.

15.16 (a) 0.082, **(b)** 0.29.

15.18 0.105; 2.05×10^{-3}.

15.20 7.09×10^{-3}.

15.22 3.3.

15.24 0.0356.

15.28 The net reaction will proceed from right to left to reach equilibrium.

15.30 0.020 atm.

15.32 $[I] = 8.58 \times 10^{-4}\ M$: $[I_2] = 0.0194\ M$.

15.34 $[CO_2] = 0.48\ M$, $[H_2] = 0.020\ M$, $[CO] = 0.075\ M$, $[H_2O] = 0.065\ M$.

15.36 $[H_2] = [CO_2] = 0.05\ M$, $[H_2O] = [CO] = 0.11\ M$.

15.42 (a) Shift to the right, **(b)** no effect, **(c)** no effect.

15.44 **(a)** No effect, **(b)** no effect, **(c)** shift to the left, **(d)** no effect, **(e)** shift to the left.
15.46 **(a)** Shift to the right, **(b)** shift to the left, **(c)** shift to the right, **(d)** shift to the left, **(e)** no effect.
15.48 No change.
15.50 **(a)** Shift to the right, **(b)** no effect, **(c)** no effect, **(d)** shift to the left, **(e)** shift to the right, **(f)** shift to the left, **(g)** shift to the right.
15.52 **(a)** NO: 0.24 atm; Cl_2: 0.12 atm, **(b)** 0.017.
15.54 **(a)** No, **(b)** yes.
15.56 **(a)** 8×10^{-44}, **(b)** The reaction needs an input of energy to get started.
15.58 **(a)** 1.7, **(b)** $P_A = 0.69$ atm, $P_B = 0.81$ atm.
15.60 1.5×10^5.
15.62 H_2: 0.28 atm, Cl_2: 0.049 atm, HCl: 1.67 atm.
15.64 49.8 atm.
15.66 3.84×10^{-2}.
15.68 3.13.
15.70 N_2: 0.860 atm; H_2: 0.366 atm; NH_3: 4.40×10^{-3} atm.
15.72 **(a)** 1.16, **(b)** 53.7%.
15.74 **(a)** 0.49 atm, **(b)** 0.23, **(c)** 0.037, **(d)** 0.0368 mol.
15.76 $[H_2] = 0.070\ M$, $[I_2] = 0.182\ M$, $[HI] = 0.825\ M$.
15.78 (c).
15.80 **(a)** 2.3×10^{-4}, **(b)** 1.5, **(c)** 0.81, **(d)** 4.2×10^3; 0.015.
15.82 0.0231; 9.60×10^{-4}.
15.84 NO_2: 0.80 atm; N_2O_4: 0.50 atm.

CHAPTER 16

16.4 **(a)** Nitrite ion: NO_2^-; **(b)** hydrogen sulfate ion: HSO_4^-; **(c)** hydrogen sulfide ion: HS^-; **(d)** cyanide ion: CN^-; **(e)** formate ion: $HCOO^-$.
16.6 **(a)** H_2S, **(b)** H_2CO_3, **(c)** HCO_3^-, **(d)** H_3PO_4, **(e)** $H_2PO_4^-$, **(f)** HPO_4^{2-}, **(g)** H_2SO_4, **(h)** HSO_4^-, **(i)** H_2SO_3, **(j)** HSO_3^-.
16.8 **(a)**
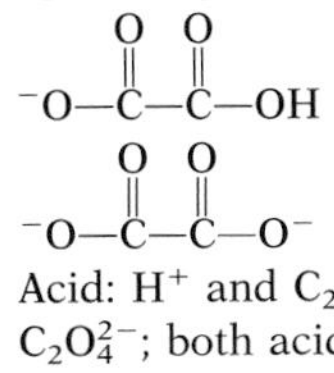
(b) Acid: H^+ and $C_2H_2O_4$; base: $C_2O_4^{2-}$; both acid and base: $C_2HO_4^-$.
16.16 **(a)** $6.3 \times 10^{-6}\ M$, **(b)** $1.0 \times 10^{-16}\ M$, **(c)** $2.7 \times 10^{-6}\ M$.
16.18 6.72.
16.20 **(a)** Acidic, **(b)** neutral, **(c)** basic.
16.22 1.98×10^{-3} mol; 0.444.
16.24 2.2×10^{-3} g.
16.30 **(a)** Strong base, **(b)** weak base, **(c)** weak base, **(d)** weak base, **(e)** strong base.
16.32 **(a)** False, **(b)** true, **(c)** true, **(d)** false.
16.34 To the left.
16.36 Only the anion of phenol can be stabilized by resonance.
16.42 $[H^+] = [CH_3COO^-] = 5.8 \times 10^{-4}\ M$; $[CH_3COOH] = 0.0181\ M$.
16.44 $2.3 \times 10^{-3}\ M$.
16.46 **(a)** 3.5%, **(b)** 9.0%, **(c)** 33%, **(d)** 79%.
16.48 **(a)** 3.9%. **(b)** 0.30%.
16.52 $[H^+] = [HCO_3^-] = 1.0 \times 10^{-4}\ M$; $[CO_3^{2-}] = 4.8 \times 10^{-11}\ M$.
16.56 7.1×10^{-7}.
16.58 1.5%.
16.66 HZ < HY < HX.
16.68 4.82.
16.70 HCl: 1.40; H_2SO_4: 1.31.
16.72 Stronger at 40°C.
16.76 $AlCl_3$ is a Lewis acid, Cl^- is a Lewis base.
16.78 CO_2 and BF_3.
16.80 0.106 L.
16.82 No.
16.84 No.
16.86 CrO is ionic and basic; CrO_3 is covalent and acidic.
16.88 4.0×10^{-2}.
16.90 0.028.

CHAPTER 17

17.6 (a) and (c).
17.8 4.74; (a).
17.10 7.03.
17.12 10; more effective against an added acid.
17.14 **(a)** 4.81. **(b)** 4.64.
17.16 HC.
17.18 89 g/mol.
17.20 **(a)** 110 g/mol. **(b)** 1.6×10^{-6}.
17.26 Red.
17.34 **(a)** 7.8×10^{-10}. **(b)** 1.8×10^{-18}.
17.36 1.8×10^{-10}.
17.38 $2.2 \times 10^{-4}\ M$.
17.40 2.3×10^{-9}.
17.42 $[Na^+] = 0.045\ M$; $[NO_3^-] = 0.076\ M$; $[Sr^{2+}] = 1.5 \times 10^{-2}$ M; $[F^-] = 1.2 \times 10^{-4}\ M$.
17.46 **(a)** $0.013\ M$. **(b)** $2.2 \times 10^{-4}\ M$. **(c)** $3.4 \times 10^{-3}\ M$.
17.48 **(a)** $1.0 \times 10^{-5}\ M$. **(b)** $1.1 \times 10^{-10}\ M$.
17.52 Complex ion formations. **(a)** $[Cu(NH_3)_4]^{2+}$. **(b)** $[Ag(CN)_2]^-$. **(c)** $[HgCl_4]^{2-}$. For equations for the formation of the complex ions see Table 17.5
17.54 $[Cd(CN)_4^{2-}] = 4.2 \times 10^{-3}\ M$; $[CN^-] = 0.48\ M$; $[Cd^{2+}] = 1.1 \times 10^{-18}\ M$.
17.56 $3.5 \times 10^{-5}\ M$.
17.60 $0.011\ M$.
17.62 Chloride ions will only precipitate Ag^+ ions or a flame test for Cu^{2+} ions.
17.64 **(a)** 1.37. **(b)** 5.28. **(c)** 8.85.
17.66 2.51-4.41.
17.68 $1.27\ M$.
17.70 $[Na^+] = 0.0835\ M$; $[OH^-] = 0.0335\ M$; $[H^+] = 3.0 \times 10^{-13}\ M$; $[CH_3COO^-] = 0.0500\ M$; $[CH_3COOH] = 8.3 \times 10^{-10}\ M$.
17.72 9.25; 9.18.
17.74 9.97 g; 13.03
17.76 $[Ag^+] = 2.0 \times 10^{-9}\ M$. $[Cl^-] = 0.080\ M$. $[Zn^{2+}] = 0.070\ M$. $[NO_3^-] = 0.060\ M$.
17.78 pH is greater than 2.68 but less than 8.11.
17.80 All except (a).
17.82 2.4×10^{-13}.
17.84 A precipitate of $Fe(OH)_2$ will form.
17.86 The original precipitate is HgI_2. At higher concentration of KI, the complex ion HgI_4^{2-} forms so mass of HgI_2 decreases.
17.88 7.82-10.38.

CHAPTER 18

18.12 **(a)** +3. **(b)** 6. **(c)** Oxalate.
18.14 **(a)** Na: +1; Mo: +6. **(b)** Mg: +2; W: +6. **(c)** K: +1; Fe: +2.
18.16 **(a)** cis-dichorobis-(ethylenediamine)cobalt(III), **(b)** pentaamminechloro-platinum(IV) chloride,

(c) hexaaminecobalt(III) chloride, **(d)** pentaamminechlorocobalt(III) chloride, **(e)** trans-diamminedichloroplatinum(II).

18.18 **(a)** $[Cr(en)_2Cl_2]^+$. **(b)** $Fe(CO)_5$. **(c)** $K_2[Cu(CN)_4]$. **(d)** $[Co(NH_3)_4(H_2O)Cl]Cl_2$.

18.20 **(a)** 2. **(b)** 2.

18.22

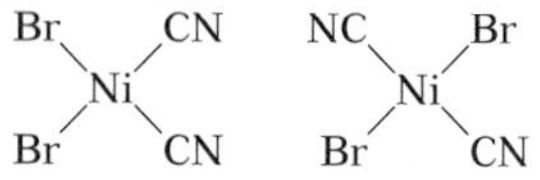

cis isomer trans isomer

18.28 CN^- is a strong-field ligand. Absorption occurs near the high-energy end of the spectrum (blue) and the complex ion looks yellow. H_2O is a weaker field ligand and absorption occurs in the orange or red part of the spectrum. Consequently, the complex ion appears green or blue.

18.30 **(a)** 2 unpaired spins. **(b)** 4 unpared spins.

18.32 $[Co(NH_3)_4Cl_2]Cl$.

18.34 6.7×10^{13}.

18.36 The $Mn(H_2O)_6^{2+}$ ion is a high-spin complex, containing 5 unpaired spins.

18.38 The following will absorb at longer wavelength: **(a)** $[Co(H_2O)_6]^{2+}$; **(b)** $[FeF_6]^{3-}$; **(c)** $[CuCl_4]^{2-}$.

18.40 **(a)** $4Au(s) + 8CN^-(aq) + O_2(g) + 2H_2O(l) \longrightarrow 4[Au(CN)_2]^-(aq) + 4OH^-(aq)$. **(b)** $Zn(s) + 2[Au(CN)_2]^-(aq) \longrightarrow [Zn(CN)_4]^{2-}(aq) + 2Au(s)$. **(c)** Linear (Au is *sp*-hybridized).

CHAPTER 19

19.8 (c) < (d) < (a) < (e) < (b).

19.10 **(a)** 47.5 J/K. **(b)** −12.5 J/K. **(c)** −242.8 J/K.

19.12 **(a)** $\Delta S < 0$. **(b)** $\Delta S > 0$. **(c)** $\Delta S > 0$. **(d)** $\Delta S < 0$.

19.16 **(a)** −1139 kJ. **(b)** −140 kJ. **(c)** −2935.0 kJ.

19.18 **(a)** Spontaneous at all temperatures. **(b)** Below 111 K.

19.22 8.0×10^1 kJ/mol.

19.24 457.2 kJ; 7×10^{-81}.

19.26 **(a)** −24.6 kJ. **(b)** −1.3 kJ.

19.28 −341 kJ/mol.

19.30 −2.87 kJ/mol.

19.32 **(a)** 8.0×10^4 J. **(b)** 4.0×10^4 J. **(c)** -3.2×10^4 J. **(d)** 6.4×10^4 J.

19.34 Free energy of reaction is positive or the rate is too slow.

19.36 **(a)** $\Delta H > 0$, $\Delta S > 0$, $\Delta G < 0$. **(b)** $\Delta H > 0$, $\Delta S > 0$, $\Delta G = 0$. **(c)** $\Delta H > 0$, $\Delta S > 0$, $\Delta G > 0$.

19.38 $\Delta S > 0$.

19.40 **(b)** Both water and ethanol have larger ΔS_{vap} because the liquids are more ordered due to hydrogen bonding.

19.42 **(b)** 0.0739.

19.44 2.6×10^{-9}.

19.46 703°C.

19.48 $\Delta S < 0$; $\Delta H < 0$.

19.50 55 J/K.

19.52 $\Delta S_{sys} < 0$ but $\Delta S_{univ} > 0$.

19.54 56 J/K.

19.56 4.9×10^{-31}; oxygen does not react with nitrogen in air.

19.58 4.8×10^{-75} atm.

19.60 (a) and (b).

19.62 4.5×10^5.

19.64 **(a)** Pentane: 3%; 2-methylbutane: 43%; 2,2-dimethylpropane: 54%. **(b)** Stability increases from pentane to 2,2-dimethylpropane (as the extent of branching increases).

19.66 $C(s) + Cu_2O(s) \longrightarrow CO(g) + 2Cu(s)$; $K = 8$.

CHAPTER 20

20.2 **(a)** $Mn^{2+} + H_2O_2 + 2OH^- \longrightarrow MnO_2 + 2H_2O$. **(b)** $2Bi(OH)_3 + 3SnO_2^{2-} \longrightarrow 2Bi + 3H_2O + 3SnO_3^{2-}$. **(c)** $Cr_2O_7^{2-} + 14H^+ + 3C_2O_4^{2-} \longrightarrow 2Cr^{3+} + 6CO_2 + 7H_2O$. **(d)** $2Cl^- + 2ClO_3^- + 4H^+ \longrightarrow Cl_2 + 2ClO_2 + 2H_2O$.

20.8 2.46 V; $3Ag^+(aq) + Al(s) \longrightarrow 3Ag(s) + Al^{3+}(aq)$.

20.10 Cl_2, MnO_4^-, F_2.

20.12 **(a)** Spontaneous, **(b)** nonspontaneous, **(c)** nonspontaneous, **(d)** spontaneous.

20.14 **(a)** Li, **(b)** H_2, **(c)** Fe^{2+}, **(d)** Br^-.

20.18 0.367 V.

20.20 **(a)** −432 kJ; 6×10^{75}. **(b)** −104 kJ; 2×10^{18}. **(c)** −178 kJ; 1×10^{31}. **(d)** -1.27×10^3 kJ; 1×10^{222}.

20.22 0.37 V; −36 kJ; 2×10^6.

20.26 **(a)** 2.23 V; 2.23 V; −430 kJ. **(b)** 0.02 V; 0.04 V; −23 kJ.

20.28 0.083 V.

20.30 0.010 V.

20.34 1.09 V.

20.42 **(a)** Anode: $2Cl^- \longrightarrow Cl_2 + 2e^-$; cathode:$Ca^{2+} + 2e^- \longrightarrow Ca$. **(b)** 0.19 g.

20.44 **(a)** $\$2.09 \times 10^3$. **(b)** $\$2.46 \times 10^3$. **(c)** $\$4.70 \times 10^3$.

20.46 **(a)** 0.14 F. **(b)** 0.123 F. **(c)** 0.10 F.

20.48 **(a)** $Ag^+ + e^- \longrightarrow Ag$. **(b)** $2H_2O \longrightarrow O_2 + 4H^+ + 4e^-$. **(c)** 6.0×10^2 C.

20.50 **(a)** 0.589 g Cu. **(b)** 0.133 A.

20.52 2.2 h.

20.54 9.66×10^4 C.

20.56 0.0710 F.

20.58 $Cr_2O_7^{2-} + 6Fe^{2+} + 14H^+ \longrightarrow 2Cr^{3+} + 6Fe^{3+} + 7H_2O$; 0.156 *M*.

20.60 45.1%.

20.62 **(a)** $2MnO_4^- + 16H^+ + 5C_2O_4^{2-} \longrightarrow 2Mn^{2+} + 10CO_2 + 8H_2O$. **(b)** 5.40%.

20.64 0.231 mg/mL Blood.

20.66 **(a)** 0.80 V. **(b)** $2Ag^+ + H_2 \longrightarrow 2Ag + 2H^+$. **(c)** (i) 0.92 V. (ii) 1.10 V. **(d)** The cell functions as a pH meter; its potential is a sensitive function of the hydrogen ion concentration.

20.68 Water would be oxidized before the F^- ion.

20.70 2.5×10^2 h.

20.72 Hg(I) exists as Hg_2^{2+} ions.

20.74 $[Mg^{2+}] = 0.0500$ *M*; $[Ag^+] = 2 \times 10^{-55}$ *M*; 1.44 g Mg.

20.76 **(a)** H_2; 0.206 L. **(b)** 6.09×10^{23}/mol.

20.78 **(a)** −1356.8 kJ. **(b)** 1.17 V.

20.80 +3.

20.82 14 kJ; 4×10^{-3}.

20.84 1.4 A.

20.86 +4.

20.88 1.60×10^{-19} C/e^-.

CHAPTER 21

21.6 **(a)** 0.049 *M*/s. **(b)** 0.025 *M*/s.
21.16 **(a)** rate = $k[F_2][ClO_2]$.
(b) 1.2/*M* · s.
(c) 2.4×10^{-4} *M*/s.
21.18 **(a)** rate = $k[X]^2[Y]^3$.
(b) 0.13 *M*/s.
21.20 **(a)** 0.046 s^{-1}.
(b) 0.13 $M^{-1}\ s^{-1}$.
21.22 3 half-lives or 2.08/k.
21.24 **(a)** 0.0198 s^{-1}. **(b)** 151 s.
21.32 135 kJ/mol.
21.34 371°C.
21.36 51.0 kJ/mol.
21.46 **(a)** rate = $k[X_2][Y]$.
(b) Z does not appear in the rate-determining step.
(c) $X_2 + Y \longrightarrow XY + X$ (slow); $X + Z \longrightarrow XZ$ (fast).
21.54 rate = $(k_1k_2/k_{-1})[E][S]$.
21.56 **(a)** Temperature,
(b) activation energy,
(c) concentration of reactant,
(d) catalyst.
21.58 Area of large sphere: 22.6 cm^2; area of 8 small spheres: 44.9 cm^2. The spheres with the larger surface area is the more effective catalyst.
21.60 $[H_2O]$ can be treated as a constant.
21.62 **(a)** rate = $k[H^+][CH_3COCH_3]$;
(b) $3.8 \times 10^{-3}/M \cdot s$.
21.64 2.63 atm.
21.66 $M^{-2}\ s^{-1}$.
21.68 56.4 min.
21.70 (b), (d), and (e).
21.72 0.098%.
21.74 **(a)** Increased, **(b)** decreased, **(c)** decreased, **(d)** increased.
21.76 0.0896 min^{-1}.
21.78 1.12×10^3 min.
21.80 **(a)** rate = $k[X][Y]^2$.
(b) 1.9 X $10^{-2}\ M^{-2}\ s^{-1}$.
21.82 3.1.
21.84 During the first ten minutes or so the engine is relatively cold so the reaction will be slow.

CHAPTER 22

22.6 **(a)** ${}_{-1}^{0}\beta$. **(b)** ${}_{20}^{40}Ca$. **(c)** α.
(d) ${}_{0}^{1}n$.
22.8 **(a)** ${}_{34}^{80}Se + {}_{1}^{2}d \longrightarrow {}_{34}^{81}Se + {}_{1}^{1}p$.
(b) ${}_{4}^{9}Be + {}_{1}^{2}d \longrightarrow {}_{3}^{9}Li + 2{}_{1}^{1}p$.
(c) ${}_{5}^{10}B + {}_{0}^{1}n \longrightarrow {}_{3}^{7}Li + {}_{2}^{4}He$.
22.10 ${}_{80}^{198}Hg + {}_{0}^{1}n \longrightarrow {}_{79}^{198}Au + {}_{1}^{1}p$.
22.18 **(a)** Li-9. **(b)** Na-22.
(c) Sc-48.
22.20 **(a)** Ne-17. **(b)** Ca-45.
(c) Tc-92. **(d)** Hg-195.
(e) Cm-242.
22.22 6×10^9 kg.
22.24 **(a)** 4.54×10^{-12} J; 1.14×10^{-12} J/nucleon.
(b) 2.363×10^{-10} J; 1.284×10^{-12} J/nucleon.
22.26 0.250 d^{-1}; 2.77 days.
22.28 2.7 day.
22.30 ${}_{82}^{208}Pb$.
22.32 A: none left; B: 0.25 mol; C: none left; D: 0.75 mol.
22.44 IO_3^- is formed only from IO_4^-.
22.46 After a few days of food intake isolate hemoglobin from red blood cells and monitor radioactivity from Fe-59.
22.48 An analogous Pauli principle for nucleons as for electrons.
22.50 **(a)** 0.343 millicurie.
(b) ${}_{93}^{237}Np \longrightarrow {}_{2}^{4}He + {}_{91}^{233}Pa$.
22.52 **(a)** 1.040×10^{-12} J/nucleon.
(b) 1.111×10^{-12} J/nucleon.
(c) 1.199×10^{-12} J/nucleon.
(d) 1.410×10^{-12} J/nucleon.
22.54 ${}_{7}^{18}N \longrightarrow {}_{8}^{18}O + {}_{-1}^{0}\beta$.
22.56 Use radioactive dating technique.
22.58 **(a)** ${}_{83}^{209}Bi + {}_{2}^{4}He \longrightarrow {}_{85}^{211}At + 2{}_{0}^{1}n$.
(b) ${}_{83}^{209}Bi(\alpha,2n){}_{85}^{211}At$.
22.60 The sun has a much greater gravitational pull which favors the fusion process.
22.62 2.8×10^3 yr.
22.64 **(a)** ${}_{19}^{40}K \longrightarrow {}_{18}^{40}Ar + {}_{+1}^{0}\beta$;
(b) 3.0×10^9 yr.
22.66 **(a)** For Sr-90 decay: $\Delta E = -5.59 \times 10^{-15}$ J: for Y-90 decay: $\Delta E = -2.84 \times 10^{-13}$ J.
(b) 0.024 mole.
(c) -4.19×10^6 kJ.
22.68 Age of shroud: 1.3×10^3 yr. It cannot be the burial cloth of Jesus Christ.

INDEX

PHOTO CREDITS

CHAPTER 1

Page 3: The Bettmann Archive. Figure 1.1: Fritz Goro/Life Magazine © Time, Inc. Figure 1.2: McGraw-Hill photos by Ken Karp. Figure 1.8: Courtesy of Mettler. Page 8: McGraw-Hill photo by Ken Karp. Page 10: NASA. Page 11: Comstock. Page 12: McGraw-Hill photo by Ken Karp.

CHAPTER 2

Page 27: left, Association Joliot-Gurie Institut du Radium, courtesy Archives Pierre et Marie Curie; right, The Bettmann Archive. Figure 2.3: McGraw-Hill photo by Ken Karp. Page 29: Niels Bohr Library. Page 31: Burndy Library; Courtesy AIP Emilio Segré Visual Archives. Page 32: Cambridge University Library, Rutherford Collection; Courtesy AIP Emilio Segré Visual Archives. Page 39: McGraw-Hill photo by Ken Karp. Page 42: McGraw-Hill photo by Ken Karp.

CHAPTER 3

Page 53: The Granger Collection. Figure 3.1: McGraw-Hill photo by Ken Karp. Page 55: Andrew Popper/Picture Group. Page 56: Chemical Heritage Foundation. Page 57: Dr. E. R. Degginger, FPSA. Page 58: Dr. E. R. Degginger, FPSA. Page 58: L. V. Bergman & Associates. Page 59: American Gas. Page 60: McGraw-Hill photo by Ken Karp. Page 62: McGraw-Hill photo by Ken Karp. Page 63: McGraw-Hill photo by Ken Karp. Page 64: Ward's Natural Science Establishment. Page 67: McGraw-Hill photo by Ken Karp. Page 68: McGraw-Hill photo by Ken Karp. Page 69: McGraw-Hill photo by Ken Karp. Page 70: McGraw-Hill photo by Ken Karp. Page 71: Raymond Chang. Page 73: McGraw-Hill photo by Ken Karp. Page 77: McGraw-Hill photo by Ken Karp.

CHAPTER 4

Page 87: Dirck Halstead/Gamma-Liaison. Figure 4.3: McGraw-Hill photo by Ken Karp. Figure 4.4: McGraw-Hill photo by Ken Karp. Figure 4.5: McGraw-Hill photo by Ken Karp. Figure 4.6: McGraw-Hill photo by Ken Karp. Figure 4.7: McGraw-Hill photo by Ken Karp. Figure 4.9: McGraw-Hill photo by Ken Karp. Figure 4.10: Joel Gordon. Figure 4.12: McGraw-Hill photos by Ken Karp. Figure 4.15: McGraw-Hill photos by Ken Karp. Figure 4.16: McGraw-Hill photos by Ken Karp. Page 93: McGraw-Hill photo by Ken Karp. Page 94: Chemical Heritage Foundation. Page 95: McGraw-Hill photo by Ken Karp. Page 96: McGraw-Hill photo by Ken Karp. Page 99: McGraw-Hill photo by Ken Karp. Page 105: McGraw-Hill photo by Ken Karp. Page 109: McGraw-Hill photo by Ken Karp.

CHAPTER 5

Page 117: The Granger Collection. Figure 5.10: McGraw-Hill photo by Ken Karp. Page 118: McGraw-Hill photo by Ken Karp. Page 120: Peter Gridley/FPG International. Page 122: AIP Emilio Serge Visual Archive. Page 125: The Bettmann Archive. Page 138: NASA. Page 140: Chemical Heritage Foundation.

CHAPTER 6

Page 149: left, UPI/Bettmann Newsphotos; right, NASA. Figure 6.2: Dr. E. R. Degginger, FPSA. Page 152: McGraw-Hill photo by Ken Karp. Page 153: McGraw-Hill photo by Ken Karp. Page 160: top, McGraw-Hill photo by Ken Karp; bottom, courtesy of the Diamond Information Center. Page 161: Dr. E. R. Degginger, FPSA. Page 165: McGraw-Hill photo by Ken Karp. Page 169: McGraw-Hill photo by Ken Karp. Page 170: Raymond Chang.

CHAPTER 7

Page 179: above, Deutsches Museum; below, Royal Observatory, Edinburgh/AATB/Science Photo Library/Photo Researchers. Figure 7.2: © 1994 B.S.I.P./Custom Medical Stock Photos. Figure 7.5: Joel Gordon. Figure 7.12: By permission of Sargent-Welch Scientific Company, Skokie, Ill. Page 180: AIP Emilio Segré Visual Archives, Gift of Jost Lemmerich. Page 183: Courtesy of the Archives, California Institute of Technology. Page 186: AIP Emilio Segré Visual Archives. Page 189: AIP Niels Bohr Library. Page 192: top, AIP Niels Bohr Library, W. F. Meggers Collection; bottom, AIP Niels Bohr Library, photo by Francis Simon. Page 200: AIP Niels Bohr Library, Goudsmit Collection.

CHAPTER 8

Page 215: Special Collections Van Pelt Library, University of Pennsylvania. Figure 8.12: Lithium and sodium, McGraw-Hill photo by Ken Karp; potassium, Life Science Library/MATTER, photo Albert Fenn © Time-Life Books, Inc.; rubidium and cesium, L. V. Bergman & Associates. Beryllium, magnesium, calcium, strontium and barium, L. V. Bergman & Associates; radium, Life Science Library/MATTER, photo by Phil Brodatz © Time-Life Books, Inc. Figure 8.14: L. V. Bergman & Associates. Figure 8.15: Graphite, McGraw-Hill photo by Ken Karp; diamond, Courtesy of Diamond Information Center; silicon, Frank Wing/Stock, Boston; germanium and lead, McGraw-Hill photo by Ken Karp; tin, L. V. Bergman & Associates. Figure 8.16: Nitrogen, Joe McNally/Wheeler Pictures; white and red phosphorus, Life Science Library/MATTER, photo Albert Fenn © Time-Life Books, Inc.; arsenic, antimony, and bismuth, L. V. Bergman & Associates. Figure 8:17: Sulfur and selenium, L. V. Bergman & Associates; tellurium, McGraw-Hill photo by Ken Karp. Figure 8.18: Joel Gordon. Figure 8.19: McGraw-Hill photos by Ken Karp. Figure 8.20: Neil Bartlett. Figure 8.21: Courtesy of Argonne National Laboratory.

CHAPTER 9

Page 249: Above, Bancroft Library, University of California; below, from G. N. Lewis, *Valence,* Dover Publications, Inc., New York 1966. Page 253: Linus Pauling Institute of Science and Medicine. Page 258: McGraw-Hill photo by Ken Karp. Page 261: Courtesy of James O. Schreck, Professor of Chemistry, University of Northern Colorado.

CHAPTER 10

Page 275: Raymond Chang. Page 286: AIP Emilio Segré Visual Archives, Fankuchen Collection. Page 300: McGraw-Hill photo by Ken Karp.

CHAPTER 11

Page 307: Left, Courtesy of Edmund Catalogs; right, courtesy Railway Technical Research Institute, Tokyo, Japan. Figure 11.3: McGraw-Hill photo by Ken Karp. Figure 11.9: McGraw-Hill photo by Ken Karp. Figure 11.11: McGraw-Hill photo by Ken Karp. Figure 11.28: McGraw-Hill photos by Ken Karp. Figure 11.32: Ken Karp. Page 324: L. V. Bergman & Associates. Page 325: Grant Heilman. Page 327: McGraw-Hill photo by Ken Karp. Page 330: McGraw-Hill photo by Ken Karp.

CHAPTER 12

Page 339: The Bettmann Archive. Figure 12.1: McGraw-Hill photo by Ken Karp. Figure 12.6: McGraw-Hill photo by Ken Karp. Page 353: Frank A. Cezus/FPG International. Page 354: McGraw-Hill photo by Ken Karp. Page 357: John Mead/Science Photo Library/Photo Researchers.

CHAPTER 13

Page 367: Courtesy Johnson & Johnson. Figure 13.5: McGraw-Hill photo by Ken Karp. Figure 13.13: Joel Gorden. Page 369: J. H. Robinson/Photo Researchers. Page 370: Courtesy of Chevron Corp. & American Petroleum Institute. Page 375: Courtesy Nabisco Foods, Inc. Page 376: Dr. E. R. Degginger, FPSA. Page 377: IBM Photo-Almaden Research Center. Page 381: McGraw-Hill photos by Ken Karp. Page 384: McGraw-Hill photo by Ken Karp.

CHAPTER 14

Page 397: National Gallery of Art, Washington. Figure 14.3: McGraw-Hill photo by Ken Karp. Figure 14.4: McGraw-Hill photo by Ken Karp. Figure 14.5: McGraw-Hill photo by Ken Karp. Figure 14.7: Charles Weckler The Image Bank. Figure 14.10: Donald Klegg. Figure 14.11: Courtesy Monstano. Page 399: Tom Tracy/The Stock Shop. Page 400: Courtesy Reynolds Metal Company. Page 401: UPI/Bettmann Newsphotos. Page 402: Richard Hutchings/Photo Researchers. Page 404: Hagley Museum and Library. Page 410: Lawrence Berkeley Laboratory.

CHAPTER 15

Page 417: Left, German Information Center; right, Grant Heilman. Figure 15.4: McGraw-Hill photo by Ken Karp. Figure 15.6: McGraw-Hill photo by Ken Karp. Figure 15.7: McGraw-Hill photo by Ken Karp. Page 418: McGraw-Hill photo by Ken Karp. Page 424: Collection Varin-Visage Jacana/Photo Researchers. Page 432: Chemical Heritage Foundation.

CHAPTER 16

Figure 16.1: McGraw-Hill photo by Ken Karp. Figure 16.6: McGraw-Hill photo by Ken Karp. Figure 16.7: McGraw-Hill photo by Ken Karp. Page 450: McGraw-Hill photo by Ken Karp. Page 472: Michael Melford Wheeler Pictures.

CHAPTER 17

Page 481: Left, Dole/The Image Bank; right, photo by A. R. Terepka, © *Scientific American,* March 1970, Vol. 222, No. 3, p. 88. Figure 17.1: McGraw-Hill photo by Ken Karp. Figure 17.2: McGraw-Hill photo by Ken Karp. Figure 17.5: McGraw-Hill photo by Ken Karp. Figure 17.7: McGraw-Hill photo by Ken Karp. Figure 17.8: McGraw-Hill photo by Ken Karp. Figure 17.9: McGraw-Hill photo by Ken Karp. Figure 17.10: Fujimoto/Kodansha. Page 482: McGraw-Hill photo by Ken Karp. Page 493: Above, Dr. E. R. Degginger, FPSA; below, Bjorn Bolstad/Peter Arnold. Page 495: McGraw-Hill photo by Ken Karp. Page 496: McGraw-Hill photo by Ken Karp. Page 497: McGraw-Hill photo by Ken Karp.

CHAPTER 18

Page 513: Stephen J. Lippard, Chemistry Department, Massachusetts Institute of Technology. Figure 18.2: Scandium, titanium, vanadium, chromium, manganese, iron, cobalt and nickel, McGraw-Hill photo by Ken Karp; copper, L. V. Bergman & Associates. Figure 18.9: McGraw-Hill photo by Ken Karp. Figure 18.15: McGraw-Hill photo by Ken Karp.

CHAPTER 19

Page 535: Farrell Grehan/Photo Researchers. Figure 19.4: Courtesy National Lime Association. Page 537: McGraw-Hill photo by Ken Karp. Page 540: McGraw-Hill photo by Ken Karp. Page 543: Yale University Library. Page 549: McGraw-Hill photo by Ken Karp.

CHAPTER 20

Page 561: The Bettmann Archive. Figure 10.1: McGraw-Hill photo by Ken Karp. Figure 20.5: McGraw-Hill photo by Ken Karp. Figure 20.10: NASA. Figure 20.11 a, Dr. E. R. Degginger, FPSA; b, McGraw-Hill photo by Ken Karp; c, Donald Dietz/Stock, Boston. Figure 20.13: McGraw-Hill photo by Ken Karp. Figure 20.16: McGraw-Hill photo by Ken Karp. Page 573: McGraw-Hill photo by Ken Karp. Page 590: Special Collections Van Pelt Library, University of Pennsylvania.

CHAPTER 21

Figure 21.3: McGraw-Hill photo by Ken Karp. Figure 21.5: McGraw-Hill photo by Ken Karp. Figure 21.15: McGraw-Hill photo by Ken Karp. Figure 21.17: Courtesy of Johnson Matthey. Figure 21.19: Courtesy GM. Page 609: McGraw-Hill photo by Ken Karp. Page 622: McGraw-Hill photo by Ken Karp.

CHAPTER 22

Page 635: AIP Emilio Segré Visual Archives, Max-Planck-Gesellschaft Pressestelle. Figure 22.5: Fermilab Visual Media Services. Figure 22.10: Pierre Kopp/West Light. Figure 22.11: M. Lazarus/Photo Researchers. Figure 22.12: Los Alamos National Laboratory. Figure 22.13: NASA. Figure 22.15: Lawrence Livermore National Laboratory. Figure 22.16: Department of Defense, Still Media Records Center, UA Navy Photo. Figure 22.17: B. Leonard Holman, MD, Brigham & Women's Hospital, Harvard Medical Center. Page 644: Francois Lochon/Gamma Liaison. Page 653: U.S. Department of Energy/Science Photo Library/Photo Researchers. Page 657: Thomas R. Fletcher/Stock, Boston.

Fundamental Constants

Avogadro's number	6.022×10^{23}
Electron charge (e)	1.6022×10^{-19} C
Electron mass	9.1095×10^{-28} g
Faraday constant (F)	96,487 C/mol electron
Gas constant (R)	8.314 J/K · mol (0.08206 L · atm/K · mol)
Planck's constant (h)	6.6256×10^{-34} J s
Proton mass	1.67252×10^{-24} g
Neutron mass	1.67495×10^{-24} g
Speed of light in vacuum	2.99792458×10^{8} m/s

Useful Conversion Factors and Relationships

1 lb = 453.6 g

1 in = 2.54 cm (exactly)

1 mi = 1.609 km

1 km = 0.6215 mi

1 pm = 1×10^{-12} m = 1×10^{-10} cm

1 atm = 760 mmHg = 760 torr = 101,325 N/m^2 = 101,325 Pa

1 cal = 4.184 J (exactly)

1 L atm = 101.325 J

1 J = 1 C × 1 V

$$?^\circ C = (^\circ F - 32^\circ F) \times \frac{5^\circ C}{9^\circ F}$$

$$?^\circ F = \frac{9^\circ F}{5^\circ C} \times (^\circ C) + 32^\circ F$$

$$K = (^\circ C + 273.15^\circ C)\left(\frac{1\ K}{1^\circ C}\right)$$